Vitagenes in avian biology and poultry health

Vitagenes
in avian biology and poultry health

Peter F. Surai

Wageningen Academic
Publishers

EAN: 9789086863532
e-EAN: 9789086869060
ISBN: 978-90-8686-353-2
e-ISBN: 978-90-8686-906-0
DOI: 10.3920/978-90-8686-906-0

First published, 2020

Dedication

To my wife Helen, my daughter Katie, my son Anton,
my grandsons Oscar, Arthur and Henry
and my granddaughter Aiste
who gave me inspiration for writing this book.

Preface

Commercial poultry production is associated with various stresses leading to decrease of productive and reproductive performance and compromised health of growing chickens, parent birds as well as commercial layers. Excess of reactive oxygen and nitrogen species (RONS) production, disturbance of redox homeostasis, oxidative stress and damages to proteins, lipids and DNA/RNA are considered to be major molecular mechanisms of the detrimental consequences of various stresses. However, recently a pleasant new face of some RONS, especially H_2O_2, has emerged and their role in cell signalling and stress adaptation has received a lot of attention. Therefore, new insight in the role of free radicals as signalling molecules, understanding the role of nutrients in gene expression and maternal programming, tremendous progress in human and animal genome work created new demands for further research related to understanding molecular mechanisms of stress development and adaptation. In fact. stress adaptation is associated with various signalling pathways and executed at the gene level. The term vitagenes refers to a group of redox-sensitive genes that are involved in stress sensing and preserving cellular adaptive homeostasis and the vitagene family includes heat shock proteins, superoxide dismutase, glutathione and thioredoxin systems and sirtuins. The vitagenes are key players in redox signalling and redox homeostasis maintenance in birds including poultry under commercial stress conditions of egg and meat production. Development of the vitagene concept become an important milestone in understanding molecular mechanisms of stresses adaptation.

A range of comprehensive reviews have been published addressing various vitagene-related issues, including their protective roles in neurodegenerative disorders, neuroprotection, aging and longevity, dermatology, free radical-related diseases, osteoporosis, Alzheimer pathology, etc. We suggested that the vitagene concept can also be useful in animal/poultry sciences and this concept in relation to poultry production was further developed in our previous publications. It seems likely that by upregulating the vitagenes and improving adaptive ability of animals/poultry to stress it is possible to decrease negative consequences of the four main types of stresses in poultry and farm animal production, including environmental, technological, nutritional and internal/biological stresses. Furthermore, there is an opportunity to nutritionally modulate the vitagene network by using various nutrients such as carnitine, taurine, betaine, vitamins A, E, D and C, phytochemicals, including silymarin, etc. In fact, activation of the vitagene network by nutritional means is considered as a fundamental mechanism for improving animal/poultry resistance to various stresses. Therefore, the goal of this volume is to provide up to date information about the roles of vitagenes in avian biology and poultry health with a special emphasis to stress adaptation.

The book is divided into 4 parts and includes 17 chapters. The first part deals with stress and antioxidant defences and includes two chapters. In Chapter 1 an analysis of main stresses in poultry production is presented indicating that it is practically impossible to avoid various stresses in commercial meat and egg production systems.

Chapter 2 is devoted to the integrated antioxidant system of the body with regulatory functions providing necessary connections between different antioxidants. A special emphasis is given to oxidative stress and redox biology, stress-response pathways and involvement of transcription factors in stress adaptation. An Nrf2 and NF-κB interplay in oxidative stress is also emphasised.

The second part of the book is devoted to roles of vitagenes in avian biology and consists of 6 chapters. The vitagene concept development is described in Chapter 3 showing how the vitagenes can be incorporated into the general scheme of antioxidant defence network and describing their roles as major players in redox signalling and stress adaptation. The next 5 chapters are dealing with individual vitagenes. In particular, Chapter 4 is devoted to superoxide dismutases (SOD), newcomers into the vitagene family. Special emphasis is given to biochemical features and protective roles of SOD in avian species, including their regulation by environmental and nutritional stimuli with a main conclusion that SODs as important AO enzymes of the first level of AO defence are an integral part of the vitagene family. Chapter 5 is devoted to heat shock proteins (HSP) with a specific emphasis to HSP70 and HSP32 called heme oxygenase 1 (HO-1). Again, important biological features of avian HSPs are described and their protective role in proteostasis maintenance is described. Protective antioxidant roles of HO-1 are also described in detail. The role of various nutrients, including vitamins E, D and C, carnitine, betaine, selenium and phytochemicals in HSP expression and activity modulation is considered. Thioredoxin system, including thioredoxin (Trx), thioredoxin reductase (TR), peroxiredoxins (Prx) and sulfiredoxin (Srx) in relation to avian biology is describe in Chapter 6. A special attention is paid to the role of Trx system in redox homeostasis maintenance in stress adaptation. Chapter 7 is devoted to the glutathione (GSH) system, including protective roles of GSH, glutathione reductase (GR), glutaredoxins (Grx) and glutathione peroxidases (GPx) in avian biology and poultry health. Biochemical features of avian elements of GSH system are analysed and their modulation by nutritional and environmental means are described. Special attention is given to non-Se-GPx as important players in redox balance maintenance. Protective roles of sirtuins in stress adaptation are described in Chapter 8. Recent data related to sirtuins in avian biology are presented. Detailed analysis of interactions of sirtuins with transcription factors (Nrf2, NF-κB, FOXO, p53, HSF1) is presented.

Part III of the book deals with nutritional modulation of vitagenes. In particular, Chapter 9 is devoted to carnitine as an effective modulator of the AO defences and vitagene network. Detailed information on carnitine absorption, assimilation and metabolism in avian species is presented with a special emphasis to the usage of this nutrient in the antistress technology. Chapter 10 deals with well-known amino acid taurine, describing in detail its protective roles in various *in vitro* and *in vivo* systems, showing its potential in vitagene modulation and stress adaptation. A possibility of taurine being a semi-essential amino acid in modern poultry production is considered. In Chapter 11 protective roles of silymarin in biological systems are described with a special emphasis to its role in the antioxidant defence network, in vitagene and transcription factor modulation. Protective roles of silymarin in poultry production are also described. Chapter 12 is devoted to vitagene modulation by natural antioxidants, including vitamins A, E, D3, C, selenium, betaine, polyphenols and synergistic combinations of various antioxidants.

The fourth part of the book is devoted to practical applications of the vitagene concept in commercial poultry production. In particular, Chapter 13 describes details of the PerforMax concept development leading to the first commercial vitagene-regulating product in the poultry/animal nutrition. In general, it is shown that supplying vitagene-regulating mixtures via drinking water could be considered as a fast-response system to deal with various stresses in poultry production. Results of successful research work and commercial trials presented in the chapter clearly indicate that the vitagene concept found its way to the commercial poultry production including broiler production, rearing birds, parent stock and commercial layers management. Chapter 14 showing ShellBone concept development as a way for the next step of the vitagene concept application for improvement of eggshell and bone quality in poultry. The development of the third vitagene-related concept called VitaTonic in poultry production dealing with liver problems is described in Chapter 15. Protective roles of vitagenes in gut health maintenance and immunocompetence are considered in Chapter 16. A relationship between nutrition, gut microbiota and redox homeostasis in the gut is characterised. A special emphasis is given to a new understanding of the role of vitagenes in protection of immunoreceptors in stress conditions. The final Chapter 17 combined all information provided in the previous chapters to emphasise an essential role of vitagene network as an integral part of the antioxidant defence mechanisms providing redox homeostasis and stress adaptation.

I understand that my views on the role of vitagenes in avian biology and poultry health are sometimes different from those of other scientists and therefore I would appreciate very much receiving any comments from readers which will help me in my future research. I would like to thank my colleagues with whom I have had the pleasure to collaborate and share my ideas related to natural antioxidants and vitagenes in particular, who helped me at various stages of this research by providing essential information and advise. I am also indebted to the World's Poultry Science Association for the Research Award and a grant of the Government of Russian Federation (Contract No. 14.W03.31.0013) supporting my research.

Peter F. Surai

About the author

Dr Peter Surai started his studies at Kharkov University, Ukraine, where he obtained his PhD and DSc in biochemistry studying effects of antioxidants on poultry. Later he became Professor of Human Physiology. In 1994 he moved to Scotland to continue his antioxidant related research in poultry and in 2000 he was promoted to a full Professor of Nutritional Biochemistry at the Scottish Agricultural College. Recently he was awarded Honorary Professorships in 7 universities in various countries, including UK, Hungary, Bulgaria, Russia and Ukraine. In 2010 he was elected to the Russian Academy of Sciences as a foreign member. He has more than 850 research publications, including 360 papers in peer-reviewed journals, 14 books and 44 chapters in various books. In 1999 he received the prestigious John Logie Baird Award for Innovation for the development of 'super-eggs' and, in 2000, The World's Poultry Science Association Award for Research in recognition of an outstanding contribution to the development of the poultry industry. In 2017 he became a member of the team at the Moscow State Academy of Veterinary Medicine and Biotechnology named after K.I. Skryabin to conduct a research under a mega-grant of the Government of Russian Federation (contract no. 14.W03.31.0013). He successfully transferred the vitagene concept of stress adaptation from medical sciences into poultry and animal sciences. For the last 20 years he has been lecturing all over the world visiting more than 70 countries.

Table of contents

Abbreviations

4-HNE	4-hydroxyalkenal
AA	ascorbic acid
Ab	antibody
AFB1	aflatoxin B1
Akt	serine/threonine kinase
AMPK	adenosine monophosphate protein kinase
AO	antioxidant
AP1	transcription factor
Ape-1	apurinic/apyrimidinic endonuclease 1
ARE	antioxidant response element
ASK1	apoptosis signal-regulating kinase 1
ATF6	activating transcription factor 6
ATM	ataxia-telangiectasia-mutated
ATR	ataxia-telangiectasia and Rad 3-related
APR	acute phase response
BD	basal diet
CAT	catalase
Cd36	scavenger receptor
CoQ	coenzyme Q
COX-2	cyclooxygenase-2
CPS1	carbamoyl phosphate synthetase 1
CREB	cAMP responsive element binding protein
CRTC2	CREB regulated transcription coactivator 2
CUL3	cullin 3 protein
DAA	dehydroascorbic acid
DC	dendritic cells
DHA	docosahexaenoic acid
DHLA	dihydrolipoic acid
DON	deoxynivalenol
EC-SOD	extracellular superoxide dismutase
ESR	energy stress response
EPR	electron paramagnetic resonance
ER	endoplasmic reticulum
ETC	electron transport chain
FB1	fumonisin B1
FCR	feed conversion ratio
FcγR	phagocytic Fcγ receptors
FOXO	forkhead box O, transcription factors
G6PD	glucose-6-phosphate dehydrogenase
γ-GCS	gamma-glutamylcysteine synthetase
GCL	glutamate cysteine ligase
GDH	glutamate dehydrogenase
GI-GSH-Px	gastrointestinal glutathione peroxidase
GIT	gastrointestinal tract

GLUT1	glucose transporter1
GPx	glutathione peroxidase
GR	glutathione reductase
Grx	glutaredoxin
GSH	reduced glutathione
GSH-syst.	glutathione system
GSSG	oxidised glutathione
GST	glutathione S-transferase
H_2O_2	hydrogen peroxide
hCPCs	human cardiac progenitor cells
HIF-1α	hypoxia-inducible factor 1α
HIF	hypoxia-inducible transcription factor
HIR	hypoxia-induced response
HO-1	heme oxygenase 1
HS	heat stress
HSF1	heat shock factor 1
HSP	heat shock protein
HSP70	heat shock protein 70
HSR	heat shock response
HISR	hypoxia-induced stress response
IBD	infectious bursal disease
IDE	insulin degrading enzyme
IDH	isocitrate dehydrogenase
IFN	interferon
Ig	immunoglobulin
IGF-1	insulin-like growth factor 1
IKK	IκB kinase
IL-1	interleukin 1
IL-2R	interleukin 2 receptor
IL-6	interleukin 6
ISR	inflammatory stress response
iNOS	inducible nitric oxide synthase
IRE1	inositol-requiring enzyme 1
IκB	inhibitor of kappa B
Keap1	Kelch-like-ECH-associated protein 1
LCAD	long-chain acyl-CoA dehydrogenase
LDH	lactate dehydrogenase
LOOH	lipid hydroperoxide
LOX	lipoxygenase
LP	lipid peroxidation
LPS	lipopolysaccharide
MAPK	mitogen-activated protein kinase
MCD1	mitotic chromosome determinant
MDA	malondialdehyde
Met	methionine
MHC	major histocompatibility complex

MIF	macrophage inflammatory protein 2
Msr	methionine sulphoxide reductase
NF-κB	nuclear factor kappa-light-chain-enhancer of activated B cells
NK cells	natural killer cells
NKT cells	natural killer T cells
NO	nitric oxide
NPGPx	non-selenocysteine containing phospholipid hydroperoxide glutathione peroxidase
NQO1	NAD(P)H:quinone acceptor oxidoreductase 1
Nrf2	nuclear factor erythroid-2 related factor 2
NSR	nutritional stress response
ONOO	peroxynitrite
OSR	oxidative stress response
OTA	ochratoxin A
oxPTM	oxidative post-translational modifications
p53	tumour protein
p65	transcription factor
PAMP	pathogen-associated molecular patterns
PARP1	poly-ADP-ribose polymerase
PCB	polychlorinated biphenyls
PDH	pyruvate dehydrogenase
PDI	protein disulphide isomerase
PFK-1	phosphofructokinase-1
PGC-1α	peroxisome proliferator-activated receptor-γ coactivator
PGE2	prostaglandin E2
PGK1	phosphoglycerate kinase 1
pGSH-Px	plasma glutathione peroxidase
PHA	phytohemagglutin
PH-GSH-Px	phospholipid glutathione peroxidase
PI3K	phosphatidylinositol 3-kinase
PLA2	phospholipase A2
PMN	polymorphonuclear leukocytes
POP	persistent organic pollutant
PPAR	peroxisome proliferator-activated receptor
PPRE	peroxisome proliferator response element
PRDX1	peroxiredoxin 1
PRR	pattern recognition receptor
Prx	peroxiredoxin
PTEN	phosphatase and tensin homolog on chromosome 10
PTP1B	protein tyrosine phosphatase 1B
PUFA	polyunsaturated fatty acid
Ref-1	redox effector factor 1
RNA Pol	RNA polymerase
RNR	ribonucleotide reductase
RNS	reactive nitrogen species
RONS	reactive oxygen and nitrogen species

ROS	reactive oxygen species
RXR	retinoid-X receptor
SD	stocking density
SeCys	selenocysteine
SelH	selenoprotein H
SelM	selenoprotein M
SelN	selenoprotein N
SelT	selenoprotein T
SelV	selenoprotein V
SeMet	selenomethionine
Sep15	selenoprotein 15
SIRT	sirtuin
SM	silymarin
SOD	superoxide dismutase
Srx	sulfiredoxin
SREBP	sterol regulatory element binding protein
T-AOC	total antioxidant capacity
TBA	thiobarbituric acid
TBARS	thiobarbituric acid reactive substances
TCR	T-cell receptor
Th cells	T helper cells
TLR	Toll-like receptors
TNF-α	tumour necrosis factor alpha
Toc	tocopherol
TR4/TAK1	nuclear receptor
Trx	thioredoxin
TrxR	thioredoxin reductase
Trx-Syst	thioredoxin system
UCP2	uncoupling protein 2
UPR	unfolded protein response

Part I.
Stresses and antioxidant defences

Necessity is the mother of invention

Chapter 1
Stresses in poultry production

Don't count your chickens before they are hatched

1.1 Introduction

Commercial poultry production is associated with various stresses leading to decrease of productive and reproductive performance of growing chickens, parent birds as well as commercial layers. Growing body of evidence indicates that most stresses in poultry production at the cellular level are associated with oxidative stress due to excess of reactive oxygen and nitrogen species (RONS) production or inadequate antioxidant protection (Surai, 2002, 2006, 2018, 2020; Surai and Fisinin, 2012b, 2015; Surai *et al.*, 2019). In animals/birds, redox-signalling pathways use RONS as signalling molecules to activate genes responsible for regulation of various functions including immunity, growth, differentiation, proliferation and apoptosis. This chapter is devoted to major stresses in poultry production with special emphasis to oxidative stress as main mechanism of detrimental consequences of stresses.

1.2 Classification of stresses in poultry production

From a physiological point of view, stress is related to a deviation from optimal internal and external conditions. Under stressful conditions, the hypothalamic-pituitary-adrenal axis, the autonomic nervous system and the immune system are responsible for re-establishing homeostasis. Therefore, a cascade of regulatory mechanisms is involved, resulting in a mobilisation of energy and a shift in metabolism with detrimental effects on growth performance and feed efficiency (Bureau *et al.*, 2009). In modern commercial poultry production oxidative stress-related nutritional metabolic diseases (e.g. encephalomalacia, exudative diathesis, muscular dystrophy, etc.) practically disappeared (Surai, 2002, 2006, 2018, 2020), however, various disorders of the biological antioxidant defence system still causing substantial problems. For example, the amount of a particular nutrient in the diet may be insufficient to meet the requirements, the diet may contain substances that inactivate the nutrient or inhibit its absorption/utilisation, or metabolism may be upset by the interaction of dietary and environmental factors causing oxidative stress (Mezes *et al.*, 1997; Surai *et al.*, 2019a). Domestication and genetic selection based on rapid growth rates, better feed conversion, and heavier BW of broilers has made domestic birds, including broilers and turkey, particularly susceptible to oxidative stress (Soleimani *et al.*, 2011). In general, there are four major types of stress in poultry industry: technological, environmental, nutritional and internal stresses (Fisinin *et al.*, 2009, 2009a; Surai,

Peter F. Surai **Vitagenes in avian biology and poultry health**
DOI 10.3920/978-90-8686-906-0_1, © Wageningen Academic Publishers 2020

2002, 2006, 2019; Surai and Borodai, 2010; Surai and Fisinin, 2012, 2012a, 2016a,b; Surai and Fotina, 2010; Surai *et al.*, 2019a; Table 1.1).

According to the recent literature review, heat and diet are among main means causing oxidative stress in domestic birds that may lead to serious health disorders, lower growth rates, and, hence, economic losses (Estevez, 2015). Therefore, dietary antioxidants are considered to be the main protective means to deal with various stresses in poultry production (Estevez, 2015; Fellenberg and Speisky, 2006; Mishra and Jha, 2019; Rehman *et al.*, 2018; Surai, 2002, 2006; Surai and Kochish, 2019; Surai *et al.*, 2019b)

1.3 Technological stresses

1.3.1 Chick placement

Chick viability is an important factor in determining profitability and, from fertilisation to placement at the broiler farm, factors such as egg quality, egg storage conditions, incubation conditions and post-hatch environment will all affect chick quality (Decuypere *et al.*, 2001). It has been proven that the first 24 hours of the chick's life are the most important (Fisinin and Surai, 2012, 2012a; Noy and Uni, 2010). It

Table 1.1. Stresses in poultry production (adapted from Surai *et al.*, 2019a).

- Technological stresses
  - Chick placement
  - Increased stocking density
  - Weighing, grading, group formation, catching, transferring to breeder houses
  - Prolonged egg storage, egg transportation, inadequate egg storage conditions, incorrect incubation regimes
- Environmental stresses
  - Inadequate temperature
  - Inadequate ventilation and increased dust
  - Inadequate lightning
- Nutritional stresses
  - Mycotoxins
  - Oxidised fat
  - Toxic metals (lead, cadmium, mercury, etc.)
  - Imbalance of minerals (Se, Zn, Cu, etc.) and other nutrients
  - Low water quality
  - Usage of coccidiostats and other drugs via feed or water
- Internal stresses
  - Vaccinations
  - Microbial or virus challenges
  - Gut dis-bacteriosis
  - Pipping and hatching

is believed that a chick should have access to the feed and water as soon as possible after hatching to stimulate the development of the digestive and immune systems. In fact, time between chick hatch and placement is stressful due to dehydration and yolk sac reserve depletion. Indeed, when putting together hatching time inside the hatcher, time of chick processing and transportation, and finally, placement at the farm, it could take up to 36-48 h before a newly hatched chick has access to feed and water and during this time body weight decreases quickly (Noy and Sklan, 1999). It has been shown that in the hatching chick the most dramatic changes in the small intestine occur within the first 24 h post-hatch (Geyra *et al.*, 2001). There is an inverse relationship between duration of post-hatching holding time and subsequent chick performance (Fisinin and Surai, 2012, 2012a; Hager and Beane, 1983; Pinchasov and Noy, 1993). Therefore, immediate access to feed and water help achieving an increased body weight of the growing chick at 3 weeks of age (Sklan *et al.*, 2000) or at market age of broilers (Vieira and Moran, 1999). It should be also mentioned that there is the hatch window (24-36 hours) or the spread between late and early hatchers which depends on the homogeneity/heterogeneity of the incubating eggs which is dependent on breeder age (Fisinin and Surai, 2012, 2012a). A spread in the hatching period will increase the numbers of chicks sitting extra hours in stressful conditions of the hatcher without food or water. Furthermore, any delay in accessing food (Bigot *et al.*, 2003; Noy *et al.*, 2001) and/or water intake after hatching as well as hatchery treatments such as vaccination, sexing and transport to the farm can result in additional stress (Geyra *et al.*, 2001a). Indeed, extended time in the hatcher (36 h) was associated with decreasing antioxidant defences indicative by decreased vitamin E and coenzyme Q concentrations in chicken tissues (Karadas *et al.*, 2011). Given the relatively high temperature and humidity in the hatcher, it is possible to make the argument that the chick may be under chronic oxidative stress during this holding time (Fisinin and Surai, 2012, 2012a; Surai and Fisinin, 2012, 2012a). Therefore, antioxidant protection at hatching time is considered to be an important determinant of chick viability during first post-hatch days (Surai, 2000, 2002; Surai *et al.*, 1998, 1999, 1999a, 2016). During chick embryo development there is an antioxidant/prooxidant (redox) balance in the tissues which supports normal embryonic development and post-hatch chick viability (Surai and Fisinin, 2015; Surai *et al.*, 1996). It has been suggested that an accumulation of the natural antioxidants like vitamins A, E and carotenoids as well as an increase in GPx activity in the embryonic liver may have an adaptive significance, evolving to protect unsaturated lipids against peroxidation during the stress imposed by hatching (Surai, 2002; Surai *et al.*, 2016).

Postnatal nutritional exposures are considered to be critical for the developmental maturation of many organ systems and optimal physiological functions. There is a growing body of evidence indicating that environmental exposures including nutritional exposures during these critical and sensitive periods of life can cause permanent changes in many physiological processes, which is known as 'programming' (Amarasekera *et al.*, 2013). Our previous investigations indicate that low quality neonatal nutrition resulted in long-term impairment in the capacity to assimilate dietary antioxidants (Blount *et al.*, 2003). It seems likely that early programming associated with epigenetic mechanisms plays a key role in chicken growth and development at time of chicken placement. Furthermore, scientific evidence is accumulating that the programming effects of conditions during early development can be transmitted to the offspring (Champagne and Rissman, 2011). Therefore, transgenerational effects of stress are potentially mediated via modulation of the hypothalamic-pituitary-adrenal axis as well as epigenetic mechanisms causing heritable changes in gene expression and it was suggested that early experiences may shape phenotypes of chickens in a long-term way (Goerlich *et al.*, 2012). In addition to the aforementioned stresses chicks are exposed to such stresses as hatching without maternal contact, transportation and social isolation. Indeed, the early life social isolation stress resulted in a dampened corticosterone response to restraint stress in affected birds and in their male offspring. Furthermore, stress-specific genes, such as early growth response 1 and corticotropin releasing hormone receptor were upregulated immediately after restraint stress (Goerlich *et al.*, 2012).

Research data are accumulating to support the hypothesis that the vitamin E status of chickens and turkey poults and probably chickens may be inadequate during the first weeks after hatching (Sell, 1996). A variety of approaches aimed at improving the vitamin E status of turkey poults have, in fact, been investigated including dietary supplementation of the poults with high levels of α-tocopherol (Applegate and Sell, 1996; Surai, 2002), bile salts (Marusich *et al.*, 1975) and fat (Soto-Salanova and Sell, 1995), as well as vitamin E injection (Soto-Salanova and Sell, 1996) and alterations in provision of n-6 and n-3 polyunsaturated fatty acids (Applegate and Sell, 1996). When vitamin E was added in the drinking water, there was an increase of α-tocopherol in tissues and a decreased susceptibility of red blood cells to haemolysis (Soto-Salanova, 1998). Moreover, day-old chickens were treated with 3.25 mg vitamin E/bird/day per os, via the drinking water, for two weeks. The vitamin E content of both the liver and the blood plasma was significantly higher in the treated chickens than in the untreated controls (Mezes, 1994). It seems likely, that provision of vitamin E and other fat-soluble vitamins (A, E and D3) with water at time of chicken placement can solve the problem of their low availability for newly hatched chicks (Surai and Borodai, 2010: Surai and Fisinin, 2012, 2012a; Surai and Fotina, 2010). Such a supplementation helps chickens overcome stress of placement and has positive effects on chicken growth and development.

When chicks are placed in winter while outside temperature is quite low there is always a temptation to decrease ventilation to keep energy usage to the minimum. However, it is very important to provide good quality, warm, fresh air that is rich in oxygen for the recently hatched chicks. Indeed, the chick's trachea is very often irritated from being boxed and shipped in the chick trays, often for many hours. Furthermore, chicks can be exposed to formaldehyde gas and contaminated air during hatch (Fisinin and Surai, 2012, 2012a; Fisinin *et al.*, 2009, 209a). Excessive amounts of irritants such as carbon dioxide and ammonia can cause depression, dehydration, emaciation as well as various problems with the respiratory system of the chick (Surai and Fisinin, 2012, 2012a; Surai and Fotina, 2010). The increased lipid peroxidation and reduced activities of antioxidant enzymes in healthy chickens reared under unfavourable microclimatic conditions such as higher air temperature and humidity, higher ammonia concentrations, and lower light intensity were indicative about an induced oxidative stress (Georgieva *et al.*, 2011). It should also be mentioned that poor ventilation is often associated with toxic carbon monoxide accumulation. Toxicity causes an irreversible physiological and biochemical changes that cannot be corrected with successive additional ventilation (Fisinin *et al.*, 2009, 2009a).

Therefore, to deal with oxidative stress at chicken placement there are several important options. They include: (1) electrolyte supplementation via drinking system to increase water consumption by chicks and keep optimal electrolyte balance in the body (Balnave and Gorman, 1993; Fisinin *et al.*, 2009; Surai and Fisinin, 2012); (2) fat-soluble vitamin supplementation via drinking water to overcome low efficacy of vitamin assimilation from the diet (Surai, 2002; Surai and Fisinin, 2012, 2012a); (3) organic acid supplementation to maintain gut health (Bourassa *et al.*, 2018); (4) other protective nutrients (ascorbic acid, Se, carnitine, betaine, lysine, methionine, etc.) supplementation with water to decrease oxidative stress related to chick placement and gut adaptation to a new type of feed (Fisinin *et al.*, 2009; Surai and Fisinin, 2012, 2012a). Improved antioxidant defences during first days of postnatal life are suggested to help immune system development in this critical period of time (Fisinin and Surai, 2013, 2013a).

1.3.2 Other technological stresses

Stocking density (SD) is a management factor which has critical implications for the poultry industry. Current recommended densities are rather variable and depend on breed, countries, and husbandry systems (Estvez, 2007). In fact, high stocking density has been reported to be a stressful condition (Puron *et al.*, 1995) affecting unfavourably the welfare and gut health of broiler chicks, predisposing them to various gut disorders including necrotic enteritis (Tsiouris *et al.*, 2015). Furthermore, high stocking density is associated with decreased locomotor activity and increased physiological (H:L ratio and bursa weight) and oxidative (GSH concentrations and GSH/GSSG ratios) stress indicators (Simitzis *et al.*, 2012) causing decrease performance, increase mortality and prevalence of leg weakness (Sørensen *et al.*, 2000) and had a negative effect on some aspects of bone quality (tibia curvature and shear strength; Buijs *et al.*, 2012). Furthermore, high SD is associated with decreased relative weights of lymphoid organs (spleen and bursa; Ravindran *et al.* 2006), reduced feed intake and weight gain with poor feed conversion ratio (Cengiz *et al.*, 2015), decreased breast muscle yield, tibial development, whereas increasing the scores of gait, footpad and hock burn, and abdominal plumage damage (Sun *et al.*, 2013) and decreased the final body weight (Tong *et al.*, 2012), and decreased carcass quality (Feddes *et al.*, 2002) of broiler chickens. In fact, under appropriate environments high SD was shown to reduce the growth performance of broilers associated with decreased growth of muscle and bone (Li *et al.*, 2019).

In addition to less efficient growth, birds in the more crowded pens had depressed immune response (Casteel *et al.*, 1994). Similarly, hens in cages with higher stocking density had lower hen-day egg production, egg mass, and feed intake compared with those in normal density cages (Mirfendereski and Jahanian, 2015). In native Chinese chickens high SD (8 hens/m^2) was shown to have an adverse effect on the performance and welfare status during 22 to 38 weeks of age (Geng *et al.*, 2020). High stocking density is physiologically stressful to broiler chickens, as indicated by serum corticosterone, ovotransferrin, ceruloplasmin and brain heat shock protein (HSP) 70 expression (Najafi *et al.*, 2015) or by increased serum MDA level and reduced GPx (Simsek *et al.*, 2009). Therefore, natural antioxidants, including vitamin E and capsaicin were shown to have beneficial effects on growth performance and lipid peroxidation in broilers reared under high-stocking-density condition (Thiamhirunsopit *et al.*, 2014). However, effects of stocking density on bird productive and reproductive indexes depend on many factors, including breed, sex, age and results are not always consistent. For example, Buijs *et al.* (2009) reported that stocking density did not affect bursa weight, mortality, or concentrations of corticosterone metabolites in droppings. Similarly, stoking density did not affect weights of the liver, spleen, bursa, and thymus, and there were no significant differences in the organ to BW ratios as density increased (Tong *et al.*, 2012) and feed conversion ratio was not affected by stocking density (Cravener *et al.*, 1992; Feddes *et al.*, 2002).

Chicken weighing, grading and group formation in rearing houses, as well as chicken catching are shown to be stressful conditions (Fisinin *et al.*, 2009, 2009a; Surai and Fisinin, 2012, 2012a) and the transferring chickens to breeder houses is always associated with increased stress and sometimes causing feather picking and cannibalism (Gunnarsson *et al.*, 1999). Therefore, antioxidant dietary supplementation could be considered as a technological measure to deal with oxidative stress in chickens caused by increased stocking density (Fisinin and Surai, 2012, 2012a).

1.4 Environmental stresses

Environmental stresses started from the moment when egg is laid, since temperature variation could cause embryo to start developing (high environmental temperature) or die (low temperature or fast temperature change; Fisinin *et al.*, 2009). It is well-known that temperature and other conditions of egg storage between egg laying and its placement into the hatchery negatively affect embryonic development. In fact, hatchability of fertile eggs declines with length of storage and there is an increase in percentages of early and late embryonic mortality with length of storage period (Elibol *et al.*, 2002; Fasenko, 2007) and most likely could affect chickens in later life. Furthermore, additional time in hatchery during hatching is also considered to be a stress causing detrimental changes in antioxidant defences of the chick (Karadas *et al.*, 2011).

1.4.1 Heat stress

Heat stress is one of the most common environmental stressors in poultry industry worldwide (Altan *et al.*, 2003). It seems likely that increased metabolic activity of modern poultry genotypes is responsible for the reduction in heat resistance

(Soleimani *et al.*, 2011). Therefore, today's chickens are shown to suffer from immune dysregulation and gut barrier dysfunction due to heat stress, leading to decreased productive and reproductive performance, immunosuppression and increased susceptibility to infectious diseases and mortality (Fisinin and Surai, 2013, 2013a; Lara and Rostagno, 2013; Quinteiro-Filho *et al.*, 2010). Indeed, the gastrointestinal tract is particularly sensitive to stressors (Surai and Fisinin, 2015), which can cause a variety of changes, including alteration of the normal, protective microbiota (Burkholder *et al.*, 2008) and decreased integrity of the intestinal epithelium (Quinteiro-Filho *et al.*, 2010) leading to gut leakage in broiler chickens (Ruff *et al.*, 2020). In fact, heat stress alters the jejunal glucose and lipid transport in chickens (Sun *et al.*, 2015). Furthermore, heat stress can inhibit the activity of digestive enzymes and reduce absorption and immune functions of intestinal mucosa (Chen *et al.*, 2014). Broilers subjected to the heat stress were characterised by reduced average daily gain and feed intake; lower viable counts of *Lactobacillus* and *Bifidobacterium* and increased viable counts of *coliforms* and *Clostridium* in small intestinal contents; shorter jejunal villus height, deeper crypt depth, and lower ratio of villus height to crypt depth (Song *et al.*, 2014). Indeed, intestinal integrity disruption is reported to be an important consequence of heat stress (Lian *et al.*, 2020). Heat stress was indicated to have immunosupressive effects and causing multiple immune abnormalities in broiler chickens by impairing the development and functional maturation of T and B cells in both primary and secondary lymphoid tissues (Hirakawa *et al.*, 2020). Furthermore, detrimental consequences of the heat stress on gut immunity (Fisinin and Surai, 2013; 2013a; Surai and Fotina, 2013) warrants further investigations. Heat stress was also shown to have detrimental effects on chicken meat quality (Awad *et al.*, 2020; Gonzalez-Rivas *et al.*, 2020; Zaboli *et al.*, 2019; Zhang *et al.*, 2020).

The negative effect of high temperature on hatching eggs could be very substantial during summer heat stress. Indeed, high environmental temperature is one of the most serious factors adversely affecting the laying performance in poultry. Egg production (De Andrade *et al.*, 1977; Mack *et al.*, 2013), egg weight (Ebeid *et al.*, 2012; Mack *et al.*, 2013; Mashaly *et al.*, 2004; Sahin *et al.*, 2007), eggshell thickness (De Andrade *et al.*, 1977; Ebeid *et al.*, 2012; Franco-Jimenez *et al.*, 2007; Lin *et al.*, 2004), eggshell percentage (Ebeid *et al.*, 2012), eggshell density (De Andrade *et al.*, 1977), eggshell breakage (Lin *et al.*, 2004) and egg freshness (Barrett *et al.*, 2019) were negatively affected by high ambient temperature. The calbindin concentration was prominently decreased in ileum, cecum, colon, and eggshell gland under heat stress conditions, which could be related to the deterioration of eggshell quality characteristics under heat stress conditions (Ebeid *et al.*, 2012). Elevated temperatures also increase mortality in both layers (Mashaly *et al.*, 2004) and broilers (Quinteiro-Filho *et al.*, 2010). In fact, at the molecular level, oxidative stress is considered to be a driving force of the negative consequences of the heat stress (Habashy *et al.*, 2019; Lambert *et al.*, 2002; Lin *et al.*, 2006; Surai and Fotina, 2013; Wen and Zhou, 2019). Therefore, natural antioxidants, including vitamin E (Liu *et al.*, 2009; Surai *et al.*, 2019b), selenium (Habibian *et al.*, 2015; Surai and Kochish, 2019), ascorbic acid (Pardue *et al.*, 1985) carnitine (Celik *et al.*, 2004), betaine (Ratriyanto and Mosenthin, 2018; Sayed and Downing, 2011; Zhao *et al.*, 2019), taurine (Lu *et al.*, 2019; Surai *et al.*, 2020), electrolytes (Ahmad and Sarvar, 2006), synbiotics (Jiang *et al.*, 2020), as well as polyphenols (Hu *et al.*, 2019) are shown to be protective in heat stressed birds (Nawab *et al.*, 2018; Saeed *et al.*, 2019). It is interesting to note that cold stress in chickens is also associated with oxidative stress and triggers a response via the Nrf2/ARE signalling pathway (Chen *et al.*, 2015). Furthermore, the differentially expressed genes due to cold stress in partridge hypothalamus included 334 down-regulated genes and 543 up-regulated genes (Chen *et al.*, 2014a). Protective and regulatory roles of heat shock proteins in poultry exposed to heat stress has been described in detail in our recent reviews (Surai, 2015e; Surai and Kochish, 2017).

1.4.2 Other environmental stresses

Chronical exposure to high levels of dust and ammonia within a broiler rearing house was shown to cause oxidative stress (Bottje *et al.*, 1998; Bottje and Wideman, 1995). This could be the case during periods of cool weather, when poultry producers often decrease ventilation to reduce heating, thereby allowing dust and gaseous pollutants to accumulate in the air. Changing lightning programs and light sources could also be a stress for poultry (Fisinin *et al.* 2009, 2009a; Huth and Archer, 2015;). For example, in the growing chickens, dimming was shown to be lower environmental stress than the abrupt light-dark transition (Van der Pol *et al.*, 2015).

1.5 Nutritional stresses

1.5.1 Mycotoxins

Silent killers', 'invisible thieves', 'unavoidable contaminants', and 'natural toxicants' – all these names have been given to the fungal secondary metabolites, mycotoxins. In general mycotoxins are considered to be unavoidable contaminants in foods and feeds and are a major problem all over the world. Mycotoxins are considered to be among major feed-related stressors in poultry production (Awad *et al.*, 2013; Heussnel and Bingle, 2015; Schwartz-Zimmermann *et al.*, 2015; Surai and Dvorska, 2005). In fact, aflatoxins (AF), zearalenone, ochratoxin A (OTA), fumonisins, trichothecenes such as deoxynivalenol (DON) and T-2 toxin, are considered to be the most common mycotoxins that can significantly impact the health and performance of poultry species (Murugesan *et al.*, 2015). Indeed, the aforementioned mycotoxins can severely affect the immune system and gut health leading to high economic losses to poultry producers (Cimbalo *et al.*, 2020; Yang *et al.*, 2020a).

Immunosuppressive effects of mycotoxins reviewed previously (Surai and Dvorska, 2005; Surai and Mezes, 2005) are associated with their prooxidant and pro-apoptotic actions (Surai *et al.*, 2008, 2010) with negative effects on transcription factors, including Nrf2 (Katika *et al.*, 2015; Limonciel *et al.*, 2014) and NF-κB (Kumar *et al.*, 2013; Ramyaa *et al.*, 2014; Figure 1.1).

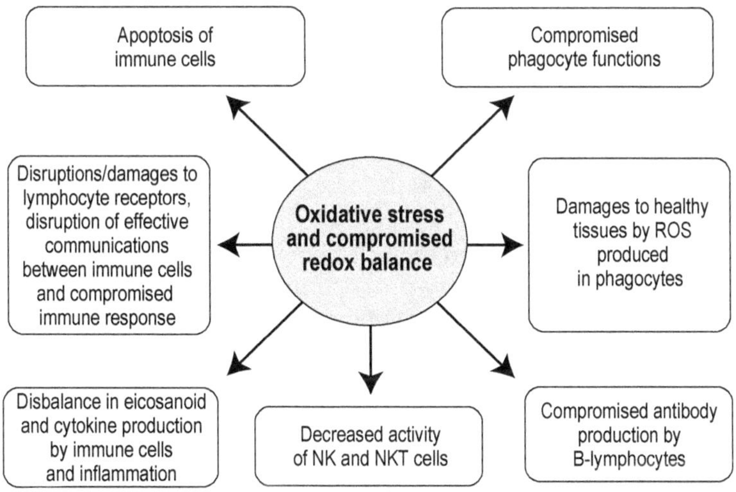

Figure 1.1. Oxidative stress caused by mycotoxins and immunity (Surai, 2006; Surai and Mezes, 2005).

The antioxidant defence systems are under regulation by various transcription factors. In recent years great attention has been paid to a basic leucine zipper transcription factor, Nuclear factor-erythroid-2 (NF-E2-) related factor 2 (Nrf2) and nuclear factor-kappa B (NF-κB). Indeed, Nrf2 has a significant role in adaptive responses to oxidative stress being involved in the induction of the expression of various antioxidant molecules to combat oxidative and electrophilic stress. This includes enzymes of the first line of the antioxidant defence, namely SOD, GPx and Catalase, detoxification enzymes (HO-1, NQO1, and GST), GSH-related proteins (γ-GCS), NADPH-producing enzymes and others stress-response proteins contributing to preventing oxidative and inflammatory damages. Furthermore, NF-κB is an inducible transcription factor that regulates many cellular processes including immunity, inflammation, apoptosis, cell proliferation and differentiation. In many cases, NF-κB activation is associated with synthesis of pro-inflammatory cytokines (for review see Surai, 2015c,d; Surai *et al.*, 2019a). Therefore, inhibition of Nrf2 and activation of NF-κB by mycotoxins are considered to be fundamental mechanisms of their toxic effects (for more information on transcription factors see Chapter 2).

It was shown that AFB1 in the broilers diet could reduce the percentages of T-cell subsets and the expression level of cytokine mRNA in the small intestine affecting the immune function of the intestinal mucosa (Jiang *et al.*, 2015). Furthermore, DON is shown to suppress the antibody response to infectious bronchitis vaccine and to Newcastle disease virus in broilers and laying hens and to decrease tumour necrosis factor alpha (TNF-α) in the plasma of broilers (Awad *et al.*, 2013). Furthermore, the feeding of OTA+T-2 toxin diets decreased the relative weight of spleen, thymus, and bursa of Fabricius and serum concentrations of total protein, albumin, and globulin, elevated the activities of serum γ-glutamyl transferase, aspartate aminotransferase, and alanine aminotransferase and impaired chick immune function (Wang *et al.*, 2009). Exposure to Fusarium mycotoxins generally exacerbates infections with parasites, bacteria and viruses including coccidiosis, necrotic enteritis and aspergillosis in poultry (Antonissen *et al.*, 2014). In particular, it has been shown that DON can compromise several crucial intestinal functions leading to increases in the susceptibility to enteric infectious diseases, being a predisposing factor for other general diseases (Ghareeb *et al.*, 2015).

It seems likely that gut health is severely affected by the mycotoxin toxicity leading to detrimental consequences in poultry production. It is interesting to note that it was hypothesised that the intestinal mucosa of birds is subject to a higher oxidative stress than is the intestines in mammals (Maurice *et al.*, 1991) and antioxidant-pro-oxidant (redox) balance in the chicken gut is considered as an important determinant of bird's health (Surai and Fisinin, 2015). In fact, the oxidative stress caused by DON toxicity in the intestinal cells could lead to DNA and cell membrane alterations and consequently induces apoptosis, atrophy, and massive death of the intestinal cells (Ghareeb *et al.*, 2015). Similarly, T-2 toxin causes oxidative stress and disturbance in energy metabolism and gut microbiome with following impaired spleen function, inhibited protein and DNA biosynthesis and immunotoxicity (Wan *et al.*, 2015). Furthermore, T-2 toxin modifies feeding behaviour by interfering with central neuronal networks devoted to central energy balance (Gaige *et al.*, 2014). Direct toxic effects of fumonisin B1 on intestinal structure, including villus architecture and enzyme activities are reported by Lessard *et al.* (2009) and increase in the trans-cellular and para-cellular permeability of pig small intestine due to fumonisin B1 were also reported (Lalles *et al.*, 2009). Similarly, zearalenone and its metabolites affected porcine intestinal cell viability, transepithelial resistance and cytokine synthesis with important implication for gut health (Liu *et al.*, 2014; Marin *et al*, 2015; Taranu *et al.*, 2014). It has been shown that aflatoxins can have a direct or indirect effect, or both, on functionality of the gastrointestinal tract in laying hens (Applegate *et al.*, 2009) and it seems likely that detrimental effects of aflatoxins on the gut could be mediated via increased apoptosis. Indeed, in AFB1 treated broilers a significant increase in the number of apoptotic cells and in the expression of Bax (an apoptosis promoter) and Caspase-3 mRNA was observed, while the expression of Bcl-2 (an apoptosis inhibitor) and the Bcl-2/Bax ratio were significantly decreased (Peng *et al.*, 2014). Indeed, the cellular Bcl-2/Bax ratio is a key regulator of apoptosis; a high Bcl-2/Bax ratio makes cells resistant to apoptotic stimuli, while a low ratio induces cell death. It is interesting to note that OTA can also cause compositional and functional changes of gut microbiota. In particular, OTA treatment decreased the within-subject diversity of the gut microbiota associated with changes in functional genes of gut microbiota including signal transduction, carbohydrate transport, amino acid transport system, and mismatch repair (Guo *et al.*, 2014). It should be mentioned that AFB_1 could also modify the gut microbiota in a dose-dependent manner (Wang *et al.*, 2016). Therefore, mycotoxins are able to compromise several key functions of the gastrointestinal tract, including decreased surface area available for nutrient absorption, modulation of nutrient transporters, or loss of barrier function with intestinal inflammation (Grenier and Applegate, 2013) and consequences of microbiota changes in the gut due to mycotoxins await further investigation.

It has been shown that, OTA, T-2 toxin and DON impose an oxidative stress and have a stimulating effect on lipid peroxidation (Fisinin and Surai, 2012a,b,c,d,e,f; Surai, 2002, 2006; Surai *et al.*, 2008). Similarly, aflatoxins (Ma *et al.*, 2015), fumonisins (Poersch *et al.*, 2014) and zearalenone (Lautert *et al.*, 2014) cause oxidative stress in poultry. In most cases, thiobarbituric acid reactive substances (TBARS) accumulation was used as a measurement of lipid peroxidation. Furthermore, ethane exhalation, EPR registered free radicals, hydroxyl radical formation, single-strand cleavage DNA, DNA adduct formation as well as LDH release were also used to confirm pro-oxidant properties of mycotoxins. Various *in vitro* and *in vivo* systems were also used including liver microsomes, phospholipid vesicles, primary cell cultures, whole organs and whole body. TBARS accumulation was substantially increased and at the same time vitamin E and GSH concentrations and activities of antioxidant enzymes significantly declined as a result of mycotoxicosis (Surai, 2006; Surai *et al.*, 2008, 2010).

A variety of physical, chemical, and biological methods have been developed for decontamination and/or detoxification of mycotoxins from contaminated foods and feeds. The most applied method for protecting poultry/animals against mycotoxicosis is the utilisation of adsorbents mixed with the feed, which are supposed to bind the mycotoxins in the gastro-intestinal tract. The efficiency of mycotoxin binders, however, differs considerably depending mainly on the chemical structure of both the adsorbent and the toxin (Huwig *et al.*, 2001). In fact, our recent analysis indicates that mycotoxin binders are not able to solve the problem of mycotoxins in poultry production (Fisinin and Surai, 2012a,b,c,d,e,f). Indeed, it is impossible to bind 100% mycotoxins during short period of time when feed is moving in the intestine. In most of cases, only about 30-50% mycotoxins are bound. In many cases, *in vitro* data on the efficacy of mycotoxin binders are not reflecting the situation in the gut and such a mycotoxin as T-2 toxin is very poorly absorbed by various adsorbents. There is a chance that unspecific binding ability of adsorbents could be associated with binding some nutrients, including vitamins and minerals making them unavailable for nutritional purposes. In our opinion, too much attention has been paid in recent years to mycotoxin binders and alternative ways of dealing with mycotoxin toxicity need to be also considered (Surai and Fisinin, 2012a,b). To conclude, mycotoxins impose oxidative stress, stimulate apoptosis and involved in gene expression regulation. In particular, these changes are responsible for immunosuppressive action of mycotoxins. Indeed, damages to receptors on the surface of macrophages, neutrophils and lymphocytes could cause miscommunication between the cells leading to immunosuppression (Surai and Mezes, 2005). Since oxidative stress and lipid peroxidation are important determinants in mycotoxin toxicity (Dai *et al.*, 2019; Surai and Dvorska, 2005), a protective effect of antioxidants is expected (Galvano *et al.*, 2001). Indeed, in several experiments with various animal species, including poultry, protective effects of antioxidants (vitamins E and C, selenium, ascorbic acid, carnitine, etc.) against the toxic effects of mycotoxins were observed (Ren *et al.*, 2019; Surai and Dvorska, 2005; Surai and Mezes, 2005; Surai *et al.*, 2019b). Therefore, there is a need to develop a nutritional support strategy for the liver and gut, main sites of mycotoxin detoxification in animal/chicken body.

1.5.2 Oxidised fat

A diet prone to oxidation is likely to cause oxidative stress and potentially induce an inflammatory response. Oxidised fatty acids are absorbed from the intestine mainly in the form of unsaturated compounds and initiate lipid peroxidation in the tissues (Mezes *et al.*, 1997). Feeding oxidised fats has been clearly shown to have detrimental effects on poultry, including increased incidence of encephalomalacia (L'estrange *et al.*, 1966), increased peroxidation of cell membranes (Asghar *et al.*, 1989; Lin *et al.*, 1989), increased plasma TBARS (Sheehy *et al.*, 1993, 1994), decreased plasma vitamin E (Tavarez *et al.*, 2011) and decreased growth and/or feed efficiency (Cabel and Waldroup, 1989; Nakamura *et al.*, 1972; Tavarez *et al.*, 2011). The intake of oxidised oil caused a growth depression after 2 weeks and the retention of fat, energy and alpha-tocopherol was lower in the group fed oxidised fat. Furthermore, these animals showed significantly higher plasma concentrations of TBARS, and lower concentrations of tocopherols, lutein, beta-carotene, and retinol in plasma and tissues (Enberg *et al.*, 1996). Indeed, chickens consumed oxidised fat exhibited lower feed efficiency in the starter period and decreased gains during the starter and grower periods in comparison to birds fed the control diet (Wang *et al.*, 1997).

It is interesting that oxidised oils also decreased feed efficiency in laying hens (Yue *et al.*, 2011) and the authors suggested that oxidised oil might affect the performance of laying hens through the regulation of apolipoproteins and oestradiol. It has been shown that dietary oxidised oils suppressed gene expression of lipogenic enzymes in rats (Eder *et al.* 2003). Oxidised oil in the chicken diet may increase the susceptibility of the gastrointestinal tract and other tissues to lipid and protein oxidation (Sheehy *et al.*, 1994; Zhang *et al.*, 2011). Oxidation of the dietary oil lowered lipid stability significantly in both raw and precooked chicken meats during chill storage (Galvin *et al.*, 1997; Jensen *et al.*, 1997). Furthermore, chickens fed on high oxidised (HO) diet were characterised by decreased concentrations of PUFAs (18:2, 18:3, 20:4, 20:5, 22:5, and 22:6) on day 42, resulting from increased PUFA oxidability in stress conditions of the oxidised diet (Lu *et al.*, 2014) and natural antioxidants in the diet were protective. In fact, HO diet caused hepatocellular necrosis by oxidative stress (Lu *et al.*, 2014). Diets with high-oxidised oil reduced stearic, linoleic and linolenic acid content in chicken breast muscles compared to low-oxidised oil samples (Delles *et al.*, 2015). Indeed, feeding diets with high oxidised oil not only increased the vulnerability of lipids and proteins to oxidation, but also reduced the activities of tissue antioxidant defence enzymes (SOD, GPx and Catalase; Delles *et al.*, 2014). It was shown that the presence of an antioxidant in the feed protects lipids from further oxidising, therefore increasing broiler performance and improving shelf life when using oxidised oil (Tavarez *et al.*, 2011). Since stress caused by feeding oxidised oils to chickens would depend on many factors related to experimental design, including diet composition, level of oil oxidation, bird age and sex, presence of dietary antioxidants, etc., the outcome of such experiments is not always consistent. For example, no significant treatment differences were observed among oxidised oil supplemented birds for BW gain, feed intake, feed efficiency, or abdominal fat pad weight (Billek, 2000; Bou *et al.*,

2006; Ehr *et al.*, 2015; Lopez-Ferrer *et al.*, 1999). Furthermore, hatching egg weight and egg production were not affected by dietary oxidised oil (Leeson *et al.*, 2008).

1.5.3 Dietary toxicants

Physicochemical environment of the gastrointestinal tract depends on many factors with diet, bacterial metabolites and body secretion being major determinants (Sanderson, 1999). There is a delicate balance between the environment of the lumen and epithelial cell functionality and dietary factors are responsible for gene expression in the intestine and its adaptation. In this regard, oxidative stress could cause changes in this balance affecting absorption of nutrients (Surai and Fisinin, 2015). Even if each of those lipid peroxidation promoters (oxidised polyunsaturated fatty acids, nitrites, nitrates, heavy metals, mycotoxins, etc.) are present at a very low concentration, their combination could be much more powerful. For example, heavy metal (lead, cadmium, mercury) concentrations in major feed sources are quite low; however, in combination with other prooxidants they potentially can be involved in generation of free radicals and cause oxidative stress in the gut (Pappas *et al.*, 2010; Surai and Fisinin, 2015).

1.5.4 Nutrient imbalances

Selenium is an essential trace element involved in regulation of many different metabolic pathways including antioxidant defences. There are at least 25 selenoproteins in human and animal/poultry bodies with tissue-specific Se-dependent expression (Pappas *et al.*, 2008; Surai, 2018). Therefore, both Se deficiency and excess could cause oxidative stress in poultry (Surai, 2006; Surai and Kochish, 2019). Recent data clarified this statement. For example, in the low-Se group chicken oxidative stress occurred in the liver and gut (Yao *et al.*, 2015). Se deficiency in chickens decreased the muscle expressions of 19 selenoproteins, 11 of which were antioxidative selenoproteins (Yao *et al.*, 2014) and influenced the expressions of 24 selenoproteins and 10 cytokines in chicken erythrocytes (Luan *et al.*, 2016). Furthermore, the mRNA levels of 19 selenoprotein genes in the layer chicken liver were decreased by dietary Se deficiency (Liu *et al.*, 2014a). It seems likely that Se excess also causes an oxidative stress in chickens (Mezes *et al.*, 1997; Surai, 2006; Xu *et al.*, 2014).

Zinc is required for the activity of over 300 enzymes and participates in many enzymatic and metabolic functions in the body, including antioxidant defences. Therefore, Zinc deficiency causes oxidative stress and loss of appetite and reduced efficiency of feed utilisation with growth retardation, bone deformities and skeletal abnormalities, decreased egg production and hatchability and increased mortality (Sahin *et al.*, 2009). It is interesting to note that chicks are quite resistant to Zn excess, but high concentrations of supplemental Cu depressed chick weight gain (Persia *et al.*, 1994), affected feed efficiency associated with oxidative stress as evidenced by MDA accumulation and SOD decrease (Cinar *et al.*, 2014) and NF-κB activation leading to inflammation (Yang *et al.*, 2020b). The authors also showed protective effects of antioxidants (vitamins E and C) in such stress conditions. In general, imbalance of most nutrients can cause oxidative stress. For example, ascorbic acid, a well-known antioxidant, in high doses in the chicken diet can cause oxidative stress (Berzina *et al.*, 2013).

1.6 Internal/biological stresses

The biggest stress for commercial layers/breeders is coming at the peak of egg production. Indeed, major compounds of the egg yolk are synthesised in the liver and it is working to its maximal ability and any stress can cause a drop in egg production which very often is not coming up after the stress is removed. Finally, eggshell quality during a second part of egg laying is considered to be a problem, especially when layer age past 80 weeks (Safaa *et al.*, 2008). Indeed, most losses are related to the poor shell quality of eggs produced at the end of the production cycle. For example, Grobas *et al.* (1999) found that the percentage of broken eggs from Brown egg-laying hens on the farm increased from 0.43% at 22 weeks to 1.81% at 74 weeks of age.

Among various stress factors/conditions vaccinations have a special place. Indeed, vaccinations are absolutely necessary to maintain chicken protection against various diseases, but vaccination itself imposes stress (Janmohammadi *et al.*, 2020; Li *et al.*, 2020) and activates the immune system with negative consequences for productive parameters, since immunity is quite expensive for the body in terms of usage of nutrients and energy (Fisinin and Surai, 2013a,b; Surai, 2006). It is generally assumed by immunologists that providing immunological defences to minimise such risks to the host is costly in terms of necessitating trade-offs with other nutrient-demanding processes such as growth, reproduction, and thermoregulation (Lochmiller and Deerenberg, 2000). It has been shown that lipopolysaccharide injection decreased feed intake and body weight gain (Lai *et al.*, 2011) and reduced ileal protein digestibility (Yang *et al.*, 2011). It is well appreciated that efficacy of vaccination is very much dependent on the immunocompetence of the birds, which could be compromised in stress conditions (Surai, 2002, 2006, 2018). It should be mentioned that environmental stresses (temperature, light, air quality, infective agents, and environmental contaminants; Dietert *et al.*, 1994), nutritional stressors (mycotoxins, nutrient deficiencies; Klassing *et al.*, 1998; Surai, 2002; Surai and Dvorska, 2005) and immunosuppressive diseases (bursal disease, infectious chicken anaemia and Marek's disease; Fussell, 1998; Hoerr, 2010) dramatically affect immune responses of poultry. In general, a relationship between stress and immunity is quite complex (Dhadhar, 2014) but detail mechanisms of it is beyond the scope of this chapter (for more details see Chapter 16). Indeed, the immune system is considered to be the most sensitive to various stresses (Dohms and Metz, 2001; Hoerr *et al.*, 2010; Lauridsen, 2019; Surai, 2006, 2018; Surai and Dvorska, 2005). In fact, vaccination was shown to induce immune stress in layer pullets (Song *et al.*, 2020) and stress-related dysfunction of the immune system weakens natural resistance to diseases (Antonissen *et al.*, 2014) and reduces efficacy of vaccinations (Ingrao *et al.*, 2013) leading to significant losses in profits. It should also be mentioned that microbial and virus challenges are considered to be main internal stresses causing detrimental consequences for productive and reproductive parameters of birds (Fisinin *et al.*, 2013, 2013a).

1.7 Conclusions

In modern commercial poultry production oxidative stress-related metabolic diseases (e.g. encephalomalacia, exudative diathesis, muscular dystrophy, etc.) practically disappeared however, various disorders of the biological antioxidant defence system still causing substantial problems. In general, there are four major types of stress in poultry industry: technological, environmental, nutritional and internal stresses. In particular, our analysis of stresses indicates that they cause detrimental consequences for chicken growth and development, decreasing productive and reproductive performance of breeders and commercial layers. In fact, a list of commercially relevant stresses in poultry production could be quite long, but the main point is most stresses suppress reproductive performance of parent birds including reduced fertility and hatchability. Furthermore, stresses are associated with impaired feed conversion, reduced average daily weight gain, immunosuppression and increased mortality in growing birds. It seems likely that oxidative stress is a driving force of the detrimental consequences of the major aforementioned stresses (Surai, 2020; Surai and Fisinin, 2016a,b; Surai and Kochish, 2019; Surai *et al.*, 2019a,b). There is a need for the development of an effective strategy to deal with stresses in poultry production and vitagene concept could be an important element of such a strategy (See next chapters). It seems likely that the vitagene concept could help understanding molecular mechanisms responsible for cell/organism adaptation to stresses and developing effective means of decreasing negative consequences of stresses. and vitagene regulation by nutritional means (Fisinin and Surai, 2011, 2011a; Surai, 2015a,b,c,d, 2016, 2020; Surai and Fisinin, 2012, 2012a, 2016c,d,e; Surai and Kochish, 2017; Surai *et al.*, 2017, 2019a) appeared as a new approach to realise a full potential of the body for adaptation to stress conditions in poultry/animal production. Details of this approach will be considered in next chapters of the book.

References

Ahmad, T. and Sarwar, M., 2006. Dietary electrolyte balance: implications in heat stressed broilers. World's Poultry Science Journal 62: 638-653.

Altan, O., Pabuçcuoğlu, A., Altan, A., Konyalioğlu, S. and Bayraktar H., 2003. Effect of heat stress on oxidative stress, lipid peroxidation and some stress parameters in broilers. British Poultry Science 44: 545-550.

Amarasekera, M., Prescott, S.L. and Palmer, D.J., 2013. Nutrition in early life, immune-programming and allergies: the role of epigenetics. Asian Pacific Journal of Allergy and Immunology 31:175-182.

Antonissen, G., Martel, A., Pasmans, F., Ducatelle, R., Verbrugghe, E., Vandenbroucke, V., Li, S., Haesebrouck, F., Van Immerseel, F. and Croubels, S., 2014. The impact of Fusarium mycotoxins on human and animal host susceptibility to infectious diseases. Toxins 6: 430-452.

Applegate, T.J., Schatzmayr, G., Prickel, K., Troche, C. and Jiang, Z., 2009. Effect of aflatoxin culture on intestinal function and nutrient loss in laying hens. Poultry Science 88: 1235-1241.

Applegate, T.J. and Sell, J.L., 1996. Effect of dietary linoleic to linolenic acid ratio and vitamin E supplementation on vitamin E status of poults. Poultry Science 75: 881-890.

Asghar, A., Lin, C.F., Gray, J.I., Buckley, D.J., Booren, A.M., Crackel, R.L. and Flegal, C.J., 1989. Influence of oxidised dietary oil and antioxidant supplementation on membrane-bound lipid stability in broiler meat. British Poultry Science 30: 815-823.

Awad, W., Ghareeb, K., Böhm, J. and Zentek, J., 2013. The toxicological impacts of the Fusarium mycotoxin, deoxynivalenol, in poultry flocks with special reference to immunotoxicity. Toxins 5: 912-925.

Awad, E.A., Najaa, M., Zulaikha, Z.A., Zulkifli, I. and Soleimani, A.F., 2020. Effects of heat stress on growth performance, selected physiological and immunological parameters, caecal microflora, and meat quality in two broiler strains. Asian-Australasian Journal of Animal Sciences 33: 778-787.

Barrett, N.W., Rowland, K., Schmidt, C.J., Lamont, S.J., Rothschild, M.F., Ashwell, C.M. and Persia, M.E., 2019. Effects of acute and chronic heat stress on the performance, egg quality, body temperature, and blood gas parameters of laying hens. Poultry Science 98: 6684-6692.

Berzina, N., Markovs, J., Dizhbite, T., Apsite, M., Vasilyeva, S., Basova, N., Smirnova, G. and Isajevs, S., 2013. Oxidative stress and innate immunity status in chickens exposed to high dose of ascorbic acid. Cell Biochemistry and Function 31: 551-559.

Bigot, K., Mignon-Grasteau, S., Picard, M. and Tesseraud, S., 2003. Effects of delayed feed intake on body, intestine, and muscle development in neonate broilers. Poultry Science 82: 781-788.

Billek, G., 2000. Health aspects of thermoxidized oils and fats. European Journal of Lipid Science and Technology 120: 587-593.

Blount, J.D., Metcalfe. N.B., Arnold, K.E., Surai, P.F., Devevey, G.L. and Monaghan, P., 2003. Neonatal nutrition, adult antioxidant defences and sexual attractiveness in the zebra finch. Proceedings of the Royal Society of London. Series B, Biological sciences 270: 1691-1696.

Bottje, W.G. and Wideman Jr., R.F., 1995. Potential role of free radicals in the pathogenesis of pulmonary hypertension syndrome. Poultry and Avian Biology Reviews 6: 211-231.

Bottje, W.G., Wang, S., Kelly, F.J., Dunster, C., Williams, A. and Mudway, I., 1998. Antioxidant defenses in lung lining fluid of broilers: impact of poor ventilation conditions. Poultry Science 77: 516-522.

Bou, R., Grimpa, S., Guardiola, F., Barroeta, A.C. and Codony, R., 2006. Effects of various fat sources, alpha-tocopheryl acetate, and ascorbic acid supplements on fatty acid composition and alpha-tocopherol content in raw and vacuum-packed, cooked dark chicken meat. Poultry Science 85: 1472-1481.

Bourassa, D.V., Wilson, K.M., Ritz, C.R., Kiepper, B.K. and Buhr, R.J., 2018. Evaluation of the addition of organic acids in the feed and/or water for broilers and the subsequent recovery of Salmonella Typhimurium from litter and ceca. Poultry Science 97: 64-73.

Buijs, S., Keeling, L., Rettenbacher, S., Van Poucke, E. and Tuyttens, F.A., 2009. Stocking density effects on broiler welfare: identifying sensitive ranges for different indicators. Poultry Science 88:1536-1543.

Buijs, S., Van Poucke, E., Van Dongen, S., Lens, L., Baert, J. and Tuyttens, F.A., 2012. The influence of stocking density on broiler chicken bone quality and fluctuating asymmetry. Poultry Science 91: 1759-1767.

Bureau, C., Hennequet-Antier, C., Couty, M. and Guémené, D., 2009. Gene array analysis of adrenal glands in broiler chickens following ACTH treatment. BMC Genomics 10: 430.

Burkholder, K.M., Thompson, K.L., Einstein, M.E., Applegate, T.J. and Patterson, J.A., 2008. Influence of stressors on normal intestinal microbiota, intestinal morphology, and susceptibility to Salmonella Enteritidis colonization in broilers. Poultry Science 87: 1734-1741.

Cabel, M.C. and Waldroup, P.W., 1989. Ethoxyquin and ethylenediaminetetraacetic acid for the prevention of rancidity in rice bran stored at elevated temperature and humidity for various lengths of time. Poultry Science 68: 438-442.

Casteel, E.T., Wilson, J.L., Buhr, R.J. and Sander, J.E., 1994. The influence of extended posthatch holding time and placement density on broiler performance. Poultry Science 73:1679-1684.

Celik, L.B., Tekeli, A. and Oztürkcan, O., 2004. Effects of supplemental L-carnitine in drinking water on performance and egg quality of laying hens exposed to a high ambient temperature. Journal of Animal Physiology and Animal Nutrition 88: 229-233.

Cengiz, Ö., Köksal, B.H., Tatlı, O., Sevim, Ö., Ahsan, U., Üner, A.G., Ulutaş, P.A., Beyaz, D.,Büyükyörük, S., Yakan, A. and Önol, A.G., 2015. Effect of dietary probiotic and high stocking density on the performance, carcass yield, gut microflora, and stress indicators of broilers. Poultry Science 94: 2395-2403.

Champagne, F.A. and Rissman, E.F., 2011. Behavioral epigenetics: a new frontier in the study of hormones and behavior. Hormones and Behavior 59: 277-278.

Chen, X.Y., Li, R. and Geng, Z.Y., 2015. Cold stress initiates the Nrf2/UGT1A1/L-FABP signaling pathway in chickens. Poultry Science 94: 2597-2603.

Chen, X.Y., Li, R., Wang, M. and Geng, Z.Y., 2014a. Identification of differentially expressed genes in hypothalamus of chicken during cold stress. Molecular Biology Reports 41: 2243-2248.

Chen, Z., Xie, J., Wang, B. and Tang, J., 2014. Effect of γ-aminobutyric acid on digestive enzymes, absorption function, and immune function of intestinal mucosa in heat-stressed chicken. Poultry Science 93: 2490-2500.

Cimbalo, A., Alonso-Garrido, M., Font, G. and Manyes, L., 2020. Toxicity of mycotoxins *in vivo* on vertebrate organisms: A review. Food and Chemical Toxicology 137: 111161.

Cinar, M., Yildirim, E., Yigit, A.A., Yalcinkaya, I., Duru, O., Kisa, U. and Atmaca, N., 2014. Effects of dietary supplementation with vitamin C and vitamin E and their combination on growth performance, some biochemical parameters, and oxidative stress induced by copper toxicity in broilers. Biological Trace Element Research 158: 186-196.

Cravener, T.L., Roush, W.B. and Mashaly, M.M., 1992. Broiler production under varying population densities. Poultry Science 71: 427-433.

Dai, C., Xiao, X., Sun, F., Zhang, Y., Hoyer, D., Shen, J., Tang, S. and Velkov, T., 2019. T-2 toxin neurotoxicity: role of oxidative stress and mitochondrial dysfunction. Archives of Toxicology 93: 3041-3056.

De Andrade, A., Rogler, N., Featherson, J.C. and Allison, W.R.C.W., 1977. Interrelationships between diet and elevated temperature (cyclic and constant) on egg production and shell quality. Poultry Science 56: 1178-1188.

Decuypere, E., Tona, K., Bruggeman, V. and Bamelis, F., 2001. The day-old chick: a crucial hinge between breeders and broilers. World's Poultry Science Journal 57: 127-138.

Delles, R.M., Xiong, Y.L., True, A.D., Ao, T. and Dawson, K.A., 2014. Dietary antioxidant supplementation enhances lipid and protein oxidative stability of chicken broiler meat through promotion of antioxidant enzyme activity. Poultry Science 93: 1561-1570.

Delles, R.M., Xiong, Y.L., True, A.D., Ao, T. and Dawson, K.A., 2015. Augmentation of water-holding and textural properties of breast meat from oxidatively stressed broilers by dietary antioxidant regimens. British Poultry Science 56: 304-314.

Dhabhar, F.S., 2014. Effects of stress on immune function: the good, the bad, and the beautiful. Immunologic Research 58: 193-210.

Dietert, R.R., Golemboski, K.A. and Austic, R.E., 1994. Environment-immune interactions. Poultry Science 73: 1062-1076.

Dohms, J.E. and Metz, A., 1991. Stress-mechanisms of immunosuppression. Veterinary Immunology and Immunopathology 30: 89-109.

Ebeid, T.A., Suzuki, T. and Sugiyama, T., 2012. High ambient and quality temperature influences eggshell quality and calbindin-D28k localization of eggshell gland and all intestinal segments of laying hens. Poultry Science 91: 2282-2287.

Eder, K., Suelzle, A., Skufca, P., Brandsch, C. and Hirche, F., 2003. Effects of dietary thermoxidized fats on expression and activities of hepatic lipogenic enzymes in rats. Lipids 38: 31-38.

Ehr, I.J., Kerr, B.J. and Persia, M.E., 2015. Effects of peroxidized corn oil on performance, AMEn, and abdominal fat pad weight in broiler chicks. Poultry Science 94: 1629-1634.

Elibol, O., Peak, S.D. and Brake, J., 2002. Effect of flock age, length of egg storage, and frequency of turning during storage on hatchability. Poultry Science 81: 945-950.

Engberg, R.M., Lauridsen, C., Jensen, S.K. and Jakobsen, K., 1996. Inclusion of oxidized vegetable oil in broiler diets. Its influence on nutrient balance and on the antioxidative status of broilers. Poultry Science 75: 1003-1011.

Estevez, I., 2007. Density allowances for broilers: where to set the limits? Poultry Science 86: 1265-1272.

Estévez, M., 2015. Oxidative damage to poultry: from farm to fork. Poultry Science 94: 1368-1378.

Fasenko, G.M., 2007. Egg storage and the embryo. Poultry Science 86: 1020-1024.

Feddes, J.J.R., Emmanuel, E.J. and Zuidhof, M.J., 2002. Broiler performance, body weight variance, feed and water intake, carcass quality at different stocking densities. Poultry Science 81: 774-779.

Fellenberg, M.A. and Speisky, H., 2006. Antioxidants: Their effects on broiler oxidative stress and its meat oxidative stability. World's Poultry Science Journal 62: 53-70.

Fisinin, V.I. and Surai, P.F., 2011. Effective protection from stresses in poultry production: from vitamins to vitagenes. Part 1. Poultry and Poultry Products (Ptitza I Ptitzeproducti, Moscow) 5: 23-26.

Fisinin, V.I. and Surai, P.F., 2011a. Effective protection from stresses in poultry production: from vitamins to vitagenes. Part 2. Poultry and Poultry Products (Ptitza I Ptitzeproducti, Moscow) 6: 10-13.

Fisinin, V.I. and Surai, P.F., 2012. First days of chicken life: from a protection against stresses to an effective adaptation. Russian Poultry Science (Ptitsevodstvo, Russia) 2: 11-15.

Fisinin, V.I. and Surai, P.F., 2012a. Early chicken nutrition and muscle tissue development. Russian Poultry Science (Ptitsevodstvo, Russia) 3: 9-12.

Fisinin, V.I. and Surai, P.F., 2012a. Properties and toxicity of DON. Mycotoxins and antioxidants: Uncompromising fighting. Part 1. Animal Production of Russia (Zhivotnovodstvo Rossii) 5: 11-14.

Fisinin, V.I. and Surai, P.F., 2012b. Properties and toxicity of DON. Mycotoxins and antioxidants: Uncompromising fighting. Part 2. Animal Production of Russia (Zhivotnovodstvo Rossii, Russia) 6: 3-5.

Fisinin, V.I. and Surai, P.F., 2012c. Mycotoxins and antioxidants: Uncompromising fighting. Ochratoxin A. Part 1. Compounded Feed (Kombikorma, Russia) 3: 55-60.

Fisinin, V.I. and Surai, P.F., 2012d. Mycotoxins and antioxidants: Uncompromising fighting. Ochratoxin A. Part 2. Compounded Feed (Kombikorma, Russia) 5: 59-60.

Fisinin, V.I. and Surai, P.F., 2012e. Mycotoxins and antioxidants: Uncompromising fighting. T-toxin – metabolism and toxicity. Part 1. Poultry and Poultry Products (Ptiza i Ptizeproducti, Russia) 3: 38-41.

Fisinin, V.I. and Surai, P.F., 2012f Mycotoxins and antioxidants: Uncompromising fighting. T-toxin – mechanisms of toxicity and protection. Part 2. Poultry and Poultry Products (Ptizai Ptizeproducti, Russia) 4: 36-39.

Fisinin, V.I. and Surai, P.F., 2013. Immunity in modern animal and poultry production: from theory to practical aspects of immunomodulation. Russian Poultry Science (Ptitsevodstvo, Russia) 5: 4-10.

Fisinin, V.I. and Surai, P.F., 2013a. Gut immunity in birds: facts and thinking. Agricultural Biology (Selskokhozaistvennaya Biologia, Russia) 4: 1-25.

Fisinin, V.I. and Surai, P.F., 2013a. Immunity in modern animal and poultry production: from theory to practical aspects of immunomodulation. Russian Poultry Science (Ptitsevodstvo, Russia) 5: 4-10.

Fisinin, V.I. and Surai, P.F., 2013b. Gut immunity in birds: facts and thinking. Agricultural Biology (Selskokhozaistvennaya Biologia, Russia) 4: 1-25.

Fisinin, V.I., Papazyan, T. and Surai, P.F., 2009. Innovative methods of fighting stresses in poultry production. Russian Poultry Science (Ptitzevodstvo, Russia) 8: 10-14.

Fisinin, V.I., Papazyan, T. and Surai, P.F., 2009a. Modern methods of stress prevention in poultry production. Today's Animal Production (Ukraine) 2: 56-61.

Franco-Jimenez, D.J., Scheideler, S.E., Kittok, R.J., Brown-Brandl, T.M., Robeson, L.R., Taira, H. and Beck, M.M., 2007. Differential effects of heat stress in three strains of laying hens. The Journal of Applied Poultry Research 16: 628-634.

Fussell, L.W., 1998. Poultry industry strategies for control of immunosuppressive diseases. Poultry Science 77: 1193-1196.

Gaigé, S., Djelloul, M., Tardivel, C., Airault, C., Félix, B., Jean, A., Lebrun, B., Troadec, J.D. and Dallaporta, M., 2014. Modification of energy balance induced by the food contaminant T-2 toxin: a multimodal gut-to-brain connection. Brain, Behaviour and Immunology 37: 54-72.

Galvano, F., Piva, A., Ritieni, A. and Galvano, G., 2001. Dietary strategies to counteract the effects of mycotoxins: a review. Journal of Food Protection 64: 120-131.

Galvin, K., Morrissey, P.A. and Buckley, D.J., 1997. Influence of dietary vitamin E and oxidised sunflower oil on the storage stability of cooked chicken muscle. British Poultry Science 38: 499-504.

Geng, A.L., Liu, H.G., Zhang, Y., Zhang, J., Wang, H.H., Chu, Q. and Yan, Z.X., 2020. Effects of indoor stocking density on performance, egg quality, and welfare status of a native chicken during 22 to 38 weeks. Poultry Science 99: 163-171.

Georgieva, N.V., Stoyanchev, K., Bozakova, N. and Jotova, I., 2011. Combined effects of muscular dystrophy, ecological stress, and selenium on blood antioxidant status in broiler chickens. Biological Trace Element Research 142: 532-545.

Geyra, A., Uni, Z. and Sklan, D., 2001. Enterocyte dynamics and mucosal development in the posthatch chick. Poultry Science 80: 776-782.

Geyra, A., Uni, Z. and Sklan, D., 2001a. The effect of fasting at different ages on growth and tissue dynamics in the small intestine of the young chick. British Journal of Nutrition 86: 53-61.

Ghareeb, K., Awad, W.A., Böhm, J. and Zebeli, Q., 2015. Impacts of the feed contaminant deoxynivalenol on the intestine of monogastric animals: poultry and swine. Journal of Applied Toxicology 35: 327-337.

Goerlich, V.C., Nätt, D., Elfwing, M., Macdonald, B. and Jensen, P., 2012. Transgenerational effects of early experience on behavioral, hormonal and gene expression responses to acute stress in the precocial chicken. Hormones and Behavior 61: 711-718.

Gonzalez-Rivas, P.A., Chauhan, S.S., Ha, M., Fegan, N., Dunshea, F.R. and Warner, R.D., 2020. Effects of heat stress on animal physiology, metabolism, and meat quality: a review. Meat Science 162: 108025.

Grenier, B. and Applegate, T.J., 2013. Modulation of intestinal functions following mycotoxin ingestion: meta-analysis of published experiments in animals. Toxins 5: 396-430.

Grobas, S., Mendez, J., De Blas, C. and Mateos, G.G., 1999. Influence of dietary energy, supplemental fat and linoleic acid concentration on performance of laying hens at two ages. British Poultry Science 40: 681-687.

Gunnarsson, S., Keeling, L.J. and Svedberg, J., 1999. Effect of rearing factors on the prevalence of floor eggs, cloacal cannibalism and feather pecking in commercial flocks of loose housed laying hens. British Poultry Science 40: 12-18.

Guo, M., Huang, K., Chen, S., Qi, X., He, X., Cheng, W.H., Luo, Y., Xia, K. and Xu W., 2014. Combination of metagenomics and culture-based methods to study the interaction between ochratoxin A and gut microbiota. Toxicological Sciences 141: 314-323.

Habashy, W.S., Milfort, M.C., Rekaya, R. and Aggrey, S.E., 2019. Cellular antioxidant enzyme activity and biomarkers for oxidative stress are affected by heat stress. International Journal of Biometeorology 63: 1569-1584.

Habibian, M., Sadeghi, G., Ghazi, S., Moeini, M.M., 2015. Selenium as a feed supplement for heat-stressed poultry: a review. Biological Trace Element Research 165: 183-193.

Hager, J.E. and Beane. W.L., 1983. Posthatch incubation time and early growth of broiler chickens. Poultry Science 62: 247-254.

Heussner, A.H. and Bingle, L.E., 2015. Comparative ochratoxin toxicity: a review of the available data. Toxins 7: 4253-4282.

Hirakawa, R., Nurjanah, S., Furukawa, K., Murai, A., Kikusato, M., Nochi, T. and Toyomizu, M., 2020. Heat stress causes immune abnormalities via massive damage to effect proliferation and differentiation of lymphocytes in broiler chickens. Frontiers in Veterinary Science 7: 46.

Hoerr, F.J., 2010. Clinical aspects of immunosuppression in poultry. Avian Diseases 54: 2-15.

Hu, R., He, Y., Arowolo, M.A., Wu, S. and He, J., 2019. Polyphenols as potential attenuators of heat stress in poultry production. Antioxidants 8: 67.

Huth, J.C. and Archer, G.S., 2015. Comparison of two LED light bulbs to a dimmable CFL and their effects on broiler chicken growth, stress, and fear. Poultry Science 94: 2027-2036.

Huwig, A., Freimund, S., Käppeli, O. and Dutler, H., 2001. Mycotoxin detoxication of animal feed by different adsorbents. Toxicology Letters 122:179-188.

Ingrao, F., Rauw, F., Lambrecht, B. and Van den Berg, T., 2013. Infectious Bursal disease: a complex host-pathogen interaction. Developmental and Comparative Immunology 41: 429-438.

Janmohammadi, A., Sheikhi, N., Nazarpak, H.H. and Nikbakht Brujeni, G., 2020. Effects of vaccination on acute-phase protein response in broiler chicken. PloS One 15, 2: e0229009.

Jensen, C., Engberg, R., Jakobsen, K., Skibsted, L.H. and Bertelsen, G., 1997. Influence of the oxidative quality of dietary oil on broiler meat storage stability. Meat Science 47: 211-222.

Jiang, M., Peng, X., Fang, J., Cui, H., Yu, Z. and Chen, Z., 2015. Effects of aflatoxin b1 on T-cell subsets and mRNA expression of cytokines in the intestine of broilers. International Journal of Molecular Sciences 16: 6945-6959.

Jiang, S., Mohammed, A.A., Jacobs, J.A., Cramer, T.A. and Cheng, H.W., 2020. Effect of synbiotics on thyroid hormones, intestinal histomorphology, and heat shock protein 70 expression in broiler chickens reared under cyclic heat stress. Poultry Science 99: 142-150.

Karadas, F., Surai. P.F. and Sparks. N.H., 2011. Changes in broiler chick tissue concentrations of lipid-soluble antioxidants immediately post-hatch. Comparative Biochemistry and Physiology. Part A, Molecular & Integrative Physiology 160: 68-71.

Katika, M.R., Hendriksen, P.J., Van Loveren H. and Peijnenburg, A., 2015. Characterization of the modes of action of deoxynivalenol (DON) in the human Jurkat T-cell line. Journal of Immunotoxicology 12: 206-216.

Klasing, K.C., 1998. Nutritional modulation of resistance to infectious diseases. Poultry Science 77: 1119-1125.

Kumar, R., Alam, S., Chaudhari, B.P., Dwivedi, P.D., Jain, S.K., Ansari, K.M. and Das, M., 2013. Ochratoxin A-induced cell proliferation and tumor promotion in mouse skin by activating the expression of cyclin-D1 and cyclooxygenase-2 through nuclear factor-kappa B and activator protein-1. Carcinogenesis 34: 647-657.

L'Estrange, J. L., Carpenter, K.J., Lea, C. H. and Parr, L.J., 1966. Nutritional effects of autoxidized fats in animal diets. 2. Beef fat in the diet of broiler chickens. British Journal of Nutrition 20: 113-122.

Lai, H.T.L., Nieuwland, M.G.B., Kemp, B.A., Aarnink, J.A. and Parmentier, H.K., 2011. Effects of repeated intratracheally administered lipopolysaccharide on primary and secondary specific antibody responses and on body weight gain of broilers. Poultry Science 90: 337-351.

Lallès, J.P., Lessard, M. and Boudry, G., 2009. Intestinal barrier function is modulated by short-term exposure to fumonisin B_1 in Using chambers. Veterinary Research Communications 33: 1039-1043.

Lambert, G.P., Gisolfi, C.V., Berg, D.J., Moseley, P.L., Oberley, L.W. and Kregel, K.C., 2002. Selected contribution: Hyperthermia induced intestinal permeability and the role of oxidative and nitrosative stress. Journal of Applied Physiology 92: 1750-1761.

Lara, L. and Rostagno, M., 2013. Impact of heat stress on poultry production. Animals 3: 356-369.

Lauridsen, C., 2019. From oxidative stress to inflammation: redox balance and immune system. Poultry Science 98: 4240-4246.

Lautert, C., Ferreiro, L., Wolkmer, P., Paim, F.C., da Silva, C.B., Jaques, J.A., Lopes, S.T. and Santurio, J.M., 2014. Individual *in vitro* effects of ochratoxin A, deoxynivalenol and zearalenone on oxidative stress and acetylcholinesterase in lymphocytes of broiler chickens. Springerplus 3: 506.

Leeson, S., Namkung, H., Caston, L., Durosoy, S. and Schlegel, P., 2008. Comparison of selenium levels and sources and dietary fat quality in diets for broiler breeders and layer hens. Poultry Science 87: 2605-2612.

Lessard, M., Boudry, G., Sève, B., Oswald, I.P. and Lallès, J.P., 2009. Intestinal physiology and peptidase activity in male pigs are modulated by consumption of corn culture extracts containing fumonisins. Journal of Nutrition 139: 1303-1307.

Li, R.F., Liu, S.P., Yuan, Z.H., Yi, J.E., Tian, Y.N., Wu, J. and Wen, L.X., 2020. Effects of induced stress from the live LaSota Newcastle disease vaccination on the growth performance and immune function in broiler chickens. Poultry Science 99: 1896-1905.

Li, X.M., Zhang, M.H., Liu, S.M., Feng, J.H., Ma, D.D., Liu, Q.X., Zhou, Y., Wang, X.J. and Xing, S., 2019. Effects of stocking density on growth performance, growth regulatory factors, and endocrine hormones in broilers under appropriate environments. Poultry Science 98: 6611-6617.

Lian, P., Braber, S., Garssen, J., Wichers, H.J., Folkerts, G., Fink-Gremmels, J. and Varasteh, S., 2020. Beyond heat stress: intestinal integrity disruption and mechanism based intervention strategies. Nutrients 12: 734.

Limonciel, A. and Jennings, P., 2014. A review of the evidence that ochratoxin A is an Nrf2 inhibitor: implications for nephrotoxicity and renal carcinogenicity. Toxins 6: 371-379.

Lin, C.F., Asghar, A., Gray, J.I., Buckley, D.J., Booren, A.M., Crackel, R.L. and Flegal, C.J., 1989. Effects of oxidized dietary oil and antioxidant supplementation on broiler growth and meat stability. British Poultry Science 30: 855-864.

Lin, H., Decuypere, E. and Buyse, J., 2006. Acute heat stress induces oxidative stress in broiler chickens. Comparative Biochemistry and Physiology. Part A, Molecular & Integrative Physiology 144: 11-17.

Lin, H., Mertens, K., Kemps, B., Govaerts, T., De Ketelaere, B., De Baerdemaeker, J., Decuypere, E. and Buyse, J., 2004. New approach of testing the effect of heat stress on eggshell quality: mechanical and material properties of eggshell and membrane. British Poultry Science 45: 476-482.

Liu C.P, Fu J., Lin S.L., Wang X.S. and Li, S., 2014a. Effects of dietary selenium deficiency on mRNA levels of twenty-one selenoprotein genes in the liver of layer chicken. Biological Trace Element Research 159: 192-198.

Liu, M., Gao, R., Meng, Q., Zhang, Y., Bi, C. and Shan, A., 2014. Toxic effects of maternal zearalenone exposure on intestinal oxidative stress, barrier function, immunological and morphological changes in rats. PLoS One 9, 9: e106412.

Lochmiller, R.L. and Deerenberg, C., 2000. Trade-offs in evolutionary immunology: just what is the cost of immunity? OIKOS 88: 87-98.

Lopez-Ferrer, S., Baucells, M.D., Barroeta, A.C. and Grashorn, M.A., 1999. Influence of vegetable oil sources on quality parameters of broiler meat. Archiv Fur Gefluegelkunde 63: 29-35.

Lu, T., Harper, A.F., Zhao, J., Corl, B.A., LeRoith, T. and Dalloul, R.A., 2014. Effects of a dietary antioxidant blend and vitamin E on fatty acid profile, liver function, and inflammatory response in broiler chickens fed a diet high in oxidants. Poultry Science 93:1658-1666.

Lu, Z., He, X.F., Ma, B.B., Zhang, L., Li, J.L., Jiang, Y., Zhou, G.H. and Gao, F., 2019. The alleviative effects and related mechanisms of taurine supplementation on growth performance and carcass characteristics in broilers exposed to chronic heat stress. Poultry Science 98: 878-886.

Luan, Y., Zhao, J., Yao, H., Zhao, X., Fan, R., Zhao, W., Zhang, Z. and Xu, S., 2016. Selenium deficiency influences the mRNA expression of selenoproteins and cytokines in chicken erythrocytes. Biological Trace Element Research 171: 427-436.

Ma, Q., Li, Y., Fan, Y., Zhao, L., Wei, H., Ji, C. and Zhang, J., 2015. Molecular mechanisms of lipoic acid protection against aflatoxin B_1-induced liver oxidative damage and inflammatory responses in broilers. Toxins 7: 5435-5447.

Mack, L.A., Felver-Gant, J.N., Dennis, R.L. and Cheng, H.W., 2013. Genetic variations alter production and behavioral responses following heat stress in 2 strains of laying hens. Poultry Science 92: 285-294.

Marin, D.E., Motiu, M. and Taranu, I., 2015. Food contaminant zearalenone and its metabolites affect cytokine synthesis and intestinal epithelial integrity of porcine cells. Toxins 7: 1979-1988.

Marusich, W.L., DeRitter, E., Ogrinz, E.F., Keating, J., Mitrovic, M. and Bunnell, R.H., 1975. Effect of supplemental vitamin E on control of rancidity in poultry meat. Poultry Science 54: 831-844.

Mashaly, M.M., Hendricks, G.L., Kalama, M.A., Gehad. A.E., Abbas, A.O. and Patterson, P.H., 2004. Effect of heat stress on production parameters and immune responses of commercial laying hens. Poultry Science 83: 889-894.

Maurice, D.V., Lightsey, S.F., Kuo-Tung, H. and Rhoades, J.F., 1991. Comparison of GSH s-transferase activity in the rat and birds: Tissue distribution and rhythmicity in chicken (*Gallus domesticus*) liver. Comparative Biochemistry and Physiology 100B: 471-474.

Mezes, M., 1994. Effect of vitamin E treatment on early postnatal changes of vitamin E status of chicken. Acta Veterinaria Hungarica 42: 477-480.

Mezes, M., Surai, P.F., Salyi, G., Speake, B.K., Gaal, T. and Maldjian, A., 1997. Nutritional metabolic diseases in poultry and disorders of the biological antioxidant defence system. Acta Veterinaria Hungarica 45: 349-360.

Mirfendereski, E. and Jahanian, R., 2015. Effects of dietary organic chromium and vitamin C supplementation on performance, immune responses, blood metabolites, and stress status of laying hens subjected to high stocking density. Poultry Science 94: 281-288.

Mishra, B. and Jha, R., 2019. Oxidative stress in the poultry gut: potential challenges and interventions. Frontiers in Veterinary Science 6: 60.

Murugesan, G.R., Ledoux, D.R., Naehrer, K., Berthiller, F., Applegate, T.J., Grenier, B., Phillips, T.D. and Schatzmayr, G., 2015. Prevalence and effects of mycotoxins on poultry health and performance, and recent development in mycotoxin counteracting strategies. Poultry Science 94: 1298-1315.

Najafi, P., Zulkifli, I., Soleimani, A.F. and Kashiani, P., 2015. The effect of different degrees of feed restriction on heat shock protein 70, acute phase proteins, and other blood parameters in female broiler breeders. Poultry Science 94: 2322-2329.

Nakamura, M., Tanaka, H., Hattori, Y. and Watanabe, M., 1972. Biological effects of autoxidized safflower oils. Lipids 8: 566-572.

Nawab, A., Ibtisham, F., Li, G., Kieser, B., Wu, J., Liu, W., Zhao, Y., Nawab, Y., Li, K., Xiao, M. and An, L., 2018. Heat stress in poultry production: mitigation strategies to overcome the future challenges facing the global poultry industry. Journal of Thermal Biology 78: 131-139.

Niu, Z.Y., Liu, F.Z., Yan, Q.L., and Li, W.C., 2009. Effects of different levels of vitamin E on growth performance and immune responses of broilers under heat stress. Poultry Science 88: 2101-2107.

Noy, Y. and Sklan, D., 1999. Different types of early feeding and performance in chicks and poults. The Journal of Applied Poultry Research 8: 16-24.

Noy, Y. and Uni, Z., 2010. Early nutrition strategy. World's Poultry Science Journal 66: 639-646.

Noy, Y., Gyra, A., and Sklan, D., 2001. The effect of early feeding on growth and small intestinal development in the posthatch poult. Poultry Science 80: 912-919.

Pappas, A.C., Zoidis, E., Fegeros, K., Zervas, G. and Surai, P.F., 2010. Cadmium, toxicity and the antioxidant system. Nova Science Publishers, Inc., New York.

Pappas, A.C., Zoidis, E., Surai, P.F. and Zervas, G., 2008. Selenoproteins and maternal nutrition. Comparative Biochemistry and Physiol. Part B Biochemistry & Molecular Biology 151: 361-372.

Pardue, S.L., Thaxton, J.P. and Brake J., 1985. Role of ascorbic acid in chicks exposed to high environmental temperature. Journal of Applied Physiology 58: 1511-1516.

Peng, X., Zhang, S., Fang, J., Cui, H., Zuo, Z. and Deng, J., 2014. Protective roles of sodium selenite against aflatoxin B1-induced apoptosis of jejunum in broilers. International Journal of Environmental Research and Public Health 11: 13130-13143.

Persia, M.E., Baker, D.H. and Parsons, C.M., 1994. Tolerance for excess basic zinc chloride and basic copper chloride in chicks. British Poultry Science 45: 672-676.

Pinchasov, Y. and Noy, Y., 1993. Comparison of post-hatch holding time and subsequent early performance of broiler chicks and Turkey poults. British Poultry Science 34: 111-120.

Poersch, A.B., Trombetta, F., Braga, A.C., Boeira, S.P., Oliveira, M.S., Dilkin, P., Mallmann, C.A., Fighera, M.R., Royes, L.F., Oliveira, M.S. and Furian, A.F., 2014. Involvement of oxidative stress in subacute toxicity induced by fumonisin B1 in broiler chicks. Veterinary Microbiology 174: 180-185.

Puron, D., Santamaria, R., Segura, J.C. and Alamilla, J.L., 1995. Broiler performance at different stocking densities. The Journal of Applied Poultry Research 4: 55-60.

Quinteiro-Filho, W.M., Ribeiro, A., Ferraz-de-Paula, V., Pinheiro, M.L., Sakai, M., Sá, L.R.M., Ferreira, A.J. and Palermo-Neto, J., 2010. Heat stress impairs performance parameters, induces intestinal injury, and decreases macrophage activity in broiler chickens. Poultry Science 89: 1905-1914.

Ramyaa, P., Krishnaswamy, R. and Padma, V.V., 2014. Quercetin modulates OTA-induced oxidative stress and redox signalling in HepG2 cells – up regulation of Nrf2 expression and down regulation of NF-κB and COX-2. Biochimica et Biophysica Acta 1840: 681-692.

Ratriyanto, A. and Mosenthin, R., 2018. Osmoregulatory function of betaine in alleviating heat stress in poultry. Journal of Animal Physiology and Animal Nutrition 102: 1634-1650.

Ravindran, V., Thomas, D.V., Thomas, D.G. and Morel, P.C.H., 2006. Performance and welfare of broilers as affected by stocking density and zinc bacitracin supplementation. Animal Science Journal 77: 110-116.

Rehman, Z.U., Meng, C., Sun, Y., Safdar, A., Pasha, R.H., Munir, M. and Ding, C., 2018. Oxidative stress in poultry: lessons from the viral infections. Oxidative Medicine and Cellular Longevity 2018: 5123147.

Ren, Z., He, H., Fan, Y., Chen, C., Zuo, Z. and Deng, J., 2019. Research progress on the toxic antagonism of selenium against mycotoxins. Biological Trace Element Research, 190: 273-280.

Ruff, J., Barros, T.L., Tellez, G., Jr, Blankenship, J., Lester, H., Graham, B.D., Selby, C., Vuong, C.N., Dridi, S., Greene, E.S., Hernandez-Velasco, X., Hargis, B.M. and Tellez-Isaias, G., 2020. Research note: evaluation of a heat stress model to induce gastrointestinal leakage in broiler chickens. Poultry Science 99: 1687-1692.

Saeed, M., Abbas, G., Alagawany, M., Kamboh, A.A., Abd El-Hack, M.E., Khafaga, A.F. and Chao, S., 2019. Heat stress management in poultry farms: a comprehensive overview. Journal of Thermal Biology 84: 414-425.

Safaa, H.M., Serrano, M.P., Valencia, D.G., Frikha, M., Jiménez-Moreno, E. and Mateos, G.G., 2008. Productive performance and egg quality of brown egg-laying hens in the late phase of production as influenced by level and source of calcium in the diet. Poultry Science 87: 2043-2051.

Sahin, K., Sahin, N., Kucuk, O., Hayirli, A. and Prasad, A.S., 2009. Role of dietary zinc in heat-stressed poultry: a review. Poultry Science 88: 2176-2183.

Sahin, N., Onderci, M., Balci, T.A., Cikim, G., Sahin, K. and Kucuk, O., 2007. The effect of soy isoflavones on egg quality and bone mineralization during the late laying period of quail. British Poultry and Science 48: 363-369.

Sanderson, I.R., 1999. The physicochemical environment of the neonatal intestine. American Journal of Clinical Nutrition 69: 1028S-1034S.

Sayed, M.A. and Downing, J., 2011. The effects of water replacement by oral rehydration fluids with or without betaine supplementation on performance, acid-base balance, and water retention of heat-stressed broiler chickens. Poultry Science 90: 157-167.

Schwartz-Zimmermann, H.E., Fruhmann, P., Dänicke, S., Wiesenberger, G., Caha, S., Weber, J. and Berthiller, F., 2015. Metabolism of deoxynivalenol and deepoxy-deoxynivalenol in broiler chickens, pullets, roosters and turkeys. Toxins 7: 4706-4729.

Sell, J.L., 1996. Recent developments in vitamin E nutrition of turkeys. Animal Feed Science and Technology 60: 229-240.

Sheehy, P.J.A., Morrissey, P.A. and Flynn A., 1994. Consumption of thermally-oxidized sunflower oil by chicks reduces α-tocopherol status and increased susceptibility of tissues to lipid oxidation. British Journal of Nutrition 71: 53-65.

Sheehy, P.J.A., Morrissey, P.A. and Flynn, A., 1993. Influence of heated vegetable oils and α-tocopheryl acetate supplementation on α-tocopherol, fatty acids and lipid peroxidation in chicken muscle. British Poultry Science 34: 367-381.

Simitzis, P.E., Kalogeraki, E., Goliomytis, M., Charismiadou, M.A., Triantaphyllopoulos, K., Ayoutanti. A., Niforou. K., Hager-Theodorides, A.L. and Deligeorgis, S.G., 2012. Impact of stocking density on broiler growth performance, meat characteristics, behavioural components and indicators of physiological and oxidative stress. British Poultry Science 53: 721-730.

Simsek, U.G., Dalkilic, B., Ciftci, M. and Yuce, A., 2009.The influences of different stocking densities on some welfare indicators, lipid peroxidation (MDA) and antioxidant enzyme activities (GSH, GSH-Px, CAT) in Broiler chickens. Journal of Animal and Veterinary Advances 8: 1568-1572.

Sklan, D., Noy, Y., Hoyzman, A. and Rozenboim, I., 2000. Decreasing weight loss in the hatchery by feeding chicks and poults in hatching trays. The Journal of Applied Poultry Research 9: 142-148.

Soleimani, A.F., Zulkifli, I., Omar, A.R. and Raha, A.R., 2011. Physiological responses of 3 chicken breeds to acute heat stress. Poultry Science 90: 1435-1440.

Song, J., Xiao, K., Ke, Y.L., Jiao, L.F., Hu, C.H., Diao, Q.Y., Shi. B. and Zou, X.T., 2014. Effect of a probiotic mixture on intestinal microflora, morphology, and barrier integrity of broilers subjected to heat stress. Poultry Science 93: 581-588.

Song, M., Lin, X., Zhao, J., Wang, X., Jiao, H., Li, H., Sun, S. and Lin, H., 2020. High frequency vaccination-induced immune stress reduces bone strength with the involvement of activated osteoclastogenesis in layer pullets. Poultry Science 99: 734-743.

Sørensen, P., Su, G. and Kestin, S.C., 2000. Effects of age and stocking density on leg weakness in broiler chickens. Poultry Science 79: 864-870.

Soto-Salanova, M.F., 1998. Vitamin E in young turkeys: a reassessment of the requirement. Retrospective Theses and Dissertations, Digital Repository @ Iowa State University.

Soto-Salanova, M.F. and Sell, J.L., 1995. Influence of supplemental dietary fat on changes in vitamin E concentration in livers of poults. Poultry Science 74: 201-204.

Soto-Salanova, M.F. and Sell, J.L., 1996. Efficacy of dietary and injected vitamin E for poults. Poultry Science 75: 1393-1403.

Sun, X., Zhang, H., Sheikhahmadi, A., Wang, Y., Jiao, H., Lin, H. and Song, Z., 2015. Effects of heat stress on the gene expression of nutrient transporters in the jejunum of broiler chickens (*Gallus gallus domesticus*). International Journal of Biometeorology 59: 127-135.

Sun, Z.W., Yan, L., G, Y.Y., Zhao, J.P., Lin, H. and Guo, Y.M., 2013. Increasing dietary vitamin D3 improves the walking ability and welfare status of broiler chickens reared at high stocking densities. Poultry Science 92: 3071-3079.

Surai, P.F., 2002. Natural antioxidants in avian nutrition and reproduction. Nottingham University Press, Nottingham, UK.

Surai, P.F., 2006. Selenium in nutrition and health. Nottingham University Press, Nottingham, UK.

Surai, P.F., 2015a. Antioxidant action of carnitine: molecular mechanisms and practical applications. EC Veterinary Science 2: 66-84

Surai, P.F., 2015b. Carnitine enigma: from antioxidant action to vitagene regulation. Part 1. Absorption, metabolism and antioxidant activities. Journal of Veterinary Science & Medicine 3, 2: 14.

Surai, P.F., 2015c. Carnitine enigma: from antioxidant action to vitagene regulation. Part 2. Transcription Factors and Practical Applications. Journal of Veterinary Science & Medicine 3, 2: 17.

Surai, P.F., 2015d. Silymarin as a natural antioxidant: an overview of the current evidence and perspectives. Antioxidants 4: 204-247.

Surai, P.F., 2015e. Antioxidant systems in poultry biology: Heat Shock Proteins. Journal of Science 5: 1188-1222.

Surai, P.F., 2016. Antioxidant systems in poultry biology: superoxide dismutase. Journal of Animal Nutrition 1, 1: 8.

Surai, P.F., 2018. Selenium in poultry nutrition and health. Wageningen Academic Publishers, Wageningen, the Netherlands.

Surai P.F., 2019. Vitagenes in poultry production: stresses and antioxidants. Suchasne Ptakhivnitstvo (Ukraine) 9-10: 24-27.

Surai P. F., 2020. Antioxidants in poultry nutrition and reproduction: an update. Antioxidants 9(2): 105.

Surai, P.F. and Borodai, V.P., 2010. Stresses in poultry production: from understanding molecular mechanisms to the development of protection methods. Modern Poultry Science (Ukraine): 7-8: 31-36.

Surai, P.F. and Dvorska, J.E., 2005. Effects of mycotoxins on antioxidant status and immunity. In: Diaz, D.E. (ed.) The mycotoxin blue book. Nottingham University Press, Nottingham, UK, pp. 93-137.

Surai, P.F. and Fisinin, V.I., 2012. Innovative methods of fighting stresses in poultry production: from vitamins to sirtuins and vitagenes. Effective Poultry Production (Ukraine), 8: 8-13.

Surai, P.F. and Fisinin, V.I., 2012a. Modern methods of fighting stresses in poultry production: from antioxidants to vitagenes. Agricultural Biology (Selskokhozaistvennaya Biologia, Russia) 4: 3-13.

Surai, P.F. and Fisinin, V.I., 2015. Antioxidant-prooxidant balance in the intestine: applications in chick placement and pig weaning. Journal of Veterinary Science & Medicine 3, 2: 16.

Surai, P.F. and Fisinin, V.I., 2016a. Vitagenes in poultry production. Part 1. Technological and environmental stresses. World's Poultry Science Journal 72: 721-733.

Surai, P.F. and Fisinin, V.I., 2016b. Vitagenes in poultry production. Part 2. Nutritional and internal stresses. World's Poultry Science Journal 72: 761-772.

Surai, P.F. and Fisinin, V.I., 2016c. Vitagenes in poultry production. Part 3. Vitagene concept development. World's Poultry Science Journal 72, 793-804.

Surai, P.F. and Fisinin, V.I., 2016d. Natural antioxidants and stresses in poultry production: from vitamins to vitagenes. Proc. XXV World Poultry Congress, Invited Lecture Papers, Beijing, China, 5-9 September 2016, pp. 116-121

Surai, P.F. and Fisinin, V.I., 2016e. Antioxidant system regulation: from vitamins to vitagenes. In: Watson, R.R. and De Meester, F. (eds) Handbook of cholesterol. Wageningen Academic Publishers, Wageningen, the Netherlands, pp. 451-481.

Surai, P.F., Fisinin, V.I. and Karadas, F., 2016. Antioxidant systems in chick embryo development. Part 1. Vitamin E, carotenoids and selenium. Animal Nutrition *2:* 1-11.

Surai, P.F. and Fotina, T.I., 2010. Once more about stresses: from changes in gene expression to inclusion of anti-stress premix into drinking water. Effective Poultry Science (Ukraine) 8: 20-25.

Surai, P.F. and Fotina, T.I., 2013. Physiological mechanisms of the heat stress development in poultry. Today's Animal Production Science (Ukraine) 6: 54-60.

Surai, P.F., Ionov, I.A., Kuchmistova, E.F., Noble, R.C. and Speake, B.K.,1998. The relationship between the levels of a-tocopherol and carotenoids in the maternal feed, yolk and neonatal tissues: comparison between the chicken, turkey, duck and goose. Journal of the Science of Food and Agriculture 76: 593-598.

Surai, P.F. and Kochish, I.I., 2017. Antioxidant systems and vitagenes in poultry biology: heat shock proteins. In: Asea, A.A.A. and Kaur, P. (ed.) Heat shock proteins in veterinary. Springer, Switzerland, pp. 123-177.

Surai, P.F. and Kochish, I.I., 2019. Nutritional modulation of the antioxidant capacities in poultry: the case of selenium. Poultry Science 98: 4231-4239.

Surai, P.F., Kochish, I.I. and Fisinin, V.I., 2017. Antioxidant systems in poultry biology: nutritional modulation of vitagenes. European Journal of Poultry Science 81: 1612-9199.

Surai, P.F., Kochish, I.I., Fisinin, V.I. and Kidd, M.T., 2019a. Antioxidant defence systems and oxidative stress in poultry biology: an update. Antioxidants 8(7): 235.

Surai, P.F., Kochish, I.I., Romanov, M.N. and Griffin, D.K., 2019b. Nutritional modulation of the antioxidant capacities in poultry: the case of vitamin E. Poultry Science 98: 4030-4041.

Surai, P.F., Kochish, I.I. and Kidd, M.T., 2020. Taurine in poultry nutrition. Animal Feed Science and Technology 260: 114339.

Surai, P.F. and Mezes, M., 2005. Mycotoxins and immunity: theoretical consideration and practical applications. Praxis Veterinaria 53: 71-88.

Surai, P.F., Mezes, M., Fotina, T. and Denev, S.D., 2010. Mycotoxins in human diet: a hidden danger. In: De Meester, F., Zibadi, S. and Watson, R.R. (eds) Modern dietary fat intakes in disease promotion. Humana Press, Springer New York London, pp.275-303.

Surai, P.F., Mezes, M., Melnichuk, S.D. and Fotina T.I., 2008. Mycotoxins and animal health: from oxidative stress to gene expression. Krmiva 50: 35-43.

Surai, P.F., Noble, R. and Speake, B., 1999. Relationship between vitamin E content and susceptibility to lipid peroxidation in tissues of the newly hatched chick. British Poultry Science 40: 406-410.

Surai, P.F., Noble, R.C. and Speake, B.K., 1996. Tissue-specific differences in antioxidant distribution and susceptibility to lipid peroxidation during development of the chick embryo. Biochimica et Biophysica Acta 1304: 1-10.

Surai, P.F., Speake, B.K., Noble, R.C. and Sparks, N.H., 1999a. Tissue-specific antioxidant profiles and susceptibility to lipid peroxidation of the newly hatched chick. Biological Trace Element Research 68: 63-78.

Taranu, I., Braicu, C., Marin, D.E., Pistol, G.C., Motiu, M., Balacescu, L., Beridan Neagoe, I. and Burlacu, R., 2014. Exposure to zearalenone mycotoxin alters *in vitro* porcine intestinal epithelial cells by differential gene expression. Toxicology Letters 232: 310-325.

Tavárez, M.A., Boler, D.D., Bess, K.N., Zhao, J., Yan, F., Dilger, A.C., McKeith, F.K. and Killefer, J., 2011. Effect of antioxidant inclusion and oil quality on broiler performance, meat quality, and lipid oxidation. Poultry Science 90: 922-930.

Thiamhirunsopit, K., Phisalaphong, C., Boonkird, S. and Kijparkorn, S., 2014. Effect of chili meal (*Capsicum frutescens* L.) on growth performance, stress index, lipid peroxidation and ileal nutrient digestibility in broilers reared under high stocking density condition. Animal Feed Science and Technology 192: 90-100.

Tong, H.B., Lu, J., Zou, J.M., Wang, Q. and Shi, S.R., 2012. Effects of stocking density on growth performance, carcass yield, and immune status of a local chicken breed. Poultry Science 91: 667-673.

Tsiouris, V., Georgopoulou, I., Batzios, C., Pappaioannou, N., Ducatelle, R. and Fortomaris, P., 2015. High stocking density as a predisposing factor for necrotic enteritis in broiler chicks. Avian Pathology 44: 59-66.

Van der Pol, C.W., Molenaar, R., Buitink, C.J., van Roovert-Reijrink, I.A., Maatjens, C.M., Van den Brand, H. and Kemp, B., 2015. Lighting schedule and dimming period in early life: consequences for broiler chicken leg bone development. Poultry Science 94: 2980-2988.

Vieira, S.L. and Moran Jr., E.T., 1999. Effects of delayed placement and used litter on broiler yields. The Journal of Applied Poultry Research 8: 75-81.

Wan, Q., Wu, G., He, Q., Tang, H. and Wang, Y., 2015. The toxicity of acute exposure to T-2 toxin evaluated by the metabolomics technique. Molecular BioSystems 11: 882-891.

Wang, G.H., Xue, C.Y., Chen, F., Ma, Y.L., Zhang, X.B., Bi, Y.Z. and Cao, Y.C., 2009. Effects of combinations of ochratoxin A and T-2 toxin on immune function of yellow-feathered broiler chickens. Poultry Science 88: 504-510.

Wang, J., Tang, L., Glenn, T.C. and Wang, J.S., 2016. Aflatoxin B1 induced compositional changes in gut microbial communities of male F344 rats. Toxicological Sciences 50: 54-63.

Wang, S.Y., Bottje, W., Maynard, P., Dibner, J. and Shermer W., 1997. Effect of Santoquin and oxidized fat on liver and intestinal glutathione in broilers. Poultry Science 76: 961-967.

Wen, C. and Zhou, Y. M., 2019. Dietary mannan oligosaccharide ameliorates cyclic heat stress-induced damages on intestinal oxidative status and barrier integrity of broilers. Poultry Science 98: 4767-4776.

Xu, J.X., Cao, C.Y., Sun, Y.C., Wang, L.L., Li, N., Xu, S.W. and. Li, J.L., 2014. Effects on liver hydrogen peroxide metabolism induced by dietary selenium deficiency or excess in chickens. Biological Trace Element Research 159: 174-182.

Yang, C., Song, G. and Lim, W., 2020a. Effects of mycotoxin-contaminated feed on farm animals. Journal of Hazardous Materials 389: 122087.

Yang, F., Liao, J., Yu, W., Pei, R., Qiao, N., Han, Q., Hu, L., Li, Y., Guo, J., Pan, J. and Tang, Z., 2020b. Copper induces oxidative stress with triggered NF-κB pathway leading to inflammatory responses in immune organs of chicken. Ecotoxicology and Environmental Safety 200: 110715.

Yang, X.J., Li, W.L., Feng, Y. and Yao, J.H., 2011. Effects of immune stress on growth performance, immunity, and cecal microflora in chickens. Poultry Science 90: 2740-2746.

Yao, H., Zhao, W., Zhao, X., Fan, R., Khoso, P.A., Zhang, Z., Liu, W. and Xu, S., 2014. Selenium deficiency mainly influences the gene expressions of antioxidative selenoproteins in chicken muscles. Biological Trace Element Research 161: 318-327.

Yao, L., Du, Q., Yao, H., Chen, X., Zhang, Z. and Xu, S., 2015. Roles of oxidative stress and endoplasmic reticulum stress in selenium deficiency-induced apoptosis in chicken liver. Biometals 28: 255-265.

Yue, H.Y., Wang, J., Qi, X.L., Ji, F., Liu, M.F., Wu, S.G., Zhang, H.J. and Qi, G.H., 2011. Effects of dietary oxidized oil on laying performance, lipid metabolism, and apolipoprotein gene expression in laying hens. Poultry Science 90: 1728-1736.

Zaboli, G., Huang, X., Feng, X. and Ahn, D.U., 2019. How can heat stress affect chicken meat quality? – a review. Poultry Science 98: 1551-1556.

Zhang, M., Dunshea, F.R., Warner, R.D., DiGiacomo, K., Osei-Amponsah, R. and Chauhan, S.S., 2020. Impacts of heat stress on meat quality and strategies for amelioration: a review. International Journal of Biometeorology 64: 1613–1628. https://doi.org/10.1007/s00484-020-01929-6

Zhang, W., Xiao, S., Lee, E.J. and Ahn, D.U., 2011. Consumption of oxidized oil increases oxidative stress in broilers and affects the quality of breast meat. Journal of Agricultural and Food Chemistry 59: 969-974.

Zhao, W., Le, H.H., McQuade, R., Furness, J.B. and Dunshea, F.R., 2019. Dietary betaine improves intestinal barrier function and ameliorates the impact of heat stress in multiple vital organs as measured by evans blue dye in broiler chickens. Animals 10: 38.

Chapter 2
Antioxidant systems in animal body

Adapt the remedy to the disease

2.1 Introduction

For the majority of organisms on Earth, life without oxygen is impossible. Animals, plants and many microorganisms rely on oxygen for the efficient production of energy. However, the high oxygen concentration in the atmosphere is potentially toxic for living organisms. It is interesting that oxygen toxicity was first described in laboratory animals in 1878 (Knight, 1998). For the last three decades free radical research has generated valuable information for further understanding not only detrimental, but also beneficial role of free radicals in cell signalling and other physiological processes. The benefit or harm of free radicals ultimately depend on the level of their production and efficiency of antioxidant defence.

2.2 Free radicals and reactive oxygen and nitrogen species

Free radicals are atoms or molecules containing one or more unpaired electrons. Free radicals are highly unstable and reactive and are capable of damaging biologically relevant molecules such as DNA, proteins or lipids. The animal body is under constant attack from free radicals, formed as a natural consequence of the body's normal metabolic activity and as part of the immune system's strategy for destroying invading microorganisms. The internal and external sources of free radicals are shown in Table 2.1.

Table 2.1. Internal and external sources of free radicals (adapted from Surai, 2006, 2018).

Internally generated	Factors promoted ROS formation
Mitochondria (ETC)	Cigarette smoke
Phagocytes (NADPH-Oxidase)	Radiation
Xanthine oxidase	UV light
Reactions with Fe^{2+} or Cu^{+}	Pollution
Arachidonate pathways	Certain drugs
Peroxisomes	Chemical reagents
Inflammation	Industrial solvents
Biomolecule oxidation (adrenaline, dopamine, tetrahydrofolates, etc.)	High level of ammonia
	Mycotoxins

DOI 10.3920/978-90-8686-906-0_2, © Wageningen Academic Publishers 2020

Collective terms reactive oxygen species (ROS) and reactive nitrogen species (RNS) have been introduced (Halliwell and Gutteridge, 2015) and recently were modified to RONS, include not only the oxygen or nitrogen radicals, but also some non-radical reactive derivatives of oxygen and nitrogen (Table 2.2).

Superoxide (O_2^{*-}) is the main free radical produced in biological systems during normal respiration in mitochondria and by autoxidation reactions with half-life at 37 °C in the range of 1×10^{-6} second. Superoxide can inactivate some enzymes due to formation of unstable complexes with transition metals of enzyme prosthetic groups, followed by oxidative self-destruction of the active site (Chaudiere and Ferrari-Iliou, 1999). Depending on condition, superoxide can act as oxidising or a reducing agent. It is necessary to mention that superoxide, by itself, is not extremely dangerous and does not rapidly cross lipid membrane bilayer (Kruidenier and Verspaget, 2002). However, superoxide is a precursor of other, more powerful RONS. For example, it reacts with nitric oxide with a formation of peroxynitrite ($ONOO^-$), a strong oxidant, which lead to formation of reactive intermediates due to spontaneous decomposition (Kontos, 2001; Mruk *et al.*, 2002). In fact, peroxynitrite was shown to damage a wide variety of biomolecules, including proteins (via nitration of tyrosine or tryptophan residues or oxidation of methionine or selenocysteine residues), DNA and lipids (Groves, 1999). Superoxide can also participate in the production of more powerful radicals by donating an electron, and thereby reducing Fe^{3+} and Cu^{2+} to Fe^{2+} and Cu^+, as follows:

$$O_2^- + Fe^{3+}/Cu^{2+} \longrightarrow Fe^{2+}/Cu^+ + O_2$$

Further reactions of Fe^{2+} and Cu^+ with H_2O_2 are a source of the hydroxyl radical ($^*OH$) in the Fenton reaction:

$$H_2O_2 + Fe^{2+}/Cu^+ \longrightarrow {}^*OH + OH^- + Fe^{3+}/Cu^{2+}$$

The sum of reactions of superoxide radical with transition metals and transition metals with hydrogen peroxide is known as the Haber-Weiss reaction.

Table 2.2. Reactive oxygen and nitrogen species (adapted from Surai, 2006, 2018).

Radicals		Non-radicals	
Alkoxyl	RO^*	Hydrogen peroxide	H_2O_2
Hydroperoxyl	HOO^*	Hypochlorous acid	HOCl
Hydroxyl	$^*OH$	Ozone	O_3
Peroxyl	ROO^*	Singlet oxygen	$^1O_2$
Superoxide	O_2^*	Peroxynitrite	$ONOO^-$
Nitric oxide	NO^*	Nitroxyl anion	NO^-
Nitrogen dioxide	NO_2^*	Nitrous acid	HNO_2

It is necessary to underline that superoxide radical is a 'double-edged sword'. It is beneficial when produced by activated polymorphonuclear leukocytes and other phagocytes as an essential component of their bactericidal activities but in excess it may result in tissue damage associated with inflammation.

Hydroxyl radical is the most reactive species with an estimated half-life of only about 10^{-9} second. It can damage any biological molecule it touches; however, its diffusion capability is restricted to only about two molecular diameters before reacting (Yu, 1994). Therefore, in most cases damaging effect of hydroxyl radical is restricted to the site of its formation. In general, hydroxyl radical can be generated in human/animal body as a result of radiation exposure from natural sources (radon gas, cosmic radiation) and from man-made sources (electromagnetic radiation and radionuclide contamination). In fact, in many cases hydroxyl radical is a trigger of chain reaction in lipid peroxidation. Therefore, RONS (Table 2.2) are constantly produced *in vivo* in the course of the physiological metabolism in tissues.

It is generally accepted that the electron-transport chain in the mitochondria is responsible for major part of superoxide production in the body (Halliwell and Gutteridge, 2015). Mitochondria are shown to contain up to 12 sites for ROS production associated with nutrient oxidation and respiration. In fact, mitochondria exhibit a highly dynamic and complicated ROS release profile that varies depending on physiological conditions and carbon source, type, and availability and cell type. However, it seems likely that complex III has consistently the highest capacity of ROS production in all tissue and cell types examined so far (Young *et al.*, 2019). Mitochondrial electron transport system consumes more than 85% of all oxygen used by the cell and, because the efficiency of electron transport is not 100%, about 1-3% of electrons escape from the chain and the univalent reduction of molecular oxygen results in superoxide anion formation (Chow *et al.*, 1999; Halliwell, 1994; Singal *et al.*, 1998). About 10^{12} O_2 molecules processed by each rat cell daily and if the leakage of partially reduced oxygen molecules is about 2%, this will yield about 2×10^{10} molecules of ROS per cell per day (Chance *et al.*, 1979). An interesting calculation has been made by Halliwell (1994), showing that in the human body about 1.72 kg/year of superoxide radical is produced. In stress condition it would be substantially increased. Clearly, these calculations showed that free radical production in the body is substantial and many thousand biological molecules can be easily damaged if are not protected. The activation of macrophages in stress conditions is another important source of free radical generation. Immune cells produce ROS/RNS and use them as an important weapon to destroy pathogens (Kettle and Winterbourn, 1997; Schwarz, 1996; for more details see Chapter 16).

The most important effect of free radicals on the cellular metabolism is due to their participation in lipid peroxidation reactions (Surai, 2006, 2018). The first step of this process is called the initiation phase, during which carbon-centred free radicals are produced from a precursor molecule, for example a polyunsaturated fatty acid (PUFA):

$$LH \xrightarrow{\text{Initiator}} L^*$$

The initiator in this reaction could by the hydroxyl radical, radiation or some other events or compounds. In presence of oxygen these radicals (L*) react with oxygen producing peroxyl radicals starting the next stage of lipid peroxidation called the propagation phase:

$$L^* + O_2 \longrightarrow LOO^*$$

At this stage, a relatively unreactive carbon-centred radical (L*) is converted to a highly reactive peroxyl radical. In fact, the resulted peroxyl radical can attack any available peroxidazable material producing hydroperoxide (LOOH) and new carbon-centred radical (L*):

$$LOO^* + LH \longrightarrow LOOH + L^*$$

Therefore, lipid peroxidation is a chain reaction and potentially large number of cycles of peroxidation could cause substantial damage to cells. In membranes the peroxidazable material is represented by PUFAs. It is generally accepted that PUFA susceptibility to peroxidation is proportional to amount double bounds in the molecules. In fact, docosahexaenoic acid (DHA, 22:6n-3) and arachidonic acid (AA, 20:4n-6) are among major substrates of the peroxidation in the biological membrane. It is necessary to underline that the same PUFAs are responsible for maintenance of physiologically important membrane properties including fluidity and permeability. Therefore, as a result of lipid peroxidation within the biological membranes their structure and functions are compromised.

Lipid peroxidation produces a wide variety of oxidation products. The main primary products of lipid peroxidation are lipid hydroperoxides (LOOH). Among them, malondialdehyde (MDA) appears to be the most mutagenic product of lipid peroxidation, whereas 4-hydroxyalkenal (4-HNE) is the most toxic one (Ayala *et al.*, 2014). Indeed, 4-HNE is highly reactive toward nucleophilic thiol and amino groups and could form covalent adducts with various cellular (macro)molecules, including lipids, proteins, and nucleic acids. This leads to various detrimental consequences of cellular structure and metabolism, including inhibition of protein and DNA synthesis, dysregulation of enzyme activities, alteration in mitochondrial coupling, etc. (Hu *et al.*, 2017). Therefore, major systems of the animal body, including cardiovascular system, reproductive system, immune system, liver and kidney are affected due to lipid peroxidation.

Lipid peroxides are shown to exert their toxic effects through two important mechanisms. Firstly, lipid peroxidation is associated with alterations in the assembly, composition, structure, and dynamics of lipid membranes leading to detrimental consequences in cell functions. Secondly, lipid peroxides could promote further generation of new RONS with formation of new reactive compounds capable of damaging/crosslinking DNA and proteins (Gaschler and Stockwell, 2017). Lipid peroxidation also plays a role in regulated cell death called apoptosis. For example, the lipid degradation product 4-HNE has been shown to induce apoptosis in specific contexts and lipid peroxidation is the primary driver of ferroptosis, a type of regulated necrotic cell death (Gaschler and Stockwell, 2017). Furthermore, 4-HNE can be found at low concentrations in human tissues and plasma and participates in the control of biological processes, such as signal transduction, cell proliferation, and differentiation (Pizzimenti *et al.*, 2013). From an analytical point of view, quantitative determination of MDA in plasma, urine and other biological samples is easier than that of HNE and MDA and HNE are shown to correlate closely with each other. However, reliable measurement of MDA in biological samples is quite challenging and requires special precautions at the pre-analysis stage (Tsikas *et al.*, 2017).

Proteins and DNA are also important targets for RONS. The complex structure of proteins and a variety of oxidisable functional groups of the amino acids make them susceptible to oxidative damage. Indeed, proteins exposure to RONS causes modification of amino acid side chains and alteration of the protein structure leading to functional changes disturbing cellular metabolism associated with several pathological states (Ahmad *et al.*, 2017). In fact, increased side-chain hydrophilicity, side-chain and backbone fragmentation, aggregation via covalent crosslinking or hydrophobic interactions, protein unfolding and altered conformation, altered interactions with biological partners and modified turnover are observed due to protein oxidation (Davis, 2016). In the case of protein oxidation and depending on the amount of oxidative modification, proteins undergo a transition from slight functional changes to a completely dysfunctional, unfolded and insoluble structures (Korovila *et al.*, 2017).

The accumulation of oxidised proteins has been implicated in a range of age-related pathologies and a range of oxidised proteins and amino acids has been characterised in biological systems (Kehrer, 2000). In general, the accumulation of oxidised proteins depends on the balance between antioxidants, prooxidants and removal/repair mechanisms and leads to the formation of reversible disulphide bridges. More severe protein oxidation causes a formation of chemically modified derivatives e.g. Shiff's base (Tirosh and Reznick, 2000). Interestingly, protein peroxides can oxidise both proteins and other targets (Davies, 2016). Nitric oxide, hydroxyl radical, alkoxyl and peroxyl radicals as well as carbon-centred radicals, hydrogen peroxide, aldehydes or other products of lipid peroxidation can attack protein molecules. Usually oxidative modification of proteins occurs by two different mechanisms: a site-specific formation of ROS via redox-active transition metals and non-metal-dependent RONS-induced oxidation of amino acids (Tirosh and Reznick, 2000). The modification of a protein occurs by either a direct oxidation of a specific amino acid in the protein molecule or cleavage of the protein backbone. In both cases biological activity of the modified proteins would be compromised. The degree of protein damage depends on many different factors (Grune *et al.*, 1997):

- the nature and relative location of the oxidant or free radical source;
- nature and structure of protein;
- the proximity of RONS to protein target;

- the nature and concentrations of available antioxidants.

It seems likely that direct oxidation of cysteine and methionine residues in proteins are major reactions due to oxidative stress and this results in altered protein activity and function (Davies, 2016). In spite of important roles of protein oxidation in pathogenesis of the development of various diseases, mechanisms for the control of protein oxidation and their repair have not been well studied and this has been a topic of great interest for the last few years. The oxidative damage to proteins is associated with alteration of transport proteins and ion dis-balance, disruption to the receptors and impair signal transduction, enzyme inactivation, etc. It is believed that conversion of –SH groups into disulphides and other oxidised species (e.g. oxyradicals) is one of the earliest events during the radical-mediated oxidation of proteins. Therefore, thioredoxin system, in particular, thioredoxin plus thioredoxin reductase deals with these changes by reducing protein disulphides to thiols and regulating redox-sensitive transcription factors (Dean *et al.*, 1997).

It is worth mentioning that the main cellular mechanisms effectively controlling protein homeostasis in the cell include (Goloubinoff *et al.*, 2016):
- the molecular chaperones, including HSPs, acting as aggregate unfolding and refolding enzymes;
- the chaperone-gated proteases, acting as aggregate unfolding and degrading enzymes;
- the aggresomes, acting as aggregate compacting machineries;
- the autophagosomes, acting as aggregate degrading organelles.

It is interesting that reversible oxidation of cysteine could be an important cellular redox sensor in some proteins (Finkel, 2000). Methionine residues in proteins are also very susceptible to oxidation with methionine sulphoxide formation, which was detected in native proteins (Gao *et al.*, 1998). This could affect activity of various proteins. In fact, many forms of RONS oxidise methionine residues of proteins to a mixture of the R- and S-isomers of methionine sulphoxide (Stadtman *et al.*, 2002). Methionine is known to be one of the most easily oxidised amino acids and Msr is responsible for reversing this oxidation and restoration of protein function, with MsrA and MsrB reducing different stereoisomers (Jiang *et al.*, 2020; Reiterer *et al.*, 2019).

Therefore, Msr can reduce either the free or the protein-bound methionine sulphoxide back to methionine. In fact, Msr is considered a repair mechanism for dealing with the product of reaction of oxidants with methionine residues (Levine *et al.*, 1996). The authors hypothesised that methionine residues function as a 'last chance' antioxidant defence system for proteins. It was shown that in bacterial glutamine synthetase surface-exposed methionine residues surrounding the entrance to the active site are preferentially oxidised and other residues (e.g. cysteine) within the critical regions of the protein are protected without loss of catalytic activity of the protein (Levine *et al.*, 1996). Indeed, due to Msr activity the methionine-methionine sulphoxide pair can function catalytically. MsrA is present in most living organisms, is encoded by a single gene and the mammalian enzyme has been detected in all tissues studied. In particular, it is found in the cytosol and mitochondria of rat liver cells (Vougier *et al.*, 2003). Msr is considered to have at least three important function in cellular metabolism including antioxidant defence, repair enzyme and a regulator of certain enzyme function and possibly participation in signal transduction (Bar-Noy and Moskovitz, 2002; Stadtman *et al.*, 2002). Interestingly, mouse that lacks the MsrA gene (Moskovitz *et al.*, 2001):

- exhibits enhanced sensitivity to oxidative stress;
- has a shorter lifespan;
- develops an atypical walking pattern;
- accumulates higher tissue levels of oxidised protein under oxidative stress;
- is less able to up-regulate expression of thioredoxin reductase under oxidative stress.

MsrA has been known for a long time, and its repairing function is well characterised, however, recently, a new methionine sulphoxide reductase was characterised (Grimaud *et al.*, 2001). It was referred to as MsrB and it was shown that the gene of MsrB is present in genomes of eubacteria, archaebacteria, and eukaryotes. Therefore, in mammals two methionine sulphoxide reductases, MsrA and MsrB, are expressed with different substrate specificity (Grimaud *et al.*, 2001).

The major mammalian MsrB has been identified as a selenoprotein (Kryukov *et al.*, 2002; Moskovitz *et al.*, 2002) and called selenoprotein R. It is a zinc-containing stereo-specific Msr (Kryukov *et al.*, 2002). Furthermore, it has been shown that there was a loss of MsrB activity in the *MsrA*$^{-/-}$ mouse in parallel with losses in the levels of MsrB mRNA and MsrB protein (Moskovitz and Stadtman, 2003). Therefore, the author suggested that MsrA might have a role in MsrB transcription. Moreover, Se deficiency in mouse was associated with a substantial decrease in the levels of MsrB-catalytic activity, MsrB protein, and MsrB mRNA in liver and kidney tissues (Moskovitz and Stadtman, 2003). It has been reported that human and mouse genomes possess three MsrB genes responsible for synthesis of the following protein products: MsrB1, MsrB2 and MsrB3 (Kim and Gladyshev, 2004). In particular, MsrB1 (Selenoprotein R) was present in the cytosol and nucleus and exhibited the highest methionine-R-sulphoxide reductase activity due to presence of selenocysteine (Sec) in its active site. Other mammalian MsrBs are not selenoproteins and contain cysteine in place of Sec and were less catalytically efficient (Kim and Gladyshev, 2004). The reduced glutathione itself can also participate in maintenance of protein –SH groups. At the same time, the thioredoxin system has alkyl hydroperoxide reductase activity. Protein disulphide isomerase is also involved in re pairing of –SH groups in proteins (Dean *et al.*, 1997). Furthermore, the cells can generally remove oxidised proteins by proteolysis. In fact, damaged proteins are degraded by the proteasome, multicatalytic proteinase (an intracellular, nonlysosomal threonine type protease, EC 3.4.99.46), which is responsible for degradation of the majority cytosolic proteins (Rock *et al.*, 1994).

It is well recognised now that the proteasome is the major enzymatic system in charge of cellular 'cleansing' and plays a key role in the degradation of damaged proteins controlling the level of altered proteins in eukaryotic cells (Friguet *et al.*, 2000). It is suggested that enhanced susceptibility to degradation by proteinases is employed as a criterion of unfolding (Dean *et al.*, 1997), however, heavily oxidised proteins are characterised by an increased resistance to proteolytic attack by most proteinases. The proteasome complex recognises hydrophobic amino acid residues, aromatic residues, and bulky aliphatic residues that are modified during the oxidative stress and catalyse the selective removal of oxidatively modified cell proteins (Grune *et al.*, 1997). By minimising protein aggregation and cross-linking and by removing potentially toxic protein fragments proteasome is an active part of the cellular defence system against oxidative stress. The selective degradation of oxidatively damaged proteins enables cells to restore vital proteins including enzymes during physiological metabolism and during moderate stress conditions (Grune *et al.*, 1997). Oxidised proteins may also be recognised as 'foreign' by the immune system with corresponding antibody formation (Halliwell and Gutteridge, 2015). Understanding molecular mechanisms of protein oxidation could have important applications in meat producing industries. Indeed, water-holding capacity of intact proteins could be substantially affected due to protein oxidation and conformation changes in protein structure (Bao and Ertbjerg, 2019; Estevez *et al.*, 2020), especially in stress conditions (Gonzalez-Rivas *et al.*, 2020). Clearly, further work is needed to clarify molecular mechanisms of the protein oxidation and its effects on egg, meat and milk quality.

It has been shown that the DNA in each cell of a rat is hit by about 100,000 free radicals a day and each cell sustains as many as 10,000 potentially mutagenic (if not repaired) lesions per day arising from endogenous sources of DNA damage (Ames and Gold, 1997, 2003; Diplock, 1994; Helbock *et al.*, 1998). Therefore, some oxidative lesions escape repair and the steady state level of oxidative lesions increased with age, and an old rat has accumulated about 66,000 oxidative DNA lesions per cell (Ames, 2003). Oxidation, methylation, deamination and depurination are four main endogenous processes leading to significant DNA damage with oxidation to be most significant one and approximately 20 types of oxidatively altered DNA molecules have been identified. The chemistry of attack by ROS on DNA is very complex and lesions in chromatin include damage to bases, sugar lesions, single strand-breaks, basic lesions and DNA-nucleoprotein cross-links (Diplock, 1994).

Since maintaining the integrity of the genome is of the vital importance, living organisms have evolved a DNA damage response (DRR) consisting of several important steps associated with damage sensing, signalling cascades and congruent damage repair (Kciuk *et al.*, 2020). In fact, DNA repair is one of the fundamental processes of life (130 human DNA repair genes have been identified; Wood *et al.*, 2001) and if the systems are compromised devastating consequences would be expected. In order to deal with the deleterious effects of such lesions, leading to genomic instability, cells have evolved a number of DNA repair mechanisms (Kciuk *et al.*, 2020). They include the direct reversal of the lesion, mismatch repair, the base excision repair, nucleotide excision repair, nucleotide incision repair, transcription-coupled repair, global genome repair, homologous recombination and non-homologous end-joining (Karakaidos *et al.*, 2020; Slupphaug *et al.*, 2003). These repair pathways are universally present in living cells and extremely well conserved. Therefore, DNA repair systems include a number of enzymatic processes ranging from base recognition and excision to ligation of DNA strands. In particular, the DNA glycosylases recognise a damaged purines and pyrimidines and hydrolyse the bond linking the abnormal base to the sugar-phosphate backbone (Halliwell and Gutteridge, 2015); the 5^{I}-apurinic endonucleases process strand breaks, sites of base loss, and the products of DNA glycosylase/apurinic lyase action. DNA polymerase fills in the one-nucleotide gap with the correct base. DNA ligases complete the repair process by sealing the 3^{I} end of the newly synthesised stretch of DNA to the original portion of the DNA chain (Cardozo-Pelaez *et al.*, 2000; Croteau and Bohr, 1997; Wallace *et al.*, 1997).

It is believed that most damaged or inappropriate bases in DNA are removed by excision repair, while a minority are repaired by direct damage reversal (Krokan *et al.*, 2000). The importance of these DNA repair systems is confirmed by the fact that defects in these can result in cell death and hypersensitivity to endogenous or environmental mutagens (Jackson, 1999). Therefore, removing mutagenic lesions in DNA is a vital task for repair systems. In general, the repair DNA damage mechanisms in bacteria are well defined, whereas in higher eukaryotes the genes and proteins responsible for repair await further investigation (Croteau and Bohr, 1997; Karakaidos *et al.*, 2020). It seems likely that DNA repair is integrated with cell cycle regulation, transcription and replication and use some common factors (Slupphaug *et al.*, 2003). However, the repair enzymes do not achieve complete repair or removal of damaged DNA molecules and this could lead to arrest of cell cycle and cell death. In fact, programmed cell death (apoptosis) is involved in maintenance of the genetic integrity by removing genetically altered cells. There are also various mechanisms of the elimination of cells bearing DNA damage including apoptosis and the activation of innate immunity by DNA injuries (Ragu *et al.*, 2020). As mentioned in the Cahper 1, in poultry production, overproduction of free radicals and oxidative stress are considered to be related to various type of stresses, including, nutritional, technological, environmental and internal stresses (Surai and Fisinin, 2016, 2016a,b,c,d). In general, it is widely believed that most human and animal/poultry diseases at different stages of their development are associated with free radical production and metabolism (Surai, 2002, 2006, 2018). Normally, there is a delicate balance between the amount of free radicals generated in the body and the antioxidants to protect against them. For the majority organisms on Earth, life without oxygen is impossible, animals, plants and many micro-organisms relying on oxygen for the efficient production of energy. However, they pay a high price for pleasure of living in an oxygenated atmosphere since high oxygen concentration in the atmosphere is potentially toxic for living organisms.

2.3 Three levels of antioxidant defence

During evolution living organisms have developed specific antioxidant protective mechanisms to deal with ROS and RNS (Halliwell and Gutteridge, 2015). Therefore, it is only the presence of natural antioxidants in living organisms which enable them to survive in an oxygen-rich environment (Halliwell, 1994, 2012). These mechanisms are described by the general term 'antioxidant system'. It is diverse and responsible for the protection of cells from the actions of free radicals. This system includes:

- natural fat-soluble antioxidants (vitamin E, carotenoids, ubiquinones, etc.);
- water-soluble antioxidants (ascorbic acid, uric acid, carnitine, taurine, etc.);
- antioxidant enzymes: glutathione peroxidase (GPx), catalase (CAT) and superoxide dismutase (SOD);
- thiol redox system consisting of the glutathione system (glutathione/glutathione reductase/glutaredoxin/glutathione peroxidase and the thioredoxin system (thioredoxin/thioredoxin reductase/thioredoxin peroxidase (peroxiredoxins)/sulfiredoxin).

The antioxidant capacity of a compound is determined by multiple factors including the chemical reactivity toward free radicals; fate of antioxidant-derived radicals; interaction with other antioxidants; concentration, distribution, mobility, and metabolism at the micro-environment (Niki, 2014, 2016). The protective antioxidant compounds are located in organelles, subcellular compartments or the extracellular space enabling maximum cellular protection to occur. Thus, the antioxidant system of the living cell includes three major levels of defence (Niki, 1996; Surai, 1999, 2002, 2006, 2018).

The first level of defence is responsible for prevention of radical formation, maintaining redox balance and cell signalling includes SOD, GPx, CAT and metal binding proteins. Recently, thioredoxin system, glutathione system as well as vitagenes and such transcription factors as Nrf2, NF-κB and HSF have also been included into the first level of the antioxidant defence (Figure 2.1).

Since the superoxide radical is the main free radical produced in physiological conditions in the cell (Halliwell, 1994) superoxide dismutase (EC 1.15.1.1) is considered to be the main element of the first level of antioxidant defence in the cell (Surai, 1999, 2016, 2020a). This enzyme dismutates the superoxide radical in the following reaction:

$$2O_2^{*} + 2H^{+} \xrightarrow{SOD} H_2O_2 + O_2$$

More detailed information on the roles and regelation of SOD is presented in Chapter 4.

Since H_2O_2 is still toxic, there is a range of enzymes, involved in its detoxification, including Catalase, GPx and peroxiredoxins as follows:

$$2H_2O_2 \xrightarrow{GPx,\ Prx,\ Catalase} 2H_2O + O_2$$

Catalase (EC 1.11.1.6) is a tetrameric enzyme consisting of four identical subunits of 60 kDa containing a single ferriprotoporphyrin group per subunit. It plays an important role in the acquisition of tolerance to oxidative stress in the adaptive response of cells (Mates and Sanchez-Jimenez, 1999). In mammalian cells, NADPH is bound to

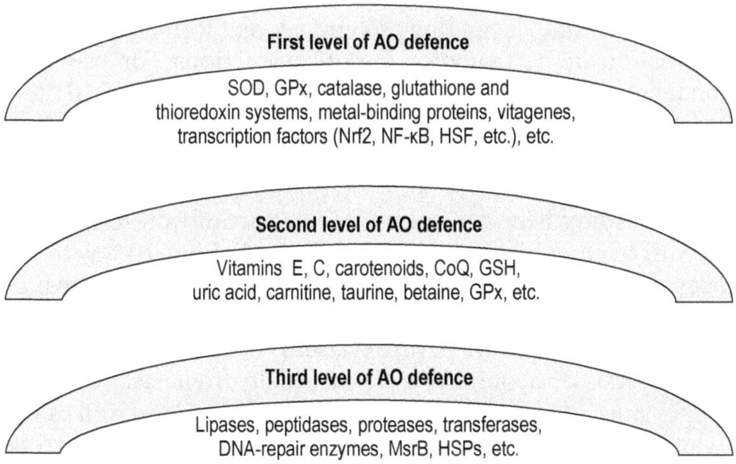

Figure 2.1. Three lines of antioxidant defence in animal cells (adapted from Surai, 1999, 2018).

catalase protecting it from inactivation by H_2O_2 (Chaudiere and Ferrari-Illiou, 1999). Since GPx has a much higher affinity for H_2O_2 than CAT (Jones *et al.*, 1981) and wider distribution in the cell (catalase is located mainly in peroxisomes), the H_2O_2 removal from the cell is very much dependent on GPx (Surai *et al.*, 2018a,b). Recently it has been shown that thioredoxin peroxidases called peroxiredoxins (Prx) are also capable of directly reducing hydrogen peroxide (Nordberg and Arner, 2001). It is interesting that the levels of antioxidant enzymes are regulated by gene expression as well as by post-translational modifications (Fujii and Taniguchi, 1999). Mammalian cells express six Prx isoforms, including Prx3 and Prx5 in the mitochondria. Prxs function by undergoing oxidation by H_2O_2 at an active site cysteine and then subsequent reduction by thioredoxin, thioredoxin reductase, and NADPH. There are eight GPx, which are oxidised by H_2O_2 and reduced by glutathione (GSH) and catalase is found in peroxisomes (Sena and Chandel, 2012). More details on the role of thioredoxin system in antioxidant defences are shown in Chapter 6.

Transition metal ions also accelerate the decomposition of lipid hydroperoxides into cytotoxic products such as aldehydes, alkoxyl radicals and peroxyl radicals:

$$LOOH + Fe^{2+} \longrightarrow LO^{*} + Fe^{3+} + OH^{-}$$

$$LOOH + Fe^{3+} \longrightarrow LOO^{*} + Fe^{2+} + H^{+}$$

Therefore, metal-binding proteins (transferrin, lactoferrin, haptoglobin, hemopexin, metallothionenin, ceruloplasmin, ferritin, albumin, myoglobin, etc.) also belong to the first level of defence. It is necessary to take into account that iron and copper are powerful promoters of free radical reactions and therefore their availability in 'catalytic' forms is carefully regulated *in vivo* (Halliwell, 1999). Indeed, organisms have evolved to keep transition metal ions safely sequestered in storage or transport proteins. In this way the metal-binding proteins prevent formation of hydroxyl radical by preventing them from participation in radical reactions. For example, transferrin binds the iron (about 0.1% of the total body reserves), transports it in the plasma pool and attaches it to the transferrin receptor. The important point is that iron associated with transferrin will not catalyse free radical reaction. Ferritin is considered to be involved in iron storage (about 30% of total body reserves) within the cytosol in various tissues including liver and spleen. Major part of iron in the body (55-60%) is associated with haemoglobin within red cells and about 10% with myoglobin in muscles (Galey, 1997). A range of other iron-containing proteins (mainly enzymes) can be found in the body including NADH dehydrogenase, cytochrome P450, ribonucleotide reductase, proline hydroxylase, tyrosine hydroxylase, peroxidases, catalase, cyclooxygenase, aconitase, succinate dehydrogenase, etc. (Galey, 1997). Despite an importance of iron in various biochemical reactions, iron can be extremely dangerous when not carefully handled by proteins. In fact, in many stress conditions a release of free iron from its normal sites and its participation in Fenton chemistry mediate damages to cells. For example, superoxide radical can release iron from ferritin and H_2O_2 degrades the heme of haemoglobin to liberate iron ions (Halliwell, 1987).

Ceruloplasmin is another major protein mediating free radical metabolism being a copper-binding protein. Under physiological conditions it binds six or seven copper ions per molecule preventing their participation in free radical generation. About 5% of human plasma copper is bound to albumin or to amino acids and the rest is bound to ceruloplasmin. Furthermore, ceruloplasmin possess antioxidant properties itself being able to scavenge superoxide radical (Yu, 1994). Therefore, it is now quite clear that metal sequestration is an important part of extracellular antioxidant defence.

Detailed information on other members of the first level of antioxidant defence, including thioredoxin system, glutathione system, transcription factors and vitagenes will be presented in the next chapters.

Unfortunately, this first level of antioxidant defence in the cell is not sufficient to completely prevent free radical formation and some radicals do escape through the preventive first level of antioxidant safety screen initiating lipid peroxidation and causing damage to DNA and proteins. Therefore, the second level of defence consists of chain-breaking antioxidants – vitamin E, ubiquinol, carotenoids, ascorbic acid, uric acid, carnitine, taurine and some other antioxidants. Glutathione and thioredoxin systems also play a substantial role in the second level of antioxidant defence. Chain-breaking antioxidants inhibit peroxidation by keeping the chain length of the propagation reaction as small as possible. Therefore, they prevent the propagation step of lipid peroxidation by scavenging peroxyl radical intermediates in the chain reaction:

$$LOO^* + Toc \longrightarrow Toc^* + LOOH$$

(LOO* is lipid peroxyl radical; Toc – tocopherol, Toc* – tocopheroxyl radical, LOOH – lipid hydroperoxide)

Vitamin E, the most effective natural free radical scavenger identified to date, is the main chain breaking antioxidant in the cell. However, hydroperoxides, produced in the reaction of vitamin E with the peroxyl radical, are toxic and if not removed, impair membrane structure and functions (Gutteridge and Halliwell, 1990). In facts, lipid hydroperoxides are not stable and in the presence of transition metal ions can decompose producing new free radicals and cytotoxic aldehydes (Diplock, 1994). Therefore, hydroperoxides have to be removed from the cell in the same way as H_2O_2, but catalase is not able to detoxify these compounds and Se-dependent GPx can deal with them converting hydroperoxides into non-reactive products (Surai *et al.*, 2018a,b) as follows:

$$\text{LOOH} + 2\text{GSH} \xrightarrow{\text{GPx}} \text{LOH (non-toxic)} + H_2O + \text{GSSG}$$

Thus, vitamin E and GPx are working in tandem providing effective antioxidant defence. Two recent reviews address this issue in relation to poultry production (Surai *et al.*, 2019a; Surai and Kochish, 2019). A summary of antioxidant system modulation by dietary vitamin E in poultry is shown in Figure 2.2. Indeed, vitamin E plays a vital role in poultry nutrition by regulating various branches of the antioxidant defence network in breeders, cockerels and growing chickens. On the one hand, vitamin E directly prevents lipid peroxidation in the egg yolk, chicken tissues and semen in stress conditions. On the other hand, vitamin E can affect other antioxidant protection mechanisms, including vitagenes (Chapter 12) and transcription factors, e.g. activation of Nrf2 (He *et al.*, 2019) and inhibition of NF-κB (Zhan *et al.*, 2020).

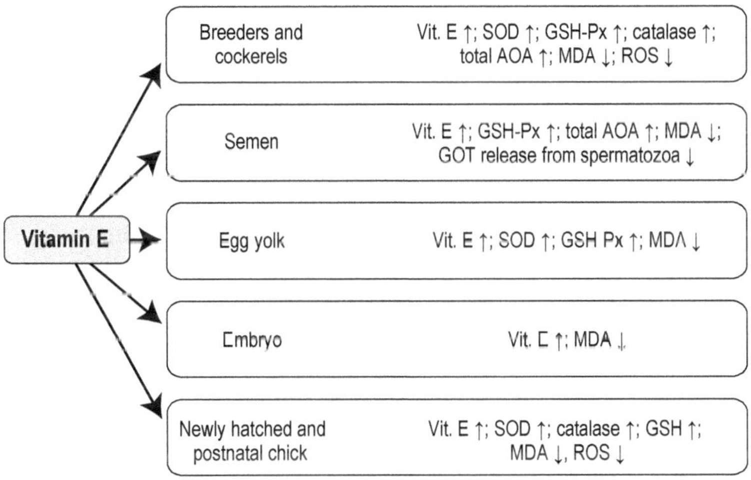

Figure 2.2. Antioxidant system modulation by dietary vitamin E in poultry (adapted from Surai *et al.*, 2019a).

Coenzyme Q, known also as ubiquinone, was discovered in 1957. The name ubiquinone is related to its 'ubiquitous' presence in all cells and the name coenzyme Q reflects the chemical structure of the compound containing one quinone group and 10 isoprenyl units. Coenzyme Q_{10} (CoQ_{10}) exists both in an oxidised and a reduced form, ubiquinone and ubiquinol, respectively (Overvad *et al.*, 1999). Importantly, ubiquinone is considered to be an important fat-soluble antioxidant and electron carrier synthesised in mitochondria (Stefely and Pagliarini, 2017). In general, dietary supplementation of CoQ does not affect the endogenous synthesis of CoQ in tissues. However, oxidative stress (physical exercise, thyroid hormone treatment, cold adaptation, vitamin A deficiency, etc.) is associated with increased CoQ synthesis reflecting a cellular adaptation (Ernster and Dallner, 1995). Therefore, CoQ synthesis is considered to be an adaptive mechanism in response to stress conditions when other antioxidants are depleted. For example, in vitamin E and Se deficient rats CoQ concentration elevated and CoQ-dependent reductase system is activated (Navarro *et al.*, 1998).

Antioxidant properties of CoQ are directly related to the protection in the gastrointestinal tract. For example, in rats treated per os with sodium nitrite increases TBARS in small intestinal mucosa and liver were observed. Pre-treatment of nitrite-poisoned rats with CoQ_{10} mitigated lipid peroxidation and increased total antioxidant status in animal blood (Grudzinski and Frankiewicz-Jozko, 2001). It directly involves in protection of biological molecules (lipids, proteins and DNA) from oxidative damage by quenching free radicals, regenerating other antioxidants (vitamins E and C) and regulating mitochondrial integrity (Varela-López *et al.*, 2016). It was suggested that Se inadequacy could compromise the cells ability to synthesise/obtain the optimal concentrations of coenzyme Q10, while optimal function of Se depends on the levels of coenzyme Q10 (Alehagen and Aaseth, 2015). It seems likely that additional synthesis of CoQ in stress conditions could be considered as an adaptive mechanism to deal with overproduction of free radicals.

Carotenoids comprise a family of more than 1,100 compounds responsible for a variety of bright colours in fall leaves, flowers (narcissus, marigold), fruits (pineapple, citrus fruits, paprika), vegetables (carrots, tomatoes), insects (ladybird), bird plumage (flamingo, cock of the rock, ibis, canary) and marine animals (crustaceans, salmon) (Maoka, 2009; Pfander, 1992; Yabuzaki *et al.* 2017). These pigments provide different colours from light yellow to dark red and when complexed with proteins they can produce green and blue colorations (Ong and Tee, 1992). Carotenoids – important elements of the antioxidant system, possessing antioxidant activities and participating directly or indirectly (for example, by recycling vitamin E or regulating expression of various genes) in antioxidant defences (Surai *et al.*, 2001a,b). An important role of canthaxanthin with a special emphasis to carotenoid antioxidant activities in breeder nutrition has been described (Surai, 2012a,b). Biological functions of these natural pigments in relation to animals or humans are not well defined but their antioxidant properties seem to be of major importance. Therefore, antioxidant interactions including their recycling provide an effective and reliable system of defence from free radicals and toxic products of their metabolism. Among many important biological functions of carotenoids, there participation in building an effective antioxidant defence network could be of vital importance. Indeed, biological value of direct AO activity of carotenoids associated with scavenging ROS is probably not very high (Costantini and Moller 2008). It seems likely that indirect effects of carotenoids on the antioxidant defences via upregulation of Nrf2 (Chang *et al.* 2018; Li *et al.* 2018; Xue *et al.* 2019; Yu *et al.* 2018; Zhao *et al.* 2017) and downregulation of NF-κB (Chang *et al.* 2018; Icel *et al.* 2019; Li *et al.* 2019a; Sahin *et al.* 2017) are a driving force of their beneficial effect in avian species in general and in poultry production in particular.

Vitamin C is a hydrophilic antioxidant functioning in an aqueous environment and possessing high free-radical-scavenging activity (Yu, 1994). It directly reacts with O_2^- and OH* and various lipid hydroperoxides and is taking part in the vitamin E recycling (Halliwell, 1996; Yu, 1994). Ascorbic acid is protective against a number of ROS (Carr and Frei, 1999; Halliwell, 1996, 1999). The major advantages of ascorbate as an antioxidant have been described as follows (Carr and Frei, 1999):

- Both ascorbate and ascorbyl radical have low reduction potentials and can react with most other biologically relevant radicals and oxidants.
- Ascorbyl radical has a low reactivity as a result of resonance stabilisation of unpaired electron and readily dismutates to ascorbate and dehydroascorbic acid (DAA).
- Ascorbyl radical and DAA can be converted into active ascorbate form by enzyme-dependent or independent pathways. In particular, ascorbyl radical can be reduced by NADH-dependent semidehydroascorbate reductase or by thioredoxin reductase. At the same time DAA can be reduced to AA by GSH, lipoic acid or glutaredoxin.
- Recent data from the epigenetics field indicate that ascorbate could play an important role in the demethylation of DNA and histone. In fact, by regulating the epigenome, ascorbate can be involved in embryonic development, postnatal development and in health maintenance in general (Camarena and Wang, 2016).

Glutathione (GSH) is the most abundant non-protein thiol in avian and mammalian cells and considered to be an active antioxidant in biological systems providing cells with their reducing milieu (Meister, 1992). Indeed, GSH is shown to be one of the most important non-enzymatic antioxidants in animals/poultry participating in redox balance maintenance and signalling, regulation of transcription factors and gene expression and many other important pathways/processes including epigenetic mechanisms (García-Giménez *et al.*, 2017; see Chapter 7 for more information).

Carnitine, taurine and silymarin are shown to be important antioxidants (see Chapters 9, 10 and 11, respectively). Betaine also can be included into an antioxidant family. Indeed, the evidence is accumulating to show that betaine also possess antioxidant properties (Alirezaei *et al.*, 2012; 2015; Hasanzadeh-Moghadam *et al.*, 2018; Li *et al.*, 2019b; Tsai *et al.*, 2015). Uric acid (UA) is traditionally considered to be a metabolically inert end-product of purine metabolism in man, without any physiological value. However, this ubiquitous compound has proven to be a selective antioxidant (Becker, 1993; Maples and Mason, 1988) which can:

- react with hydroxyl radicals and hypochlorous acid, itself being converted to innocuous products;
- serve as an oxidisable cosubstrate for the enzyme cyclooxygenase;
- protect against reperfusion damage induced by activated granulocytes;
- prevents oxidative inactivation of endothelial enzymes in stress conditions;
- chelate transition metal ions and scavenging ROS.

Recently it has been shown that the antioxidant and neuroprotective effects afforded by UA treatment involved the modulation of Nrf2-mediated oxidative stress and regulation of BDNF and NGF expression levels (Ya *et al.*, 2018). Interestingly, it has been suggested that uric acid could be an important biomarker for cell death rather than an antioxidant for neural protection (Liu *et al.*, 2019).

Polyphenolic compounds, including flavonoids has received tremendous attention as natural antioxidant. However, their direct involvement in antioxidant defences as free-radical scavengers has been questioned (Surai, 2014). In fact, polyphenolic concentrations in target tissues (except gut) are too low to show direct antioxidant activities. However, there are other mechanisms of polyphenolics involvement in antioxidant defences, including activation of Nrf2, inhibition of NF-κB (Di Meo *et al.*, 2020) and vitagene modulation (see Chapter 12 for details).

Some specific enzymes which hydrolyse oxidised bases preventing their incorporation into DNA can also be considered as a part of the second level of antioxidant defence (Slupphaug *et al.*, 2003). However, even the second level of antioxidant defence in the cell is not able to prevent damaging effects of ROS and RNS on lipids, proteins and DNA. In this case, the third level of defence is based on systems that eliminate damaged molecules or repair them. This level of antioxidant defence includes lipolytic (lipases), proteolytic (peptidases or proteases) and other enzymes (Msr, DNA repair enzymes, ligases, nucleases, polymerases, proteinases, phospholipases, various transferases, etc.) as well as protein chaperones, including HSPs.

All the antioxidants are operating in the body in association with each other forming an integrated antioxidant system. The co-operative interactions between antioxidants in the cell are vital for maximum protection from the deleterious effects of free radicals and toxic products of their metabolism. For-example, it is well established that vitamin E is the major antioxidant in biological membranes, a 'head quarter' of antioxidant network. However, it is usually present there in low molar ratios (one molecule per 2,000-3,000 phospholipids) but vitamin E deficiency is difficult to induce in adult animals. It is probably due to the fact that oxidised vitamin E can be converted back into the active reduced form by reacting with other antioxidants: ascorbic acid, glutathione, ubiquinols or carotenoids (Figure 2.3).

As a result of antioxidant action of vitamin E, tocopheroxyl radical is formed. This radical can be reduced back to an active form of α-tocopherol by coupling with ascorbic acid oxidation. Ascorbic acid can be regenerated back from the oxidised form by recycling with glutathione which can receive a reducing potential from

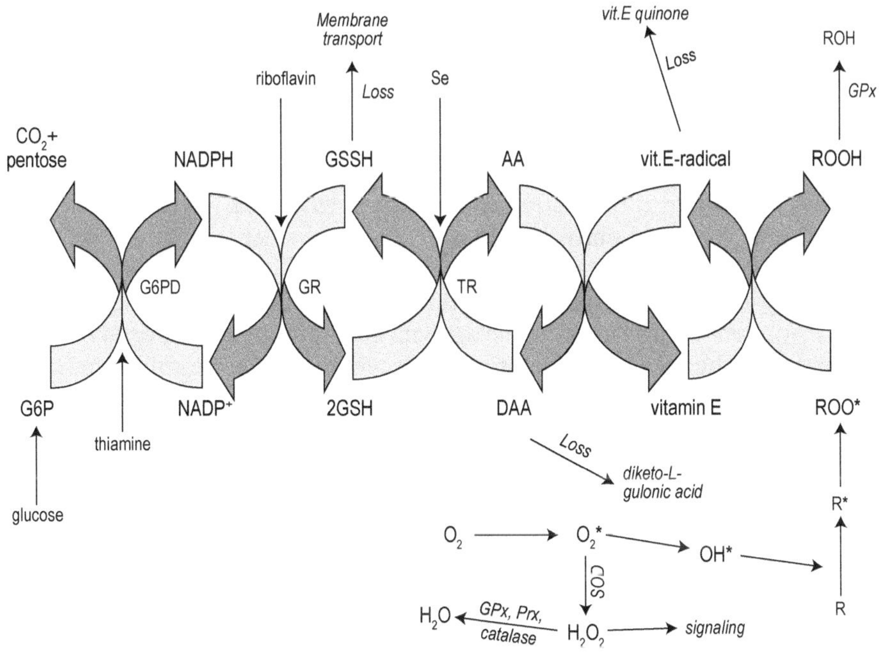

Figure 2.3. Redox cycle of vitamin E (adapted from Surai, 1999; Winkler *et al.*, 1994).

NADPH, synthesised in the pentose phosphate cycle of carbohydrate metabolism. Enzymes involved in vitamin E recycling are as follows: (1) thioredoxin reductase; (2) glutathione reductase; (3) glucose-6-phosphate dehydrogenase. Due to incomplete regeneration (the efficiency of recycling is usually less than 100%) in biological systems, the antioxidants have to be obtained from the diet (vitamin E and carotenoids) or synthesised in the tissues (ascorbic acid and glutathione).

This figure demonstrates a connection of antioxidant defence to the general body metabolism (the pentose phosphate cycle is the major producer of reducing equivalents in the form of NADPH) and shows involvement of other nutrients in this process. For example, dietary protein is a source of essential amino acids for glutathione synthesis, riboflavin is an essential part of glutathione reductase, niacin is a part of NADPH and Se is an integral part of thioredoxin reductase. At the same time thiamine is required for transketolase in the pentose phosphate pathway. Thus, a major finding in recent years is the possibility of direct or indirect vitamin E recycling (Surai, 2002, 2006, 2014; Surai and Fisinin, 2014). The rate of regeneration, or recycling, of the vitamin E radicals that form during their antioxidant action may affect both its efficiency in antioxidant action and its lifetime in biological systems and the greater recycling activity is associated with increased efficiency of inhibition of lipid peroxidation (Packer, 1995). It seems likely that vitamin E efficacy is very often more dependent on its recycling efficiency than on its concentration *per se*. Therefore, the antioxidant protection in the cell depends not only on vitamin E concentration and location, but also relies on the effective recycling. Indeed, if the recycling is effective then even low vitamin E concentrations are able to maintain high antioxidant protection in physiological conditions. For example, this could be demonstrated using chicken brain as a model system. Indeed, our data (Surai, 2002) indicate that the brain is characterised by extremely high concentrations of long chain polyunsaturated fatty acids predisposing this tissue to lipid peroxidation. Furthermore, brain contains much lower levels of vitamin E than other body tissues. However, in fresh chicken brain, levels of products of lipid peroxidation are very low, which could be a reflection of an effective vitamin E recycling by ascorbic acid which is present in this tissue in comparatively high concentrations. Antioxidant recycling is the most important element in understanding mechanisms involved in antioxidant protection against oxidative stress. The rate of regeneration, or recycling, of the vitamin E radicals may affect both its antioxidant efficiency and its lifetime in biological systems.

2.4 Antioxidant defence network

Living cells permanently balance the process of formation and inactivation of ROS and as a result ROS level remains low but still above zero. Adverse environmental conditions initiate attempts of organisms to resist the environment that became more aggressive (Skulachev, 1998). Cells can usually tolerate mild oxidative stress by additional synthesis of various antioxidants (glutathione, antioxidant enzymes, etc.) in an attempt to restore antioxidant/oxidant balance. At the same time, energy expenditures are increased, and respiration is activated leading to the increased yield of ROS (Skulachev, 1998). However, these adaptive mechanisms have limited ability. Once the free radical production exceeds the ability of antioxidant system to neutralise them, lipid peroxidation develops and causes damage to unsaturated lipids in cell membranes, amino acids in proteins and nucleotides in DNA and as a result, membrane and cell integrity is disrupted. Membrane damage is associated with a decreased efficiency of absorption of different nutrients and leads to an imbalance of vitamins, amino acids, inorganic elements and other nutrients in the organism. All these events result in decreased productive and reproductive performances of animals. Immunity incompetence and unfavourable changes in the cardio-vascular system, brain and neurones and muscle system due to increased lipid peroxidation make the situation even worse.

Therefore, the antioxidant defence includes several options (Surai, 2015, 2015a,b,c,d, 2016, 2017, 2018, 2020; Surai and Fisinin, 2014; 2015; Surai *et al.*, 2019):

- decrease localised oxygen concentration;
- decrease activity of pro-oxidant enzymes (carnitine, silymarin);
- improve efficiency of electron chain in the mitochondria and decreasing electron leakage leading to superoxide production (carnitine);
- induction of various transcription factors (e.g. NF-E2-related factor 2 [Nrf2], nuclear factor-κB [NF-κB] and others) and ARE-related synthesis of AO enzymes (SOD, GPx, CAT, glutathione reductase [GR], glutathione S-transferase [GST], etc.);
- binding metal ions (metal-binding proteins) and metal chelating;

- decomposition of peroxides by converting them to non-radical, nontoxic products (Se-GPx);
- chain breaking by scavenging intermediate radicals such as peroxyl and alkoxyl radicals (vitamins E, C, GSH, uric acid, carnitine, ubiquinol, bilirubin, etc.);
- repair and removal of damaged molecules (methionine sulfoxide reductase, DNA-repair enzymes, HSPs and other chaperons, etc.);
- redox-signalling and vitagene activation with synthesis and increased expression of protective molecules (GSH, thioredoxins, SOD, HSPs, sirtuins, etc.);
- antioxidant recycling mechanisms, including vitamin E recycling;
- protein glutathionylation is a way to prevent its irreversible oxidation;
- apoptosis activation and removal terminally damaged cells and restriction of mutagenesis (Figure 2.4).

As it was shown above all antioxidants in the body are working as a 'team' responsible for antioxidant defence and we call this team the antioxidant system. In this team one member helps another one working efficiently. In general vitamin E and coenzyme Q are considered to be a 'head-quarter' of the antioxidant defences (Surai *et al.*, 2019a), while Se is a 'chief executive' of antioxidant defence, since from 25 known selenoproteins, more than half participate in antioxidant defences (Surai, 2018; Surai and Kochish, 2019). Furthermore, a central role in antioxidant system regulation belongs to vitagene expression and additional synthesis of protective molecules in stress conditions ('ministry of defence') to improve adaptive ability to stress (Surai, 2018). Therefore, if relationships in this team are effective, which happens only in the case of balanced diet and sufficient provision of dietary antioxidant nutrients, then even low doses of such antioxidants as vitamin E could be effective. On the

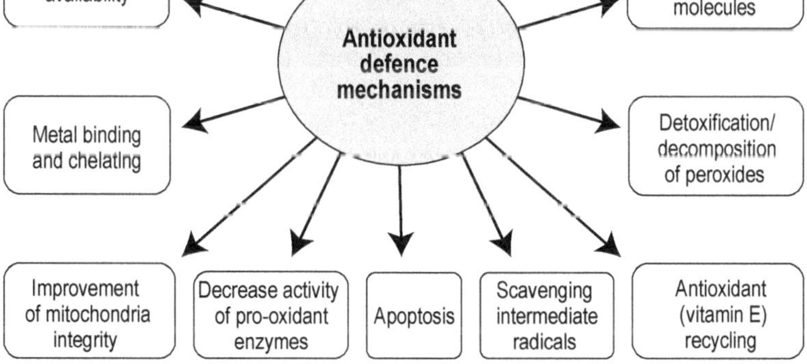

Figure 2.4. Antioxidant defence mechanisms (adapted from Surai *et al.*, 2019).

other hand, when this team is subjected to high stress conditions, free radical production is increased dramatically. During these times, without external help it is difficult to prevent damage to major organs and systems. This 'external help' is dietary supplementation with increased concentrations of natural antioxidants. For nutritionist or feed formulator it is a great challenge to understand when the internal antioxidant team in the body requires help, how much of this help to provide and what the economic return will be. Again, it is necessary to remember about essentiality of keeping right balance between free radical production and antioxidant defence. Indeed, ROS and RNS have another more attractive face participating in a regulation of varieties of physiological functions.

2.5 Oxidative stress and redox biology

The concept of oxidative stress as an imbalance between oxidants and antioxidants and oxidative stress responses was formulated in 1985 by Sies (Sies, 1985) and later it was updated (Sies, 2015, 2018, 2019; Sies and Jones, 2020; Sies *et al.*, 2017) to include current development and understanding of the topic. In particular, low-level physiological oxidative stress is called 'oxidative eustress' while high level oxidative stress is named as 'oxidative distress'. (Sies, 2019). Indeed, to deal with the oxidative challenge and to maintain redox homeostasis a stress response is initiated including the activation of gene expression of defence systems. There are two major 'master regulators' of the stress response, including the Nrf2/Keap1 and the NF-κB/IκB systems. Booth transcription factors are translocated to the nucleus and create protective, but in many cases opposite responses. While Nrf2 activates genes responsible for synthesis of an array of protective antioxidant molecules, NF-κB activates the expression of genes involved in inflammatory, immune, and acute phase responses. The stress response also includes other important factors such as the hypoxia induced response, the heat shock response, the unfolded protein response, and various repair programs as well as removal programs including autophagy, mitophagy, apoptosis, necroptosis, ferroptosis, etc. (Sies, 2019). Oxidative stress and its relationship to redox signalling is shown in Figure 2.5.

Indeed, redox signalling is shown to be integrated with main homeostatic mechanisms at the molecular, organellar, cellular, tissue and organismic levels (Sies and Jones, 2020) and associated with the vitagene network and various transcription factors (Surai, 2020).

2.6 Stress-response pathways

Avian species manage stress via adopting various mechanisms generally called 'stress response' (SR) associated with induction of various genes responsible for synthesis of various cyto-protective molecules. Depending on conditions, SR can be immediate, lasting from a few seconds to several hours and associated with receptor-mediated intracellular signalling or SR may be delayed with involvement of various modulators

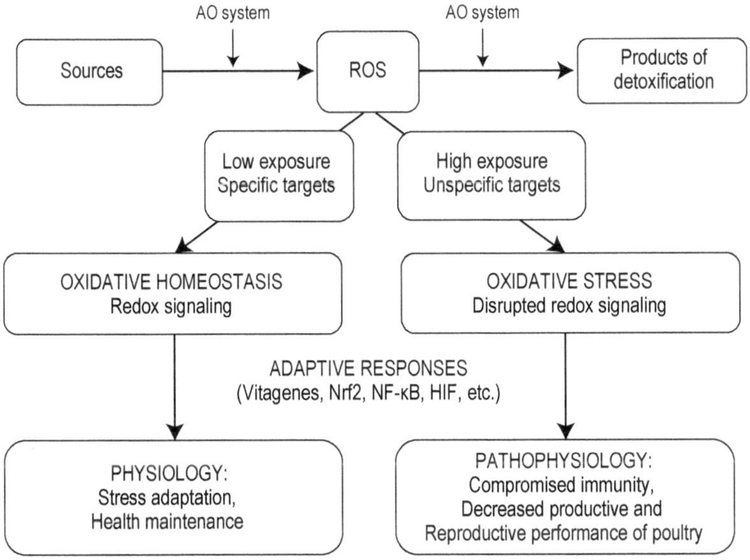

Figure 2.5. Oxidative stress and its relationship to redox signalling (adapted from Sies, 2018; Surai, 2018, 2020; Surai *et al.*, 2019).

and downstream effectors (Bhattacharya and Rattan, 2019). There are at least 8 stress-response pathways responsible for stress sensing and development of the adequate response (Figure 2.6).

For example, various stresses in poultry production caused by increased/reduced temperature, dust in air, mycotoxins, etc. can activation of HSF1 with following HSP activation (Surai and Kochish, 2017). Oxidative stress can cause an accumulation of unfolded proteins in the ER lumen triggering an SR called the unfolded protein response (UPR) or ER stress response (Bhattacharya and Rattan, 2019). Low oxygen levels, some metals and various chemicals can activate hypoxia-induced stress response (HISR) associated with activation of HIF inducible genes responsible for synthesis of protective molecules, including erythropoietin, HO-1, etc. helping deal with homeostasis disturbance (Hirota, 2020). Radiation, various pro-oxidants (pesticides, mycotoxins, etc.), as well as RONS can cause DNA damage and activate DNA damage response (DDR) associated with activation of a key serine/threonine protein signalling kinase (ATM) and Rad3-related (ATR). In the next step, ATM and ATR are recruited to double-strand and single strand breaks with following activation of DNA repair enzymes to repair or remove and replace damaged parts by intact once (Bhattacharya and Rattan, 2019; Li *et al.*, 2016).

Various pathogens, damaged macromolecules, allergens and various chemicals can activate inflammatory stress response (ISR) associated with activation/translocation to nucleus of NF-κB and synthesis of various pro-inflammatory cytokines (Bhattacharya and Rattan, 2019; Fairaq *et al.*, 2015). Low ATP/AMP or NAD+/NADH ratio due to lack

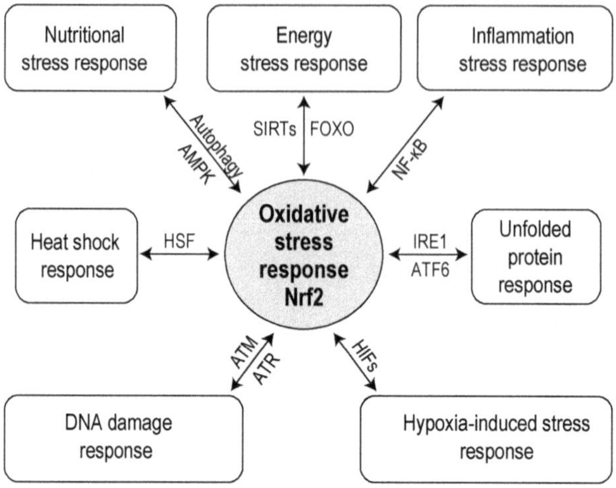

Figure 2.6. Hypothetical stress-response creation scheme (adapted from Bhattacharya and Rattan, 2019; Sies and Jones, 2020; Surai, 2020; Surai *et al.*, 2019). AMPK: AMP-activated protein kinase; ATF6: activating transcription factor 6; ATM: Ataxia-telangiectasia-mutated; ATR: Ataxia-telangiectasia and Rad 3-related; FOXO: forkhead box protein; HIF: hypoxia inducible factor; HSF: heat shock factor; IRE1: inositol-requiring enzyme 1; NF-κβ: nuclear factor kappa-light-chain-enhancer of activated B cells; Nrf2: nuclear factor erythroid-2 related factor 2; SIRTs-sirtuins.

of energy could cause energy stress response (ESR) mediated mainly via sirtuin system activation, AMPK activation and deacetylation of PGC-1α, FOXO and other important molecules (Bhattacharya and Rattan, 2019; Lin *et al.*, 2014). Nutritional disbalance/inadequacy, hypoxia and damaged cellular organelles can lead to nutritional stress response (NSR) associated with AMPK activation and autophagosome formation with following lysosomal digestion of damaged mitochondria (Bhattacharya and Rattan, 2019; Oh *et al.*, 2018; Stroeve *et al.*, 2015). Finally, oxidative stress response is related to Nrf2 activation and synthesis of a range protective molecules (Surai *et al.*, 2019). Interestingly, all 8 stress-responses are interrelated, and oxidative stress response can be placed into the centre of stress-response creation (Figure 2.6).

2.7 Oxidative stress and transcription factors

It is important to mention that ROS are no longer viewed as just toxic by-products of mitochondrial respiration but are now appreciated for their role in regulating a myriad of cellular signalling pathways (Reczek and Chandel, 2015). It has been suggested that the signalling ROS are produced in a subtly regulated manner, while many deleterious ROS are produced and react randomly (Niki, 2014). Therefore, it is unlikely that nutritional antioxidants detrimentally affect physiologically important signalling functions, since the antioxidants do not scavenge signalling ROS/RNS nor do they inhibit the formation of signalling molecules (Niki, 2012, 2016). Recent evidence suggests that several selenoproteins could participate in cell signalling. In particular, selenoprotein W and six other small thioredoxin-like mammalian selenoproteins (SelH, SelM, SelN, SelT, SelV and Sep15) may serve to transduce hydrogen peroxide signals into regulatory disulphide bonds in specific target proteins (Hawkes and Alkan, 2010). Similarly, GPx and TrxR are also involved in cellular redox balance regulation (Labunskyy *et al.*, 2014). Oxidation-reduction (redox) based regulation of gene expression is a fundamental regulatory mechanism in cell biology acting at the cell-signalling level. In fact, redox signalling is the overlap of signal transduction with redox biology. Redox signalling is essential in physiological homeostasis and alterations in redox signalling are observed in stress conditions and aging; sustained deviation from redox homeostasis results in disease (Forman, 2016), and decreased productive and reproductive performance of poultry.

Since ROS are damaging to many biological molecules, the antioxidant systems are responsible for the prevention of this damage. However, a basal level of oxidative stress is essential for cell adaptation and survival. Therefore, a moderate level of oxidative stress can create adaptive responses and improve the adaptive ability to stressful challenges/conditions (Yan, 2014). Indeed, in animals, redox-signalling pathways use ROS as signalling molecules to activate genes responsible for regulation of various functions, including growth, differentiation, proliferation and apoptosis. Furthermore, the antioxidant defence systems are also under regulation by various transcription factors (Kweider *et al.*, 2014; Ma and He, 2012; Majzunova *et al.*, 2013; Song and Zou, 2014). In fact, the redox balance is controlled by a battery of transcriptional factors, including Nrf2, NF-κB, PPARs, PGC-1a, p53, FoxO, MAPK, AP-1, etc. (Lushchak, 2011; Wang and Hai, 2016). They regulate redox status by modulating ROS-generating enzymes and antioxidant enzymes in a cooperative and interactive way. In recent years great attention has been paid to basic leucine zipper transcription factor, Nrf2 and NF-κB.

2.7.1 Transcription factor Nrf2

It is known that Nrf2 is the redox-sensitive master regulator of oxidative stress signalling and stress response, and critical for cell survival under stressful conditions (Itoh *et al.*, 2010). It has been shown that the Nrf2 antioxidant response pathway is an important player in the cellular antioxidant defence. Indeed, it is responsible for activation of a variety of genes involved in early defence reactions of higher organisms (Ma, 2013; Van der Wijst *et al.*, 2014). High expression of Nrf2 in organs that face environmental stress, including lungs and the small intestine (Itoh *et al.*, 2015), is a confirmation of its importance in stress adaptation processes. Clearly, Nrf2 has a significant role in adaptive responses to oxidative stress, being involved in the induction of the expression of various antioxidant molecules to combat oxidative and electrophilic stress (Howden, 2013; Keum and Choi, 2014; Tang *et al.*, 2014; Vriend and Reiter, 2015). It is suggestive that under normal physiological conditions, Nrf2 is kept in the cytoplasm as an inactive complex with the negative regulator, Kelch-like-ECH-associated protein 1 (Keap1), which is anchored to the actin cytoskeleton. In fact, Keap1 sequesters Nrf2 in the cytoplasm and forwards it to a Cul3-based E3 ligase which is followed by rapid ubiquitin-proteasome degradation leading to a short (about 20 min) half-life of Nrf-2 under physiological conditions (for review see Choi *et al.*, 2014).

It seems likely that, Keap-1 is an important cellular redox sensor and upon exposure to oxidative or electrophilic stress, critical cysteine thiols of Keap1 are modified/oxidised and Keap1 loses its ability to ubiquitinate Nrf2 resulting in preventing its degradation. There are also other ways of Nrf2 activation. For example, phosphorylation of Nrf2 at specific serine and/or tyrosine residues also causes an Nfr2-Keap1 dissociation resulting in Nrf2 release and translocation to nucleus, where it combines with a small musculoaponeurotic fibrosarcoma protein called Maf to form a heterodimer (Bhakkiyalakshmi *et al.*, 2015). Indeed, by binding to ARE in the upstream promoter region of genes encoding various antioxidant molecules, Nrf2 regulates the expression of antioxidant proteins, thiol molecules and other protective molecules. This includes enzymes of the first line of the antioxidant defence, namely SOD, GPx and catalase, detoxification enzymes (HO-1, NQO1, and GST), GSH related proteins (γ-GCS), NADPH-producing enzymes and others stress-response proteins contributing to the prevention of oxidative and inflammatory damage (Lee *et al.*, 2013; Zhou *et al.*, 2014; Figure 2.7).

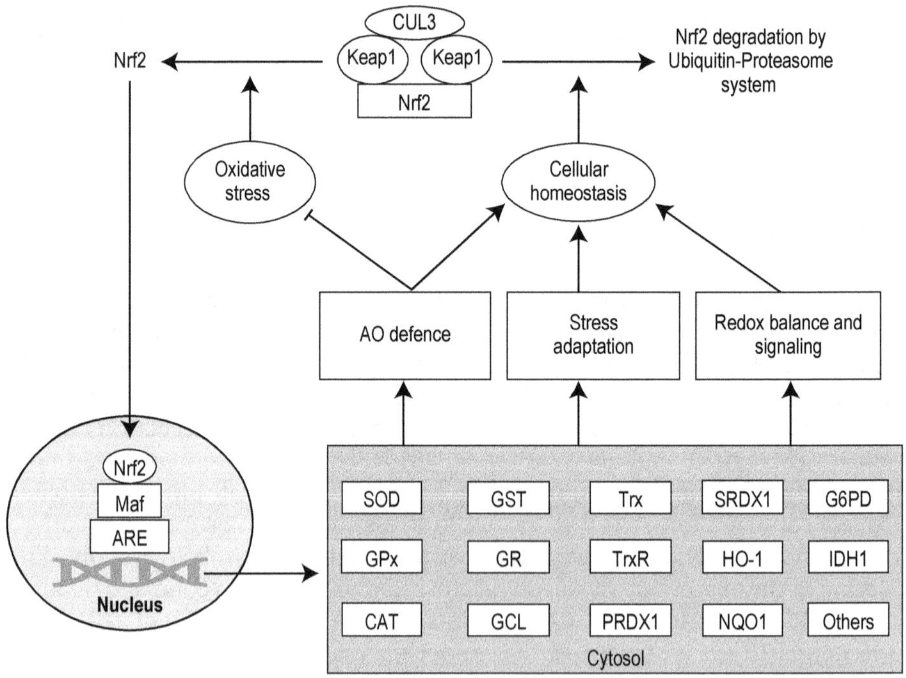

Figure 2.7. Participation of Nrf2 in the AO defence network (adapted from Surai *et al.*, 2019).

In cells under physiological homeostatic conditions, cytosolic transcription factor Nrf2 is kept at low levels being bound to Keap1 by the ubiquitin ligase complex Cullin (Cul)3-RING-box protein (Rbx)1 (Cul3). This complex ubiquitinates Nrf2, triggering its constant proteasomal degradation. Under oxidative stress, ROS modify/oxidise SH-groups within Keap1 leading to conformational changes inducing the Nrf2 release from Keap1. This prevents Nrf2 proteasomal degradation and Nrf2 translocates to the nucleus. In the nucleus, Nfr2 binds to the ARE and initiates the transcription of an array of direct or indirect antioxidant enzymes including SOD, GPx, CAT, GST, GR, GCL, Trx, TrxR, PRDX1, SRDX1, HO-1, NQO1. G6PD, IDH2, etc. These enzymes contribute to the improvement of the antioxidant defence network and reduce the cellular oxidative stress.

The Nrf-2 induced synthesis of AO enzymes also participates in regulation of stress adaptation and redox signalling. The restoration of cellular homeostasis leads to Nrf2-Keap-1 complex formation and activation of Nrf2 degradation by ubiquitin-proteosome system and decreases the Nrf-2 mediated synthesis of AO enzymes. In fact, hundreds of cytoprotective genes are regulated by Nrf2 (Itoh *et al.*, 2015) and gene products (proteins) are involved in the maintenance and responsiveness of the cellular antioxidant systems. Indeed, an orchestrated change in gene expression via Nrf2 and ARE is a key mechanism of the protective effect against oxidative stress (Lee *et al.*, 2003).

It is suggestive that Nrf2 is controlled through a complex transcriptional/epigenetic and post-translational network that provides regulatory mechanisms ensuring Nrf2 activity increases in response to redox disturbances, inflammation, growth factor stimulation and nutrient/energy fluxes orchestrating adaptive responses to diverse forms of stress (Hayes and Dinkova-Kostova, 2014). As mentioned above, there is a range of Nrf2 activating mechanisms, including stabilisation of Nrf2 via Keap1 cysteine thiol modification and phosphorylation of Nrf2 by upstream kinases (Surh, 2008; Surh *et al.*, 2008). It is proven that effects of Nrf2 on the adaptive ability of cells is quite broad and goes beyond activation of synthesis of antioxidant molecules. Indeed, Nrf2 also contributes to homeostasis by up-regulating the repair and degradation of damaged macromolecules, and by modulating intermediary metabolism conducting direct metabolic reprogramming during stress (Zhou *et al.*, 2014). Recently molecular mechanisms of regulating roles of transcription factors in cellular adaptation to stress have been extensively studied. In particular, it has been suggested that low intensity oxidative stress is predominantly sensed by the Keap1/Nrf2 system (Lushchak, 2011) followed by downstream up-regulation of the protective AO genes. It is interesting to note that intermediate oxidative stress also increases the activity of antioxidant enzymes, but mainly via NF-κB and AP-1 pathways (Lushchak, 2011). Furthermore, at both, low and intermediate intensity oxidative stresses, MAP-kinases and other kinases seem to be involved in signal sensing and cellular response, leading to enhanced antioxidant potential (Zhou *et al.*, 2014). Emerging evidence clearly indicates that Nrf2 can interact with other transcription factors, including heat shock factor (Hsf1; Dayalan Naidu *et al.*, 2015) to provide additional options for AO system regulation. As mentioned above, the Nrf2 stress pathway intimately communicates with mitochondria to maintain cellular homeostasis during oxidative stress (Itoh *et al.*, 2015). Recent findings have shown that KEAP1 contains multiple stress sensors which allow Nrf2 effective response to diverse cellular signals, from oxidative stress and cellular metabolites to dysregulated autophagy (Baird and Yamamoto, 2020). Therefore, pharmaceutical/nutritional modulation of NRF2's cytoprotective activity is of great importance in stress adaptation and resistance. Indeed, there is great interest in modulation of Nrf2 expression and activity by various natural products (Hassanein *et al.*, 2020; Zhang and Chapman, 2020), including silymarin (Vargas-Mendoza *et al.*, 2020).

2.7.2 Transcription factor nuclear factor-kappa B

NF-κB is an inducible transcription factor that regulates many cellular processes including immunity and inflammation. NF-κB consists of a group of five related proteins that are capable of binding to DNA. This transcription factor is activated by a wide range of stimuli including oxidative stress. It has been shown that NF-κB regulates the transcription of many different genes, including pro-inflammatory cytokines and leukocyte adhesion molecules, acute phase proteins and anti-microbial peptides (Buelna-Chontal and Zazueta, 2013; Pedruzzi *et al.*, 2012; Tkach *et al.*,2014). There are some similarities in regulation of Nrf2 and NF-κB. For example, in physiological conditions, NF-κB is found in cytoplasm in an inactive state associated with the inhibitory IκB (inhibitor of kappa B) protein preventing its binding to target sites. It has been proven that activation of NF-κB is an effective mechanism of host defence against infection and stress (Pal *et al.*, 2014). As a result of action of cytokines and other stressors, IκB proteins are rapidly phosphorylated by IκB kinase on specific serine residues, followed by ubiquitination, and degradation by the 26S proteasome. The following release of NF-κB and its translocation to the nucleus is responsible for the transcription of target genes, for cell survival, and involved with inflammation, apoptosis, cell proliferation and differentiation (Hayden and Ghosh, 2014).

NF-κB transcription factors, such as p65, can combine to form hetero- and homodimers of different composition, providing a tool for effective regulation of different sets of gene targets (Grilli and Memo, 1997). There is a range of additional stimuli implicated into the NF-κB activation including, cell-surface receptors, inhibitory κB kinases, IB proteins, and factors that are involved in the posttranslational modification of the Rel proteins, etc. (Buelna-Chontal and Zazueta, 2013; Hayden and Ghosh, 2014; Pal *et al.*, 2014; Pedruzzi *et al.*, 2012; Tkach *et al.*, 2014). Accumulating evidence indicates that there is a complex interplay/crosstalk between Nrf2 and NF-κB pathways. For example, several Nrf2 activators can inhibit NF-κB pathway. NF-κB may also directly antagonise the transcriptional activity of Nrf2 (for review see Tkach *et al.*, 2014). In recent years, several compounds, including LC, have been shown to have inhibitory activities against multiple components of NF-κB activation pathway. It is interesting to note that SIRT1 and NF-κB show an antagonistic relationship in controlling inflammation (de Gregorio *et al.*, 2020). It seems likely that NF-κB has dual roles in development and immunity of various organisms being an important element of the adaptive response to environmental challenges (Williams and Gilmore, 2020). In fact, various chaperones, including vitagene HSP70, controlling the NF-κB pathway, are important regulators of the transduction of cytokine-mediated signals, modulating systemic inflammatory responses (Fusella *et al.*, 2020). Indeed, NF-κB signalling is responsible for context-dependent transcriptional control in immune cells (Mulero *et al.*, 2019). It seems likely that interaction/interplay between Nrf2 and NF-κB is an important evolutionary conserved mechanisms of stress adaptation and resistance.

2.7.3 Nrf2 and NF-κB interplay in oxidative stress

It seems likely that under oxidative stress the transcription factors NF-κB and Nrf2 antagonise each other by impairing activation of the other (Zhang *et al.*, 2019). This substantially complicates the evaluation of the relative impact of each pathway into regulation of the oxidative stress and stress adaptation. For example, some antioxidant enzymes are controlled by booth, Nrf2 and NF-κB. Indeed, expression HO-1 is regulated by of Nrf2, NF-κB and HIF-1α signalling (Zhang *et al.*, 2019). Accumulating evidence indicates that there is a complex interplay/crosstalk between Nrf2 and NF-κB pathways under stress and a variety of pathophysiological conditions (Sivandzade *et al.*, 2019). The authors reviewed existing evidence proving the point that deletion of Nrf2 (Nrf2 knock out mice) was associated with enhanced inflammation, while its upregulation decreases pro-inflammatory and immune responses transcriptionally regulated by NF-κB. For example, several Nrf2 activators can inhibit NF-κB pathway. NF-κB may also directly antagonise the transcriptional activity of Nrf2 (for review see Tkach *et al.*, 2014). The details of Nrf2-NF-κB interactions, provided by (Sivandzade *et al.*, 2019) can be summarised as follows:

- Nrf2 can inhibit the activation of NF-κB pathway by increasing antioxidant defences neutralising ROS, thus reduces ROS-mediated NF-κB activation.
- Nrf2 can also prevent the degradation of IκB-α leading to blockage of NF-κB nuclear translocation and prevention of transcription of pro-inflammatory genes.
- NF-κB can inhibit Nrf2 activity through stimulation of the recruitment of histone deacetylase3 (HDAC3) to the ARE region associated with prevention of ARE gene transcription.
- NF-κB is able to decrease expression of free CREB binding protein (CBP) by competing for CH1-KIX domain of CBP with Nrf2 leading to decreased Nrf2 expression.

It has been shown that the activation of the Nrf2/HO-1 signal transduction pathway can inhibit NF-κB mediated effects in various model systems (Jiang *et al.*, 2014; Wang *et al.*, 2018). Very often, various protective nutrients possessing AO activities affect both Nrf2 and NF-κB. Most research in this area was conducted with various model systems using plant extracts and individual plant polyphenolics to prove this point. Furthermore, various toxic compounds were also used showing deferent direction of activation of Nrf2 and NF-κB. In fact, the oxidative stress accepted to activate various transcription factors, including Nrf2, NF-κB, AP-1, HIF-1α, p53, PPAR-γ, and *β*-catenin/Wnt (Reuter *et al.*, 2010). The main results are summarised below.

It is proven that Nrf2 and NF-κB affect each other expression and activity to coordinate anti-oxidative and inflammatory responses, but it is not yet known how this interconnection takes place (Sivandzade *et al.*, 2019). Thus, stress-associated changed in redox balance and in activities of transcription factors such as Nrf2/ Keap1 and NF-κB/ IκB/IKK provide adaptive cell responses to oxidants and variety of stress stimuli through regulation of gene expression under both physiological and pathological conditions (Moldogazieva *et al.*, 2018) Despite the accepted concept of physiological ROS/RNS signalling there is still no complete consensus on molecular mechanisms explaining beneficial or deleterious effects of RONS on biomolecules and cellular functions (Moldogazieva *et al.*, 2018). Hypothetical scheme of Nrf2-NF-κB cross-talk is shown in the Figure 2.8.

There is a delicate balance between Nrf2 and NF-κB expression in various tissues and in physiological conditions the balance is well maintained. It seems likely that increased NF-κB expression due to various stresses can cause simultaneous increase in expression of Nrf2 leading to improved antioxidant defences and decreased NF-κB expression as a feedback mechanism. Other transcription factors and vitagenes are also involved in regulation of the balance. Once the ability to balance AO defences against ROS production is overwhelmed due to extremely high stress, redox status would be changed, Nrf2/NF-κB balance would be broken leading to detrimental consequences in terms of health, productive and reproductive performance maintenance in poultry and farm animals.

Indeed, in the body a delicate critical balance exists between antioxidant defence and repair systems and free radical generation. In physiological conditions the right and left parts of the so-called 'balances' are in equilibrium i.e. free radical generation

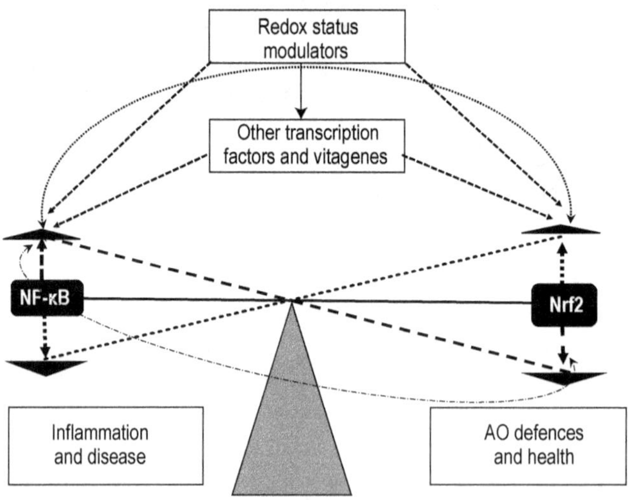

Figure 2.8. Hypothetical Nrf2-NF-κB crosstalk (adapted from Surai *et al.*, 2019).

is neutralised by the antioxidant system. Exogenous factors are among the most important elements, which increase an efficiency of the antioxidant system of the organism. Natural and synthetic antioxidants in the feed as well as optimal levels of Mn, Cu, Zn and Se help to maintain the efficient levels of endogenous antioxidants in the tissues. Optimal diet composition allows the antioxidants of the food to be efficiently absorbed and metabolised. Optimal temperature, humidity and other environmental conditions are also required for the effective protection against free radical production. The prevention of different diseases by antibiotics and other drugs is an integral part of the efficient antioxidant defence as well.

2.8 Conclusions

Antioxidant-prooxidant balance in the cell is an important determinant of various physiological functions. Indeed, oxidative stress occurs when this balance is disturbed due to overproduction of free radicals or compromised antioxidant defences. Free radical overproduction and oxidative stress are considered as a pathobiochemical mechanism involved in the initiation or progression phase of various diseases. In poultry production free radial generation, lipid peroxidation and protein oxidation are responsible for the decrease of productive and reproductive performance as well as for decreased product quality. Dietary antioxidants are important players in protecting against the development of the oxidative stress in stress conditions of commercial egg and meat production. However, recent evidence suggests that oxidative stress can induce changes in gene expression. In fact, some free radicals, such as H_2O_2, are now considered to be signal molecules taking part in signal transduction in the cell, affecting redox homeostasis and stress adaptation. In fact, there is a range of redox-sensitive transcription factors, including Nrf2, NF-κB, FOXO, p53, PGC-1α, HIF-1 and HSF1 (Figure 2.9), which participate in regulation of various cellular processes including adaptation to stress. Indeed, activation of the aforementioned transcription factors in stress conditions would lead to additional synthesis of an array of protective molecules to deal with oxidative stress and to re-establish adaptive homeostasis.

The regulation of gene expression by oxidants, antioxidants, and redox state has emerged as a novel subdiscipline in molecular biology that has promising implications for the feed industry and poultry production. Thus, the redox state/homeostasis of the cell, which reflects antioxidant/prooxidant balance, can be considered as an important element of gene regulation. Therefore, the effect of antioxidants on animal health is much deeper than one could expect several years ago. The mechanisms by which natural antioxidants act at the molecular and cellular level include roles in gene expression and regulation, apoptosis, and signal transduction. Antioxidants are involved in fundamental metabolic and homeostatic processes. However, there are still many gaps in our knowledge of the basic molecular mechanisms of oxidative damage and antioxidant defences.

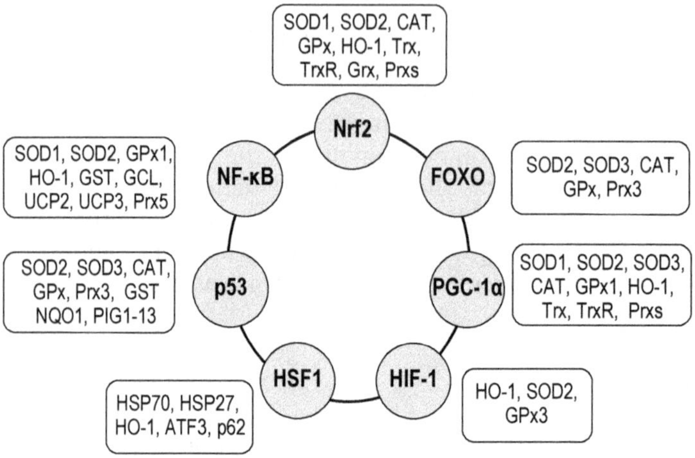

Figure 2.9. Transcription factors and their clients involved in redox status regulation (adapted from Dayalan Naidu *et al.*, 2015; Dengler *et al.*, 2014; Surai *et al.*, 2019; Wang and Hai, 2016).

The transcription factors interact with each other and with other important signalling pathways in a cooperative and interactive way to stimulate additional synthesis of various antioxidants to deal with oxidative stress and to re-establish adaptive homeostasis under various stress conditions. Molecular mechanisms of interactions between antioxidants, transcription factors and vitagenes and their participation in stress adaptation and establishment of adaptive homeostasis will be considered in the next chapters of this book.

References

Ahmad, S., Khan, H., Shahab, U., Rehman, S., Rafi, Z., Khan, M.Y., Ansari, A., Siddiqui, Z., Ashraf, J.M., Abdullah, S.M., Habib, S. and Uddin, M., 2017. Protein oxidation: an overview of metabolism of sulphur containing amino acid, cysteine. Frontiers in Bioscience 9: 71-87.

Alehagen, U. and Aaseth, J., 2015. Selenium and coenzyme Q10 interrelationship in cardiovascular diseases - a clinician's point of view. Journal of Trace Elements in Medicine and Biology 31: 157-162.

Alirezaei, M., Khoshdel, Z., Dezfoulian, O., Rashidipour, M. and Taghadosi, V., 2015. Beneficial antioxidant properties of betaine against oxidative stress mediated by levodopa/benserazide in the brain of rats. The journal of Physiological Sciences 65: 243-252.

Alirezaei, M., Reza Gheisari, H., Reza Ranjbar, V. and Hajibemani, A., 2012. Betaine: a promising antioxidant agent for enhancement of broiler meat quality. British Poultry Science 53: 699-707.

Ames, B.N., 2003. An Enthusiasm for Metabolism. Journal of Biological Chemistry 278: 4369-4380.

Ames, B.N. and Gold, L.S., 1997. The causes and prevention of cancer: gaining perspective. Environmental Health Perspectives 105 (Suppl 4): 865-873

Ayala, A., Muñoz, M.F. and Argüelles, S., 2014. Lipid peroxidation: production, metabolism, and signaling mechanisms of malondialdehyde and 4-hydroxy-2-nonenal. Oxidative Medicine and Cellular Longevity 2014: 360438.

Baird, L. and Yamamoto, M., 2020. The molecular mechanisms regulating the KEAP1-NRF2 pathway. Molecular and Cellular Biology 40: e00099-20. https://doi.org/10.1128/MCB.00099-20

Bao, Y. and Ertbjerg, P., 2019. Effects of protein oxidation on the texture and water-holding of meat: a review. Critical Reviews in Food Science and Nutrition 59: 3564-3578.

Bar-Noy, S. and Moskovitz, J., 2002. Mouse methionine sulfoxide reductase B: effect of selenocysteine incorporation on its activity and expression of the seleno-containing enzyme in bacterial and mammalian cells. Biochemical and Biophysical Research Communications 297: 956-961.

Becker, B.F., 1993. Towards the physiological function of uric acid. Free Radical Biology and Medicine 14: 615-631

Bhakkiyalakshmi, E., Sireesh, D., Rajaguru, P., Paulmurugan, R. and Ramkumar, K.M., 2015. The emerging role of redox-sensitive Nrf2-Keap1 pathway in diabetes. Pharmacological Research 91: 104-114.

Bhattacharya, S. and Rattan, S.I., 2019. Primary stress response pathways for preconditioning and physiological hormesis. In: Rattan, S.I.S. and Kyriazi, M. (eds) The science of hormesis in health and longevity. Academic Press, London, UK, pp. 35-51.

Buelna-Chontal, M. and Zazueta, C., 2013. Redox activation of Nrf2 & NF-κB: a double end sword? Cellular Signalling 25: 2548-2557.

Camarena, V. and Wang, G., 2016. The epigenetic role of vitamin C in health and disease. Cellular and Molecular Life Sciences 73: 1645-1658.

Cardozo-Pelaez, F., Brooks, P.J., Stedeford, T., Song, S. and Sanchez-Ramos, J., 2000. DNA damage, repair, and antioxidant systems in brain regions: a correlative study. Free Radical Biology and Medicine 28: 779-785.

Carr, A. and Frei, B., 1999. Does vitamin C act as a pro-oxidant under physiological conditions? FASEB Journal 13: 1007-1024.

Chance, B., Sies, H. and Boveries, A., 1979. Hydroperoxide metabolism in mammalian organs. Physiological Reviews 59: 527-605.

Chang, J., Zhang, Y., Li, Y., Lu, K., Shen, Y., Guo, Y., Qi, Q., Wang, M. and Zhang, S., 2018. NrF2/ARE and NF-κB pathway regulation may be the mechanism for lutein inhibition of human breast cancer cell. Future Oncology 14: 719-726.

Chaudiere, J. and Ferrari-Iliou, R., 1999. Intracellular antioxidants: from chemical to biochemical mechanisms. Food and Chemical Toxicology 37: 949-962.

Choi, B.H., Kang, K.S. and Kwak, M.K., 2014. Effect of redox modulating NRF2 activators on chronic kidney disease. Molecules 19: 12727-12759.

Chow, C.K., Ibrahim, W., Wei, Z. and Chan, A.C., 1999. Vitamin E regulates mitochondrial hydrogen peroxide generation. Free Radical Biology and Medicine 27: 580-587.

Costantini, D. and Møller, A.P., 2008. Carotenoids are minor antioxidants for birds. Functional Ecology 22, 367-370.

Croteau, D.L. and Bohr, V.A., 1997. Repair of oxidative damage to nuclear and mitochondrial DNA in mammalian cells. Journal of Biological Chemistry 272: 25409-25412.

Davies, M.J. 2016. Protein oxidation and peroxidation. Biochem Journal 473: 805-825.

Dayalan Naidu, S., Kostov, R.V. and Dinkova-Kostova, A.T., 2015. Transcription factors Hsf1 and Nrf2 engage in crosstalk for cytoprotection. Trends in Pharmacological Sciences 36: 6-14.

De Gregorio, E., Colell, A., Morales, A. and Marí, M., 2020. Relevance of SIRT1-NF-κB axis as therapeutic target to ameliorate inflammation in liver disease. International Journal of Molecular Sciences 21: E3858.

Dean, R.T., Fu, S., Stocker, R. and Davies, M.J., 1997. Biochemistry and pathology of radical-mediated protein oxidation. Biochemical Journal 324: 1-18.

Dengler, V.L., Galbraith, M. and Espinosa, J.M., 2014. Transcriptional regulation by hypoxia inducible factors. Critical Reviews in Biochemistry and Molecular Biology 49: 1-15.

Di Meo, F., Valentino, A., Petillo, O., Peluso, G., Filosa, S. and Crispi, S., 2020. Bioactive polyphenols and neuromodulation: molecular mechanisms in neurodegeneration. International Journal of Molecular Sciences 21: 2564.

Diplock, A.T., 1994. Antioxidants and disease prevention. Molecular Aspects of Medicine 15: 295-376.

Ernster, L. and Dallner, G., 1995. Biochemical, physiological and medical aspects of ubiquinone function. Biochimica et Biophysica Acta 1271: 195-204.

Estévez, M., Geraert, P.A., Liu, R., Delgado, J., Mercier, Y. and Zhang, W., 2020. Sulphur amino acids, muscle redox status and meat quality: more than building blocks – invited review. Meat Science 163: 108087.

Fairaq, A., Goc, A., Artham, S., Sabbineni, H. and Somanath, P.R., 2015. TNFα induces inflammatory stress response in microvascular endothelial cells via Akt- and P38 MAP kinase-mediated thrombospondin-1 expression. Molecular and Cellular Biochemistry 406: 227-236.

Finkel, T., 2000. Redox-dependent signal transduction. FEBS Letters 476: 52-54.

Forman, H.J., 2016. Redox signaling: an evolution from free radicals to aging. Free Radical Biology and Medicine 97: 398-407.

Friguet, B., Bulteau, A.L., Chondrogianni, N., Conconi, M. and Petropoulos, I., 2000. Protein degradation by the proteasome and its implications in aging. Annals of the New York Academy of Sciences 908:143-154.

Fujii, J. and Taniguchi, N., 1999. Down regulation of superoxide dismutases and glutathione peroxidase by reactive oxygen and nitrogen species. Free Radical Research 31: 301-308.

Fusella, F., Seclì, L., Cannata, C. and Brancaccio, M., 2020. The one thousand and one chaperones of the NF-κB pathway. Cellular and Molecular Life Sciences 77: 2275-2288.

Galey, J-B., 1997. Potential use of iron chelators against oxidative damage. In: Sies, H. (ed.) Antioxidants in disease mechanisms and therapy. Academic Press, San Diego, CA, USA, pp. 167-203.

Gao, J., Yin, D.H., Yao, Y., Sun, H., Qin, Z., Schoneich, C., Williams, T.D. and Squier, T.C., 1998. Loss of conformational stability in calmodulin upon methionine oxidation. Biophysical Journal 74: 1115-1134.

García-Giménez, J.L., Romá-Mateo, C., Pérez-Machado, G., Peiró-Chova, L. and Pallardó, F.V., 2017. Role of glutathione in the regulation of epigenetic mechanisms in disease. Free Radical biology & Medicine 112: 36-48.

Gaschler, M.M. and Stockwell, B.R., 2017. Lipid peroxidation in cell death. Biochemical and Biophysical Research communications 482: 419-425.

Goloubinoff, P., 2016. Mechanisms of protein homeostasis in health, aging and disease. Swiss Medical Weekly 146: w14306.

Gonzalez-Rivas, P.A., Chauhan, S.S., Ha, M., Fegan, N., Dunshea, F.R. and Warner, R.D., 2020. Effects of heat stress on animal physiology, metabolism, and meat quality: a review. Meat Science 162: 108025.

Grilli, M. and Memo, M., 1997. Transcriptional pharmacology of neurodegenerative disorders: novel venue towards neuroprotection against excitotoxicity? Molecular Psychiatry 2: 192-194.

Grimaud, R., Ezraty, B., Mitchell, J.K., Lafitte, D., Briand, C., Derrick, P.J. and Barras, F., 2001. Repair of oxidized proteins. Identification of a new methionine sulfoxide reductase. Journal of Biological Chemistry 276: 48915-48920.

Groves, J.T., 1999. Peroxynitrite: reactive, invasive and enigmatic. Current Opinion in Chemical Biology 3: 226-235.

Grudziński, I.P. and Frankiewicz-Jóźko, A., 2001. Effects of oral coenzyme Q10 supplementation on sodium nitrite-induced lipid peroxidation in rats. Roczniki Panstwowego Zakladu Higieny 52: 213-218.

Grune, T., Reinheckel, T. and Davies, K.J., 1997. Degradation of oxidized proteins in mammalian cells. FASEB Journal 11: 526-534.

Gutteridge, J.M. and Halliwell, B., 1990. The measurement and mechanism of lipid peroxidation in biological systems. Trends in Biochemical Sciences 15: 129-135.

Halliwell, B., 2012. Free radicals and antioxidants: updating a personal view. Nutrition Reviews 70: 257-265.

Halliwell, B., 1987. Oxidants and human disease: some new concepts. FASEB Journal 1: 358-364.

Halliwell, B., 1994. Free radicals and antioxidants: A personal view. Nutrition Reviews 52: 253-265.

Halliwell, B., 1996. Vitamin C: antioxidant or pro-oxidant *in vivo*? Free Radical Research 25: 439-454

Halliwell, B., 1999. Vitamin C: poison, prophylactic or panacea? Trends in Biochemical Sciences 24: 255-259.

Halliwell, B. and Gutteridge, J.M.C. (2015). Free radicals in biology and medicine. Oxford University Press, USA.

Hasanzadeh-Moghadam, M., Khadem-Ansari, M.H., Farjah, G.H. and Rasmi, Y., 2018. Hepatoprotective effects of betaine on liver damages followed by myocardial infarction. Veterinary Research Forum 9: 129-135.

Hassanein, E., Sayed, A.M., Hussein, O.E. and Mahmoud, A.M., 2020. Coumarins as modulators of the Keap1/Nrf2/ARE signaling pathway. Oxidative Medicine and Cellular Longevity 2020: 1675957.

Hawkes, W.C. and Alkan, Z., 2010. Regulation of redox signaling by selenoproteins. Biological Trace Element Research 134: 235-251.

Hayden, M.S. and Ghosh, S., 2014. Regulation of NF-κB by TNF family cytokines. Seminars in Immunology 26: 253-266.

Hayes, J.D. and Dinkova-Kostova, A.T., 2014. The Nrf2 regulatory network provides an interface between redox and intermediary metabolism. Trends in Biochemical Sciences 39: 199-218.

He, W., Xu, Y., Ren, X., Xiang, D., Lei, K., Zhang, C. and Liu, D., 2019. Vitamin E ameliorates lipid metabolism in mice with nonalcoholic fatty liver disease via Nrf2/CES1 signaling pathway. Digestive Diseases and Sciences 64: 3182-3191.

Helbock, H.J., Beckman,K.B., Shigenaga, M.K., Walter, P.B., Woodall, A.A., Yeo, H.C. and Ames, B.N., 1998. DNA oxidation matters: the HPLC-electrochemical detection assay of 8-oxo-deoxyguanosine and 8-oxo-guanine. Proceedings of the National Academy of Science of the USA 95: 288-293.

Hirota, K., 2020. Basic biology of hypoxic responses mediated by the transcription factor HIFs and its implication for medicine. Biomedicines 8: 2.

Howden, R., 2013. Nrf2 and cardiovascular defense. Oxidative Medicine and Cellular Longevity 2013: 104308.

Hu, C., Wang, M. and Han, X., 2017. Shotgun lipidomics in substantiating lipid peroxidation in redox biology: Methods and applications. Redox Biology 12: 946-955.

Icel, E., Icel, A., Uçak, T., Karakurt, Y., Elpeze, B., Keskin Çimen, F. and Süleyman, H., 2019. The effects of lycopene on alloxan induced diabetic optic neuropathy. Cutaneous and Ocular Toxicology 38: 88-92.

Itoh, K., Mimura, J. and Yamamoto, M., 2010. Discovery of the negative regulator of Nrf2, Keap1: a historical overview. Antioxidants and Redox Signaling 13: 1665-1678.

Itoh, K., Ye, P., Matsumiya, T., Tanji, K. and Ozaki, T., 2015. Emerging functional cross-talk between the Keap1-Nrf2 system and mitochondria. Journal of Clinical Biochemistry and Nutrition 56: 91-97.

Jackson, S.P., 1999. Colworth medal lecture. Detection, repair and signalling of DNA double-strand breaks. Biochemical Society Transactions 27: 1-13.

Jiang, T., Tian, F., Zheng, H., Whitman, S.A., Lin, Y., Zhang, Z., Zhang, N. and Zhang, D.D., 2014. Nrf2 suppresses lupus nephritis through inhibition of oxidative injury and the NF-κB-mediated inflammatory response. Kidney International 85: 333-343.

Jiang, B., Adams, Z., Moonah, S., Shi, H., Maupin-Furlow, J. and Moskovitz, J., 2020. The antioxidant enzyme methionine sulfoxide reductase A (MsrA) interacts with Jab1/CSN5 and regulates its function. Antioxidants 9: E452.

Jones, D.P., Eklow, L., Thor, H. and Orrenius, S., 1981. Metabolism of hydrogen peroxide in isolated hepatocytes: relative contributions of catalase and glutathione peroxidase in decomposition of endogenously generated H2O2. Archives of Biochemistry and Biophysics 210: 505-516.

Karakaidos, P., Karagiannis, D. and Rampias, T., 2020. Resolving DNA damage: epigenetic regulation of DNA repair. Molecules 25: E2496.

Kciuk, M., Bukowski, K., Marciniak, B. and Kontek, R., 2020. Advances in DNA repair-emerging players in the arena of eukaryotic DNA repair. International Journal of Molecular Sciences 21: E3934.

Kehrer, J.P., 2000. The Harber-Weiss reaction and mechanism of toxicity. Toxicology 149: 43-50.

Kettle, A.J. and Winterbourn, C.C., 1997. Myeloperoxidase: a key regulator of neutrophil oxidant production. Redox Report 3: 3-15.

Keum, Y.S. and Choi, B.Y., 2014. Molecular and chemical regulation of the Keap1-Nrf2 signaling pathway. Molecules 19: 10074-10089.

Kim, H.Y. and Gladyshev, V.N., 2004. Methionine sulfoxide reduction in mammals: characterization of methionine-R-sulfoxide reductases. Molecular Biology of the Cell, 15: 1055-1064.

Knight, J.A., 1998. Free radicals: their history and current status in aging and disease. Annals of the Clinical and Laboratory Sciences 28: 331-346.

Kontos, H.A., 2001. Oxygen radicals in cerebral ischemia: the 2001 Willis lecture. Stroke 32: 2712-2716.

Korovila, I., Hugo, M., Castro, J.P., Weber, D., Höhn, A., Grune, T. and Jung T., 2017. Proteostasis, oxidative stress and aging. Redox Biology 13:550-567.

Krokan, H.E., Nilsen, H., Skorpen, F., Otterlei, M. and Slupphaug, G., 2000. Base excision repair of DNA in mammalian cells. FEBS Letters 476: 73-77.

Kruidenier, L. and Verspaget, H.W., 2002. Review article: oxidative stress as a pathogenic factor in inflammatory bowel disease – radicals or ridiculous? Alimentary Pharmacology & Therapeutics 16: 1997-2015.

Kryukov, G.V., Kumar, R.A., Koc, A., Sun, Z. and Gladyshev, V.N., 2002. Selenoprotein R is a zinc-containing stereo-specific methionine sulfoxide reductase. Proceedings of the National Academy of Science of the USA 99: 4245-4250.

Kweider, N., Huppertz, B., Kadyrov, M., Rath, W. and Pufe, T., 2014. A possible protective role of Nrf2 in preeclampsia. Annals of Anatomy 196: 268-277.

Labunskyy, V.M., Hatfield, D.L. and Gladyshev, V.N., 2014. Selenoproteins: molecular pathways and physiological roles. Physiological Reviews 94: 739-777.

Lee, B.H., Hsu, W.H., Hsu, Y.W. and Pan, T.M., 2013. Dimerumic acid attenuates receptor for advanced glycation end-products signal to inhibit inflammation and diabetes mediated by Nrf2 activation and promotes methylglyoxal metabolism into d-lactic acid. Free Radical Biology and Medicine 60: 7-16.

Lee, J.M., Calkins, M.J., Chan, K., Kan, Y.W. and Johnson, J.A., 2003. Identification of the NF-E2-related factor-2-dependent genes conferring protection against oxidative stress in primary cortical astrocytes using oligonucleotide microarray analysis. Journal of Biological Chemistry 278: 12029-12038.

Levine, R.L., Mosoni, L., Berlett, B.S. and Stadtman, E.R., 1996. Methionine residues as endogenous antioxidants in proteins. Proceedings of the National Academy of Science of the USA 93: 15036-15040.

Li, C., Wang, Y., Li, L., Han, Z., Mao, S. and Wang, G., 2019. Betaine protects against heat exposure-induced oxidative stress and apoptosis in bovine mammary epithelial cells via

Li, Z., Pearlman, A.H. and Hsieh, P., 2016. DNA mismatch repair and the DNA damage response. DNA Repair 38: 94-101.

Li, H., Huang, C., Zhu, J., Gao, K., Fang, J. and Li, H., 2018. Lutein suppresses oxidative stress and inflammation by Nrf2 activation in an osteoporosis rat model. Medical Science Monitor 24: 5071-5075.

Li, R., Hong, P. and Zheng, X., 2019a. β-carotene attenuates lipopolysaccharide-induced inflammation via inhibition of the NF-κB, JAK2/STAT3 and JNK/p38 MAPK signaling pathways in macrophages. Animal Science Journal; 90: 140-148.

Lin, A., Yao, J., Zhuang, L., Wang, D., Han, J., Lam, E.W. and Gan, B., 2014. The FoxO-BNIP3 axis exerts a unique regulation of mTORC1 and cell survival under energy stress. Oncogene 33: 3183-3194.

Liu, D., Yun, Y., Yang, D., Hu, X., Dong, X., Zhang, N., Zhang, L., Yin, H. and Duan, W,. 2019. What is the biological function of uric acid? An antioxidant for neural protection or a biomarker for cell death. Disease Markers 2019: 4081962.

Lushchak, V.I., 2011. Adaptive response to oxidative stress: bacteria, fungi, plants and animals. Comparative Biochemistry and Physiology Part C: Pharmacology, Toxicology and Endocrinology 153: 175-190.

Ma, Q. and He, X., 2012. Molecular basis of electrophilic and oxidative defense: promises and perils of Nrf2. Pharmacological Reviews 64: 1055-1081.

Ma, Q., 2013. Role of Nrf2 in oxidative stress and toxicity. Annual Review of Pharmacology and Toxicology 53: 401-426.

Majzunova, M., Dovinova, I., Barancik, M. and Chan, J.Y., 2013. Redox signalling in pathophysiology of hypertension. Journal of Biomedical Science 20: 69.

Maoka, T., 2009. Recent progress in structural studies of carotenoids in animals and plants. Archives of Biochemistry and Biophysics 483: 191-195.

Maples, K.R. and Mason, R.P., 1988. Free radical metabolite of uric acid. Journal of Biological Chemistry 263: 1709-1712.

Mates, J.M. and Sanchez-Jimenez, F., 1999. Antioxidant enzymes and their implications in pathophysiologic processes. Frontiers in Bioscience 4: D339-D345.

Moldogazieva, N.T., Mokhosoev, I.M., Feldman, N.B. and Lutsenko, S.V., 2018. ROS and RNS signalling: adaptive redox switches through oxidative/nitrosative protein modifications. Free Radical Research 52: 507-543.

Moskovitz, J. and Stadtman, E.R., 2003. Selenium-deficient diet enhances protein oxidation and affects methionine sulfoxide reductase (MsrB) protein level in certain mouse tissues. Proceedings of the National Academy of Science of the USA 100: 7486-7490.

Moskovitz, J., Bar-Noy, S., Williams, W.M., Requena, J., Berlett, B.S. and Stadtman, E.R., 2001. Methionine sulfoxide reductase (MsrA) is a regulator of antioxidant defense and lifespan in mammals. Proceedings of the National Academy of Science of the USA 98: 12920-12925.

Moskovitz, J., Singh, V.K., Requena, J., Wilkinson, B.J., Jayaswal, R.K. and Stadtman, E.R., 2002. Purification and characterization of methionine sulfoxide reductases from mouse and *Staphylococcus aureus* and their substrate stereospecificity. Biochemical and Biophysical Research communications 290: 62-65.

Mruk, D.D., Silvestrini, B., Mo, M.Y. and Cheng, C.Y., 2002. Antioxidant superoxide dismutase – a review: its function, regulation in the testis, and role in male fertility. Contraception 65: 305-311.

Mulero, M. C., Huxford, T. and Ghosh, G., 2019. NF-κB, IκB, and IKK: Integral components of immune system signaling. Advances in Experimental Medicine and Biology 1172: 207-226.

Navarro, F., Navas, P., Burgess, J.R., Bello, R.I., De Cabo, R., Arroyo, A. and Villalba, J.M., 1998. Vitamin E and selenium deficiency induces expression of the ubiquinone-dependent antioxidant system at the plasma membrane. FASEB Journal 12:1665-1673.

Niki, E., 1996. α-tocopherol. In: Cadenas E. and Packer L. (eds) Handbook of antioxidants. Marcel Dekker, New York, NY, USA, pp. 3-25.

Niki, E., 2012. Do antioxidants impair signaling by reactive oxygen species and lipid oxidation products? FEBS Letters 586: 3767-3770.

Niki, E., 2014. Antioxidants: basic principles, emerging concepts, and problems. Biomedical Journal 37: 106-111.

Niki, E., 2016. Oxidative stress and antioxidants: distress or eustress? Archives of Biochemistry and Biophysics 595: 19-24.

Nordberg, J. and Arner, E.S., 2001. Reactive oxygen species, antioxidants, and the mammalian thioredoxin system. Free Radical Biology and Medicine 31: 1287-1312.

Oh, S., Suganuma, T., Gogol, M. M. and Workman, J. L., 2018. Histone H3 threonine 11 phosphorylation by Sch9 and CK2 regulates chronological lifespan by controlling the nutritional stress response. eLife 7: e36157.

Ong, A.S.H. and Tee, E.S., 1992. Natural sources of carotenoids from plants and oils. Methods in Enzymology 213: 142-167.

Overvad, K., Diamant, B., Holm, L., Holmer, G., Mortensen, S. A. and Stender, S., 1999. Coenzyme Q10 in health and disease. European Journal of Clinical Nutrition 53: 764-770.

Packer, L., 1995. Nutrition and biochemistry of the lipophilic antioxidants vitamin E and carotenoids. In: Ong, A.S.H., Niki, E. and Packer, L. (eds) Nutrition, lipids, health, and disease. AOCS Press, Champaign, IL, USA, pp. 8-35.

Pal, S., Bhattacharjee, A., Ali, A., Mandal, N.C. and Mandal, S.C., 2014. Chronic inflammation and cancer: potential chemoprevention through nuclear factor kappa B and p53 mutual antagonism. Journal of Inflammation 11: 23.

Pedruzzi, L.M., Stockler-Pinto, M.B., Leite Jr., M. and Mafra, D., 2012. Nrf2-keap1 system versus NF-κB: the good and the evil in chronic kidney disease? Biochimie 94: 2461-2466.

Pfander, H., 1992. Carotenoids: an overview. Methods in Enzymology. 213: 3-13.

Pizzimenti, S., Ciamporcero, E., Daga, M., Pettazzoni, P., Arcaro, A., Cetrangolo, G., Minelli, R., Dianzani, C., Lepore, A., Gentile, F. and Barrera, G., 2013. Interaction of aldehydes derived from lipid peroxidation and membrane proteins. Frontiers in Physiology 4: 242.

Ragu, S., Matos-Rodrigues, G. and Lopez, B. S., 2020. Replication stress, DNA damage, inflammatory cytokines and innate immune response. Genes 11: 409.

Reczek, C.R. and Chandel, N.S., 2015. ROS-dependent signal transduction. Current Opinion in Cell Biology 33: 8-13.

Reiterer, M., Schmidt-Kastner, R. and Milton, S.L., 2019. Methionine sulfoxide reductase (Msr) dysfunction in human brain disease. Free Radical Research 53: 1144-1154.

Reuter, S., Gupta, S.C., Chaturvedi, M.M. and Aggarwal, B.B., 2010. Oxidative stress, inflammation, and cancer: how are they linked? Free Radical Biology & Medicine 49: 1603-1616.

Rock, K.L., Gramm, C., Rothstein, L., Clark, K., Stein, R., Dick, L., Hwang, D. and Goldberg, A.L., 1994. Inhibitors of the proteasome block the degradation of most cell proteins and the generation of peptides presented on MHC class I molecules. Cell 78: 761-771.

Sahin, K., Orhan, C., Akdemir, F., Tuzcu, M., Sahin, N., Yılmaz, I. and Juturu, V., 2017. β-cryptoxanthin ameliorates metabolic risk factors by regulating NF-κB and Nrf2 pathways in insulin resistance induced by high-fat diet in rodents. Food and chemical toxicology: an international journal published for the British Industrial Biological Research Association 107: 270-279.

Schwarz, K.B., 1996. Oxidative stress during viral infection: a review. Free Radical Biology and Medicine 21: 641-649.

Sena, L.A. and Chandel, N.S., 2012. Physiological roles of mitochondrial reactive oxygen species. Molecular Cell 48: 158-167.

Sies, H., 1985. Oxidative stress: introductory remarks. In: Sies, H. (ed.) Oxidative stress. Academic Press, London, UK, p.1-8.

Sies, H., 2015. Oxidative stress: a concept in redox biology and medicine. Redox Biology 24: 180-183.

Sies, H., 2018. On the history of oxidative stress: concept and some aspects of current development. Current Opinion in Toxicology 7: 122-126.

Sies, H., 2019. Oxidative stress: eustress and distress in redox homeostasis. In: Fink, G. (ed.) Stress: physiology, biochemistry, and pathology. Academic Press, London, UK, pp. 153-163.

Sies, H., Berndt, C. and Jones, D.P., 2017. Oxidative stress. Annual Review of Biochemistry 86: 715-748.

Sies, H. and Jones, D.P., 2020. Reactive oxygen species (ROS) as pleiotropic physiological signalling agents. Nature Reviews. Molecular Cell Biology 21: 363-383. https://doi.org/10.1038/s41580-020-0230-3

Singal, P.K., Khaper, N., Palace, V. and Kumar, D., 1998. The role of oxidative stress in the genesis of hart disease. Cardiovascular Research 40: 426-432.

Sivandzade, F., Prasad, S., Bhalerao, A. and Cucullo, L., 2019. NRF2 and NF-κB interplay in cerebrovascular and neurodegenerative disorders: Molecular mechanisms and possible therapeutic approaches. Redox Biology 21: 101059.

Skulachev, V.P., 1998. Biochemical mechanisms of evolution and the role of oxygen. Biochemistry (Moscow) 63: 1335-1343.

Slupphaug, G., Kavli, B. and Krokan, H.E., 2003. The interacting pathways for prevention and repair of oxidative DNA damage. Mutation Research 531: 231-251.

Song, P. and Zou, M.H., 2014. Redox regulation of endothelial cell fate. Cellular and Molecular Life Sciences 71: 3219-3239.

Stadtman, E.R., Moskovitz, J., Berlett, B.S. and Levine, R.L., 2002. Cyclic oxidation and reduction of protein methionine residues is an important antioxidant mechanism. Molecular and Cellular Biochemistry 234-235: 3-9.

Stefely, J.A. and Pagliarini, D.J., 2017. Biochemistry of mitochondrial coenzyme Q biosynthesis. Trends in Biochemical Sciences 42: 824-843.

Stroeve, J., Van Wietmarschen, H., Kremer, B., Van Ommen, B. and Wopereis, S., 2015. Phenotypic flexibility as a measure of health: the optimal nutritional stress response test. Genes & Nutrition 10, 3: 13.

Surai, P.F., 1999. Vitamin E in avian reproduction. Poultry and Avian Biology Review 10: 1-60.

Surai, P.F., 2002. Natural antioxidants in avian nutrition and reproduction. Nottingham University Press, Nottingham, UK.

Surai, P.F., 2006. Selenium in nutrition and health. Nottingham University Press, Nottingham, UK.
Surai, P.F., 2012a. The antioxidant properties of canthaxanthin and its potential effects in the poultry eggs and on embryonic development of the chick. Part 1. Worlds Poultry Science Journal 68: 465-475.
Surai, P.F., 2012b. The antioxidant properties of canthaxanthin and its potential effects in the poultry eggs and on embryonic development of the chick. Part 2. Worlds Poultry Science Journal 68: 717-726.
Surai, P.F., 2014. Polyphenol compounds in the chicken/animal diet: from the past to the future. Journal of Animal Physiology and Animal Nutrition 98: 19-31.
Surai, P.F., 2015. Antioxidant action of carnitine: molecular mechanisms and practical applications. EC Veterinary Science 2.1: 66-84.
Surai, P.F., 2015a. Carnitine enigma: from antioxidant action to vitagene regulation. Part 1. Absorption, metabolism and antioxidant activities. Journal of Veterinary Science & Medicine 3, 2: 14.
Surai P.F., 2015b. Carnitine enigma: from antioxidant action to vitagene regulation. Part 2. Transcription factors and practical applications. Journal of Veterinary Science & Medicine 3, 2: 17.
Surai, P.F., 2015c. Silymarin as a natural antioxidant: an overview of the current evidence and perspectives. Antioxidants 4: 204-247.
Surai, P.F., 2015d. Antioxidant systems in poultry biology: heat shock proteins. Journal of Science 5: 1188-1222.
Surai, P.F., 2016. Antioxidant systems in poultry biology: superoxide dismutase. Journal of Animal Research and Nutrition 1: 1, 8.
Surai, P.F., 2017. Antioxidant defences: food for thoughts. EC Nutrition 10(2): 65-66.
Surai, P.F., 2018. Selenium in poultry nutrition and health. Wageningen Academic Publishers, Wageningen, The Netherlands.
Surai, P.F., 2020. Antioxidants in poultry nutrition and reproduction: an update. Antioxidants 9(2): 105.
Surai, P.F., 2020a. Superoxide Dismutase as a New Entrant into the Vitagene Family in Animals/Poultry. EC Nutrition 15.3: 01-03.
Surai, P.F. and Fisinin, V.I., 2014. Antioxidant systems of the body: from vitamin E to polyphenols and beyond. Proceedings of the 35th Western Nutrition Conference. September 24-25, 2014, Edmonton, Alberta, Canada.
Surai, P.F. and Fisinin, V.I., 2015. Antioxidant-prooxidant balance in the intestine: applications in chick placement and pig Weaning. Journal of Veterinary Science & Medicine 3, 1: 16.
Surai, P.F. and Fisinin, V.I., 2016. Antioxidant system regulation: from vitamins to vitagenes. In: Watson, R.R. and De Meester, F. (eds) Handbook of cholesterol. Wageningen Academic Publishers, Wageningen, the Netherlands, pp. 451-481.
Surai, P.F. and Fisinin, V.I., 2016a. Vitagenes in poultry production. Part 1. Technological and environmental stresses. World's Poultry Science Journal 72: 721-733.
Surai, P.F. and Fisinin, V.I., 2016b. Vitagenes in poultry production. Part 2. Nutritional and internal stresses. World's Poultry Science Journal 72: 761-772.
Surai, P.F. and Fisinin, V.I., 2016c. Vitagenes in poultry production. Part 3. Vitagene concept development. World's Poultry Science Journal 72: 793-804.
Surai, P.F. and Fisinin, V.I., 2016d. Natural antioxidants and stresses in poultry production: from vitamins to vitagenes. Proc. XXV World Poultry Congress, Invited Lecture Papers, Beijing, China, 5-9 September, 2016, pp. 116-121.
Surai, P.F. and Kochish, I.I., 2017. Antioxidant systems and vitagenes in poultry biology: heat shock proteins. In: Asea, A.A.A. and Kaur, P. (ed.) Heat shock proteins in veterinary. Springer, Switzerland, pp. 123-177.

Surai, P.F. and Kochish, I.I., 2019. Nutritional modulation of the antioxidant capacities in poultry: the case of selenium. Poultry Science 98: 4231-4239.

Surai, P.F., Kochish, I.I. and Fisinin, V.I., 2018a. Glutathione peroxidases in poultry biology: Part 1. Classification and mechanisms of action. World's Poultry Science Journal 73: 185-197.

Surai, P.F., Kochish, I.I. and Fisinin V.I., 2018b. Glutathione peroxidases in poultry biology: Part 2. Modulation of enzymatic activities. World's Poultry Science Journal 73: 239-250.

Surai, P.F., Kochish, I.I., Fisinin, V.I. and Kidd, M.T., 2019. Antioxidant defence systems and oxidative stress in poultry biology: an update. Antioxidants 8(7): 235.

Surai, P.F., Kochish, I.I., Romanov, M.N. and Griffin, D.K., 2019a. Nutritional modulation of the antioxidant capacities in poultry: the case of vitamin E. Poultry Science 98: 4030-4041.

Surai, P.F., Speake, B.K. and Sparks, N.H.C., 2001a. Carotenoids in avian nutrition and embryonic development. 2. Antioxidant properties and discrimination in embryonic tissues. Journal of Poultry Science 38: 117-145.

Surai, P.F., Speake, B.K. and Sparks, N.H.C., 2001b. Carotenoids in avian nutrition and embryonic development. 1. Absorption, availability and levels in plasma and egg yolk Journal of Poultry Science 38: 1-27.

Surh, Y.J., 2008. NF-kappa B and Nrf2 as potential chemopreventive targets of some anti-inflammatory and antioxidative phytonutrients with anti-inflammatory and antioxidative activities. Asia Pacific Journal of Clinical Nutrition 17, Suppl. 1: 269-272.

Surh, Y.J., Kundu, J.K. and Na, H.K., 2008. Nrf2 as a master redox switch in turning on the cellular signaling involved in the induction of cytoprotective genes by some chemopreventive phytochemicals. Planta Medica 74: 1526-1539.

Tang, W., Jiang, Y.F., Ponnusamy, M. and Diallo, M., 2014. Role of Nrf2 in chronic liver disease. World Journal of Gastroenterology 20: 13079-13087.

Tirosh, O. and Reznick. A.Z., 2000. Chemical bases and biological relevance of protein oxidation. In: Sen, C.K., Packer, L. and Hanninen, O.O.P., Handbook of oxidants and antioxidants in exercise. Elsevier, Amsterdam, the Netherlands, pp. 89-114.

Tkach, K.E., Oyler, J.E. and Altan-Bonnet, G., 2014. Cracking the NF-κB code. Science Signalling 7: pe5.

Tsai, M.T., Chen, C.Y., Pan, Y.H., Wang, S.H., Mersmann, H.J. and Ding, S.T., 2015. Alleviation of carbon-tetrachloride-induced liver injury and fibrosis by betaine supplementation in chickens. Evidence-Based Complementary and Alternative Medicine 2015: 725379.

Tsikas, D., 2017. Assessment of lipid peroxidation by measuring malondialdehyde (MDA) and relatives in biological samples: Analytical and biological challenges. Analitical Biochemistry 524: 13-30.

Van der Wijst, M.G., Brown, R. and Rots, M.G., 2014. Nrf2, the master redox switch: the Achilles' heel of ovarian cancer? Biochimica et Biophysica Acta 1846: 494-509.

Varela López, A., Giampieri, F., Battino, M. and Quiles, J.L., 2016. Coenzyme Q and its role in the dietary therapy against aging. Molecules 21, 3: 373.

Vargas-Mendoza, N., Morales-González, Á., Morales-Martínez, M., Soriano-Ursúa, M.A., Delgado-Olivares, L., Sandoval-Gallegos, E.M., Madrigal-Bujaidar, E., Álvarez-González, I., Madrigal-Santillán, E. and Morales-Gonzalez, J.A., 2020. Flavolignans from silymarin as Nrf2 bioactivators and their therapeutic applications. Biomedicines 8: E122.

Vougier, S., Mary, J. and Friguet, B., 2003. Subcellular localization of methionine sulphoxide reductase A (MsrA): evidence for mitochondrial and cytosolic isoforms in rat liver cells. Biochemical Journal 373: 531-537.

Vriend, J. and Reiter, R.J., 2015. The Keap1-Nrf2-antioxidant response element pathway: a review of its regulation by melatonin and the proteasome. Molecular and Cellular Endocrinology 401: 213-220.

Wallace, S.S., 1997. Oxidative damage to DNA and its repair. In: Scandalios, J.D. (ed.) Oxidative stress and the molecular biology of antioxidant defences. Cold Spring Harbor Laboratory Press, Cold Harbor, NY, USA, pp. 49-89.

Wang, X. and Hai, C., 2016. Novel insights into redox system and the mechanism of redox regulation. Molecular Biology Reports 43: 607-628.

Wang, Y., Chen, Y., Chen, Y., Zhou, B., Shan, X. and Yang, G., 2018. Eriodictyol inhibits IL-1β-induced inflammatory response in human osteoarthritis chondrocytes. Biomedicine & Pharmacotherapy 107: 1128-1134.

Williams, L.M. and Gilmore, T.D., 2020. Looking down on NF-κB. Molecular and Cellular Biology 40: e00104-20. https://doi.org/10.1128/MCB.00104-20

Winkler, B.S., Orselli, S.M. and Rex, T.S., 1994. The redox couple between glutathione and ascorbic acid: a chemical and physiological perspective. Free Radical Biology & Medicine 17: 333-349.

Wood, R.D., Mitchell, M., Sgouros, J. and Lindahl, T., 2001. Human DNA repair genes. Science 291:1284-1289.

Xue, Y., Sun, C., Hao, Q. and Cheng, J., 2019. Astaxanthin ameliorates cardiomyocyte apoptosis after coronary microembolization by inhibiting oxidative stress via Nrf2/HO-1 pathway in rats. Naunyn-Schmiedeberg's Archives of Pharmacology 392: 341-348.

Ya, B.L., Liu, Q., Li, H.F., Cheng, H.J., Yu, T., Chen, L., Wang, Y., Yuan, L.L., Li, W.J., Liu, W.Y. and Bai, B., 2018. Uric acid protects against focal cerebral ischemia/reperfusion-induced oxidative stress via activating Nrf2 and regulating neurotrophic factor expression. Oxidative Medicine and Cellular Longevity 2018: 6069150.

Yabuzaki, J., 2017. Carotenoids Database: structures, chemical fingerprints and distribution among organisms. Database 2017, 1: bax004.

Yan, L.J., 2014. Positive oxidative stress in aging and aging-related disease tolerance. Redox Biology 2C: 165-169.

Young, A., Gill, R. and Mailloux, R.J., 2019. Protein S-glutathionylation: The linchpin for the transmission of regulatory information on redox buffering capacity in mitochondria. Chemico-Biological Interactions 299: 151-162.

Yu, B.P., 1994. Cellular defences against damage from reactive oxygen species. Physiological Reviews 74: 139-162.

Yu, M., Yan, W., Beight, C., 2018. Lutein and zeaxanthin isomers protect against light-induced retinopathy via decreasing oxidative and endoplasmic reticulum stress in BALB/cJ Mice. Nutrients 10, 7.

Zhan, T., Han, Y., Tang, C., Zhao, Q., Sun, D., Li, Y., Jia, X., Zhou, L. and Zhang, J., 2020. Metabolism and biological activity of α-tocopherol derived from vitamin E-enriched transgenic maize in broilers. Journal of the Science of Food and Agriculture 100: 4319-4328. https://doi.org/10.1002/jsfa.10480

Zhang, L., Yousefzadeh, M.J., Suh, Y., Niedernhofer, L.J. and Robbins, P.D., 2019. Signal transduction, ageing and disease. Sub-Cellular Biochemistry 91: 227-247.

Zhang, D.D. and Chapman, E., 2020. The role of natural products in revealing NRF2 function. Natural Product Reports 37: 797-826. https://doi.org/10.1039/c9np00061e

Zhao, B., Ren, B., Guo, R., Zhang, W., Ma, S., Yao, Y., Yuan, T., Liu, Z. and Liu, X., 2017. Supplementation of lycopene attenuates oxidative stress induced neuroinflammation and cognitive impairment via Nrf2/NF-κB transcriptional pathway. Food and Chemical Toxicology 109: 505-516.

Zhou, S., Sun, W., Zhang, Z. and Zheng, Y., 2014. The role of Nrf2-mediated pathway in cardiac remodelling and heart failure. Oxidative Medicine and Cellular Longevity 2014: 260429.

Part II.
Vitagenes in avian biology

All fair in love and war

Chapter 3
Vitagene concept development

The chain is no stronger than its weakest link

3.1 Introduction

The term 'vitagene' was introduced in 1998 by Rattan who wrote 'Our survival and the physical quality of life depends upon an efficient functioning of various maintenance and repair processes. This complex network of the so-called longevity assurance processes is composed of several genes, which may be called *vitagenes*'. Later the vitagene concept has been further developed in medical sciences by Calabrese and colleagues (Calabrese *et al.*, 2004, 2007, 2009a, 2014) and major prosurvival mechanisms controlled by homodynamic vitagene network are shown in Figure 3.1.

3.2 Vitagene family

In accordance with Calabrese *et al.* (2007, 2009, 2014), Surai and Fisinin (2016a,b) and Surai (2016, 2020a,b) the term vitagenes refers to a group of redox-sensitive genes that

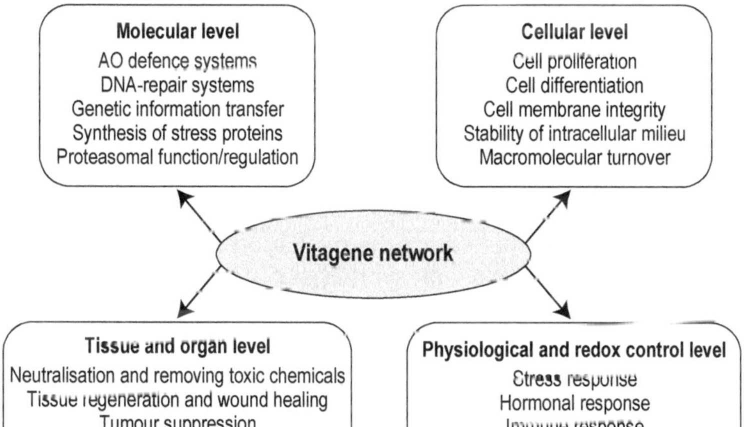

Figure 3.1. Major components of the vitagene network (adapted from Calabrese *et al.*, 2007; Rattan, 1998; Surai, 2018a, 2019; Surai and Fisinin, 2016).

DOI 10.3920/978-90-8686-906-0_3, © Wageningen Academic Publishers 2020

are strictly involved in stress sensing and preserving cellular adaptive homeostasis and the vitagene family includes:

- Heat shock proteins (HSP);
  - HO-1;
  - HSP70;
- SOD;
  - SOD1;
  - SOD2;
  - SOD3;
- Thioredoxin system;
  - Trx;
  - TR;
  - Prx;
  - Srx;
- Glutathione system;
  - GSH;
  - GR;
  - GPx;
  - Grx;
- Sirtuins;
  - SIRT1;
  - SIRT2;
  - SIRT3;
  - SIRT4;
  - SIRT5;
  - SIRT6;
  - SIRT7.

The products of the above-mentioned genes actively operate in detecting and controlling diverse forms of stress and cell injuries by regulation of synthesis of an array of protective molecules. The cooperative mechanisms of the vitagene network are reviewed in recently published comprehensive reviews (Calabrese *et al.*, 2014; Trovato Salinaro *et al.*, 2014) with a major conclusion indicating an essential regulatory role of the vitagene network in cell and whole organism adaptation to various stresses. Indeed, cellular stress response is mediated via the regulation of pro-survival pathways and vitagene activation with the following synthesis of a range of protective antioxidant molecules is the central event in such an adaptation. The vitagene concept found its acceptance in medical sciences, including neurodegenerative disorders (Calabrese *et al.*, 2004), neuroprotection (Calabrese *et al.*, 2009), aging and longevity (Calabrese *et al.*, 2007, 2011, 2012, 2014), dermatology (Calabrese *et al.*, 2008), free radical-related diseases (Calabrese *et al.*, 2010), osteoporosis and Alzheimer pathology (Cornelius *et al.*, 2013, 2014; Dattilo *et al.*, 2015; Surai and Fisinin, 2016).

We suggested that the vitagene concept can also be useful in poultry production (Fisinin and Surai, 2011, 2011a; Surai and Fisinin, 2012, 2012a). The vitagene concept in relation to poultry production was further developed in our previous publications (Fisinin and Surai, 2011a,b; Surai, 2015a,b,c, 2016, 2020a,b; Surai and Fisinin, 2012a,b; Surai *et al.*, 2019). It seems likely that by upregulating the vitagenes and improving adaptive ability of animals to stress it is possible to decrease negative consequences of various stresses in poultry and farm animal production. Furthermore, there is an opportunity to nutritionally modulate the vitagene network by using various natural antioxidants: carnitine (Calabrese *et al.*, 2009; Surai, 2015a,b), betaine, vitamins A, E, D, C (Surai *et al.*, 2017), taurine (Surai, 2018b; Surai *et al.*, 2018, 2020), phytochemicals (Calabrese *et al.*, 2012), including silymarin (Surai, 2015c) and other nutrients. In fact, activation of the vitagene network by nutritional means is considered as a fundamental mechanism for improving animal/poultry resistance to various stresses (Surai and Fisinin, 2016a,b; Surai *et al.*, 2017).

As can be seen from the Figure 3.2 vitagenes provide optimal conditions for redox signalling, redox homeostasis being major players in cell/organism adaptation to various stresses.

3.3 Conclusions

Development of the vitagene concept and its transfer from medical sciences to poultry/animal sciences become an important milestone in understanding molecular mechanisms of stress development and stress adaptation. The vitagenes are considered to be key players in redox signalling and redox homeostasis maintenance under commercial stress conditions of egg and meat production. In fact, interactions

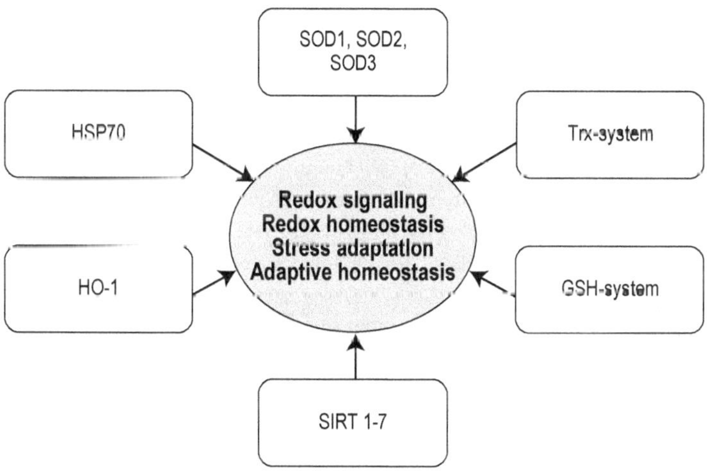

Figure 3.2. Protective roles of vitagenes in adaptive homeostasis.

between vitagenes, transcription factors and a range of signalling mechanisms become key steps in poultry/animal adaptation to stresses and development of adaptive homeostasis. In the next chapters details of vitagene actions, mechanisms of their nutritional modulation and examples of the usage of the vitagene concept in commercial poultry production will be presented.

References

Calabrese, V., Boyd-Kimball, D., Scapagnini, G. and Butterfield, D.A., 2004. Nitric oxide and cellular stress response in brain aging and neurodegenerative disorders: the role of vitagenes. *In vivo* 18: 245-267.

Calabrese, V., Guagliano, E., Sapienza, M., Panebianco, M., Calafato, S., Puleo, E., Pennisi, G., Mancuso, C., Butterfield, D. A. and Stella, A. G., 2007. Redox regulation of cellular stress response in aging and neurodegenerative disorders: role of vitagenes. Neurochemical Research 32: 757-773.

Calabrese, V., Calafato, S., Puleo, E., Cornelius, C., Sapienza, M., Morganti, P. and Mancuso, C., 2008. Redox regulation of cellular stress response by ferulic acid ethyl ester in human dermal fibroblasts: role of vitagenes. Clinics in Dermatology 26: 358-363.

Calabrese, V., Cornelius, C., Mancuso, C., Barone, E., Calafato, S., Bates, T., Rizzarelli, E. and Kostova, A. T., 2009. Vitagenes, dietary antioxidants and neuroprotection in neurodegenerative diseases. Frontiers in Bioscience 14: 376-397.

Calabrese, V., Cornelius, C., Dinkova-Kostova, A.T. and Calabrese, E.J., 2009. Vitagenes, cellular stress response, and acetylcarnitine: Relevance to hormesis. Biofactors 35: 146-160.

Calabrese, V., Cornelius, C., Trovato, A., Cavallaro, M., Mancuso, C., Di Rienzo, L., Condorelli, D., De Lorenzo, A. and Calabrese, E.J., 2010. The hormetic role of dietary antioxidants in free radical-related diseases. Current Pharmaceutical Design 16: 877-883.

Calabrese, V., Cornelius, C., Cuzzocrea, S., Iavicoli, I., Rizzarelli, E. and Calabrese, E.J., 2011. Hormesis, cellular stress response and vitagenes as critical determinants in aging and longevity. Molecular Aspects of Medicine 32: 279-304.

Calabrese, V., Cornelius, C., Dinkova-Kostova, A.T., Iavicoli, I., Di Paola, R., Koverech, A., Cuzzocrea, S., Rizzarelli, E. and Calabrese, E.J., 2012. Cellular stress responses, hormetic phytochemicals and vitagenes in aging and longevity. Biochimica et Biophysica Acta 1822: 753-783.

Calabrese, V., Scapagnini, G., Davinelli, S., Koverech, G., Koverech, A., De Pasquale, C., Salinaro, A.T., Scuto, M., Calabrese, E.J. and Genazzani, A.R., 2014. Sex hormonal regulation and hormesis in aging and longevity: role of vitagenes. Journal of Cell Communication and Signaling 8: 369-384.

Calabrese, V., Dattilo, S., Petralia, A., Parenti, R., Pennisi, M., Koverech, G., Calabrese, V., Graziano, A., Monte, I., Maiolino, L., Ferreri, T. and Calabrese, E J., 2015. Analytical approaches to the diagnosis and treatment of aging and aging-related disease: redox status and proteomics. Free Radical Research 49: 511-524.

Calabrese, V., Giordano, J., Signorile, A., Laura Ontario, M., Castorina, S., De Pasquale, C., Eckert, G. and Calabrese, E.J., 2016. Major pathogenic mechanisms in vascular dementia: Roles of cellular stress response and hormesis in neuroprotection. *Journal of Neuroscience Research* 94: 1588-1603.

Calabrese, V., Giordano, J., Ruggieri, M., Berritta, D., Trovato, A., Ontario, M.L., Bianchini, R. and Calabrese, E.J., 2016. Hormesis, cellular stress response, and redox homeostasis in autism spectrum disorders. Journal of Neuroscience Research 94: 1488-1498.

Calabrese, V., Giordano, J., Crupi, R., Di Paola, R., Ruggieri, M., Bianchini, R., Ontario, M.L., Cuzzocrea, S. and Calabrese, E.J., 2017. Hormesis, cellular stress response and neuroinflammation in schizophrenia: Early onset versus late onset state. Journal of Neuroscience Research 95: 1182-1193.

Cornelius, C., Trovato Salinaro, A., Scuto, M., Fronte, V., Cambria, M.T., Pennisi, M., Bella, R., Milone, P., Graziano, A., Crupi, R., Cuzzocrea, S., Pennisi, G. and Calabrese, V., 2013. Cellular stress response, sirtuins and UCP proteins in Alzheimer disease: role of vitagenes. Immunity & Ageing 10,1: 41.

Cornelius, C., Koverech, G., Crupi, R., Di Paola, R., Koverech, A., Lodato, F., Scuto, M., Salinaro, A.T., Cuzzocrea, S., Calabrese, E.J. and Calabrese, V., 2014. Osteoporosis and alzheimer pathology: Role of cellular stress response and hormetic redox signaling in aging and bone remodeling. Frontiers in Pharmacology 5: 120.

Dattilo, S., Mancuso, C., Koverech, G., Di Mauro, P., Ontario, M.L., Petralia, C.C., Petralia, A., Maiolino, L., Serra, A., Calabrese, E.J. and Calabrese, V., 2015. Heat shock proteins nd hormesis in the diagnosis and treatment of neurodegenerative diseases. Immunity & Ageing 12: 20.

Fisinin V.I. and Surai P.F., 2011a. Effective protection against stresses in poultry production: from vitamins to vitagenes. Poultry and Poultry Products (Moscow) 5: 23-26.

Fisinin V.I. and Surai P.F., 2011b. Effective protection against stresses in poultry production: from vitamins to vitagenes. Poultry and Poultry Products (Moscow) 6: 10-13.

Rattan, S.I., 1998. The nature of gerontogenes and vitagenes. Antiaging effects of repeated heat shock on human fibroblasts. Annals of the New York Academy of Sciences 854: 54-60.

Surai, P.F., 2015a. Carnitine enigma: from antioxidant action to vitagene regulation. Part 2. Transcription factors and practical applications. Journal of Veterinary Science & Medicine 3, 2: 17.

Surai, P.F., 2015b. Antioxidant action of carnitine: molecular mechanisms and practical applications. EC Veterinary Science 2.1: 66-84.

Surai, P.F., 2015c. Silymarin as a natural antioxidant: an overview of the current evidence and perspectives. Antioxidants 4: 204-247.

Surai, P.F., 2016. Antioxidant systems in poultry biology: superoxide dismutase. Journal of Animal Research and Nutrition 1, 1: 8.

Surai, P.F., 2018a. Selenium in poultry nutrition and health. Wageningen Academic Publishers, Wageningen, the Netherlands.

Surai, P.F., 2018b. Taurine and carnitine in poultry production: from vitagene activation to chicken health maintenance. Ptakhivnitstvo.ua (Ukrainian Poultry Science) 1-2: 12-17.

Surai P.F., 2019. Vitagenes in poultry production: adaptation to commercially relevant stresses. Suchasne Ptakhivnitstvo (Ukraine) 7-8: 28-32.

Surai, P.F., 2020a. Antioxidants in poultry nutrition and reproduction: an update. Antioxidants 9, 2: 105.

Surai, P.F., 2020b. Superoxide dismutase as a new entrant into the vitagene family in animals/poultry. EC Nutrition 15.3: 01-03.

Surai, P.F. and Fisinin, V.I., 2012a. The modern antistress technologies in poultry: from antioxidants to vitagenes. Agricultural Biology (Moscow) 4: 3-13.

Surai, P.F. and Fisinin, V.I., 2012b. Modern methods for fighting stresses in poultry production: from antioxidants to sirtuins and vita-genes. Effectivne Ptakhivnitstvo, Ukraine (Effective Poultry Production) 8: 8-13.

Surai, P.F. and Fisinin, V.I., 2016a. Vitagenes in poultry production. Part 3. Vitagene concept development. Worlds Poultry Science Journal 72: 793-804.

Surai, P.F. and Fisinin, V.I., 2016b. Antioxidant system regulation: from vitamins to vitagenes. In: Watson, R.R. and De Meester, F. (eds) Handbook of cholesterol. Wageningen Academic Publishers, Wageningen, the Netherlands, pp. 451-481.

Surai, P.F., Kochish, I.I. and Fisinin, V.I., 2017. Antioxidant systems in poultry biology: nutritional modulation of vitagenes. European Journal of Poultry Science 81: 1612-9199.

Surai, P.F., Kochish, I.I., Fisinin, V.I. and Kidd, M.T., 2019. Antioxidant defence systems and oxidative stress in poultry biology: an update. Antioxidants 8, 7: 235.

Surai P.F., Kochish I.I., Kidd M.T., 2020. Taurine in poultry nutrition. Animal Feed Science and Technology 260: 114339

Surai, P.F., Kochish, I.I., Fisinin, V.I., Grozina, A.A. and Shatskikh, E.V., 2018. Molecular mechanisms of chicken gut health maintenance: role of microbiota. Agricultural Technologies, Moscow, Russia.

Trovato Salinaro, A., Cornelius, C., Koverech, G., Koverech, A., Scuto, M., Lodato, F., Fronte, V., Muccilli, V., Reibaldi, M., Longo, A., Uva, M.G. and Calabrese, V., 2014. Cellular stress response, redox status, and vitagenes in glaucoma: a systemic oxidant disorder linked to Alzheimer's disease. Frontiers in Pharmacology 5: 129.

Trovato, A., Siracusa, R., Di Paola, R., Scuto, M., Ontario, M.L., Bua, O., Di Mauro, P., Toscano, M.A., Petralia, C., Maiolino, L., Serra, A., Cuzzocrea, S. and Calabrese, V., 2016. Redox modulation of cellular stress response and lipoxin A4 expression by *Hericium erinaceus* in rat brain: relevance to Alzheimer's disease pathogenesis. Immunity & Ageing 13: 23.

Trovato, A., Siracusa, R., Di Paola, R., Scuto, M., Fronte, V., Koverech, G., Luca, M., Serra, A., Toscano, M.A., Petralia, A., Cuzzocrea, S. and Calabrese, V., 2016. Redox modulation of cellular stress response and lipoxin A4 expression by *Coriolus versicolor* in rat brain: *Relevance* to Alzheimer's disease pathogenesis. Neurotoxicology 53: 350-358.

Chapter 4
Superoxide dismutases (SODs)

The first blow is half the battle

4.1 Introduction

A growing body of evidence indicates that most stresses in poultry production at the cellular level are associated with oxidative stress. Recently, a concept of the cellular antioxidant defence has been revised with special attention paid to cellular redox status/balance maintenance and cell signalling. In fact, antioxidant systems of the living cell are based on three major levels of defence and superoxide dismutases (SODs) are shown to belong to the first level of the antioxidant defence network (Surai, 2018, 2020b). Furthermore, cellular antioxidant defences are shown to include several options and vitagene activation in stress conditions is considered as a fundamental adaptive mechanism (Surai, 2020a,b). The vitagene family includes various genes regulating synthesis of protective molecules including elements of thioredoxin and glutathione systems, sirtuins, heat shock proteins and SODs (Surai, 2020b; Surai *et al.*, 2019). On one hand, SODs are the main cellular antioxidant mechanism dealing with overproduction of free radicals in stress conditions. On the other hand, in biological systems SODs are important source of H_2O_2, main signalling molecule participating in stress adaptation (Surai, 2020a; Surai *et al.*, 2019). Interest in SOD among scientists has been very high and Medline search for Superoxide dismutase or SOD in paper title (between 1973-2020) conducted on June 15th, 2020 gave 11,344 hits/publications (2,675 hits for the last 10 years), including 315 review papers. Therefore, the aim of this chapter is to present updated information related to roles of SOD in avian biology and poultry production as an important part of the vitagene network.

4.2 Superoxide dismutase in biological systems

SOD was discovered by McCord and Fridovich in 1969 as an enzymatic activity in preparations of carbonic anhydrase or myoglobin that inhibited the aerobic reduction of cytochrome C by xanthine oxidase (McCord and Fridovich, 1969). Therefore, haemocuprein, which was discovered much earlier, became Cu,Zn-SOD (Bannister, 1988). This discovery opened new era in free radical research. At present, three distinct isoforms of SOD have been identified in mammals, and their genomic structure, cDNA, and proteins have been described (Zelko, *et al.*, 2002). The fourth form of the enzyme Fe-SOD was isolated from various bacteria but not found in animal. Furthermore, a novel type of nickel-containing SOD was purified to apparent homogeneity from the cytosolic fractions of *Streptomyces sp.* (Youn *et al.*, 1996). The biosynthesis of SODs, in most biological systems, is well controlled. In fact, exposure to increased pO_2,

Peter F. Surai **Vitagenes in avian biology and poultry health**
DOI 10.3920/978-90-8686-906-0_4, © Wageningen Academic Publishers 2020

increased intracellular fluxes of O_2^-, metal ions perturbation, and exposures to several environmental oxidants have been shown to influence the rate of SOD synthesis in both prokaryotic and eukaryotic organisms (Hassan,1988). A range of transcriptional factors, including NF-κB, AP-1, AP-2, and Sp1, as well as CCAAT-enhancer-binding protein (C/EBP), have been shown to regulate the constitutive or inducible expression levels of all three SODs (Miao and St Clair, 2009). Furthermore, it seems likely that in addition to transcriptional control, epigenetic regulation and posttranscriptional modifications are responsible for a regulation of the SOD functional activity (Miao and St Clair, 2009). Comparative characteristics of SOD1, SOD2 and SOD3 are summarised in Table 4.1 (Miao and St Clair, 2009; Huang *et al.*, 2012).

SOD1, or Cu,Zn-SOD, was the first enzyme of this family to be characterised and is a copper and zinc-containing homodimer that is found almost exclusively in intracellular cytoplasmic spaces. It exists as a 32 kDa homodimer and is present in the cytoplasm and nucleus of every cell type examined (Zelko *et al.*, 2002). The chromosomal localisation and characteristics of the sod1 gene have been identified in rodents, bovines, and humans and the human sod1 gene is shown to be localised on chromosome 21q22. Furthermore, sod1 gene consists of five exons interrupted by four introns, which is significantly similar in different species in terms of the size

Table 4.1. Biochemical properties of mammalian superoxide dismutase (adapted from Huang *et al.*, 2012; Miao and St Clair, 2009).

Enzymes	Cu,Zn-SOD	Mn-SOD	EC-SOD
Gene designation (human/ mouse)	SOD1/Sod1	SOD2/Sod2	SOD3/Sod3
Chromosome location man/ mouse)	HAS21/MMU16	HAS6/MMU17	HAS4/MMU5
Disease caused by enzyme defects	Amyotrophic lateral sclerosis (ALS)	None	None
Metal co-factor(s)	Cu^{2+} – catalytically active Zn^{2+} – maintains enzyme stability	Mn^{2+} – catalytically active	Cu^{2+} – catalytically active Zn^{2+} – maintains enzyme stability
Active form	Dimer	Tetramer	Tetramer
Molecular Mass, kDa	88	32	135
Subcellular locations	Cytosol, intermembrane space of mitochondria, nucleus	Mitochondria matrix	Extracellular matrix and circulation
Tissue distribution (from high to low)	Liver, kidney, brain, heart	Heart, brain, skeletal muscle	Blood vessels, lung, kidney, uterus
Post-translational modification	Nitration, phosphorylation, glutathiolation, glycation	Acetylation, nitration, phosphorylation	Glycosylation
Inducibility	Not inducible	Inducible	Induced by antioxidants and regulated through NRF

of exons, particularly the coding regions (Miao and St Clair, 2009). The sequence and structure of Cu,Zn-SOD is highly conserved from prokaryotes to eukaryote and mammalian SOD1 is highly expressed in the liver and kidney (Culotta *et al.*, 2006). Enzymatic activity of SOD1 depends on the presence of the Cu and Zn. While copper is needed for SOD1 catalytic activity, Zn participates in proper protein folding and stability. Over 100 mutations in the human gene SOD1 are described to lead to some inherited diseases, but their mechanisms remain unclear (Fukai and Ushio-Fukai, 2011). Recently it has been shown that SOD1 can acts as a nuclear transcription factor to regulate oxidative stress resistance (Figure 4.1; Tsang *et al.*, 2014). SOD1 is known to be the major cytosolic superoxide dismutase responsible for dismutating superoxide. In response to increased level of H_2O_2 SOD1 can be phosphorylated and translocated to the nucleus.

At the next step of SOD1 action, it becomes associated with the promoters of the target 'oxidative response' genes to regulate gene expression at the transcriptional level. In particular this includes genes involved in AO defence, ROS-induced DNA replication stress and DNA damage responses, general cellular stress and maintenance of cellular redox homeostasis (Tsang *et al.*, 2014). In this way SOD1 improved cell/tissue adaptability to stress. It should be also noted that H_2O_2 can directly diffuse into the nucleus and cause genomic DNA damage. Therefore, activation of the aforementioned genes and synthesis of protective molecules can deal with this problem. Four post-translational modifications (PTMs) are shown to contribute to the

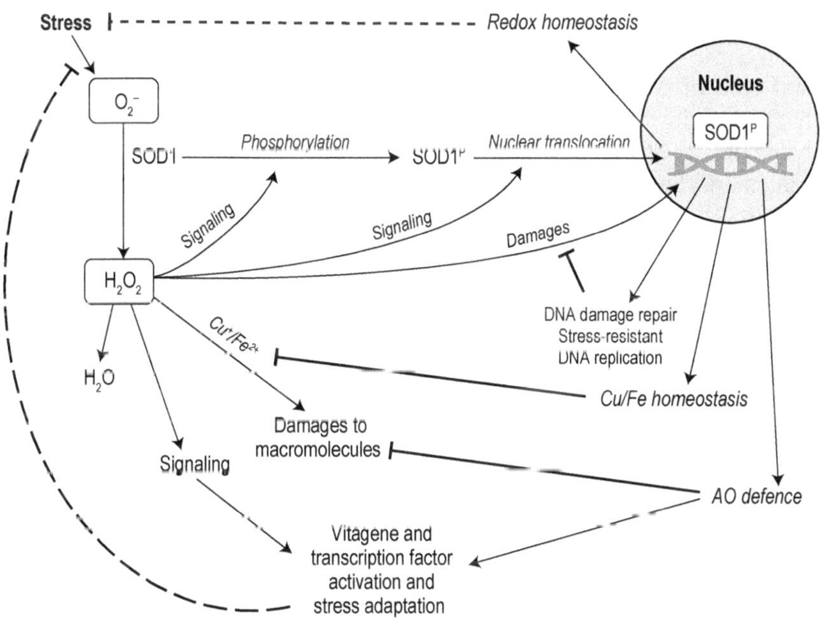

Figure 4.1. A suggested working model for superoxide dismutase (SOD)1 to act as a nuclear transcription factor to regulate oxidative stress resistance (adapted from Surai, 2016, 2020b; Tsang *et al.*, 2014).

stability and function of SOD1 including zinc and copper acquisition, intra-subunit disulphide bond formation and dimerisation. Furthermore, SOD1 is also shown to undergo a range of secondary PTMs including acetylation, glycation, glucosylation, deamidation, palmitoylation. oxidation, acylation, phosphorylation, deamidation, etc. (Wright *et al.*, 2019). Such PTMs are able to change the chemical and biophysical properties and activity of SOD1.

The second member of the family (SOD2) has manganese (Mn) as a cofactor and therefore called Mn-SOD. SOD2 is shown to have a unique genetic organisation and little similarity with SOD1 and SOD3 (Miao and St Clair, 2009). The primary structure of SOD2 genes is sown to be highly conserved and it shares more than 90% sequence homology in the coding region in mouse, rat, bovine and human and the human sod2 is located on chromosome 6q25.3 (Miao and St Clair, 2009). It was shown to be a 96 kDa homotetramer and located exclusively in the mitochondrial matrix, a prime site of superoxide radical production (Halliwell and Gutteridge, 2015). Therefore, the expression of Mn-SOD is considered to be essential for the survival of all aerobic organisms from bacteria to humans and it participates in the development of cellular resistance to oxygen radical-mediated toxicity (Fridovich, 1995). Indeed, Mn-SOD is shown to play a critical role in the defence against oxidant-induced injury and apoptosis in various cells. In fact, Mn-SOD is inducible enzyme and its activity is affected by cytokines and oxidative stress. Therefore, Mn-SOD has been shown to play a major role in promoting cellular differentiation and in protecting against hyperoxia-induced pulmonary toxicity (Fridovich,1995) being a crucial determinant of redox status of the cell. Furthermore, Mn-SOD influences the activity of transcription factors (such as HIF-1α, AP-1, NF-κB and p53) and affects DNA stability (Miriyala *et al.*, 2012).

A critical role of Mn-SOD under physiological and pathological conditions has recently been reviewed in details and the following findings of Mn-SOD confirm the critical role of Mn-SOD in the survival of aerobic life (Holley *et al.*, 2012; Indo *et al.*, 2015; Mates and Sanchez-Jimenez, 1999; Miriyala *et al.*, 2011; Nguyen *et al.*, 2020; Sah *et al.*, 2020):

- *Escherichia coli* and yeasts lacking the Mn-SOD gene are highly sensitive to oxidative stress.
- Mn-SOD gene knockout mice can only survive few days after birth, with pathological findings of many various diseases due to mitochondrial disorder, suggesting a critical role of the enzyme.
- Cells transfected with Mn-SOD cDNAs have increased resistance to various free-radical-generating toxicants (paraquat, tumour necrosis factor, doxorubicin, mitomycin C, irradiation, ischemic reperfusion, smoking, etc.).
- Human Mn-SOD gene transgenic mice show reduced severity of free-radical-induced pulmonary damage and adriamycin-induced myocardial damage.
- Overexpression of Mn-SOD was shown to protect against tert-butyl hydroperoxide induced apoptosis, radiation-induced intestinal syndrome, and lung injury, reduce inflammation and improved mitochondrial respiration in stress conditions.
- Ablation of Sod2 was shown to increase sensitivity to oxygen toxicity and induces multiple organ failure and early neonatal death.

It is important to mention that MnSOD is of great importance for detoxification of the major ROS in biological systems (e.g. superoxide) and at the same time MnSOD is the main source of H_2O_2, major signalling molecule in the same systems. Therefore, MnSOD activity is tightly regulated at the transcriptional, translational, and post-translational levels, depending on the intracellular signals or environmental triggers (Zou *et al.*, 2016). In this regard, it is interesting to note that SOD, as a vitagene, interacts with other member of the vitagene family to build an effective system of stress resistance and adaptability. For example, SIRT3, another member of vitagene family located in mitochondria, can regulate the activity of MnSOD through deacetylation. In fact, MnSOD contains reversible acetyl lysines and deacetylation of lysines 68 and 122 can significantly increases the MnSOD enzymatic activity (Figure 4.2). Furthermore, loss of SIRT3 in different cell lines lead to increased intracellular and mitochondrial superoxide levels. In contrast, increased SIRT3 gene expression was associated with decreased cellular ROS and mitochondrial superoxide levels (Zou *et al.*, 2016).

There is a range of post-translation modification of Mn-SOD and Cu,Zn-SOD leading to reduced their activity, including acetylation, phosphorylation, glutathionylation, nitration and glycation. They represent an important mechanism of SOD activity regulation in various stressful conditions.

In 1982, a third SOD isozyme was discovered by Marklund and co-workers and called extracellular superoxide dismutase (EC-SOD), due to its exclusive extracellular location. EC-SOD is a glycoprotein with a molecular weight of 135,000 kDa and high affinity for heparin (Marklund *et al.*, 1982). However, there are some species-specific variations in molecular weight. The human EC-SOD gene has been mapped to chromosome 4q21 and consists of three exons and two introns (Nozik-Grayck *et al.*, 2005). The full-length mouse EC-SOD cDNA is sown to be 82% identical to that of rat and 60% identical to the human EC-SOD (Miao and St Clair, 2009).

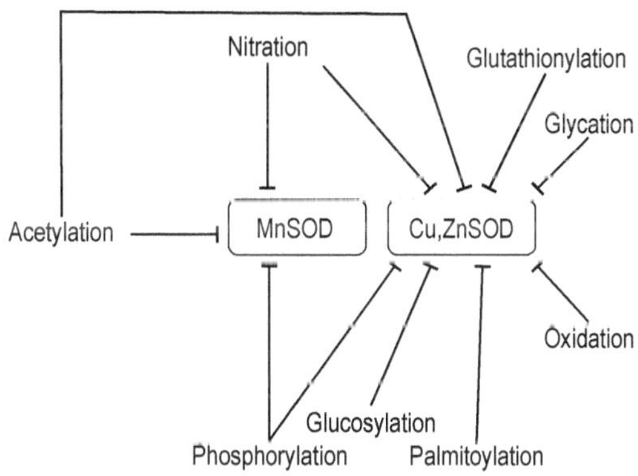

Figure 4.2. Post-translational modifications as negative regulators of superoxide dismutase (SOD) activity (adapted from Wright *et al.*, 2019; Yamakura and Kawasaki, 2010; Zou *et al.*, 2016).

Mature form of EC-SOD consists of three functional domains: the glycosylation domain (1-95 amino acid) at amino-terminal end (responsible for increased solubility of the protein), a catalytic domain (96-193 amino acids) containing the active site and a heparin-binding domain (194-222 amino acids) responsible for binding to heparin sulphate proteoglycans (Sah *et al.*, 2020). EC-SOD is the only antioxidant enzyme that scavenges superoxide specifically in the extracellular space (Yan and Spaulding, 2020). EC-SOD is present in various organisms as a tetramer or, less commonly, as a dimer and contains one copper and one zinc atom per subunit, which are required for enzymatic activity (Fattman *et al.*, 2003). The expression pattern of EC-SOD is highly restricted to the specific cell type and tissues (e.g. lung and kidney, Yan and Spaulding, 2020) where its activity can exceed that of Cu,Zn-SOD or Mn-SOD. As a copper-containing enzyme, the activity of EC-SOD is regulated by copper availability (Nozik-Grayck *et al.*, 2005). Interestingly, EC-SOD was shown to act not only on the cell surface and in the extracellular matrix of cells in a paracrine manner but can also be distributed to other tissues in an endocrine manner (Yan and Spaulding, 2020). EC-SOD is comparatively resistant to high temperatures, extreme pH, and high urea concentrations, it can be inhibited by various agents including azide and cyanide and inactivated by diethyldithiocarbamate and hydrogen peroxide. Oxidative stress and post-translational modification of EC-SOD are shown to cause loss of EC-SOD activity (Miao and St Clair, 2009). Interestingly, EC-SOD was reported to have a comparatively long half-life (~ 20 h) in circulation, whereas Cu,Zn-SOD and Mn-SOD are characterised by a very short half-life at ~ 20 min and 5-6 h, respectively (Nguyen *et al.*, 2020). Genetic evidence supports a causal protective role of EC-SOD activity in virus pathologies and the detrimental effects of reduced EC-SOD levels/activities in disease development as well as the protective effects of enhanced EC-SOD leading to reduced ROS and oxidative damage under disease/stress conditions have been clearly shown (Yan and Spaulding, 2020). Interestingly, *Sod3−/−* mouse was shown to have a normal phenotype under physiological condition but was characterised by increased stress (high O_2 tension) susceptibility and showed earlier onset of severe lung oedema, and exhibited increased susceptibility to stress-induced pulmonary hypertension (Nguyen *et al.*, 2020). Overexpression of EC-SOD was found to suppress the release of inflammatory mediators and adhesion molecules associated with restriction of the inflammation during tissue damage (Sah *et al.*, 2020). Indeed, enhanced EC-SOD expression is considered to be an effective mechanism for protection against oxidative damage. Interestingly, EC-SOD is shown to have immunomodulatory action: downregulate receptors such as TLR2, TLR4, TLR7, histamine receptor 4 (H4R) and IL-4Rα, interact with TLR4, H4R and IL-4R inhibit dendritic cells maturation and T cell activation and differentiation (Sah *et al.*, 2020). EC-SOD was shown to control receptor-ligand complexes formation and interaction between signalling molecules leading to inhibition of inflammation through multiple mechanisms (Nguyen *et al.*, 2020):

- elimination of ROS products;
- modulation of immune cell (T cells, macrophages, NK cells, DCs) function;
- suppression of inflammatory mediators;
- regulation of cellular signalling cascades (TLRs, NF-κB, MAPKs, and JAK-STAT, etc.).

The aforementioned data clearly showed an importance of all three forms SOD in maintaining cellular homeostasis, redox balance, adaptation to various stresses and increasing resistance to a range of oxidative stress-associated diseases/disorders in humans (Lewandowski *et al.*, 2019) and animals, including poultry (Surai, 2016).

4.3 Superoxide dismutase in avian biology

4.3.1 Chicken superoxide dismutase

Chicken SOD was described and purified in early 1970th. Indeed, in chicken liver two types of SOD were identified, one of which was localised in the mitochondria while the other was found in the cytosol (Weisiger and Fridovich, 1973). The cytosol SOD was inhibited by cyanide, whereas the mitochondrial enzyme was not. Later this feature was used to distinguish between two forms of enzymes during assays. The cytosol SOD was purified to homogeneity with apparent molecular weight in presence of mercaptoethanol to be 30,600 Da and to contain copper and zinc, being similar to the other Cu, Zn-SOD which have been isolated from diverse eukaryotes. In fact, purified cytosol SOD from chicken liver contained 0.30% copper and 0.25% zinc. This corresponds to 0.9 atom of copper and 0.8 atom of zinc per subunit. It was also shown that this chicken liver Cu, Zn-SOD had a tendency to polymerise (Weisiger and Fridovich, 1973). In contrast, the mitochondrial SOD was found in chicken liver to be a manganoprotein which has a molecular weight of 80,000 Da. It is composed of four subunits of equal size, which are not covalently joined. It contains 2.3 atoms of manganese per molecule and is strikingly similar to the SOD previously isolated from bacteria. This supports the theory that mitochondria have evolved from aerobic prokaryotes. In fact, Mn-SOD was first isolated from the chicken liver (Weisiger and Fridovich, 1973). The Mn-SOD was found primarily in the mitochondrial matrix space whereas the Cu,Zn-SOD, previously isolated from the cytosol, was found in the intermembrane space (Weisiger and Fridovich, 1973a).

Cu,Zn-SOD was purified from chicken liver with a subunit Mr of 16900 (Dameron and Harris, 1987). Low dietary copper was associated with a decrease in SOD activity and when the 10 day old deficient chicks were injected with 0.5 mol of $CuSO_4$ intraperitoneally, SOD activity in aorta was restored to control levels in about 8 h. Indeed, dietary copper regulates SOD activity in the tissues of young developing animals. The authors also suggested that a copper deficiency suppresses Cu,Zn-SOD activity without inhibiting synthesis or accumulation of the Cu,Zn-SOD protein in this tissue (Dameron and Harris, 1987). Interestingly, molecular properties (amino acid composition, molecular mass and subunit composition) of the chicken enzyme was shown to be similar to those of a bovine erythrocyte Cu,Zn SOD (Michalski and Prowse, 1991). Purified chicken liver Cu,Zn-SOD was confirmed to contain two subunits having Cu and Zn elements with a molecular weight of 16,000±500 for each (Oztürk-Urek and Tarhan, 2001). The optimum pH of purified Cu,Zn-SOD was determined to be 8.9. The enzyme was found to have fair thermal stability up to 45 °C at pH 7.4 over a 1-h incubation period. The SOD enzyme was not inhibited by

DTT and beta-mercaptoethanol but inhibited by CN(-) and H_2O_2 (Oztürk-Urek and Tarhan, 2001). SOD purified from chicken heart has a molecular weight 31.0±1.0 kDa and is composed of two equally sized subunits each having 1.1±0.03 and 0.97±0.02 atoms of Cu and Zn elements, respectively (Demirel and Tarhan, 2004). Cu,Zn-SOD from egg yolk of hens was shown to be cytoplasmic, homodimeric enzyme with a mass of 33.38 kDa and pI of 6.3 (Wawrzykowski and Kankofer, 2017). The authors described similar Cu,Zn-SOD dimer with a molecular weight of 31.77 kDa and the two monomers with molecular weight of 15.59 kDa in erythrocytes of hens. The Mn-SOD cDNA in chicken heart was shown to be 1,108 bp in length. The molecular weight of the mature peptide was 22 kDa. A comparison of the deduced amino acid sequence with those of the human, rat, C. elegans and D. melanogaster showed that the amino acid homology rates were 82.4, 84.7, 62.4 and 59.3%, respectively (Bu *et al.*, 2001). Interestingly, Cu,Zn-SOD activity in the gg yolk (98.5 U/g) was 15-fold higher than that in the egg white (6.1 U/g; Wawrzykowski and Kankofer, 2017). Interestingly, the authors were not able to confirm the occurrence of Cu, Zn-SOD in the egg white by using MALDI-TOF-MS and the question remains if SOD activity in egg white was due to other compounds. SOD activity in avian tissues depends on many different factors including genetics, age, nutrition and various stress-related factors. For example, SOD activity in the Jungle Fowl feather melanocytes was shown to be 2- and 4-fold higher than that in Barred Plymouth Rock and White Leghorn tissue respectively (Bowers *et al.*, 1994). There were lower activities of total SOD along with an elevation in MDA content in the ileum of laying hens in the late phase of production as compared with those at peak production (Wang *et al.*, 2019). Indeed, understanding the molecular mechanisms of the regulation of SOD gene expression and the factors involved in tissue- and cell-specific expression of the SOD genes are of great importance for a developing novel strategies for preventing negative consequences of various stresses in poultry production.

4.3.2 Superoxide dismutase in chicken embryo

Chick embryo tissues contain a high proportion of highly polyunsaturated fatty acids in the lipid fraction (Speake *et al.*, 1998) and therefore need antioxidant defence (Surai, 1999). The antioxidant system of the newly hatched chick includes the antioxidant enzymes SOD, GPx, catalase (Surai, 1999a), fat-soluble antioxidants vitamin E and carotenoids (Surai *et al.*, 1996), water-soluble antioxidants ascorbic acid (Surai *et al.*, 1996) and glutathione (Surai, 1999a) as well as selenium (Surai, 2000, 2002, 2002a; Surai and Fisinin, 2014). Vitamin E (Surai and Speake, 1998), carotenoids (Surai, 2012, 2012a; Surai and Speake, 1998a; Surai *et al.*, 2001; 2001a, 2003) and selenium (Surai, 2000, 2002, 2002a; Surai and Fisinin, 2014) are transferred from feed into egg and further to embryonic tissues. Glutathione and antioxidant enzymes GPx, SOD and catalase are also expressed in the embryonic tissues at various stages of their development (Surai, 1999a; Surai *et al.*, 1999). Our results indicate that there are tissue-specific features in antioxidant defence strategy during embryonic development of the chicken and SOD plays a crucial role as an integral part of the antioxidant network.

In the embryonic liver, SOD specific activity was maximal at day 11 but decreased sharply by day 15 and remained relatively constant thereafter. By contrast, the specific activity of SOD in the brain from day 15 onwards was approximately 2 times higher than that in the liver. In the YSM SOD specific activity increased gradually between days 10 and 15 and then decreased gradually between day 15 and hatching (Surai, 1999a). The specific activities of SOD in kidney, lung, heart and skeletal muscle all showed a gradual decrease between day 15 and hatching. The tissues displayed a considerable variation in the Mn-SOD activity, with the heart having the highest value and lung the lowest (Surai *et al.*, 1999). By contrast, the lung was characterised by high Cu,Zn-SOD activity; in the heart, activity of Cu,Zn-SOD was comparable to the other tissues. Based on the total SOD activity the tissues could be placed in the following descending order: heart >muscle>YSM>kidney>lung>liver. Mn-SOD is the main enzymatic form in the liver and heart comprising 73.2 and 68% of the total SOD activity respectively. In great contrast, in the lung, YSM and thigh muscle, SOD is exclusively represented by Cu,Zn-SOD comprising 98.5, 98.3 and 84.7% of the total SOD activity respectively. In various parts of the brain (cerebrum, cerebellum, brain stem and optic lobes) of the newly hatched chick the Cu,Zn-SOD activity is also almost 2-fold higher than that of Mn-SOD (Surai *et al.*, 1999). Notably, in the kidney both SOD forms are equally represented. Furthermore, the tissues differed markedly in the GPx activities. In all the tissues, Se-dependent GPx was the main enzymatic form, comprising from 65% (lung) up to 90% (heart) of the total enzyme activity. The liver and kidney displayed the highest total GPx activity and the muscle the lowest. As in the case of GPx, catalase activity was also maximal in the liver and kidney.

4.3.3 Superoxide dismutase in avian semen

Despite the importance of SOD in the protection of cells against lipid peroxidation, its activity in avian semen has received only limited attention. A comparison of SOD activity in sperm from various species including boar, rabbit, stallion, donkey, ram, bull, man and chicken indicated that donkey sperm had the highest and fowl the lowest SOD activity (Mannella and Jones, 1980). Furthermore, turkey spermatozoa were found to contain even less SOD activity than fowl spermatozoa (Froman and Thurston, 1981). Our data indicate that in seminal plasma of 5 avian species, KCN inhibited 100% of SOD activity, an observation reflecting the presence of only Cu, Zn-SOD (Surai *et al.*, 1998). In the seminal plasma, the highest SOD activity was recorded in turkey and guinea fowl while the lowest activity was found in duck. Overall, avian species classified in accordance with decreasing SOD activity (expressed per mg seminal plasma protein) can be placed in the following order: guinea fowl>chicken>goose>duck>turkey. Similarly, in seminal plasma, the activity of GPx was two times greater in the ganders than in chickens, whereas SOD activity was lower than in chickens (Partyka *et al.*, 2012). In contrast, the SOD activity in spermatozoa, from pre-cited species is classified in an opposite order to that observed in seminal plasma (goose>duck>chicken=guinea fowl>turkey (Surai *et al.*, 1998). In chicken semen, the SOD activity significantly increased in cryopreserved seminal plasma with simultaneous decrease of its activity in cells (Partyka *et al.*, 2012). In sperm both forms of SOD are expressed with significant species-specific differences.

For example in goose, Cu,Zn-SOD appears twice higher than Mn-SOD and an opposite distribution between different forms of SOD was recorded in guinea fowl where Mn-SOD was more than two-fold higher compared to Cu,Zn-SOD (Surai *et al.*, 1998). In chicken, about 67% of total SOD activity was detected in spermatozoa as compared to 33% in seminal plasma (Surai *et al.*, 1998a). The biological meaning and physiological consequences of such species-specific differences in SOD activity and distribution remain to be established. Notably, in laying hens, SOD activity in the utero-vaginal junction was shown to be increased compared to other regions of the lower oviduct (vagina, uterus; Breque *et al.*, 2003, 2006).

4.4 Superoxide dismutase up- and down-regulation in stress conditions

4.4.1 Heat stress

High environmental temperature is one of the most important stressors causing economic losses to the poultry industry, including poor growth performance, immunosuppression, high mortality, decreased reproductive performance and deterioration of meat quality (Lin *et al.*, 2006). Since SOD is an inducible enzyme, depending on conditions, stresses can tissue-specifically increase or decrease SOD activity in various avian species. For example, acute heat stress (34 °C) in chickens was shown to induce a significant production of ROS, and antioxidant enzymes, including SOD, CAT and GPx (Yang *et al.*, 2010). On exposure to chronic heat stress, GPx activity remained relatively constant, though a temperature-dependent elevation in Cu,Zn-SOD activity was observed in skeletal muscle of broiler chickens (Azad *et al.*, 2010). Chicken exposure to heat stress increased SOD activity and MDA levels in skeletal muscle and vitamin E or vitamin E + Se dietary supplementation further enhanced SOD activity in muscles in heat-stressed birds (Ghazi Harsini *et al.*, 2012). In broiler chickens, plasma activity of SOD was increased, whereas GPx was suppressed by heat stress (32±1 °C). Furthermore, heat exposure increased SOD and catalase activities in breast muscle but the reverse was true in thigh muscle. On the other hand, heat stress increased SOD and decreased GPx activities of mitochondria regardless of muscle types (Huang *et al.*, 2015). Interestingly, in restrictedly fed broiler breeders plasma MDA, protein carbonyl content, activity of SOD and corticosterone content were not altered after acute (33 °C) and prolonged heat challenges (Xie *et al.*, 2015). Probably the stress intensity was not high enough to upregulate SOD. On the other hand, if stress is too high adaptive functions of SOD can be overwhelmed with the following SOD decrease. For example, heat stress in black-boned chickens reduced daily feed intake and BW gain; decreased serum GSH and inhibited GPx, SOD and CAT activities compared with birds subjected to thermo-neutral circumstances (Liu *et al.*, 2014). Similarly, in chickens heat stress induced higher levels of TNF-α, IL-4, HSP27, HSP70, and MDA levels but lower level of IFN-γ, IL-2, GPx, and SOD in spleen (Xu *et al.*, 2014, 2015). These responses were ameliorated by the treatment of Se, polysaccharide of *Atractylodes macrocephala* Koidz alone or in combination (Xu *et al.*, 2014). Similarly, chronic thermal stress (36 °C) was shown to increase the expression of SOD and HSP70 genes in the liver of Ross 308 and Cobb 500 broilers (Roushdy *et al.*, 2018). It seems likely that heat stress response depends on the conditions, including temperature, duration of stress, age of birds, tissue analysed, etc. For example, heat stress in chickens (32.1 °C and 55-65% RH for 6 h/day from day 28 to day 42) was associated with a significant decrease in SOD, GPx and CAT activity in the testes (Xiong *et al.*, 2020). It seems likely that stress duration is an important variable in terms of AO response. For example, at the end of the first week (d35) heat exposure (36±2 °C; 8 h/day) a clear picture of oxidative stress was evident with a significant decrease in SOD, GPx, GR activity and GSH concentration simultaneously with increased MDA in the chicken serum and increased H_2O_2 production in pectoralis muscles, duodenum, jejunum and ileum. However, at the end of the second week of the heat stress (d42) chickens adapted to the stress as evidenced by increased AO activities (SOD, GPx, GR and CAT) in chicken serum, alleviation of increased MDA in serum and H_2O_2 production in pectoralis muscles, duodenum, jejunum and ileum (Wang *et al.*, 2019). In another study it was confirmed that SOD activity changes due to heat stress in chickens are tissue-specific and stress duration-dependent (Habashy *et al.*, 2019). In fact, male (Cobb500) broilers were grown at high temperature (35 °C; 40-50% humidity) from day 14 until day 26 of age. After first day of heat stress SOD activity in the chicken liver and pectoralis muscle did not change while after 12-day treatment it was significantly increased in the liver but did not change in the muscle (Habashy *et al.*, 2019).

4.4.2 Cold stress

Environmental temperature either below or above the comfort zone causes discomfort in avian species. In fact, the increase in metabolic rate at temperatures below the comfort zone (cold stress) is a significant cause of increased mortality from the pulmonary hypertension syndrome (ascites) in broilers (Julian *et al.*, 2005). Initially, it was shown that when broilers were exposed to a cool environment for 3 weeks, plasma SOD activity was decreased (Pan *et al.*, 2005). Similarly, cold exposure reduced chicken plasma SOD and supplemental L-carnitine (100 mg/kg) was shown to restore the SOD activity in cold-stressed birds (Tan *et al.*, 2008). Broilers with cold-induced ascites were characterised by a significantly decreased SOD activity in the liver (Wang *et al.*, 2012). Opposite results were also reported. In fact, during acute cold stress, the SOD activity of the lung increased compared with their control group at each stress time point (Jia *et al.*, 2009). Similarly, there was a significant decrease in CAT and SOD in blood, but increased SOD activity was evident in the liver (Ramnath and Rekha, 2009). A complexity of the SOD response to various stresses is also illustrated in the next two papers. In chick duodenum, under acute cold stress MDA level increased and the activity of SOD and iNOS first increased and then decreased. In contrast, under chronic cold stress the activity of SOD, NO, and NOS in duodenum first decreased and then increased, whereas the MDA level increased (Zhang *et al.*, 2011). In immune organs, the activities of SOD and GPx were first increased then decreased, and activity of total antioxidant capacity was significantly decreased at the acute cold stress in chicks (Zhao *et al.*, 2014).

4.4.3 Other environmental stresses

Effects of environmental stresses on SOD activity is, probably, tissue-specific and depend on many factors, including strength and duration of the stress. For example, in broilers corticosterone administration caused decreases in serum SOD activity as well as in the apparent digestibility of energy, relative weight of bursa and thymus, total antioxidant capacity, and antibody titres to Newcastle disease virus (Zeng *et al.*, 2014). In contrast, there was an increase in SOD activity in the chicken heart during short-term corticosterone administration (Lin *et al.*, 2004). In growing chickens exposed to high ammonia and low humidity blood antioxidative capacities and pectoral muscle SOD and GPx activities were significantly reduced (Wei *et al.*, 2014). Hepatic mitochondrial SOD activity decreased at 14 d in feed-restricted broiler chicks (Yang *et al.*, 2010). However, the plasma SOD activity of feed-restricted birds was markedly higher than those fed *ad libitum* on d 35 and d 42 (Pan *et al.*, 2005).

4.4.4 Toxicological stresses

Administration of cadmium to chickens decreased SOD activities in various tissues, including liver (Gupta and Kar, 1999; Li *et al.*, 2013), kidney (Liu *et al.*, 2015), blood (Erdogan *et al.*, 2005), ovary (Yang *et al.*, 2012), testes (Li *et al.*, 2010) and splenic lymphocytes *in vitro* (Liu *et al.*, 2014). Usually, decreased SOD activity was accompanied by decreased GPx activity and increased lipid peroxidation in the same tissues. In contrast to the aforementioned results, Cd oral administration produced peroxidative damage in chickens, as indicated by increase in TBARS, reduction in GSH concentration in liver and kidney, but increased CAT and SOD activities were observed in erythrocytes (Bharavi *et al.*, 2010). Dietary nickel chloride is also shown to have a negative effect on SOD and other antioxidant enzymes (GPx and CAT) in the intestine (Wu *et al.*, 2013), caecal tonsil (Wu *et al.*, 2014) or splenocytes (Huang *et al.*, 2013). Similarly, vanadium inhibited SOD activity in chicken liver and kidney (Liu *et al.*, 2012). The list of chicken SOD inhibitors includes aluminium (Swain and Chainy, 1997, 1998), fluorine (Chen *et al.*, 2011), polychlorinated biphenyls (Zhang, 2005; Zhou and Zhang, 2005), 4-nitrophenol (Mi *et al.*, 2010), dioxin (Lim *et al.*, 2007), organophosphate (Zhang *et al.*, 2007), thiram (Li *et al.*, 2007), furazolidone (Sas, 1993), florfenicol (Han *et al.*, 2020), valproic acid (Hsieh *et al.*, 2013), oxidised oil (Açıkgöz *et al.*, 2011). It seems likely that mycotoxins can also decrease SOD activity in various chicken tissues. In particular, DON decreased SOD activity in embryo fibroblast DF-1 cells (Li *et al.*, 2014) and AFB1 feed contamination was associated with decreased SOD in the chicken liver (Cao and Wang, 2014; Yarru *et al.*, 2009) and erythrocytes (Sirajudeen *et al.*, 2011). However, the activities of SOD, GST and non-protein thiol levels in the chicken liver were not altered by the FB1-containing (100 mg/kg) diet fed for 21 days (Poersch *et al.*, 2014). Furthermore. gene expression of Nrf2 and its target genes (HO-1, GPx, MnSOD, and CAT) was downregulated in chicken kidney following OTA exposure (50 μg/kg OTA, Li *et al.*, 2020).

4.4.5 Diseases and gut health

Various avian diseases also negatively affect antioxidant defences including decrease SOD activity in jejunal and ileal parts of the gut challenged with *Salmonella pullorum* (Wang *et al.*, 2012), brain and liver of Newcastle disease virus-infected chickens (Subbaiah *et al.*, 2011), erythrocytes of the *Eimeria acervulina* infected birds (Georgieva *et al.*, 2011) and plasma of *Eimeria tenella* challenged birds (Wang *et al.*, 2008). Since antioxidant-pro-oxidant balance in the gut plays an important role in chicken health and immunity (Surai and Fisinin, 2015), special emphasis should be given to this area of research. For example, in vitamin-D-replete chicks, Cu,Zn-SOD was shown to be associated with the apical border (microvilli) of the duodenal absorptive cells (Davis *et al.*, 1989). Furthermore, inclusion of γ-aminobutyric acid (GABA) in laying hen diet was associated with significant increasing the activity of SOD and GPx and decreasing MDA levels in serum (Zhang *et al.*, 2012). Similarly, serum SOD and catalase activities were significantly increased, and MDA was decreased by dietary sodium butyrate at 0.5 or 1.0 g/kg feeding to chickens from hatch for 21 days (Zhang *et al.*, 2011). Broilers fed a diet supplemented with 1×10^9 cfu *Clostridium butyricum*/kg diet had greater SOD activity in the ileal mucosa on d21 and in jejunal mucosa on d42 than those in the other groups fed antibiotic aureomycin or lower doses of the probiotic (Liao *et al.*, 2015).

Recently, the impaired redox status and activated Nrf2/ARE pathway in wooden breast myopathy in broiler chickens have been shown. In fact, increased SOD expression together with other vitagenes (HO-1, GPx1, etc.) were not potent enough to prevent mitochondrial damage and lipid and protein oxidation (Pan *et al.*, in press).

4.5 Clinical significance of superoxide dismutase activity in different tissues

When studying SOD, results interpretation could be a challenging task. First of all, plasma is easily obtained material, however, the meaning of increased or decreased total SOD in plasma sometimes could be misleading. Indeed, in normal human plasma three forms of SOD are found with the lowest amount of SOD1 (5.6-35.5 ng/ml), somehow higher amount of SOD2 (47-150 ng/ml) and even more SOD3 (79-230 ng/ml; Saitoh *et al.*, 2001). Therefore, ideally individual SODs should be determined in plasma to have maximum information to analyse. However, practically in all studies related to SOD in avian plasma only total SOD was determined. Secondly, in tissues Mn-SOD and Cu,Zn-SOD should be distinguished. However, similar to plasma SOD, in most of poultry-related studies only total SOD was analysed. Thirdly, since Mn-SOD is an inducible enzyme, an increased SOD activity in tissues could mean an adaptive response to stress situation or could indicate a potential of the antioxidant defence in the stress conditions. Indeed, when natural antioxidants are supplemented with diets there could be upregulation of SOD indicating an increase in antioxidant defences or downregulation of SOD reflecting a decreased need for SOD because of other antioxidant mechanisms are increased. However, as mentioned above SOD is the main enzyme dealing with superoxide production in mitochondria, a primary site of ROS formation, and most likely it cannot be replaced by other antioxidants. Furthermore, when stress is too strong there is a decrease in SOD activity indicating that the antioxidant defence network was overwhelmed by increased production of free radicals and the body is not able to adequately adapt to the situation. Clearly, there is a need for additional research on individual forms of SOD in avian species with specific emphasis to various transcription factors, including NF-κB and Nrf2, responsible for or involved in SOD activation in stress conditions.

In general, the free radical-initiated oxidative damage of lipids, proteins, and DNA as part of the unspecific immune response caused by some viral (Marek's disease, Newcastle diseases, or infectious bursal disease), bacterial diseases (*Salmonella*, *Staphylococcus*, *Clostridium*, or *E. coli*), or parasitic infections (coccidiosis) has been recently reviewed (Mezes and Balogh, 2011). Indeed, roles of superoxide production and SOD activity in many of those diseases in poultry await investigations. In fact, it has been suggested that oxidative damage may regulate the occurrence and development of avian infectious bronchitis and SOD activity in the serum of chickens inoculated with infectious bronchitis virus significantly decreased (Wang *et al.*, 2011). Similarly, blood SOD was shown to be significantly decreased in broiler birds infected with *E. tenella* (Georgieva *et al.*, 2006).

4.6 Dietary modulation of superoxide dismutase

4.6.1 Mn and Cu in the diet

Mn-SOD is shown to be highly expressed in various organs containing a large number of mitochondria such as the heart, liver, and kidneys. Indeed, in comparison to other tissues, the heart has the highest steady state mRNA Mn-SOD expression level in chickens (Kong *et al.*, 2003). It has been proven that Mn availability is a regulating factor of Mn-SOD activity. For example, in primary cultured broiler myocardial cells Mn-SOD mRNA, Mn-SOD protein, and Mn-SOD activity were induced by manganese in dose- and time-dependent manner. Manganese regulates Mn-SOD expression not only at transcriptional level but also at translational and/or posttranslational levels (Gao *et al.*, 2011). In both heart and kidney, Mn-SOD activity was significantly depressed by decreased dietary manganese; greatest reduction occurred in the heart (Paynter, 1980). Decreased heart Mn-SOD and Cu,Zn-SOD activities, resulting from dietary Mn and Cu deficiencies, were both associated with increased peroxidation (Paynter, 1980a). It seems likely that Mn-SOD activity is very sensitive to dietary Mn levels in commercial corn-soybean meal diets. In fact, Mn deficiency in growing chickens caused the reductions of Mn concentrations of the liver and heart as well as Mn-SOD activity of the heart (Luo *et al.*, 1992). In chickens, dietary Mn contents required to reach the plateau of Mn concentrations of the liver, pancreas, kidney, heart, spleen and muscle and to obtain the maximum Mn-SOD activity of heart were calculated to be 110, 111, 141, 123, 109, 99 and 121 mg/kg respectively. Interestingly, Mn-SOD of liver and pancreas were not affected. Therefore, for broilers fed the basal corn-soybean meal diet, 120 mg/kg Mn was suggested as the required level (Luo *et al.*, 1991) which corresponds to the presently recommended levels of Mn supplementation. Chickens fed a Mn-deficient diet from hatching had significantly lower levels of Mn-SOD activity in liver than did controls. However, activity of the Cu,Zn-SOD in the liver was higher in Mn-deficient chickens than in controls (De Rosa *et al.*, 1980). The activity of both forms of SOD reached normal levels when a Mn-supplemented (1000 mg/kg) diet was fed to deficient chickens, but the activity of the manganese enzyme was not affected by feeding the supplemented diet to manganese sufficient chickens. It was shown that heart Mn-SOD activity and heart Mn-SOD mRNA levels increased linearly as dietary Mn levels increased, confirming that dietary Mn significantly affected heart Mn-SOD gene transcription (Li *et al.*, 2004). Furthermore, birds fed supplemental Mn had lower MDA content in leg muscle and greater Mn-SOD activities and Mn-SOD mRNA level in breast or leg muscle than those fed the control diet (Lu *et al.*, 2007). Compared with control chickens fed on a diet without Mn supplementation, chickens fed Mn-supplemented diets had higher Mn concentrations, Mn-SOD mRNA levels, Mn-SOD protein concentrations, and Mn-SOD activities within heart tissue (Li *et al.*, 2011, 2011a). Therefore, dietary Mn can activate Mn-SOD gene expression at both the transcriptional and translational levels (Li *et al.*, 2011). However, Mn excess can be toxic for birds. In fact, the activities of SOD and GPx in chicken serum and immune organs (spleen, thymus, and bursa of Fabricius; Liu *et al.*, 2013) and testes (Liu *et al.*, 2013a) were decreased due to Mn dietary excess.

It seems likely that dietary Cu is involved in regulation of the SOD activity and in the case low Cu levels in the basic diet, it is possible to upregulate Cu,Zn-SOD in chickens by dietary Cu supplementation. For example, in the basal low-Cu group, Cu, Zn-SOD activity decreased in the liver, RBC and heart to 14, 25, and 61%, respectively, of control activities after 6 weeks' depletion (Paynter *et al.*, 1979). On the other hand, Cu,Zn-SOD activity in chicken erythrocytes from the Cu- and vitamin C-supplemented birds was increased by 39 and 20% respectively (Aydemir *et al.*, 2000). Similarly, in the Cu-supplemented chickens, Cu,Zn-SOD activity in the liver, erythrocyte, kidney and heart significantly increased by 75, 40, 12, 12% respectively. Furthermore, Mn-SOD activity in the heart, liver, kidney and brain of the vitamin C –supplemented chickens was increased. In addition, in the heart of Cu-supplemented chickens Mn-SOD was found to be increased by approximately 15%, while in liver tissue of the Cu-supplemented group it was reduced by 19% (Oztürk-Urek *et al.*, 2001). However, in an earlier study, hepatic Mn SOD and Cu,Zn-SOD were not influenced by dietary Cu level or source or LPS in broiler chicks (Koh *et al.*, 1996) probably reflecting differences in the background Cu levels. However, excessive Cu intake was shown to cause oxidative stress associated with decrease in activities of SOD, CAT and GPx, but increase contents/expression of malondialdehyde MDA, proinflammatory cytokines, NF-κB in immune organs of chicken (Yang *et al.*, 2020). In conclusion, excessive Cu could cause pathologic changes and induce oxidative stress with triggered NF-κB pathway and might further regulate the inflammatory response in immune organs of chicken.

4.6.2 Vitamins, carnitine and amino acids

Dietary vitamin A excess was shown to decrease SOD activity in the chicken liver and brain (Surai *et al.*, 2000). Similarly, increased vitamin E supplementation (40-60 mg/kg) or CCl_4 injection decreased the activity of SOD in the chicken blood (Mahmoud and Hiajazi, 2007). However, in a recent study a higher vitamin E level (60 vs 30 mg/kg) significantly increased alpha-tocopherol concentrations and SOD activity in serum of laying hens (Zduńczyk *et al.*, 2013). In heat stressed (34 °C for 8 hours/day for 42 days) chickens SOD, GPx, CAT and GST activities in liver, heart and kidney tissues were decreased while expression of HSP60, HSP70, HSP90, HSF1 and HSF3 were significantly increased. Dietary supplementation of vitamin C (1 g/kg) was found to correct those parameters towards the normal control value (Albokhadaim *et al.*, 2019). Recently, betaine has been recognised as a natural anti-heat stress agent able to mitigate heat-induce oxidative stress in poultry industry (Saeed *et al.*, 2017). Initially, L-carnitine dietary supplementation was shown to increase blood SOD activity in chickens (Geng *et al.*, 2004). Furthermore, when chicken fed corn-soybean diets supplemented with different doses of lipoic acid SOD activity in serum (300 mg/kg), liver (100, 200 and 300 mg/kg) and leg muscle (200 or 300 mg/kg) was significantly increased (Chen *et al.*, 2011). It was shown that increased lipoic acid (LA) or acetyl-L-carnitine (ALC) resulted in increased total antioxidant capacity and SOD and GPx activities and decreased levels of MDA in serum and liver of birds (Jia *et al.*, 2014). Notably, birds fed diets containing 50 mg/kg of LA and 50 mg/kg of ALC had higher serum and liver SOD activities than those fed diets containing 100 mg/kg of LA or ALC alone. In laying hens reared in a hot and humid climate L-threonine supplementation at 0.2% maximised the SOD activity in both serum and liver (Azzam *et al.*, 2012). Serum SOD increased linearly and quadratically in laying hens receiving excess dietary tryptophan (0·4 g/kg) (Dong *et al.*, 2012). Broilers given a diet containing 5.9 g/kg methionine had enhanced serum SOD activity and decreased hepatic MDA content at day 7 (Chen *et al.*, 2013). Dietary taurine supplementation was shown to enhance antioxidative capacity, including increased SOD activity in breast muscles of broiler chickens (Xu *et al.*, 2020).

4.6.3 Selenium

Low-Se diet caused a significant decrease in the activities of SOD and GPx, and an increase MDA content in thymus, spleen, Bursa of Fabricius and serum (Zhang *et al.*, 2012). Interestingly, not only Se deficiency (0.03 mg Se per kg of diet) but also Se excess (3 mg/kg) in chickens significantly lowered SOD and CAT activities in the liver and serum (Xu *et al.*, 2014). It seems likely that SOD in adult birds is also affected by Se status. For example, laying hens fed the Se-supplemented diet showed higher SOD and GPx activity and lower MDA content in plasma compared with those fed the control (non-supplemented) diet (Jing *et al.*, 2015). Positive effects of dietary Se on SOD activities in avian species depend not only on Se concentration, but also on the form of Se used, with organic Se being more effective than sodium selenite. In fact, the activities of serum GPx, SOD and total antioxidant capacity were significantly higher in selenium yeast than sodium selenite-fed chickens (Chen *et al.*, 2014). Similarly, dietary Se-Met significantly elevated T-AOC, GPX, T-SOD, CAT activities, contents of GSH and reduced carbonyl protein content in chicken breast muscle (Jiang *et al.*, 2009). It was shown that dietary organic Se significantly increased the Se content and the activities of CAT and SOD but decreased the MDA content in chicken breast muscle at 42 days of age (Ahmad *et al.*, 2012). In fact, Cd/Pb poisoning and heat stress were shown to impose oxidative stress as evidenced by increase oxidation, mRNA levels of inflammatory proteins, and apoptotic proteins. At the same time, Se was reported to enhance the antioxidant status and alleviates those effects via upregulation of antioxidant proteins and other molecular effects (Seremelis *et al.*, 2019; Surai, 2018).

4.6.4 Phytochemicals

Polyphenolic compounds and various plant extracts have received substantial attention as an important means of decreasing oxidative stress *in vitro* and *in vivo*. For example, in cultured muscle cells of embryonic broilers, pretreatment with low-dosage phytoestrogen equol (1 μM) restored altered (decreased) by H_2O_2 intracellular SOD activity. However, pretreatment with high-dosage equol (10 and 100 μM) showed a synergistic effect with H_2O_2 in inducing cell damage, but had no effect on MDA content, SOD or GPx activity (Wei *et al.*, 2011). Similarly, in chicken HD11 macrophages challenged with LPS activity of SOD increased in cells treated with the higher concentration of equol (80 or 160 μmol/l, but not in 10, 20 or 40 μmol/l groups; Gou *et al.*, 2015). In a chicken erythrocyte model both curcumin and cyanidin-3-rutinoside were shown to significantly attenuate apoptosis and haemolysis, decreasing MDA content, and increasing SOD activity in a time- and dose-dependent manner (Zhang *et al.*, 2014). Similarly, feeding diets with added flavonoids (hesperidin and naringenin) to laying hens increased the blood serum SOD activity (Lien *et al.*, 2008). There was a significant increase in the activities of SOD chicken blood due to Brahma Rasayana supplementation (Ramnath *et al.*, 2008). Dietary xanthophyll (lutein+zeaxanthin) supplementation (20 or 40 mg/kg) for 3 or 4 weeks was shown to increase serum SOD activity in chickens (Gao *et al.*, 2013). However, the SOD activity was not affected in the chicken liver or jejunal mucosa. Inclusion into the chicken diet of polysavone (1·5 g/kg), a natural extract from alfalfa, for 6 weeks increased serum and liver SOD activity, while breast muscle SOD activity at 6 weeks of age were significantly higher and MDA content was significantly lower in 1·0 and 1·5 g/kg polysavone groups than in the control group (Dong *et al.*, 2011). Notably, effects of plant extracts added to chicken diets on the SOD activity would depend on many factors including polyphenol composition, concentration and bioavailability. In fact, low availability of polyphenolic compounds for growing chickens, breeders and layers (Surai, 2014) is an important limiting factor of their biological efficacy and nutritive value. For example, there was no effect of dietary turmeric rhizome powder (0.25-0.75%) on the activities of GPx and SOD in thigh muscle (Daneshyar, 2012) or serum (Daneshyar *et al.*, 2012). Feeding to broiler chicks diets enriched with selected herbal supplements failed to affect the growth performance of chickens at 42 days of age. In addition, this supplementation had no influence on the activities of SOD and GPx, concentration of vitamin A and selected lipid metabolism indices (Petrovic *et al.*, 2012).

Dihydromyricetin (a flavonoid component of herbal medicines) was shown to attenuate Escherichia Coli LPS-induced ileum injury in chickens by inhibiting NLRP3 inflammasome and TLR4/NF-κB signalling pathway associated with prevention of alteration in SOD and GPx activity and GSH concentration in chicken plasma and ileum (Chang *et al.*, 2020). Similarly, quercetin was found to attenuate the LPS-induced inhibition of Nrf2 activation, translocation, and downstream gene expression, including Mn-SOD and HO-1 (Sun *et al.*, 2020). In general, nutritional strategies to deal with oxidative stress during associated with increased environmental temperature is on agenda of many research groups (Nawab *et al.*, 2018; Zaboli *et al.*, 2019) and vitagene, including SOD, activation is an important and effective approach in this area (Surai, 2020; Surai *et al.*, 2017, 2018, 2019a).

4.7 Conclusions

Protective roles of SOD in animal/poultry physiology are shown in Figure 4.3.

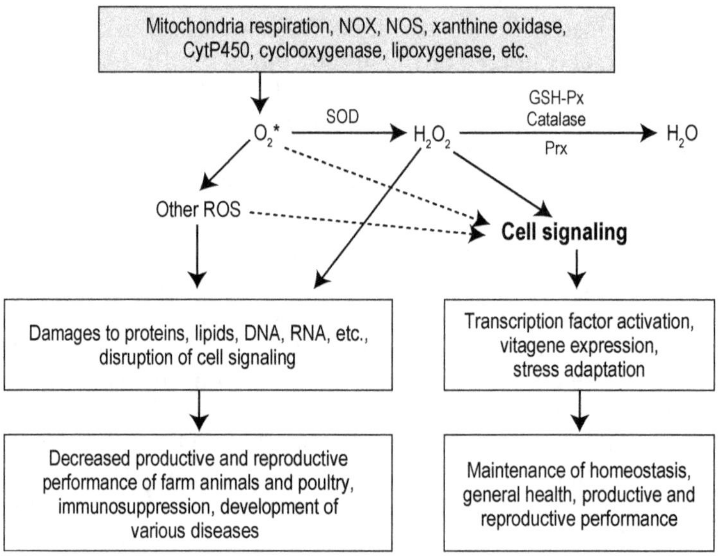

Figure 4.3. Protective roles of superoxide dismutase in animal/poultry physiology (adapted from Surai, 2018, 2020b).

From the aforementioned analysis of the data related to SOD in avian/poultry physiology and adaptation to stresses it is possible to conclude:

- SOD as important vitagene is the main driving force in cell/body adaptation to various stress conditions. Indeed, in stress conditions additional synthesis of SOD is an adaptive mechanism to decrease ROS formation.
- SOD is the main regulator of production of H_2O_2, an important signalling molecule, and, therefore, the enzyme expression and activity are tightly regulated at transcription and post-transcription levels and they are regulated by an array of transcription factors.
- If the stress is too high SOD activity is decreased and apoptosis is activated.
- There are tissue-specific differences in SOD expression which also depends on the strength of such stress-factors as heat, heavy metals, mycotoxins and other toxicants.
- In most studies related to SOD in avian species mainly total activity of the enzyme was studied and molecular characterisation of individual (e.g. SOD1, SOD2 and SOD3) forms of enzymes in various avian species awaits investigation.
- SOD is shown to provide an effective protection against lipid peroxidation in chicken embryonic tissues and in semen.
- SOD is proven to be protective in heat and cold stress, toxicity stress as well as in other oxidative stress-related conditions in poultry production.
- There are complex interactions inside the antioxidant network of the cell/body to ensure an effective maintenance of homeostasis in stress conditions. Indeed, in many cases nutritional antioxidants (vitamin E, selenium, phytochemicals, etc.) in the feed can increase SOD expression in chicken tissues.
- Regulating effects of various phytochemicals on SOD need further investigation.
- Nutritional means of additional SOD upregulation in stress conditions of poultry production and physiological and commercial consequences await investigation. Indeed, in medical sciences manipulation of SOD expression and usage of SOD mimics are considered as an important approach in disease prevention and treatment.
- SOD upregulation in stress conditions is emerging as an effective means for stress management.
- Transcription factor-like activity of SOD deserves more attention and investigation.
- SODs are important elements of the vitagene protective network in avian species as well as in human and animals in general.

References

Açıkgöz, Z., Bayraktar, H., Altan, O., Akhisaroglu, S.T., Kırkpınar, F. and Altun, Z., 2011. The effects of moderately oxidised dietary oil with or without vitamin E supplementation on performance, nutrient digestibility, some blood traits, lipid peroxidation and antioxidant defence of male broilers. Journal of the Science of Food and Agriculture 91: 1277-1282.

Ahmad, H., Tian, J., Wang, J., Khan, M.A., Wang, Y., Zhang, L. and Wang, T., 2012. Effects of dietary sodium selenite and selenium yeast on antioxidant enzyme activities and oxidative stability of chicken breast meat. Journal of Agricultural and Food Chemistry 60: 7111-7120.

Albokhadaim, I.F., Althnaian, T.A. and El-Bahr, S.M., 2019. Gene expression of heat shoc kproteins/factors (HSP60, HSP70, HSP90, HSF-1, HSF-3) and antioxidant enzyme activities in heat stressed broilers treated with vitamin C. Polish Journal of Veterinary Sciences 22, 3: 565-572.

Aydemir, T., Oztürk, R., Bozkaya, L.A. and Tarhan, L., 2000. Effects of antioxidant vitamins A, C, E and trace elements Cu, Se on CuZn SOD, GSH-Px, CAT and LPO levels in chicken erythrocytes. Cell Biochemistry and Function 18: 109-115.

Azad, M.A., Kikusato, M., Maekawa, T., Shirakawa, H. and Toyomizu, M., 2010. Metabolic characteristics and oxidative damage to skeletal muscle in broiler chickens exposed to chronic heat stress. Comparative Biochemistry and Physiology. Part A, Molecular & Integrative Physiology 155: 401-406.

Azzam, M.M., Dong, X.Y., Xie, P. and Zou, X.T., 2012. Influence of L-threonine supplementation on goblet cell numbers, histological structure and antioxidant enzyme activities of laying hens reared in a hot and humid climate. British Poultry Science 53: 640-645.

Bannister, W.H., 1988. From haemocuprein to copper-zinc superoxide dismutase: a history on the fiftieth anniversary of the discovery of haemocuprein and the twentieth anniversary of the discovery of superoxide dismutase. Free Radical Research Communications 5: 35-42.

Bharavi, K., Reddy, A.G., Rao, G.S., Reddy, A.R. and Rao, S.V., 2010. Reversal of cadmium-induced oxidative stress in chicken by herbal adaptogens *Withania somnifera* and *Ocimum sanctum*. Toxicology International 17: 59-63.

Bowers, R.R., Lujan, J., Biboso, A., Kridel, S. and Varkey, C., 1994. Premature avian melanocyte death due to low antioxidant levels of protection: fowl model for vitiligo. Pigment Cell Research 7: 409-418.

Breque, C., Surai, P. and Brillard, J-P., 2006. Antioxidant status of the lower oviduct in fowl varies with age and vitamin E diet supplementation. Molecular Reproduction and Development 73: 1045-1051.

Breque, C., Surai, P. and Brillard, J-P., 2003. Roles of antioxidants in prolonged storage of avian spermatozoa *in vivo* and *in vitro*. Molecular Reproduction and Development 66: 314-323.

Bu, Y., Luo, X., Li, S., Lu, C., Li, Y., Kuang, X., Liu, B., Li, J. and Yu, S., 2001. Cloning and sequence analysis of manganese-containing superoxide dismutase (MnSOD) cDNA of chickens. Chinese Journal of Biochemistry and Molecular Biology 17: 463-467.

Cao, J. and Wang, W., 2014. Effects of astaxanthin and esterified glucomannan on hematological and serum parameters, and liver pathological changes in broilers fed aflatoxin-B1-contaminated feed. Animal Science Journal 85: 150-157.

Chang, Y., Yuan, L., Liu, J., Muhammad, I., Cao, C., Shi, C., Zhang, Y., Li, R., Li, C. and Liu, F., 2020. Dihydromyricetin attenuates Escherichia coli lipopolysaccharide-induced ileum injury in chickens by inhibiting NLRP3 inflammasome and TLR4/NF-κB signalling pathway. Veterinary Research 51, 1: 72.

Chen, G., Wu, J. and Li, C., 2014. Effect of different selenium sources on production performance and biochemical parameters of broilers. Journal of Animal Physiology and Animal Nutrition 98: 747-754.

Chen, P., Ma, Q.G., Ji, C., Zhang, J.Y., Zhao, L.H., Zhang, Y. and Jie, Y. Z., 2011. Dietary lipoic acid influences antioxidant capability and oxidative status of broilers. International Journal of Molecular Sciences 12: 8476-8488.

Chen, T., Cui, H., Cui, Y., Bai, C. and Gong, T., 2011. Decreased antioxidase activities and oxidative stress in the spleen of chickens fed on high-fluorine diets. Human & Experimental Toxicology 30: 1282-1286.

Chen, Y.P., Chen, X., Zhang, H. and Zhou, Y.M., 2013. Effects of dietary concentrations of methionine on growth performance and oxidative status of broiler chickens with different hatching weight. British Poultry Science 54: 531-537.

Culotta, V.C., Yang, M. and O'Halloran, T.V., 2006. Activation of superoxide dismutases: putting the metal to the pedal. Biochimica et Biophysica Acta 1763: 747-758.

Dameron, C.T. and Harris, E.D., 1987. Regulation of aortic CuZn-superoxide dismutase with copper. Effects *in vivo*. Biochemical Journal 248: 663-668.

Daneshyar, M., 2012. Effect of dietary turmeric on antioxidant properties of thigh meat in broiler chickens after slaughter. Animal Science Journal 83: 599-604.

Daneshyar, M., Kermanshahi, H. and Golian, A., 2012. The effects of turmeric supplementation on antioxidant status, blood gas indices and mortality in broiler chickens with T(3)-induced ascites. British Poultry Science 53: 379-385.

Davis, W.L., Matthews, J.L., Shibata, K., Kipnis, M., Farmer, G.R., Cortinas, E., Meiyr, J.C. and Goodman, D.B., 1989. The immunocytochemical localization of superoxide dismutase in the enterocytes of the avian intestine: the effect of vitamin D3. The Histochemical Journal 21: 194-202.

De Rosa, G., Keen, C.L., Leach, R.M. and Hurley, L.S., 1980. Regulation of superoxide dismutase activity by dietary manganese. Journal of Nutrition 110: 795-804.

Demirel, L.A. and Tarhan, L., 2004. Dismutation properties of purified and GDA modified CuZnSOD from chicken heart. Artificial Cells, Blood Substitutes, and Immobilization Biotechnology 32: 609-624.

Dong, X.F., Gao, W.W., Su, J.L., Tong, J.M. and Zhang, Q., 2011. Effects of dietary polysavone (alfalfa extract) and chlortetracycline supplementation on antioxidation and meat quality in broiler chickens. British Poultry Science 52: 302-309.

Dong, X.Y., Azzam, M.M., Rao, W., Yu, D.Y. and Zou, X.T., 2012. Evaluating the impact of excess dietary tryptophan on laying performance and immune function of laying hens reared under hot and humid summer conditions. British Poultry Science 53: 491-496.

Erdogan, Z., Erdogan, S., Celik, S. and Unlu, A., 2005. Effects of ascorbic acid on cadmium-induced oxidative stress and performance of broilers. Biological Trace Element Research 104: 19-32.

Fattman, C.L., Schaefer, L.M. and Oury, T.D., 2003. Extracellular superoxide dismutase in biology and medicine. Free Radical Biology and Medicine 35: 236-256.

Fridovich, I., 1995. Superoxide radical and superoxide dismutases. Annual Review of Biochemistry 64: 97-112.

Froman, D.P. and Thurston, R.J., 1981. Chicken and turkey spermatozoal superoxide dismutase: a comparative study. Biology of Reproduction 24: 193-200.

Fukai, T. and Ushio-Fukai, M., 2011. Superoxide dismutases: role in redox signaling, vascular function, and diseases. Antioxidants & Redox Signaling 15: 1583-1606.

Gao, T., Wang, F., Li, S., Luo, X. and Zhang, K., 2011. Manganese regulates manganese-containing superoxide dismutase (MnSOD) expression in the primary broiler myocardial cells. Biological Trace Element Research 144: 695-704.

Gao, Y.Y., Xie, Q.M., Ma, J.Y., Zhang, X.B., Zhu, J.M., Shu, D.M., Sun, B.L., Jin, L. and Bi, Y.Z., 2013. Supplementation of xanthophylls increased antioxidant capacity and decreased lipid peroxidation in hens and chicks. The British Journal of Nutrition 109: 977-983.

Geng, A., Guo, Y. and Yuan, J., 2004. Effects of dietary L-carnitine and coenzyme Q10 supplementation on performance and ascites mortality of broilers. Archives of Animal Nutrition 58: 473-482.

Georgieva, N. V., Gabrashanska, M., Koinarski, V. and Yaneva, Z., 2011. Zinc Supplementation against Eimeria acervulina-Induced Oxidative Damage in Broiler Chickens. Veterinary Medicine International 2011: 647124.

Georgieva, N.V., Koinarski, V. and Gadjeva, V., 2006. Antioxidant status during the course of *Eimeria tenella* infection in broiler chickens. Veterinary Journal 172: 488-492.

Ghazi Harsini, S., Habibiyan, M., Moeini, M.M. and Abdolmohammadi, A.R., 2012. Effects of dietary selenium, vitamin E, and their combination on growth, serum metabolites, and antioxidant defense system in skeletal muscle of broilers under heat stress. Biological Trace Element Research 148: 322-330.

Gou, Z., Jiang, S., Zheng, C., Tian, Z. and Lin, X., 2015. Equol inhibits LPS-induced oxidative stress and enhances the immune response in chicken HD11 macrophages. Cellular Physiology and Biochemistry 36: 611-621.

Gupta, P. and Kar, A., 1999. Cadmium induced thyroid dysfunction in chicken: hepatic type I iodothyronine 5'-monodeiodinase activity and role of lipid peroxidation. Comparative biochemistry and physiology. Part C, Pharmacology, Toxicology & Endocrinology 123: 39-44.

Habashy, W.S., Milfort, M.C., Rekaya, R. and Aggrey, S.E., 2019. Cellular antioxidant enzyme activity and biomarkers for oxidative stress are affected by heat stress. International Journal of Biometeorology 63: 1569-1584.

Halliwell, B. and Gutteridge, J.M.C., 2015. Free radicals in biology and medicine. Oxford University Press, USA.

Han, C., Wei, Y., Cui, Y., Geng, Y., Bao, Y. and Shi, W., 2020. Florfenicol induces oxidative stress and hepatocyte apoptosis in broilers via Nrf2 pathway. Ecotoxicology and Environmental Safety 191: 110239.

Hassan, H.M., 1988. Biosynthesis and regulation of superoxide dismutases. Free Radical Biology and Medicine 5: 377-385.

Holley, A.K., Dhar, S.K., Xu, Y. and St Clair, D.K., 2012. Manganese superoxide dismutase: beyond life and death. Amino Acids 42:139-158.

Hsieh, C. L., Chen, K. C., Ding, C. Y., Tsai, W. J., Wu, J. F. and Peng, C. C., 2013. Valproic acid substantially downregulated genes folr1, IGF2R, RGS2, COL6A3, EDNRB, KLF6, and pax-3, N-acetylcysteine alleviated most of the induced gene alterations in chicken embryo model. Romanian Journal of Morphology and Embryology 54: 993-1004.

Huang, C., Jiao, H., Song, Z., Zhao, J., Wang, X. and Lin, H., 2015. Heat stress impairs mitochondria functions and induces oxidative injury in broiler chickens. Journal of Animal Science 93: 2144-2153.

Huang, J., Cui, H., Peng, X., Fang, J., Zuo, Z., Deng, J. and Wu, B., 2013. The association between splenocyte apoptosis and alterations of Bax, Bcl-2 and caspase-3 mRNA expression, and oxidative stress induced by dietary nickel chloride in broilers. International Journal of Environmental Research and Public Health 10: 7310-7326.

Huang, T.T., Zou, Y. and Corniola, R., 2012. Oxidative stress and adult neurogenesis--effects of radiation and superoxide dismutase deficiency. Seminars in Cell & Developmental Biology 23: 738-744.

Indo, H.P., Yen, H.C., Nakanishi, I., Matsumoto, K., Tamura, M., Nagano, Y., Matsui, H., Gusev, O., Cornette, R., Okuda, T., Minamiyama, Y., Ichikawa, H., Suenaga, S., Oki, M., Sato, T., Ozawa, T., Clair, D.K. and Majima, H.J., 2015. A mitochondrial superoxide theory for oxidative stress diseases and aging. Journal of Clinical Biochemistry and Nutrition 56: 1-7.

Jia, H.Y., Li, J.M., Yu, Q., Wang, J.J. and Li, S., 2009. The effect of cold stress on DNA oxidative damage of lung in chicken. Zhongguo Yingyong Shenglixue Zazhi 25: 373-376.

Jia, R., Bao, Y.H., Zhang, Y., Ji, C., Zhao, L.H., Zhang, J.Y., Gao, C.Q. and Ma, Q.G., 2014. Effects of dietary α-lipoic acid, acetyl-L-carnitine, and sex on antioxidative ability, energy, and lipid metabolism in broilers. Poultry Science 93: 2809-2817.

Jiang, Z., Lin, Y., Zhou, G., Luo, L., Jiang, S. and Chen, F., 2009. Effects of dietary selenomethionine supplementation on growth performance, meat quality and antioxidant property in yellow broilers. Journal of Agricultural and Food Chemistry 57: 9769-9772.

Jing, C.L., Dong, X.F., Wang, Z.M., Liu, S. and Tong, J.M., 2015. Comparative study of DL-selenomethionine vs sodium selenite and seleno-yeast on antioxidant activity and selenium status in laying hens. Poultry Science 94: 965-975.

Julian, R.J., 2005. Production and growth-related disorders and other metabolic diseases of poultry – a review. The Veterinary Journal 169: 350-369.

Koh, T.S., Peng, R.K. and Klasing, K.C., 1996. Dietary copper level affects copper metabolism during lipopolysaccharide-induced immunological stress in chicks. Poultry Science 75: 867-872.

Kong, B.W., Kim, H. and Foster, D.N., 2003. Expression analysis and mitochondrial targeting properties of the chicken manganese-containing superoxide dismutase. Biochimica et Biophysica Acta 1625: 98-108.

Lewandowski, Ł., Kepinska, M. and Milnerowicz, H., 2019. The copper-zinc superoxide dismutase activity in selected diseases. European Journal of Clinical Investigation, 49, 1: e13036.

Li, D., Ye, Y., Lin, S., Deng, L., Fan, X., Zhang, Y., Deng, X., Li, Y., Yan, H. and Ma, Y., 2014. Evaluation of deoxynivalenol-induced toxic effects on DF-1 cells *in vitro*: cell-cycle arrest, oxidative stress, and apoptosis. Environmental Toxicology and Pharmacology 37: 141-149.

Li, J., Bi, D., Pan, S. and Zhang, Y., 2007. Effect of diet with thiram on liver antioxidant capacity and tibial dyschondroplasia in broilers. British Poultry Science 48: 724-728.

Li, J.L., Gao, R., Li, S., Wang, J.T., Tang, Z.X. and Xu, S.W., 2010. Testicular toxicity induced by dietary cadmium in cocks and ameliorative effect by selenium. Biometals 23: 695-705.

Li, J.L., Jiang, C.Y., Li, S. and Xu, S.W., 2013. Cadmium induced hepatotoxicity in chickens (*Gallus domesticus*) and ameliorative effect by selenium. Ecotoxicology and Environmental Safety 96: 103-109.

Li, K., Cao, Z., Guo, Y., Tong, C., Yang, S., Long, M., Li, P. and He, J., 2020. Selenium yeast alleviates ochratoxin a-induced apoptosis and oxidative stress via modulation of the PI3K/AKT and Nrf2/Keap1 signaling pathways in the kidneys of chickens. Oxidative Medicine and Cellular Longevity 2020: 4048706.

Li, S., Lin, Y., Lu, L., Xi, L., Wang, Z., Hao, S., Zhang, L., Li, K. and Luo, X., 2011. An estimation of the manganese requirement for broilers from 1 to 21 days of age. Biological Trace Element Research 143: 939-948.

Li, S., Lu, L., Hao, S., Wang, Y., Zhang, L., Liu, S., Liu, B., Li, K. and Luo, X., 2011. Dietary manganese modulates expression of the manganese-containing superoxide dismutase gene in chickens. The Journal of Nutrition 141: 189-194.

Li, S., Luo, X., Liu, B., Crenshaw, T.D., Kuang, X., Shao, G. and Yu, S., 2004. Use of chemical characteristics to predict the relative bioavailability of supplemental organic manganese sources for broilers. Journal of Animal Science 82: 2352-2363.

Liao, X.D., Ma, G., Cai, J., Fu, Y., Yan, X.Y., Wei, X.B. and Zhang, R.J., 2015. Effects of *Clostridium butyricum* on growth performance, antioxidation, and immune function of broilers. Poultry Science 94: 662-667.

Lien, T.F., Yeh, H.S. and Su, W.T., 2008. Effect of adding extracted hesperetin, naringenin and pectin on egg cholesterol, serum traits and antioxidant activity in laying hens. Archives of Animal Nutrition 62: 33-43.

Lim, J., DeWitt, J.C., Sanders, R.A., Watkins, J.B. and Henshel, D.S., 2007. Suppression of endogenous antioxidant enzymes by 2,3,7,8-tetrachlorodibenzo-p-dioxin-induced oxidative stress in chicken liver during development. Archives of Environmental Contamination and Toxicology 52: 590-595.

Lin, H., Decuypere, E. and Buyse, J., 2004. Oxidative stress induced by corticosterone administration in broiler chickens (*Gallus gallus domesticus*) 2. Short-term effect. Comparative Biochemistry and Physiology. Part B, Biochemistry & Molecular Biology 139: 745-751.

Lin, H., Jiao, H.C., Buyse, J. and Decuypere, E., 2006. Strategies for preventing heat stress in poultry. World's Poultry Science Journal 62: 71-86.

Liu, J., Cui, H., Liu, X., Peng, X., Deng, J., Zuo, Z., Cui, W., Deng, Y. and Wang, K., 2012. Dietary high vanadium causes oxidative damage-induced renal and hepatic toxicity in broilers. Biological Trace Element Research 145: 189-200.

Liu, L., Yang, B., Cheng, Y. and Lin, H., 2015. Ameliorative effects of selenium on cadmium-induced oxidative stress and endoplasmic reticulum stress in the chicken kidney. Biological Trace Element Research 167: 308-319.

Liu, L.L., He, J.H., Xie, H.B., Yang, Y.S., Li, J.C. and Zou, Y., 2014. Resveratrol induces antioxidant and heat shock protein mRNA expression in response to heat stress in black-boned chickens. Poultry Science 93: 54-62.

Liu, S., Xu, F.P., Yang, Z.J., Li, M., Min, Y.H. and Li, S., 2014. Cadmium-induced injury and the ameliorative effects of selenium on chicken splenic lymphocytes: mechanisms of oxidative stress and apoptosis. Biological Trace Element Research 160: 340-351.

Liu, X.F., Li, Z.P., Tie, F., Liu, N., Zhang, Z.W. and Xu, S.W., 2013. Effects of manganese-toxicity on immune-related organs of cocks. Chemosphere 90: 2085-2100.

Liu, X.F., Zhang, L.M., Guan, H.N., Zhang, Z.W. and Xu, S.W., 2013. Effects of oxidative stress on apoptosis in manganese-induced testicular toxicity in cocks. Food and Chemical Toxicology 60: 168-176.

Lu, L., Luo, X.G., Ji, C., Liu, B. and Yu, S.X., 2007. Effect of manganese supplementation and source on carcass traits, meat quality, and lipid oxidation in broilers. Journal of Animal Science 85: 812-822.

Luo, X.G., Su, Q., Huang, J.C. and Liu, J.X., 1991. A study on the optimal manganese (Mn) level in a practical diet of broiler chicks. Chinese Journal of Animal and Veterinary Sciences 22: 313-317.

Luo, X.G., Su, Q., Huang, J.C. and Liu, J.X., 1992. Effects of manganese (Mn) deficiency on tissue Mn-containing superoxide dismutase (MnSOD) activity and its mitochondrial ultrastructures of broiler chicks fed a practical diet. Chinese Journal of Animal and Veterinary Sciences 23: 97-101.

Mahmoud, K.Z. and Hijazi, A.A., 2007. Effect of vitamin A and/or E on plasma enzymatic antioxidant systems and total antioxidant capacity of broiler chickens challenged with carbon tetrachloride. Journal of Animal Physiology and Animal Nutrition 91: 333-340.

Mannella, M.R.T. and Jones, R., 1980. Properties of spermatozoal superoxide dismutase and lack of involvement of superoxides in metal-ion-catalysed lipid peroxidation reactions in semen. Biochemical Journal 191: 289-297.

Marklund, S.L., Holme, E. and Hellner, L., 1982. Superoxide dismutase in extracellular fluids. Clinica Chimica Acta 126: 41-51.

Mates, J.M. and Sanchez-Jimenez, F., 1999. Antioxidant enzymes and their implications in pathophysiologic processes. Frontiers in Bioscience 4: D339-D345.

McCord, J.M. and Fridovich, I., 1969. Superoxide dismutase: an enzymatic function for erythrocuprein (hemocuprein). Journal of Biological Chemistry 244: 6049-6055.

Mezes, M. and Balogh, K., 2011. Free radicals and antioxidants in avian diseases. In: Mandelker, L. and Vajdovich, P. (eds) Studies on veterinary medicine. Humana Press, Springer New York, NY, USA, pp. 175-190.

Mi, Y., Zhang, C., Li, C.M., Taneda, S., Watanabe, G., Suzuki, A.K. and Taya, K., 2010. Protective effect of quercetin on the reproductive toxicity of 4-nitrophenol in diesel exhaust particles on male embryonic chickens. The Journal of Reproduction and Development 56: 195-199.

Miao, L. and St Clair, D.K., 2009. Regulation of superoxide dismutase genes: implications in disease. Free Radical Biology & Medicine 47: 344-356.

Michalski, W.P. and Prowse, S.J., 1991. Cu,Zn superoxide dismutase from chicken erythrocytes. Comparative Biochemistry and Physiology. B, Comparative Biochemistry 100: 371-375.

Miriyala, S., Holley, A.K. and St Clair, D.K., 2011. Mitochondrial superoxide dismutase – signals of distinction. Anti-cancer Agents in Medicinal Chemistry 11: 181-190.

Miriyala, S., Spasojevic, I., Tovmasyan, A., Salvemini, D., Vujaskovic, Z., St Clair, D. and Batinic-Haberle, I., 2012. Manganese superoxide dismutase, MnSOD and its mimics. Biochimica et Biophysica Acta 1822: 794-814.

Nawab, A., Ibtisham, F., Li, G., Kieser, B., Wu, J., Liu, W., Zhao, Y., Nawab, Y., Li, K., Xiao, M. and An, L., 2018. Heat stress in poultry production: mitigation strategies to overcome the future challenges facing the global poultry industry. Journal of Thermal Biology 78: 131-139.

Nguyen, N.H., Tran, G.B. and Nguyen, C.T., 2020. Anti-oxidative effects of superoxide dismutase 3 on inflammatory diseases. Journal of Molecular Medicine 98: 59-69.

Nozik-Grayck, E., Suliman, H.B. and Piantadosi, C.A., 2005. Extracellular superoxide dismutase. The International Journal of Biochemistry & Cell Biology 37: 2466-2471.

Oztürk-Urek, R., Bozkaya, L.A. and Tarhan, L., 2001. The effects of some antioxidant vitamin and trace element-supplemented diets on activities of SOD, CAT, GSH-Px and LPO levels in chicken tissues. Cell Biochemistry and Function 19: 125-132

Oztürk-Urek, R. and Tarhan, L., 2001. Purification and characterization of superoxide dismutase from chicken liver. Comparative Biochemistry and Physiology. Part B, Biochemistry & Molecular Biology 128: 205-212.

Pan, J.Q., Tan, X., Li, J.C., Sun, W.D. and Wang, X.L., 2005. Effects of early feed restriction and cold temperature on lipid peroxidation, pulmonary vascular remodelling and ascites morbidity in broilers under normal and cold temperature. British Poultry Science 46: 374-381.

Pan, X., Zhang, L., Xing, T., Li, J. and Gao, F., in press. The impaired redox status and activated Nrf2/ARE pathway in wooden breast myopathy in broiler chickens. Asian-Australasian Journal of Animal Sciences. https://doi.org/10.5713/ajas.19.0953

Partyka, A., Lukaszewicz, E. and Niżański, W., 2012. Lipid peroxidation and antioxidant enzymes activity in avian semen. Animal Reproduction Science 134: 184-190.

Paynter, D.I., 1980. Changes in activity of the manganese superoxide dismutase enzyme in tissues of the rat with changes in dietary manganese. The Journal of Nutrition 110: 437-447.

Paynter, D.I., 1980. The role of dietary copper, manganese, selenium, and vitamin E in lipid peroxidation in tissues of the rat. Biological Trace Element Research 2: 121-135.

Paynter, D.I., Moir, R.J. and Underwood, E.J., 1979. Changes in activity of the Cu-Zn superoxide dismutase enzyme in tissues of the rat with changes in dietary copper. The Journal of Nutrition 109: 1570-1576.

Petrovic, V., Marcincak, S., Popelka, P., Simkova, J., Martonova, M., Buleca, J., Marcincakova, D., Tuckova, M., Molnar, L. and Kovac, G., 2012. The effect of supplementation of clove and agrimony or clove and lemon balm on growth performance, antioxidant status and selected indices of lipid profile of broiler chickens. Journal of Animal Physiology and Animal Nutrition 96: 970-977.

Poersch, A.B., Trombetta, F., Braga, A.C., Boeira, S.P., Oliveira, M.S., Dilkin, P., Mallmann, C.A., Fighera, M.R., Royes, L.F., Oliveira, M.S. and Furian, A.F., 2014. Involvement of oxidative stress in subacute toxicity induced by fumonisin B1 in broiler chicks. Veterinary Microbiology 174: 180-185.

Ramnath, V. and Rekha, P.S., 2009. Brahma rasayana enhances *in vivo* antioxidant status in cold-stressed chickens (*Gallus gallus domesticus*). Indian Journal of Pharmacology 41: 115-119.

Ramnath, V., Rekha, P.S. and Sujatha, K.S., 2008. Amelioration of heat stress induced disturbances of antioxidant defense system in chicken by brahma rasayana. Evidence-based Complementary and Alternative Medicine 5: 77-84.

Roushdy, E.M., Zaglool, A.W. and El-Tarabany, M.S., 2018. Effects of chronic thermal stress on growth performance, carcass traits, antioxidant indices and the expression of HSP70, growth hormone and superoxide dismutase genes in two broiler strains. Journal of Thermal Biology 74: 337-343.

Saeed, M., Babazadeh, D., Naveed, M., Arain, M.A., Hassan, F.U. and Chao, S., 2017. Reconsidering betaine as a natural anti-heat stress agent in poultry industry: a review. Tropical Animal Health and Production 49: 1329-1338.

Sah, S.K., Agrahari, G. and Kim, T.Y., 2020. Insights into superoxide dismutase 3 in regulating biological and functional properties of mesenchymal stem cells. Cell & Bioscience 10: 22.

Saitoh, D., Ookawara, T., Fukuzuka, K., Kawakami, M., Sakamoto, T., Ohno, H. and Okada, Y., 2001. Characteristics of plasma extracellular SOD in burned patients. Burns 27: 577-581.

Sas, B., 1993. Contribution to the pathobiochemistry of furazolidone-induced oxidative toxicity in chickens. Acta Veterinaria Hungarica 41: 103-121.

Seremelis, I., Danezis, G.P., Pappas, A.C., Zoidis, E. and Fegeros, K., 2019. Avian stress-related transcriptome and selenotranscriptome: role during exposure to heavy metals and heat stress. Antioxidants 8, 7: 216.

Sirajudeen, M., Gopi, K., Tyagi, J.S., Moudgal, R.P., Mohan, J. and Singh, R., 2011. Protective effects of melatonin in reduction of oxidative damage and immunosuppression induced by aflatoxin B1-contaminated diets in young chicks. Environmental Toxicology 26: 153-160.

Speake, B.K., Murray, A.M.B. and Noble, R.C., 1998. Transport and transformation of yolk lipids during development of the avian embryo. Progress in Lipid Research 37: 1-32.

Subbaiah, K.C., Raniprameela, D., Visweswari, G., Rajendra, W. and Lokanatha, V., 2011. Perturbations in the antioxidant metabolism during Newcastle disease virus (NDV) infection in chicken: protective role of vitamin E. Naturwissenschaften 98: 1019-1026.

Sun, L., Xu, G., Dong, Y., Li, M., Yang, L. and Lu, W., 2020. Quercetin protects against lipopolysaccharide-induced intestinal oxidative stress in broiler chickens through activation of Nrf2 pathway. Molecules 25, 5:1053.

Surai, A.P., Surai, P.F., Steinberg, W., Wakeman, W.G., Speake, B.K. and Sparks, N.H., 2003. Effect of canthaxanthin content of the maternal diet on the antioxidant system of the developing chick. British Poultry Science 44: 612-619.

Surai, P.F., 1999. Tissue-specific changes in the activities of antioxidant enzymes during the development of the chicken embryo. British Poultry Science 40: 397-405.

Surai, P.F., 2000. Effect of the selenium and vitamin E content of the maternal diet on the antioxidant system of the yolk and the developing chick. British Poultry Science 2000; 41: 235-243.

Surai, P.F., 2002a. Selenium in poultry nutrition: a new look at an old element. 1. Antioxidant properties, deficiency and toxicity. World's Poultry Science Journal 58: 333-347.

Surai, P.F., 2002b. Selenium in poultry nutrition: a new look at an old element. 2. Reproduction, egg and meat quality and practical applications. World's Poultry Science Journal 58: 431-450.

Surai, P.F., 2012a. The antioxidant properties of canthaxanthin and its potential effects in the poultry eggs and on embryonic development of the chick. Part 1. World's Poultry Science Journal 68: 465-475.

Surai, P.F., 2012b. The antioxidant properties of canthaxanthin and its potential effects in the poultry eggs and on embryonic development of the chick. Part 2. World's Poultry Science Journal 68: 717-726.

Surai, P.F., 2014. Polyphenol compounds in the chicken/animal diet: from the past to the future. Journal of Animal Physiology and Animal Nutrition 98: 19-31

Surai, P.F., 2016. Antioxidant systems in poultry biology: superoxide dismutase. Journal of Animal Nutrition 1, 1: 8.

Surai, P.F., 2018. Selenium in poultry nutrition and health. Wageningen Academic Publishers, Wageningen, the Netherlands.

Surai P.F., 2020a. Antioxidants in poultry nutrition and reproduction: an update. Antioxidants 9, 2: 105.

Surai, P.F., 2020b. Superoxide dismutase as a new entrant into the vitagene family in animals/poultry. EC Nutrition 15.3: 01-03.

Surai, P.F., Blesbois, E., Grasseau, I., Chalah, T., Brillard, J.P., Wishart, G.J., Cerolini, S. and Sparks, N.H., 1998. Fatty acid composition, glutathione peroxidase and superoxide dismutase activity and total antioxidant activity of avian semen. Comparative Biochemistry and Physiology. Part B, Biochemistry & Molecular Biology 120: 527-533.

Surai, P.F., Cerolini, S., Wishart, G.J., Speake, B.K., Noble, R.C. and Sparks, N.H., 1998a. Lipid and antioxidant composition of chicken semen and its susceptibility to peroxidation. Poultry and Avian Biology Reviews 9: 11-23.

Surai, P.F. and Fisinin, V.I., 2014. Selenium in poultry breeder nutrition: an update. Animal Feed Science and Technology 191: 1-15.

Surai, P.F., Kochish, I.I. and Fisinin, V.I., 2017. Antioxidant systems in poultry biology: nutritional modulation of vitagenes. European Journal of Poultry Science 81: 1612-9199.

Surai P.F., Kochish I.I., Fisinin V.I., Grozina A.A. and Shatskikh E.V., 2018. Molecular mechanisms of chicken gut health maintenance: role of microbiota. Agricultural Technologies, Moscow, Russia.

Surai, P.F., Kochish, I.I., Fisinin, V.I. and Kidd, M.T., 2019. Antioxidant defence systems and oxidative stress in poultry biology: an update. Antioxidants 8, 7: 235.

Surai, P.F., Kuklenko, T.V., Ionov, I.A., Noble, R.C. and Sparks, N.H., 2000. Effect of vitamin A on the antioxidant system of the chick during early postnatal development. British Poultry Science 41: 454-458.

Surai, P.F., Noble, R.C. and Speake, B.K., 1996. Tissue-specific differences in antioxidant distribution and susceptibility to lipid peroxidation during development of the chick embryo. Biochimica et Biophysica Acta 1304, 1-10.

Surai, P. F. and Speake, B.K., 1998. Selective excretion of yolk-derived tocotrienols into the bile of chick embryo. Comparative Biochemistry and Physiology 121B: 393-396.

Surai, P.F. and Speake, B.K., 1998a. Distribution of carotenoids from the yolk to the tissues of the chick embryo. The Journal of Nutritional Biochemistry 9: 645-651.

Surai, P.F., Speake, B.K., Noble, R.C. and Sparks, N.H.C., 1999. Tissue-specific antioxidant profiles and susceptibility to lipid peroxidation of the newly hatched chick. Biological Trace Element Research 68: 63-78.

Surai, P.F., Speake, B.K. and Sparks, N.H.C., 2001. Carotenoids in avian nutrition and embryonic development. 1. Absorption, availability and levels in plasma and egg yolk. Journal of Poultry Science 38: 1-27.

Surai, P.F., Speake, B.K. and Sparks, N.H.C., 2001a. Carotenoids in avian nutrition and embryonic development. 2. Antioxidant properties and discrimination in embryonic tissues. Journal of Poultry Science 38:117-145.

Swain, C. and Chainy, G.B., 1997. Aluminum effect on lipid peroxidation and on the activities of superoxide dismutase and catalase in the cerebral hemisphere and liver of young chicks. Journal of Trace Elements in Medicine and Biology 11: 77-82.

Swain, C. and Chainy, G.B., 1998. Effects of aluminum sulphate and citric acid ingestion on lipid peroxidation and on activities of superoxide dismutase and catalase in cerebral hemisphere and liver of developing young chicks. Molecular and Cellular Biochemistry 187: 163-172.

Tan, X., Hu, S.H. and Wang, X.L., 2008. The effect of dietary L-carnitine supplementation on pulmonary hypertension syndrome mortality in broilers exposed to low temperatures. Journal of Animal Physiology and Animal Nutrition 92: 203-210.

Wang H., Zhong X., Shi W. and Guo, B., 2011. Study of malondialdehyde (MDA) content, superoxide dismutase (SOD) and glutathione peroxidase (GSH-Px) activities in chickens infected with avian infectious bronchitis virus. African Journal of Biotechnology 10: 9213-9217.

Wang, J., Xue, X., Liu, Q., Zhang, S., Peng, M., Zhou, J., Chen, L. and Fang, F., 2019. Effects of duration of thermal stress on growth performance, serum oxidative stress indices, the expression and localization of ABCG2 and mitochondria ROS production of skeletal muscle, small intestine and immune organs in broilers. Journal of Thermal Biology 85: 102420.

Wang, L.C., Zhang, T.T., Wen, C., Jiang, Z.Y., Wang, T. and Zhou, Y.M., 2012. Protective effects of zinc-bearing clinoptilolite on broilers challenged with *Salmonella pullorum*. Poultry Science 91: 1838-1845.

Wang, M.L., Suo, X., Gu, J.H., Zhang, W.W., Fang, Q. and Wang, X., 2008. Influence of grape seed proanthocyanidin extract in broiler chickens: effect on chicken coccidiosis and antioxidant status. Poultry Science 87: 2273-2280.

Wang, W.W., Wang, J., Zhang, H.J., Wu, S.G. and Qi, G.H., 2019. Transcriptome analysis reveals mechanism underlying the differential intestinal functionality of laying hens in the late phase and peak phase of production. BMC Genomics 20, 1: 970.

Wang, Y., Guo, Y., Ning, D., Peng, Y., Cai, H., Tan, J., Yang, Y. and Liu, D., 2012. Changes of hepatic biochemical parameters and proteomics in broilers with cold-induced ascites. Journal of Animal Science and Biotechnology 3,1: 41.

Wawrzykowski, J. and Kankofer, M., 2017. Partial biochemical characterization of Cu,Zn-superoxide dismutase extracted from eggs of hens (*Gallus gallus domesticus*). Food Chemistry 227: 390-396.

Wei, F.X., Hu, X.F., Sa, R.N., Liu, F.Z., Li, S.Y. and Sun, Q.Y., 2014. Antioxidant capacity and meat quality of broilers exposed to different ambient humidity and ammonia concentrations. Genetics and Molecular Research 13: 3117-3127.

Wei, X.J., Wu, J., Ni, Y.D., Lu, L.Z. and Zhao, R.Q., 2011. Antioxidant effect of a phytoestrogen equol on cultured muscle cells of embryonic broilers. *In vitro* Cellular & Developmental Biology. Animal 47: 735-741.

Weisiger, R.A. and Fridovich, I., 1973. Mitochondrial superoxide simutase. Site of synthesis and intramitochondrial localization. The Journal of Biological Chemistry 248: 4793-4796.

Weisiger, R.A. and Fridovich, I., 1973a. Superoxide dismutase. Organelle specificity. The Journal of Biological Chemistry 248: 3582-3592.

Wright, G., Antonyuk, S.V. and Hasnain, S.S., 2019. The biophysics of superoxide dismutase-1 and amyotrophic lateral sclerosis. Quarterly Reviews of Biophysics, 52, e12.

Wu, B., Cui, H., Peng, X., Fang, J., Zuo, Z., Deng, J. and Huang, J., 2013. Dietary nickel chloride induces oxidative intestinal damage in broilers. International Journal of Environmental Research and Public Health 10: 2109-2119.

Wu, B., Cui, H., Peng, X., Fang, J., Zuo, Z., Deng, J. and Huang, J., 2014. Dietary nickel chloride induces oxidative stress, apoptosis and alters Bax/Bcl-2 and caspase-3 mRNA expression in the cecal tonsil of broilers. Food and Chemical Toxicology 63: 18-29.

Xie, J., Tang, L., Lu, L., Zhang, L., Lin, X., Liu, H. C., Odle, J. and Luo, X., 2015. Effects of acute and chronic heat stress on plasma metabolites, hormones and oxidant status in restrictedly fed broiler breeders. Poultry Science 94: 1635-1644.

Xiong, Y., Yin, Q., Li, J. and He, S., 2020. Oxidative stress and endoplasmic reticulum stress are involved in the protective effect of alpha lipoic acid against heat damage in chicken testes. Animals 10, 3: 384.

Xu, D., Li, W., Huang, Y., He, J. and Tian, Y., 2014. The effect of selenium and polysaccharide of *Atractylodes macrocephala* Koidz. (PAMK) on immune response in chicken spleen under heat stress. Biological Trace Element Research 160: 232-237.

Xu, D. and Tian, Y., 2015. Selenium and polysaccharides of *Atractylodes macrocephala* Koidz play different roles in improving the immune response induced by heat stress in chickens. Biological Trace Element Research 168: 235-241.

Xu, J.X., Cao, C.Y., Sun, Y.C., Wang, L.L., Li, N., Xu, S.W. and Li, J.L., 2014. Effects on liver hydrogen peroxide metabolism induced by dietary selenium deficiency or excess in chickens. Biological Trace Element Research 159: 174-182.

Xu, S.W., Lu, Z., Ma, B.B., Xing, T., Li, J.L., Zhang, L., Jiang, Y. and Gao, F., 2020. Dietary taurine supplementation enhances antioxidative capacity and improves breast meat quality of broiler chickens. British Poultry Science 61, 2: 140-145.

Yamakura, F. and Kawasaki, H., 2010. Post-translational modifications of superoxide dismutase. Biochimica et Biophysica Acta 1804: 318-325.

Yan, Z. and Spaulding, H.R., 2020. Extracellular superoxide dismutase, a molecular transducer of health benefits of exercise. Redox Biology 32: 101508.

Yang, F., Liao, J., Yu, W., Pei, R., Qiao, N., Han, Q., Hu, L., Li, Y., Guo, J., Pan, J. and Tang, Z., 2020. Copper induces oxidative stress with triggered NF-κB pathway leading to inflammatory responses in immune organs of chicken. Ecotoxicology and Environmental Safety 200: 110715.

Yang, L., Tan, G.Y., Fu, Y.Q., Feng, J.H. and Zhang, M.H., 2010. Effects of acute heat stress and subsequent stress removal on function of hepatic mitochondrial respiration, ROS production and lipid peroxidation in broiler chickens. Comparative biochemistry and physiology. Toxicology & Pharmacology 151: 204-208.

Yang, S., Zhang, Z., He, J., Li, J., Zhang, J., Xing, H. and Xu, S., 2012. Ovarian toxicity induced by dietary cadmium in hen. Biological Trace Element Research 148: 53-60.

Yang, X., Zhuang, J., Rao, K., Li, X. and Zhao, R., 2010. Effect of early feed restriction on hepatic lipid metabolism and expression of lipogenic genes in broiler chickens. Research in Veterinary Science 89: 438-444

Yarru, L.P., Settivari, R.S., Gowda, N.K., Antoniou, E., Ledoux, D.R. and Rottinghaus, G.E., 2009. Effects of turmeric (*Curcuma longa*) on the expression of hepatic genes associated with biotransformation, antioxidant, and immune systems in broiler chicks fed aflatoxin. Poultry Science 88: 2620-2627.

Youn, H.D., Kim, E.J., Roe, J.H., Hah, Y.C. and Kang, S.O., 1996. A novel nickel-containing superoxide dismutase from *Streptomyces* spp. Biochemical Journal 318: 889-896.

Zaboli, G., Huang, X., Feng, X. and Ahn, D., 2019. How can heat stress affect chicken meat quality? – a review. Poultry Science 98: 1551-1556.

Zduńczyk, Z., Drazbo, A., Jankowski, J., Juśkiewicz, J., Czech, A. and Antoszkiewicz, Z., 2013. The effect of different dietary levels of vitamin E and selenium on antioxidant status and immunological markers in serum of laying hens. Polish Journal of Veterinary Sciences 16: 333-339.

Zelko, I.N., Mariani, T.J. and Folz, R.J., 2002. Superoxide dismutase multigene family: a comparison of the CuZn-SOD (SOD1), Mn-SOD (SOD2), and EC-SOD (SOD3) gene structures, evolution, and expression. Free Radical Biology and Medicine 33: 337-349.

Zeng, Z.K., Li, Q.Y., Piao, X.S., Liu, J.D., Zhao, P.F., Xu, X., Zhang, S. and Niu, S., 2014. *Forsythia suspensa* extract attenuates corticosterone-induced growth inhibition, oxidative injury, and immune depression in broilers. Poultry Science 93: 1774-1781

Zhang, J., Hou, X., Ahmad, H., Zhang, H., Zhang, L. and Wang, T., 2014. Assessment of free radicals scavenging activity of seven natural pigments and protective effects in AAPH-challenged chicken erythrocytes. Food Chemistry 145: 57-65.

Zhang, L.P., Wang, Q.S., Guo, X., Zhu, Y.J., Zhou, G.Z. and Xie, K.Q., 2007. Time-dependent changes of lipid peroxidation and antioxidative status in nerve tissues of hens treated with tri-ortho-cresyl phosphate (TOCP). Toxicology 239: 45-52.

Zhang, M., Zou, X.T., Li, H., Dong, X.Y. and Zhao, W., 2012. Effect of dietary γ-aminobutyric acid on laying performance, egg quality, immune activity and endocrine hormone in heat-stressed Roman hens. Animal Science Journal 83: 141-147.

Zhang, W.H., Jiang, Y., Zhu, Q.F., Gao, F., Dai, S.F., Chen, J. and Zhou, G.H., 2011. Sodium butyrate maintains growth performance by regulating the immune response in broiler chickens. British Poultry Science 52: 292-301.

Zhang, Y.M., 2005. Protective effect of quercetin on aroclor 1254-induced oxidative damage in cultured chicken spermatogonial cells. Toxicological sciences: an official journal of the Society of Toxicology 88: 545-550.

Zhang, Z.W., Lv, Z.H., Li, J.L., Li, S., Xu, S.W. and Wang, X.L., 2011. Effects of cold stress on nitric oxide in duodenum of chicks. Poultry Science 90: 1555-1561.

Zhang, Z.W., Wang, Q.H., Zhang, J.L., Li, S., Wang, X.L. and Xu, S.W., 2012. Effects of oxidative stress on immunosuppression induced by selenium deficiency in chickens. Biological Trace Element Research 149: 352-361.

Zhao, F.Q., Zhang, Z.W., Qu, J.P., Yao, H.D., Li, M., Li, S. and Xu, S.W., 2014. Cold stress induces antioxidants and Hsps in chicken immune organs. Cell Stress & Chaperones 19: 635-648.

Zhou, C. and Zhang, C., 2005. Protective effects of antioxidant vitamins on Aroclor 1254-induced toxicity in cultured chicken embryo hepatocytes. Toxicology in Vitro 19: 665-673.

Zou, X., Santa-Maria, C.A., O'Brien, J., Gius, D. and Zhu, Y., 2016. Manganese superoxide dismutase acetylation and dysregulation, due to loss of SIRT3 activity, promote a luminal B-like breast carcinogenic-permissive phenotype. Antioxidants & Redox Signaling 25: 326-336.

Chapter 5
Heat shock proteins

A stich in time saves nine

5.1 Introduction

Understanding roles of vitagenes in stress resistance of poultry as a background for the development of effective strategies to deal with stresses is an emerging topic of research (Surai, 2015; 2015a; 2015b; 2014; Shatskikh *et al.*, 2015; Surai and Fisinin, 2015; Surai *et al.*, 2019; Surai, 2020). It is known that vitagenes are responsible for synthesis of various protective molecules and HSP70 and HSP32 (HO-1) synthesis is under vitagene control.

Therefore, the aim of this chapter is a critical analysis of the role of HSPs in poultry biology with special emphasis to the HSP70 and HO-1 functions as an essential part of the vitagene network, responsible for adaptive ability of the cells or whole organisms to various stress conditions.

5.2 Heat shock response and heat shock factors

The heat shock response (HSR) is one of the main adaptive stress responses of the cell, restoring cellular homeostasis upon exposure to proteotoxic stress, including heat shock, cold, oxidative stress, hypoxia, toxins, chemicals, pathogen, etc. (Meijering *et al.*, 2015; Pockley and Multhoff, 2008; Velichko *et al.*, 2013). In fact, cooperative interactions between the transcription factors and various homeostatic mechanisms are responsible for effective adaptation to stressful conditions (Fujimoto and Nakai, 2010; Sakurai and Enoki, 2010; Takii *et al.*, 2015). Indeed, to maintain vital life function it is imperative that organisms preserve the integrity of their proteins. Therefore, HSR in vertebrates is characterised by the induction of HSPs and related elements, such as the ubiquitin-proteasome system (Velichko *et al.*, 2013). Because HSPs act as molecular chaperones that facilitate protein folding and suppress protein aggregation, this response plays a major role in maintaining protein homeostasis. Generally, HSR is regulated mainly at the level of transcription by four heat shock transcription factors (HSFs), including HSF1, HSF2, HSF3, and HSF4, which bind to HSE (Fujimoto and Nakai, 2010), thus resulting in stimulation of HSPs expression.

Among other heat shock factors, HSF1 has received tremendous attention as the main factor governing the HSR by coordinating stress-induced transcription (Richter *et al.*, 2010). Although originally discovered as a response to thermal stress, HSR can be triggered by a variety of stress conditions that interfere with protein folding and

DOI 10.3920/978-90-8686-906-0_5,

result in accumulation of misfolded or aggregated proteins (Liu and Chang, 2008). HSF1 activation is a multistep process that is negatively regulated by chaperones, including HSP90 and HSP70 (de Thonel *et al.*, 2012), which sets the stage for rapid induction of gene expression within minutes of cellular stress (Stetler *et al.*, 2010). In physiological conditions the majority of HSFs form a complex with HSP70 or HSP90 interacting with the HSF1 activation domain. In stress conditions, HSP70 and HSP90 form complexes with denatured proteins, which releases HSFs (Kantidze *et al.*, 2015). Furthermore, in unstressed state, HSF1 is present in the cytoplasm as a latent monomeric molecule. Upon heat shock, monomeric HSF1 is hyperphosphorylated and converts to a trimer with the capacity to bind DNA that accumulates in the nucleus and subsequently binds to the heat shock element within the promoter region of HSP genes. In addition, extensive posttranslational modifications such as phosphorylation, acetylation, and sumoylation are thought to fine-tune HSF1 activity (Meijering *et al*, 2015, Takii *et al.*, 2015). The increased expression of HSPs continues until the amount of HSP70 and HSP90 reaches the level sufficient to block the activation domain of the HSFs (Kantidze *et al.*, 2015).

Therefore, there are specialised adaptive mechanisms in different cellular compartments, leading to the transcriptional activation of target gene expression upon stress exposure. In fact, in the cytosol, the HSF1 is kept inactive. As a result of proteotoxic stress, HSF1 forms an active trimeric complex that drives target gene expression in the nucleus, called the cytosolic heat-shock response (HSR^{Cyt}). On the other hand, proteostatic imbalance in the endoplasmic reticulum (ER) is shown to activate a transcriptional program called the unfolded protein response in the ER (UPR^{ER}) with XBP1 (X box binding protein 1) being a key transcription factor responsible for regulation of the UPR^{ER}. It is believed that XBP1 is activated as a result of alternative splicing of its mRNA by the ER transmembrane sensor IRE1 (inositol requiring enzyme 1). Furthermore, cellular response in response to proteotoxic stress in mitochondria is associated with the transcription factor ATFS-1 (activating transcription factor associated with stress) triggering the UPR in the mitochondria (UPR^{MIT}) (Franz and Hoppe, 2018). Under normal conditions, ATFS-1 is imported into mitochondria, where it is degraded. However, in stress conditions ATFS-1 translocates to the nucleus to induce a broad transcriptional response including the upregulation of mitochondrial chaperones, antioxidant genes, glycolysis genes, and amino acid catabolism pathways (Tian *et al.*, 2016).

5.3 Chicken heat shock factors

Avian cells express at least three HSFs (HSFs 1-3). Initially, three avian HSF genes corresponding to a novel factor, HSF3, and the avian homologs of mammalian HSF1 and HSF2 have been cloned (Nakai and Morimoto, 1993). The predicted amino acid sequence of HSF3 is approximately 40% related to the sequence of HSF1 and HSF2. Similar to HSF1 and HSF2, the HSF3 message, is coexpressed during development and in most tissues, which suggests a general role for the regulatory pathway involving HSF3 (Nakai and Morimoto, 1993). It was shown that the regulatory domain is located between the transcriptional activation domains and the DNA binding domain of HSF1 and is conserved between mammalian and chicken HSF1 but is not found in HSF2 or HSF3 (Green *et al.*, 1995). Indeed, the regulatory domain was found to be functionally homologous between chicken and human HSF1. In fact, HSF3 is negatively regulated in avian cells and acquires DNA-binding activity in certain cells upon heat shock (Nakai *et al.*, 1995). Induction of HSF3 DNA-binding activity is delayed compared with that of HSF1 and heat shock leads to the translocation of HSF3 to the nucleus (Nakai *et al.*, 1995). It has been shown that HSF1 is rapidly activated by even mild heat shock, while HSF3 is activated only by severe heat shock. In contrast, HSF2 is not activated by heat stress and has been speculated to have developmental functions (Tanabe *et al.*, 1997).

Indeed, cHSF3 (chick HSF3) was activated at higher temperatures than the cHSF1. In fact, at a mild heat shock, such as 41 °C, only cHSF1 was activated, whereas both cHSF1 and cHSF3 were activated following a severe heat shock at 45 °C. Similarly, cHSF3 was activated by treating cells with higher concentrations of sodium arsenite compared to cHSF1. Furthermore, the DNA binding activity of cHSF3 by severe heat shock lasted for a longer period than that of cHSF1. In addition, the total amount of cHSF3 increased only upon severe heat shock, whereas that of HSF1 decreased. Indeed, cHSF3 is involved in the persistent and burst activation of stress genes upon severe stress in chicken cells (Tanabe *et al.*, 1997). It seems likely that denaturation of nascent polypeptides could be the first trigger for the activation of cHSF1 and cHSF3 (Tanabe *et al.*, 1997). It has been suggested that HSF3 has a dominant role in the regulation of the heat shock response and directly influences HSF1 activity. Thus, disruption of the HSF3 gene results in the severe reduction of heat shock gene expression and loss of thermotolerance (Tanabe *et al.*, 1998). In addition, null cells lacking HSF3, yet expressing normal levels of HSF1, exhibited a severe reduction in the heat shock response, as measured by inducible expression of heat shock genes, and did not exhibit thermotolerance.

Important information related to HSFs in avian species has been obtained in experiments with chick embryos. In fact, it was shown that HSF3 was almost constantly expressed in various tissues during early to late chicken embryonic development (Kawazoe *et al.*, 1999). The expression of HSF1 was equally high in most tissues early in development and thereafter declined to different levels in a tissue-dependent manner and HSF3 became the dominant heat-responsive factor mediating stress signals to heat shock gene expression in the chicken. Furthermore, the high-level and ubiquitous expression of HSF2 as well as HSF1 and HSF3 in early embryogenesis suggest the involvement of these factors in all developmental processes (Kawazoe *et al.*, 1999). It is interesting to note that in avian, HSF1 and HSF3 are maintained in a cryptic monomer and dimer form, respectively, in the cytoplasm in the absence of stress. Upon heat stress, they undergo conformational change associated with the formation of a trimer and nuclear translocation and the nuclear localisation signal acts positively on the trimer formation of cHSF3 upon stress conditions (Nakai and Ishikawa, 2000). Indeed, avian cells express two redundant heat-shock responsive factors, HSF1 and HSF3, which differ in their activation kinetics and threshold induction temperature.

For example, in birds, HSF1 only slightly induces HSP70 expression during heat shock and indeed HSF3 is a master regulator of the heat shock genes in avian cells, as is HSF1 in mammalian cells (Inouye *et al.*, 2003). Avian cells lacking two heat-inducible HSFs, HSF1 and HSF3 were generated (Nakai and Ishikawa, 2001). In addition to complete loss of activation of heat shock genes under stress conditions, these cells exhibited a marked reduction in HSP90α expression under normal growth conditions. Reduction in HSP90α expression caused instability of a cyclin-dependent kinase, Cdc2, and cell cycle progression was blocked mainly at the G2 phase, but also at G1 phase even at mild heat shock temperatures. Restoration of HSP90α expression rescued the temperature sensitivity without induction of HSPs (Nakai and Ishikawa, 2001). Whereas HSF1 mediates transcriptional activity only in the brain upon severe heat shock, HSF3 is exclusively activated in blood cells upon light, moderate, and severe heat shock, promoting induction of heat-shock genes (Shabtay and Arad, 2006). Although not activated, HSF1 is expressed in blood cell nuclei in a granular appearance, suggesting regulation of genes other than heat-shock genes. It was shown that HSF1 and HSF3 mediate transcriptional activity of adult tissues and differentiated cells in a nonredundant manner. Instead, an exclusive, tissue-specific activation is observed, implying that redundancy may be developmentally related (Shabtay and Arad, 2006). The heat shock response regulated by the HSF family should consist of the induction of classical as well as of nonclassical heat shock genes, both of which might be required to maintain protein homeostasis (Fujimoto and Nakai, 2010). Recently, additional information on the roles of HSF2 has been obtained. In particular it has been shown that vertebrate HSF2 is activated during heat shock in the physiological range (Shinkawa *et al.*, 2011). HSF2 deficiency reduces threshold for chicken HSF3 or mouse HSF1 activation, resulting in increased HSP expression during mild heat shock. HSF2-null cells are more sensitive to sustained mild heat shock than wild-type cells, associated with the accumulation of ubiquitylated misfolded proteins. Furthermore, loss of HSF2 function increases the accumulation of aggregated polyglutamine protein and shortens the lifespan of R6/2 Huntington's disease mice, partly through αB-crystallin expression (Shinkawa *et al.*, 2011). In fact, HSF2 was identified as a major regulator of proteostasis capacity against febrile-range thermal stress (Shinkawa *et al.*, 2011). It was also shown that chicken HSF3, but not chicken HSF1, also induces the expression of the major avian pyrogenic cytokine IL-6 during heat shock (Prakasam *et al.*, 2013). In general, important roles of HSFs in adaptation of poultry to various stress conditions are difficult to overestimate. However, recent genome-wide studies have revealed that HSF1 is capable of reprogramming transcription more extensively than previously assumed; it is also involved in a multitude of processes in stressed and non-stressed cells (Vihervaara and Sistonen, 2014).

5.4 Heat shock proteins

Heat shock proteins (HSPs) are highly conserved families of proteins discovered in 1962 (Ritossa, 1962). Interestingly, the discovery of HSP was associated with an observation associated with increased thermostat temperature (by a colleague of Dr Ritossa not related to his experiment; it was just a chance created a new unexpected pattern) where Dr Ritossa kept his *Drosophila* experimental samples (Ritossa, 1996). Remarkably, the first draft of the paper describing the discovery was rejected by editors of highly reputable journal as a new finding was considered to be 'irrelevant to scientific community'. Later, it has been realised that most HSPs have strong cytoprotective effects and are molecular chaperones for other cellular proteins. Taking into account current knowledge of the mode of action of HSPs, the name of "stress proteins" would be more appropriate for them but due to historical reasons they are still called HSPs. Indeed, in the case of oxidative stress, HSP network participates in detecting intracellular changes, protecting against protein misfolding and preventing activation of downstream events related to inflammation and apoptosis (Figure 5.1; Kalmar and Greensmith, 2009).

Since oxidative stress plays a major role in a number of diseases and disease mechanisms in human (Kalmar and Greensmith, 2009) and decreases productive and reproductive performance in farm animals (Surai, 2006, 2018), it is likely that any medication/treatment that is able to reduce levels of oxidative stress will make a significant impact on human health and animal performance. Some HSPs are constitutively expressed, whereas others are strictly stress inducible. Under physiologic conditions, HSPs play an important role as molecular chaperones by promoting the correct protein folding and participating in the transportation of proteins across intracellular membranes and repair of denatured proteins. Therefore, HSPs participate in the regulation of essential cell functions, such as protein translocation, refolding, assembly and the recognition, prevention of protein aggregation, renaturation of misfolded proteins, degradation

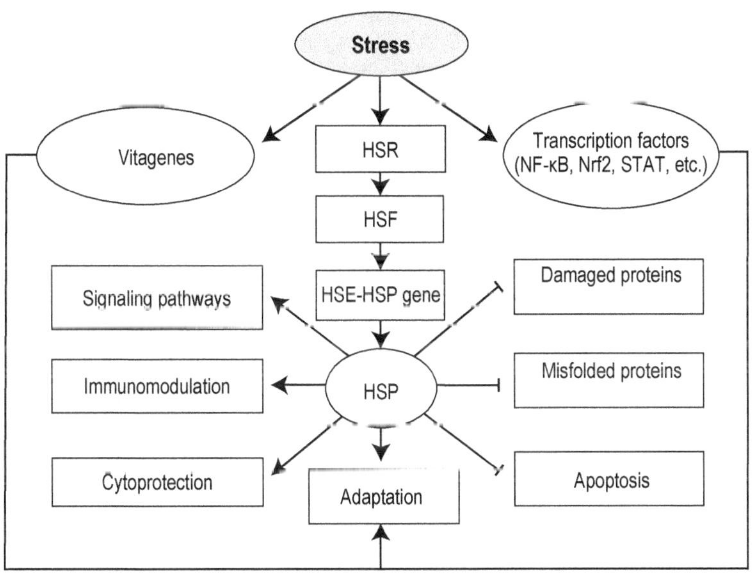

Figure 5.1. Functions of heat shock proteins (HSP) under stress conditions (adapted from Khalil *et al.*, 2011; Surai, 2015b; Surai and Kochish, 2017).

of unstable proteins, etc. (Zilaee *et al.*, 2014). It should be mentioned that the events of cell stress and cell death are linked and HSPs induced in response to stress appear to function at key regulatory points in the control of apoptosis (Garrido *et al.*, 2001; Kennedy *et al.*, 2014). A key feature of HSPs is their ability to provide cytoprotection. Synthesis of these proteins under stress conditions is a highly conserved mechanism of the cell response and adaptation is common among all living organisms.

In fact, HSPs are synthesised in response to a great variety of cellular stresses, including heat stress, hypoxia, ischemia, hypothermia, virus infections as well as the effects of various toxicants, including mycotoxins (Velichko *et al.*, 2013). It is important to note that upregulation of the synthesis of HSPs is considered an endogenous adaptive phenomenon leading to improved tolerance to various stress conditions/factors. In mammals and birds, the HSP superfamily includes five broadly conserved families of proteins (Table 5.1). Among them HSP70, HSP90 and HSP32 (HO-1) are considered as vitagenes.

Table 5.1. Mammalian heat shock proteins (adapted from Bozaykut *et al.*, 2014; O'Neill *et al.*, 2014; Surai and Kochish, 2017).

HSP family	Location	Summary of structural features and domains	Main established functions
HSP90	cytosol	homodimer with two cytosolic isoforms α and β, dimerisation occurs at the C-terminal and nucleotide exchange at the N-terminal	chaperone for a multitude of client proteins and regulator of protein complex formation; mainly responsible for cell viability, keeps proteins in folded state
HSP70	cytosol/nucleus/ mitochondria	consists of a N-terminal (ATPase domain) and a C-terminal substrate-binding domain connected by a short flexible linker	protein trafficking and degradation, refolding of denatured proteins; during stress anti-apoptotic properties; protein quality control and turnover
HSP60	mitochondria	arranged as two stacked heptameric rings with three domains (apical, intermediate and equatorial)	mitochondrial protein folding and assembly
HSP40	cytosol/nucleus	J-domain that stimulates the ATPase activity of HSP70 and C-terminal that loads polypeptides to HSP70	regulates activity of HSP70; binds non-native proteins; processes pro-collagen; substrate delivery to HSP70, targets non-native proteins to ERAD
sHSP	cytosol	conserved C-terminal and highly variable N-terminal (WDPF domain)	preventing unfolded protein aggregation; prevent the accumulation of aggregated proteins

5.4.1 HSP70

Among the HSPs, HSP70 is one of the most conserved and important protein family and has been extensively reviewed (Albakova *et al.*, 2020; Duncan *et al.*, 2015; Mayer, 2013; Mayer and Gierash, 2019; Moram Luengo *et al.*, 2019; Qu *et al.*, 2015; Rosenzweig *et al.*, 2019; Shiber and Ravid, 2014) and will be briefly dealt with here. HSP70 refers to a family of 70 kDa chaperone proteins. Some of the important house-keeping functions attributed to HSP70 include (Garrido *et al.*, 2006): import of proteins into cellular compartments; folding of proteins in the cytosol, endoplasmic reticulum and mitochondria; degradation of unstable proteins; dissolution of protein complexes; control of regulatory proteins; refolding of misfolded proteins and translocation of precursor proteins into mitochondria. These molecular chaperones are implicated in a wide variety of cellular processes, including protein biogenesis, protection of the proteome from negative consequences of stress, recovery of proteins from aggregates, facilitation of protein translocation across membranes, as well as disassembly of particular protein complexes and cell signalling for growth, differentiation, and apoptosis (Clerico *et al.*, 2015). In particular, HSP70 can inhibit apoptosis by interfering with target proteins (Ravagnan *et al.*, 2001). In eukaryotic cells, HSP70s are subject to a large number of post-translational modifications (Mayer, 2013). These ATP-dependent chaperones represent central components of the cellular protein surveillance network and are involved in a large variety of protein-folding processes. In fact, they effectively interact with practically all proteins in their unfolded, misfolded, or aggregated states but do not interact with their folded counterparts (Mayer, 2013). A number of eukaryotic proteins are regulated through transient association with HSP70, including steroid hormone receptors, kinases and transcription factors.

Thirteen different and unique HSP70 have been identified in eukaryote/human cells being distributed in different subcellular compartments, including cytosol, nucleus, endoplasmic reticulum, and mitochondria (Daugaard *et al.*, 2007; Mahalka *et al.*, 2014; Rosenzweig *et al.*, 2019). The two most important members of the HSP70 family are the constitutively expressed 73 kDa heat shock cognate (HSC73, HSC70, HSPA8) and stress-inducible 72 kDa heat shock protein (HSP72, HSP1A) (Meimaridou *et al.*, 2009). Indeed, under normal conditions HSP70 proteins function as ATP-dependent molecular chaperones maintaining important cell functions related to proteostasis (Mayer, 2013; Figure 5.2).

Under various stress conditions additional synthesis of stress inducible HSP70 enhances the ability of stressed cells to deal with increased concentrations of unfolded or denatured proteins (Clerico *et al.*, 2015; Figure 5.2). HSP70 expression is associated with a reduction in JNK1 phosphorylation and/or an increase in oxidative capacity consequential of improvements in mitochondrial homeostasis (Henstridge *et al.*, 2014). It seems likely that HSP70s do not work alone but with a team of cochaperones. Recently it has been found that the organelle distribution of HSP70 is determined by their specific lipid compositions. In particular, HSP70 attach to lipids by extended phospholipid anchorage, with specific acidic phospholipids associating with HSP70

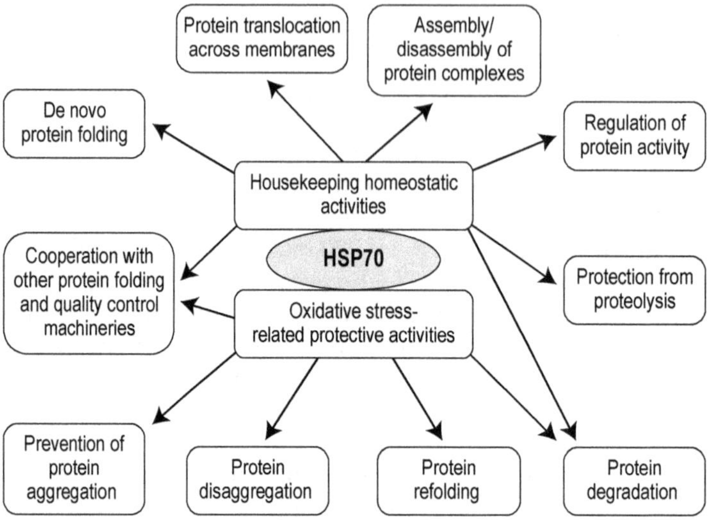

Figure 5.2. The important housekeeping and stress-related activities of heat shock proteins (HSP)70s (adapted from Albakova *et al.*, 2020; Rosenzweig *et al.*, 2019; Surai and Kochish, 2017).

in the extended conformation with acyl chains inserting into hydrophobic crevices within HSP70, and other chains remaining in the bilayer (Mahalka *et al.*, 2014). It seems likely that this could represent an important connection between HSPs and lipid quality control in the cell and the HSP90/HSP70-based chaperone machinery may function as a comprehensive protein management system for quality control of damaged proteins. Actually in a recently developed model, it was proposed that the heat shock protein HSP90/HSP70-based chaperone machinery played a major role in determining the selection of proteins that have undergone oxidative or other toxic damage for ubiquitination and proteasomal degradation (Pratt *et al.*, 2010). Indeed, HSP70s were reported to have a large set of substrates, including nascent polypeptide chains at the ribosome or translocation pore, misfolded, aggregated, or amyloidic proteins, oligomeric protein complexes, as well as some native proteins (Mayer and Gierasch, 2019; Figure 5.3).

HSP70 is shown to be involved in regulation of the inflammation process, mitochondrial function, and ER stress being a promising target for nutritional/drug modulation (Mulyani *et al.*, 2020) as a therapeutic approach for disease prevention (Konstantinova *et al.*, 2019).

5.4.2 Chicken HSP70

In 1978 it was shown that the pattern of proteins synthesised by chicken embryo fibroblasts changes dramatically after heat treatment (45 °C for a few hours). In fact, three proteins (Mr = 22,000, 76,000, and 95,000) accounted for almost 50% of the cell's protein synthetic capacity immediately after the heat-shock (Kelley and Schlesinger,

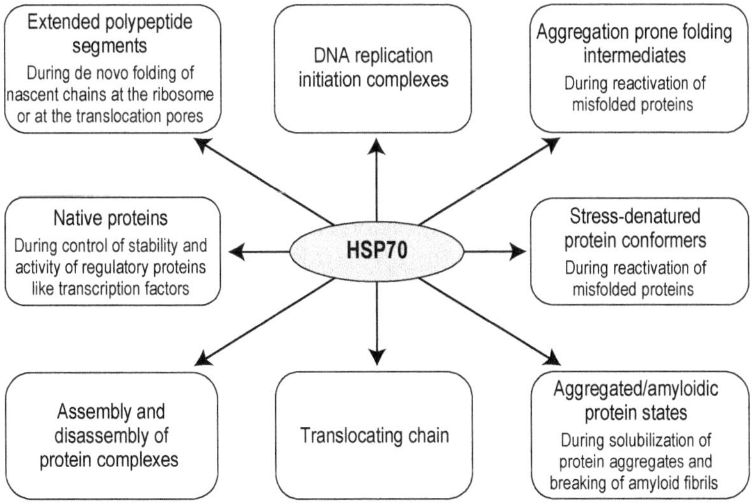

Figure 5.3. Some important substrates for heat shock proteins (HSP)70 (adapted from Mayer and Gierasch, 2019; Rosenzweig *et al.*, 2019; Surai and Kochish, 2017).

1978). The universality of the heat shock response and conservation of proteins induced by this type of stress was proven in different experimental conditions. In particular, antibodies to chicken HSPs, cHSP89 and cHSP70, cross-reacted with proteins of similar molecular weights in embryonic and adult chicken tissues and in extracts from widely different organisms ranging from yeast to mammals (Kelley and Schlesinger, 1982). Heat-shock polypeptides of identical sizes of 85,000, 70,000, and 25,000 Da were synthesised predominantly in chicken embryo fibroblasts and in many different organs of 18-day-old embryos at 42.5-44 °C (Voellmy and Bromley, 1982). Effects of heat treatments on chick embryo fibroblasts, Drosophila embryonic cells, and human lymphoblastoid cells have been compared (Voellmy *et al.*, 1983). Cells from all three species synthesise large HSPs with Mr=70,000 and 84,000-85,000. Different small HSPs with Mr between 22,000 and 27,000 are made at high rates in heat-treated chicken and Drosophila cells but could not be observed in human cells. It was found that chicken reticulocytes respond to elevated temperatures by the induction of only one heat shock protein, HSP70, whereas lymphocytes induce the synthesis of all four heat shock proteins (HSP89, HSP70, HSP23 and HSP22). The synthesis of HSP70 in lymphocytes was rapidly induced by small increases in temperature (2-3 °C) and blocked by preincubation with actinomycin D (Morimoto and Fodor, 1984). Furthermore, incubation of chicken reticulocytes at elevated temperatures (43-45 °C) resulted in a rapid change in the pattern of protein synthesis, characterised by the decreased synthesis of normal proteins, e.g. alpha and beta globin, and the preferential and increased synthesis of HSP70 (Banerji *et al.*, 1984). Indeed, the rapid 20-fold increase in the synthesis of HSP70 was observed after heat shock and preincubation of reticulocytes with the transcription inhibitor actinomycin D or 5,6-dichloro-1-beta-D-ribofuranosylbenzimidazole blocked the heat shock-induced synthesis of HSP70.

In 1986 Morimoto and co-authors studied organisation, nucleotide sequence, and transcription of the chicken HSP70 gene. They isolated a gene encoding a 70,000-Da heat shock protein (HSP70) from a chicken genomic library and showed that the order and spacing of the sequences share many features in common with the promoter for the human HSP70 gene (Morimoto *et al.*, 1986). The expression of HSP70 during maturation of avian erythroid cells was also studied (Banerji *et al.*, 1987). It was shown that definitive red cells respond to heat shock by a 10- to 20-fold increase in HSP70 protein synthesis with little change in HSP70 mRNA levels. Therefore, the increased expression of HSP70 in cells was due to increased translatability of HSP70 mRNA. Furthermore, the authors showed that HSP70 expression in erythroid cells is lineage specific and although HSP70 was constitutively expressed, neither HSP70 synthesis nor HSP70 mRNA levels were heat shock inducible in primitive red cells.

HSP70 was shown to constitutively expressed in the embryonic chicken lens. In fact, HSP70 mRNA in the embryonic chicken lens was associated primarily with cells in the early stages of fibre formation, and increased transcription of this gene was part of the differentiation process (Dash *et al.*, 1994). It was shown that the heat induced increase in HSP70 mRNA and protein in broiler liver, *in vivo*, are time dependent, similar to that in mammals (Gabriel *et al.*, 1996). An increase in the amount of HSP70 was detected from the first up to the fifth hour of acute heat exposure (35 °C for 5 h), while an increase in HSP70 mRNA peaked at 3 h. It seems likely that heat shock response in avian species is related to temperatures above 41 °C. For example, the spatial expression of HSP70 transcripts was detected in chicken embryos under normal incubation conditions and moderate heat stress (41 °C) did not induce enhancements on HSP70 mRNA levels (Gabriel *et al.*, 2002). At the same time, acute exposure to severe heat stress (44 °C) for one hour resulted in a fifteen-fold increase in HSP70 mRNA levels. It is interesting to note that the return of stressed embryos to normal incubation temperature resulted in increased HSP70 mRNA levels for three hours which was normalised after six hours. The increased expression of HSP70 in broiler chicken embryos was shown to be affected not only by heat (40 °C) but also by cold (32 °C) stress and is tissue- and age-dependent (Leandro *et al.*, 2004). In fact, HSP70 was detected in the liver, heart, breast muscle, and lungs and the brain contained 2- to 5-times more HSP70 protein compared to the other embryonic tissues. These data are in agreement with our observations indicating low level of vitamin E and high levels of PUFAs in chicken embryonic brain (Surai *et al.*, 1996). Therefore, increased HSP70 expression is an adaptive mechanism of increasing antioxidant defences. Younger embryos had higher HSP70 synthesis than older embryos, irrespective of the type of thermal stressor (Leandro *et al.*, 2004). Again, these data confirm our finding about maturation of the antioxidant defences during chicken embryonic development (Surai, 1999).

It was shown that HSP70 expression in postnatal chickens is tissue- and allele-dependent (Zhen *et al.*, 2006). Indeed, the expression of HSP70 gene in the liver was significantly (more than 2-fold) higher than that in the muscle under normal growth conditions. This could reflect an importance of HSP70 chaperone functions, since the liver is the major site of synthesis of many important proteins. However, during acute heat stress (44 °C for 4 hours) the expression of HSP70 gene in the brain was the highest being significantly different from those in the liver and muscle. This adaptive response by HSP70 also is an important mechanism to compensate for relatively low levels of antioxidants in the brain tissue of the chicken (Surai, 2002). Long-term, moderate heat stress (30-32 °C) was associated with significantly increased HSP70 levels in mononuclear blood cells of laying hens (Maak *et al.*, 2003). However, the age-dependent responses of different genotypes were not uniform. HSP70 gene expression was gender-dependent with significantly higher levels in male than in female chickens (Figueiredo *et al.*, 2007) and tissue-dependent heat induction of HSP70 expression may correlate with DNA methylation pattern in the HSP70 promoter (Gan *et al.*, 2013a). During the exposure to heat stress (37±1 °C), the heart, liver and kidney of broiler chickens exhibited increased amounts of HSP70 protein and mRNA. The expression of HSP70 mRNA in the heart, liver and kidney of heat-stressed broilers increased significantly and attained the highest level after a 2-h exposure to elevated temperatures. Significant elevations in HSP70 protein occurred after 2, 5, and 3 h of heat stressing, respectively, indicating that the stress-induced responses vary among different tissues (Yu *et al.*, 2008). Furthermore, the expression of HSF3 and HSP70 mRNA in Lingshan chickens (LSC) and White Recessive Rock (WRR) exhibited species-specific and tissue-specific differences during heat treatment (Zhang *et al.*, 2014). For example, after 2 h of heat treatment, HSP70 expression was significantly higher in the liver and leg muscle of WRR compared to LSC. Recent analysis of genetic diversity of the HSP70 gene in 8 native Chinese chicken breeds indicates presence of 36 variations, which included 34 single nucleotide polymorphisms and 2 indel mutations (Gan *et al.*, 2015). Furthermore, 57 haplotypes were observed, of which, 43 were breed-specific and 14 were shared.

HSP expression in the gut could be considered as an important mechanism of the antioxidant protection (Surai and Fisinin, 2015). However, there were no effects of HSP70 overexpression on intestinal morphology under heat stress, but there was a strong correlation between HSP70 expression and the digestive enzyme activity in broilers (Hao *et al.*, 2012). In another study from the same department, HSP70 induction was shown to protect the intestinal mucosa from heat-stress injury by improving antioxidant capacity of broilers and inhibiting the lipid peroxidation (Gu *et al.*, 2012). In fact, HSP70 significantly protected the integrity of the intestinal mucosa from heat stress (36±1 °C) by significantly elevating antioxidant enzyme activities (SOD, GPx and total antioxidant capacity) and inhibiting lipid peroxidation to relieve intestinal mucosal oxidative injury.

To investigate the alterations introduced by domestication and selective breeding in heat stress response, two experiments were conducted using Red Jungle Fowl (RJF), village fowl (VF), and commercial broilers (CB). Birds of similar age (30 d old) or common body weight (930±15 g) were exposed to 36±1 °C for 3 h (Soleimani *et al.*, 2011). The RJF at a common age and common BW showed significantly higher levels of basal HSP70 and cortisone compared with VF and CB. Heat treatment was shown to significantly increase body temperature, heterophil:lymphocyte ratio, and plasma corticosterone concentration in CB but not in VF and RJF. Irrespective of stage of heat treatment, RJF showed lower heterophil:lymphocyte ratio and higher plasma corticosterone concentration than VF and CB. It was concluded that domestication and selective breeding are leading to individuals that are more susceptible to stress rather than resistant (Soleimani *et al.*, 2011). Furthermore, laying hens exposed to HS (32.6 °C) showed higher concentrations of HSP70 in the liver (Felver-Gant *et al.*, 2012). In addition, kind gentle hens (a line of group-selected hens for high productivity and survivability) had higher concentrations of HSP70 than DeKalb XL hens (commercial line of individually selected hens for high egg production) regardless of treatment. Interestingly, feed restriction in broiler chickens was associated with a significant 3-fold increase in HSP70 expression in their brain (Najafi *et al.*, 2018). However, reduced protein level in the chicken diet did not affect HSP70 expression in their plasma (Zulkifli *et al.*, 2018). HSP70 in chickens can also be upregulated by various toxicants as an adaptive response. For example, dietary Cd was found to upregulate HSP70 as well as NF-κB and TNF-α in the chicken liver (Wang *et al.*, in press). Similar changes in HSP70 expression were observed in the chicken rectum tissue due to the single use or combined exposure to chronic arsenite and Cu^{2+} (Yang *et al.*, 2020).

HSP70 is also shown to be expressed in other avian species. Notably, quail HSP70 showed 98% homology with HSP70 stress protein in *Gallus gallus* and 99% homology with *Numida meleageris* (Gaviol *et al.*, 2008). Duck HSP70 gene was also identified and characterised (GenBank: EU678246) and shown to contain no introns (Xia *et al.*, 2013). Fifteen variations were identified within the open reading frame. The expression of duck HSP70 gene was tissue-specific and the highest expression level was seen in pectoral muscle (Xia *et al.*, 2013).

To sum up, the results from the aforementioned studies consistently demonstrate that increased HSP70 expression in chicken tissues is one of the most important protective responses to prevent or deal with, detrimental changes in protein structure and functions due to various stresses. However, there is a need for further research to understand molecular mechanisms of HSP70 regulation in avian species.

5.4.3 HSP90

HSP90, the major soluble protein of the cell, has recently received great attention and a range of reviews described its structure, functions and regulation (Erlejman *et al.*, 2014; Karagöz and Rüdiger, 2015; Khurana and Bhattacharyya, 2015; Mayer and Le Breton, 2015). In fact, in the cell, HSP90 is known to comprise 1-2% of total proteins under non-stress conditions and it is further upregulated under stress (Csermely *et al.*, 1998). For example, heat shock (37-42 °C) have been reported to induce HSP90 levels by as much as twofold (Bagatell *et al.*, 2000). Furthermore, fish naturally living in a hot spring with relatively high water temperature (34.4±0.6 °C) is characterised by increased levels of all the studied HSPs (HSP70, HSP60, HSP90, HSC70 and GRP75) compared with fish living in normal river water temperature (Oksala *et al.*, 2014).

HSP90 is expressed as a 90 kDa protein and its functional molecule is a homodimer (α/α or β/β) and each monomer consists of three domains. They are NH_2-terminal nucleotide binding domain (a binding site for ATP/ADP), the middle domain (the binding site for nuclear localisation signal and client proteins) and the C-terminal domain (the site of dimerisation and co-chaperone binding) (Li *et al.*, 2012, 2012a). The HSP90 family in mammalian cells consists of four major homologs including two cytoplasmic isoforms HSP90α (inducible form) and HSP90β (constitutive form) (Sreedhar *et al.*, 2004), HSP90B located in endoplasmic reticulum and tumour necrosis factor receptor-associated protein (TRAP) found in mitochondria and the inner membrane space (Revathi and Prashanth, 2015; Table 5.2). It is interesting to note that HSP90α and HSP90β share 86% amino acid identity and are expressed in all nucleated cells.

HSP90 is a highly efficient, ATP-dependent molecular chaperone involved in the maturation and stabilisation of a wide-range of proteins in both physiological and stress conditions being an important hub in the protein network that maintains cellular homeostasis and function (Jackson, 2013). HSP90 belongs to a family of proteins known as 'chaperones,' which are solely dedicated to helping other proteins (client proteins) correct folding, function and stability. Indeed, cellular stress causes protein denaturation, and they cannot function properly and must be repaired or eliminated with the help of chaperones (Garcia-Carbonero *et al.*, 2013). HSP90 deals with more than 200 important clients which are involved in signal transduction, including many steroid hormone receptors, receptor tyrosine kinases, Src family members, serine-threonine kinases, cell cycle regulators, telomerase and many other proteins (Li *et al.*, 2012a; Wayne *et al.*, 2011; Zhang and Burrows, 2004).

It is difficult to overestimate chaperoning functions of HSP90 related to various nuclear proteins regulating DNA replication, DNA repair, DNA metabolism, RNA transcription and RNA processing (Li *et al.*, 2012a) and the protective action of HSP90 is related to posttranslational modifications of soluble nuclear factors as well as histones (Erlejman *et al.*, 2014). It was suggested that HSP90 clients are associated with major physiological events including signal transduction, cell cycle progression, transcriptional regulation, natural and acquired immunity and intracellular movement of proteins (Li *et al.*, 2012a; Taipale *et al.*, 2010; Figure 5.4). In fact, HSP90

Table 5.2. Isoforms of heat shock proteins 90 (HSP90) (adapted from Revathi and Prashanth, 2015; Surai and Kochish, 2017).

Family	Subcellular localisation	Subfamily	Gene	Protein
HSP90A	Cytosolic	HSP90AA (inducible)	HSP90AA1	Hsp90-α1
			HSP90AA2	Hsp90-α2
		HSP90AB (constitutively expressed)	HSP90AB1	Hsp90-β
HSP90B	Endoplasmic reticulum		HSP90B1	Endoplasmic/GRP-94
TRAP	Mitochondrial		TRAP1	TNF receptor-associated protein 1

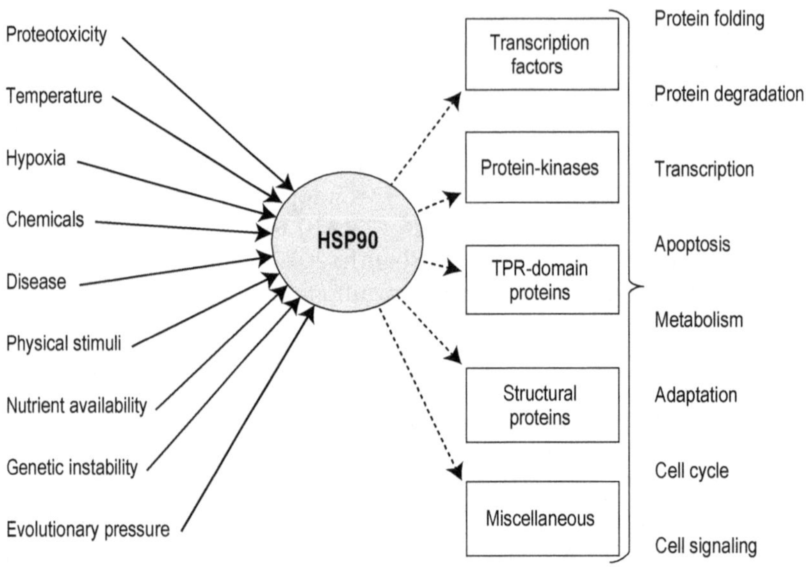

Figure 5.4. Regulatory roles of heat shock proteins (HSP)90 (adapted from Erlejman *et al.*, 2014; Hoter *et al.*, 2018; Surai and Kochish, 2017).

participates in many cellular processes including cell cycle control, cell survival, hormone and other signalling transduction pathways, often acting as hormone receptors and is considered to be key player in maintaining cellular homeostasis and adaptive response to stress (Jackson, 2013). In many cases, HSP90-associated stress response is orchestrated via HSF1, which under stress conditions upregulates several hundred genes including HSP90. It is known that under physiological condition, as a client protein, HSF1 is kept in an inactive monomeric form through the transient interaction with Hsp90 (Li *et al.*, 2012a). During stress, HSF1 dissociates from HSP90, homotrimerises, undergoes phosphorylation and translocates to the nucleus to perform its gene-expression regulatory functions (Li *et al.*, 2012a). As a matter of fact, HSP90 is regulated transcriptionally through direct interactions with the transcription factor HSF (Trinklein *et al.*, 2004). Generally, HSP90 is present in cells in equilibrium between a low chaperoning activity 'latent state' in physiological conditions and an 'activated state', with increased chaperoning efficiency in stress conditions (Chiosis *et al.*, 2004).

HSP90 protects cellular homeostasis against various stresses and preserves cellular homeostasis by modulating the functions of hundreds of client factors leading to involvement in major signalling and homeostatic events. HSP90 usually works as a complex with other chaperones and over 20 co-chaperones (Hong *et al.*, 2013) and increased expression of HSP90 have been shown to be associated with the tolerance of hypothermia, cell proliferation, and cell cycle control (Herring and Gawlik, 2007). In fact, co-chaperones assist HSP90 in its conformational cycling, act as substrate recognition proteins and provide additional enzymatic activity (Barrott and Haystead,

2013). It seems likely that under heat stress conditions, co-chaperones allow HSP90 to prevent aggregation of unfolded proteins (Richter *et al.*, 2010). Indeed, HSP90 involves in the folding, stabilisation, activation and assembly of its client proteins through the formation of complexes with co-chaperones such as HSP70, HSP40, Hop, Hip and p23 (Whitesell and Lindquist, 2005). The molecular chaperones HSP90 and HSP70 form a multichaperone complex, in which both are connected by a third protein called Hop. Indeed, Hop (HSP70/HSP90 organising protein) facilitates interaction between HSP90 and HSP70 helping substrate to be efficiently transferred from HSP70 to HSP90 (Daniel *et al.*, 2008). It seems likely that the interplay between the two chaperone machineries affecting the trafficking and turnover of several hundred signalling proteins as well as removal of damaged and aberrant proteins via the ubiquitin-proteasome pathway is of great importance for cell viability and adaptability. HSP90 is shown to possess an ATPase activity, which is known to be essential to modulate the conformational dynamics of the protein. In fact, ATP hydrolysis is associated with the HSP90 dimer transitioning into its "open" conformation and releasing the client protein (Taipale *et al.*, 2010). The system is regulated by post-translational modifications including phosphorylation, acetylation, nitrosylation and methylation and uses a range of co-chaperones mediating interactions with HSP90 client proteins (Jackson, 2013; Li *et al.*, 2012a). Therefore, HSP90 has been considered to be a key factor at the crossroads of genetics and epigenetics (Erlejman *et al.*, 2014).

5.4.4 Chicken HSP90

A cDNA clone for the 90 kDa heat-shock protein was isolated by direct immunological screening of a chicken smooth muscle cDNA expression library (Catelli *et al.*, 1985). It was shown that HSP90 is increased in heat-shocked chick embryo fibroblasts (Catelli *et al.*, 1985a). Furthermore, HSP90 from chicken liver has been purified and physically characterised (Iannotti *et al.*, 1988). The protein was shown to be an elongated dimer with a molecular weight of 160,000 and a frictional ratio of 1.6, extensively phosphorylated and partitioned totally into the aqueous phase. A comparison of the amino acid sequence of the chick HSP90 to that of the homologous HSP90 from yeast to man, reveals 64-96% identity respectively (Binart *et al.*, 1989). The authors suggested that two hydrophilic regions A and B may play a role in the interaction of HSP90 with other proteins such as steroid hormone receptors. In fact, the dimeric form of the HSP90 was confirmed and its structure was shown to be stabilised by hydrogen bonds (Radanyi *et al.*, 1989). Furthermore, the cDNA-derived amino acid sequence of chick HSP90 revealed a 'DNA like' structure: potential site of interaction with steroid receptors (Binart *et al.*, 1989). The nucleotide sequence of a 2652 bp derived from a chicken HSP90 genomic clone was reported and two introns have been identified (Vourch *et al.*, 1989). It was proven that HSP90 gene expression is constitutive and heat inducible. In the chick oviduct cells, HSP90 was located in the cytoplasm as aggregates, often inside small vesicles, while in the apical part of the cell, HSP90 was located at the Golgi complex (Pekki, 1991). The epithelium also exhibited some cells with high levels of HSP90. It is interesting to note that HSP90 is associated with both microtubules and microfilaments (Czar *et al.*, 1996). In fact, C-terminal half of HSP90 contains a sequence which is responsible for the cytoplasmic localisation of the protein and the cytoplasmic anchoring signal is located between amino acids 333 and 664 (Passinen *et al.*, 2001). It was shown that in contrast to HSP70, the 35S metabolically-labelled HSP90, which accumulates in the cytosoluble fraction 6-8 h after serum treatment, is not preferentially translocated to the nuclear compartment, although a small fraction is always present in the nucleus (Jérôme *et al.*, 1993). It was also demonstrated that serum- or insulin-induced accumulation of HSP90α mRNA results from an activation of gene transcription and that hsp90α promotor activity is induced approximately fivefold after serum stimulation. Therefore, chicken HSP90 constitutively expressed in most cells, is up-regulated by thermal stress and by developmental and mitogenic stimuli. Indeed, a transient induced expression of the HSP90α gene takes place at both the messenger RNA and the protein synthesis level. This response is protein synthesis dependent and DNA synthesis independent. A possible link between cell cycle and HSP90α regulation was suggested (Jérôme *et al.*, 1991).

It seems likely that the HSP90 alpha and beta genes are the result of a gene duplication event that occurred at the time of the emergence of vertebrates (Meng *et al.*, 1993). Furthermore, avian HSP90β mRNA is not inducible by thermal stress or mitogenic stimuli, contrary to the mouse and human HSP90 alpha and beta mRNAs. Indeed, chicken HSP90β is the only vertebrate HSP90 insensitive to heat shock and there are some specific features of HSP90 beta gene structure and location explaining why chicken HSP90 beta mRNA is generally less abundant than alpha and is not inducible by heat shock or serum/growth factor stimulation (Meng *et al.*, 1995).

The importance of ATP binding and hydrolysis by HSP90 in formation and function of protein heterocomplexes was shown (Grenert *et al.*, 1999). Chicken HSP90 hydrolysing ATP activity was found to be 10-100-fold lower than that in yeast HSP90 and TRAP1, an HSP90 homologue found in mitochondria (Owen *et al.*, 2002). The authors showed that sequences within the last one-fourth of HSP90 regulate ATP hydrolysis. The N-terminal ATP binding domain of HSP90 is necessary and sufficient for interaction with oestrogen receptor (Bouhouche-Chatelier *et al.*, 2001). There are two sites in HSP90 binding ATP. In fact, HSP90 N-terminal domain has a nonconventional nucleotide binding site and HSP90 possesses a second ATP-binding site located on the C-terminal part of the protein (Garnier *et al.*, 2002). HSP90 chaperone activity was shown to require the full-length protein and interaction among its multiple domains, indicating that the cooperation of multiple functional domains is essential for active, chaperone-mediated folding (Johnson *et al.*, 2000).

The expression of HSP90 increased in the heart, liver and kidney of broilers after exposure to increased temperature for 2 h (Lei *et al.*, 2009). In the heart and kidney, HSP90 mRNA transcription levels exhibited the same trend as the protein expression of HSP90. Induction of HSP90 mRNA and HSP90 protein at an early stressing stage indicated that heat stress can directly stimulate and quickly initiate the transcription of HSP90 mRNA and translation of HSP90 protein to protect cells. The HSP90α gene is shown to play an evolutionarily conserved role during somitogenesis in vertebrates in addition to providing protection to all cells of the embryo following stress (Sass and Krone, 1997).

It was suggested that HSP90 can participate directly in the function of a broad range of cellular signal transduction components, including retinoid receptor signal transduction (Holley and Yamamoto, 1995). In eukaryotic cells, HSP90 is associated with several protein kinases and regulates their activities. HSP90 was also reported to possess an autophosphorylase activity (Kim *et al.*, 1999). In fact, chicken HSP90 participates in folding and stabilisation of signal-transducing molecules including steroid hormone receptors and protein kinases and both amino- and carboxyl-terminal domains of HSP90 interact to modulate chaperone activity (Marcu *et al.*, 2000). Depletion of HSP90β induces multiple defects in B cell receptor signalling (Shinozaki *et al.*, 2006). Indeed, inhibition of HSP90 with geldanamycin resulted in the inactivation of MAPK/ERK and PI3K/AKT pathways leading to significantly reduced levels of IFN-γ, IL-6 and NO mRNAs in avian macrophages (Bhat *et al.*, 2010). Therefore, in contrast to mammals, HSP90α but not HSP90β may play a major role in CpG ODN(2007) induced immunoactivation in avian macrophage cells. Collectively, these observations strongly suggest that signalling roles of HSP90 in avian species need further investigation. Recently, four novel members of the 90 kDa heat shock protein (HSP90) family expressed in Japanese quail, *Coturnix japonica* have been described (Nagahori *et al.*, 2010). The coding regions of the genes, CjHSP90AA1, CjHSP90AB1, CjHSP90B1 and CjTRAP1, exhibited more than 94% similarity to their related genes in chicken. Furthermore, CjHSP90AA1 exhibited a robust response to heat shock treatment.

5.4.5 HSP32 (HO-1)

HO-1 is the stress-inducible isoform of the three HO isoforms described to date, serving as a critical protective mechanism in vertebrate systems responsible for adaptation to oxidative, inflammatory, and cytotoxic stress (Fredenburgh *et al.*, 2015; Wu *et al.*, 2011). In fact, HO-1 (32 kDa), also known as heat shock protein-32 (HSP32), is shown to be expressed at a relatively low level in most tissues. It is proven that HO-1 is endoplasmic reticulum phase II enzyme catalysing the rate-limiting step in heme degradation, producing free iron (Fe^{2+}), carbon monoxide (CO) and biliverdin (Soares and Bach, 2009). Biliverdin is subsequently reduced to bilirubin by biliverdin reductase. It is interesting to mention that the products of the aforementioned reaction can trigger signalling cascades leading to improvement of antioxidant defences and protection against oxidative stress. In particular, CO can modulate the production of proinflammatory or anti-inflammatory cytokines and mediators having immunomodulatory effects with respect to regulating the functions of antigen-presenting cells, dendritic cells, and regulatory T cells (Ryter and Choi, 2016). It seems likely that products of the HO-1 reaction namely CO and biliverdin have also cytoprotective, anti-inflammatory and anti-apoptotic properties in stress conditions (Durante *et al.*, 2010; Haines *et al.*, 2012; Zahir *et al.*, 2015). Cells exposed to low concentrations of CO were shown to respond by an increase in ROS formation (e.g. oxidative conditioning) with important consequences for inflammation, proliferation, mitochondria biogenesis, and apoptosis (Bilban *et al*, 2008). Actually, the degradation of heme by HO-1, the signalling actions of CO, the antioxidant protective action of biliverdin/bilirubin, and the sequestration of Fe^{2+} by ferritin are suggested to contribute to the anti-inflammatory effects of HO-1 (Pae and Chung, 2009) and increase stress resistance. Furthermore, recent studies have demonstrated that HO-1 inhibits stress-induced extrinsic and intrinsic apoptotic pathways *in vitro* (Morse *et al.*, 2009). The vital importance of HO-1 in stress adaptation have been confirmed in HO-1-deficient mice models showing atypical proinflammatory immune response (Kapturczak *et al.*, 2004) with increased vulnerability to endotoxin sepsis (Poss and Tonegawa, 1997), defective expression of interferon-β (Tzima *et al.*, 2009) and increased susceptibility to apoptosis (True *et al.*, 2007; Vachharajani *et al.*, 2000). Moreover, HO-1 knockout mice were characterised by very low survival (~1-5% of litters) and high levels of oxidative stress with a shortened life span (Wegiel *et al.*, 2014, 2014a). In fact, HO-1 knockout mice were shown to be extremely sensitive to oxidative stress caused by ischemia and reperfusion (Liu *et al.*, 2005; Yet *et al.*, 1999) and to develop anaemia associated with hepatic and renal iron overload leading to oxidative tissue injury and chronic inflammation (Poss and Tonegawa, 1997). The aforementioned observations provide substantial evidence to support the implication of HO-1 in stress response.

The half-lives of HO-1 mRNA and protein are shown to be approximately 3 hours and 15-21 hours, respectively (Dennery, 2000). In humans, the HO-1 gene (*Hmox1*) is located on chromosome 22q12 and consists of four introns and five exons. The regulatory region of the mammalian HO-1 gene has a promoter, a proximal enhancer, and two or more distal enhancers (for review see Schipper and Song, 2015). The *Hmox1* promoter is shown to exhibit a range of binding sites (for AP-1, AP-2, NF-κB, and HIF-1), as well as HSE sequences, metal response elements and stress-response elements. Therefore, the complex gene structure explains its high sensitivity to induction by diverse pro-oxidant and inflammatory stimuli including heme, dopamine, TNF-α, IL-1β, cysteamine, β-amyloid, H_2O_2, hyperoxia, UV light, heavy metals, lipopolysaccharide, etc. (Schipper and Song, 2015). In vertebrates HO-1 is shown to be upregulated by its substrate heme as well as by a wide variety of stressors including heavy metals, heat shock, ischemia, ROS, RNS, bacterial endotoxins, radiation, hypoxia, H_2O_2, nitric oxide, etc. (Chang *et al.*, 2009; Wegiel *et al.*, 2014). Furthermore, inflammatory mediators such IL-1, TNF-α, LPS are also shown to upregulate HO-1 *in vitro* (Niess *et al.*, 1999; Terry *et al.*, 1998). At the cellular level, HO-1 is highly expressed in the organs participating in degrading senescent red blood cells, including spleen, reticuloendothelial cells of the liver and bone marrow (Immenschuh *et al.*, 1999) as well as in macrophages (Bissell *et al.*, 1972) and dendritic cells (Chauveau *et al.*, 2005). In fact, HO-1 upregulation in various cells is shown to attenuate the expression of various proinflammatory genes (Lee and Chau, 2002; Wijayanti *et al.*, 2004). Furthermore, HO-1 is of great importance for building immunocompetence. Indeed, induction of HO-1 in dendritic cells alters their maturation state and interaction with other cells (Chauveau *et al.*, 2005; Remy *et al.*, 2009), including T lymphocytes (George *et al.*, 2008; Moreau *et al.*, 2009) and macrophages (Choi *et al.*, 2010; Nakamichi *et al.*, 2005; Wegiel *et al.*, 2014, 2014a).

Regulation of HO-1 activity as an adaptive response to stress is mediated via several key initiator and feedback control processes. In particular, the transcriptional regulation of the HO-1 gene is shown to be attributed to several transcription factors including Nrf2, Bach1 (Igarashi and Sun, 2006; Jang *et al.*, 2009), HIF-1 (Semenza, 2010) and PPARs (Ndisang, 2014). It seems likely that MAPK signalling is involved in HO-1 induction (De Backer *et al.*, 2009). In particular, the anti-inflammatory cytokine IL-10 was shown to induce HO-1 expression via a p38 MAPK-dependent pathway (Lee and Chau, 2002). Indeed, the antiapoptotic effect of CO was shown to be mediated by the activation of the p38 MAPK signal transduction pathway and required the activation of the transcription factor NF-κB (Soares *et al.*, 2002). Furthermore, the phosphatidylinositol-3 kinase (PI3K)/Akt signalling also modulates HO-1 activity (Salinas *et al.*, 2004). In addition, HO-1 is involved in suppression of the expression of the pro-inflammatory cytokine TNF-α, while an HO-1 inhibitor (zinc protoporphyrin) attenuated this effect (Lee and Chau, 2002). Furthermore, HO-1 is an important regulator of cellular metabolism, and its activity may affect NADPH- and oxygen-consuming pathways, including fatty acid synthesis, oxidative metabolism of cytochrome p450, or modulation of ROS generation in phagocytes (Węgiel *et al.*, 2014).

Cytoprotective action of HO-1 is summarised in Figure 5.5. In fact, there is a range of stress-related factors which could activate HO-1 expression, mainly via Nrf2 pathway. Products of HO-1 action on heme are involved in cytoprotection.

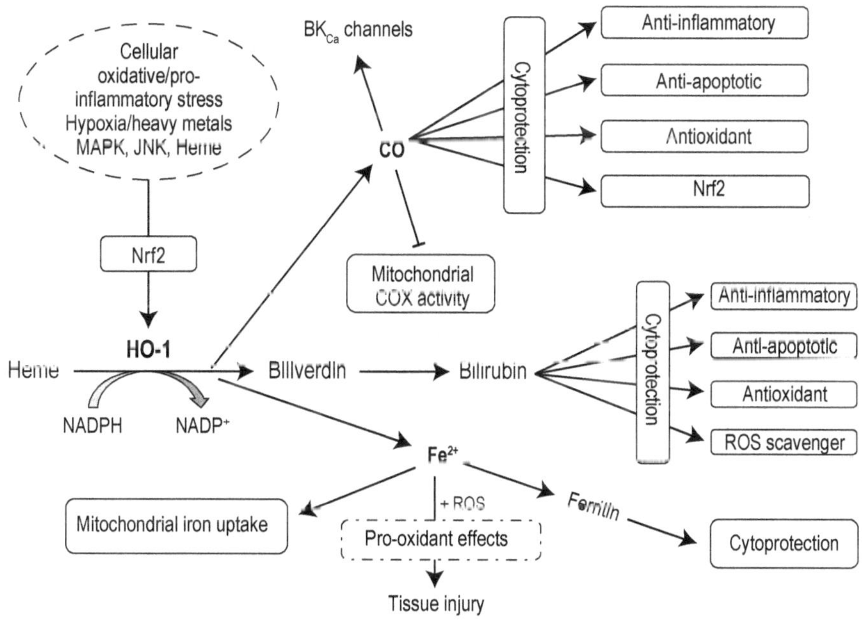

Figure 5.5. Cytoprotective action of heme oxygenase 1 (HO-1) (adapted from Duvigneau *et al.*, 2019; Liu *et al.*, 2019; Ryter, 2019; Surai and Kochish, 2017).

5.4.6 Chicken heme oxygenase 1

Data on HO-1 expression and its protective actions in poultry production are very limited. In early 1990th, HO-1 was purified from liver microsomes of chicks pre-treated with cadmium chloride (Bonkovsky *et al.*, 1990). The molecular weight of the enzyme was shown to be 33,000 Da and the pH optimum of the reaction was 7.4. It was also shown that Hg^{2+} inhibited HO-1 activity by 67% at 10 μM and totally at 15 μM. Comparison of sequences to those derived from cDNA sequences for the major inducible rat and human HO-1 showed 69% and 76% similarities, respectively (Bonkovsky *et al.*, 1990). Next year, a cDNA from a chick liver library that encodes for HO-1 has been cloned and sequenced (Evans *et al.*, 1991). The protein corresponding to this fragment of DNA was found to compose of 296 amino acid residues and has a molecular mass of 33,509 Da. The similarity of chick HO-1 to rat and human HO-1 (nucleotides 66% and amino acids 62%) was confirmed to be moderately high. It was also shown that Cd-dependent induction of HO-1 was due to increased transcription of the gene or stabilisation of its message (Evans *et al.*, 1991). Similar to mammalian HO-1, chicken HO-1 has five exons and four introns (Lu *et al.*, 1998). In the DNA sequence there are consensus sequences corresponding to numerous transcription factor recognition elements, including AP-1, AP-2, NF-κB, C/EBP, c-Myc and a metal-responding element identified in the promoter region (Lu *et al.*, 1998). Furthermore, chick HO-1 promoter region responded to sodium arsenite, H_2O_2 and transition metals, but not to heme. The chick HO-1 promoter region also contains a unique sequence that localised at -3.7 kb upstream of the transcription start site of the chick HO-1 gene and serves up-regulation of the gene by metalloporphyrins (Shan *et al.*, 2002, 2004). Furthermore, the chick HO-1 promoter region was shown to contain 'expanded' by three base pairs AP-1 sites that are important for up-regulation of the gene by heme and cobalt protoporphyrin, but not other metalloporphyrins (Shan *et al.*, 2004).

HO-1 could be detected in microsomes from all chick or rat organs studied, including spleen, testis and brain (Greene *et al.*, 1991). The effects of heme on the induction of mRNA and protein synthesis for HO-1 have been studied in primary cultures of chick embryo liver cells (Cable *et al.*, 1993). It was shown that heme increased (up to 20-fold) the amount of mRNA and the rate of HO-1 gene transcription in a dose-dependent fashion. In fact, 7-15 h after heme addition, the half-life of HO-1 mRNA was 3.5 h in the presence or absence of actinomycin D, while the half-life of heme-induced HO-1 protein was 15 h (Cable *et al.*, 1993). Similarities were observed with respect to regulation of HO-1 expression in primary chick embryo hepatocytes and chicken hepatoma cells (Gabis *et al.*, 1996). It seems likely that HO-1 synthesis is under hormonal control. For example, the effects of various hormones on the induction of HO-1 in monolayer cultures in chick embryo hepatocytes were examined (Sardana *et al.*, 1985). Indeed, insulin is shown to suppress the activity of basal as well as Co^{2+}-induced HO-1, while hydrocortisone suppressed the basal enzyme activity and slightly enhanced Co^{2+}-induced enzyme activity. In contrast, triiodothyronine caused a slight increase of both uninduced and induced levels of the enzyme (Sardana *et al.*, 1985).

There is a range of *in vitro* studies, mainly with embryonic chick cells, to address possible mechanisms of HO-1 induction by various metals. For example, in primary cultures of embryonic chick liver cells HO-1 activity was shown to be upregulated by inorganic cobalt (Maines and Sinclair, 1977). Treatment of isolated chick embryo liver cells *in vitro* with sodium arsenite or melarsoprol also showed a potent induction of HO-1 (Sardana *et al.*, 1981). In monolayer cultures of chick embryo liver cells the most potent HO-1 inducing action was exhibited by Co^{2+}, Cd^{2+}, Sb^{3+}, As^{3+}, and Au^{1+} followed by lower induction observed with Cu^{2+}, Fe^{2+}, and Fe^{3+}(Sardana *et al.*, 1982). In contrast, adding Zn^{2+} (20 μM), Mn^{2+} (50 μM) or cysteine (400 μM) to Co^{2+}-treated cells blocked/inhibited the HO-1 induction. Cycloheximide also blocked the HO-1 induction, indicating that HO-1 activation is dependent on fresh RNA and protein synthesis (Sardana *et al.*, 1982). The activity of HO-1 in chick embryo is shown to be enhanced by cadmium chloride treatment (Prasad and Datta, 1984). It has been suggested that induction of HO-1 by drugs and metals occurs by different mechanisms. For example, a drug phenobarbitone induced HO-1 by increasing hepatic haem formation, while increases in HO activity by metals (cobalt, cadmium or iron) were not dependent on increased haem synthesis and were not inhibited by 4,6-dioxoheptanoic acid (Lincoln *et al.*, 1988). In cultured chick embryo liver cells, synergistic induction of HO-1 by iron, added with the phenobarbital-like drug, glutethimide was heme-dependent (Cable *et al.*, 1990). Addition of an inhibitor of heme biosynthesis abolished the synergistic induction of heme oxygenase providing evidence for the heme-dependent mechanism of induction. Both HO-1 mRNA and protein levels were shown to correlate with changes in HO-1 activity indicating that glutethimide and iron induce HO-1 at the transcriptional level. Induction of the HO-1 gene by heme is shown to be fundamentally different from that produced by transition metals or sodium arsenite and expression of the HO-1 gene is highly conserved across species (Lu *et al.*, 1997). Notably, in chick embryo liver cell cultures, HO-1 responded to sodium arsenite treatment in a dose-dependent fashion, and the response was rapid and transient. Although 2.5 μM arsenite is shown to induce HO-1 four- to six-fold, this had no effect on degradation of exogenous heme (Jacobs *et al.*, 1999).

It seems likely that similar to mammals, in birds HO-1 induction in stress conditions is mediated by various signalling pathways. For example, in chicken hepatoma cells, MAP kinases ERK and p38 are shown to be involved in the induction of HO-1, and at least one AP-1 element is involved in this response (Elbirt *et al.*, 1998). In particular, it was shown that the phenylarsine oxide (PAO), an inhibitor of protein tyrosine phosphatases, upregulated HO-1 gene activity in dose- and time-dependent fashion and both an AP-1 element and a metal responsive element were involved in the PAO mediated induction of the HO-1 activity (Shan *et al.*, 1999). Indeed, a short (1-15 min) exposure of normal hepatocytes to low concentrations (0.5-3 μM) of PAO are shown to produce a marked increase in mRNA and protein of HO-1, which occur without producing changes in cellular glutathione levels or stabilisation of HO-1 message (Gildemeister *et al.*, 2001). Furthermore, preincubation of cells with inhibitors of protein synthesis decreased the ability of PAO to increase levels of HO-1 mRNA, suggesting that the inductive effect requires *de novo* protein synthesis. Addition of thiol donors abrogated the PAO-mediated induction of HO-1 in a dose-dependent fashion. Addition of genistein, a tyrosine kinase inhibitor, blunted the induction produced by both PAO and heme (Gildemeister *et al.*, 2001). It was shown that induction of the chicken HO-1 gene by sodium arsenite or cobalt chloride is mediated through oxidative stress pathway(s) by activation of AP-1 proteins (Lu *et al.*, 2000). It seems likely that vascular endothelial growth factor upregulates HO-1 protein expression *in vivo* in chicken embryo chorioallantoic membranes by a mechanism dependent on an increase in cytosolic calcium levels and activation of protein kinase C (Fernandez and Bonkovsky, 2003).

In chick embryo hepatocytes heme breakdown occurred predominantly, if not solely, by heme oxygenase (Lincoln *et al.*, 1989). It seems likely, that increased HO-1 expression in chicken embryos between internal (day 19) and external pipping (day 20) (Druyan *et al.*, 2007) is an adaptive mechanism responsible for increased protection of tissues during this stressful period of the ontogenesis. Similarly, increased concentrations of vitamin E and carotenoids were observed in chicken embryonic tissues at the same period of time (Surai, 2002), providing an effective protection at hatching. It is well-known that various phytochemicals can affect HO-1 activity (Barbagallo *et al.*, 2013; Murakami, 2014), however, more research is needed to understand molecular mechanisms of their interactions. For example, sulpharaphane containing broccoli extract and four different essential oils were tested in the 2-week-old broilers as feed additives for 3 weeks. The phytogenic feed additives increased HO-1 activity in the jejunum, but decreased it in the liver (Mueller *et al.*, 2012). It is interesting to note that relative mRNA expression of HIF-1 (heart) was increased and HO-1 (heart and liver) was decreased at week 4 in broilers fed with high ME and protein diet (Peng *et al.*, 2013). From the aforementioned data it is clear that HO-1 is well characterised in avian species, however, its response to different stresses in commercial and wild birds are still not fully characterised. Thus, an analysis of the published data leads to the conclusion that HSPs play a significant role in cell/organism protection against various stresses being an integral part of the antioxidant network responsible for proteostasis maintenance.

5.5 Practical applications of heat shock proteins expression in poultry production

5.5.1 Heat stress and heat shock proteins in avian species

The universality of the HSR and conservation of proteins induced by heat stress were shown in experiments with various species. As mentioned above, effect of heat stress on the expression of HSPs in avian species started in early 1980th (Kelley and Schlesinger, 1982; Voellmy and Bromley, 1982; Voellmy *et al.*, 1983). Similarly, exposure of chick myotube cultures to an increased temperature (45 °C) caused extensive synthesis of three major HSPs (25 kDa, 65 kDa and 81 kDa). When experimental cells were allowed to recover from heat-shock treatment at 37 °C for 6-8 h, HSP synthesis declined to levels comparable to those in control cultures maintained at 37 °C (Bag *et al.*, 1983, 1983a). Therefore, four major chicken stress mRNAs coding HSPs with apparent molecular weights of 88 kDa, 71 kDa, 35 kDa and 23 kDa were separated and their properties were studied (White and Hightower, 1984). Exposure of the 11-day embryonic chicken lens to elevated temperature (45 °C) dramatically increased the synthesis of three HSPs with subunit molecular weights of 89,000, 70,000 and 24,000 Da. Furthermore, the functional half-lives at 37 °C of the mRNAs encoding the lens HSPs were about 3-5 h (Collier and Schlesinger, 1986).

The intracellular distributions of the major heat shock proteins, HSP89, HSP70, and HSP24 were studied in chicken embryo fibroblasts stressed by heat shock, allowed to recover and then restressed (Collier and Schlesinger, 1986a). It was shown that HSP89 was localised primarily to the cytoplasm and during the restress a portion of this protein was associated with the nuclear region. In contrast, significant amount of HSP70 was shown to move to the nucleus during stress. In general, the nuclear HSPs reappeared in the cytoplasm in cells allowed to recover at normal temperatures. It is interesting to note that, sodium arsenite also induces HSPs and their distributions were similar to that observed after heat shock, except for HSP89, which remained cytoplasmic (Collier and Schlesinger, 1986a). Reticulocytes, purified from the blood of quail and chickens responded to heat shock by the synthesis of HSP90, HSP70 and HSP26 (quail) or HSP24 (chicken) and the depressed synthesis of many other proteins normally produced at a physiological temperature (Atkinson *et al.*, 1986). It was shown that the expression of each protein depended upon the particular temperature and duration of heat exposure. It was noted that HSP70 was constitutively synthesised and selectively partitioned between cellular compartments. Furthermore, heat shock induced synthesis of the HSP90, HSP70 and HSP26 in quail was prevented by actinomycin D (Atkinson *et al.*, 1986).

Heat shock response is a universal biological protective mechanism in stress conditions. Indeed, cultured bovine, equine, ovine and chicken lymphocytes responded to heat stress by the increased synthesis of HSPs. In particular, HSP70 and HSP90 were synthesised in all species and induction time of the HSPs synthesis comprised 30-60 minutes (Guerriero and Raynes, 1990). Heat shock response is an important mechanism of immune cells protection. Actually, heat-induced chicken macrophages synthesised HSP23, HSP70 and HSP90. The optimal temperature and time for induction of these HSPs was 45-46 °C for 1 h, with a variable recovery period for each HSP (Miller and Qureshi, 1992). A comparison of HSP synthesis among peritoneal macrophages (PM) from chickens, turkeys, quail, and ducks shows the highly conserved nature of heat-shock response within birds. In fact, macrophage cultures from each avian species expressed the three major HSPs (HSP23, HSP70 and HSP90) following heat-shock exposure (1 h heat shock at 45 °C) (Miller and Qureshi, 1992a). There was also increased expression of a new HSP called P32, which probably was HSP32 (known as HO-1) in all 4 species. The authors also showed that the duck P32 and HSP23 were lower in molecular mass than their respective homologues expressed in chickens, turkeys, and quail macrophage cultures indicating some species-specific differences between HSPs in avian species (Miller and Qureshi, 1992b). Chicken macrophages (mononuclear phagocytic cell line MQ-NCSU) exposed to LPS under control (41 °C) temperatures expressed enhanced synthesis of classical HSP23, HSP70, and HSP90, as well as heat-inducible 32-kDa protein (P32), and a novel LPS-induced 120-kDa protein (P120). In comparison to LPS treatment, MQ-NCSU cells exposed to 45 °C (HS) expressed HSP23, HSP70, HSP90, and P32 but not P120 (Miller and Qureshi, 1992b). It is interesting to note that lead acetate caused similar upregulation of the same four HSPs (HSP23, HSP70, HSP90 and P32) previously expressed by macrophages after *in vitro* and *in vivo* heat treatment (Miller and Qureshi, 1992c). It seems likely that various nutritional deficiencies could affect HSP response. For example, during acute *in vivo* heat stress, a HSP response was not inducible in chickens deficient in inorganic phosphorus (Edens *et al.*, 1992) and they were more susceptible heat stress. Increased HSPs expression in response to various stresses, including heat stress, is shown to be a universal mechanism in various chicken tissues. For example, both the amount and polyadenylation of HSP70 and ubiquitin transcripts increased when male germ cells from adult chicken testis were exposed to elevated (46 °C) temperatures (Mezquita *et al.*, 1998). Similarly, there was a marked increase in HSP70 expression in the brains of female broiler chickens after 4 days (from d35 to d38) of heat treatment (38±1 °C for 2 h/d; Zulkifli *et al.*, 2002). In addition, in chicken pineal cells several heat shock proteins (HSPs 25, 70, and 90) are shown to be synthesised under temperature conditions (Wolfe and Zatz, 1994). Thermal stress (41 °C) caused induction of HSP90α and HSP90β in chicken heart, liver and spleen, but HSP90α and HSP90β mRNA levels were stable in brain. Transcription of HSP70 also increased in all organs from chickens in heat stress groups when compared to chickens in control groups (Mahmoud *et al.*, 2004, 2004a). The elevation of the three HSPs in heart, may act as protective mechanism in adverse environments. For example, three main chicken HSPs (HSP60, HSP70, HSP90), and their corresponding mRNAs in the heart tissue of heat-stressed (37 °C for 2-10 hours) broilers, elevated significantly after 2 h of heat exposure and decreased quickly with continued heat stress. However, the level of HSP60 protein in the heart increased and maintained throughout heat exposure (Yu *et al.*, 2008). Indeed, there is a great diversity in heat shock response in different tissues. For example, thirty-two-week-old broiler breeders were subjected either to acute (step-wisely increasing temperature from 21 to 35 °C within 24 hours) or chronic (32 °C for 8 weeks) high temperature exposure. There was a tissue specificity in the response to acute and chronic stress (Xie *et al.*, 2014). For example, in the heart, acute heat challenge increased lipid peroxidation and upregulated gene expression of all four HSFs. Furthermore, during chronic heat treatment, the HSP 70 mRNA level was increased and HSP 90 mRNA was decreased. At the same time, in the liver, protein oxidation was alleviated during acute heat challenge and gene expression of HSF2, 3 and 4 and HSP70 were highly induced. In addition, HSP90 expression was increased by chronic thermal treatment. In the muscle, both types of heat stress increased protein oxidation, but HSFs and HSPs gene expression remained unaltered and only tendencies to increase were observed in HSP70 and HSP90 gene expression after acute heat stress (Xie *et al.*, 2014). The expression of HSP27, HSP70, and HSP90 mRNA in the bursa of Fabricius and spleen of 42-d old chickens were increased due to heat stress (37±2 °C for 15 d; Liu *et al.*, 2014). However, under the same stress conditions the expression of HSP27 and HSP90 mRNA in thymus were decreased. In testis of heat-stressed cockerels (38 °C for 4 hours) the heat shock proteins, chaperonin containing t-complex, and proteasome subunits were downregulated (Wang *et al.*, 2014). Therefore, acute heat stress impairs the processes of translation, protein folding, and protein degradation resulting in apoptosis and spermatogenesis disturbance. Heat stress in 21-day-old broilers was associated with up-regulation of the rectal temperature and the mRNA expression of HSP70 in the liver (Zuo *et al.*, 2015). Heat stress (40 °C for 2 h) in the growing chickens (41 day old) caused significant increases in sera corticosterone, LDH, MDA and SOD, the expression of HSP90 and HSP70 in the pectoralis major. Furthermore, HSP90 was shown to positively correlate with corticosterone and SOD activities (Hao and Gu, 2014). In chicken hypothalamus the transcripts of HSP90 decreased while HSP40 increased in response to thermal stress (34 °C for 24 h; Sun *et al.*, 2015).

It seems likely that gene expression changes due to heat stress are of great importance for cell adaptation to stress. For example, heat stress (38 °C for 4 hours) was associated with upregulation of 169 genes and downregulation of 140 genes in rooster testis (Wang *et al.*, 2013). Differentially expressed genes were mainly related to response to stress, transport, signal transduction, and metabolism. Indeed, HSP genes (HSP25, HSP70 and HSP90AA1) and related chaperones were the major upregulated groups in chicken testes after acute heat stress. Heat stress in chickens was associated with 166 differentially expressed genes in the brain, 219 in the leg muscle and 317 in the liver (Luo *et al.*, 2014). Six of these genes were differentially expressed in all three tissues and included heat shock protein genes (HSPH1-heat shock 105/110 kDa protein 1 and HSP25), apoptosis-related genes (RB1CC1, BAG3), a cell proliferation and differentiation-related gene and the hunger and energy metabolism related gene. Various functional clusters were related to the effects of heat stress, including those for cytoskeleton, extracellular space, ion binding and energy metabolism (Luo *et al.*, 2014). It seems likely that HSP expression in response to increased temperature is a universal cellular mechanism protecting proteins against unfavourable changes, including misfolding and molecular mechanisms of HSR need further research.

5.5.2 Dietary antioxidants and heat shock proteins

Since all antioxidants in the body are working together to build the effective antioxidant defence network, the increase concentration of one antioxidant can be associated with no need for increase another antioxidant element in stress conditions.

Vitamin E

Vitamin E is considered to be a main chain-breaking antioxidant in biological systems and its roles in poultry production are difficult to overestimate (Surai, 1999a, 2002, 2014; Surai and Fisinin, 2014). It was shown that vitamin E, added to the Vero cell culture prior mycotoxins (citrinin, zearalenone and T2 toxin) was able to prevent an induction of HSP70 expression due to mycotoxins (El Golli *et al.*, 2006). In isolation-stressed quail, vitamin E or vitamin C were shown to prevent an increase in HSP70 expression in the brain and heart (Soleimani *et al.*, 2012). In crossbred cows, treatment with α-tocopherol acetate during dry period resulted in reduced oxidative stress and

HSP70 (Aggarwal *et al.*, 2013). In cultured rat hepatocytes vitamin E significantly counteracted the effect of cyclosporine A-induced increase in HSP70 (Andrés *et al.*, 2000). However, in young men, γ-tocopherol was shown to prevent the exercise-induced increase of HSP72 in skeletal muscle as well as in the circulation (Fischer *et al.*, 2006).

However, in most of cases effect of vitamin E on HO-1 expression is different from the aforementioned effects on HSP70. Indeed, recently it has been shown that vitamin E activated the HO-1 promoter via the cAMP-response element but not the ARE enhancer through the extracellular signal-regulated kinase and protein kinase A (Reed *et al.*, 2015). It was shown that α-tocopheryl succinate increases nuclear translocation and electrophile-responsive/antioxidant-responsive elements binding activity of Nrf2, resulting in up-regulation of downstream genes cystine-glutamic acid exchange transporter and HO-1, while decreasing NF-κB nuclear translocation (Bellezza *et al.*, 2012). It seems likely that α-tocopherol protects human retinal pigment epithelial cells from acrolein-induced cellular toxicity, not only as a chain-breaking antioxidant, but also as a Phase II enzyme inducer, including Nrf-2, SOD and HO-1 induction (Feng *et al.*, 2010). Similarly, in a murine prostate cancer model γ-tocopherol-enriched mixed tocopherols significantly upregulated the expression of Nrf2 and its related detoxifying and antioxidant enzymes, including SOD and HO-1 (Barve *et al*, 2009). In rats, protective effect of vitamin E against focal brain ischemia and neuronal death was shown. In fact, vitamin E induced the expression of the alpha subunit of hypoxia-inducible factor-1 (HIF-1) and its target genes, including vascular endothelial growth factor (VEGF) and heme oxygenase-1 (Zhang *et al.*, 2004).

Ascorbic acid

Ascorbic acid is main water-soluble antioxidant provided with feed and synthesised within the animal/chicken body (Chakraborthy *et al.*, 2014). It has been shown that chickens experience a less severe stress response after exposure to high temperatures when they are provided dietary ascorbic acid. In fact, heart HSP70 expression decreased in ascorbic acid-fed chickens and the HSP70 increase after heat was two-fold lower in ascorbic acid-fed birds in comparison with the control chickens. Furthermore, plasma corticosterone and heart HSP70 were positively correlated, while plasma ascorbic acid and heart HSP70 were negatively correlated (Mahmoud *et al.*, 2004a). In the ascorbic acid-fed chickens, neither the lower constitutive HSC70 nor the decreased HSP70 response to heat stress (42 °C) in the heart and liver were sex-dependent (Mahmoud *et al.*, 2003). A lower expression of HSP70 associated with lower body temperature in heat-stress conditions reflected a lower stress response in the ascorbic acid-fed birds. Indeed, ovary and brain HSP70 expression linearly decreased as dietary vitamin C or vitamin E supplementation increased in heat-stressed quail. However, HSP70 expression of ovary and brain was not affected by vitamin C or E supplementation under thermo-neutral conditions (Sahin *et al.*, 2009).

Effects of ascorbic acid on HSP70 expression were also evaluated in experiments with laboratory animals or in human trials. For example, lymphocytes from non-supplemented subjects responded to hydrogen peroxide with increased HSP60 and HSP70 content over 48 h. In fact, in vitamin C supplemented subjects, baseline HSP60 (lymphocytes) and HSP70 (muscles) content were elevated, but they did not respond to hydrogen peroxide or exercise (Khassaf *et al.*, 2003). In elderly, increased concentration of vitamins C and E was associated with a reduction in oxidative stress and leukocytes HSP72 (Simar *et al.*, 2012). Ascorbic acid was shown to attenuate increase in HSP expression due to various toxic agents or heat stress. For example, human brain astrocyte cells enriched with ascorbic acid before being exposed to ethanol, were reported to be better protected against the alcohol-mediated toxicity than non-supplemented cells, and showed significantly lower concentrations of HSP70 (Sánchez-Moreno *et al.*, 2003). Ascorbic acid significantly attenuated Cd-induced upregulation of GRP78 in mouse testes (Ji *et al.*, 2012). Cyclic heat stress (23 to 38 to 23 °C, for 2 h on each of seven consecutive days) activated hepatic HSP70, TNF-α, iNOS, and GPx genes, whereas vitamin C (0.5% in water) during heat stress ameliorated heat stress-induced cellular responses in rats (Yun *et al.*, 2012). It is interesting to note that there was a specific disappearance of HSP70 in the testes of 20-day-old ascorbic acid-deficient mice (Yazama *et al.*, 2006). It seems likely that effects of ascorbic acid on HSPs is not universal and for HO-1 is different from HSP70. Indeed, the HO-1 mRNA and protein level in rat kidney, liver, and lung were highly induced by ascorbate treatment (100 mg/kg b.w.) under normal and HS conditions. In particular, in HS the HO-1 activity in tissues was enhanced by both ascorbate pre- and post-treatment (Zhao *et al.*, 2014).

Vitamin D3

Vitamin D is known for its classical functions in calcium uptake and bone metabolism. However, recently, vitamin D has been recognised for its non-classical actions including modulation of antioxidant defences (Xu *et al.*, 2015; Zhong *et al.*, 2014) through regulating oxidant and antioxidant enzyme genes. It was shown that HO-1 was down-regulated in the livers of mice fed the vitamin D deficient diet (Zhu *et al.*, 2015). At the same time, vitamin D deficiency increases the expression of the hepatic mRNA levels of HO-1 in obese rats (Roth *et al.*, 2012). In a model of reperfusion of bilateral femoral vessels pre-treatment of rats with vitamin D3 results in a significant increase in leukocyte HO-1 expression in rat model of reperfusion (Shih *et al.*, 2011). By employing microarray technology, the effect of a single dose of 1,25-(OH)2D3 on gene expression in the intestine of vitamin D-deficient rats was shown. Indeed, at 3 h, there was a 1.9-fold increased expression of HO-1 (Kutuzova and DeLuca, 2007). The effects of 1,25-D3 treatment on HO-1 expression following focal cortical ischemia elicited by photothrombosis in glial cells were studied. Postlesional treatment with 1,25-D3 (4 μg/kg body weight) resulted in a transient, but significant upregulation of glial HO-1 immunoreactivity (Oermann *et al.*, 2004).

Carnitine and betaine

Carnitine is considered as a novel mitochondria-targeted antioxidant with a range of antioxidant actions (Surai, 2015, 2015c,d), while betaine is reported to have antioxidant properties in various oxidative stress-generating model systems (Alirezaei *et al.*, 2015). In human endothelial cells in culture carnitine was shown to increase gene and protein expression of HO-1 (Calò *et al.*, 2006). Furthermore, in humans and in an animal model it was shown that carnitine-mediated improved response to erythropoietin involves induction of HO-1 (Calò *et al.*, 2008). Indeed, L-carnitine treatment was associated with an increased level of HO-1 in the retinal ganglion cells (Cao *et al.*, 2015). L-carnitine prevented increase in HSP70 in the testes of cadmium-exposed rats (Selim *et al.*, 2012). It was shown that Acetyl-L-carnitine-induced up-regulation of heat shock proteins protects cortical neurons against amyloid-beta peptide 1-42-mediated oxidative stress and neurotoxicity (Abdul *et al.*, 2006). Acetyl-L-carnitine induces heme oxygenase (increased the amount and activity of HO) in rat astrocytes and protects against oxidative stress (Calabrese *et al.*, 2005). From the aforementioned data it is clear that carnitine can be considered as an important regulator of the vitagene network. The influence of hyperosmotic shrinkage and the osmolyte betaine on heme oxygenase HO-1 expression was studied in cultured rat hepatocytes. Hyperosmolarity transiently suppressed HO-1 induction in response to hemin or medium addition at the levels of mRNA and protein expression. Pretreatment of the cells with betaine largely restored induction of both HO-1 mRNA and protein under hyperosmotic conditions (Lordnejad *et al.*, 2001).

Selenium

Selenium is a central part of the antioxidant defence network via at least 25 selenoproteins (Surai, 2006, 2018). The protective effect of selenium against cadmium-induced cytotoxicity in chicken splenic lymphocytes was shown to be mediated via the HSP pathway (Chen *et al.*, 2012). Indeed, the mRNA expression of HSPs (HSP27, HSP40, HSP60, HSP70 and HSP90) exposed to 10^{-6} mol/l Cd showed a sustained decrease at 12-48 h exposure. In contrast, adding to the medium Se (10^{-7} mol/l) was associated with a significant increase in the mRNA expression of HSPs, as compared to the control group of chicken splenic lymphocytes. Concomitantly, treatment of chicken splenic lymphocytes with Se in combination with Cd prevented a decrease in the mRNA expression of HSPs due to Cd treatment. A different HSP response to arsenic was observed. The expression of HSPs mRNA and protein (HSP70 and HO1) in rat liver were increased by 5 and 3 folds in the arsenic-fed animals compared with the control group, and selenium prevented the occurring of oxidative damage from arsenic and significantly reduced expression of HSPs mRNA and protein (Xu *et al.*, 2013).

The HSP70 response was shown to be significantly lower in the chickens fed selenium and challenged with either enteropathogenic Escherichia coli or heat stress than in those chickens given no supplemental selenium (Mahmoud and Edens, 2005, 2003). An acute heat stress induced HSP70 in 22 d turkey embryos and the embryos from selenium-fed dams were shown to have less HSP70 after the 3 h post-heat stress recovery period (Rivera *et al.*, 2005) demonstrating that selenium had the ability to reduce the impact of heat stress. In fact, heat stress enhanced HSP70 and HSP27 expression and concentration in chicken spleen and dietary Se prevented the aforementioned increase in HSPs (Xu *et al.*, 2014). Similarly, in piglets under heat stress conditions selenium can down-regulate the mRNA levels of HSPs in various tissues (Gan *et al.*, 2013). The relative messenger RNA (mRNA) and protein expression of HSP60, HSP70, and HSP90 in PBMC was observed highest in heat-stressed goats and Se + vitamin E supplementation decreased the HSP expression (Dangi *et al.*, 2015).

In contrast, Se deficiency increased the mRNA levels of HSPs (HSP90, 70, 60, 40, and 27) in chicken neutrophils (Chen *et al.*, 2014). Indeed, HSPs played an important role in the protection of the chicken liver after oxidative stress due to Se deficiency. For example, the mRNA levels of HSPs and the protein expression of HSPs (HSP60, 70, and 90) increased significantly in the Se-deficient group compare to the corresponding control group (Liu *et al.*, 2015). In exudative diathesis (ED) broiler chicken model caused by Se deficiency, the antioxidant function was shown to decline remarkably, and most of the HSP expression levels increased significantly in the spleen, thymus, and bursa of Fabricius of the broiler chicks with ED (Yang *et al.*, 2016). Se deficiency causes defects in the chicken bursa of Fabricius associated with decreased selenoprotein expression (Khoso *et al.*, 2015). As a compensatory response to changes due to Se deficiency, the mRNA and protein expression levels of HSPs (HSP27, HSP40, HSP60, HSP70, and HSP90) were significantly increased. Similar observations with Se deficient mouse were recorded. For example, Se deficiency was shown to increase HSP70 levels in mouse testis (Kaur and Bansal, 2003). A significant increase in the stress-inducible HSP70 gene and protein expression was observed in the mice fed Se-deficient or Se-excess supplemented diet as compared with Se adequate fed group (Kaushal and Bansal, 2009). It is interesting to note that the testis-specific HSP70-2 expression significantly decreased as result of Se deficiency. It is clear that increased expression of HSPs in response to toxic metals is an adaptive mechanism to deal with oxidative stress imposed by such toxicants. Similarly, in the case of Se deficiency increased HSP expression is also an adaptive mechanism to compensate for lack of synthesis of selenoproteins and their antioxidant protective functions. As mentioned above, HSP response to various stressors and to nutritional supplements would depend on many factors, including the model used, stressor nature and strength, etc. For example, in human lens epithelial cells sodium selenite gradually increased the expression of HSP70 in a time-dependent manner (Zhu *et al.*, 2011). In rat hippocampus with ischemia-induced neuronal damage, selenium pretreatment was shown to significantly increase the level of HSP70 when compared with ischemic group (Yousuf *et al.*, 2007). In fact, a significant increase in hippocampal HSP70 expression in the ischemic group was observed but the expression was even higher in the selenium-pretreated group than ischemic group.

Phytochemicals

Regulatory and health promoting properties of various phytochemicals and their effects on HSPs have received substantial attention and there is a range of comprehensive reviews covering the subject (Calabrese *et al.*, 2011; De Roos and Duthie, 2015; Mattson and Cheng, 2006; Murakami *et al.*, 2013). They are beyond the scope of the present review. Therefore, only effects of silymarin, possessing a range of antioxidant-related activities (Surai, 2015a), are reviewed below.

Silymarin

It seems likely that SM, similar to other flavonoids, can affect the vitagene network. In fact, SM/silybin affects HSP32 (HO-1) activity in different model systems. For example, As-intoxicated rats showed a significant up-regulation of myocardial NADPH (NOX) oxidase sub-units such as NOX2 and NOX4 as well as Keap1 and down-regulation of Nrf2 and vitagene HO-1 protein expressions. Pre-administration of silibinin (75 mg/kg/BW) recovered all these altered parameters to near normalcy in As-induced cardiotoxic rat (Muthumani and Prabu, 2014). Similarly, in a model of liver injury caused by alcohol plus pyrazole, SM administration (50 mg/kg/BW) had a protective effect with a trend in restoring the decreased activity of HO-1 and Nrf2 (Choi *et al.*, 2013). SM (250 mg/kg/BW) possesses substantial protective effect against B(a)P-induced damages by increasing (restoring) HO-1 (vitagene) activity (Kiruthiga *et al.*, 2015). Similarly, *in vitro* SM (500 μM) reduced tBH-induced hepatocyte toxicity by activating HO-1 gene expression (Cerný *et al.*, 2009). Indeed, the enzyme HO-1 is an important regulatory molecule present in most mammalian cells. In fact, the main function of HO-1 is to break down the pro-oxidant molecule heme into three products; carbon monoxide (CO), biliverdin and free iron and actively participate in the antioxidant defence in the human/animal body (Venditti and Smith, 2014). Indeed, HO-1 is a stress-inducible protein and can be induced by various oxidative and inflammatory signals. From the data presented above it is clear that SM/silibinin can upregulate HO-1 and improve antioxidant defences. It is likely that SM/silibinin can affect other HSPs including HSP70. Indeed, in an *in vitro* system based on CHO-K1 cells treated with As, SM (5 μM) significantly decreased HSP70 expression previously elevated by As (Bongiovanni *et al.*, 2007). In another *in vitro* system based on heat-induced chicken hepatocytes, SM (259 μM) affected HSP70 expression significantly, preventing its alleviation by heat stress (Oskoueian *et al*, 2014). A similar protective effect of SM (100 mg/kg/BW) on HSP70 was seen in rats given SM for 7 days prior to mesenteric ischemia-reperfusion (I-R) compared to I-R group (Demir *et al.*, 2014). It is interesting to note that silybin was identified as a novel HSP90 inhibitor (Zhao *et al.*, 2011). Therefore, silibinin can decrease HSP70 expression in stressed cells indicating improved AO defences and decrease stress by other means (e.g. Nrf2-related increased AO synthesis). Indeed, effects of silymarin on HSPs in avian species awaits investigation, while other phytochemicals are shown to be effective. For example, resveratrol, a plant phytochemical possessing antioxidant activities, attenuated the heat stress-induced overexpression of HSP27, HSP70, and HSP90 mRNA in the bursa of Fabricius and spleen and increased the low expression of HSP27 and HSP90 mRNA in thymus in 42 d old chickens upon heat stress (Liu *et al.*, 2014). Indeed, there is a need for more detailed investigation of the relationship between nutritional antioxidants and HSP expressions in physiological and stress conditions.

5.6 Conclusions

From the aforementioned analysis of the data related to HSPs in poultry physiology and adaptation to stresses it is possible to conclude:

- HSPs as important vitagenes are main driving force in cell/body adaptation to various stress conditions.
- In physiological condition some HSP play house-keeping role and their expression is typically low. However, under stress conditions synthesis of most cellular proteins decreases while HSP expression and synthesis are usually significantly increased.
- HSPs being cellular chaperones are responsible for proteostasis and involved in protein quality control in the cell to prevent misfolding or to facilitate degradation of misfolded/damaged proteins, making sure that proteins are in optimal structure for their biological activities.
- There are tissue-specific differences in HSP expression which also depends on the strength/intensity of such stress-factors as heat, heavy metals, mycotoxins and other toxicants.
- HSP70, HSP90 and HSP32 are shown to be protective in heat stress, toxicity stress as well as in other oxidative-stress related conditions in poultry production.
- Molecular mechanisms of HSP participation in acquisition of thermotolerance need further detailed investigation.
- There are complex interactions inside the antioxidant systems of the cell/body to ensure an effective maintenance of homeostasis in stress conditions. Indeed, in many cases nutritional antioxidants (vitamin E, ascorbic acid, selenium) in the feed can decrease oxidative stress and as a result HSP expression could be decreased as well.
- Regulating effects of various phytochemicals on HSPs need further investigation.
- Protective effects of HSPs in immunity under stress conditions await practical applications in poultry production.
- Nutritional means of additional HSP upregulation in stress conditions of poultry production and its physiological and commercial consequences await investigation. Indeed, in medical sciences manipulation of HSP expression is considered as an important approach in disease prevention and treatment. It seems likely that in poultry/animal sciences nutritional manipulation of vitagenes is a new way in managing commercially-relevant stresses.

References

Abdul, H.M., Calabrese, V., Calvani, M. and Butterfield, D.A., 2006. Acetyl-L-carnitine-induced up-regulation of heat shock proteins protects cortical neurons against amyloid-beta peptide 1-42-mediated oxidative stress and neurotoxicity: implications for Alzheimer's disease. Journal of Neuroscience Research 84: 398-408.

Aggarwal, A., Ashutosh, Chandra, G. and Singh, A.K., 2013. Heat shock protein 70, oxidative stress, and antioxidant status in periparturient crossbred cows supplemented with α-tocopherol acetate. Tropical Animal Health and Production 45: 239-245.

Albakova, Z., Armeev, G.A., Kanevskiy, L.M., Kovalenko, E.I. and Sapozhnikov, A.M., 2020. HSP70 multi-functionality in cancer. Cells 9, 3: 587.

Alirezaei, M., Khoshdel, Z., Dezfoulian, O., Rashidipour, M. and Taghadosi, V., 2015. Beneficial antioxidant properties of betaine against oxidative stress mediated by levodopa/benserazide in the brain of rats. The Journal of Physiological Sciences 65: 243-252.

Andrés, D., Alvarez, A.M., Díez-Fernández, C., Zaragoza, A. and Cascales, M., 2000. HSP70 induction by cyclosporine A in cultured rat hepatocytes: effect of vitamin E succinate. Journal of Hepatology 33: 570-579.

Atkinson, B.G., Dean, R.L. and Blaker, T.W., 1986. Heat shock induced changes in the gene expression of terminally differentiating avian red blood cells. Canadian journal of genetics and cytology. Journal Canadien de Genetique et de Cytologie 28: 1053-1063.

Bag, J., 1983. Regulation of heat-shock protein synthesis in chicken muscle culture during recovery from heat shock. European Journal of Biochemistry 135: 373-378.

Bag, J., 1983a. Free messenger ribonucleoprotein complexes of chicken primary muscle cells following modification of protein synthesis by heat-shock treatment. European Journal of Biochemistry 135: 187-196.

Bagatell, R., Paine-Murrieta, G.D., Taylor, C.W., Pulcini, E.J., Akinaga, S., Benjamin, I.J. and Whitesell, L., 2000. Induction of a heat shock factor 1-dependent stress response alters the cytotoxic activity of hsp90-binding agents. Clinical Cancer Research 6: 3312-3318.

Banerji, S.S., Laing, K. and Morimoto, R.I., 1987. Erythroid lineage-specific expression and inducibility of the major heat shock protein HSP70 during avian embryogenesis. Genes 1: 946-953.

Banerji, S.S., Theodorakis, N.G. and Morimoto, R.I., 1984. Heat shock-induced translational control of HSP70 and globin synthesis in chicken reticulocytes. Molecular and Cellular Biology 4: 2437-2448.

Barbagallo, I., Galvano, F., Frigiola, A., Cappello, F., Riccioni, G., Murabito, P., D'Orazio, N., Torella, M., Gazzolo, D. and Li Volti, G., 2013. Potential therapeutic effects of natural heme oxygenase-1 inducers in cardiovascular diseases. Antioxidants & Redox Signaling 18: 507-521.

Barrott, J.J. and Haystead, T.A., 2013. Hsp90, an unlikely ally in the war on cancer. The FEBS Journal 280: 1381-1396.

Barve, A., Khor, T.O., Nair, S., Reuhl, K., Suh, N., Reddy, B., Newmark, H. and Kong, A.N., 2009. Gamma-tocopherol-enriched mixed tocopherol diet inhibits prostate carcinogenesis in TRAMP mice. International Journal of Cancer 124: 1693-1699.

Bellezza, I., Tucci, A., Galli, F., Grottelli, S., Mierla, A.L., Pilolli, F. and Minelli, A., 2012. Inhibition of NF-κB nuclear translocation via HO-1 activation underlies α-tocopheryl succinate toxicity. The Journal of Nutritional Biochemistry 23: 1583-1591.

Bhat, A., Gomis, S., Potter, A. and Tikoo, S.K., 2010. Role of Hsp90 in CpG ODN mediated immunostimulation in avian macrophages. Molecular Immunology 47: 1337-1346.

Bilban, M., Haschemi, A., Wegiel, B., Chin, B.Y., Wagner, O. and Otterbein, L.E., 2008. Heme oxygenase and carbon monoxide initiate homeostatic signaling. Journal of Molecular Medicine 86: 267-279.

Binart, N., Chambraud, B., Dumas, B., Rowlands, D.A., Bigogne, C., Levin, J.M., Garnier, J., Baulieu, E.E. and Catelli, M.G., 1989. The cDNA-derived amino acid sequence of chick heat shock protein Mr 90,000 (HSP 90) reveals a 'DNA like' structure: potential site of interaction with steroid receptors. Biochemical and Biophysical Research Communications 159: 140-147.

Bissell, D.M., Hammaker, L. and Schmid, R., 1972. Liver sinusoidal cells. Identification of a subpopulation for erythrocyte catabolism. The Journal of Cell Biology 54: 107-119.

Bongiovanni, G.A., Soria, E.A. and Eynard, A.R., 2007. Effects of the plant flavonoids silymarin and quercetin on arsenite-induced oxidative stress in CHO-K1 cells. Food and Chemical Toxicology 45: 971-976.

Bonkovsky, H.L., Healey, J.F. and Pohl, J., 1990. Purification and characterization of heme oxygenase from chick liver. Comparison of the avian and mammalian enzymes. European Journal of Biochemistry 189: 155-166.

Bouhouche-Chatelier, L., Chadli, A. and Catelli, M.G., 2001. The N-terminal adenosine triphosphate binding domain of Hsp90 is necessary and sufficient for interaction with estrogen receptor. Cell Stress Chaperones 6: 297-305.

Bozaykut, P., Ozer, N.K. and Karademir, B., 2014. Regulation of protein turnover by heat shock proteins. Free radical Biology & Medicine 77: 195-209.

Cable, E., Greene, Y., Healey, J., Evans, C.O. and Bonkovsky, H., 1990. Mechanism of synergistic induction of hepatic heme oxygenase by glutethimide and iron: studies in cultured chick embryo liver cells. Biochemical and Biophysical Research Communications 168: 176-181.

Cable, E.E., Cable, J.W. and Bonkovsky, H.L., 1993. Repression of hepatic delta-aminolevulinate synthase by heme and metalloporphyrins: relationship to inhibition of heme oxygenase. Hepatology 18: 119-127.

Calabrese, V., Cornelius, C., Dinkova-Kostova, A.T., Iavicoli, I., Di Paola, R., Koverech, A., Cuzzocrea, S., Rizzarelli, E. and Calabrese, E.J., 2012. Cellular stress responses, hormetic phytochemicals and vitagenes in aging and longevity. Biochimica et Biophysica Acta 1822: 753-783.

Calabrese, V., Ravagna, A., Colombrita, C., Scapagnini, G., Guagliano, E., Calvani, M., Butterfield, D.A. and Giuffrida Stella, A.M., 2005. Acetylcarnitine induces heme oxygenase in rat astrocytes and protects against oxidative stress: involvement of the transcription factor Nrf2. Journal of Neuroscience Research 79: 509-521.

Calò, L.A., Davis, P.A., Pagnin, E., Bertipaglia, L., Naso, A., Piccoli, A., Corradini, R., Spinello, M., Savica, V. and Dalla Libera, L., 2008. Carnitine-mediated improved response to erythropoietin involves induction of haem oxygenase-1: studies in humans and in an animal model. Nephrology, Dialysis, Transplantation 23: 890-895.

Calò, L.A., Pagnin, E., Davis, P.A., Semplicini, A., Nicolai, R., Calvani, M. and Pessina, A.C., 2006. Antioxidant effect of L-carnitine and its short chain esters: relevance for the protection from oxidative stress related cardiovascular damage. International Journal of Cardiology 107: 54-60.

Cao, Y., Li, X., Wang, C.J., Li, P., Yang, B., Wang, C.B. and Wang, L.X., 2015. Role of NF-E2-related factor 2 in neuroprotective effect of L-carnitine against high glucose-induced oxidative stress in the retinal ganglion cells. Biomedicine & Pharmacotherapy 69: 345-348.

Catelli, M.G., Binart, N., Feramisco, J.R. and Helfman, D.M., 1985. Cloning of the chick hsp 90 cDNA in expression vector. Nucleic Acids Research 13: 6035-6047.

Catelli, M.G., Binart, N., Jung-Testas, I., Renoir, J.M., Baulieu, E.E., Feramisco, J.R. and Welch, W.J., 1985. The common 90-kd protein component of non-transformed '8S' steroid receptors is a heat-shock protein. The EMBO Journal 4: 3131-3135.

Cerný, D., Canová, N.K., Martínek, J., Horínek, A., Kmonícková, E., Zídek, Z. and Farghali, H., 2009. Effects of resveratrol pretreatment on tert-butylhydroperoxide induced hepatocyte toxicity in immobilized perifused hepatocytes: involvement of inducible nitric oxide synthase and hemoxygenase-1. Nitric Oxide: Biology and Chemistry 20: 1-8.

Chakraborthy, A., Ramani, P., Sherlin, H.J., Premkumar, P. and Natesan, A., 2014. Antioxidant and pro-oxidant activity of Vitamin C in oral environment. Indian Journal of Dental Rsearch 25: 499-504.

Chang, A.Y., Chan, J.Y., Cheng, H-L., Tsai, C-Y. and Chan, S.H., 2009. Hypoxia-inducible factor 1/ heme oxygenase 1 cascade as upstream signals in the prolife role of heat shock protein 70 at rostral ventrolateral medulla during experimental brain stem death. Shock 32: 651-658.

Chauveau, C., Rémy, S., Royer, P.J., Hill, M., Tanguy-Royer, S., Hubert, F.X., Tesson, L., Brion, R., Beriou, G., Gregoire, M., Josien, R., Cuturi, M.C. and Anegon, I., 2005. Heme oxygenase-1 expression inhibits dendritic cell maturation and proinflammatory function but conserves IL-10 expression. Blood 106: 1694-1702.

Chen, X., Yao, H., Yao, L., Zhao, J., Luan, Y., Zhang, Z. and Xu, S., 2014. Selenium deficiency influences the gene expressions of heat shock proteins and nitric oxide levels in neutrophils of broilers. Biological Trace Element Research 161: 334-340.

Chen, X., Zhu, Y.H., Cheng, X.Y., Zhang, Z.W. and Xu, S.W., 2012. The protection of selenium against cadmium-induced cytotoxicity via the heat shock protein pathway in chicken splenic lymphocytes. Molecules 17: 14565-14572.

Chiosis, G., Vilenchik, M., Kim, J. and Solit, D., 2004. Hsp90: the vulnerable chaperone. Drug Discovery Today 9: 881-888.

Choi, K.M., Kashyap, P.C., Dutta, N., Stoltz, G.J., Ordog, T., Shea Donohue, T., Bauer, A.J., Linden, D.R., Szurszewski, J.H., Gibbons, S.J. and Farrugia, G., 2010. CD206-positive M2 macrophages that express heme oxygenase-1 protect against diabetic gastroparesis in mice. Gastroenterology 138: 2399-2409.

Choi, M.K., Han, J.M., Kim, H.G., Lee, J.S., Lee, J.S., Wang, J.H., Son, S.W., Park, H.J. and Son, C.G., 2013. Aqueous extract of *Artemisia capillaris* exerts hepatoprotective action in alcohol-pyrazole-fed rat model. Journal of Ethnopharmacology 147: 662-670.

Clerico, E.M., Tilitsky, J.M., Meng, W. and Gierasch, L.M., 2015. How hsp70 molecular machines interact with their substrates to mediate diverse physiological functions. Journal of Molecular Biology 427: 1575-1588.

Collier, N.C. and Schlesinger, M.J., 1986. Induction of heat-shock proteins in the embryonic chicken lens. Experimental Eye Research 43: 103-117.

Collier, N.C. and Schlesinger, M.J., 1986. The dynamic state of heat shock proteins in chicken embryo fibroblasts. The Journal of Cell Biology 103: 1495-1507.

Csermely, P., Schnaider, T., Soti, C., Prohászka, Z. and Nardai, G., 1998. The 90-kDa molecular chaperone family: structure, function, and clinical applications. A comprehensive review. Pharmacology &Therapeutics 79: 129-168.

Czar, M.J., Welsh, M.J. and Pratt, W.B., 1996. Immunofluorescence localization of the 90-kDa heat-shock protein to cytoskeleton. European Journal of Cell Biology 70: 322-330.

Dangi, S.S., Gupta, M., Dangi, S.K., Chouhan, V.S., Maurya, V.P., Kumar, P., Singh, G. and Sarkar, M., 2015. Expression of HSPs: an adaptive mechanism during long-term heat stress in goats (*Capra hircus*). International Journal of Biometeorology 59: 1095-1106.

Daniel, S., Bradley, G., Longshaw, V.M., Söti, C., Csermely, P. and Blatch, G.L., 2008. Nuclear translocation of the phosphoprotein Hop (Hsp70/Hsp90 organizing protein) occurs under heat shock, and its proposed nuclear localization signal is involved in Hsp90 binding. Biochimica et Biophysica Acta 1783: 1003-1014.

Dash, A., Chung, S. and Zelenka, P.S., 1994. Expression of HSP70 mRNA in the embryonic chicken lens: association with differentiation. Experimental Eye Research 58: 381-387.

Daugaard, M., Rohde, M. and Jäättelä, M., 2007. The heat shock protein 70 family: Highly homologous proteins with overlapping and distinct functions. FEBS Letters 581: 3702-3710.

De Backer, O., Elinck, E., Blanckaert, B., Leybaert, L., Motterlini, R. and Lefebvre, R.A., 2009. Water-soluble CO-releasing molecules Hemeoxygenase-1 in gastrointestinal diseases reduce the development of postoperative ileus via modulation of MAPK/HO-1 signalling and reduction of oxidative stress. Gut 58: 347-356.

De Roos, B. and Duthie, G.G., 2015. Role of dietary pro-oxidants in the maintenance of health and resilience to oxidative stress. Molecular Nutrition & Food Research 59: 1229-1248.

De Thonel, A., Le Mouël, A. and Mezger, V., 2012. Transcriptional regulation of small HSP-HSF1 and beyond. The International Journal of Biochemistry & Cell Biology 44: 1593-1612.

Demir, M., Amanvermez, R., Kamalı Polat, A., Karabıçak, I., Cınar, H., Kesicioğlu, T. and Polat, C., 2014. The effect of silymarin on mesenteric ischemia-reperfusion injury. Medical Principles and Practice 23: 140-144.

Dennery, P. A., 2000. Regulation and role of heme oxygenase in oxidative injury. Current Topics in Cellular Regulation 36: 181-199.

Druyan, S., Cahaner, A. and Ashwell, C.M., 2007. The expression patterns of hypoxia-inducing factor subunit alpha-1, heme oxygenase, hypoxia upregulated protein 1, and cardiac troponin T during development of the chicken heart. Poultry Science 86: 2384-2389.

Duncan, E.J., Cheetham, M.E., Chapple, J.P. and Van der Spuy, J., 2015. The role of HSP70 and its co-chaperones in protein misfolding, aggregation and disease. Sub-cellular Biochemistry 78: 243-273.

Durante, W., 2010. Targeting heme oxygenase-1 in vascular disease. Current Drug Targets 11: 1504-1516.

Duvigneau, J.C., Esterbauer, H. and Kozlov, A.V., 2019. Role of heme oxygenase as a modulator of heme-mediated pathways. Antioxidants 8, 10: 475.

Edens, F.W., Hill, C.H. and Wang, S., 1992. Heat shock protein response in phosphorus-deficient heat-stressed broiler chickens. Comparative Biochemistry and Physiology. B, Comparative Biochemistry 103: 827-831.

El Golli, E., Hassen, W., Bouslimi, A., Bouaziz, C., Ladjimi, M.M. and Bacha, H., 2006. Induction of Hsp 70 in vero cells in response to mycotoxins cytoprotection by sub-lethal heat shock and by Vitamin E. Toxicology Letters 166: 122-130.

Elbirt, K.K., Whitmarsh, A.J., Davis, R. and Bonkovsky, H.L., 1998. Mechanism of sodium arsenite-mediated induction of heme oxygenase-1 in hepatoma cells. Role of mitogen-activated protein kinases. Journal of Biological Chemistry 273: 8922-8931.

Erlejman, A.G., Lagadari, M., Toneatto, J., Piwien-Pilipuk, G. and Galigniana, M.D., 2014. Regulatory role of the 90-kDa-heat-shock protein (Hsp90) and associated factors on gene expression. Biochimica et Biophysica Acta 1839: 71-87.

Evans, C.O., Healey, J.F., Greene, Y. and Bonkovsky, H.L., 1991. Cloning, sequencing and expression of cDNA for chick liver haem oxygenase. Comparison of avian and mammalian cDNAs and deduced proteins. Biochemical Journal 273: 659-666.

Felver-Gant, J.N., Mack, L.A., Dennis, R.L., Eicher, S.D. and Cheng, H.W., 2012. Genetic variations alter physiological responses following heat stress in 2 strains of laying hens. Poultry Science 91: 1542-1551.

Feng, Z., Liu, Z., Li, X., Jia, H., Sun, L., Tian, C., Jia, L. and Liu, J., 2010. α-Tocopherol is an effective Phase II enzyme inducer: protective effects on acrolein-induced oxidative stress and mitochondrial dysfunction in human retinal pigment epithelial cells. The Journal of Nutritional Biochemistry 21: 1222-1231.

Fernandez, M. and Bonkovsky, H.L., 2003. Vascular endothelial growth factor increases heme oxygenase-1 protein expression in the chick embryo chorioallantoic membrane. British Journal Pharmacology 139: 634-640.

Figueiredo, D., Gertler, A., Cabello, G., Decuypere, E., Buyse, J. and Dridi, S., 2007. Leptin downregulates heat shock protein-70 (HSP-70) gene expression in chicken liver and hypothalamus. Cell and Tissue Research 329: 91-101.

Fischer, C.P., Hiscock, N.J., Basu, S., Vessby, B., Kallner, A., Sjöberg, L.B., Febbraio, M.A. and Pedersen, B.K., 2006. Vitamin E isoform-specific inhibition of the exercise-induced heat shock protein 72 expression in humans. Journal of Applied Physiology 100: 1679-1687.

Franz, A. and Hoppe, T., 2018. Stressed out: mitohormesis is crossing borders. Trends in Endocrinology & Metabolism 29: 525-527.

Fredenburgh, L.E., Merz, A.A. and Cheng, S., 2015. Haeme oxygenase signalling pathway: implications for cardiovascular disease. European Heart Journal 36: 1512-1518.

Fujimoto, M. and Nakai, A., 2010. The heat shock factor family and adaptation to proteotoxic stress. FEBS Journal 277: 4112-4125.

Gabis, K.K., Gildemeister, O.S., Pepe, J.A., Lambrecht, R.W. and Bonkovsky, H.L., 1996. Induction of heme oxygenase-1 in LMH cells. Comparison of LMH cells to primary cultures of chick embryo liver cells. Biochimica et Biophysica Acta 1290: 113-120.

Gabriel, J.E., da Mota, A.F., Boleli, I.C., Macari, M. and Coutinho, L.L., 2002. Effect of moderate and severe heat stress on avian embryonic hsp70 gene expression. Growth, Development, and Aging 66: 27-33.

Gabriel, J.E., Ferro, J.A., Stefani, R.M., Ferro, M.I., Gomes, S.L. and Macari, M., 1996. Effect of acute heat stress on heat shock protein 70 messenger RNA and on heat shock protein expression in the liver of broilers. British Poultry Science 37: 443-449.

Gan, F., Ren, F., Chen, X., Lv, C., Pan, C., Ye, G., Shi, J., Shi, X., Zhou, H., Shituleni, S. A. and Huang, K., 2013. Effects of selenium-enriched probiotics on heat shock protein mRNA levels in piglet under heat stress conditions. Journal of Agricultural and Food Chemistry 61: 2385-2391.

Gan, J.K., Jiang, L.Y., Kong, L.N., Zhang, X.Q. and Luo, Q.B., 2015. Analysis of genetic diversity of the heat shock protein 70 gene on the basis of abundant sequence polymorphisms in chicken breeds. Genetics and Molecular Research 14: 1538-1545.

Gan, J.K., Zhang, D.X., He, D.L., Zhang, X.Q., Chen, Z.Y. and Luo, Q.B., 2013. Promoter methylation negatively correlated with mRNA expression but not tissue differential expression after heat stress. Genetics and Molecular Research 12: 809-819.

Garcia-Carbonero, R., Carnero, A. and Paz-Ares, L., 2013. Inhibition of HSP90 molecular chaperones: moving into the clinic. Lancet Oncology 14: e358-369.

Garnier, C., Lafitte, D., Tsvetkov, P.O., Barbier, P., Leclerc-Devin, J., Millot, J.M., Briand, C., Makarov, A.A., Catelli, M.G. and Peyrot, V., 2002. Binding of ATP to heat shock protein 90: evidence for an ATP-binding site in the C-terminal domain. The Journal of Biological Chemistry 277: 12208-12214.

Garrido, C., Brunet, M., Didelot, C., Zermati, Y., Schmitt, E. and Kroemer, G., 2006. Heat shock proteins 27 and 70: anti-apoptotic proteins with tumorigenic properties. Cell Cycle 5: 2592-2601.

Garrido, C., Gurbuxani, S., Ravagnan, L. and Kroemer, G., 2001. Heat shock proteins: endogenous modulators of apoptotic cell death. Biochemical and Biophysical Research Communications 286: 433-442.

Gaviol, H.C., Gasparino, E., Prioli, A.J. and Soares, M.A., 2008. Genetic evaluation of the HSP70 protein in the Japanese quail (*Coturnix japonica*). Genetics and Molecular Research 7: 133-139.

George, J.F., Braun, A., Brusko, T.M., Joseph, R., Bolisetty, S., Wasserfall, C.H., Atkinson, M.A., Agarwal, A. and Kapturczak, M.H., 2008. Suppression by CD4+CD25+ regulatory T cells is dependent on expression of heme oxygenase-1 in antigen-presenting cells. The American Journal of Pathology 173: 154-160.

Gildemeister, O.S., Pepe, J.A., Lambrecht, R.W. and Bonkovsky, H.L., 2001. Induction of heme oxygenase-1 by phenylarsine oxide. Studies in cultured primary liver cells. Molecular and Cellular Biochemistry 226: 17-26.

Green, M., Schuetz, T.J., Sullivan, E.K. and Kingston, R.E., 1995. A heat shock-responsive domain of human HSF1 that regulates transcription activation domain function. Molecular and Cellular Biology 15: 3354-3362.

Greene, Y.J., Healey, J.F. and Bonkovsky, H.L., 1991. Immunochemical studies of haem oxygenase. Preparation and characterization of antibodies to chick liver haem oxygenase and their use in detecting and quantifying amounts of haem oxygenase protein. The Biochemical Journal 279: 849-854.

Grenert, J.P., Johnson, B.D. and Toft, D.O., 1999. The importance of ATP binding and hydrolysis by hsp90 in formation and function of protein heterocomplexes. The Journal of Biological Chemistry 274: 17525-17533.

Gu, X.H., Hao, Y. and Wang, X.L., 2012. Overexpression of heat shock protein 70 and its relationship to intestine under acute heat stress in broilers: 2. Intestinal oxidative stress. Poultry Science 91: 790-799.

Guerriero, V. and Raynes, D.A., 1990. Synthesis of heat stress proteins in lymphocytes from livestock. Journal of Animal Science 68: 2779-2783.

Haines, D.D., Lekli, I., Teissier, P., Bak, I. and Tosaki, A., 2012. Role of haeme oxygenase-1 in resolution of oxidative stress-related pathologies: focus on cardiovascular, lung, neurological and kidney disorders. Acta Physiologica 204: 487-501.

Hao, Y. and Gu, X.H., 2014. Effects of heat shock protein 90 expression on pectoralis major oxidation in broilers exposed to acute heat stress. Poultry Science 93: 2709-2717.

Hao, Y., Gu, X.H. and Wang, X.L., 2012. Overexpression of heat shock protein 70 and its relationship to intestine under acute heat stress in broilers: 1. Intestinal structure and digestive function. Poultry Science 91: 781-789.

Henstridge, D.C., Whitham, M. and Febbraio, M.A., 2014. Chaperoning to the metabolic party: the emerging therapeutic role of heat-shock proteins in obesity and type 2 diabetes. Molecular Metabolism 3: 781-793.

Herring, G. and Gawlik, D.E., 2007. The role of stress proteins in the study of allostatic overload in birds: use and applicability to current studies in avian ecology. TheScientificWorldJournal 7: 1596-1602.

Holley, S.J. and Yamamoto, K.R., 1995. A role for Hsp90 in retinoid receptor signal transduction. Molecular Biology of the Cell 6: 1833-1842.

Hong, D.S., Banerji, U., Tavana, B., George, G.C., Aaron, J. and Kurzrock, R., 2013. Targeting the molecular chaperone heat shock protein 90 (HSP90): lessons learned and future directions. Cancer Treatment Reviews 39: 375-387.

Hoter, A., El-Sabban, M.E. and Naim, H.Y., 2018. The HSP90 family: structure, regulation, function, and implications in health and disease. International Journal of Molecular Sciences 19, 9: 2560.

Iannotti, A.M., Rabideau, D.A. and Dougherty, J.J., 1988. Characterization of purified avian 90,000-Da heat shock protein. Archives of Biochemistry and Biophysics 264: 54-60.

Igarashi, K. and Sun, J., 2006. The heme-Bach1 pathway in the regulation of oxidative stress response and erythroid differentiation. Antioxidants & Redox Signaling 8: 107-118.

Immenschuh, S, Tan, M. and Ramadori, G., 1999. Nitric oxide mediates the lipopolysaccharide dependent upregulation of the heme oxygenase-1 gene expression in cultured rat Kupffer cells. Journal of Hepatology 30: 61-69.

Inouye, S., Katsuki, K., Izu, H., Fujimoto, M., Sugahara, K., Yamada, S., Shinkai, Y., Oka, Y., Katoh, Y. and Nakai, A., 2003. Activation of heat shock genes is not necessary for protection by heat shock transcription factor 1 against cell death due to a single exposure to high temperatures. Molecular and Cellular Biology 23: 5882-5895.

Jackson, S.E., 2013. Hsp90: structure and function. Topics in Current Chemistry 328: 155-240.

Jacobs, J.M., Marek, D., Walton, H.S., Sinclair, P.R. and Sinclair, J.F., 1999. Effect of sodium arsenite on heme metabolism in cultured chick embryo hepatocytes. Archives of Biochemistry and Biophysics 371: 8-14.

Jang, J.S., Piao, S., Cha, Y-N. and Kim, C., 2009. Taurine chloramine activates Nrf2, increases HO-1 expression and protects cells from death caused by hydrogen peroxide. Journal of Clinical Biochemistry and Nutrition 45: 37-43.

Jérôme, V., Léger, J., Devin, J., Baulieu, E.E. and Catelli, M.G., 1991. Growth factors acting via tyrosine kinase receptors induce HSP90 alpha gene expression. Growth Factors 4: 317-327.

Jérôme, V., Vourc'h, C., Baulieu, E.E. and Catelli, M.G., 1993. Cell cycle regulation of the chicken hsp90 alpha expression. Experimental Cell Research 205: 44-51.

Ji, Y.L., Wang, Z., Wang, H., Zhang, C., Zhang, Y., Zhao, M., Chen, Y.H., Meng, X.H. and Xu, D.X., 2012. Ascorbic acid protects against cadmium-induced endoplasmic reticulum stress and germ cell apoptosis in testes. Reproductive Toxicology 34: 357-363.

Johnson, B.D., Chadli, A., Felts, S.J., Bouhouche, I., Catelli, M.G. and Toft, D.O., 2000. Hsp90 chaperone activity requires the full-length protein and interaction among its multiple domains. The Journal of Biological Chemistry 275: 32499-32507.

Kalmar, B. and Greensmith, L., 2009. Induction of heat shock proteins for protection against oxidative stress. Advanced Drug Delivery Reviews 61: 310-318.

Kantidze, O.L., Velichko, A.K. and Razin, S.V., 2015. Heat stress-induced transcriptional repression. Biochemistry (Moscow) 80: 990-993.

Kapturczak, M.H., Wasserfall, C., Brusko, T., Campbell-Thompson, M., Ellis, T.M., Atkinson, M.A. and Agarwal, A., 2004. Heme oxygenase-1 modulates early inflammatory responses: evidence from the heme oxygenase-1-deficient mouse. The American Journal of Pathology 165: 1045-1053.

Karagöz, G.E. and Rüdiger, S.G., 2015. Hsp90 interaction with clients. Trends in Biochemical Sciences 40: 117-125.

Kaur, P. and Bansal, M.P., 2003. Effect of oxidative stress on the spermatogenic process and hsp70 expressions in mice testes. Indian Journal of Biochemistry & Biophysics 40: 246-251.

Kaushal, N. and Bansal, M.P., 2009. Diminished reproductive potential of male mice in response to selenium-induced oxidative stress: involvement of HSP70, HSP70-2, and MSJ-1. Journal of Biochemical and Molecular Toxicology 23:125-136.

Kawazoe, Y., Tanabe, M., Sasai, N., Nagata, K. and Nakai, A., 1999. HSF3 is a major heat shock responsive factor duringchicken embryonic development. European Journal of Biochemistry 265: 688-697.

Kelley, P.M. and Schlesinger, M.J., 1978. The effect of amino acid analogues and heat shock on gene expression in chicken embryo fibroblasts. Cell 15: 1277-1286.

Kelley, P.M. and Schlesinger, M.J., 1982. Antibodies to two major chicken heat shock proteins cross-react with similar proteins in widely divergent species. Molecular and Cellular Biology 2: 267-274.

Kennedy, D., Jäger, R., Mosser, D.D. and Samali, A., 2014. Regulation of apoptosis by heat shock proteins. IUBMB Life 66: 327-338.

Khalil, A.A., Kabapy, N.F., Deraz, S.F. and Smith, C., 2011. Heat shock proteins in oncology: diagnostic biomarkers or therapeutic targets? Biochimica et Biophysica Acta 1816: 89-104.

Khassaf, M., McArdle, A., Esanu, C., Vasilaki, A., McArdle, F., Griffiths, R.D., Brodie, D.A. and Jackson, M.J., 2003. Effect of vitamin C supplements on antioxidant defence and stress proteins in human lymphocytes and skeletal muscle. The Journal of Physiology, 549: 645-652.

Khoso, P.A., Yang, Z., Liu, C. and Li, S., 2015. Selenoproteins and heat shock proteins play important roles in immunosuppression in the bursa of Fabricius of chickens with selenium deficiency. Cell Stress Chaperones 20: 967-978.

Khurana, N. and Bhattacharyya, S., 2015. Hsp90, the concertmaster: tuning transcription. Frontiers in Oncology 5: 100.

Kim, H.R., Kang, K.I., Kang, H.S. and Kim, H.D., 1999. Identification of heat shock protein 90-associated 84-kDa phosphoprotein. Journal of Biochemistry 126:1025-1032.

Kiruthiga, P.V., Karthikeyan, K., Archunan, G., Pandian, S.K. and Devi, K.P., 2015. Silymarin prevents benzo(a)pyrene-induced toxicity in Wistar rats by modulating xenobiotic-metabolizing enzymes. Toxicology and Industrial Health 31: 523-541.

Konstantinova, E.V., Chipigina, N.S., Shurdumova, M.H., Kovalenko, E.I. and Sapozhnikov, A.M., 2019. Heat shock protein 70 kDa as a target for diagnostics and therapy of cardiovascular and cerebrovascular diseases. Current Pharmaceutical Design 25: 710-714.

Kutuzova, G.D. and DeLuca, H.F., 2007. 1,25-Dihydroxyvitamin D3 regulates genes responsible for detoxification in intestine. Toxicology and Applied Pharmacology 218: 37-44.

Leandro, N.S., Gonzales, E., Ferro, J.A., Ferro, M.I., Givisiez, P.E. and Macari, M., 2004. Expression of heat shock protein in broiler embryo tissues after acute cold or heat stress. Molecular Reproduction and Development 67: 172-177.

Lee, T.S. and Chau, L.Y., 2002. Heme oxygenase-1 mediates the anti-inflammatory effect of interleukin-10 in mice. Nature Medicine 8: 240-246.

Lei, L., Yu, J. and Bao, E., 2009. Expression of heat shock protein 90 (Hsp90) and transcription of its corresponding mRNA in broilers exposed to high temperature. British Poultry Science 50: 504-511.

Li, J., Soroka, J. and Buchner, J., 2012a. The Hsp90 chaperone machinery: conformational dynamics and regulation by co-chaperones. Biochimica et Biophysica Acta 1823: 624-635.

Li, J., Sun, L., Xu, C., Yu, F., Zhou, H., Zhao, Y., Zhang, J., Cai, J., Mao, C., Tang, L., Xu, Y., and He, J., 2012. Structure insights into mechanisms of ATP hydrolysis and the activation of human heat-shock protein 90. Acta Biochimica et Biophysica Sinica 44: 300-306.

Lincoln, B.C., Aw, T.Y. and Bonkovsky, H.L., 1989. Heme catabolism in cultured hepatocytes: evidence that heme oxygenase is the predominant pathway and that a proportion of synthesized heme is converted rapidly to biliverdin. Biochimica et Biophysica Acta 992: 49-58.

Lincoln, B.C., Healey, J.F. and Bonkovsky, H.L., 1988. Regulation of hepatic haem metabolism. Disparate mechanisms of induction of haem oxygenase by drugs and metals. The Biochemical Journal 250: 189-196.

Liu, Y.T., Lin, Z.M., He, S.J. and Zuo, J.P., 2019. Heme oxygenase-1 as a potential therapeutic target in rheumatic diseases. Life Sciences 218: 205-212.

Liu, C.P., Fu, J., Xu, F.P., Wang, X.S. and Li, S., 2015. The role of heat shock proteins in oxidative stress damage induced by Se deficiency in chicken livers. Biometals 28:163-173.

Liu, L.L., He, J.H., Xie, H.B., Yang, Y.S., Li, J.C. and Zou, Y., 2014. Resveratrol induces antioxidant and heat shock protein mRNA expression in response to heat stress in black-boned chickens. Poultry Science 93: 54-62.

Liu, X., Wei, J., Peng, D.H., Layne, M.D. and Yet, S.F., 2005. Absence of heme oxygenase-1 exacerbates myocardial ischemia/reperfusion injury in diabetic mice. Diabetes 54: 778-784.

Liu, Y. and Chang, A., 2008. Heat shock response relieves ER stress. EMBO Journal 27: 1049-1059.

Lordnejad, M.R., Schliess, F., Wettstein, M. amd Häussinger, D., 2001. Modulation of the heme oxygenase HO-1 expression by hyperosmolarity and betaine in primary rat hepatocytes. Archives of Biochemistry and Biophysics 388: 285-292.

Lu, T.H., Lambrecht, R.W., Pepe, J., Shan, Y., Kim, T. and Bonkovsky, H.L., 1998. Molecular cloning, characterization, and expression of the chicken heme oxygenase-1 gene in transfected primary cultures of chick embryo liver cells. Gene 207: 177-186.

Lu, T.H., Pepe, J.A., Gildemeister, O.S., Tyrrell, R.M. and Bonkovsky, H.L., 1997. Regulation of expression of the human heme oxygenase-1 gene in transfected chick embryo liver cell cultures. Biochimica et Biophysica Acta 1352: 293-302.

Lu, T.H., Shan, Y., Pepe, J., Lambrecht, R.W. and Bonkovsky, H.L., 2000. Upstream regulatory elements in chick heme oxygenase-1 promoter: a study in primary cultures of chick embryo liver cells. Molecular and Cellular Biochemistry 209: 17-27.

Luo, Q.B., Song, X.Y., Ji, C.L., Zhang, X.Q. and Zhang, D.X., 2014. Exploring the molecular mechanism of acute heat stress exposure in broiler chickens using gene expression profiling. Gene 546: 200-205.

Maak, S., Melesse, A., Schmidt, R., Schneider, F. and Von Lengerken, G., 2003. Effect of long-term heat exposure on peripheral concentrations of heat shock protein 70 (Hsp70) and hormones in laying hens with different genotypes. British Poultry Science 44: 133-138.

Mahalka, A.K., Kirkegaard, T., Jukola, L.T., Jäättelä, M. and Kinnunen, P.K., 2014. Human heat shock protein 70 (Hsp70) as a peripheral membrane protein. Biochimica et Biophysica Acta 1838: 1344-1361.

Mahmoud, K.Z., Edens, F.W., Eisen, E.J. and Havenstein, G.B., 2004. The effect of dietary phosphorus on heat shock protein mRNAs during acute heat stress in male broiler chickens (*Gallus gallus*). Comparative Biochemistry and Physiology. Toxicology & Pharmacology 137: 11-18.

Mahmoud, K.Z. and Edens, F.W., 2003. Influence of selenium sources on age-related and mild heat stress-related changes of blood and liver glutathione redox cycle in broiler chickens (*Gallus domesticus*). Comparative Biochemistry and Physiology. Part B, Biochemistry & Molecular Biology 136: 921-934.

Mahmoud, K.Z. and Edens, F.W., 2005. Influence of organic selenium on hsp70 response of heat-stressed and enteropathogenic Escherichia coli-challenged broiler chickens (*Gallus gallus*). Comparative Biochemistry and Physiology. Toxicology & Pharmacology 141: 69-75.

Mahmoud, K.Z., Edens, F.W., Eisen, E.J. and Havenstein, G.B., 2004a. Ascorbic acid decreases heat shock protein 70 and plasma corticosterone response in broilers (*Gallus gallus domesticus*) subjected to cyclic heat. Comparative Biochemistry and Physiology. Part B, Biochemistry & Molecular Biology 137: 35-42.

Mahmoud, K.Z., Edens, F.W., Eisen, E.J. and Havenstein, G.B., 2003. Effect of ascorbic acid and acute heat exposure on heat shock protein 70 expression by young white Leghorn chickens. Comparative Biochemistry and Physiology. Toxicology & Pharmacology 136: 329-335.

Maines, M.D. and Sinclair, P., 1977. Cobalt regulation of heme synthesis and degradation in avian embryo liver cell culture. The Journal of Biological Chemistry 252: 219-223.

Marcu, M.G., Chadli, A., Bouhouche, I., Catelli, M. and Neckers, L.M., 2000. The heat shock protein 90 antagonist novobiocin interacts with a previously unrecognized ATP-binding domain in the carboxyl terminus of the chaperone. The Journal of Biological Chemistry 275: 37181-37186.

Mattson, M.P. and Cheng, A., 2006. Neurohormetic phytochemicals: Low-dose toxins that induce adaptive neuronal stress responses. Trends in Neurosciences 29: 632-639.

Mayer, M.P. and Gierasch, L.M., 2019. Recent advances in the structural and mechanistic aspects of Hsp70 molecular chaperones. The Journal of Biological Chemistry 294: 2085-2097.

Mayer, M.P., 2013. Hsp70 chaperone dynamics and molecular mechanism. Trends in Biochemical Sciences 38: 507-514.

Mayer, M.P. and Le Breton, L., 2015. Hsp90: breaking the symmetry. Molecular Cell 58: 8-20.

Meijering, R.A., Henning, R.H. and Brundel, B.J., 2015. Reviving the protein quality control system: therapeutic target for cardiac disease in the elderly. Trends in Cardiovascular Medicine 25: 243-247.

Meimaridou, E., Gooljar, S.B. and Chapple, J.P., 2009. From hatching to dispatching: the multiple cellular roles of the Hsp70 molecular chaperone machinery. Journal of Molecular Endocrinology 42: 1-9.

Meng, X., Baulieu, E.E. and Catelli, M.G., 1995. Isolation of chicken hsp90 beta gene promoter. Biochemical and Biophysical Research Communications 206: 644-651.

Meng, X., Jérôme, V., Devin, J., Baulieu, E.E. and Catelli, M.G., 1993. Cloning of chicken hsp90 beta: the only vertebrate hsp90 insensitive to heat shock. Biochemical and Biophysical Research Communications 190: 630-636.

Mezquita, B., Mezquita, C. and Mezquita, J., 1998. Marked differences between avian and mammalian testicular cells in the heat shock induction and polyadenylation of Hsp70 and ubiquitin transcripts. FEBS Letters 436: 382-386.

Miller, L. and Qureshi, M.A., 1992. Heat-shock protein synthesis in chicken macrophages: influence of and *in vitro* heat shock, lead acetate, and lipopolysaccharide. Poultry Science 71: 988-998.

Miller, L. and Qureshi, M.A., 1992a. Comparison of heat-shock-induced and lipopolysaccharide-induced protein changes and tumoricidal activity in a chicken mononuclear cell line. Poultry Science 71: 979-987.

Miller, L. and Qureshi, M.A., 1992c. Induction of heat-shock proteins and phagocytic function of chicken macrophage following *in vitro* heat exposure. Veterinary Immunology and Immunopathology 30: 179-191.

Miller, L. and Qureshi. M.A., 1992b. Molecular changes associated with heat-shock treatment in avian mononuclear and lymphoid lineage cells. Poultry Science 71: 473-481.

Mulyani, W., Sanjiwani, M., Sandra, Prabawa, I., Lestari, A., Wihandani, D.M., Suastika, K., Saraswati, M.R., Bhargah, A. and Manuaba, I., 2020. Chaperone-based therapeutic target innovation: heat shock protein 70 (HSP70) for type 2 diabetes mellitus. Diabetes, metabolic Syndrome and Obesity: Targets and Therapy 13: 559-568.

Morán Luengo, T., Mayer, M.P. and Rüdiger, S., 2019. The Hsp70-Hsp90 chaperone cascade in protein folding. Trends in Cell Biology 29: 164-177.

Moreau, A., Hill, M., Thebault, P., 2009. Tolerogenic dendritic cells actively inhibit T cells through heme oxygenase-1 in rodents and in nonhuman primates. FASEB Journal 23: 3070-3077.

Morimoto, R. and Fodor, E., 1984. Cell-specific expression of heat shock proteins in chicken reticulocytes and lymphocytes. The Journal of Cell Biology 99: 1316-1323.

Morimoto, R.I., Hunt, C., Huang, S.Y., Berg, K.L. and Banerji, S.S., 1986. Organization, nucleotide sequence, and transcription of the chicken HSP70 gene. The Journal of Biological Chemistry 261: 12692-12699.

Morse, D., Lin, L., Choi, A.M. and Ryter, S.W., 2009. Heme oxygenase-1, a critical arbitrator of cell death pathways in lung injury and disease. Free Radical Biology & Medicine, 47: 1-12.

Mueller, K., Blum, N.M., Kluge, H. and Mueller, A.S., 2012. Influence of broccoli extract and various essential oils on performance and expression of xenobiotic and antioxidant enzymes in broiler chickens. The British Journal of Nutrition 108: 588-602.

Murakami, A., 2013. Modulation of protein quality control systems by food phytochemicals. Journal of Clinical Biochemistry and Nutrition 52: 215-227.

Murakami, A., 2014. Dose-dependent functionality and toxicity of green tea polyphenols in experimental rodents. Archives of Biochemistry and Biophysics 557: 3-10.

Muthumani, M. and Prabu, S.M., 2014. Silibinin potentially attenuates arsenic-induced oxidative stress mediated cardiotoxicity and dyslipidemia in rats. Cardiovascular Toxicology 14: 83-97.

Nagahori, K., Iwamoto, S., Maruyama, A., Suzuki, S., Hosomichi, K., Shiina, T., Hara, H., Yoshida, Y. and Hanzawa, K., 2010. Basic characterization of 90 kDa heat shock protein genes HSP90AA1, HSP90AB1, HSP90B1 and TRAP1 expressed in Japanese quail (*Coturnix japonica*). Animal Science Journal 81: 513-518.

Najafi, P., Zulkifli, I. and Soleimani, A. F., 2018. Inhibition of corticosterone synthesis and its effect on acute phase proteins, heat shock protein 70, and interleukin-6 in broiler chickens subjected to feed restriction. Poultry Science 97: 1441-1447.

Nakai, A. and Ishikawa, T., 2000. A nuclear localization signal is essential for stress-induced dimer-to-trimer transition of heat shock transcription factor 3. The Journal of Biological Chemistry 275: 34665-34671.

Nakai, A. and Ishikawa, T., 2001. Cell cycle transition under stress conditions controlled by vertebrate heat shock factors. EMBO Journal 20: 2885-2895.

Nakai, A. and Morimoto, R.I., 1993. Characterization of a novel chicken heat shock transcription factor, heat shock factor 3, suggests a new regulatory pathway. Molecular and Cellular Biology 13: 1983-1997.

Nakai, A., Kawazoe, Y., Tanabe, M., Nagata, K. and Morimoto, R.I., 1995. The DNA-binding properties of two heat shock factors, HSF1 and HSF3, are induced in the avian erythroblast cell line HD6. Molecular and Cellular Biology 15: 5268-5278.

Nakamichi, I., Habtezion, A., Zhong, B., Contag, C.H., Butcher, E.C. and Omary, M.B., 2005. Hemin-activated macrophages home to the pancreas and protect from acute pancreatitis via heme oxygenase-1 induction. The Journal of Clinical Investigation 115: 3007-3014.

Ndisang, J.F., 2014. Cross-talk between heme oxygenase and peroxisome proliferator-activated receptors in the regulation of physiological functions. Frontiers in Bioscience 19: 916-935.

Niess, A.M., Passek, F., Lorenz, I., Schneider, E.M., Dickhuth, H.H., Northoff, H. and Fehrenbach, E., 1999. Expression of the antioxidant stress protein heme oxygenase-1 (HO-1) in human leukocytes: acute and adaptational responses to endurance exercise. Free Radical Biology and Medicine 26: 184-192.

Oermann, E., Bidmon, H.J., Witte, O.W. and Zilles, K., 2004. Effects of 1alpha,25 dihydroxyvitamin D3 on the expression of HO-1 and GFAP in glial cells of the photothrombotically lesioned cerebral cortex. Journal of Chemical Neuroanatomy 28: 225-238.

Oksala, N.K., Ekmekçi, F.G., Ozsoy, E., Kirankaya, S., Kokkola, T., Emecen, G., Lappalainen, J., Kaarniranta, K. and Atalay, M., 2014. Natural thermal adaptation increases heat shock protein levels and decreases oxidative stress. Redox Biology 3: 25-28.

O'Neill, S., Harrison, E.M., Ross, J.A., Wigmore, S.J. and Hughes, J., 2014. Heat-shock proteins and acute ischaemic kidney injury. Nephron. Experimental Nephrology 126: 167-174.

Oskoueian, E., Abdullah, N., Idrus, Z., Ebrahimi, M., Goh, Y.M., Shakeri, M. and Oskoueian, A., 2014. Palm kernel cake extract exerts hepatoprotective activity in heat-induced oxidative stress in chicken hepatocytes. BMC Complementary and Alternative Medicine 14: 368.

Owen, B.A., Sullivan, W.P., Felts, S.J. and Toft, D.O., 2002. Regulation of heat shock protein 90 ATPase activity by sequences in the carboxyl terminus. The Journal of Biological Chemistry 277: 7086-7091.

Pae, H.O. and Chung. H.T., 2009. Heme oxygenase-1: its therapeutic roles in inflammatory diseases. Immune Network 9, 12-19.

Passinen, S., Valkila, J., Manninen, T., Syvälä, H. and Ylikomi, T., 2001. The C-terminal half of Hsp90 is responsible for its cytoplasmic localization. European Journal of Biochemistry 268: 5337-5342.

Pekki, A.K., 1991. Different immunoelectron microscopic locations of progesterone receptor and HSP90 in chick oviduct epithelial cells. The Journal of Histochemistry and Cytochemistry 39: 1095-1101.

Peng, Y.Z., Wang, Y.W., Ning, D. and Guo, Y.M., 2013. Changes of haematic parameters, redox status and mitochondrial complex activity in the heart and liver of broilers fed with different density diets under low ambient temperature. Avian Pathology 42: 327-334.

Pockley, A.G. and Multhoff, G., 2008. Cell stress proteins in extracellular fluids: friend or foe? Novartis Foundation Symposium 291: 86-95.

Poss, K.D. and Tonegawa, S., 1997. Reduced stress defense in heme oxygenase 1-deficient cells. Proceedings of the National Academy of Sciences of the United States of America, 94: 10925-10930.

Prakasam, R., Fujimoto, M., Takii, R., Hayashida, N., Takaki, E., Tan, K., Wu, F., Inouye, S., and Nakai, A., 2013. Chicken IL-6 is a heat-shock gene. FEBS letters 587: 3541-3547.

Prasad, A.R. and Datta, K., 1984. Altered regulation of hepatic heme metabolism in cadmium exposed chick embryo. Biochemistry International 8: 289-298.

Pratt, W.B., Morishima, Y., Peng, H.M. and Osawa, Y., 2010. Proposal for a role of the Hsp90/Hsp70-based chaperone machinery in making triage decisions when proteins undergo oxidative and toxic damage. Experimental Biology and Medicine 235: 278-289.

Qu, B., Jia, Y., Liu, Y., Wang, H., Ren, G. and Wang, H., 2015. The detection and role of heat shock protein 70 in various nondisease conditions and disease conditions: a literature review. Cell Stress & Chaperones 20: 885-892.

Radanyi, C., Renoir, J.M., Sabbah, M. and Baulieu, E.E., 1989. Chick heat-shock protein of Mr =90,000, free or released from progesterone receptor, is in a dimeric form. The Journal of Biological Chemistry 264: 2568-2573.

Ravagnan, L., Gurbuxani, S., Susin, S.A., Maisse, C., Daugas, E., Zamzami, N., Mak, T., Jäättelä, M., Penninger, J.M., Garrido, C. and Kroemer, G., 2001. Heat-shock protein 70 antagonizes apoptosis-inducing factor. Nature Cell Biology 3: 839-843.

Reed, D.K., Hall, S. and Arany, I., 2015. α-tocopherol protects renal cells from nicotine- or oleic acid-provoked oxidative stress via inducing heme oxygenase-1. Journal of Physiology and Biochemistry 71: 1-7.

Remy, S., Blancou, P. and Tesson, L., 2009. Carbon monoxide inhibits TLR-induced dendritic cell immunogenicity. Journal of Immunology 182. 1877-1884.

Revathi, B. and Prashanth, K., 2015. Potential Hsp90 inhibitors: a novel target for cancer therapy. Chemotherapy 4: 146.

Richter, K., Haslbeck, M. and Buchner, J., 2010. The heat shock response: life on the verge of death. Molecular Cell 40: 253-266.

Ritossa, F., 1962. A new puffing pattern induced by temperature shock and DNP in Drosophila. Experientia 18: 571-573.

Ritossa F., 1996. Discovery of the heat shock response. Cell Stress & Chaperones 1: 97-98.

Rivera, R.E., Christensen, V.L., Edens, F.W. and Wineland, M.J., 2005. Influence of selenium on heat shock protein 70 expression in heat stressed turkey embryos (*Meleagris gallopavo*). Comparative Biochemistry and Physiology. Part A, Molecular & Integrative Physiology 142: 427-432.

Rosenzweig, R., Nillegoda, N.B., Mayer, M.P. and Bukau, B., 2019. The Hsp70 chaperone network. Nature Reviews. Molecular cell biology 20: 665-680.

Roth, C.L., Elfers, C.T., Figlewicz, D.P., Melhorn, S.J., Morton, G.J., Hoofnagle, A., Yeh, M.M., Nelson, J.E. and Kowdley, K.V., 2012. Vitamin D deficiency in obese rats exacerbates nonalcoholic fatty liver disease and increases hepatic resistin and Toll-like receptor activation. Hepatology 55: 1103-1111.

Ryter, S.W., 2019. Heme oxygenase-1/carbon monoxide as modulators of autophagy and inflammation. Archives of Biochemistry and Biophysics 678: 108186.

Ryter, S.W. and Choi, A.M., 2016. Targeting heme oxygenase-1 and carbon monoxide for therapeutic modulation of inflammation. Translation Research 167: 7-34.

Sahin, N., Tuzcu, M., Orhan, C., Onderci, M., Eroksuz, Y. and Sahin, K., 2009. The effects of vitamin C and E supplementation on heat shock protein 70 response of ovary and brain in heat-stressed quail. British Poultкн Science 50: 259-265.

Sakurai, H. and Enoki, Y., 2010. Novel aspects of heat shock factors: DNA recognition, chromatin modulation and gene expression. FEBS Journal 277: 4140-4149.

Salinas, M., Wang, J., Rosa de Sagarra, M., Martín, D., Rojo, A.I., Martin-Perez, J., Ortiz de Montellano, P.R. and Cuadrado, A., 2004. Protein kinase Akt/PKB phosphorylates heme oxygenase-1 *in vitro* and *in vivo*. FEBS Letters 578: 90-94.

Sánchez-Moreno, C., Paniagua, M., Madrid, A. and Martín, A., 2003. Protective effect of vitamin C against the ethanol mediated toxic effects on human brain glial cells. The Journal of Nutritional Biochemistry 14: 606-613.

Sardana, M.K., Drummond, G.S., Sassa, S. and Kappas, A., 1981. The potent heme oxygenase inducing action of arsenic and parasiticidal arsenicals. Pharmacology 23: 247-253.

Sardana, M.K., Sassa, S. and Kappas, A., 1982. Metal ion-mediated regulation of heme oxygenase induction in cultured avian liver cells. The Journal of Biological Chemistry 257: 4806-4811.

Sardana, M.K., Sassa, S. and Kappas, A., 1985. Hormonal regulation of heme oxygenase induction in avian hepatocyte culture. Biochemical Pharmacology 34: 2937-2944.

Sass, J.B. and Krone, P.H., 1997. HSP90alpha gene expression may be a conserved feature of vertebrate somitogenesis. Experimental Cell Research 233: 391-394.

Schipper, H.M. and Song, W., 2015. A heme oxygenase-1 transducer model of degenerative and developmental brain disorders. International Journal of Jolecular Sciences 16: 5400-5419.

Selim, M.E., Rashedel, H.A., Aleisa, N.A. and Daghestani, M.H., 2012. The protection role of heat shock protein 70 (HSP-70) in the testes of cadmium-exposed rats. Bioinformation 8: 58-64.

Semenza, G.L., 2010. Hypoxia-inducible factor 1: regulator of mitochondrial metabolism and mediator of ischemic preconditioning. Biochimica et Biophysica Acta 1813: 1263-1268.

Shabtay, A. and Arad, Z., 2006. Reciprocal activation of HSF1 and HSF3 in brain and blood tissues: is redundancy developmentally related? American journal of physiology. Regulatory, Integrative and Comparative Physiology 291: R566-R572.

Shan, Y., Lambrecht, R.W. and Bonkovsky, H.L., 2004. Identification of key elements that are responsible for heme-mediated induction of the avian heme oxygenase-1 gene. Biochimica et Biophysica Acta 1679: 87-94.

Shan, Y., Lambrecht, R.W., Hong Lu, T. and Bonkovsky, H.L., 1999. Effects of phenylarsine oxide on expression of heme oxygenase-1 reporter constructs in transiently transfected cultures of chick embryo liver cells. Archives of Biochemistry and Biophysics 372: 224-229.

Shan, Y., Pepe, J., Lambrecht, R.W. and Bonkovsky, H.L., 2002. Mapping of the chick heme oxygenase-1 proximal promoter for responsiveness to metalloporphyrins. Archives of Biochemistry and Biophysics 399: 159-166.

Shiber, A. and Ravid, T., 2014. Chaperoning proteins for destruction: diverse roles of Hsp70 chaperones and their co-chaperones in targeting misfolded proteins to the proteasome. Biomolecules 4: 704-724.

Shih, P.K., Chen, Y.C., Huang, Y.C., Chang, Y.T., Chen, J.X. and Cheng, C.M., 2011. Pretreatment of vitamin D3 ameliorates lung and muscle injury induced by reperfusion of bilateral femoral vessels in a rat model. The Journal of Surgical Research 171: 323-328.

Shinkawa, T., Tan, K., Fujimoto, M., Hayashida, N., Yamamoto, K., Takaki, E., Takii, R., Prakasam, R., Inouye, S., Mezger, V. and Nakai, A., 2011. Heat shock factor 2 is required for maintaining proteostasis against febrile-range thermal stress and polyglutamine aggregation. Molecular Biology of the Cell 22: 3571-3583.

Shinozaki, F., Minami, M., Chiba, T., Suzuki, M., Yoshimatsu, K., Ichikawa, Y., Terasawa, K., Emori, Y., Matsumoto, K., Kurosaki, T., Nakai, A., Tanaka, K. and Minami, Y., 2006. Depletion of hsp90beta induces multiple defects in B cell receptor signaling. The Journal of Biological Chemistry 281: 16361-16369.

Simar, D., Malatesta, D., Mas, E., Delage, M. and Caillaud, C., 2012. Effect of an 8-weeks aerobic training program in elderly on oxidative stress and HSP72 expression in leukocytes during antioxidant supplementation. J. Nutr. Health Aging 16, 155-161.

Soares, M.P. and Bach, F.H., 2009. Heme oxygenase-1: from biology to therapeutic potential. Trends in Molecular Medicine 15: 50-58.

Soares, M.P., Usheva, A., Brouard, S., Berberat, P.O., Gunther, L., Tobiasch, E. and Bach, F.H., 2002. Modulation of endothelial cell apoptosis by heme oxygenase-1-derived carbon monoxide. Antioxidants & Redox Signaling 4: 321-329.

Soleimani, A.F., Zulkifli, I., Omar, A.R. and Raha, A.R., 2011. Physiological responses of 3 chicken breeds to acute heat stress. Poultry Science 90: 1435-1440.

Soleimani, A.F., Zulkifli, I., Omar, A.R. and Raha, A.R., 2012. The relationship between adrenocortical function and Hsp70 expression in socially isolated Japanese quail. Comparative Biochemistry and Physiology. Part A, Molecular & Integrative Physiology 161: 140-144.

Sreedhar, A.S., Kalmár, E., Csermely, P. and Shen, Y.F., 2004. Hsp90 isoforms: functions, expression and clinical importance. FEBS Letters 562: 11-15.

Stetler, R.A., Gan, Y., Zhang, W., Liou, A.K., Gao, Y., Cao, G. and Chen, J., 2010. Heat shock proteins: cellular and molecular mechanisms in the central nervous system. Progress in Neurobiology 92: 184-211

Sun, H., Jiang, R., Xu, S., Zhang, Z., Xu, G., Zheng, J. and Qu, L., 2015. Transcriptome responses to heat stress in hypothalamus of a meat-type chicken. Journal of Animal Science and Biotechnology 6,1: 6.

Surai, P.F., 1999. Tissue-specific changes in the activities of antioxidant enzymes during the development of the chicken embryo. British Poultry Science 40: 397-405.

Surai, P.F., 1999a. Vitamin E in avian reproduction. Poultry and Avian Biology Reviews 10: 1-60.

Surai, P.F., 2002. Natural antioxidants in avian nutrition and reproduction. Nottingham University Press, Nottingham, UK.

Surai, P.F., 2006. Selenium in nutrition and health. Nottingham University Press, Nottingham, UK.

Surai, P. F., 2014. Polyphenol compounds in the chicken/animal diet: from the past to the future. Journal of Animal Physiology and Animal Nutrition 98: 19-31.

Surai, P.F., 2015. Antioxidant action of carnitine: Molecular mechanisms and practical applications. EC Veterinary Science 2.1: 66-84.

Surai, P.F., 2015a. Silymarin as a natural antioxidant: an overview of the current evidence and perspectives. Antioxidants 4: 204-247.

Surai, P.F., 2015b. Antioxidant systems in poultry biology: heat shock proteins. Journal of Science 5: 1188-1222.

Surai, P.F., 2015c. Carnitine enigma: from antioxidant action to vitagene regulation. Part 1. Absorption, metabolism and antioxidant activities. Journal of Veterinary Science & Medicine 3, 2: 14.

Surai P.F., 2015d. Carnitine enigma: from antioxidant action to vitagene regulation. Part 2. Transcription factors and practical applications. Journal of Veterinary Science & Medicine 3, 2: 17.

Surai, P.F., 2018. selenium in poultry nutrition and health. Wageningen Academic Publishers, Wageningen, The Netherlands.

Surai, P.F. and Fisinin, V.I., 2014. Antioxidant systems of the body: from vitamin E to polyphenols and beyond. Proceedings of the 35th Western Nutrition Conference, Edmonton, Alberta, 265-277.

Surai, P.F. and Fisinin, V.I., 2015. Antioxidant-prooxidant balance in the intestine: applications in chick placement and pig Weaning. Journal of Veterinary Science & Medicine 3, 1: 16.

Surai, P.F. and Kochish, I.I., 2017. Antioxidant systems and vitagenes in poultry biology: heat shock proteins. In: Asea, A.A.A. and Kaur, P. (ed.) Heat shock proteins in veterinary. Springer, Switzerland, pp. 123-177.

Surai, P.F., Noble, R.C. and Speake, B.K., 1996. Tissue-specific differences in antioxidant distribution and susceptibility to lipid peroxidation during development of the chick embryo. Biochimica et Biophysica Acta 1304: 1-10.

Taipale, M., Jarosz, D.F. and Lindquist, S., 2010. HSP90 at the hub of protein homeostasis: emerging mechanistic insights. Nature Reviews. Molecular Cell Biology 11: 515-528.

Takii, R., Fujimoto, M., Tan, K., Takaki, E., Hayashida, N., Nakato, R., Shirahige, K. and Nakai, A., 2015. ATF1 modulates the heat shock response by regulating the stress-inducible heat shock factor 1 transcription complex. Molecular and Cellular Biology 35: 11-25.

Tanabe, M., Kawazoe, Y., Takeda, S., Morimoto, R.I., Nagata, K. and Nakai, A., 1998. Disruption of the HSF3 gene results in the severe reduction of heat shock gene expression and loss of thermotolerance. EMBO Journal 17: 1750-1758.

Tanabe, M., Nakai, A., Kawazoe, Y. and Nagata, K., 1997. Different thresholds in the responses of two heat shock transcription factors, HSF1 and HSF3. The Journal of Biological Chemistry 272: 15389-15395.

Terry, C.M., Clikeman, J.A., Hoidal, J.R. and Callahan, K.S., 1998. Effect of tumor necrosis factor-alpha and interleukin-1 alpha on heme oxygenase-1 expression in human endothelial cells. The American Journal of Physiology 274: H883-H891.

Tian, Y., Merkwirth, C. and Dillin, A., 2016. Mitochondrial UPR: a double-edged sword. Trends in Cell Biology 26: 563-565.

Trinklein, N.D., Chen, W.C., Kingston, R.E. and Myers, R.M., 2004. Transcriptional regulation and binding of heat shock factor 1 and heat shock factor 2 to 32 human heat shock genes during thermal stress and differentiation. Cell Stress Chaperones 9: 21-28.

True, A.L., Olive, M., Boehm, M., San, H., Westrick, R.J., Raghavachari, N., Xu, X., Lynn, E.G., Sack, M.N., Munson, P.J., Gladwin, M.T. and Nabel, E.G., 2007. Heme oxygenase-1 deficiency accelerates formation of arterial thrombosis through oxidative damage to the endothelium, which is rescued by inhaled carbon monoxide. Circulation Research 101: 893-901.

Tzima, S., Victoratos, P., Kranidioti, K., Alexiou, M. and Kollias, G., 2009. Myeloid heme oxygenase-1 regulates innate immunity and autoimmunity by modulating IFN-beta production. The Journal of Experimental Medicine 206: 1167-1179.

Vachharajani, T.J., Work, J., Issekutz, A.C. and Granger, D.N., 2000. Heme oxygenase modulates selectin expression in different regional vascular beds. American Journal of Physiology. Heart and Circulatory Physiology 278: H1613-H1617.

Velichko, A.K., Markova, E.N., Petrova, N.V., Razin, S.V. and Kantidze, O.L., 2013. Mechanisms of heat shock response in mammals. Cellular and Molecular Life Sciences 70: 4229-4241.

Venditti, C.C. and Smith, G.N., 2014. Involvement of the heme oxygenase system in the development of preeclampsia and as a possible therapeutic target. Women's Health 10: 623-643.

Vihervaara, A. and Sistonen, L., 2014. HSF1 at a glance. Journal of Cell Science 127: 261-266.

Voellmy, R. and Bromley, P.A., 1982. Massive heat-shock polypeptide synthesis in late chicken embryos: convenient system for study of protein synthesis in highly differentiated organisms. Molecular and Cellular Biology 2: 479-483.

Voellmy, R., Bromley, P. and Kocher, H.P., 1983. Structural similarities between corresponding heat-shock proteins from different eukaryotic cells. The Journal of Biological Chemistry 258: 3516-3522.

Vourch, C., Binart, N., Chambraud, B., David, J. P., Jérôme, V., Baulieu, E. E. and Catelli, M. G., 1989. Isolation and functional analysis of chicken 90-kDa heat shock protein gene promoter. Nucleic Acids Research 17: 5259-5272.

Wang, Y., Liu, J., Chen, R., Qi, M., Tao, D. and Xu, S., in press. The antagonistic effects of selenium yeast (SeY) on cadmium-induced inflammatory factors and the heat shock protein expression levels in chicken livers. Biological Trace Element Research. https://doi.org/10.1007/s12011-020-02039-5

Wang, S.H., Cheng, C.Y., Chen, C.J., Chen, H.H., Tang, P.C., Chen, C.F., Lee, Y.P. and Huang, S.Y., 2014. Changes in protein expression in testes of L2 strain Taiwan country chickens in response to acute heat stress. Theriogenology 82: 80-94.

Wang, S.H., Cheng, C.Y., Tang, P.C., Chen, C.F., Chen, H.H., Lee, Y.P. and Huang, S.Y., 2013. Differential gene expressions in testes of L2 strain Taiwan country chicken in response to acute heat stress. Theriogenology 79: 374-282.

Wayne, N., Mishra, P. and Bolon, D.N., 2011. Hsp90 and client protein maturation. Methods in Molecular Biology 787: 33-44.

Wegiel, B., Hedblom, A., Li, M., Gallo, D., Csizmadia, E., Harris, C., Nemeth, Z., Zuckerbraun, B.S., Soares, M., Persson, J.L. and Otterbein, L.E., 2014. Heme oxygenase-1 derived carbon monoxide permits maturation of myeloid cells. Cell Death & Disease 5, 3: e1139.

Wegiel, B., Nemeth, Z., Correa-Costa, M., Bulmer, A.C. and Otterbein, L.E., 2014. Heme oxygenase-1: a metabolic nike. Antioxidants & Redox Signaling 20: 1709-1722.

White, C.N. and Hightower, L.E., 1984. Stress mRNA metabolism in canavanine-treated chicken embryo cells. Molecular and Cellular Biology 4: 1534-1541.

Whitesell, L. and Lindquist, S.L., 2005. HSP90 and the chaperoning of cancer. Nature reviews. Cancer 5: 761-772.

Wijayanti, N., Huber, S., Samoylenko, A., Kietzmann, T. and Immenschuh, S., 2004. Role of NF-kappaB and p38 MAP kinase signaling pathways in the lipopolysaccharide-dependent activation of heme oxygenase 1 gene expression. Antioxidants & Redox Signaling 6: 802-810.

Wolfe, M.S. and Zatz, M., 1994. Synthesis of heat shock proteins in cultured chick pineal cells. Brain Research 662: 273-277.

Wu, M.L., Ho, Y.C., Lin, C.Y. and Yet, S.F., 2011. Heme oxygenase-1 in inflammation and cardiovascular disease. American Journal of Cardiovascular Disease 1: 150-158.

Xia, M., Gan, J., Luo, Q., Zhang, X. and Yang, G., 2013. Identification of duck HSP70 gene, polymorphism analysis and tissue expression under control and heat stress conditions. British Poultry Science 54: 562-566.

Xie, J., Tang, L., Lu, L., Zhang, L., Xi, L., Liu, H.C., Odle, J. and Luo, X., 2014. Differential expression of heat shock transcription factors and heat shock proteins after acute and chronic heat stress in laying chickens (*Gallus gallus*). PloS One 9, 7: e102204.

Xu, D., Li, W., Huang, Y., He, J. and Tian, Y., 2014. The effect of selenium and polysaccharide of *Atractylodes macrocephala* Koidz. (PAMK) on immune response in chicken spleen under heat stress. Biological Trace Element Research 160: 232-237.

Xu, S., Chen, Y.H., Tan, Z.X., Xie, D.D., Zhang, C., Xia, M.Z., Wang, H., Zhao, H., Xu, D.X. and Yu, D.X., 2015. Vitamin D3 pretreatment alleviates renal oxidative stress in lipopolysaccharide-induced acute kidney injury. The Journal of Steroid Biochemistry and Molecular Biology 152: 133-141.

Xu, Z., Wang, Z., Li, J.J., Chen, C., Zhang, P.C., Dong, L., Chen, J.H., Chen, Q., Zhang, X.T. and Wang, Z.L., 2013. Protective effects of selenium on oxidative damage and oxidative stress related gene expression in rat liver under chronic poisoning of arsenic. Food and chemical toxicology 58: 1-7.

Yang, Z., Liu, C., Zheng, W., Teng, X. and Li, S., 2016. The functions of antioxidants and heat shock proteins are altered in the immune organs of selenium-deficient broiler chickens. Biological Trace Element Research 169: 341-351.

Yang, X., Zhao, H., Wang, Y., Liu, J., Guo, M., Fei, D., Mu, M. and Xing, M., 2020. The activation of heat-shock protein after copper(II) and/or arsenic(III)-induced imbalance of homeostasis, inflammatory response in chicken rectum. Biological Trace Element Research 195: 613-623.

Yazama, F., Furuta, K., Fujimoto, M., Sonoda, T., Shigetomi, H., Horiuchi, T., Yamada, M., Nagao, N. and Maeda, N., 2006. Abnormal spermatogenesis in mice unable to synthesize ascorbic acid. Anatomical Science International 81: 115-125.

Yet, S.F., Perrella, M.A., Layne, M.D., Hsieh, C.M., Maemura, K., Kobzik, L., Wiesel, P., Christou, H., Kourembanas, S. and Lee, M.E., 1999. Hypoxia induces severe right ventricular dilatation and infarction in heme oxygenase-1 null mice. The Journal of Clinical Investigation 103: R23-R29.

Yousuf, S., Atif, F., Ahmad, M., Hoda, M.N., Khan, M.B., Ishrat, T. and Islam, F., 2007. Selenium plays a modulatory role against cerebral ischemia-induced neuronal damage in rat hippocampus. Brain Research 1147: 218-225.

Yu, J., Bao, E., Yan, J. and Lei, L., 2008. Expression and localization of Hsps in the heart and blood vessel of heat-stressed broilers. Cell Stress Chaperones 13: 327-335.

Yun, S.H., Moon, Y.S., Sohn, S.H. and Jang, I.S., 2012. Effects of cyclic heat stress or vitamin C supplementation during cyclic heat stress on HSP70, inflammatory cytokines, and the antioxidant defense system in Sprague Dawley rats. Experimental Animals 61: 543-553.

Zahir, F., Rabbani, G., Khan, R.H., Rizvi, S.J., Jamal, M.S. and Abuzenadah, A.M., 2015. The pharmacological features of bilirubin: the question of the century. Cellular & Molecular Biology Letters 20: 418-447.

Zhang, B., Tanaka, J., Yang, L., Yang, L., Sakanaka, M., Hata, R., Maeda, N. and Mitsuda, N., 2004. Protective effect of vitamin E against focal brain ischemia and neuronal death through induction of target genes of hypoxia-inducible factor-1. Neuroscience 126: 433-440.

Zhang, H. and Burrows, F., 2004. Targeting multiple signal transduction pathways through inhibition of Hsp90. Journal of Molecular Medicine 82: 488-499.

Zhang, W.W., Kong, L.N., Zhang, X.Q. and Luo, Q.B., 2014. Alteration of HSF3 and HSP70 mRNA expression in the tissues of two chicken breeds during acute heat stress. Genetics and Molecular Research 13: 9787-9794.

Zhao, B., Fei, J., Chen, Y., Ying, Y.L., Ma, L., Song, X.Q., Huang, J., Chen, E.Z. and Mao, E.Q., 2014. Vitamin C treatment attenuates hemorrhagic shock related multi-organ injuries through the induction of heme oxygenase-1. BMC Complementary and Alternative Medicine 14: 442.

Zhao, H., Brandt, G.E., Galam, L., Matts, R.L. and Blagg, B.S., 2011. Identification and initial SAR of silybin: an Hsp90 inhibitor. Bioorganic & Medicinal Chemistry Letters 21: 2659-2664.

Zhen, F.S., Du, H.L., Xu, H.P., Luo, Q.B. and Zhang, X.Q., 2006. Tissue and allelic-specific expression of hsp70 gene in chickens: basal and heat-stress-induced mRNA level quantified with real-time reverse transcriptase polymerase chain reaction. British Poultry Science 47: 449-455.

Zhong, W., Gu, B., Gu, Y., Groome, L.J., Sun, J. and Wang, Y., 2014. Activation of vitamin D receptor promotes VEGF and CuZn-SOD expression in endothelial cells. The Journal of Steroid Biochemistry and Molecular Biology 140: 56-62.

Zhu, L., Kong, M., Han, Y.P., Bai, L., Zhang, X., Chen, Y., Zheng, S., Yuan, H. and Duan, Z., 2015. Spontaneous liver fibrosis induced by long term dietary vitamin D deficiency in adult mice is related to chronic inflammation and enhanced apoptosis. Canadian Journal of Physiology and Pharmacology 93: 385-394.

Zhu, X., Guo, K. and Lu, Y., 2011. Selenium effectively inhibits 1,2-dihydroxynaphthalene-induced apoptosis in human lens epithelial cells through activation of PI3-K/Akt pathway. Molecular Vision 17: 2019-2027.

Zilaee, M., Ferns, G.A. and Ghayour-Mobarhan, M., 2014. Heat shock proteins and cardiovascular disease. Advances in Clinical Chemistry 64: 73-115.

Zulkifli, I., Che Norma, M.T., Israf. D.A. and Omar, A.R., 2002. The effect of early-age food restriction on heat shock protein 70 response in heat-stressed female broiler chickens. British Poultry Science 43: 141-145.

Zulkifli, I., Akmal, A.F., Soleimani, A.F., Hossain, M.A. and Awad, E.A., 2018. Effects of low-protein diets on acute phase proteins and heat shock protein 70 responses, and growth performance in broiler chickens under heat stress condition. Poultry Science 97: 1306-1314.

Zuo, J., Xu, M., Abdullahi, Y.A., Ma, L., Zhang, Z. and Feng, D., 2015. Constant heat stress reduces skeletal muscle protein deposition in broilers. Journal of the Science of Food and Agriculture 95: 429-436.

Chapter 6
Thioredoxin system

Never trouble trouble till trouble troubles you

6.1 Introduction

A fine balance between oxidising and reducing conditions called redox status/homeostasis is essential for the normal function and survival of cells being a major determinant of many different pathways including cell signalling and gene regulation (Hunyadi*et al.*, 2019; Miyazawa *et al.*, 2019; Schmidlin*et al.*, 2019; Sies, 2019; Sies and Jones, 2020). It was suggested that reversible oxidations of protein thiols could be a coordinated metabolic response to hydrogen peroxide regulating both redox signalling and protecting cells/tissues from oxidative stress (Foley *et al.*, 2020). In fact, H_2O_2 is shown to be produced by almost every cell in the body and participates in many important cellular processes including membrane signal transduction, gene expression, cell differentiation, insulin metabolism, cell shape determination and other signalling cascades (Pravda, 2020). There are three major redox couples in cells determining redox equilibrium including NADP+/NADPH, GSSG/2GSH, and Trx(ox)/Trx(red). These redox couples are thermodynamically connected to each other in the maintenance of redox status of cells. with NADPH being major source of reducing equivalents (Penney and Roy, 2013). Furthermore, pentose phosphate cycle is the major source of NADPH connecting redox balance in the cell to carbohydrate metabolisms. A thiol redox system consisting of the thioredoxin system (thioredoxin/thioredoxin reductase/thioredoxin peroxidase (peroxiredoxins)/sulfiredoxin; Zhang *et al.*, 2013) and glutathione system (glutathione/glutathione reductase/glutaredoxin/glutathione peroxidase) are believed to be the major players in the redox equilibrium regulation and maintenance in biological systems (Gromer *et al.*, 2004; Holmgren, 2000). Together they supply electrons for deoxyribonucleotide formation, antioxidant defence, protein and DNA synthesis and repair and redox regulation of signal transduction, transcription, cell growth, differentiation and apoptosis (Mustacich and Powis, 2000). Indeed, thioredoxin system not only plays a crucial role as thiol/disulphide redox controller, it is also essential for certain organisms as the only system ensuring the redox homeostasis (Koháryová and Kollárová, 2015). In sulfhydryl-containing proteins, their thiol groups (PSH) play crucial roles in modulating their respective functions. Indeed, depending on the oxidative stimulus, protein SH (PSH) oxidation can lead to the formation of (Farina and Aschner, 2019):

- disulphide linkages (–S–S–);
- cysteinyl radical (P-S–);
- sulfenic acid (PSOH);
- sulfinic acid (PSO_2H);
- sulfonic acid (PSO_3H), etc.

DOI 10.3920/978-90-8686-906-0_6, © Wageningen Academic Publishers 2020

Importantly, under normal physiological conditions PSH oxidation to disulphides is reversible in cells and the reduction of disulphides is proceeded at the expense of electrons initially derived from NADPH in events with glutaredoxins or thioredoxins being intermediary reducing agents (Farina and Aschner, 2019). Therefore, biological roles of the thioredoxin system are diverse and include (Balsera and Buchanan, 2019; Gromer *et al.*, 2004; Lu and Holmgren, 2014):

- Antioxidant defence by direct catalysis of several antioxidant reactions and by regeneration of other antioxidant enzymes such as 2-Cys peroxiredoxins or methionine sulphoxide reductase inactivated by oxidative stress as well as by recycling dehydroascorbate to ascorbate and reduction of ubiquinone to ubiquinol.
- Cell signalling pathways where Trx participates in adaptive regulation of enzymes in response to environmental signals.
- General metabolism: being a substrate for ribonucleotide reductase in DNA synthesis and 3'-phosphoadenylylsulphate reductase in sulphur assimilation.
- Chaperone function: dealing with unfolded and denatured proteins.
- Other functions including participation in protein biosynthesis, hormone and cytokine action, apoptosis, etc.

This chapter is devoted to description of molecular mechanisms of the thioredoxin system action as an integrated part of the vitagene network with a special emphasis to data from avian species/poultry.

6.2 Thioredoxins

Thioredoxin (Trx), an approximately 12 kDa thiol/disulphide oxidoreductase, was first characterised in 1964 in *E. coli* and three years later it was described in rat hepatoma cells (see Powis *et al.*, 2000 for review). Thioredoxin with a redox-active dithiol/disulphide is an electron donor for essential enzymes including ribonucleotide reductase and a general protein disulphide reductase (Holmgren, 2001). Furthermore, Trx represents an intracellular redox regulator that has been shown to be important for the regulation of redox-sensitive transcription factors and maintaining them in active form during oxidative stress. Indeed, Trx a broad specificity oxidoreductase is considered to be an essential cofactor for many redox-dependent enzymes. Furthermore, Trx is involved in reduction of disulphides in signalling proteins, transcription factors, and oxidatively 'damaged' proteins under oxidative stress conditions (Veal *et al.*, 2018). In fact, most, if not all, of the functions of Trx depend on the activity of TrxR.

The cDNA sequence for chicken Trx predicts a protein of 105 amino acids with a molecular weight of 11,700 (Jones and Luk, 1988). The authors showed that the sequence of the chicken Trx is very similar to the sequences of other thioredoxins. Comparison of the chicken Trx protein sequence with those from bacteria and plants indicates structural features that appear to be essential for activity. To investigate the biological significance of Trx2, chicken Trx2 cDNA was cloned and clones of the conditional Trx2-deficient cells were generated using chicken B-cell line, DT40. It was shown that chicken Trx2 is an essential gene and that Trx2-deficient cells undergo apoptosis upon repression of the Trx2 transgene, showing an accumulation of intracellular ROS (Tanaka *et al.*, 2002; Wang *et al.*, 2006). Increased Trx expression in chicken ovarian follicles was associated with high rates of egg production (Yang *et al.*, 2008). Trx was found to be expressed in chicken jejunum (Xiao *et al.*, 2012) and was shown to be important protein of the chicken seminal plasma (Marzoni *et al.*, 2013). Furthermore, chicken mitochondrial Trx2 was found to possess disulphide reductase activity in concentration-dependent manner showing protective effects on LPS-induced oxidative stress in chicken hepatocytes (Hu *et al.*, 2015). Trx expression in the chicken liver was shown to be age dependent (Del Vesco *et al.*, 2017). In chicken, based on the amino acid combinations, which are important for forming binding hot spots, Trx was suggested to interact with a range of selenoproteins, including TR1, TR2, TR3, SPS1, Sep15, SelN, SelM, SelI, Gpx1, Gpx2, Gpx3, Gpx4, Dio1, Dio3, SelH, SelT, SelW, and Sepx1 (Liu *et al.*, 2017). Interesting, gene expression of selenoproteins was reported to be regulated by Txn silence in chicken cardiomyocytes. In fact, low expression of Txn was shown to significantly decrease the mRNA levels of Dio1, Dio2, GPx1, GPx2, GPx3, GPx4, TR1, TR2, TR3, SelT, SelW, SelK and MsrB whereas the mRNA levels of the rest of selenoproteins were increased (Yang *et al.*, 2017a). Furthermore, Se deficiency was reported to cause Trx suppression and thioredoxin knock down was found to disbalance insulin responsiveness in chicken cardiomyocytes through PI3K/Akt pathway inhibition (Yang *et al.*, 2017b). In addition, Txn knockdown in chicken cardiomyocytes was indicated to lead to cytosolic Ca^{2+} overload through upregulated gene expression of Ca^{2+} channel-related genes in the cytoplasmic and ER membranes (Yang *et al.*, 2018). In fact, Txn-deficient chicken cardiomyocytes were characterised by oxidative stress and activated autophagy with severe inflammation and damages to cardiomyocytes (Yang *et al.*, 2020). It was demonstrated that heat stress significantly downregulated Trx2 expression in the hepatic mitochondria of broiler chicks (Zhang *et al.*, 2018). Interestingly, broilers fed methionine in the form of DL-2-hydroxy-4-methylthio-butanoic acid was shown to have increased *Trx* gene expression in the duodenum and ileum, but decreased glutaredoxin, glutathione reductase, and glutathione synthetase genes expression (Wang *et al.*, 2019). It seems likely that Trx plays an important role in maintaining activities of various immune receptors (Yarana *et al.*, 2019) which could be a mechanism of prevention of immunosuppression in stress conditions (Surai, 2018) and Trx role in avian immunity deserves more attention.

6.3 Thioredoxin reductase

Thioredoxin reductase (TrxR) was first characterised by Holmgren in 1977 from calf liver and thymus and 5 years later it was purified from rat liver cytosol by Luthman and Holmgren (1982). It was shown that TrxR had a subunit molecular weight of 58,000 and a native molecular weight of 116,000. The enzyme was highly specific for NADPH with a Km of 6 μM. It contained an FAD prosthetic group and was sensitive to inhibition by arsenic. Fourteen years later it was shown that human TrxR is a selenoenzyme (Gladyshev *et al.*, 1996; Tamura and Stadtman, 1996). Selenocysteine is required for the activity of this enzyme, since the Cys mutant enzyme is inactive.

Whereas H_2O_2 was a substrate for the wild-type enzyme, all mutant enzymes lacked hydroperoxidase activity (Zhong and Holmgren, 2000). Furthermore, radiolabelling of proteins by incubation of the cDNA-transfected cells with sodium [^{75}Se] selenite showed that ^{75}Se was incorporated into the expressed TrxR protein (Fujiwara *et al.*, 1999) confirming a requirement for Se for the formation of functional TrxR. Therefore, mammalian TrxRs are a class of flavoproteins that use NADPH as an electron donor and belong to the family of oxidoreductases (Ganther, 1999) that share sequence identity and mechanistic similarity with glutathione reductases (Gasdaska *et al.*, 1995; Mustacich and Powis, 2000). These enzymes are involved in linking the thioredoxin system to reduced glutathione and the nucleotide cofactors (Holmgren and Bjornstedt, 1995). Therefore, TrxRs are involved in protein folding and critical protein-protein and protein-DNA interactions and mammalian TrxRs show increased activity with Se supplementation in the nutritional to supranutritional ranges (Surai, 2006, 2018). TrxR activity in cells is modulated by an intricate interplay, involving regulation by Se availability, posttranscriptional regulation and posttranslational inactivation by ROS. Both *in vivo* and *in vitro* studies demonstrated that Trx and TrxR have protective roles against cytotoxicity mediated by the generation of ROS (Calabrese *et al*, 2009b).

There are at least three forms of this enzyme (Table 6.1). TrxR1 is located predominantly in the cytosol; TrxR2 is found in mitochondria (Miranda-Vizuete *et al.*, 2000; Powis *et al.*, 2000). In fact, human mitochondrial TrxR consists of 521 amino acid residues with a calculated molecular mass of 56.2 KDa. It is also highly homologous to the

Table 6.1 Classification of human thioredoxins and thioredoxin reductases (adapted from Miranda-Vizuete *et al.*, 2004 and Surai, 2006, 2018).

Name	Chromosomal localisation	Size, kDa	Tissue specificity	Subcellular localisation
Thioredoxins				
Trx-1	9q31	11.71	ubiquitous	mainly cytosolic, nuclear upon certain stimuli
Trx-2	22q13.1	11.87	ubiquitous	mitochondrial
Txl-1/Trp32	18q21.2	32.25	ubiquitous	cytosolic
Erdj5/JDPI	2p22.1-23.1	91.08	ubiquitous	endoplasmic reticulum
Sptrx-1	18p11.2-11.31	53.27	testis/spermatid	sperm fibrous sheath
Sptrx-2	7p14.1	67.27	testis/spermatid	sperm fibrous sheath
Sptrx-3	not determined	14.57	testis/spermatid	Golgi
Txl-2	3q22.3-23	36.85	ubiquitous, especially in testis and lung	associated with microtubules in cilia and flagella
Thioredoxin reductases				
TrxR1	12q23-24.1	54.71	ubiquitous	cytosolic
TrxR2	22q11.21	53.06	ubiquitous	mitochondrial
TGR	3p13-q13.33	63.63	ubiquitous, but highly expressed in testis	cytosolic

previously described cytosolic TrxR1. It is interesting that TrxR2 has extra 33 amino acids in its molecule at the N-termins. It was shown that mRNA for TrxR2 is highly expressed in prostate, testis and liver. TrxR2 gene consists of 18 exons spanning about 67 kb with a chromosomal localisation at position 22q11.2 (Miranda-Vizuele *et al.*, 2000). The third member of the family called TrxR3 is located in the testes (Sun *et al.*, 1999). In fact, Sun *et al.* (2001) demonstrated that testes TrxR has broad substrate specificity and can reduce several components of the thioredoxin and glutathione systems (Mustacich and Powis, 2000). Therefore, it is also called thioredoxin and glutathione reductase (TGR). It has been shown that TrxR1 and TrxR2 are essential for embryonic development in mice (Gladishev, 2016).

It has been established that TrxR is a homodimer and a selenenyl sulphide was identified as the active site of TrxR and a structural model and mechanisms for the enzyme were proposed (Zhong *et al.*, 2000). The most striking feature of TrxR enzymes is their sensitivity to oxidising conditions that cause changes in conformation (Gorlatov and Stadtman, 1998). Such conformational changes are suggested as important with regard to triggering cell signalling in response to oxidative stress (Ganther, 1999). In addition to participation of TrxR in cell signalling and redox regulation of transcription factors, reactivation of oxidatively inactivated proteins (Ganther, 1999) could be of great importance in antioxidant defence in the cell. Therefore, TrxRs are involved in protein folding and critical protein-protein and protein-DNA interactions (Ebert-Duming *et al.*, 1999).

TrxR can also directly reduce thioredoxin, hydrogen peroxide, lipid hydroperoxides, ascorbyl free radical, dehydroascorbic acid, lipoic acid and selenite (Holmgren, 2001) and may have a role in detoxification reactions (Holmgren and Bjornstedt, 1995). The ability of mammalian TrxR to reduce dehydroascorbic acid (May *et al.*, 1997) could be an important link between Se activity with Se supplementation in the nutritional to supranutritional ranges (Ganther, 1999; Holmgren, 2000). An additional unique property of TrxR is its hydroperoxidase activity, which provides self-protection from inactivation by hydroxyl radical (Zhong and Holmgren, 2000). A general scheme of reactions and functions of thioredoxin reductase in the cell are shown in Figure 6.1 and detailed information on this enzyme is presented by Nordberg and Arner (2001). TrxR activity in cells is modulated by an intricate interplay, involving:

- Regulation by Se availability: In rat liver and kidney TrxR activity increased several fold as a result Se supplementation of the deficient diet (Berggren *et al.*, 1999). However, there is a tissue-specificity in this regulation. For example, after 12 months low Se diet consumption by rats TrxR activity decreased in the heart, liver, and kidney, but increased in the arterial wall (Wu and Huang, 2004).
- Regulation of the promoter of TrxR: a housekeeping type promoter in combination with alternative splice variants and transcriptional start sites.
- Posttranscriptional regulation through AU-rich elements. Mammalian TrxR1 and TrxR2 exhibit alternative splicing around the first exon. Regulation via Au-rich elements enables quick expression responses to various stimuli.
- Posttranslational inactivation by ROS and electrophilic agents (prostaglandin derivatives, lipid aldehydes, iodoacetic acid, arsenicals, gold compounds, quinines, nitrosoureas, cisplantin, dinitrohalobenzenes).

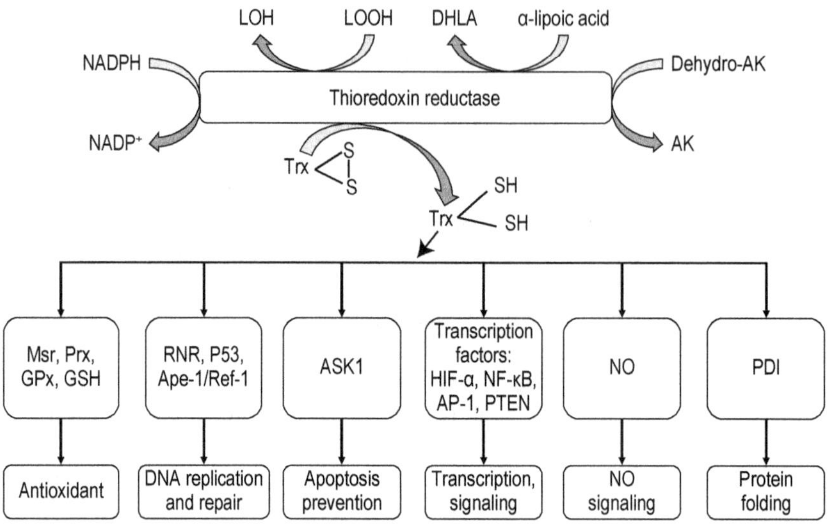

Figure 6.1. Thioredoxin reductase functions (adapted from Surai, 2006, 2018; Zhang *et al.*, 2020).

TrxRs are involved in protein folding and critical protein-protein and protein-DNA interactions and mammalian TrxRs show increased activity with Se supplementation in the nutritional to supranutritional ranges (Surai, 2006, 2018). TrxR activity in cells is modulated by an intricate interplay, involving regulation by Se availability, posttranscriptional regulation and posttranslational inactivation by ROS. Biological roles of the thioredoxin system are diverse and include (Das, 2004; Gromer *et al.*, 2004; Rundlof and Arner, 2004):

- Antioxidant defence: by direct catalysis of several antioxidant reactions and by regeneration of other antioxidant enzymes such as peroxiredoxins or methionine sulphoxide reductase inactivated by oxidative stress; recycling dehydroascorbate to ascorbate and reduction of ubiquinone to ubiquinol. In fact, the thioredoxin system is a major line of cellular defence against oxygen damage (Hirt *et al.*, 2002). Indeed, cytochrome c is a substrate for both TrxR1 and TrxR2 and cells overexpressing TrxR2 are more resistant to impairment of complex III in the mitochondrial respiratory chain upon both antimycin A and myxothiazol treatments, suggesting a complex III bypassing function of TrxR2 (Nalvarte *et al.*, 2004).
- Redox regulation: by reducing oxidised Trx back to an active form and being involved in regulation of various transcription factors.
- Gene regulation by modulating several transcriptional factors, including nuclear factor-κB, FOS, Jun, Ref-1 and p53. The reducing activity of Trx for transcriptional factors more than 100-fold higher than that of GSH (Nakamura, 2004).
- Modulation of protein phosphorylation: by affecting activity of mitogen activating protein kinases and phosphoprotein phosphatases.
- Regulation of apoptosis: by controlling apoptosis signal-regulating kinase 1.
- Redox regulation of various cellular functions including cell proliferation, differentiation and maintenance of viability.

- Regulation of the synthesis of deoxyribonucleotides (DNA synthesis and repair) by providing reducing equivalents to ribonucleotide reductase.
- Involvement in hormone action and cytokine function: Trx can act as an autocrine growth-factor synergising with IL-1 and IL-2; there is evidence that Trx can act as ID activator.
- Protein biosynthesis: the Trx system is important to maintain high activity of protein biosynthesis machinery in the cell.

Interactions between TR and Trx are depicted in Figure 6.1.

Recently, it has been proposed that TrxR1 is a potent regulator of Nrf2 playing a central role in redox homeostasis, defence against oxidative stress, and regulation of redox signalling pathways (Cebula *et al.*, 2015). Indeed, disruption of TrxR1 protects mice from acute acetaminophen-induced hepatotoxicity through enhanced NRF2 activity (Patterson *et al.*, 2013). It seems likely that TrxR1 reduces the disulphide bonds in Keap1 to arrests Nrf2 in the cytoplasm. On the other hand, inactivation or decreased activity of TrxR1 is associated with disulphide bond formation in Keap1, leading to Nrf2 release and its transfer into the nucleus to drive the transcription of many cytoprotective genes (Cebula *et al.*, 2015). Furthermore, TrxR1 is shown to be an Nrf2 target gene. In fact, Nrf2 has been reported to bind to an ARE in gene promoters in Trx, TrxR, PRDX1 and PRDX6 (Hawkes *et al.*, 2014). Therefore, interactions within the antioxidant system are key factors regulating many physiological functions. The catalytic circle of avian thioredoxin system is shown in Figure 6.2.

Oxidised cysteine in protein is reduced back to the reduced form due to action of Trx which is oxidised in the reaction. TR is responsible for reduction of the Trx into the active form due to reducing potential provided by NAPPH produced in the pentose phosphate cycle. Data on TrxR activity in various tissues obtained mainly with mammals, including laboratory animals and humans and much less information is available on avian TRs. For example, Smith *et al.* (2001) compared TrxR activity in mammals and chickens, finding chickens to have extremely low TrxR activities probably reflecting low TrxR protein expression or being a result of differences

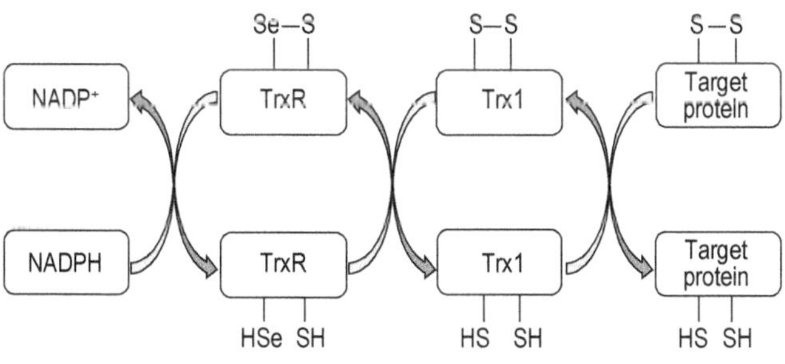

Figure 6.2. Avian thioredoxin system (adapted from Matsuzawa, 2017; Mohammadi *et al.*, 2019; Zhang *et al.*, 2017).

between mammalian and chicken TrxR. In fact, Gowdy (2004) used Western blots and found TrxR protein expression at relatively low levels as well as some differences in molecular weight of the chicken TrxR in comparison to the mammalian enzyme.

Data on TrxR activity in chicken tissues have been presented by Edens and Gowdy (2004). They showed that when Se supplementation was low, the highest TrxR activity was found in kidney and brain and lowest in the liver. After Se supplementation TrxR activity increased practically in all tissues studied. Furthermore, organic Se supplemented at 0.3 mg/kg increased TrxR activity significantly more than selenite at the same dose in heart and thymus. There was similar tendency of increased Se availability from organic Se for activation of TrxR in the brain, breast muscle, bursa, thymus and spleen. The authors also showed that the highest TrxR activity was found in nuclear pellet and mitochondrial lysates and the lowest activity was seen in mitochondrial pellets (Edens and Gowdy, 2004). Recently, TrxR activity has been detected in a range of tissues (liver, lung, heart, kidney, brain, breast muscle, bursa, thymus, spleen, RBC and plasma) in broiler chickens (Gowdy *et al.*, 2015). Similar to mammalian species, activity of chicken TrxR is shown to be selenium dependent. Subcellular distribution of TrxR activity was found in association with the cytosolic, nuclear pellet and mitochondrial fractions. Compared with sodium selenite, Se-Yeast or selenomethionine (SM) significantly increased the activity and TrxR1 mRNA in the liver and kidney of broiler breeders and their offspring (Yuan *et al.*, 2012). Selenium dietary supplementation (0.4 mg/kg diet) increased TrxR activity in duodenal mucosa, liver and in the kidney in chickens (Placha *et al.*, 2014). Se deficiency was associated with a decreased expression of TrxR2 in chicken thyroids (Lin *et al.*, 2014). Similarly, Se deficiency in chickens was associated with a significant decrease in activity of TrxR1 (by 50%), TrxR2 (by 83%) and TrxR3 (by 36%) in pancreas by 55th day of the experiment (Zhao *et al.*, 2014).

Furthermore, TrxR expression decreased in chicken adipose tissues due to Se deficiency (Liang *et al.*, 2014). Low Se diet (0.028 mg/kg) or high Se diet (3 mg/kg) significantly reduced TrxR activity in chicken kidney with changes in their mRNA levels. In particular, low Se diet downregulated the mRNA expression of TrxR3 (Xu *et al.*, 2016). Se deficiency was also shown to downregulate TrxR1, GPx3, GPx4, and selenoprotein S, but upregulated SELENOT and SELENOU in spleen in AFB1 administered chickens (Zhao *et al.*, 2019). Interestingly, Cd toxicity (100 mg/kg) in chicken was associated with a significant increase TrxR1 gene expression in the liver (Zoidis *et al.*, 2019).

6.4 Peroxiredoxins

First peroxiredoxin (Prx) was described in *Saccharomyces cerevisiae* in 1988 as a specific 'protector' protein inhibiting enzyme inactivation by a thiol/Fe(III)/O_2 oxidation system (Kim *et al.*, 1988). The discovered protein did not possess catalase, glutathione peroxidase, superoxide dismutase, or iron chelation activities and the authors suggested that its function is related to a sulphur radical scavenging. For

40 years of research these proteins were given multiple names, including 'protector protein', 'thiol-specific antioxidants', 'thioredoxin-linked thiol peroxidase' and 'thioredoxin peroxidase' and now the name 'peroxiredoxins' is generally accepted (Mishra *et al.*, 2015). Peroxiredoxins (Prxs) are a family of hydrogen peroxide scavengers that reduce peroxides via the oxidation of a catalytic (or peroxidatic) cysteine to sulfenic acid (Hopkins and Neumann, 2019). They are highly conserved antioxidative proteins (non-seleno peroxidases), currently comprising six members (Prx1, Prx2, Prx3, Prx4, Prx5, and Prx6) in mammals and located in different parts of the cell (Figure 6.3). Indeed, Prxs have a wide subcellular distribution and perform divergent biological functions (Poynton and Hampton, 2014).

They also subcategorised into three subfamilies including typical 2-Cys Prx (Prx1–4), atypical 2-Cys Prx (Prx5) and 1-Cys Prx (Prx6) (Rhee and Kil, 2016). Typical 2-Cys Prxs have a conserved N-terminal (peroxidatic) and C-terminal (resolving) Cys residues that are located in different subunits in the obligate homodimer and involved in the peroxidase catalytic activity. Atypical 2-Cys Prx is characterised by only one conserved N-terminal Cys residue and one additional but less conserved Cys residue in the same polypeptide. Finally, 1-Cys Prx has only one N-terminal conserved Cys residue to be used for catalysis. Prx activity is based on a redox-active cysteine that is oxidised to a sulfenic acid by hydroperoxides including hydrogen peroxide, organic peroxides, peptide and protein hydroperoxides, and peroxynitrite being the most important thiol-dependent non-selenium peroxidases in biological systems. It was suggested that Prxs are responsible for a reduction of 90% of cellular peroxides such as hydrogen peroxide, peroxynitrite and hydroperoxides (Shahnaj *et al.*, 2019).

It was calculated that because the rate constant of Prx-thiol oxidation is substantially higher than most of other thiol-based proteins, Prxs are approximately 10^5-10^7 times more efficient than other thiol-based antioxidants including GSH and Thioredoxin and Prxs are able to reduce the ROS present even in minute amounts that cannot

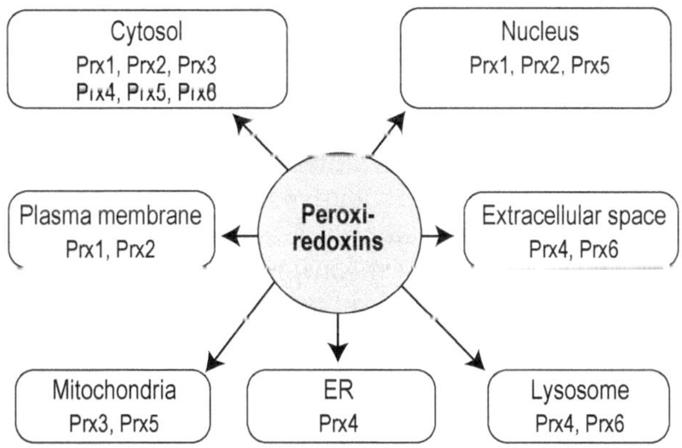

Figure 6.3. Subcellular distribution of peroxiredoxins (adapted from Heo *et al.*, 2020; Sharapov and Novoselov, 2019).

be eliminated by other antioxidants (Mishra *et al.*, 2015). These cysteine-dependent peroxidases play major roles not only in peroxide detoxification, but also in regulating peroxide-mediated cell signalling being critical regulators of biological functions through their ability to control the redox status, allowing either the promotion or dampening of signals from H_2O_2 or related oxidants with signalling abilities (Elko *et al.*, 2019). Prxs are considered to have a bifunctional activity profile; with thioredoxin-dependent peroxidase and signalling activities at low H_2O_2 concentrations and catalase and chaperone/other signalling activities at higher H_2O_2 concentrations (Veal *et al.*, 2018). Interestingly, hyperoxidised Prx were only detected once the cellular H_2O_2-buffering capacity was breached (Veal *et al.*, 2018). Therefore, Prxs have been suggested to regulate redox signalling by acting as peroxide transducers to initiate the oxidation of redox regulated proteins as well as by affecting the oxidation of thioredoxin family proteins (Veal *et al.*, 2018). Prxs have been discovered to be multifunctional proteins, with a chaperone activity similar to that in HSP (e.g. HSP70) that protects against protein aggregation (Veal *et al.*, 2018). Therefore, Prx importance is unarguable, as knockouts of the most highly expressed Prxs are associated with increased oxidative stress and reduced genome stability. Prx1 is the most ubiquitously expressed member of the peroxiredoxin family involving in antioxidant defence, cell differentiation and proliferation, immune responses, regulation of apoptosis, and chaperone actions (Daly *et al.*, 2008). Due to high affinity toward H_2O_2, 2-Cys Prxs can efficiently reduce H_2O_2 at low concentration. Interestingly, Prxs exhibit 24 h rhythms in their redox state in all kingdoms of life (Del Olmo *et al.*, 2019). It is important to mention that Prxs are working in close concert with other antioxidants since they require the Trx/TR/NADPH, Srx redox system, and in some cases Grx/GSH, for their reduction and Prx chaperone functions are controlled by the redox status (Elko *et al.*, 2019). Members of the typical 2-Cys Prx subfamily of Prxs (Prx1 to Prx4 in mammals) are shown to be inactivated via hyperoxidation of the active-site cysteine to sulfinic acid (Cys–SO_2H) during catalysis and can be reactivated via an ATP-consuming reaction catalysed by sulfiredoxin (Srx, Jeong *et al.*, 2012).

At least 4 different classes of Prx protein have been evolutionary conserved in chickens (Han *et al.*, 2005). Interestingly, recently converging evidence supporting loss of PRDX5 in aves has been presented, while PRDX5 appears to be conserved in non-avian species (Pirson *et al.*, 2018). Chicken Prx proteins demonstrate antioxidant activity similar to those of the mammalian enzymes and Prx expression in chickens are not tissue specific, indicating their essential role as a housekeeping gene in all tissues to protect against oxidative damage (Han *et al.*, 2005). Prx1 was shown to be expressed in chicken jejunum (Xiao *et al.*, 2012), chicken embryonic kidney (Cao *et al.*, 2011) and chicken macrophages (Lavric *et al.*, 2008), while Prx6 was indicated to be expressed in chicken liver (Huang *et al.*, 2011) and chicken gut (Lee *et al.*, 2014). It seems likely that in chickens Prxs are adaptive antioxidants and depending on conditions they can be activated or inhibited by stress. Recently, global gene and protein expression in the small yellow follicle (SYF; 6-8 mm in diameter) tissues of chickens in response to acute heat stress were investigated and upregulated expression of peroxiredoxin family was considered as an adaptive mechanism of dealing with heat-stress induced oxidative stress (Cheng *et al.*, 2018). Furthermore, it was shown that heat stress significantly downregulated Prx3 expression in the hepatic mitochondria of broiler chicks (Zhang *et al.*, 2018). Proteomic analysis of chick retina during early recovery from lens-induced myopia revealed that Prx4 was upregulated in the recovery retinas compared with the control eye retinas (Zhou *et al.*, 2018). Acute heat stress was shown to induce differential protein expression in the hypothalamus of the L2 strain Taiwan country chickens, including upregulated expressions of Prx1 (Tu *et al.*, 2018). Recently it has been demonstrated that *chPRDX3* is required for cell proliferation in chicken fibroblast cells and the knockdown of chicken *PRDX3* was reported to suppress cell proliferation through an increase in oxidative stress (Choi *et al.*, 2020). Therefore, Trxs, Prdxs and TrxRs can function as signal transduction proteins regulating stress-induced signalling cascades. They are important antioxidants participating in cellular/organismal adaptation to stress and their upregulation is considered to be an important approach to improve stress resistance of poultry.

6.5 Sulfiredoxin

Sulfiredoxin (Srx) was initially identified in 2003 in yeast as a protein of relative molecular mass M(r) = 13,000, which was named sulphiredoxin (identified by the US spelling 'sulfiredoxin'), that is conserved in higher eukaryotes and reduces the hyperoxidised cysteine-sulphinic acid (Cys-SO_2H) in the yeast peroxiredoxin Tsa1 (Biteau *et al.*, 2003). Human Srx was shown to have a length of 137 amino acids (Findlay *et al.*, 2006). In fact, Srx is present in mammals, birds and multiple other eukaryotic organisms and few prokaryotes (Perkins *et al.*, 2014). In normal human tissues, Srx is present in kidney, lungs, and pancreas (Chang *et al.*, 2004) and in mice Srx is expressed in adrenal gland, heart, lung, liver, kidney, pancreas, spleen, skin and brain (Rhee and Kil, 2016). Srx is mainly a cytosolic protein that can be translocated into mitochondria under oxidative stress conditions (Noh *et al.*, 2009). Indeed, cytosolic Srx was found to be imported into mitochondria via a mechanism that requires formation of a disulphide-linked complex with HSP90, which is likely promoted by H_2O_2 released from mitochondria. Furthermore, the imported Srx was found to be degraded by Lon protease in a manner dependent on Prx3 hyperoxidation state (Rhee and Kil, 2016). The authors described an elegant model of Srx action as follows. In the cytosol, H_2O_2 released from mitochondria can promote formation of a disulphide-linked complex between Srx and HSP90, and the resulting Srx–S–S–HSP90 complex is imported into mitochondria by the TOM complex. A cochaperone of HSP90 called FKBP, is also involved in the import process. The imported Srx can bind tightly to PrxIII-SO_2H and reduce it to Prx3–SH leading to decreased concentration of Prx3–SO_2H and at this point Srx becomes vulnerable to degradation by Lon. This leads to Srx down-regulation to basal levels and the consequent Prx3–SO_2H accumulation and H_2O_2 release (Rhee and Kil, 2016).

It is believed that the cellular level of H_2O_2, main signalling molecule in biological systems, is strictly regulated by a battery of redox enzymes including members of the Prxs family. Under oxidative stress conditions, the 2-Cys site of Prxs can be further oxidised to generate sulfinic acid and sulfonic acid, a process known as hyperoxidation or overoxidation (Chawsheen *et al.*, 2019). Therefore, a recycling reaction responsible for reduction of hyperoxidised Cys residue of Prxs from sulfinic to sulfenic acid catalysed by Srx is considered to be an adaptive evolutionary mechanism to deal with ROS overproduction in stress conditions (Chang *et al.*, 2004; Chawsheen *et al.*, 2019). Trx and GSH are suggested to be important partners in the aforementioned recycling. Interestingly, Srx is discovered to be highly specific to 2-Cys containing Prxs, including Prx1, 2, 3 and 4 (Woo *et al.*, 2005) and reduction of hyperoxidised Prx by Srx is considered to be a rate limiting step in reduction of hyperoxidised Prx (Mishra *et al.*, 2015). The reversible hyperoxidation of Prxs has been proposed to protect H_2O_2 signalling molecules from premature removal by 2-Cys Prxs or/and to upregulate the chaperone function of these enzymes. In addition to its sulfinic acid reductase activity, Srx catalyses the removal of glutathione (deglutathionylation) from modified proteins. (Jeong *et al.*, 2012). It is proven that sulfiredoxin is important for the reduction of hyperoxidised Prx and there is a range of transcriptional, posttranscriptional, and post-translational mechanisms involved in maintenance of very low basal levels of sulfiredoxin (with undetectably low levels of its mRNA expression) in specific cellular compartments (the cytoplasm) under optimal physiological conditions (Veal *et al.*, 2018). However, in mammalian cells, sulfiredoxin gene expression was shown to be rapidly induced in response to various stress conditions with its possible transport into the mitochondria, a place of superoxide radical production, and Prx hyperoxidation, to manage mitochondria redox balance (Veal *et al.*, 2018). The expression level of Srx is under the coordinated control of transcription factors including Nrf2 (Kim *et al.*, 2010; Soriano *et al.*, 2008), AP-1 (Wei *et al.*, 2008) and NF-κB (Jeong *et al.*, 2012).

In mammalian cells Srx expression is shown to be rapidly induced under a variety of stressful conditions, such as in metabolically stimulated pancreatic β cells, in immunostimulated macrophages, in neuronal cells engaged in synaptic communication, in lung cells exposed to hyperoxia or cigarette smoke, in hepatocytes of ethanol-fed animals, and in several types of cells exposed to chemopreventive agents (Jeong *et al.*, 2012). Similarly, in mouse macrophages, treatment with lipopolysaccharide strongly induces Srx expression in an Nrf2 and AP1 dependent manner, and the absence of either significantly affect the levels of Srx induction (Kim *et al.*, 2010). Srx can regulate the chaperone function of Prx1 by controlling its levels of glutathionylation. In fact. the glutathionylation of Cys83 of Prx1 is shown to favour formation of dimer over decamer, resulting in the loss of chaperone activity (Chae *et al.*, 2012).

Overexpressing Srx1 in human cardiac progenitor cells (hCPCs) was shown to lead to a significant increase in cell survival in response to H_2O_2 challenge. At the same time, silencing of Srx1 increases cell death upon treatment of hCPCs with H_2O_2 (Li *et al.*, 2018). It was also found that overexpressing Srx1 in hCPCs was associated with activating survival signalling molecules, including ERK and Nrf2, as well as mediating the expression of anti-oxidant genes (SOD2, CAT, TrxR1, Prx1, and Prx3) and anti-apoptotic genes (BCL-2 and BCL-xL), leading to inhibition of apoptosis under oxidative stress (Li *et al.*, 2018). It seems likely that Srx is involved in the maintenance of ER homeostasis, since knockdown of Srx sensitises human lung cancer cells to ER-stress induced cell death (Chawsheen *et al.*, 2019). The authors showed that Srx can form a complex with the ER-resident protein thioredoxin domain-containing protein 5 (TXNDC5) and in response to ER-stress Srx exhibits an increased association with TXNDC5, facilitating the retention of Srx in the ER (Chawsheen *et al.*, 2019).

It was shown that exposure of A549 or wild-type mouse embryonic fibroblast (MEF) cells to low steady-state levels of H_2O_2 (10-20 μm) did not cause any significant oxidative injury due to the maintenance of balance between H_2O_2 production and elimination (Baek *et al.*, 2012). In contrast, in Srx-depleted A549 and $Srx^{-/-}$ MEF cells a dramatic increase in extra- and intracellular H_2O_2, sulfinic 2-Cys Prxs, and apoptosis were clearly demonstrated. At the same time, re-expression of Srx in Srx-depleted A549 or $Srx^{-/-}$ MEF cells was shown to promote the reactivation of sulfinic 2-Cys Prxs and lead to cellular resistance to apoptosis (Baek *et al.*, 2012). These results indicate that Srx functions as an important component of the antioxidant defence system maintaining redox status by balancing between H_2O_2 production and elimination and thus helping survival of cells exposed to low, steady state levels of H_2O_2.

It was shown that silencing of Srxn1 expression in astrocytes was associated with upregulation of inflammatory cytokine levels, promotion of inflammatory responses, and aggravation of H_2O_2-induced cells apoptosis (Yu *et al.*, 2015; Zhou *et al.*, 2015). Overexpression of Srxn1 was indicated to inhibit the expression of apoptosis-related proteins and cytochrome C release, affect the P13K/AKT signalling pathway and alleviate myocardial cell injury induced by ischemia-reperfusion (Zhang *et al.*, 2016). Srxn1 was found to increase the proliferation and differentiation of cardiac stem cells and to reduce the apoptosis of cardiomyocytes caused by oxidative stress by reducing the production of ROS and maintaining the balance of mitochondrial membrane potential via upregulation of the ERK/Nrf2-signal pathway (Li *et al.*, 2018). Srxn1 was also shown to protect astrocytes from oxidative stress injury induced by H_2O_2 by activation of Notch signalling pathway. In fact, knockdown of Srxn1 was found to promote the secretion of LDH and MDA (indexed of oxidative stress), decrease the activity of SOD (main AO enzyme) and aggravate apoptosis of astrocytes induced by H_2O_2. At the same time, activation of the Notch signalling pathway attenuated the effect of Srxn1 on H_2O_2-induced oxidative damage and apoptosis of astrocytes (Li *et al.*, 2019). Some effectors of Srx expression are shown in Figure 6.4.

There is no data available on Srx expression and activity in tissues of avian spices and this topic deserves more attention.

6.6 Conclusions

Redox status of the cell is considered to be a major determinant of many different pathways including regulation of gene expression. A thiol redox system consisting of the thioredoxin system (thioredoxin/thioredoxin reductase/thioredoxin peroxidase (peroxiredoxins)/ sulfiredoxin and glutathione system (glutathione/glutathione

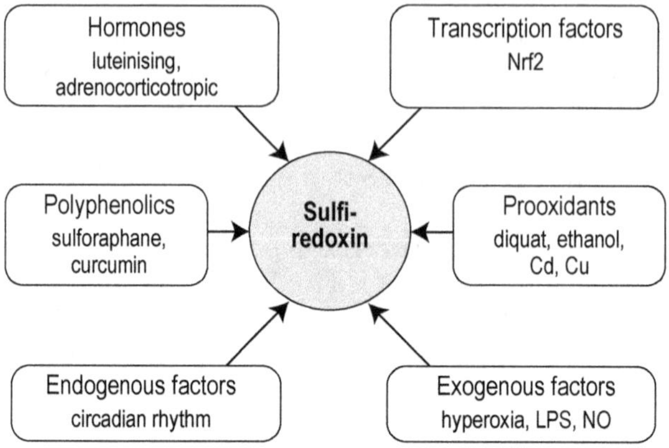

Figure 6.4. Positive effectors of Srx expression (adapted from Ramesh *et al.*, 2014).

reductase/glutaredoxin/glutathione peroxidase) is considered to be the major player in this regulation. There is only limited information available on the elements of Trx system expression and activity in avian species and most publication in this area came for the last 5 years. However, it seems likely that Trx system is a universal regulatory system responsible for redox homeostasis maintenance in various stress conditions, including those occurring in poultry production. Indeed, Trxs, TRs, Prxs and Srx are important part of the vitagene family and they interact with other vitagenes, namely SOD, GSH-system, HSP and sirtuins, and participate in regulation of many important physiological functions via maintenance of redox homeostasis associated with activation of an array of transcription factors and stress adaptation. Nutritional modulation of thioredoxin system to improve antioxidant defences and maintain redox homeostasis under various stress conditions awaits further investigations.

References

Balsera, M. and Buchanan, B.B., 2019. Evolution of the thioredoxin system as a step enabling adaptation to oxidative stress. Free Radical Biology & Medicine 140: 28-35.

Baek, J.Y., Han, S.H., Sung, S.H., Lee, H.E., Kim, Y.M., Noh, Y.H., Bae, S.H., Rhee, S.G. and Chang, T.S., 2012. Sulfiredoxin protein is critical for redox balance and survival of cells exposed to low steady-state levels of H2O2. The Journal of Biological Chemistry 287: 81-89.

Berggren, M.M., Mangin, J.F., Gasdaka, J.R. and Powis, G., 1999. Effect of selenium on rat thioredoxin reductase activity: increase by supranutritional selenium and decrease by selenium deficiency. Biochemical Pharmacology 57: 187-193.

Biteau, B., Labarre, J. and Toledano, M.B., 2003. ATP-dependent reduction of cysteine-sulphinic acid by *S. cerevisiae* sulphiredoxin. Nature 425: 980-984.

Calabrese, V., Cornelius, C., Dinkova-Kostova, A.T. and Calabrese, E.J., 2009. Vitagenes, cellular stress response, and acetylcarnitine: relevance to hormesis. Biofactors 35: 146-160.

Cao, Z., Han, Z., Shao, Y., Geng, H., Kong, X. and Liu, S., 2011. Proteomic analysis of chicken embryonic trachea and kidney tissues after infection in ovo by avian infectious bronchitis coronavirus. Proteome Science 9: 11.

Cebula, M., Schmidt, E.E. amd Arnér, E.S., 2015. TrxR1 as a potent regulator of the Nrf2-Keap1 response system. Antioxidants & Redox Signaling 23: 823-853.

Chae, H.Z., Oubrahim, H., Park, J.W., Rhee, S.G. and Chock, P.B., 2012. Protein glutathionylation in the regulation of peroxiredoxins: a family of thiol-specific peroxidases that function as antioxidants, molecular chaperones, and signal modulators. Antioxidants & Redox Signaling 16:506-523.

Chang, T.S., Jeong, W., Woo, H.A., Lee, S.M., Park, S., and Rhee, S.G., 2004. Characterization of mammalian sulfiredoxin and its reactivation of hyperoxidized peroxiredoxin through reduction of cysteine sulfinic acid in the active site to cysteine. The Journal of biological chemistry 279: 50994-51001

Chawsheen, H.A., Jiang, H., Ying, Q., Ding, N., Thapa, P. amd Wei, Q., 2019. The redox regulator sulfiredoxin forms a complex with thioredoxin domain-containing 5 protein in response to ER stress in lung cancer cells. The Journal of Biological Chemistry 294: 8991-9006.

Cheng, C.Y., Tu, W.L., Chen, C.J., Chan, H.L., Chen, C.F., Chen, H.H., Tang, P.C., Lee, Y.P., Chen, S.E. and Huang, S.Y., 2018. Functional genomics study of acute heat stress response in the small yellow follicles of layer-type chickens. Scientific Reports 8: 1320.

Choi, H., Jung, S.K., Kim, J.S., Oh, K.B., Yang, H., Lee, G., Lee, H.C., Woo, J.S. and Byun, S.J., 2020. Chicken PRDX3 is required for proliferation of chicken embryo fibroblast cells. British Poultry Science 61: 22-25.

Daly, K.A., Lefévre, C., Nicholas, K., Deane, E. and Williamson, P., 2008. Characterization and expression of Peroxiredoxin 1 in the neonatal tammar wallaby (*Macropus eugenii*). Comparative Biochemistry and Physiology. Part B, Biochemistry & Molecular Biology 149: 108-119.

Das, K.C., 2004. Thioredoxin system in premature and newborn biology. Antioxidants and Redox Signalling 6: 177-184.

Del Olmo, M., Kramer, A. and Herzel, H., 2019. A robust model for circadian redox oscillations. International Journal of Molecular Sciences 20: 2368.

Del Vesco, A.P., Khatlab, A.S., Goes, E.S.R., Utsunomiya, K.S., Vieira, J.S., Oliveira Neto, A.R. and Gasparino, E., 2017. Age-related oxidative stress and antioxidant capacity in heat-stressed broilers. Animal 11: 1783-1790.

Ebert-Dumig, R., Seufert, J., Schneider, D., Kohrle, J., Schutze, N. and Jakob, F., 1999. Expression of selenoproteins in monocytes and macrophages-implications for the immune system. Medizinische Klinik 9494: 29-34.

Edens, F.W. and Gowdy K.M., 2004. Selenium sources and selenoproteins in practical poultry production. In: Lyons, T.P. and. Jacques, K.A. (eds) Nutritional biotechnology in the feed and food industries. Proceedings of 20th Alltech's Annual Symposium. Nottingham University Press, Nottingham, UK, pp. 35-55.

Elko, E.A., Cunniff, B., Seward, D.J., Chia, S.B., Aboushousha, R., Van de Wetering, C., Van der Velden, J., Manuel, A., Shukla, A., Heintz, N.H., Anathy, V., Van der Vliet, A., and Janssen-Heininger, Y., 2019. Peroxiredoxins and beyond; redox systems regulating lung physiology and disease. Antioxidants & Redox Signaling 31: 1070-1091.

Farina, M. and Aschner, M., 2019. Glutathione antioxidant system and methylmercury-induced neurotoxicity: an intriguing interplay. Biochimica et Biophysica Acta. General Subjects 1863: 129285.

Findlay, V.J., Townsend, D.M., Morris, T.E., Fraser, J.P., He, L. and Tew, K.D., 2006. A novel role for human sulfiredoxin in the reversal of glutathionylation. Cancer Research 66: 6800-6806.

Foley, T.D., Montovano, G. and Camacho Ayala, M., 2020. The reducible disulfide proteome of synaptosomes supports a role for reversible oxidations of protein thiols in the maintenance of neuronal redox homeostasis. Neurochemical Research 45: 1825–1838. https://doi.org/10.1007/s11064-020-03046-7

Fujiwara, N., Fujii, T., Fujii, J. and Taniguchi, N., 1999. Functional expression of rat thioredoxin reductase: selenocysteine insertion sequence element is essential for the active enzyme. The Biochemical Journal 340: 439-444.

Ganther, H.E., 1999. Selenium metabolism, selenoproteins and mechanisms of cancer prevention: complexities with thioredoxin reductase. Carcinogenesis 20: 1657-1666.

Gasdaska, P.Y., Gasdaska, J.R., Cochran, S. and Powis, G., 1995. Cloning and sequencing of a human thioredoxin reductase. FEBS Letters 373: 5-9

Gladyshev, V.N., 2016. Eukaryotic proteomes. In: Hatfield D. L., Schweizer, U., Tsui P.A. and Gladyshev V.N. (eds) Selenium. Its molecular biology and role in human health. Fourth Edition, Springer, New York, NY, USA, pp.127-139.

Gladyshev, V.N., Jeang, K.T. and Stadtman, T.C., 1996. Selenocysteine, identified as the penultimate C-terminal residue in human T-cell thioredoxin reductase, corresponds to TGA in the human placental gene. Proceedings of the National Academy of Sciences of the USA 93: 6146-6151.

Gorlatov, S.N. and Stadtman, T.C., 1998. Human thioredoxin reductase from HeLa cells: selective alkylation of selenocysteine in the protein inhibits enzyme activity and reduction with NADPH influences affinity to heparin. Proceedings of the National Academy of Sciences of the USA 95: 520-8525.

Gowdy, K.M., 2004. Selenium supplementation and antioxidant protection in broiler chickens. M.Sc. Thesis, The Graduate School, North Carolina State University, Raleigh, NC, USA.

Gowdy, K.M., Edens, F.W. and Mahmoud, K.Z., 2015. Comparative effects of various forms of selenium on thioredoxin reductase activity in broiler chickens. International Journal of Poultry Science 14: 376-382.

Gromer, S., Urig, S. and Becker, K., 2004. The thioredoxin system-from science to clinic. Medical Research Reviews 24: 40-89.

Han, J.Y., Song, K.D., Shin, J.H., Han, B.K., Park, T.S., Park, H.J., Kim, J.K., Lillehoj, H.S., Lim, J.M. and Kim, H., 2005. Identification and characterization of the peroxiredoxin gene family in chickens. Poultry Science 84: 1432-1438.

Hawkes, H.J., Karlenius, T.C. and Tonissen, K.F., 2014. Regulation of the human thioredoxin gene promoter and its key substrates: a study of functional and putative regulatory elements. Biochimica et Biophysica Acta 1840: 303-314.

Heo, S., Kim, S. and Kang, D., 2020. The role of hydrogen peroxide and peroxiredoxins throughout the cell cycle. Antioxidants 9: 280.

Hirt, R.P., Muller, S., Embley, T.M. and Coombs, G.H., 2002. The diversity and evolution of thioredoxin reductase: new perspectives. Trends in Parasitology 18: 302-308.

Holmgren, A., 1977. Bovine thioredoxin system. Purification of thioredoxin reductase from calf liver and thymus and studies of its function in disulfide reduction. The Journal of Biological Chemistry 252: 4600-4606.

Holmgren, A., 2000. Redox regulation by thioredoxin and thioredoxin reductase. Biofactors 11: 63-64.

Holmgren, A., 2001. Selenoproteins of the thioredoxin system. In: Hatfield D.L. (ed.) Selenium. Its molecular biology and role in human health. Kluwer Academic Publishers, Boston, MA, USA, pp. 179-188.

Holmgren, A. and Bjornstedt, M., 1995. Thioredoxin and thioredoxin reductase. Methods in Enzymology 252: 199-208.

Hopkins, B.L. and Neumann, C.A., 2019. Redoxins as gatekeepers of the transcriptional oxidative stress response. Redox Biology 21: 101104.

Hu, L., Yu, W., Li, Y., Li, Y., Guo, J. and Tang, Z., 2015. Prokaryotic expression and antioxidant properties of mitochondrial thioredoxin-2 from broiler chicken. Chinese Veterinary Science 4: S831.

Huang, J., Ruan, J., Tang, X., Zhang, W., Ma, H. and Zou, S., 2011. Comparative proteomics and phosphoproteomics analyses of DHEA-induced on hepatic lipid metabolism in broiler chickens. Steroids 76: 1566-1574.

Hunyadi, A., 2019. The mechanism(s) of action of antioxidants: from scavenging reactive oxygen/nitrogen species to redox signaling and the generation of bioactive secondary metabolites. Medicinal Research Reviews 39: 2505-2533.

Jeong, W., Bae, S.H., Toledano, M.B. and Rhee, S.G., 2012. Role of sulfiredoxin as a regulator of peroxiredoxin function and regulation of its expression. Free Radical Biology & Medicine 53: 447-456.

Jones, S.W. and Luk, K.C., 1988. Isolation of a chicken thioredoxin cDNA clone. Thioredoxin mRNA is differentially expressed in normal and Rous sarcoma virus-transformed chicken embryo fibroblasts. The Journal of Biological Chemistry 263: 9607-9611.

Kim, K., Kim, I.H., Lee, K.Y., Rhee, S. G. and Stadtman, E.R., 1988. The isolation and purification of a specific 'protector' protein which inhibits enzyme inactivation by a thiol/Fe(III)/O2 mixed-function oxidation system. The Journal of Biological Chemistry 263: 4704-4711.

Kim, H., Jung, Y., Shin, B.S., Song, H., Bae, S.H., Rhee, S.G. and Jeong, W., 2010. Redox regulation of lipopolysaccharide-mediated sulfiredoxin induction, which depends on both AP-1 and Nrf2. The Journal of Biological Chemistry 285: 34419-34428.

Koháryová, M. and Kollárová, M., 2015. Thioredoxin system – a novel therapeutic target. General Physiology and Biophysics 34: 221-233.

Lavric, M., Maughan, M.N., Bliss, T.W., Dohms, J.E., Bencina, D., Keeler, C.L., Jr. and Narat, M., 2008. Gene expression modulation in chicken macrophages exposed to *Mycoplasma synoviae* or *Escherichia coli*. Veterinary Microbiology 126: 111-121.

Lee, S.H., Lillehoj, H.S., Jang, S.I., Jeong, M., Kim, D.K., Xu, S., Lee, S.K., Kim, J.B., Park, H.J., Kim, H.R. and Bravo, D.M., 2014. Immune and antioxidant effects of in ovo selenium proteinate on post-hatch experimental avian necrotic enteritis. Veterinary Parasitology 206: 115-122.

Li, L., Lin, G., Gu, H., Yu, L. and Ni, C., 2019. Effects of Srxn1 on growth and Notch signalling of astrocyte induced by hydrogen peroxide. Artificial Cells, Nanomedicine, and Biotechnology 47: 1917-1923.

Li, X., He, P., Wang, X.L., Zhang, S., Devejian, N., Bennett, E. and Cai, C., 2018. Sulfiredoxin-1 enhances cardiac progenitor cell survival against oxidative stress via the upregulation of the ERK/NRF2 signal pathway. Free Radical Biology and Medicine 123: 8-19.

Liang, Y., Lin, S.L., Wang, C.W., Yao, H.D., Zhang, Z.W. and Xu, S.W., 2014. Effect of selenium on selenoprotein expression in the adipose tissue of chickens. Biological Trace Element Research 160: 41-48.

Lin, S.L., Wang, C.W., Tan, S.R., Liang, Y., Yao, H.D., Zhang, Z.W. and Xu, S.W., 2014. Selenium deficiency inhibits the conversion of thyroidal thyroxine (T4) to triiodothyronine (T3) in chicken thyroids. Biological Trace Element Research 161: 263-271.

Liu Q, Yang J, Cai J, Luan Y, Sattar H, Liu M, Xu S, Zhang Z. 2017. Analysis of the interactions between thioredoxin and 20 selenoproteins in chicken. Biological Trace Element Research 179: 304-317.

Lu, J. and Holmgren, A., 2014. The thioredoxin antioxidant system. Free Radical Biology & Medicine 66: 75-87.

Luthman, M. and Holmgren, A., 1982. Rat liver thioredoxin and thioredoxin reductase: purification and characterization. Biochemistry 21: 6628-6633.

Marzoni, M., Castillo, A., Sagona, S., Citti, L., Rocchiccioli, S., Romboli, I. and Felicioli, A., 2013. A proteomic approach to identify seminal plasma proteins in roosters (*Gallus gallus domesticus*). Animal Reproduction Science 140: 216-223.

Matsuzawa, A., 2017. Thioredoxin and redox signaling: roles of the thioredoxin system in control of cell fate. Archives of Biochemistry and Biophysics 617: 101-105.

May, J.M., Mendiratta, S., Hill, K.E. and Burk, R.F., 1997. Reduction of dehydroascorbate to ascorbate by the selenoenzyme thioredoxin reductase. The Journal of Biological Chemistry 272: 22607-22610.

Miranda-Vizuete, A., Damdimopoulos, A.E. and Spyrou, G., 2000. The mitochondrial thioredoxin system. Antioxidants and Redox Signalling 2: 801-810.

Miranda-Vizuete, A., Sadek, C.M., Jimenez, A., Krause, W.J., Sutovsky, P. and Oko, R., 2004. The mammalian testis-specific thioredoxin system. Antioxidants and Redox Signalling 6: 25-40.

Mishra, M., Jiang, H., Wu, L., Chawsheen, H.A., and Wei, Q., 2015. The sulfiredoxin-peroxiredoxin (Srx-Prx) axis in cell signal transduction and cancer development. Cancer Letters 366: 150-159.

Miyazawa, T., Burdeos, G.C., Itaya, M., Nakagawa, K. and Miyazawa, T., 2019. Vitamin E: regulatory redox interactions. IUBMB Life 71: 430-441.

Mohammadi, F., Soltani, A., Ghahremanloo, A., Javid, H. and Hashemy, S.I., 2019. The thioredoxin system and cancer therapy: a review. Cancer Chemotherapy and Pharmacology 84: 925-935.

Mustacich, D. and Powis, G., 2000. Thioredoxin reductase. The Biochemical Journal 346: 1-8.

Nakamura, H., 2004. Thioredoxin as a key molecule in redox signalling. Antioxidants and Redox Signalling 6: 15-17.

Nalvarte, I., Damdimopoulos, A.E. and Spyrou, G., 2004. Human mitochondrial thioredoxin reductase reduces cytochrome c and confers resistance to complex III inhibition. Free Radical Biology & Medicine 36: 1270-1278.

Noh, Y.H., Baek, J.Y., Jeong, W., Rhee, S.G. and Chang, T.S., 2009. Sulfiredoxin translocation into mitochondria plays a crucial role in reducing hyperoxidized peroxiredoxin III. The Journal of Biological Chemistry 284: 8470-8477.

Nordberg, J. and Arner, E.S., 2001. Reactive oxygen species, antioxidants, and the mammalian thioredoxin system. Free Radical Biology and Medicine 31: 1287-1312.

Patterson, A.D., Carlson, B.A., Li, F., Bonzo, J.A., Yoo, M.H., Krausz, K.W., Conrad, M., Chen, C., Gonzalez, F.J. and Hatfield, D.L., 2013. Disruption of thioredoxin reductase 1 protects mice from acute acetaminophen-induced hepatotoxicity through enhanced NRF2 activity. Chemical Research in Toxicology 26: 1088-1096.

Penney, R.B. and Roy, D., 2013. Thioredoxin-mediated redox regulation of resistance to endocrine therapy in breast cancer. Biochimica et Biophysica Acta (BBA)-Reviews on Cancer 1836: 60-79.

Perkins, A., Poole, L.B. and Karplus, P.A., 2014. Tuning of peroxiredoxin catalysis for various physiological roles. Biochemistry 53: 7693-7705.

Pirson, M., Clippe, A. and Knoops, B., 2018. The curious case of peroxiredoxin-5: what its absence in aves can tell us and how it can be used. BMC Evolutionary Biology 18: 18.

Placha, I., Takacova, J., Ryzner, M., Cobanova, K., Laukova, A., Strompfova, V., Venglovska, K. and Faix, S., 2014. Effect of thyme essential oil and selenium on intestine integrity and antioxidant status of broilers. British Poultry Science 55: 105-114.

Powis, G., Mustacich, D. and Coon, A., 2000. The role of the redox protein thioredoxin in cell growth and cancer. Free Radical Biology and Medicine 29: 312-322.

Poynton, R.A. and Hampton, M.B., 2014. Peroxiredoxins as biomarkers of oxidative stress. Biochimica et Biophysica Acta 1840: 906-912.

Pravda, J., 2020. Hydrogen peroxide and disease: towards a unified system of pathogenesis and therapeutics. Molecular Medicine 26: 41.

Ramesh, A., Varghese, S.S., Doraiswamy, J. and Malaiappan, S., 2014. Role of sulfiredoxin in systemic diseases influenced by oxidative stress. Redox Biology 2: 1023-1028.

Rhee, S.G. and Kil, I.S., 2016. Mitochondrial H_2O_2 signaling is controlled by the concerted action of peroxiredoxin III and sulfiredoxin: Linking mitochondrial function to circadian rhythm. Free Radical Biology & Medicine 100: 73-80.

Rundlof, A.K. and Arner, E.S., 2004. Regulation of the mammalian selenoprotein thioredoxin reductase 1 in relation to cellular phenotype, growth, and signalling events. Antioxidants and Redox Signalling 6: 41-52.

Schmidlin, C.J., Dodson, M.B., Madhavan, L. and Zhang, D.D., 2019. Redox regulation by NRF2 in aging and disease. Free Radical Biology & Medicine 134: 702-707.

Shahnaj, S., Chowhan, R.K., Meetei, P.A., Kakchingtabam, P., Herojit Singh, K., Rajendrakumar Singh, L., Nongdam, P., Fisher, A.B. and Rahaman, H., 2019. Hyperoxidation of peroxiredoxin 6 induces alteration from dimeric to oligomeric state. Antioxidants 8: 33.

Sharapov, M.G. and Novoselov, V.I., 2019. Catalytic and signaling role of peroxiredoxins in carcinogenesis. Biochemistry 84: 79-100.

Sies, H., 2019. Oxidative stress: eustress and distress in redox homeostasis. In: Fink, G. (ed.) Stress: physiology, biochemistry, and pathology. Academic Press, London, UK, pp. 153-163.

Sies, H. and Jones, D.P., 2020. Reactive oxygen species (ROS) as pleiotropic physiological signalling agents. Nature Reviews. Molecular Cell Biology 21: 363-383. https://doi.org/10.1038/s41580-020-0230-3

Smith, A.D., Morris, V.C. and Levander, O.A., 2001. Rapid determination of glutathione peroxidase and thioredoxin reductase activities using a 96-well microplate format: comparison to standard cuvette-based assays. International Journal for Vitamin and Nutrition Research 71: 87-92.

Soriano, F.X., Leveille, F., Papadia, S., Higgins, L.G., Varley, J., Baxter, P., Hayes, J.D., and Hardingham, G.E., 2008. Induction of sulfiredoxin expression and reduction of peroxiredoxin hyperoxidation by the neuroprotective Nrf2 activator 3H-1,2-dithiole-3-thione. Journal of Neurochemistry 107: 533-543.

Sun, Q.A., Kirnarsky, L., Sherman, S. and Gladyshev, V.N., 2001. Selenoprotein oxidoreductase with specificity for thioredoxin and glutathione systems. Proceedings of the National Academy of Sciences of the USA 98: 3673-3678.

Sun, Q.A., Wu, Y., Zappacosta, F., Jeang, K.T., Lee, B.J., Hatfield, D.L. and Gladyshev, V.N., 1999. Redox regulation of cell signalling by selenocysteine in mammalian thioredoxin reductases. The Journal of Biological Chemistry 274: 24522-24530.

Surai, P.F., 2006. Selenium in nutrition and health. Nottingham University Press, Nottingham, UK.

Surai, P.F., 2018. Selenium in poultry nutrition and health. Wageningen Academic Publishers, Wageningen, the Netherlands.

Tamura, T. and Stadtman, T.C., 1996. A new selenoprotein from human lung adenocarcinoma cells: purification, properties, and thioredoxin reductase activity. Proceedings of the National Academy of Sciences of the USA 93: 1006-1011.

Tanaka, Y., Tran, P.O., Harmon, J. and Robertson, R.P., 2002. A role for glutathione peroxidase in protecting pancreatic beta cells against oxidative stress in a model of glucose toxicity. Proceedings of the National Academy of Sciences of the USA 99: 12363-12368.

Tu, W.L., Cheng, C.Y., Chen, C.J., Chan, H.L., Wang, S.H., Tang, P.C., Chen, C.F., Lee, Y.P., Chen, S.E. and Huang, S.Y., 2018. Annotation of differential protein expression in the hypothalami of layer-type Taiwan country chickens in response to acute heat stress. Journal of Thermal Biology 77: 157-172.

Veal, E.A., Underwood, Z.E., Tomalin, L.E., Morgan, B.A. and Pillay, C.S., 2018. Hyperoxidation of peroxiredoxins: gain or loss of function? Antioxidants & Redox Signaling 28: 574-590.

Wang, D., Masutani, H., Oka, S., Tanaka, T., Yamaguchi-Iwai, Y., Nakamura, H. and Yodoi, J., 2006. Control of mitochondrial outer membrane permeabilization and Bcl-xL levels by thioredoxin 2 in DT40 cells. The Journal of Biological Chemistry 281: 7384-7391.

Wang, Y., Yin, X., Yin, D., Lei, Z., Mahmood, T. and Yuan, J., 2019. Antioxidant response and bioavailability of methionine hydroxy analog relative to DL-methionine in broiler chickens. Animal Nutrition 5: 241-247.

Wei, Q., Jiang, H., Matthews, C.P., and Colburn, N.H., 2008. Sulfiredoxin is an AP-1 target gene that is required for transformation and shows elevated expression in human skin malignancies. Proceedings of the National Academy of Sciences of the United States of America 105: 19738-19743.

Woo, H.A., Jeong, W., Chang, T.S., Park, K.J., Park, S.J., Yang, J.S., and Rhee, S.G., 2005. Reduction of cysteine sulfinic acid by sulfiredoxin is specific to 2-cys peroxiredoxins. The Journal of Biological Chemistry 280: 3125-3128.

Wu, Q. and Huang, K., 2004. Effect of long-term se deficiency on the antioxidant capacities of rat vascular tissue. Biological Trace Element Research 98: 73-84.

Xiao, R., Power, R.F., Mallonee, D., Routt, K., Spangler, L., Pescatore, A.J., Cantor, A.H., Ao, T., Pierce, J.L. and Dawson, K.A., 2012. Effects of yeast cell wall-derived mannan-oligosaccharides on jejunal gene expression in young broiler chickens. Poultry Science 91: 1660-1669.

Xu, J.X., Zhang, C., Cao, C.Y., Zhu, S.Y., Li, H., Sun, Y.C. and Li, J.L., 2016. Dietary selenium status regulates the transcriptions of selenoproteome and activities of selenoenzymes in chicken kidney at low or super-nutritional levels. Biological Trace Element Research 170: 438-448.

Yang, J., Gong, Y., Cai, J., Liu, Q., Zhang, Y., Zheng, Y., Yu, D. and Zhang, Z., 2020. Dysfunction of thioredoxin triggers inflammation through activation of autophagy in chicken cardiomyocytes. Biofactors 46: 579-590. https://doi.org/10.1002/biof.1625

Yang, J., Gong, Y., Liu, Q., Cai, J., Zhang, B. and Zhang, Z., 2018. Thioredoxin silencing-induced cardiac supercontraction occurs through endoplasmic reticulum stress and calcium overload in chicken. Metallomics 10: 1667-1677.

Yang, J., Hamid, S., Cai, J., Liu, Q., Xu, S. and Zhang, Z., 2017b. Selenium deficiency-induced thioredoxin suppression and thioredoxin knock down disbalanced insulin responsiveness in chicken cardiomyocytes through PI3K/Akt pathway inhibition. Cellular Signalling 38: 192-200.

Yang, J., Hamid, S., Liu, Q., Cai, J., Xu, S. and Zhang, Z., 2017a. Gene expression of selenoproteins can be regulated by thioredoxin (Txn) silence in chicken cardiomyocytes. Journal of Inorganic Biochemistry 177: 118-126.

Yang, K.T., Lin, C.Y., Huang, H.L., Liou, J.S., Chien, C.Y., Wu, C.P., Huang, C.W., Ou, B.R., Chen, C.F., Lee, Y.P., Lin, E.C., Tang, P.C., Lee, W.C., Ding, S.T., Cheng, W.T. and Huang, M.C., 2008. Expressed transcripts associated with high rates of egg production in chicken ovarian follicles. Molecular and Cellular Probes 22: 47-54.

Yarana, C., Thompson, H., Chaiswing, L., Butterfield, D.A., Weiss, H., Bondada, S., Alhakeem, S., Sukati, S. and St Clair, D.K., 2019. Extracellular vesicle-mediated macrophage activation: an insight into the mechanism of thioredoxin-mediated immune activation. Redox Biology 26: 101237.

Yu, S., Wang, X., Lei, S., Chen, X., Liu, Y., Zhou, Y., Zhou, Y., Wu, J. and Zhao, Y., 2015. Sulfiredoxin-1 protects primary cultured astrocytes from ischemia-induced damage. Neurochemistry International 82: 19-27.

Yuan, D., Zhan, X.A. and Wang, Y.X., 2012. Effect of selenium sources on the expression of cellular glutathione peroxidase and cytoplasmic thioredoxin reductase in the liver and kidney of broiler breeders and their offspring. Poultry Science 91: 936-942.

Zhang, J., Bai, K. W., He, J., Niu, Y., Lu, Y., Zhang, L. and Wang, T., 2018. Curcumin attenuates hepatic mitochondrial dysfunction through the maintenance of thiol pool, inhibition of mtDNA damage, and stimulation of the mitochondrial thioredoxin system in heat-stressed broilers. Journal of Animal Science 96: 867-879.

Zhang, J., He, Z., Guo, J., Li, Z., Wang, X., Yang, C. and Cui, X., 2016. Sulfiredoxin-1 protects against simulated ischaemia/reperfusion injury in cardiomyocyte by inhibiting PI3K/AKT-regulated mitochondrial apoptotic pathways. Bioscience Reports 36: e00325.

Zhang, M., An, C., Gao, Y., Leak, R.K., Chen, J. and Zhang, F., 2013. Emerging roles of Nrf2 and phase II antioxidant enzymes in neuroprotection. Progress in Neurobiology 100: 30-47.

Zhang, J., Zhang, B., Li, X., Han, X., Liu, R. and Fang, J., 2019. Small molecule inhibitors of mammalian thioredoxin reductase as potential anticancer agents: An update. Medicinal Research Reviews 39: 5-39.

Zhang, Y., Roh, Y.J., Han, S.J., Park, I., Lee, H.M., Ok, Y.S., Lee, B.C. and Lee, S.R., 2020. Role of selenoproteins in redox regulation of signaling and the antioxidant system: a review. Antioxidants 9: E383.

Zhao, X., Yao, H., Fan, R., Zhang, Z. and Xu, S., 2014. Selenium deficiency influences nitric oxide and selenoproteins in pancreas of chickens. Biological Trace Element Research 161: 341-349.

Zhao, L., Feng, Y., Deng, J., Zhang, N.Y., Zhang, W.P., Liu, X.L., Rajput, S.A., Qi, D.S., and Sun, L.H., 2019. Selenium deficiency aggravates aflatoxin B1-induced immunotoxicity in chick spleen by regulating 6 selenoprotein genes and redox/inflammation/apoptotic signaling. The Journal of Nutrition 149: 894-901.

Zhong, L., Arnér, E.S. and Holmgren, A., 2000. Structure and mechanism of mammalian thioredoxin reductase: the active site is a redox-active selenolthiol/selenenylsulfide formed from the conserved cysteine-selenocysteine sequence. Proceedings of the National Academy of Sciences of the United States of America 97: 5854-5859.

Zhong, L. and Holmgren, A., 2000. Essential role of selenium in the catalytic activities of mammalian thioredoxin reductase revealed by characterization of recombinant enzymes with selenocysteine mutations. The Journal of Biological Chemistry 275: 18121-18128.

Zhou, Y., Duan, S., Zhou, Y., Yu, S., Wu, J., Wu, X., Zhao, J. and Zhao, Y., 2015. Sulfiredoxin-1 attenuates oxidative stress via Nrf2/ARE pathway and 2-Cys Prdxs after oxygen-glucose deprivation in astrocytes. Journal of Molecular Neuroscience 55: 941-950.

Zhou, Y.Y., Chun, R., Wang, J.C., Zuo, B., Li, K.K., Lam, T.C., Liu, Q. and To, C.H., 2018. Proteomic analysis of chick retina during early recovery from lens-induced myopia. Molecular Medicine Reports 18: 59-66.

Zoidis, E., Papadomichelakis, G., Pappas, A.C., Theodorou, G. and Fegeros, K., 2019. Effects of selenium and cadmium on breast muscle fatty-acid composition and gene expression of liver antioxidant proteins in broilers. Antioxidants 8: 147.

Chapter 7
Glutathione system in avian biology

Danger foreseen is half avoided

7.1 Introduction

Thioredoxin and glutathione systems are the two major thiol-dependent redox systems in cells participating in antioxidant defences, DNA synthesis and repair as well as in prevention of protein oxidation and stress adaptation. The regulation of oxidative stress and prevention of its detrimental effects are key to the maintenance of aerobic life. It is well appreciated that at physiological conditions, low levels of RONS are important elements of the cell signalling via induction of discrete, reversible and site-specific protein modifications (Nikolaenko *et al.*, 2018). Therefore, RONS are suggested to act as important second messengers related to stress response. In fact, the redox signalling is based on the ability of RONS to reversely modulate protein cysteines, resulting in S-nitrosylation, S-glutathionylation, and disulphide bond formation and affecting activity of the proteins involved in different signalling cascades. Specific enzymes of the glutathione and thioredoxin systems utilising the reducing power of NADPH are responsible for reduction of the affected proteins. For example, disulphide bridges and mixed disulphides (S-glutathionylation) can be reduced by both thioredoxin and glutathione/glutaredoxin systems, while thioredoxin system can reduce S-nitrosothiols (Nikolaenko *et al.*, 2018). Most, organisms, including mammals and birds, have the glutathione system in the centre of their cellular redox control where reduced glutathione is maintained by many mechanisms, including its de novo synthesis, import and reduction of oxidised glutathione (GSSG), export and sequestration of GSSG (Couto *et al.*, 2016). The glutathione system consists of glutathione (GSH), glutathione reductase (GR), glutaredoxins (Grx) and glutathione peroxidases (GPx).

This chapter deals with recent findings related to GSH system functioning as an integrated part of the vitagene network with specific emphasis to its role in avian biology and poultry protection against various stresses.

7.2 Glutathione

Glutathione (GSH, γ-l-glutamyl-l-cysteinylglycine)) is the most abundant non-protein thiol in avian and mammalian cells and considered to be an active antioxidant in biological systems providing cells with their reducing milieu (Meister, 1992). Indeed, GSH is shown to be one of the most important non-enzymatic antioxidants in animals/poultry participating in redox balance maintenance and signalling,

Peter F. Surai **Vitagenes in avian biology and poultry health**
DOI 10.3920/978-90-8686-906-0_7, © Wageningen Academic Publishers 2020

regulation of transcription factors and gene expression and many other important pathways/processes including epigenetic mechanisms (García-Giménez *et al.*, 2017). In fact, GSH is the dominant low-molecular weight antioxidant in mammalian cells. Cellular GSH plays a key role in many biological processes (Sen and Packer, 2000):

- the synthesis of DNA and proteins;
- cell growth and proliferation;
- regulation of programmed cell death;
- immune regulation;
- the transport of amino acids;
- xenobiotic metabolism;
- redox-sensitive signal transduction.

Furthermore, GSH thiol group can react directly with (Lenzi *et al.*, 2000; Meister and Anderson, 1983):

- H_2O_2;
- superoxide anion;
- hydroxyl radicals;
- alkoxyl radicals;
- hydroperoxides.

Therefore, a crucial role for GSH is as free radical scavenger, particularly effective against the hydroxyl radical (Bains and Shaw, 1997), since there are no enzymatic defences against this species of radical. Usually decreased GSH concentration in tissues is associated with increased lipid peroxidation (Thompson *et al.*, 1992). Furthermore, in stress conditions GSH prevents the loss of protein thiols and vitamin E (Palamanda and Kehrer, 1993) and plays an important role as a key modulator of cell signalling (Elliott and Koliwad, 1997). Animals and human are able to synthesise glutathione. Indeed, the reduced glutathione itself can participate in maintenance of protein –SH groups. At the same time the thioredoxin system has alkyl hydroperoxide reductase activity. Protein disulphide isomerase is also involved in re-pairing of –SH groups in proteins (Dean *et al.*, 1997).

Interestingly in the ER, GSH is mostly oxidised, while nuclear GSH is found in the reduced form and plays a key role in preserving proteins involved in DNA repair and gene transcription. Mitochondrial GSH preserves the mitochondrial integrity participating in controlling mitochondrial ROS generation and apoptotic signalling (Conde de la Rosa *et al.*, 2014). Therefore, cellular GSH plays a key role in many biological processes, including synthesis of DNA and proteins, regulating cell growth and proliferation, apoptosis, immunity, amino acid transport, xenobiotic and endogenous oxidant metabolism/detoxification, redox-sensitive signal transduction, etc. (Aquilano *et al.*, 2014; Hansen and Harris, 2015). In fact, GSH thiolic group can react directly with a range of ROS, such as H_2O_2, superoxide anion, hydroxyl radicals, alkoxyl radicals, hydroperoxides (Ribas *et al.*, 2014). There is a range of proteins with a GSH-dependent hydroperoxidase activity. In addition to specialised GPx and Prx-isoforms, some Grx and many GST can also act as hydroperoxidases on their own (Deponte *et al.*, 2013). Furthermore, in stress conditions GSH, being a redox buffer controlling redox status of the living cells, can prevent the loss of protein thiols and plays an important role as a key modulator of cell signalling (Griffiths *et al.*, 2014), either via glutathionylation or via metabolism of hydrogen peroxide (Farina and Aschner, 2019). Therefore, under oxidative stress, a decreased GSH/GSSG ratio is associated with protein S-glutathionylation: a direct modification of protein cysteine residues by the addition of GSH leading to a mixed disulphide formation between reactive thiols and GSH resulting in altering physiological functions of affected proteins (Figure 7.1).

A notable amount of glutathione can be reversibly bound to the -SH of protein cysteinyl residues (P-Cys-SH) by a mechanism called S-glutathionylation, which generates S-glutathionylated proteins (P-Cys-SSG). Therefore, free thiols on reactive cysteinyl residues are modified by the formation of an intermediate thiol derivative, or by direct thiol-disulphide exchange. S-glutathionylation can be reversed by the action of thiol-modifying enzymes such as the thiol-disulphide oxidoreductase glutaredoxin (Grx) accounting for most of the deglutathionylating activity in mammalian cells.

In fact, cysteinyl residues in proteins are especially sensitive to oxygen and nitrogen species (RONS) which can cause a range of oxidative post-translational modifications (PTM) including S-nitrosylation, S-sulfenylation and S-glutathionylation. Indeed, oxidative post-translational modifications (oxPTM) of receptors, enzymes, ion channels and transcription factors are considered to be important contributors to

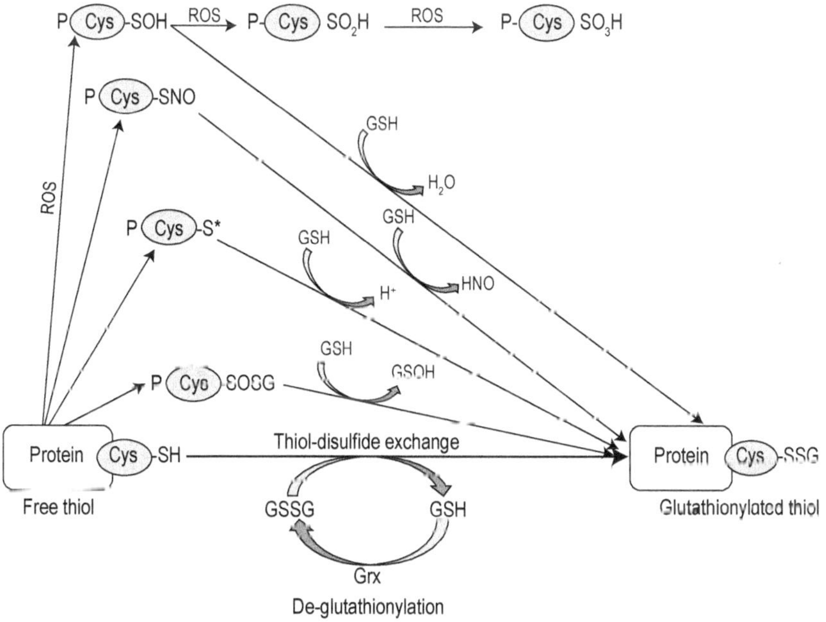

Figure 7.1. Potential reaction pathways leading to protein S-glutathionylation (adapted from Dalle-Donne *et al.*, 2009; Gallogly and Mieya, 2007; Lehrman and Murdoch, 2019; Zhang *et al.*, in press).

cell signalling. Furthermore, protein S-glutathionylation has been proven to modulate various mitochondrial functions including nutrient metabolism, ATP production, ROS release, solute import, permeability transition, protein uptake, and fission/fusion (Young *et al*, 2019) being a vital regulator of major signalling pathways in combination with other posttranslational regulatory mechanisms. In fact, oxPTMs are key cellular events affecting cell behaviour during diverse stress conditions. Among oxPTM, S-glutathionylation of reactive cysteines emerges as an important regulator of cellular homeostasis by modulating cell responses to their local redox environment (Lehrman and Murdoch, 2019). It is important to underline that protein S-glutathionylation is a reversible process, since it can be reversed by thiol modifying enzymes, predominantly glutaredoxin (Grx). Therefore, S-glutathionylation is considered as a protective mechanism against permanent protein damage following irreversible cysteine oxidation by RONS. It seems likely that S-glutathionylation can coordinate gene transcription by modulating epigenetics and transcription factors. In fact, by activation of Nrf2 and repression of NF-κB in stress conditions S-glutathionylation contribute to effective stress-adaptation. It is believed that S-glutathionylation exhibits a general inhibitory effect on enzymes by altering the structure of their catalytic site and impairing their activity. This is especially important in relation to phosphatases, GTPases and kinases, known to be key signal transducers in the cell (Lehrman and Murdoch, 2019). It seems likely that RONS production is also controlled by S-glutathionylation, since activity of many RONS producing enzymes are affected by this process. Protein S-glutathionylation is shown to be rapid, mainly enzymatically mediated, specific and reversible and these unique features make S-glutathionylation ideal for the regulation of cell functions in response to various stress stimuli/conditions (Young *et al*, 2019). In fact, abnormal protein S-glutathionylation is related to diverse cellular activities, including protein aggregation, protein degradation, apoptosis, and mitochondrial dysfunction (Ren *et al.*, 2017). At the same time, Grxs possessing high affinity and selectivity for glutathionylated proteins are known to be the major biological deglutathionylases.

Therefore, protective roles of GSH in the cells/body is of great importance for homeostasis maintenance and stress adaptation. In fact, the GSH/GSSG couple is the redox buffer responsible for maintaining appropriate redox conditions from the sub-organellar to the organismic level (Deponte *et al.*, 2013). The ratio GSH/GSSG reflects the cellular redox potential and redox balance. The expression of enzymes responsible for biosynthesis of GSH, including the catalytic and regulatory subunits of GCL and GSH synthase, is shown to be under the transcriptional control of Nrf2. Furthermore, Nrf2 also regulates the import of the GSH substrate cysteine through the cystine/glutamate antiporter, while important enzymes of GSH metabolism, including GPx2, GSTs, and GR, are synthesised under Nrf2 control (Lawerenz *et al.*, 2013).

It is well-known that GSH can be synthesised in the human/animal body from three amino acids (L-glutamate, L-cysteine and glycine) with Cys availability and concentration being a limiting factor, while γ-glutamylcysteinesynthetase (γ-GCS) is known to be rate-limiting in glutathione biogenesis (Couto *et al.*, 2016). GSH is exclusively synthesised in the cytosol and compartmentalised in different organelles, including nuclei, endoplasmic reticulum (ER), and mitochondria. In fact, in mammalian cells glutathione found mainly (>98%) in the thiol-reduced form (GSH) and it is located predominantly (about 90%) in the cytosol (1-10 mM), ~10% in the mitochondria (5-10 mM) and the rest is located in the endoplasmic reticulum and the nucleus (Dalle-Donne *et al.*, 2009). The liver is believed to be the major producer and exporter of GSH. Because in most of the functions GSH is used in its reduced form, an active enzyme mechanism exists in the form of glutathione reductase for the reduction of GSSG to GSH.

There are species-specific differences in GSH level in the liver of the different classes: mammals (6-8 µmol/g), birds (2.5-3.7 µmol/g), amphibians (0.9-2.2 µmol/g), and reptiles (except anoxia-tolerant ones) (1-1.2 µmol/g) (Storey, 1996). Hen's eggs before incubation was shown to contain small amount of glutathione, all of which was found in the yolk. The average glutathione content was reported to be about 3.05 µg/g of egg, or 11.6 µg/g of yolk comprising 134.4-193.5 µg/egg (Cazorla and Guzman Barron, 1958). The author showed that developing embryo synthesises GSH, since by the end of 15 days' incubation GSH content increased to 45 micrograms per gram of egg, a 14-fold increase in comparison to the initial value (Figure 7.2). Interestingly, the increase of GSH content in the embryo, from 68 hours up to 140 hours was shown to be due to transport of the yolk glutathione to the embryo, since there was no absolute increase in the whole egg (Cazorla and Guzman Barron, 1958).

Similarly, the glutathione concentration in skeletal muscles was found to be increasing between the 9th and 18th days of chicken embryonic development (Boldyrev *et al.*, 1988). It seems likely that there is a redistribution of GSH between different tissues during embryonic development, since GSH concentration in the chicken embryonic

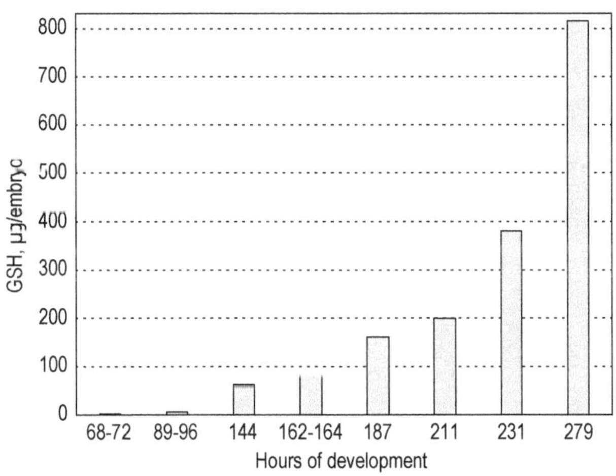

Figure 7.2. Reduced glutathione (GSH) in chicken embryo during development, µg/embryo (adapted from Cazorla and Guzman Barron, 1958).

liver was shown to gradually decrease throughout development (Figure 7.3; Surai, 1999). There is also tissue specificity in GSH concentration in the newly hatched chickens (Figure 7.4; Surai *et al.*, 1999).

Indeed, the highest GSH concentration was reported in the kidney, while its concentration in the lung and thigh muscles was more 2 times lower. The beneficial effect of organic Se supplementation of the breeders on the level of GSH in the liver

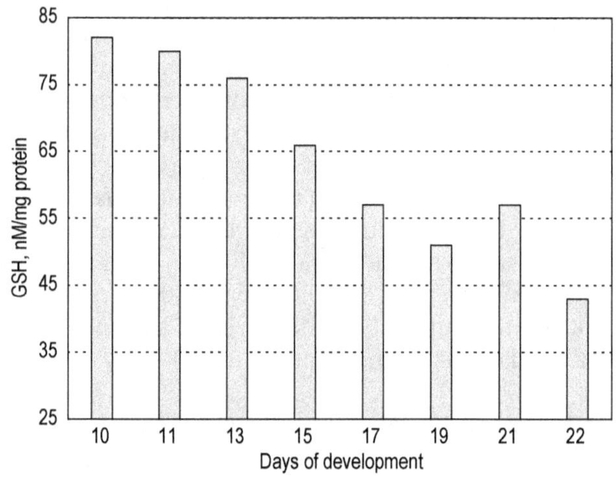

Figure 7.3. Reduced glutathione (GSH) concentration in the embryonic liver (adapted from Surai, 1999).

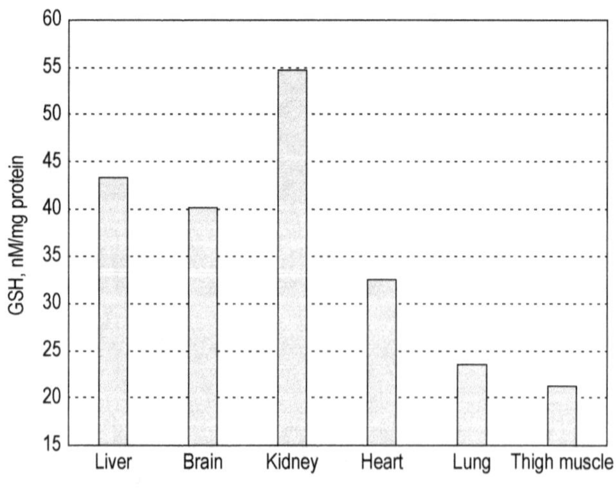

Figure 7.4. Reduced glutathione (GSH) concentration in the tissues of a newly hatched chick (adapted from Surai, 1999).

of newly hatched chicks was shown. Furthermore, vitamin E at 200 mg/kg in the maternal diet also increased the concentration of GSH in the liver of newly hatched chicks (Surai, 2000). The glutathione redox system was reported to be activated by mycotoxin exposure of chickens. In particular, shortly after starting the T-2 toxin exposure: the significantly increased GSH concentration in blood plasma at 24 and 48 h, in liver at 12, 24 and 36 h, and in kidney and spleen at 24 h were observed (Bócsai *et al.*, 2015). Similarly, GSH content in the liver of T-2/HT-2 toxin-treated chickens was significantly higher than in the control group (Nakade *et al.*, 2018). Furthermore, multi-trichothecene (T-2 toxin +DON) mycotoxin exposure was shown to activate glutathione-redox system in broiler chicken liver as evidenced by increased GSH concentration on day 3 of feeding (Pelyhe *et al.*, 2018). In addition, GSH concentration in the chicken blood plasma and liver was shown to be increased due to high (1 mg/kg) dietary ochratoxin A consumption for 14 days (Kövesi *et al.*, 2019). However, AFB1 chicken feed contamination (92.0 μg/ kg feed) was shown to decrease GSH concentration in the liver at day 14 of feeding (Kövesi *et al.*, 2020). Interestingly, higher and lower AFB1 doses as well as shorter (7 days) or longer (21 days) treatment did not affect liver GSH concentration. Interestingly, another stress due to Se excess was also able to induce GSH concentration. Indeed, excessive Se supplementation (24.5 mg Se/kg feed for 4 days) in inorganic or organic form of 3-week-old broilers was associated with elevated GSH concentration and GPx activity in plasma and liver (Balogh *et al.*, 2007).

7.3 Glutathione reductase

GR is a flavoenzyme of the pyridine nucleotide disulphide oxidoreductase family, an NADPH:GSSG oxidoreductase (EC 1.8.1.7). The enzyme has three substrates, namely NADPH, H^+ and GSSG and it plays a central role in GSH metabolism by linking the cellular NADPH pool with the thiol/disulphide-pool and helping maintain a reducing intracellular milieu. Different GR-isoforms are found in the cytosol and in the mitochondrial matrix (Deponte *et al.*, 2013). Therefore, GR is an essential enzyme that recycles oxidised glutathione back to the reduced form:

$$\text{GSSG} + \text{NADPH} + \text{H}^+ \xrightarrow{\text{GR}} 2\text{GSH} + \text{NADP}^+$$

This enzyme is highly conserved across nature and a high degree of similarity has been shown between its three-dimensional structures in various species. For example, at the level of the primary structure, *Saccharomyces cerevisiae* glutathione reductase shares 51% identity with its human homologue (Couto *et al.*, 2016). GR contains two conserved cysteines (C61, C65) at the catalytic site and these form a disulphide bond. In eukaryotes GR is found in the cytoplasm, nucleus and mitochondria, while oxidoreductase activity was also detected in the endoplasmic reticulum and in the lysosomes with a single gene expressing both the cytosolic and mitochondrial forms of GR (Couto *et al.*, 2016).

7.3.1 Avian glutathione reductase

Chicken liver GR was purified 1,714-fold to have a specific activity of 120 EU/mg of protein and characterised in 2005 (Erat *et al.*, 2005). Chiken GR has a molecular weight of 43 kDa. Because the molecular weight of GR determined by gel filtration chromatography (100 kDa) was approximately twice that by SDS-PAGE, native chicken liver GR was suggested to exist as a dimer in an active state. In general, GR of different origin were shown to have similar molecular weight, as follows: 100 kDa from human erythrocyte, 100 kDa from calf liver, 103 kDa from porcine erythrocyte, 116 kDa from sheep brain, 125 kDa from gerbil liver and 125 kDa from rat liver The optimum pH of cGR was determined as 7.0 and the stable pH of the enzyme was demonstrated to be 7.4 in Tris-HCl buffer. The enzyme's highest activity point was found to be at 50 °C. Interestingly, the *KM* for NADPH was shown to be lower than that for GSSG, suggesting a higher affinity of cGR to NADPH when compared with GSSG (Erat *et al.*, 2005). In accordance with GR activity in the liver various vertebrate species can be placed in the following descending order: rat>chick>>lizard>frog (Venditti *et al.*, 1996). Similar order was characteristic for GPx activity. GR activity in chicken liver and erythrocytes was shown to increase between 2 and 4-weeks of age (Mahmoud and Edens, 2003). There was a dramatic decrease in GR gene expression in the liver of broilers between 21 and 42 days of age (Del Vesco *et al.*, 2017). GR activity in chicken embryo was observed as early as 3 d day of incubation and there was a 2-fold increase in GR activity between days 3 and 6 of the embryo development (Figure 7.5; Cazorla and Guzman Barron, 1958).

There was also a decrease in GR activity in the chicken liver (by 32%) between 14 and 35 days of age but in the duodenal mucosa of broilers GR activity did not change

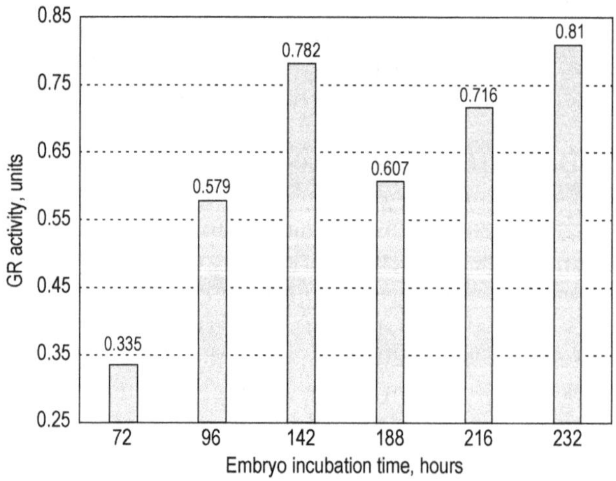

Figure 7.5. Glutathione reductase (GR) activity in chicken embryo (adapted from Cazorla and Guzman Barron, 1958).

(Liu *et al.*, 2017). The specific activities of GPx and GR, and the level of TBARS in muscles from chickens and mice with genetic muscular dystrophy were shown to be significantly increased over those of control animals (Omaye *et al.*, 1974). There were no differences in GPx or GR activities between groups but GSH levels tended to be higher and GSH/GSSG ratio tended to be lower in broilers with high feed efficiency (Ojano-Dirain *et al.*, 2005). Under extreme hypoxia conditions (14% oxygen concentration for 14 h and then 10.5% oxygen concentration for 6 h), values for the GSH content, the GSH:GSSG ratio, and the activity of GR in the liver of the Tibet chicken were higher than those of the Silky chicken, while in normoxia there was no difference between chick breeds (Bao *et al.*, 2011). It means that Tibet chickens were better adapted to hypoxia.

7.3.2 Environmental and nutritional modulation of avian glutathione reductase

Thermal stress

Under tropical summer conditions increased vitamin E supplementation (125 vs 25 mg/kg) or dietary supplementation of ascorbic acid (200 vs 0 mg/kg) were shown to increase GR activity in erythrocytes of White Leghorn layers (Panda *et al.*, 2008). In similar heat-stress conditions feeding of sprouts to chickens significantly increased the activities of GR, GPx and SOD and decreased lipid peroxidation in liver and spleen of broilers compared to the control group (Rama Rao *et al*, 2018). In the spleen of the heat stressed (33±1 °C for 10 h/day) chickens GR, GPx, MnSOD, HO-1, Nrf2 mRNA levels were decreased and resveratrol dietary supplementation (400 mg/kg diet) was shown to have protective effects (Zhang *et al.*, 2018). It seems likely that GR changes in heat stress are tissue-specific and dependent on the duration of the HS. For example, when chickens were heat stressed (35 °C for 12 days), GR activity in the liver and muscle increased at 1 day post-stress but decreased at day 12 post-stress in the liver in comparison to non-stressed birds (Habashy *et al.*, 2019). Brahma Rasayana (BR) dietary supplementation (2 g/kg daily, orally) during cold stress (4 °C for 6 h daily during 5 or 10 days) was shown to increase antioxidant enzyme activities in the chicken liver including GR, GPx, SOD and CAT (Ramnath and Rekha, 2009).

Mycotoxins

Effects of dietary contamination with various levels (3.4 and 8.2 mg/kg) of DON and zearalenone (ZEA) for 2 weeks were investigated on Ross 308 hybrid 2 weeks-old broilers. Intake of both contaminated diets resulted in a significantly decreased activity of GPx and increased level of MDA in liver tissue. Activities of TR in liver and GPx in duodenal mucosa tissues, SOD in erythrocytes as well as levels of MDA in duodenal mucosa and alpha-tocopherol in plasma were not affected by dietary mycotoxins (Borutova *et al.*, 2008). In the liver of the short-term (48 h) supplemented T-2/HT-2 toxin (3.74/1.26 mg/kg) chickens expression of the GR gene was significantly lower than in the control at 12-hour post supplementation. Similarly, 12, 24 and 48 h posts supplementation of DON (16.12 mg/kg), GR expression was shown to be reduced in the chicken liver (Nakade *et al.*, 2018). It was also shown that 0.3 mg/kg dietary

AFB1 could increase MDA contents, and decrease GSH contents, GPx, SOD, GR and CAT activities, which demonstrated an oxidative stress in spleen of broiler (Chen *et al.*, 2016; Wang *et al.*, 2013). Simultaneous supplementation with sodium selenite (0.2-0.6 mg/kg diet) was shown to restore these parameters to be close to those in control group (Wang *et al.*, 2013). Interestingly, even much lower AFB1 dose (22.5-25 μg/kg) in broiler chicken diet was shown to significantly decrease GR in the liver (Liu *et al.*, 2016). In addition to a decreased GR activity, AFB1 (40 μg/kg for 35 days) was shown to dramatically decrease GR activity in the chicken duodenal mucosa and dietary lactic acid bacteria showed a protective effect (Liu *et al.*, 2016). The same AFB1 dose was found to decrease by GR activity in the chicken serum by 35% (Liu *et al.*, 2018a). Similarly, AFB_1 (1 mg/kg contaminated corn) significantly increased lipid peroxidation (MDA) and decreased total SOD, CAT, GPx, GST activities and GSH within the liver and serum and grape seed proanthocyanidin extract was shown to have a protective effect (Ali Rajput *et al.*, 2017). In the short-term (48 h) feeding aflatoxin contaminated diet (170.3 μg/kg AFB1) to 49-week-old laying hens, expression of GR gene was significantly decreased at 24 h post-supplementation. Interestingly, GPX4 expression was significantly reduced due to AFB1 treatment at 12 and 24 h, but induced later (Erdélyi *et al.*, 2018). The gene expression of GR was significantly lower on the first day of AFB1 exposure (149.1 μg/kg feed). On the second and seventh day of AFB1 exposure there was a significant increase in the expression of GR gene compared to the control group (Balogh *et al.*, 2019).

Heavy metals

Dietary NiCl2 in excess of 300 mg/kg was shown to cause renal oxidative damage in broilers by reducing mRNA expression levels and activities of antioxidant enzymes (GR, GPx, GST, SOD) and enhancing free radicals generation, lipid peroxidation and DNA oxidation (Guo *et al.*, 2014). Dietary mercuric chloride (0.280, 3.325, 9.415, or 27.240 mg/kg) was shown to induce oxidative stress by decreasing antioxidant enzymes (GR, CAT and SOD) activities and Nrf2-Keap1 signal pathway in the ovary (Ma *et al.*, 2018a) as well as in liver and kidney of laying hens (Ma *et al.*, 2018a).

Disease challenge

The activities of GR, GPx, GST, SOD, CAT and levels of GSH were significantly decreased in brain and liver of NDV-infected chickens over controls. On the other hand, a significant decreased MDA levels and enhanced antioxidant enzyme activity levels were observed in NDV + vit. E-treated animals (75 IU/kg body wt. for 10 days) compared to NDV-infected chickens (Subbaiah *et al.*, 2011). Glutathione-related antioxidant enzyme activities (GR, GPx and GST) in liver of laying hens naturally infected with Salmonella enterica were shown to be significantly increased in comparison to uninfected layers (Buiazus *et al*, 2017). However, this increase was not able to prevent oxidative stress as indicated by increased ROS production and TBARS production. Probiotic *Bacillus subtilis* fmbJ added into the broiler diets for 42 days was shown to increase GR, GPx and SOD activity and to decrease lipid peroxidation in serum and liver (Bai *et al.*, 2017).

Dietary supplements

Dietary mushroom (*Agaricus bisporus*, 10 or 20 g of dried mushroom/kg of feed for 6 weeks) were shown to dose-dependently reduce MDA levels in liver, breast, and thigh tissues and to elevate GR, GSH, GPx, and GST compared with the control treatment (Giannenas *et al.*, 2010). An experiment was carried out to evaluate the effect of two plants belonging to Chinese herbal medicines, *Ligustrum lucidum* (LL) and *Schisandra chinensis* (SC), on the antioxidant status of hens during heat stress. The results showed that diets supplement with 1% of either LL or SC significantly increased GR activity in the chicken heart, liver and sera (Ma *et al.*, 2005), serum or kidney (Ma *et al.*, 2009). Plant extracts (a combination of extract from the crop tops of agrimony, *Agrimonia eupatoria L.*, and extract from red grape vine pomace, *Vitis vinifera L.*, administered in the drinking water to growing chickens were shown to increase GR activity in mitochondria from the liver, heart and kidney (Fejerčáková *et al.*, 2014). Supplemental yeast cell walls were shown to increase GR activity and GSH concentration (by 17.4% and 15.6% respectively) in the duodenal mucosa of chickens (Liu *et al.*, 2018). Organic chromium dietary supplementation (100, 200, 300, or 400 μg/kg diet) for 42 days was shown to increase GR and GPx activities and decrease MDA in chicken plasma in comparison to unsupplemented chickens (Rao *et al.*, 2012).

7.4 Glutaredoxins

Glutaredoxins (Grxs) are small proteins, usually around 9-15 kDa, existing in large number of isoforms in in most living organisms, from prokaryotes up to humans (Berndt *et al.*, 2008). Grxs belong to a family of GSH-dependent thiol-disulphide oxidoreductases facilitating direct reversible redox chemistry between protein thiols and the cellular GSSG/GSH (a direct electron acceptor/donor). Therefore, reduction via the Grx system takes place as follows: NADPH transfers electrons to GR, which then transfers electrons further to GSH. In this case, GSH functions as a cofactor for one of the Grx enzymes reducing target proteins via thiol exchanges (Hopkins and Neumann, 2019; Figure 7.6). In fact, they share a thioredoxin fold with a Cys-xx-Cys active site motif and belong to the thioredoxin superfamily that includes thioredoxins (Trxs), protein disulphide isomerases (PDIs) and the disulphide bond protein A (DsbA) (Xiao *et al.*, 2019). The glutaredoxin system was first described in 1976 in a mutant lacking Trx1 in *E. coli* as a dithiol hydrogen donor system for ribonucleotide reductase (Fernandes and Holmgren, 2004). In fact, Grxs can be divided into dithiol Grxs, containing two cysteine residues in their active motifs (Grx1 and Grx2), and monothiol Grxs (Grx3 and Grx5), containing a single cysteine residue in their putative motifs (Table 7.1). Therefore, Grxs are a class of important enzymes participating in cell signalling and redox homeostasis and their main characteristics are shown in Table 7.1 indicating that Grxs catalyse deglutathionylation and other types of protein thiol redox processes and playing a role in cellular iron homeostasis. However, molecular mechanisms of regulation of the cellular functions of Grxs remain poorly characterised. Therefore, reversal reduction of disulphide bonds can be mediated by a variety of thiol-redox enzymes, containing an active site with the sequence motif Cys-xx-Cys.

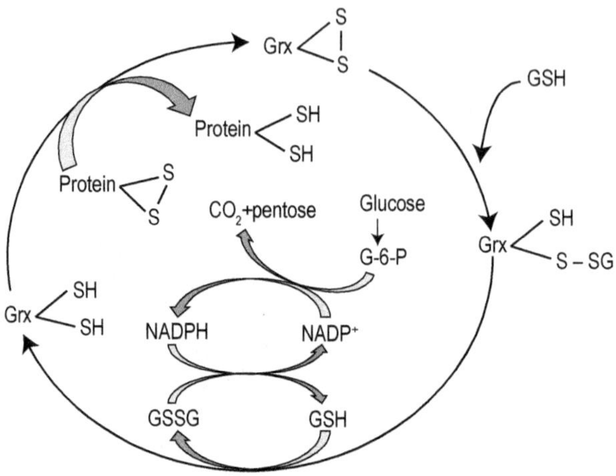

Figure 7.6. Role of glutaredoxins (Grxs) in the reduction of protein disulphides in biological systems.

Table 7.1. Main features of human glutaredoxins (Grxs) (adapted from Donelson *et al.*, 2019; Ouyang *et al.*, 2018; Xiao *et al.*, 2019).

Name	Location	Function	Additional information
Grx1	cytosol, mitochondria, nucleus	Cell signalling and protection	Mediates both oxidation and reduction of the copper metallochaperone Atox1; catalyses reduction of a protein disulphide bond in Cu,Zn-SOD and Trx1; catalyses GSH-dependent folding of reduced ribonuclease
Grx2	mitochondria	Redox sensor	Catalyses reversible oxidation and glutathionylation of mitochondrial membrane thiol proteins, and reduction of protein disulphides and deglutathionylation of mixed disulphides
Grx3(thioredoxin-like 2 (Txnl2) or PICOT	cytosol, nucleus	Redox sensor, essential for early embryonic growth and development	Grx3 deletion in cardiomyocytes alters both ROS levels and intracellular Ca^{2+} handling; modulates both cellular redox homeostasis and Ca^{2+} handling in the heart; upregulated by H_2O_2.
Grx5	mitochondria	Fe-S cluster assembly, heme synthesis	Regulates cellular iron metabolism and redox balance; mutation of the grx5 gene increases the accumulation of iron in the mitochondria, leading to mitochondrial DNA damage and respiratory metabolic disorder

These proteins are responsible for fast and reversible thiol-disulphide exchange reactions between their active-site cysteine residue and half-cystines of their disulphide substrates. Therefore, thioredoxins and glutaredoxins are abundant proteins with a number of isoforms in different species, operating in essential biosynthetic reactions and are believed to be responsible for the reduction of intracellular disulphides *in vivo* being important regulators of many biological functions. (Fernandes and Holmgren, 2004).

Many different functions have been described for Grxs, both as electron donors as well as modulators of cellular function in response to oxidative stress, including the regulation of cellular differentiation, transcription and apoptosis (Berndt *et al.*, 2008). It seems likely, that upregulation of Grx system could be an important element of adaptation to stress, since under normal conditions Grx1 inactivation in mice had no detrimental effect on development. Similarly, the knockout mice had the same sensitivity to heart hypoxia as wildtype counterparts. However, an increased glutathionylation of several proteins was observed in stress condition imposed by treating selected tissues with H_2O_2, while overexpression of Grx1 in mice induced tolerance to heart anoxia and overexpression of human Grx2 was shown to reduce myocardial cell death (Meyer *et al.*, 2009). Human glutaredoxins have been implied in several diseases (Berndt *et al.*, 2008) and Grx system could be considered as a possible backup of the thioredoxin system. Recently it has been shown that Trx and/or Grx are involved in redox modifications of targeted cysteines of several glycolytic enzymes affecting their activity being an adaptive response to environmental changes. Sequence of reactions of Grx in the reduction of protein disulphides in biological systems is shown in Figure 7.6.

Therefore, Grxs are a class of glutathione-dependent thiol-disulphide oxidoreductase enzymes facilitating reversible redox chemistry between GSH and protein thiols being versatile players in cellular redox signalling and redox homeostasis (Xiao *et al.*, 2019). Interestingly, it is known that Trx and Grx share a number of protein Cys redox targets but it was shown that down regulation of either redoxin has markedly different metabolic outcomes: silencing of Trx1 stimulates glycolytic flux while silencing of Grx1 decelerates it (López-Grueso *et al.*, 2019).

Glutaredoxins are still not described in avian species. Recently Glutaredoxin-like protein C5orf63 homolog was identified in chickens (*Gallus gallus*) by proteomics study (Likittrakulwong *et al.*, 2019). Earlier, Grx2 and Trx2 genes were cloned from skin tissue of Pucrpiano chicken and Tengchongxue chicken in China (Fang *et al.*, 2014). Furthermore, DL-2-hydroxy-4-methylthio-butanoic acid (HMTBa) dietary supplementation was shown to decrease gene expression of Grx, GSR and GSS in the chicken ileum while increased expression of Trx in the duodenum and ileum was observed (Wang *et al.*, 2019).

7.5 Glutathione peroxidases

GPx enzymes are widespread in the three domains of life. Of the eight avian GPx isoenzymes (encoded by GPX1-8 genes), four (GPx1, GPx2, GPx3, GPx4) contain a Sec residue in their active site, and four (GPx5, GPx6, GPx7, and GPx8) employ an active-site cysteine (Table 7.2). GPx are proven to belong to the first level of AO defence

Table 7.2. Main glutathione peroxidase (GPx) characteristics (adapted from Brigelius-Flohé and Mariano, 2013; Chabory *et al.*, 2010; Surai, 2018).[1]

GPx Nomenclature	Localisation	Peroxidatic residue	Substrates	Electron donors	Other characteristics
Cytosolic GPx					
GPx1	Intracellular, cytosolic, partly mitochondria	Sec	H_2O_2, t-BHP	GSH	Erythrocytes, kidney and liver
Gastrointestinal GPx					
GPx2	Intracellular, cytosolic	Sec	H_2O_2, t-BHP	GSH	Mucosal epithelial cells in GIT
Extracellular (plasma) GPx					
GPx3	Plasma	Sec	H_2O_2, t-BHP, PLOOH	GSH, Trx, Grx	Expressed in kidney, HIF target
Phospholipid hydroperoxide GPx					
GPx4	Intracellular, partly cytosolic, mitochondrial, membrane-bound	Sec	H_2O_2, PLOOH	GSH, DTT, 2-ME, L-Cys	Renal epithelial cells and testes
GPx5		Cys	n.d	n.d	Tetrameric, epididymis
GPx6		Sec in human; Cys in rodents	n.d.	n.d.	Olfactory, epithelium
GPx7	ER	Cys	H_2O_2, PLOOH	GSH, PDI	Monomeric, free in the lumen; Umbilical cord, ovary
GPx8	ER	Cys	H_2O_2, PLOOH	GSH PDI	Monomeric, an intrinsic membrane peroxidase with its active site facing the lumen, HIF target, oviduct

[1] t-BHP = tert-butyl hydroperoxide; DTT = 1,4-ditiothreitol; 2-ME = 2_mercaptoethanol; L-Cys = L-cysteine; PLOOH = phospholipid hydroperoxide.

including H_2O_2/peroxides detoxification and signalling (Surai, 2018). Members of GPx family differ in molecular weight, substrate specificity, cell distribution and perform a variety of functions. The PubMed search conducted on April 19th, 2020 on the 'glutathione peroxidase' gave 38,305 publications, including more than 2,463 references in 2019. Indeed, an interest in this subject is tremendous. Therefore, GPx properties and functions in relation to poultry biology with special emphasis to its role in chicken adaptation to various stress conditions are presented below.

The antioxidant system of the chicken is complex and well regulated. It was shown that glutathione peroxidase (GPx) belongs to the first and second levels of the antioxidant network and is involved in regulation of many important cellular pathways including maintenance of the redox balance and signalling. Indeed, since the discovery of GPx as a selenoprotein in 1973, a great body of evidence has been accumulated to confirm the importance of this vital enzyme in eukaryotes (Surai, 2018). In poultry the GPx family includes four Se-dependent forms of the enzyme, however only GPx1 and GPx4 are well characterised and received substantial attention as important enzymes participating in chicken adaptation to commercially relevant stresses. Main characteristics of 8 members of GPx family are shown in Table 7.2.

7.6 Se-dependent glutathione peroxidases

7.6.1 Cytosolic glutathione peroxidase

Cytosolic GPx (glutathione H_2O_2 oxidoreductase E.C. 1.11.1.9) was discovered by Mills in 1957, who showed that this enzyme had a protective effect in erythrocytes against H_2O_2 or ascorbate-induced haemolysis. Sixteen years later it became clear that GPx was a selenoenzyme. In fact, Rotruck *et al.* (1973) were the first to show that in rat red cells Se was tightly bound to the enzyme and demonstrated the uptake of ^{75}Se by GPx. As mentioned above, GPx is responsible for detoxification of hydroperoxides and hydrogen peroxide in the following reactions:

$$ROOH + 2GSH \xrightarrow{GPx} ROH + GSSG + H_2O$$

$$H_2O_2 + 2GSH \xrightarrow{GPx} GSSG + 2H_2O$$

These reactions employ a ping-pong mechanism. In particular, SeCys in the active centre of the enzyme is oxidised with a selenenic acid formation, which is reduced back by a reaction with 2 molecules of GSH. The Se atom in the enzyme catalytic site undergoes a redox cycle involving the selenolate anion as the active form which reduces H_2O_2 and organic peroxides (Mugesh and Singh, 2000). Recently, an elegant mathematical model and unified catalytic scheme with the incorporation of pH regulation mechanism of GPx have been developed and confirmed that GPx follows a ping pong mechanism (Pannala *et al*, 2014). GPx is characterised by high specificity for GSH as a donor of a reducing equivalent (substrate) and catalyses the reduction of a variety of hydroperoxides. It is interesting to note that thioredoxin also can be used, beside GSH, as reducing substrate, and GPx4 can also use other protein thiols as reducing equivalents (Brigelius-Flohé and Maiorino, 2013). However, GPx1 activity is related only to free peroxides and it is not able to reduce esterified fatty acid hydroperoxides. Therefore, in the biological system hydroperoxides in membranes have to be released by other enzymatic systems (e.g. phospholipases) or another member of GPx family (GPx4) can deal with them. GPx activity is dependent on the Se status of tissues. In fact, dietary Se supplementation has been shown to be effective in increasing GPx in a variety of animal species including rat, mouse, chicken, quail, sheep, cattle, horse, pig, deer, salmon, etc. (Flohé and Brigelius-Flohé, 2016). On the other hand, there is a range of nutritional means of decreasing GPx activities in various tissues including vitamin E excess, deficiencies of iron, zinc, riboflavin, vitamin B_6 or copper as well as consumption of silver, tri-o-cresyl phosphate or doxorubicin (Surai. 2006). In fact, depending on concentration and duration of exposure various chemical elements and compounds can either decrease or increase GPx activity in tissues. When Se is available, increased GPx activity could be a compensatory mechanism to deal with stress conditions and an oxygen-responsive elements (responsive to the oxygen tension in culture) were identified in the 5^{I}-flanking region of the GPx gene (Cowan *et al.*, 1993).

The earliest response in the specific activity of Se-dependent GPx occurred in chicken plasma at 8 hours and in liver at 24 hours after Se administration (Bunk and Combs, 1980). Importantly, GPx was found in chicken egg proteome (Mann and Mann, 2008). GPx activity was shown to have species- and tissue-specificity. For example, GPx in mouse leg muscle was shown to be almost 10-fold higher than that in the chicken muscle (Omaye amd Tappel, 1974). A comprehensive study of GPx in various tissues of different animals was conducted by Tappel *et al.* (1982; Table 7.3). In fact, the total GPx activities found in the study of chicken liver, heart and lung were 33, 27 and 10 nmol NADPH oxidised/min/mg protein, respectively. GPx activity in chicken, duck, turkey, ostrich and lamb muscles were measured (Daun and Akesson, 2004). It was shown that the activity of GPx varied more than 5-fold among the muscles from different species. The highest activity, found in duck muscles, was significantly higher than that in all other species. Moreover, lamb muscles had a significantly higher GPx activity than chicken and turkey breast and ostrich fillet (Daun and Akesson, 2004). It is interesting to note that GPx activity was shown to be 2.5-fold higher in duck embryo liver in comparison to chicken liver (Jin *et al.*, 2001) or 15-fold higher in the postnatal duck muscle in comparison to chicken muscle (Hoac *et al.*, 2006).

Furthermore, GPx activity in chicken meat was almost 2-fold lower than that in camel or cattle meat (Gheisari and Motamedi, 2010). In the liver of emperor penguins, GPx activities were 2-3 times higher than those in other avian species (Zenteno-Savin *et al.*, 2010). GPx activities in the liver of rat, chicken, lizard and frog were as follows 36,6; 23.2; 14.3 and 9.2 μmol NADPH/min/g respectively (Venditti *et al.*, 1999). In comparison to rats, turkey is characterised by a 10-fold decrease in GPx1 activity and increased (10-fold) GPx4 activity in the liver (Sunde and Hadley, 2010). In accordance with GPx1 activity (IU/g protein) turkey tissues can be placed in the following descending order: kidney>>gizzard>heart>liver>>thigh muscle>>breast muscle (Sunde *et al.*, 2015). Interestingly, in turkey kidney GPx3 expression was low while GPx1 expression was comparatively high (Sunde *et al.*, 2015). Whole blood GPx in chicken reduced by age while the enzymatic activity was constant in the chicken liver at 2, 4 and 6 weeks of age (Chadio *et al.*, 2015). Interestingly, there was a significant decrease in GPx activity in the utero-vaginal junction of the laying hens between 40 and 60 weeks of age (Breque *et al.*, 2006).

Studies of GPx1 knockout mice led De Haan *et al.* (2003) to a conclusion that GPx1 functions as the primary protection against acute oxidative stress, particularly in stress conditions, where high levels of ROS occur. A review of research results (Surai, 2006) indicated that overexpression of GPx is associated with an increased protection against oxidative stress created as a result of various environmental or nutritional manipulations. Indeed, GPx is well regulated enzymes and its increased activity can be considered as an important protective mechanism in stress conditions. Most of research related to GPx activity in poultry was related exclusively to GPx1 and only

Table 7.3. Total glutathione peroxidase (GPx) activity in the liver of various animals (U/mg protein) (adapted from Tappel *et al.*, 1982).

Animal	GPx activity
Hamster	920
Gerbil	683
Rabbit	496
Mouse	476
White mouse	468
Wild house mice	446
Rat	245
Carp	143
Cattle	70
Cat	67
Sheep	64
Ground squirrel	49
Chicken	33
Fence lizard	22
Dog	20
Guinea pig	12
American toad	2
Western newt	1.5
Blue gill sunfish	3.4
Rainbow trout	0.9

in a few studies GPx4 activity was measures, while GPx2 and GPx3 data are mainly based on gene expression studies.

In the newly hatched chickens, the highest GPx activity was found in the liver and kidney, with intermediate activity in the heart, lung and yolk sac membrane (YSM) and comparatively low GPx activity was shown in muscles and brain (Surai *et al.*, 1999). In all the tissues, Se-dependent GPx was the main enzymic form, comprising from 65% (lung) up to 90% (heart) of the total enzyme activity (Surai *et al.*, 1999). Similarly, in the chicken liver Se-dependent GPx comprises about a half (48%) of total activity of the enzyme (Engberg *et al.*, 1996). The specific activity of GPx in embryonic liver increases continuously during the 2nd half of the *in vivo* developmental period so that the activity at hatching was 3 times greater than that at embryonic day 10 (Surai, 1999a). The most rapid increase in GPx activity occurred between days 11 and 15 with a much more gradual increase thereafter. Interestingly, by the time of hatching, the specific activity of the enzyme in the liver was 6.1 times greater than that in the brain. Of note, GPx activity in the prenatal normoxic lung demonstrated a sharp increase between day 16 and day 18 and remained constant until hatch (Starrs *et al.*, 2001). According to GPx activity tissues of 35 days old chickens can be placed in the following descending order: liver>>kidney>plasma=erythrocytes>>femoral muscle>>pectoral muscle (Arai *et al.*, 1994).

GPx has been found to be expressed in chicken seminal plasma and spermatozoa (Surai *et al.*, 1998a,b). There are species-specific differences in activity and distribution of GPx in avian semen. For example, in seminal plasma total GPx activity was the highest in turkey and lowest in duck and goose (Surai *et al.*, 1998a). In spermatozoa, on the other hand, the highest GPx activities were found for goose and duck and much lower GPx activity was recorded for guinea fowl, turkey and chicken. In seminal plasma, the activity of GPx was two times greater in the White Koluda ganders than in chickens (Partyka *et al.*, 2012). A process of freezing and thawing fowl semen was associated with increased GPx activity in the seminal plasma (Partyka *et al.*, 2012a). It has also been shown that despite a high proportion of PUFAs and a low level of vitamin E, duck spermatozoa have the same susceptibility to lipid peroxidation as chicken spermatozoa (Surai *et al.*, 2000). It has been suggested that an increased activity of Se-GPx in duck semen compensates for the relatively low concentrations of other antioxidants. If selenium is limited in the diet (which is the case in many countries in the world), then dietary supplementation of this trace element should have a beneficial effect on the antioxidant defence in various tissues including sperm. This was confirmed in our studies. Inclusion of Se in the cockerel diet significantly increased Se-GPx activity in the liver, testes, spermatozoa and seminal plasma (Surai *et al.*, 1998c). As a result, a significant decrease in the sperm's and tissue susceptibility to lipid peroxidation was observed. It is extremely important that an inducible form of the enzyme (Se-GPx) represents more than 75% of the total enzymatic activity in chicken spermatozoa and more than 60% in the testes and liver of cockerels. In layers, increased GPx activity in the utero-vaginal glands compared to other regions of the lower oviduct (vagina, uterus) could be related to a necessity of AO defence during sperm storage in sperm-storage glands (Breque *et al.*, 2006).

7.6.2 Gastrointestinal glutathione peroxidase

Gastrointestinal GPx2 was first described in 1993 (Chu *et al.*, 1993) indicating that the enzymatic and physical properties of this enzyme to be very similar to those of cytosolic GPx. In fact, the authors showed similar substrate specificities for GPx1 and GPx2. Furthermore, GPx2 mRNA was readily detected in human liver and colon, and occasionally in human breast samples, but not in other human tissues including kidney, heart, lung, placenta, or uterus. On the other hand, in rodent tissues, GPx2 mRNA was only detected in the gastrointestinal tract, and not in other tissues including liver (Chu *et al.*, 1993). In fact, GPx2 appeared to be the major GSH-dependent peroxidase activity in rodent GI tract where at least three more selenoproteins including plasma GPx, selenoprotein P and thioredoxin reductase (TR) are found (Mork *et al.*, 1998). There are several important unique features of GPx2. First of all, GPx2 mRNA is comparatively stable in Se deficiency ranking this enzyme high in the selenoprotein hierarchy. This indicates a vital importance of this enzymes in the intestine and probably in other tissues. In fact, Se deficiency was associated with increased expression of GPx2 and decreased GPx1 expression in chicken testes (Gao *et al.*, 2017). Secondly, upon re-supplementation of Se after its deficiency, GPx-2 is shown to be synthesised first confirming its priority position among selenoproteins. Thirdly, GPx2 is exclusively located in the crypts of the intestine. Fourthly, in the gut a high expression of GPx2 is related to Paneth cells which are responsible for secretion of antimicrobial defensins upon exposure to bacteria suggesting an important role of this enzyme in gut immunity (Banning *et al.*, 2005).

The data on GPx-2 clearly indicate that this enzyme should be considered as a major antioxidant defence in the intestine. GPx2 knockout mice are shown to be viable and there is a great synergy between GPx2 and GPx1 in their participation in antioxidant defence. In fact, GPx2 could be considered as an effective barrier against hydroperoxide absorption and an important regulator of gut inflammation (Flohé and Brigelius-Flohé, 2016). GPx2 activity was shown to be detected in both the villus and crypt regions of rat mucosal epithelium and its activity nearly equalled that of GPx1 throughout the small intestine and colorectal segments (Esworthy *et al.*, 1998). It seems likely that induction of GPx2 in other tissues could be an important part of the stress response, since its gene expression is regulated by the antioxidant response element (ARE) and the Keap1/Nrf2/ARE pathway is proven to regulate the gene expression of various enzymes, including AO enzymes (Lubos *et al.*, 2011). Indeed, it is well appreciated that activation of ARE is associated with the transcription of a number of antioxidant proteins, detoxifying enzymes and transport proteins. Therefore, GPx-2 is considered to be an important oxidative stress-inducible cellular GPx isoform and its basal and inducible expression is shown to be dependent on Nrf2 (Singh *et al.*, 2006). In fact, by binding to the ARE in the upstream promoter region of genes encoding various antioxidant molecules, Nrf2 regulates the expression of hundreds of cytoprotective genes responsible for synthesis of a range of protective molecules involved in the maintenance and responsiveness of the cellular antioxidant systems (for review see Surai *et al*, 2019). This includes enzymes of the first line of the antioxidant defence (SOD, GPx and Catalase), detoxification enzymes (HO-1, NQO1, and GST), GSH-related proteins (γ-GCS), NADPH-producing enzymes and others stress-response proteins contributing to preventing oxidative and inflammatory damages. In fact, Nrf2 together with other transcription factors such as NF-κB orchestrate adaptive responses to diverse forms and levels of stress.

7.6.3 Plasma glutathione peroxidase

GPx3 from human plasma was purified to homogeneity by Takahashi and co-workers in 1987. This enzyme is shown to be a glycoprotein synthesised in the kidney. Indeed, GPx3 is extracellular enzyme found in blood plasma, chamber water of the eye or amniotic fluid. Furthermore, Maddipati and Marnett (1987) showed that the human plasma GPx3 is a tetramer of identical subunits of 21.5 kDa molecular mass. Furthermore, GPx3 is found to be a selenoprotein containing one selenium per subunit (Maddipati and Marnett, 1987). In general, the protein has a molecular weight of approximately 92,000 Da and containing four Se atoms per molecule (Cohen and Avissar, 1993).

The biological role of this enzyme remains speculative since plasma contains comparatively low concentrations of extracellular GSH or reduced thioredoxin. Indeed, GPx3 is considered to have intermediate specificity to peroxides. It can reduce lipid hydroperoxides in LDL, however, it is not active against peroxidised cholesterol esters (for review see Flohé and Brigelius-Flohé, 2016). In turkey, the highest expression of GPx3 was shown to be in liver, heart and kidney (Sunde *et al.*, 2015), while transcripts for GPx3 were shown to be highly expressed in chicken pectoral muscle at day 42 (Yao *et al.*, 2014). Plasma GPx3 activity in Se-deficient chicks was shown to decrease to 3% of Se-adequate levels (Li and Sunde, 2016). There are species-specific differences in GPx3 expression. For example, high expression of GPx3 transcript in chicken gizzard and pancreas was identified and these tissues were suggested to secrete and probably participate in regulation of this enzyme in the chicken (Li and Sunde, 2016), while in mammals kidney is shown to be the major source of plasma GPx3 (Flohé and Brigelius-Flohé, 2016). Indeed, GPx3 is considered to be a redox buffer involved in a regulation of inflammatory reactions and its more detail characterisation in avian species is a priority for future research.

7.6.4 Phospholipid glutathione peroxidase

In 1985 Ursini and co-workers reported that another form of GPx, which used a phosphatidyl choline hydroperoxide as a substrate, was Se-dependent (Ursini *et al.*, 1985). They showed that the enzyme was a monomer of 23 kDa. It contained one g-atom Se in the selenol form per 22,000 g protein. The kinetic data of GPx4 action were compatible with a ping-pong mechanism, described for the GPx1. The authors suggested that this enzyme was active at the interface of the membrane and the aqueous phase of the cell. In fact, GPx4 is distinguished from classical GPx as it is active in monomeric form and has a different amino acid composition (Sunde, 1993). There are three forms of GPx4. It is synthesised as a long form (L-form; 23 kDa) and a short form (S-form, 20 kDa) from mRNA that is transcribed from two initiation sites in exon 1a of GPx4 genomic DNA (Imai and Nakagawa, 2003). S-form GPx4 is the nonmitochondrial GPx4 and L-form GPx4 is the mitochondrial GPx4. Recently, the third form of GPx4, a 34 kDa selenoprotein, was detected in rat sperm nuclei and was called sperm nuclei GPx (snGPx). However, in chicken there is no snGPx (Bertelsmann *et al.*, 2007).

The GPx4 is unique in its capability of reducing ester lipid hydroperoxides incorporated in biomembranes or lipoproteins. It is well-known that GPx4 is widely expressed in normal tissue, and especially high in testis (Imai *et al.*, 1995), where it has an important role in spermatogenesis and sperm function. In this organ a relevant GPx4 activity is strongly linked to mitochondria of cells undergoing differentiation to spermatozoa. In testes mitochondria GPx4 is electrostatically bound to the inner surfaces of the organelle (Roveri *et al.*, 1994). Interestingly, GPx4 is found to be localised in the midpiece of spermatozoa in various species including Drosophila melanogaster, frog, fish, cock, mouse, rat, pig, bull, and human (Nayernia *et al.*, 2004). It is also important to mention that GPx4 mRNA expression in the male reproductive organs is under oestrogen control (Nam *et al.*, 2003). The most extraordinary discovery related to GPx4 is the fact of its polymerisation and conversion from active enzyme to the structural protein. Indeed, GPx4 protein was identified as the major constituent of the keratin-like material that embeds the helix of mitochondria in midpiece of mammalian spermatozoa (Ursini *et al.*, 1999).

In 1991 chick liver GPx activity was separated into three peaks by gel permeation chromatography (Miyazaki, 1991). The relative molecular weights and enzyme activities indicated that the first peak was Se-GPx1 and the second peak was related to non-Se-GPx. The third peak was the monomeric GPx, later called GPx4. The proportions of the GPx1, non-Se-GPx and GPx4 activities to total liver GPx activity were approximately 30, 42 and 28%, respectively (Mityazaki and Motoi, 1992). In the chick samples examined, the total GPx activity ranged from 15.3 nmol/min/mg in plasma to 118 nmol/min/mg in kidney. In all tissues except plasma GPx activity was separated into three peaks, while in plasma only one peak of GPx1 was detected. In terms of percentage of total GPx activity, Se-GPx activity was high in plasma and erythrocytes, intermediate in testis, brain, kidney and liver, and low in duodenum. All the organs examined contained GPx4 in different proportions. Specific GPx4 activity was high in liver, duodenum and kidney, intermediate in testis and low in brain. The high GPx4 activity in bird livers suggests that this enzyme is a major enzymatic system for reducing membrane lipid hydroperoxides in avian species. Se-GPx was the main GPx activity in rat liver while non-Se-GPx was predominant in bovine liver. In avian livers, GPx4 activity ranged from 10% of the total GPx activity in Japanese quail to 28% in chicks. In terms of specific activity toward cumene hydroperoxide, GPx4 activity of mammalian livers was below 6% of the activity of chick liver (Mityazaki and Motoi, 1992). Later, the same authors purified GPx4 to homogeneity from a broiler chick liver cytosolic fraction using 5 different column chromatographic methods (Miyazaki and Motoi, 1996). The molecular weight of the purified enzyme determined by sodium dodecyl sulphate-polyacrylamide gel electrophoresis was 19,500. Therefore, it was suggested that the enzyme protein is a single polypeptide. The isoelectric point of the enzyme was determined to be 7.0 and the optimum pH for the enzyme reaction was 7.0. The purified enzyme catalysed the reduction of hydrogen peroxide, cumene hydroperoxide, tert-butyl hydroperoxide and linoleic acid hydroperoxide. By using an antiserum against the purified enzyme, it was shown that it reacted with the 19.5 kDa polypeptide in the liver cytosol of duck and quail suggesting presence of the enzyme in these avian species (Miyazaki and Motoi, 1996). GPx4 has been shown to exist as both a 197 amino acid mitochondrial targeting protein and as a 170 amino acid non-mitochondrial protein (Kong *et al.*, 2003). The cDNA encoding the non-mitochondrial chicken GPx (cGPx4) was isolated from a chicken embryonic fibroblast cell line cDNA library. The nucleotide sequence of cGPx4 was shown to be 802 bp in length with an open reading frame that encoded 170 amino acids but lacked the N-terminal domain that encoded the mitochondrial leader sequence. Chicken non-mitochondrial GPx4 was highly expressed in brain and stromal tissues. The authors also showed that ovarian stromal tissue cGPx4 expression is regulated according to the reproductive status of the bird and its steroid hormone status, suggesting that GPx4 may play an important role in avian reproduction (Kong *et al.*, 2003). GPx4 in avian species is shown to be very sensitive to Se status. In fact, the liver had the highest GPx activity in Se-adequate poults, and GPx4 activity in Se-deficient liver decreased to 5% of Se-adequate levels (Sunde and Hadley, 2010). Based on GPx4 activity turkey tissues can be placed in the following descending order: liver>heart>kidney>>gizzard>thigh muscle>breast muscle (Sunde *et al.*, 2015). It is interesting that liver GPx4 mRNA levels could be down-regulated by excess of Se in chicken diet (Zoidis *et al.*, 2010).

7.6.5 Glutathione peroxidase ranking

As mentioned above GPx activity depends on Se provision in the diet. In an experiment, chicks produced from hens marginally deficient in Se and vitamin E were used (Kim and Combs, 1993). The hepatic activity of Se-GPx was significantly greater in Se-adequate chicks than in Se-deficient ones which was about 20% of the control level. When an experiment of the same design was conducted using chicks produced from hens that had been depleted of Se and vitamin E for a longer period of time (9 months), the hepatic activity of Se-GPx of chicks in that treatment group was about one-fifth of the activity observed for the same dietary treatment in the previous experiment (Kim and Combs, 1993). There are substantial differences among different forms of GPx with regard to response to Se deficiency (Flohé and Brigelius-Flohé, 2016). The selenoproteins retained in tissues for longer periods during progressive Se deficiency are considered to have higher physiological significance in comparison to those whose activities rapidly decline. In this respect, the main GPx forms rank as follows (Flohé and Brigelius-Flohé, 2016): GPx2>GPx4>GPx3=GPx1. However, the GPx ranking is likely species- and tissue-specific. For example, in chicken CNS the rank of GPx is as follows GPx3>GPx4>GPx2>GPx1 (Jiang *et al.*, 2017). Furthermore, liver and gizzard GPx activities in Se-deficient chicks were shown to be only 2 and 5%, respectively, of Se-adequate birds (Li and Sunder, 2016), while GPx4 activities in the same tissues in Se-deficient chicks comprised 10 and 5%, respectively, of values of Se-adequate birds. At the same time, plasma GPx3 activity in Se-deficient chicks was only 3% of Se-adequate levels (Li and Sunder, 2016). It seems likely that pancreas is more resistant to Se depletion, since GPx1 and GPx4 activities in Se-deficient chicks decreased to 39 and 25% of the physiological level (Li and Sunder, 2016). Similarly, in comparison to Se-adequate growing turkey Se deficiency is shown to decrease plasma GPx3, liver GPx1 and liver GPx4 activities to 2, 3, and 7%, respectively (Taylor and Sunde, 2016). In fact, recently it has been suggested that GPx, TrxR1, SELP, and SPS2 to play a more important role than the other selenoproteins in poultry (Luan *et al.*, 2016). Interestingly, the mRNA levels of GPx2, GPx4 and GPx3 were increased in chicken marrow due to low Se diet (0.028mg/kg; Jiang *et al.*, 2017). Similarly, in chicken kidney GPx1, GPx2 and GPx4 were upregulated in Se deficiency (Zhang *et al.*, 2016).

The aforementioned data clearly indicate that for 44 years of research Se-dependent GPx received substantial attention as a key player in the antioxidant defence system in animals and human. Indeed, GPx, being an inducible enzyme, participates in poultry adaptation to stress conditions. From one hand, it was shown that GPx1 can modulate redox-dependent cellular responses and signalling by regulating mitochondrial function (Handy *et al.*, 2009). On the other hand, as mentioned above, an oxygen-responsive element was identified in the GPx gene (Cowan *et al.*, 1993). Indeed, relationship between GPx and various transcription factors deserves more attention. It seems likely that GPx1 plays a vital role in regulating pro-inflammatory pathways, including mitogen-activated protein kinases (MAPK) and transcription factor nuclear factor-κB (NF-κB; Sharma *et al.*, 2016). In particular, NF-κB was shown to be upregulated in the GPX1$^{-/-}$ mouse (Crack *et al.*, 2006). Importantly, NF-κB is known to be a key regulator of cellular death and survival under oxidative stress conditions. Furthermore, involvement of Nrf2 in basal and inducible expression of GPx2 (Singh *et al.*, 2006) and regulation of redox-sensitive genes by GPx4 (Savaskan *et al.*, 2007) warrant further investigation.

In poultry production only two forms of Se-dependent GPx (GPx1 and GPx4) received substantial attention as important antioxidant status markers as well as indexes of Se status. Indeed, there is a need to expand research related to roles of GPx2 and GPx3 in poultry biology. Modulation of GPx activity in poultry by various factors will be considered below.

7.6.6 Effects of dietary selenium on glutathione peroxidase in poultry

Selenium deficiency

In 1980s it was proven that Se deficiency in chickens was associated with decreased GPx in various tissues (Surai, 2006). For the next 30 years this question was studied in more detail. Indeed, Se deficiency decreased GPx activity and/or expression in liver (Liu *et al.*, 2015), brain (Xu *et al.*, 2013), pancreas (Zhao *et al.*, 2014a), muscles (Yao *et al.*, 2014), thyroid (Lin *et al.*, 2014), duodenal mucosa (Liu *et al.*, 2016a), spleen and other immune organs (thymus and bursa of Fabricius; Zhang *et al.*, 2012). In Se-deficient chicks activities of GPx3, liver and gizzard GPx1, liver and gizzard GPx4 decreased dramatically to 3, 2, 5, 10 and 5%, respectively, of Se-adequate levels (Li and Sunde, 2016). Furthermore, Se deficiency in chickens decreased mRNA expression of GPx, GPx protein expression and activities in duodenum, jejunum and rectum (Yu *et al.*, 2015). Compared with the Se-supplemented chicks, the Se deficient chicks had lower muscle mRNA levels of GPx1, GPx3, GPx4 and decreased protein expression of GPx1 and GPx4 (Huang *et al.*, 2015). Similarly, Se deficiency in turkey was associated with a decrease in GPx4 mRNA levels in the liver (Sunde and Hadley, 2010).

Selenium supplementation

The addition of Se to various diets significantly elevated GPx activity in chicken plasma (Rao *et al.*, 2013), liver (Placha *et al.*, 2014), seminal plasma, spermatozoa, testes (Surai *et al.*, 1998) and egg yolk (Wang *et al.*, 2010). Correlation analysis has shown that tissue Se concentration (pooled data) was correlated to Se added to feed (r=0.529, $P<0.01$, log values) and to GPx activity (r=0.332, $P<0.05$), with the latter not being correlated with Se added to feed (Zoidis *et al.*, 2014). Information is actively accumulated to show that in comparison to traditional sodium selenite, organic Se in some cases is more effective in upregulation of GPx activity (Surai, 2006). Organic Se in the form of Se-Yeast at 0.3 mg/kg was shown to increase GPx activity in egg in comparison to sodium selenite (SS) supplemented breeders (Rajashree *et al.*, 2014). Similarly, Se-Yeast dietary supplementation increased GPx activity in the liver and breast muscle of 42-day old chickens in comparison to SS-supplemented birds (Chen *et al.*, 2014). SeMet and Se-Yeast were more effective than SS in increasing GPx in chicken plasma (Jing *et al.*, 2015). Se-Yeast supplementation in broiler diets resulted in greater tissue Se concentrations than SS and pGPx3 and tissue Se concentrations remained greater in birds previously fed a diet with Se-Yeast than in those fed SS after being fed a low-Se diet (Payne and Southern, 2005). Importantly, when the diet is adequate in Se usually there is no response to extra Se supplementation. For example, no effect of Se source (Se-Met vs Se-Yeast) or dosage (0.1; 0.3 or 0.5 mg/kg) was observed on serum GPx activities in chickens (Delezie *et al.*, 2014). Similarly, plasma GPx activity was not affected by Se source or concentration (Payne and Southern, 2005).

Selenium excess

It was shown that Se excess in chicken diet imposes oxidative stress (Surai, 2002a,b,c) which is associated with decreased GPx activity. For example, in acute Se toxicosis GPx activity of the RBC was significantly elevated at the first sampling (3 h after treatment) and decreased to the control level thereafter (Mézes and Sályi, 1994). In fact, GPx activity declined in the blood plasma and in the red blood cell hemolysate (Balogh *et al.*, 2004), spleen (Peng *et al.*, 2012a) and liver (Zoidis *et al.*, 2010) as a result of Se excess in the diet. Indeed, sodium selenite-induced (10 and 15 mg/kg diet) oxidative stress in chicken spleen was associated with decrease in GPx, SOD and catalase activities and increase in lipid peroxidation (Peng *et al.*, 2012a). Dietary Se (3 mg/kg) depressed growth performance of chicken and down-regulated nine and three selenoprotein genes in thymus and spleen, respectively, and only Sepp1 was up-regulated in the bursa of Fabricius (Tang *et al.*, 2017).

7.6.7 Glutathione peroxidase in stress conditions

Heat stress

It seems likely that GPx activation is an important adaptive mechanism to deal with oxidative stress imposed by heat stress. Indeed, heat stress increased GPx activity in the chicken liver (Tan *et al.*, 2010), serum (Yang *et al.*, 2010), erythrocytes (Aengwanich and Suttajit, 2013) and thigh muscle (Huang *et al.*, 2015a), while it was not affected in breast muscle. In contrast, Se and vitamin E supplementation increased GPx activity in skeletal muscles of heat stressed chickens, while there was no effect of such supplementation in control unstressed chickens (Ghazi Harsini *et al.*, 2012). It should also be mentioned that heart muscle is extremely sensitive to heat stress and various feed additives can help protecting it. Indeed, in heart of heat-stressed chickens at day 31 of age, both *Curcuma xanthorrhiza* and *Origanum compactum* essential oils were shown to increase GPx activity compared with unsupplemented control (Akbarian *et al.*, 2014). However, when stress is too high to be adapted to, GPx activity decreases. In fact, extreme heat stress decreased GPx activity in serum, muscle (Sahin *et al.*, 2016), spleen (Xu *et al.*, 2014) and bursa of Fabricius (Xu and Tian, 2015).

Cold stress

Data on the effect of cold stress on GPx activity are not consistent. For example, when broilers were exposed to a cool environment for 3 weeks, plasma GPx activity decreased compared with normal temperature chicks (Ramnath and Rekha, 2009). In contrast, acute cold stress was associated with increased GPx activity in spleen, thymus and bursa of Fabricius (Zhao *et al.*, 2014b). In fact, acute cold stress initially increased (1-3 hours) and then decreased (24h) GPx activity in chicken heart tissue (Zhao *et al.*, 2013). In contrast, in chronic cold stress, GPx activity in the heart initially decreased (5 days) and later (10-20 days) recovered (Zhao *et al.*, 2013).

Other stresses

Various stress-conditions could affect GPx activity in chicken blood and tissues. Among the mycotoxins studied aflatoxin B1 showed the most consistent negative effect on GPx activity in various chicken tissues including liver, spleen (Liu *et al.*, 2016b). Depending on the experimental conditions, including doses used, T-2 toxin was shown to decrease hepatic activity of GPx (by 36.8%; Dvorska *et al.*, 2007 or by 30%; Bócsai *et al.*, 2016) or increase hepatic GPx activity 3-fold (Leal *et al.*, 1999). It is interesting to note that subtoxic dietary level of deoxynivalenol (DON) increased GPx activity in chicken duodenal mucosa (Placha *et al.*, 2009) and dietary zearalenone (ZEA) increased GPx in the duodenal mucosa and kidney tissues (Grešáková *et al.*, 2012).

It was shown that GPx activity in liver and muscle was significantly higher in feed withdrawal and the darkening layer group than in control (Naziroglu *et al.*, 2000). Chicken transportation increased GPx in muscles (Wang *et al.*, 2015), while it did not affect erythrocyte GPx activity (Perai *et al.*, 2015). Increased humidity is shown to increase GPx activity in chicken pectoral muscles (Wei *et al.*, 2014). Exposure of broilers to blue light increased GPx in their breast and leg muscles (Ke *et al.*, 2011). In great contrast, in pullets, fasting resulted in a significant decrease of whole blood hemolysate GPx activity (Milinković-Tur *et al.*, 2007) and high nutrient density diet decreased GPx in the chicken heart (Peng *et al.*, 2013).

Diseases

Eimeria tenella challenge was associated with substantial (more than 2.5-fold) increase in GPx activity in chicken plasma (Bun *et al.*, 2011). Mitochondrial GPx activity in lung was elevated in broilers with pulmonary hypertension syndrome (PHS) compared to controls (Iqbal *et al.*, 2002). Similarly, GPx was shown to be upregulated in the brain of the cold-induced pulmonary hypertensive chickens (Hassanpour *et al.*, 2015). However, in most cases disease challenge was associated with a decreased GPx activity. In particular, a decreased blood GPx activity was shown in necrotic enteritis (Zhou *et al.*, 2016), infection bronchitis (Lin *et al.*, 2015) and *Ascaridia* infection (Gabrashanska *et al.*, 2007). Similarly, Newcastle disease infection decreased GPx activity in chicken brain and liver (Subbaiah *et al.*, 2011).

7.6.8 Nutritional modulation of glutathione peroxidase

Antioxidant supplements

Dietary antioxidants can affect GPx activity in chickens. However, data on the relationship between dietary vitamin E and GPx are not consistent. In fact, in most of the cases dietary vitamin E activated GPx activity in liver (Hu *et al.*, 2015) and erythrocytes (Cinar *et al.*, 2014). Of note, in different experimental conditions vitamin E supplementation decreased chicken plasma GPx (Mahmoud and Hijazi, 2007) or did not affect GPx activity in muscles (Voljc *et al.*, 2011). Ascorbic acid, a water-soluble dietary antioxidant, was shown to increase activity of the GPx activity in chicken plasma (Oztürk-Urek *et al.*, 2001) or erythrocytes (Cinar *et al.*, 2014). Similarly, GPx activity in serum, liver and breast muscle significantly increased in chickens fed with lipoic acid (Chen *et al.*, 2011).

Plant extracts

Various plant material or extracts are shown to increase GPx activity in chicken tissues. For example, dietary polysavone (1·5 g/kg), a natural extract from alfalfa rich in carotenoids and flavonoids, increased GPx in chicken serum and liver (Dong *et al.*, 2011). Furthermore, dietary lycopene (carotenoid) in the maternal diet increased GPx activity in the liver of the newly hatched chicks (Sun *et al.*, 2015). Similarly, tomato pomace, a rich source of lycopene, increases serum GPx (Hosseini-Vashan *et al.*, 2016). As mentioned above, *C. xanthorrhiza* and *O. compactum*, rich sources of flavonoids, increased GPx in chicken heart (Akbarian *et al.*, 2014), while orange peel extract and *C. xanthorrhiza* essential oil increased GPx in chicken erythrocytes (Akbarian *et al.*, 2015). Supplementation of ginger increased GPx activity in chicken serum (Zhang *et al.*, 2009), while supplementation of *Astragalus membranaceus* root powder increased activity of GPx in the serum of chickens at 21 and 42 days (Zhang *et al.*, 2013). Interestingly, garlic oil also increase GPx activity in blood plasma, erythrocytes and liver of chicken (Ancsin *et al.*, 2013).

However, plants and their extracts are not always effective in modulating GPx activity. For example, xanthophyll supplementation enhanced antioxidant capacity and reduced lipid peroxidation in different tissues of hens and chicks but did not affect serum or liver GPx activity (Gao *et al.*, 2013). Furthermore, dietary supplementation of clove and agrimony or clove and lemon balm did not affect blood GPx activity in chickens (Petrovic *et al.*, 2012).

Other dietary supplements

There is a range of feed supplements increasing GPx activity in chicken tissues. For example, live yeast probiotic, containing *S. cerevisiae* at 4.125×10^6 cfu per 100 ml, increased serum GPx activity in broiler chickens (Aluwong *et al.*, 2013). Similarly, *Lactobacillus plantarum* (probiotic) dietary supplementation increases GPx activities in chicken serum and liver at 21 days of age (Shen *et al.*, 2014). Stimulating effect of a probiotic on liver and serum GPx was confirmed later (Bai *et al.*, 2017). It has been shown that LPS can decrease GPx activity in intestinal mucosa (Wu *et al.*, 2016) and downregulates GPx mRNA expressions (Zheng *et al.*, 2016). At the same time in chicken bursal lymphocytes *Sargassum* polysaccharide increases GPx activity (Zhang *et al.*, 2011) and bursopentine, a novel pentapeptide isolated from chicken bursa of Fabricius, increases GPx activity in dendritic cells (Qin *et al.*, 2015). Furthermore, vitamin A in high doses decreased GPx activity in the chicken liver and brain (Surai *et al.*, 2000), while menadione did not affect GPx in chicken liver (Marchionatti *et al.*, 2008).

Drugs

There are some drugs used in poultry industry which are able to affect GPx. For example, it was shown that salinomycin (Kamashi *et al.*, 2004) and deltamethrin (Jayasree *et al.*, 2003) increased GPx in whole blood of chickens. However, in another paper it was shown that GPx activity in the liver decreased rapidly as a result of salinomycin toxic effects (Mezes *et al.*, 1992). In fact, in acute monensin poisoning GPx activity in liver and breast muscle initially decreased, then tended to rise (Sályi *et al.*, 1990).

7.6.9 Biological significance of selenium-glutathione peroxidase in poultry production

The various GPx are characterised by different tissue-specificity and are expressed from different genes. Initially it was thought that the major function of these peroxidases is the removal and detoxification of hydrogen peroxide and lipid hydroperoxides. However, it seems likely that emerging roles of GPx in maintenance of cellular redox state/balance have much bigger impact on cell metabolisms and stress resistance than it was expected. Indeed, GPx can affect such physiological events as differentiation, signal transduction and regulation of pro-inflammatory cytokine production, etc. Furthermore, peroxynitrite scavenging and participation of GPx enzymes in regulating biosynthesis of various eicosanoids (leukotrienes, thromboxanes and prostaglandins) can affect cell signalling and transcription factor activation responsible for the modulation of many important cellular pathways (Surai, 2006). Indeed, many different environmental factors and messengers modulate GPx in a complex manner, providing physiological regulation of antioxidant defences in the cell in stress conditions.

It is necessary to underline that, different forms of GPx perform their protective functions in concert, with each providing antioxidant protection at different sites of the body. For example, GPx2 could be considered to be a barrier against hydroperoxide resorption (Flohé and Brigelius-Flohé, 2016). In particular, it has been suggested that digestive tract is a major site of antioxidant-prooxidant interaction in the body (Surai, 2002, 2006; Surai and Fisinin, 2015). As mentioned above, GPx is an important antioxidant in plasma, which together with selenoprotein P and other antioxidant compounds, maintain antioxidant protection. On the other hand, GPx4 is an important antioxidant inside biological membranes where lipid peroxidation occurs and lipid hydroperoxides are produced. It seems likely that interactions between vitamin E and GPx4 are important elements in the antioxidant defences of biological membranes. Therefore, GPx1 and GPx4 are found in cytosol, while GPx3 and SeP are located in the extracellular fluids and working together they can provide antioxidant defence to the biological molecules inside and outside the cell (Takebe *et al.*, 2002). Clearly, the GPx family is an important part of antioxidant defences in animal/chicken bodies and specific roles of the enzymes in regulation of other important functions warrant further investigation.

7.7 Non-Se glutathione peroxidases

There are four Se-independent GPx including GPx5, GPx6, GPx7 and GPx8 respectively (Flohé and Brigelius-Flohé, 2016). From those human GPx6 is shown to be selenoprotein expressed in olfactory epithelium and probably participating in antioxidant defences. Furthermore, GPx5 is an extracellular Cys-containing GPx located in the epididymis with suggestive regulatory roles in sperm physiology. GPx7, is another Cys-containing GPx, located at the luminal site of the endoplasmic reticulum, detoxifying H_2O_2 and participating in protein folding. Interestingly, the knockout of *Gpx7* is shown to be associated with multiple organ dysfunctions and shortened life span. Finally, GPx8 is also a Cys-containing GPx located in the ER and participating in oxidative protein folding and insulin signalling (Flohé and Brigelius-Flohé, 2016). Therefore, non-Se GPx in avian species awaits further investigations. Indeed, non-Se GPx have some unique tissue-specific features of their expression. The most striking feature of these peroxidase is their comparatively efficiency of ROS, in particular H_2O_2 and hyperoxides, detoxifying ability (Herbette *et al.*, 2007). It seems likely that they participate in other important biological protective functions, including redox signalling, regulation of protein homeostasis and other functions which will be considered in the following paragraphs.

7.7.1 GPx5

The GPx5 story stated at the begging of 1990s when a specific major, androgen-dependent, mouse epididymal protein with M, 24,000 (MEP24) was shown to be actively secreted by the caput, but not by the corpus or the cauda epididymis (Ghyselinck *et al.*, 1989, 1990, 1990a). The homology of MEP24 with glutathione peroxidase (GPx) was shown (Ghyselinck *et al.*, 1991a,b) and homologous mRNA species in rat, human, rabbit, ram and boar epididymal RNA were shown (Ghyselinck *et al.*, 1990). This protein was cloned and sequencing of a partial cDNA coding for GPx5 was reported (Ghyselinck *et al.* 1993). Three years later, a secreted 24-kDa sperm-bound selenium-independent GPx protein was characterised and cloned and called GPx5 (Vernet *et al.*, 1996). It contains a 663 bp open-reading frame that, after conceptual translation, shows extensive identity with proteins belonging to the GPx family (Vernet *et al.*, 1996). The authors showed that GPx5-expressing cells can metabolise hydrogen peroxide in a manner that is consistent with a peroxidase activity. However, GPx5 is insensitive to a regular inhibitor of GPx enzymes. The expression of GPx5 was found to be restricted to the mouse caput epididymis (Schwaab *et al.*, 1998), but low levels of expression were shown in the kidney and liver (Dufaure *et al.*, 1996) and in other mouse tissues (Lahti *et al.*, 2001). It has been suggested that the Se-independent GPx5 could function as a backup system for Se-dependent GPxs (Vernet *et al.*, 1999). For example, following dietary Se deprivation it was shown that the epididymis is still efficiently protected against increasing peroxidative conditions. In fact, the caput epididymis of selenium-deficient animals showed a limited production of lipid peroxides, a total GPx activity which was not dramatically affected by the shortage in selenium availability and an increase in GPx5 mRNA and protein levels (Vernet *et al.* 1999). Therefore, in mice, GPx5 is synthesised by the caput epididymis. The protein is secreted as early as the initial segment of the caput and is found subsequently associated with the sperm plasma membrane in a sub-acrosomal localisation. In fact, GPx5 is present in the caput and cauda epididymis lumens in three different locations: either free as a soluble protein in the caput epididymal fluid, weakly bound to caput sperm membranes, or, finally, associated to lipid containing structures conferring to the protein a protective effect against proteolytic digestions. The protein was shown to be also associated to the head region of spermatozoa transiting through the epididymis to the vas deferens (Rejraji *et al.*, 2002). Within the cauda epididymis, the amount of free GPx5 is low compared to the caput and the association with sperm membranes proved to be more solid. In both caput and cauda sperm samples, the association of GPx5 with the sperm membrane protects GPx5 from proteolytic cleavages. Therefore, it was suggested that GPx5 to play an important role in sperm maturation from the early events up to the onset of fertilisation. During ontogenesis, GPx5 appeared at 20 days postnatal, before the completion of the morphological differentiation of the caput and concomitantly with the appearance of spermatozoa within the epididymis (Vernet *et al.*, 1997). Similarly, relative abundance of GPx5 mRNA was significantly greater in the stallion caput epididymis than in the remaining testicular and epididymal tissues (Koziorowska-Gilun *et al.*, 2018). In sheep, GPx5 is age-dependently expressed mainly in the epididymis (Li *et al.*, 2018). GPx5 was shown to be expressed in testes, epididymis and accessory sex glands and in seminal plasma GPx5 concentration differed among boars, among ejaculates within boar and among portions within ejaculate showing a positive relationship with sperm quality and fertility outcomes of liquid-stored semen AI-doses (Barranco *et al.*, 2016). Furthermore, GPx5 is under the hormonal control (Vernet *et al.*, 1997), since hormonal privation by castration abolished the accumulation of the GPx5 protein and an androgen-responsive element was discovered in the promoter which responded to dihydrotestosterone (Ghyselinck *et al.* 1993). Interestingly, in mouse androgens, epididymis-specific transcription factors and testicular factors are involved in regulation of GPx expression and activity (Brigelius-Flohé and Maiorino, 2013).Therefore, GPx5 is an epididymis-specific non-Se GPx in mice, rats, pigs, monkey and humans and its closest homolog is GPx3. Furthermore GPx3 + GPx5 represent more than 95% of epididymal GPx RNA and protein (Brigelius-Flohé and Maiorino, 2013). Mice that lack epididymal expression of the GPx5 was generated and they showed no obvious defects in epididymis and sperm cells and fertilisation rate was not affected (Chabory *et al.*, 2009). However, a higher incidence of miscarriages and developmental defects were observed when WT female mice were mated with Gpx5-deficient males over 1 year old compared with WT males of the same age. This was associated with DNA oxidative damages in GPx5-deficient males (Chabory *et al.*, 2009). Therefore, maintenance of sperm DNA integrity by removing excess of hydroperoxides could be an important function for GPx5 in the luminal compartment of the mouse cauda epididymis. Recently, by using a mouse transgenic model (*Gpx5*$^{-/-}$) of sperm nuclear oxidation, it has been shown that DNA damage does not disrupt chromosome organisation, but caused mild reductive DNA challenge associated with the fragility of the organisation of the oxidised sperm nucleus with possible detrimental consequences during post-fertilisation events (Champroux *et al.*, 2018). A CHO-K1 mammalian cell line expressing recombinant rat GPx5 was used to assess antioxidant protective effect of GPx5. It was shown that cells expressing GPx5 had increased resistance to oxidative challenge with decreased levels of lipid peroxidation and decreased levels of DNA oxidation (8-oxodG) compared with control cells (Taylor *et al.*, 2013). Similarly, spermatozoa of mice lacking both the sperm nucleus glutathione peroxidase 4 (snGPx4) and GPx5 activities displayed sperm nucleus structural abnormalities including delayed and defective nuclear compaction, nuclear instability and DNA damage (Noblanc *et al.*, 2012). Furthermore, GPx5 expression in the caput could be related to the regulation of disulphide formation by this antioxidant enzyme (Brigelius-Flohé and Maiorino, 2013). The activities of the purified epididymis-specific porcine GPx5 toward H_2O_2 or organic hydroperoxides were by far lower than the activity of cytosolic Se-GPx1 (less than 0.1%; Okamura *et al.*, 1997). It seems likely that, GPx5 protects spermatozoa from the premature acrosome reaction and maintains their fertilising ability in the epididymis. Indeed, from the one hand, GPx5 is shown to bind to the acrosomal region of the epididymal sperm and to disappear during the acrosome reaction. On the other hand, GPx5 significantly retarded the *in vitro* induced acrosome reaction (Okamura *et al.*, 1997). It seems likely that the sperm-associated forms of GPx4 (mitochondrial GPx4 and nuclear GPx4), the secreted GPx5 protein, and the epithelial proteins GPx1, GPx3, and cellular GPx4, are involved in maintaining optimal redox balance in the mammalian epididymis at different stages of the sperm's epididymal journey being active in different epididymis compartments (Noblanc *et al.*, 2011;Chabory *et al.*, 2010). Since avian species are different in male reproductive anatomy and physiology, not having epididymis, classical GPx5 in birds was not found.

7.7.2 GPx6

GPx6 was discovered in 1991 by molecular cloning of putative odorant binding and metabolising enzymes as a novel homologue of GPx localised in Bowman's glands, the site of olfactory-specific biotransformation enzymes (Dear *et al.*, 1991). It was shown to be a close homolog to plasma GPx3 and the gene encoding human and porcine GPx6 were cloned (Kryukov *et al.*, 2003). Indeed, it is a homotetramer selenoprotein in humans, but in rodents SeCys was replaced by Cys and therefore this enzyme is not a selenoprotein and considered to be the ancestral form of GPxs (Toppo *et al.*, 2008). Indeed, in mouse and rat, Sec was replaced by Cys in Gpx6, while chicks miss a mammal-selenoprotein gene GPX6 (Li *et al.*, 2018). It seems likely, that GPx6 is an important part of the antioxidant defence mechanisms and its gene was shown to be upregulated in the cochleae in mice with age-related hearing loss (Tanaka *et al.*, 2012). Furthermore, GPx6 was identified as a modulator of Huntington's disease since overexpression of GPx6 can dramatically alleviate both behavioural and molecular phenotypes associated with a mouse model of Huntington's disease (Shema *et al.*, 2015). However, heat stress was shown to downregulate GPx6 gene in IPEC-J2 Cells (Cao *et al.*, 2016) and nicotine treatment downregulated expression of GPx6 gene in HIV-1 transgenic rat brain (Song *et al.*, 2016). Therefore, GPx6 is an Se-independent enzyme playing an important role in regulation of AO system and decreasing the damaging effects of RONS and products of their metabolism. Indeed, the knowledge on this GPx is very limited, and there is no data available about possible existence of GPx6 in avian species.

7.7.3 GPx7

GPx7 was first described as a novel putative non-selenocysteine cysteine containing phospholipid hydroperoxide glutathione peroxidase (NPGPx) which incorporates cysteine instead of selenocysteine in the conserved catalytic centre Brca1-null mouse embryonic fibroblasts (Utomo *et al.*, 2004). The authors showed that molecular mass of the protein to be about 22 kDa with a cytoplasmic little detectable glutathione peroxidase activity *in vitro*. Expression of NPGPx was shown to suppresses toxic effects of H_2O_2 and omega-3 PUFAs. GPx7 is shown to be 187 amino acids in length, with a 19-amino-acid. It also contains N-terminal signal sequence and the C-terminal putative ER retrieval motif as well as the conserved catalytic residue (Cys57) required for GPx activity and absence of the loop required for tetramerisation and the residues required to define glutathione specificity (Nguyen *et al.*, 2011). *In vitro*, GPx7 can react with phospholipid hydroperoxides or H_2O_2 (Brigelius-Flohé and Maiorino, 2013) as well as with the reducing substrates PDI family members (Bosello-Travain *et al.*, 2013; Wang *et al.*, 2014), glutathione (Bosello-Travain *et al.*, 2013), or Grp78 (Wei *et al.*, 2012). Interestingly, with a standard GPx activity assay, GPx7 and GPx8 were found to have a very low GPx activity. Indeed, GPx7 was shown to be at least 95-fold more efficient with PDI than with GSH and GPx8 was indicated to be at least 250-fold more efficient with PDI than with GSH (Nguyen *et al.*, 2011). GPx7 is endoplasmic reticulum-resident protein disulphide isomerase peroxidase which is located in the lumen of the endoplasmic reticulum (Nguyen *et al.*, 2011) and support oxidative protein folding. In fact, In contrast to other members of GPx family, GPx7 is an unusual/atypical Cys-containing GPx which catalyses the peroxidatic cycle by using only one cysteine residue through a mechanism in which reduced GSH and protein disulphide isomerase serve as alternative substrates (Bosello-Travain *et al.*, 2013). In fact, GPx7 was shown to be a stress sensor that transmits oxidative stress signals to chaperones and deficiency of this enzyme causes accumulation of reactive oxygen species. First, a disulphide bond between its Cys57 and Cys86 residues is formed with following binding to glucose-regulated protein GRP78, causing disulphide bond formation within GRP78, and enhancing its chaperone activity to maintain physiological homeostasis (Wei *et al.*, 2012).

It is necessary to mention that folding/assembly of secretory and membrane proteins which are commonly rich in disulphide bonds also known as oxidative protein folding takes place in ER lumen. Among different enzymes catalysing this process ER sulfhydryl oxidase Ero1α has a special place catalysing de novo disulphide bond formation and generating H_2O_2 as a by-product. Interestingly, ROS production in the aforementioned reaction can be quite substantial. For example, Ero1α-mediated disulphide formation is believed to account for up to 25% of cellular peroxide/superoxide produced during protein synthesis, with the other 75% to come from mitochondria (Tu and Weissman, 2004). In fact, the major route for disulphide bond formation in the ER is believed to be via the action of Ero1 family members (Nguyen *et al.*, 2011). Therefore, GPx7 catalyses the formation of one disulphide bond for each H_2O_2 molecule consumed and simultaneously provides protection against H_2O_2-mediated cell injury. Furthermore, Wei *et al.* (2012) clearly showed that GPx7-deficient cells are characterised by increased ROS production, accumulation of misfolded proteins with impaired GRP78 chaperone activity. Furthermore, complete loss of GPx7 in GPx7-knockout animals can cause systemic oxidative stress, increases carcinogenesis, and shortens life span. Similarly, endogenous GPx7 showed protective effects in oesophageal cells against acid-mediated oxidative stress (Peng *et al.*, 2012) or in fibroblasts against ER stress (Wei *et al.*, 2012). GPx7 was shown to protect against fat accumulation in mice and human via modulating ROS (Chang *et al.*, 2013). Antioxidant protective actions of GPx7 (and GPx8) in ER were proven in an elegant study with a specific model system based on rat β-cells, where GPx7 and GPx8 were not originally expressed. Therefore, expression of GPx7 (or GPx8) attenuated FFAs-mediated H_2O_2 generation, ER stress, and apoptosis induction (Mehmeti *et al.*, 2017). GPx7 was shown to be downregulated in the mouse arteries and human endothelial cells by hyperhomocysteinaemia (HHcy) and Ero1α knockdown or GPx7 overexpression were able to protect the endothelium from HHcy-induced ER oxidative stress and inflammation (Wu *et al.*, 2019).

In accordance with the Gene database (https://www.ncbi.nlm.nih.gov/gene/424643) the GPX7 gene was shown to be conserved in human, chimpanzee, Rhesus monkey, dog, cow, mouse, rat, zebrafish, *S. cerevisiae, Kluyveromyces lactis, Eremothecium gossypii, Schizosaccharomyces pombe, Magnaporthe oryzae, Neurospora crassa*, and frog. Furthermore, 284 organisms have orthologs with human gene GPX7. Expression of GPx7 was shown in 10 chicken tissues including liver, heart, kidney, brain, lung, testes, spleen, skeletal muscles, colon and female gonad with highest expression level detected in heart. The enzyme is considered to be an important stress-response element.

Data on GPx7 expression in avian species are very limited and appeared in the last few years. For example, GPx7 expression in the muscle of broilers in the grower period almost triple due to acute heat stress and a combination of heat stress and Met supplementation caused the highest GPx7 expression (Del Vesco *et al.*, 2015). There is a dramatic decrease in GPx7 gene expression in the chicken liver between 21 and 42 days of age (Del Vesco *et al.*, 2017). Interesting, acute stress at starter period did not affect GPx7 expression in chicken muscles. Clearly, increased in GPx7 expression together with enhanced expression of Trx, TrxR1, GR, Glutathione synthetase, SOD in response to heat stress (Del Vesco *et al.*, 2017) is an adaptive response to prevent/ decrease possible damages due to oxidative stress. Recently, it has been shown that Se supplementation (1 mg/kg) increased expression of GPx5 in the chicken intestine and spleen. Interestingly, lower Se doses (0.25 or 0.5 mg/kg) significantly increased GPx7 expression in the spleen but did not affect it in the intestine (Xu *et al.*, 2015a). Since GPx7 is not a selenoprotein, activating effects of Se on this enzyme warrant further investigations. It seems likely, that GPx7 has tissue- and age-specific specific expressions in response to various stimuli. For example, in another study, vitamin E deficiency or Se deficiency in chickens were shown to upregulate muscle Gpx7 at week 2 (Huang *et al.*, 2015). The inverse relationship between the muscle Gpx7 mRNA level and Vit. E/Se status possibly indicates a compensatory function of GPx in stress conditions. It seems likely that *Eimeria/Clostridium* coinfection in the growing chickens upregulates antioxidant defences as indicated by increased gut GPx7 mRNA levels in parallel with increased CAT and SOD activities, in the infected group compared with the uninfected control group (Lee *et al.*, 2014). The authors demonstrated that in the experimental NE model system *in ovo* Se injection decreased oxidative stress in postnatally infected chickens as indicated by decreased GPx7 transcript levels, diminished CAT and SOD activities and lowered serum MDA concentrations compared with the controls. It was shown that GPx7 and GPX8 expressions were upregulated in birds affected with Wooden Breast (Mutryn *et al.*, 2015), as a response to elevated ROS levels and also to compensate for decreased expression of other antioxidants. Interestingly, in H_2O_2-damaged chicken cardiac muscle cells high level of uric acid supplementation (1,200 µM) was shown to depress SOD activity and expression of GPx7 together with GPx1, SOD1, and GCLC (Sun *et al.*, 2017).

7.7.4 GPx8

Glutathione peroxidase 8 (GPx8) is a type II transmembrane protein having a C-terminal KEDL motif localised in ER, in addition to a highly conserved N-terminal transmembrane domain (TMD; Nguyen *et al.*, 2011; Toppo *et al.*, 2008). In stress conditions associated with deregulated Ero1α activity GPx8 was indicated to limit ER hyperoxidation and H_2O_2 leakage to the cytosol (Ramming *et al.*, 2014). However, GPx8 was shown to be less efficiently reduced by PDI than GPx7 (Nguyen *et al.*, 2011). It seems likely that GPx8 can form heterotrimeric complexes with Ero1α and protein disulphide isomerase (PDI) to provide clearance of H_2O_2 coupled with disulphide bond formation (Ramming *et al.*, 2015). In addition, GPx8 was shown to be transcriptionally regulated by HIFα and as such upregulated by chemical hypoxia and fibroblast growth factor (FGF) treatment. Indeed, GPx8 depletion in cells was found to affect FGF and insulin signalling (Bosello-Travain *et al.*, 2015). Therefore, Gpx8 was suggested to be considered as a negative regulator of growth and metabolism. Furthermore, GPx8 silencing in HeLa cells, was shown to lead to H_2O_2 leakage, creation of a more oxidising environment and substantial decrease PUFA proportion in ER membrane oxidative stress s (Bosello-Travian *et al.*, 2018), probably as an adaptive response to oxidative stress. Recently, it has been shown that GPx8 is enriched in mitochondria associated membranes (MAM) and regulates Ca^{2+} signalling (Ca^{2+} storage and fluxes; Yoboue *et al.*, 2017). The authors showed that a highly conserved N-terminal transmembrane domain (TMD) is essential for the regulation of Ca^{2+} fluxes, since GPx7, which lacks a TMD, does not share these properties. Antioxidant protective actions of GPx8 and GPx7 in ER were proven in an elegant study with a specific model system based on rat β-cells, where GPx7 and GPx8 were not originally expressed. Therefore, expression of GPx7 or GPx8 attenuated FFAs-mediated H_2O_2 generation, ER stress, and apoptosis induction. (Mehmeti *et al.*, 2017). Indeed, H_2O_2 is shown to be cleared from the ER lumen with peroxidases, including peroxiredoxin 4 (Prx4), GPx7 and GPx8 (Ushioda and Nagata, 2019).

In *G. gallus* (chicken) GPx8 is expressed in 8 tissues including heart, kidney, lung, testes, spleen, skeletal muscle tissue, colon and female gonad (see https://bgee.org/?page=gene&gene_id=ENSGALG00000026704 for details) with highest expression level in female gonad. There is no yet data available on the activity and biological roles of GPx8 in avian species.

7.7.5 Physiological significance of non-selenium-glutathione peroxidase

The distribution of various forms of GPx has tissue and species-specificity. For example, Se-GPx and non-Se-GPx were determined in tissues of calves (Scholz *et al.*, 1981). Spleen, cardiac muscle, erythrocytes, brain, thymus, adipose tissue, and striated muscles of calves were found to contain only the Se-dependent GPx. Tissues having both enzymes included liver, lungs, adrenal glands, testes, kidney medulla, and kidney cortex. The liver contained the highest percentage of non-Se-dependent GPx activity of the calf tissues. In the rat liver and adrenal gland non-Se-GPx comprises about 35% of total GPx activity. However, in the heart and lung all GPx activity was related to Se-GPx (Lawrence and Burk, 1978). In contrast to other tissues, in testis non-Se-GPx was the major form of the enzyme. It seems likely that there is a species-specific expression of non-Se-GPx in animal tissues. In particular in the liver non-Se-GPx was shown to represent from 100% in guinea pig down to 35% in domestic pigs (Table 7.4, Lawrence and Burk, 1978).

As can be seen from data shown in the Table 7.4, chickens are characterised by the lowest total GPx activity in the liver in comparison to other species studied, while non-Se-GPx was the main form of the enzyme in this tissue. Selenium-dependent and non-selenium-dependent GPx in human tissue extracts were studied by Carmagnol *et al.* (1983) and they showed that:

Table 7.4. Glutathione peroxidase activity in the soluble fraction of liver of different species (adapted from Lawrence and Burk, 1978).

Species	Total U/g	Se-dependent U/g	Non-Se dependent U/g	Non-Se-dependent % of total
Hamster	45.1	26.0	19.1	43
Rat	28.0	19.6	8.4	35
Sheep	21.4	3.8	17/6	81
Pig	8.5	2.5	6.0	67
Human	9.7	1.3	8.4	84
Guinea pig	7.3	0	7.3	100
Chicken	4.9	1.5	3.4	70

- the non-Se-dependent GPx is predominant in liver, in renal cortex and skeletal muscle;
- non-Se-dependent and selenium-dependent GPx are in equal amounts in renal medulla;
- the Se-dependent GPx is predominant in adrenal glands and platelets;
- the selenium-dependent GPx represents 100% of the glutathione peroxidase activity in the other organs, including heart and brain.

Similarly, in rat tissues, only Se-dependent GPx was found in spleen, heart, lung, thymus and intestinal mucosa, while 91% GPx activity in testes was represented by non-Se-GPx. Interestingly, non-Se-GPx represented by 38, 35, 31, 26 and 21% of total GPx activity in rat adrenal, liver, kidney, brain and fat, respectively (Lawrence and Burk, 1978). In fact, in avian spermatozoa the highest total GPx was found in goose and lowest in guinea fowl respectively. The non-Se GPx represented from 15% (goose) up to 22% (chicken, Table 7.5; Surai *et al.*, 1998).

In the seminal plasma the highest GPx activity was evident in turkey, while the lowest one in the duck. In chicken, duck and goose seminal plasma non-Se-GPx

Table. 7.5. Glutathione peroxidase activity in avian spermatozoa (adapted from Surai *et al.*, 1998).

Species	Total U/10^9 spermatozoa	Se-dependent U/10^9 spermatozoa	Non-Se dependent U/10^9 spermatozoa	Non-Se-dependent % of total
Chicken	44.5	34.6	9.9	22.2
Turkey	31.8	25.5	6.3	19.8
Guinea fowl	11.8	10.3	1.5	14.4
Duck	72.6	58.7	13.9	19.1
Goose	178.7	151.2	27.5	15.4

represented about 20% of total activity, while in guinea fowl and turkey it was 62 and 29% respectively (Surai *et al.*, 1998). In cerebrum, cerebellum, brain stem and optic lobes of the newly hatched chicks non-Se-GPx represents about 15-17% total GPx activity (Surai *et al.*, 1999). It could well be that increased activity on non-Se-GPx activity in the case of Se deficiency could be a compensatory mechanism to deal with oxidative stress (Surai, 2006).

It seems likely that Se-independent GPxs are important elements of the backup system developed during evolution to deal with various stress conditions, including those related to Se inadequacy/deficiency. Furthermore, important roles of GPx7 and GPx8 in maintenance of proteostasis await further investigation.

7.8 Conclusions

All organisms are equipped with at least one thiol-dependent antioxidant systems, responsible for a control of redox homoeostasis. Interestingly, in mammals and birds, the thioredoxin system operates in concert with the glutathione system to provide effective AO defence and optimal conditions for redox signalling. In general, two main thiol-dependent redox systems can be shown as follows:

$$NADPH + H^+ + H_2O_2 \xrightarrow[GR/GSH/GPx]{TR/Trx/Prx} NADP^+ + 2H_2O$$

In general, details of interactions between Trx and GSH antioxidant systems in mammalian/avian cells are shown in Figure 7.7.

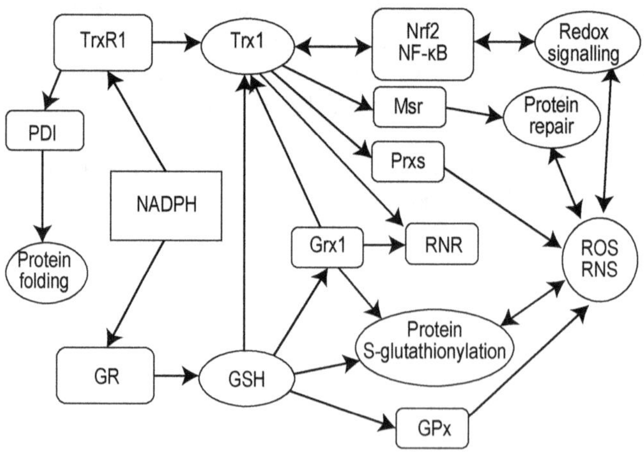

Figure 7.7. Thioredoxin (Trx) and glutathione (GSH) antioxidant systems in mammalian/avian cells (adapted from Branco *et al.*, 2012; Lu and Holmgren, 2014).

Therefore, Trx- and GSH systems are considered to be two major thiol-dependent redox systems in mammalian/avian cells operating in concert. In particular, thioredoxin system is shown to provide electrons to Trx-dependent peroxidases, which are responsible for ROS removal/detoxification. Trx is also involved in the repair of oxidised proteins by reducing Msr. Furthermore, Trx regulates the activities of many oxidative-sensitive transcription factors, including NF-κB and Nrf2, providing optimal conditions for the redox signalling. The GSH is served as a backup system to reduce Trx when the electron transfer pathway from TrxR is locked/deficient. Interestingly, protein disulphide isomerase (PDI) is known to be a substrate for TR (Lundström and Holmgren, 1990) and, therefore, protein folding and other important functions of PDI are connected to the Trx system. Indeed, it is well appreciated that ROS are important intracellular and extracellular signalling molecules (Surai *et al.*, 2019). Furthermore, complex cross talk between ROS, GSH, GSSG, other thiols and antioxidant enzymes, including GR, GPx, etc. regulates redox status of the living cells and affects their fate being the main switch between cell life (growth) and death (apoptosis; Couto *et al.*, 2016). In general, in most aerobic cells and organisms the glutathione-thioredoxin system serves as the central metabolic network responsible for redox balance maintenance and a removal or modification of endogenous electrophilic compounds and numerous xenobiotics (Couto *et al.*, 2016). An interconnections inside GSH- and Trx-systems are shown in Figure 7.8.

Hydrogen peroxide (H_2O_2), produced as a result of SOD action, is the major signalling molecule affecting various signalling pathways, including oxidation of Cys in various proteins with disulphide (S-S) bond formation. H_2O_2 can be reduced/detoxified by peroxiredoxins (Prx), glutathione peroxidases (GPx) or catalase. The reduction of H_2O_2 is conducted by the expense of oxidation of glutathione (GSH). Thioredoxin

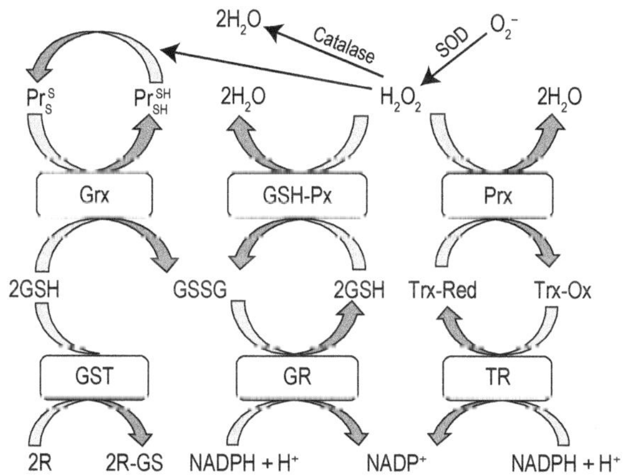

Figure 7.8. Protective roles of glutaredoxin (Grx), peroxiredoxin (Prx), thioredoxin (Trx), and glutathione (GSH) containing antioxidant systems (adapted from Bradshaw, 2019; Espinosa-Diez *et al.*, 2015; Ren *et al.*, 2017; Xie *et al.*, 2019).

(Trx) is responsible for reduction of oxidised Prx and oxidised Trx is reduced by thioredoxin reductase (TrxR) in a NADPH-dependent reaction. Oxidised glutathione (GSSG) is reduced by glutathione reductase (GR) using reducing potential off NADPH. Further, glutaredoxins (Grx) reduce disulphide (S–S) bonds in proteins (Pr), and glutathione S-transferase (GST) can detoxify reactive electrophilic compounds (R) by conjugation them with GSH. Therefore, thioredoxin and GSH-redox systems working in concert with other antioxidant systems, regulated by various transcription factors (e.g. Nrf2, NF-κB, etc.) and vitagenes are responsible for adaptation to stress and maintenance of adaptive homeostasis.

Therefore, glutathione system together with thioredoxin system are important elements of the vitagene network responsible for redox homeostasis maintenance and effective cell signalling under various stress conditions. Molecular mechanisms of the nutritional modulation of these thiol-dependent redox systems in avian species await further investigation.

References

Aengwanich, W. and Suttajit, M., 2013. Effect of polyphenols extracted from tamarind (*Tamarindus indica* L.) seed coat on pathophysiological changes and red blood cell glutathione peroxidase activity in heat-stressed broilers. International Journal of Biometeorology 57: 137-143.

Aesif, S.W., Kuipers, I., Van der Velden, J., Tully, J.E., Guala, A.S., Anathy, V., Sheely, J.I., Reynaert, N.L., Wouters, E.F., Van der Vliet, A. and Janssen-Heininger, Y.M., 2011. Activation of the glutaredoxin-1 gene by nuclear factor κB enhances signaling. Free Radical Biology and Medicine 51: 1249-1257.

Akbarian, A., Golian, A., Kermanshahi, H., De Smet, S. and Michiels, J., 2015. Antioxidant enzyme activities, plasma hormone levels and serum metabolites of finishing broiler chickens reared under high ambient temperature and fed lemon and orange peel extracts and *Curcuma xanthorrhiza* essential oil. Journal of Animal Physiology and Animal Nutrition 99: 150-162.

Akbarian, A., Michiels, J., Golian, A., Buyse, J., Wang, Y. and De Smet, S., 2014. Gene expression of heat shock protein 70 and antioxidant enzymes, oxidative status, and meat oxidative stability of cyclically heat-challenged finishing broilers fed *Origanum compactum* and *Curcuma xanthorrhiza* essential oils. Poultry Science 93: 1930-1941.

Ali Rajput, S., Sun, L., Zhang, N., Mohamed Khalil, M., Gao, X., Ling, Z., Zhu, L., Khan, F.A., Zhang, J. and Qi, D., 2017. Ameliorative effects of grape seed proanthocyanidin extract on growth performance, immune function, antioxidant capacity, biochemical constituents, liver histopathology and aflatoxin residues in broilers exposed to aflatoxin B_1. Toxins 9: 371.

Aluwong, T., Kawu, M., Raji, M., Dzenda, T., Govwang, F., Sinkalu, V. and Ayo, J., 2013. Effect of yeast probiotic on growth, antioxidant enzyme activities and malondialdehyde concentration of broiler chickens. Antioxidants 2: 326-339.

Ancsin, Z., Erdélyi, M., Balogh, K., Szabó-Fodor, J. and Mézes, M., 2013. Effect of garlic oil supplementation on the glutathione redox system of broiler chickens fed with T-2 toxin contaminated feed. World Mycotoxin Journal 6: 73-81.

Aquilano, K., Baldelli, S. and Ciriolo, M.R., 2014. Glutathione: new roles in redox signaling for an old antioxidant. Frontiers in Pharmacology 5: 196.

Arai T., Sugawara, M., Sako, T., Motoyoshi, S., Shimura, T. Tsutsui, N. and and Konno, T. (1994). Glutathione peroxidase activity in tissues of chickens supplemented with dietary selenium. Comparative Biochemistry and Physiology Part A: Physiology, 107: 245-248.

Bai, K., Huang, Q., Zhang, J., He, J., Zhang, L. and Wang, T., 2017. Supplemental effects of probiotic *Bacillus subtilis* fmbJ on growth performance, antioxidant capacity, and meat quality of broiler chickens. Poultry Science 96: 74-82.

Bains, J.S. and Shaw, C.A., 1997. Neurodegenerative disorders in human: the role of glutathione in oxidative stress-mediated neuronal death. Brain Research Reviews 25: 335-358.

Balogh, K., Kövesi, B., Zándoki, E., Kulcsár, S., Ancsin, Z., Erdélyi, M., Dobolyi, C., Bata-Vidács, I., Inotai, K., Szekeres, A., Mézes, M. and Kukolya, J., 2019. Effect of sterigmatocystin or aflatoxin contaminated feed on lipid peroxidation and glutathione redox system and expression of glutathione redox system regulatory genes in broiler chicken. Antioxidants 8: 201.

Balogh, K., Weber, M., Erdélyi, M. and Mézes, M., 2004. Effect of excess selenium supplementation on the glutathione redox system in broiler chicken. Acta Veterinaria Hungarica 52: 403-411.

Balogh, K., Weber, M., Erdélyi, M. and Mézes, M., 2007. Investigation of lipid peroxide and glutathione redox status of chicken concerning on high dietary selenium intake. Acta Biologica Hungarica 58: 269-279.

Banning, A., Deubel, S., Kluth, D., Zhou, Z. and Brigelius-Flohé, R., 2005. The GI-GPx gene is a target for Nrf2. Molecular and Cellular Biology 25: 4914-4923.

Bao, H.G., Wang, X.Y., Li, JY. and Wu, C.X., 2011. Comparison of effects of hypoxia on glutathione and activities of related enzymes in livers of Tibet chicken and Silky chicken. Poultry Science 90: 648-52.

Barranco, I., Tvarijonaviciute, A., Perez-Patiño, C., Vicente-Carrillo, A., Parrilla, I., Ceron, J.J., Martinez, E.A., Rodriguez-Martinez, H. and Roca, J., 2016. Glutathione peroxidase 5 is expressed by the entire pig male genital tract and once in the seminal plasma contributes to sperm survival and *in vivo* fertility. PloS One 11: e0162958.

Berndt, C., Lillig, C.H. and Holmgren, A., 2008. Thioredoxins and glutaredoxins as facilitators of protein folding. Biochimica et Biophysica Acta 1783: 641-50.

Bertelsmann, H., Kuehbacher, M., Weseloh, G., Kyriakopoulos, A. and Behne, D., 2007. Sperm nuclei glutathione peroxidases and their occurrence in animal species with cysteine-containing protamines. Biochimica et Biophysica Acta 1770: 1459-67.

Biazus, A.H., Da Silva, A.S., Bottari, N.B., Baldissera, M.D., do Carmo, G.M., Morsch, V.M., Schetinger, M.R., Casagrande, R., Guarda, N.S., Moresco, R.N., Stefani, L.M., Campigotto, G. and Boiago, M.M., 2017. Fowl typhoid in laying hens cause hepatic oxidative stress. Microbial Pathogenesis 103: 162-166.

Bócsai, A., Pelyhe, C., Zándoki, E., Ancsin, Z., Szabó-Fodor, J., Erdélyi, M., Mézes, M. and Balogh, K., 2016. Short term effects of T-2 toxin exposure on some lipid peroxide and glutathione redox parameters of broiler chickens. Journal of Animal Physiology and Animal Nutrition 100: 520-525.

Boldyrev, A.A., Dupin, A.M., Siambela, M. and Stvolinsky, S.L., 1988. The level of natural antioxidant glutathione and histidine-containing dipeptides in skeletal muscles of developing chick embryos. Comparative Biochemistry and Physiology. B, Comparative Biochemistry 89: 197-200.

Borutova, R., Faix, S., Placha, I., Gresakova, L., Cobanova, K. and Leng, L., 2008. Effects of deoxynivalenol and zearalenone on oxidative stress and blood phagocytic activity in broilers. Archives of Animal Nutrition 62: 303-312.

Bosello-Travain, V., Conrad, M., Cozza, G., Negro, A., Quartesan, S., Rossetto, M., Roveri, A., Toppo, S., Ursini, F., Zaccarin, M. and Maiorino, M., 2013. Protein disulfide isomerase and glutathione are alternative substrates in the one Cys catalytic cycle of glutathione peroxidase 7. Biochimica et Biophysica Acta 1830: 3846-3857.

Bosello-Travain, V., Forman, H. J., Roveri, A., Toppo, S., Ursini, F., Venerando, R., Warnecke, C., Zaccarin, M. and Maiorino, M., 2015. Glutathione peroxidase 8 is transcriptionally regulated by HIFα and modulates growth factor signaling in HeLa cells. Free Radical Biology & Medicine 81: 58-68.

Bosello-Travain, V., Miotto, G., Vučković, A.M., Venerando, R., Cozza, G., Toppo, S., Ursini, F. and Maiorino, M., 2018. Missing GPx8 activity in ER impacts on lipid composition. Free Radical Biology & Medicine 120: S130.

Bradshaw, P.C., 2019. Cytoplasmic and mitochondrial NADPH-coupled redox systems in the regulation of aging. Nutrients 11: 504.

Branco, V., Canário, J., Lu, J., Holmgren, A. and Carvalho, C., 2012. Mercury and selenium interaction *in vivo*: effects on thioredoxin reductase and glutathione peroxidase. Free Radical Biology & Medicine 52: 781-793.

Breque, C., Surai, P. and Brillard, J.P., 2006. Antioxidant status of the lower oviduct in the chicken varies with age and dietary vitamin E supplementation. Molecular Reproduction and Development 73: 1045-1051.

Brigelius-Flohé, R. and Maiorino, M., 2013. Glutathione peroxidases. Biochimica et Biophysica Acta 1830: 3289-3303.

Bun, S.D., Guo, Y.M., Guo, F.C., Ji, F.J. and Cao, H., 2011. Influence of organic zinc supplementation on the antioxidant status and immune responses of broilers challenged with *Eimeria tenella*. Poultry Science 90: 1220-1226.

Bunk, M.J. and Combs, G.F., 1980. Effect of selenium on appetite in the selenium-deficient chick. The Journal of Nutrition 110: 743-749.

Cao, L., Tang, J., Li, Q., Xu, J., Jia, G., Liu, G., Chen, X., Shang, H., Cai, J. and Zhao, H., 2016. Expression of selenoprotein genes is affected by heat stress in IPEC-J2 cells. Biological Trace Element Research 172: 354-360.

Carmagnol, F., Sinet, P.M. and Jerome, H., 1983. Selenium-dependent and non-selenium-dependent glutathione peroxidases in human tissue extracts. Biochimica et Biophysica Acta 759: 49-57.

Cazorla, A. and Barron, E.G. 1958. The formation of glutathione and the development of some enzymes for carbohydrate metabolism during the development of chicken embryo. Experimental Cell Research 14: 68-79.

Chabory, E., Damon, C., Lenoir, A., Henry-Berger, J., Vernet, P., Cadet, R., Saez, F. and Drevet, J.R., 2010. Mammalian glutathione peroxidases control acquisition and maintenance of spermatozoa integrity. Journal of Animal Science 88: 1321-1331.

Chabory, E., Damon, C., Lenoir, A., Kauselmann, G., Kern, H., Zevnik, B., Garrel, C., Saez, F., Cadet, R., Henry-Berger, J. and Schoor, M., 2009. Epididymis seleno-independent glutathione peroxidase 5 maintains sperm DNA integrity in mice. The Journal of Clinical Investigation 119: 2074-2085.

Chadio, S.E., Pappas, A.C., Papanastasatos, A., Pantelia, D., Dardamani, A., Fegeros, K. and Zervas, G., 2015. Effects of high selenium and fat supplementation on growth performance and thyroid hormones concentration of broilers. Journal of Trace Elements in Medicine and Biology 29: 202-207.

Champroux, A., Damon-Soubeyrand, C., Goubely, C., Bravard, S., Henry-Berger, J., Guiton, R., Saez, F., Drevet, J. and Kocer, A., 2018. Nuclear Integrity but not topology of mouse sperm chromosome is affected by oxidative DNA damage. Genes, 9: 501.

Chang, Y.C., Yu, Y.H., Shew, J.Y., Lee, W.J., Hwang, J.J., Chen, Y.H., Chen, Y.R., Wei, P.C., Chuang, L.M. and Lee, W.H., 2013. Deficiency of NPGPx, an oxidative stress sensor, leads to obesity in mice and human. EMBO Molecular Medicine 5: 1165-1179.

Chen, G., Wu, J. and Li, C., 2014. Effect of different selenium sources on production performance and biochemical parameters of broilers. Journal of Animal Physiology and Animal Nutrition 98: 747-754.

Chen, J., Chen, K., Yuan, S., Peng, X., Fang, J., Wang, F., Cui, H., Chen, Z., Yuan, J. and Geng, Y., 2016. Effects of aflatoxin B1 on oxidative stress markers and apoptosis of spleens in broilers. Toxicology and Industrial Health 32: 278-284.

Chen, P., Ma, Q.G., Ji, C., Zhang, J.Y., Zhao, L.H., Zhang, Y. and Jie, Y.Z., 2011. Dietary lipoic acid influences antioxidant capability and oxidative status of broilers. International Journal of Molecular Sciences 12: 8476-8488.

Chu, F.F., Doroshow, J.A. and Esworthy, R.S., 1993. Expression, characterization, and tissue distribution of a new cellular selenium-dependent glutathione peroxidase, GSHPx-GI. Journal of Biological Chemistry 268: 2571-2576.

Cinar, M., Yildirim, E., Yigit, A.A., Yalcinkaya, I., Duru, O., Kisa, U. and Atmaca, N., 2014. Effects of dietary supplementation with vitamin C and vitamin E and their combination on growth performance, some biochemical parameters, and oxidative stress induced by copper toxicity in broilers. Biological Trace Element Research 158: 186-196.

Cohen, H.J. and Avissar, N., 1993. Molecular and biochemical aspects of selenium metabolism and deficiency. Progress in Clinical and Biological Research 380: 191-202.

Conde de la Rosa L., García-Ruiz C., Fernández-Checa J.C., 2014. Glutathione in mammalian biology. In: Laher I. (ed.) Systems biology of free radicals and antioxidants. Springer, Berlin, Germany, pp.617-644.

Couto, N., Wood, J. and Barber, J., 2016. The role of glutathione reductase and related enzymes on cellular redox homoeostasis network. Free Radical Biology and Medicine 95: 27-42.

Cowan, D.B., Weisel, R.D., Williams, W.G. and Mickle, D.A., 1993. Identification of oxygen responsive elements in the 5'-flanking region of the human glutathione peroxidase gene. Journal of Biological Chemistry 268: 26904-26910.

Crack, P.J., Taylor, J.M., Ali, U., Mansell, A. and Hertzog, P.J., 2006. Potential contribution of NF-κB in neuronal cell death in the glutathione peroxidase-1 knockout mouse in response to ischemia-reperfusion injury. Stroke 37: 1533-1538.

Dalle-Donne, I., Rossi, R., Colombo, G., Giustarini, D. and Milzani, A., 2009. Protein S-glutathionylation: a regulatory device from bacteria to humans. Trends in Biochemical Sciences 34: 85-96.

Daun, C. and Åkesson, B., 2004. Glutathione peroxidase activity, and content of total and soluble selenium in five bovine and porcine organs used in meat production. Meat Science 66: 801-807.

De Haan, J.B., Crack, P.J., Flentjar, N., Iannello, R.C., Hertzog, P.J. and Kola, I., 2003. An imbalance in antioxidant defense affects cellular function: the pathophysiological consequences of a reduction in antioxidant defense in the glutathione peroxidase-1 (Gpx1) knockout mouse. Redox Report 8: 69-79.

Dean, R.T., Fu, S., Stocker, R. and Davies, M.J., 1997. Biochemistry and pathology of radical-mediated protein oxidation. Biochemical Journal 324: 1-18.

Dear, T.N., Campbell, K. and Rabbitts, T.H., 1991. Molecular cloning of putative odorant-binding and odorant-metabolizing proteins. Biochemistry 30: 10376-10382.

Del Vesco, A.P., Gasparino, E., de Oliveira Grieser, D., Zancanela, V., Soares, M.A.M. and de Oliveira Neto, A.R., 2015. Effects of methionine supplementation on the expression of oxidative stress-related genes in acute heat stress-exposed broilers. British Journal of Nutrition 113: 549-559.

Del Vesco, A.P., Khatlab, A.S., Goes, E.S.R., Utsunomiya, K.S., Vieira, J.S., Neto, A.O. and Gasparino, E., 2017. Age-related oxidative stress and antioxidant capacity in heat-stressed broilers. Animal 11:1783-1790.

Delezie, E., Rovers, M., Van der Aa, A., Ruttens, A., Wittocx, S. and Segers, L., 2014. Comparing responses to different selenium sources and dosages in laying hens. Poultry Science 93: 3083-3090.

Deponte, M., 2013. Glutathione catalysis and the reaction mechanisms of glutathione-dependent enzymes. Biochimica et Biophysica Acta 1830: 3217-3266.

Donelson, J., Wang, Q., Monroe, T.O., Jiang, X., Zhou, J., Yu, H., Mo, Q., Sun, Q., Marini, J.C., Wang, X. and Nakata, P.A., 2019. Cardiac-specific ablation of glutaredoxin 3 leads to cardiac hypertrophy and heart failure. Physiological Reports 7: e14071.

Dong, X.F., Gao, W.W., Su, J.L., Tong, J.M. and Zhang, Q., 2011. Effects of dietary polysavone (Alfalfa extract) and chlortetracycline supplementation on antioxidation and meat quality in broiler chickens. British Poultry Science 52: 302-309.

Dufaure, J.P., Lareyre, J.J., Schwaab, V., Mattei, M.G. and Drevet, J.R., 1996. Structural organization, chromosomal localization, expression and phylogenetic evaluation of mouse glutathione peroxidase encoding genes. Comptes Rendus de l'Academie des Sciences. Serie III, Sciences de la vie 319: 559-568.

Dvorska, J.E., Pappas, A.C., Karadas, F., Speake, B.K. and Surai, P.F., 2007. Protective effect of modified glucomannans and organic selenium against antioxidant depletion in the chicken liver due to T-2 toxin-contaminated feed consumption. Comparative Biochemistry and Physiology Part C: Toxicology & Pharmacology 145: 582-587.

Elliott, S.J. and Koliwad, S.K., 1997. Redox control of ion channel activity in vascular endothelial cells by glutathione. Microcirculation 4: 341-437.

Engberg, R.M., Lauridsen, C., Jensen, S.K. and Jakobsen, K., 1996. Inclusion of oxidized vegetable oil in broiler diets. Its influence on nutrient balance and on the antioxidative status of broilers. Poultry Science 75: 1003-1011.

Erdélyi, M., Balogh, K., Pelyhe, C., Kövesi, B., Nakade, M., Zándoki, E., Mézes, M. and Kovács, B., 2018. Changes in the regulation and activity of glutathione redox system, and lipid peroxidation processes in short-term aflatoxin B1 exposure in liver of laying hens. Journal of Animal Physiology and Animal Nutrition 102: 947-952.

Espinosa-Diez, C., Miguel, V., Mennerich, D., Kietzmann, T., Sánchez-Pérez, P., Cadenas, S. and Lamas, S., 2015. Antioxidant responses and cellular adjustments to oxidative stress. Redox Biology 6: 183-197.

Esworthy, R.S., Swiderek, K.M., Ho, Y.S. and Chu, F.F., 1998. Selenium-dependent glutathione peroxidase-GI is a major glutathione peroxidase activity in the mucosal epithelium of rodent intestine. Biochimica et Biophysica Acta 1381: 213-226.

Fang, F., Chunyong, Z., Yanxuan, L., Meiquan, L., Qingcong, A., Kelin, C. and Rongfu, G., 2014. Cloning Sequencing and bioinformatics analysis of GRX2 and TRX2 from Puerpiao chicken and Tengchongxue chicken. Animal Husbandry and Feed Science, 6: 91-96

Farina, M. and Aschner, M., 2019. Glutathione antioxidant system and methylmercury-induced neurotoxicity: an intriguing interplay. Biochimica et Biophysica Acta (BBA)-General Subjects, 1863: 129285.

Fejerčáková, A., Vašková, J., Bača, M., Vaško, L., Marcinčák, S., Hertelyová, Z., Petrášová, D. and Guothová, L., 2014. Effect of dietary microbially produced gamma-linolenic acid and plant extracts on enzymatic and non-enzymatic antioxidants in various broiler chicken organs. Journal of Animal Physiology and Animal Nutrition 98: 860-866.

Fernandes, A.P. and Holmgren, A., 2004. Glutaredoxins: glutathione-dependent redox enzymes with functions far beyond a simple thioredoxin backup system. Antioxidants and Redox Signaling 6: 63-74.

Flohé, L. and Brigelius-Flohé, R., 2016. Basics and news on glutathione peroxidases. In: Hatfield, D.L., Schweizer, U., Tsui, P.A. and Gladyshev, V.N. (eds) Selenium. Its molecular biology and role in human health. Fourth Edition, Springer, New York, NY, USA, pp.211-222.

Gabrashanska, M., Galvez-Morros, M., Teodorova, S.E., Ermidou-Pollet, S. and Pollet, S., 2007. Effect of selenium and *Ascaridia galli* infection on antioxidant biomarkers in broiler chickens: a mathematical model for parasite reduction and host growth. Journal of Helminthology 81: 399-408.

Gao, Y., Zhang, J., Huang, X. and Zhang, G., 2017. Glutathione peroxidase 1, selenoprotein K, and selenoprotein H may play important roles in chicken testes in response to selenium deficiency. Biological Trace Element Research 179: 271-276.

Gallogly, M.M. and Mieyal, J.J., 2007. Mechanisms of reversible protein glutathionylation in redox signaling and oxidative stress. Current Opinion in Pharmacology 7: 381-391.

Gao, Y.Y., Xie, Q.M., Ma, J.Y., Zhang, X.B., Zhu, J.M., Shu, D.M., Sun, B.L., Jin, L. and Bi, Y.Z., 2013. Supplementation of xanthophylls increased antioxidant capacity and decreased lipid peroxidation in hens and chicks. British Journal of Nutrition 109: 977-983.

García-Giménez, J.L., C. Romá-Mateo, G. Pérez-Machado, L. Peiró-Chova and F.V. Pallardó. 2017. Role of glutathione in the regulation of epigenetic mechanisms in disease. Free Radical Biology & Medicine 112: 36-48.

Ghazi Harsini, S., Habibiyan, M., Moeini, M.M. and Abdolmohammadi, A.R., 2012. Effects of dietary selenium, vitamin E, and their combination on growth, serum metabolites, and antioxidant defense system in skeletal muscle of broilers under heat stress. Biological Trace Element Research 148: 322-330.

Gheisari, H.R. and Motamedi, H., 2010. Chloride salt type/ionic strength and refrigeration effects on antioxidant enzymes and lipid oxidation in cattle, camel and chicken meat. Meat Science 86: 377-383.

Ghyselinck, N.B., Dufaure, I., Lareyre, J.J., Rigaudiere, N., Mattei, M.G. and Dufaure, J.P., 1993. Structural organization and regulation of the gene for the androgen-dependent glutathione peroxidase-like protein specific to the mouse epididymis. Molecular Endocrinology 7: 258-272.

Ghyselinck, N.B. and Dufaure, J.P., 1990a. A mouse cDNA sequence for epididymal androgen-regulated proteins related to glutathione peroxidase. Nucleic Acids Research, 18: 7144.

Ghyselinck, N.B., Jimenez, C., Courty, Y. and Dufaure, J.P., 1989. Androgen-dependent messenger RNA (s) related to secretory proteins in the mouse epididymis. Reproduction 85: 631-639.

Ghyselinck, N.B., Jimenez, C. and Dufaure, J.P., 1991b. Sequence homology of androgen-regulated epididymal proteins with glutathione peroxidase in mice. Reproduction, 93: 461-466.

Ghyselinck, N.B., Jimenez, C., Lefrancois, A.M. and Dufaure, J.P., 1990. Molecular cloning of a cDNA for androgen-regulated proteins secreted by the mouse epididymis. Journal of Molecular Endocrinology 4: 5-12.

Ghyselinck, N.B., Rigaudière, N. and Dufaure, J.P., 1991a. Androgen-dependent protein secreted by mouse caput epididymis shows high homologies with different glutathione peroxidases. Comptes rendus de l'Academie des sciences. Serie III, Sciences de la vie 313: 1-6.

Giannenas, I., Pappas, I.S., Mavridis, S., Kontopidis, G., Skoufos, J. and Kyriazakis, I., 2010. Performance and antioxidant status of broiler chickens supplemented with dried mushrooms (*Agaricus bisporus*) in their diet. Poultry Science 89: 303-311.

Grešáková, Ľ., Bořutová, R., Faix, Š., Plachá, I., Čobanová, K., Košíková, B. and Leng, Ľ., 2012. Effect of lignin on oxidative stress in chickens fed a diet contaminated with zearalenone. Acta Veterinaria Hungarica 60: 103-114.

Griffiths, H.R., Dias, I.H., Willetts, R.S. and Devitt, A., 2014. Redox regulation of protein damage in plasma. Redox Biology 2: 430-435.

Guo, H., Wu, B., Cui, H., Peng, X., Fang, J., Zuo, Z., Deng, J., Wang, X., Deng, J., Yin, S. and Li, J., 2014. NiCl 2-down-regulated antioxidant enzyme mRNA expression causes oxidative damage in the broiler's kidney. Biological Trace Element Research 162: 288-295.

Habashy, W.S., Milfort, M.C., Rekaya, R. and Aggrey, S.E., 2019. Cellular antioxidant enzyme activity and biomarkers for oxidative stress are affected by heat stress. International Journal of Biometeorology 63: 1569-1584.

Handy, D.E., Lubos, E., Yang, Y., Galbraith, J.D., Kelly, N., Zhang, Y.Y., Leopold, J.A. and Loscalzo, J., 2009. Glutathione peroxidase-1 regulates mitochondrial function to modulate redox-dependent cellular responses. Journal of Biological Chemistry 284: 11913-11921.

Hansen, J.M. and Harris, C., 2015. Glutathione during embryonic development. Biochimica et Biophysica Acta 1850: 1527-1542.

Hassanpour, H., Khalaji-Pirbalouty, V., Nasiri, L., Mohebbi, A. and Bahadoran, S., 2015. Oxidant and enzymatic antioxidant status (gene expression and activity) in the brain of chickens with cold-induced pulmonary hypertension. International Journal of Biometeorology 59: 1615-1621.

Herbette, S., Roeckel-Drevet, P. and Drevet, J.R., 2007. Seleno-independent glutathione peroxidases. More than simple antioxidant scavengers. FEBS Journal 274: 2163-2180.

Hoac, T., Daun, C., Trafikowska, U., Zackrisson, J. and Åkesson, B., 2006. Influence of heat treatment on lipid oxidation and glutathione peroxidase activity in chicken and duck meat. Innovative Food Science and Emerging Technologies 7: 88-93.

Hopkins, B.L. and Neumann, C.A., 2019. Redoxins as gatekeepers of the transcriptional oxidative stress response. Redox Biology 21:101104.

Hosseini-Vashan, S.J., Golian, A. and Yaghobfar, A., 2016. Growth, immune, antioxidant, and bone responses of heat stress-exposed broilers fed diets supplemented with tomato pomace. International Journal of Biometeorology 60: 1183-1192.

Hu, Z.P., Wang, T., Ahmad, H., Zhang, J.F., Zhang, L.L. and Zhong, X., 2015. Effects of different formulations of α-tocopherol acetate (vitamin E) on growth performance, meat quality and antioxidant capacity in broiler chickens. British Poultry Science 56: 687-695.

Huang, C., Jiao, H., Song, Z., Zhao, J., Wang, X. and Lin, H., 2015a. Heat stress impairs mitochondria functions and induces oxidative injury in broiler chickens. Journal of Animal Science 93: 2144-2153.

Huang, J.Q., Ren, F.Z., Jiang, Y.Y., Xiao, C. and Lei, X.G., 2015. Selenoproteins protect against avian nutritional muscular dystrophy by metabolizing peroxides and regulating redox/apoptotic signaling. Free Radical Biology and Medicine 83: 129-138.

Imai, H., Sumi, D., Hanamoto, A., Arai, M. and Sugiyama, A., 1995. Molecular cloning and functional expression of a cDNA for rat phospholipid hydroperoxide glutathione peroxidase: 3'-untranslated region of the gene is necessary for functional expression. Journal of Biochemistry 118: 1061-1067

Imai, H. and Nakagawa, Y., 2003. Biological significance of phospholipid hydroperoxide glutathione peroxidase (PHGPx, GPx4) in mammalian cells. Free Radical Biology & Medicine 34: 145-169.

Iqbal, M., Cawthon, D., Beers, K., Wideman Jr, R.F. and Bottje, W.G., 2002. Antioxidant enzyme activities and mitochondrial fatty acids in pulmonary hypertension syndrome (PHS) in broilers. Poultry Science 81: 252-260.

Jayasree, U., Reddy, A.G., Reddy, K.S., Anjaneyulu, Y. and Kalakumar, B., 2003. Evaluation of vitamin E against deltamothrin toxicity in broiler chicks. Indian Journal of Physiology and Pharmacology 47: 447-452.

Jiang, X.Q., Cao, C.Y., Li, Z.Y., Li, W., Zhang, C., Lin, J., Li, X.N. and Li, J.L., 2017. Delineating hierarchy of selenotranscriptome expression and their response to selenium status in chicken central nervous system. Journal of Inorganic Biochemistry 169: 13-22.

Jin, X., Kennedy, S.W., Di Muccio, T. and Moon, T.W., 2001. Role of oxidative stress and antioxidant defense in 3, 3', 4, 4', 5-pentachlorobiphenyl-induced toxicity and species-differential sensitivity in chicken and duck embryos. Toxicology and Applied Pharmacology 172: 241-248.

Jing, C.L., Dong, X.F., Wang, Z.M., Liu, S. and Tong, J.M., 2015. Comparative study of DL-selenomethionine vs sodium selenite and seleno-yeast on antioxidant activity and selenium status in laying hens. Poultry Science 94: 965-975.

Kamashi, K., Reddy, A.G., Reddy, K.S. and Reddy, V.R., 2004. Evaluation of zinc against salinomycin toxicity in broilers. Indian Journal of Physiology and Pharmacology 48: 89-95.

Ke, Y.Y., Liu, W.J., Wang, Z.X. and Chen, Y.X., 2011. Effects of monochromatic light on quality properties and antioxidation of meat in broilers. Poultry Science 90: 2632-2637.

Kim, Y.S. and Combs, G.F., 1993. Effect of dietary selenium and vitamin E on glutathione concentrations and glutathione S-transferase activities in chick liver and plasma. Nutrition Research 13: 455-463.

Kong, B.W., Kim, H. and Foster, D.N., 2003. Cloning and expression analysis of chicken phospholipid-hydroperoxide glutathione peroxidase. Animal Biotechnology 14: 19-29.

Kövesi, B., Cserháti, M., Erdélyi, M., Zándoki, E., Mézes, M. and Balogh, K., 2019. Long-term effects of ochratoxin A on the glutathione redox system and its regulation in chicken. Antioxidants 8: 178.

Kövesi, B., Cserháti, M., Erdélyi, M., Zándoki, E., Mézes, M. and Balogh, K., 2020. Lack of dose- and time-dependent effects of aflatoxin B1 on gene expression and enzymes associated with lipid peroxidation and the glutathione redox system in chicken. Toxins 12: 84.

Koziorowska-Gilun, M., Gilun, P., Mietelska, K. and Kordan, W., 2018. Determination of the activity and relative abundance of mRNA for antioxidant enzymes in stallion testicular and epididymal tissues: a comparison between two breeding seasons. Animal Reproduction Science 196: 230-238.

Kryukov, G.V., Castellano, S., Novoselov, S.V., Lobanov, A.V., Zehtab, O., Guigó, R. and Gladyshev, V.N., 2003. Characterization of mammalian selenoproteomes. Science, 300: 1439-1443.

Lahti, P.P., Shariatmadari, R., Penttinen, J.K., Drevet, J.R., Haendler, B., Vierula, M., Parvinen, M., Huhtaniemi, I.T. and Poutanen, M., 2001. Evaluation of the 5'-flanking regions of murine glutathione peroxidase five and cysteine-rich secretory protein-1 genes for directing transgene expression in mouse epididymis. Biology of Reproduction 64: 1115-1121.

Lawrence, R.A. and Burk, R.F., 1978. Species, tissue and subcellular distribution of non Se-dependent glutathione peroxidase activity. The Journal of Nutrition 108: 211-215.

Leal, M., Shimada, A., Ruíz, F. and de Mejía, E.G., 1999. Effect of lycopene on lipid peroxidation and glutathione-dependent enzymes induced by T-2 toxin *in vivo*. Toxicology Letters 109: 1-10.

Lee, S.H., Lillehoj, H.S., Jang, S.I., Jeong, M.S., Xu, S.Z., Kim, J.B., Park, H.J., Kim, H.R., Lillehoj, E.P. and Bravo, D.M., 2014. Effects of *in ovo* injection with selenium on immune and antioxidant responses during experimental necrotic enteritis in broiler chickens. Poultry Science 93: 1113-1121.

Lenzi, A., Gandini, L., Picardo, M., Tramer, F., Sandri, G. and Panfili, E., 2000. Lipoperoxidation damage of spermatozoa polyunsaturated fatty acids (PUFA): Scavenger mechanisms and possible scavenger therapies. Frontiers in Bioscience 5: 1-15.

Lermant, A. and Murdoch, C.E., 2019. Cysteine glutathionylation acts as a redox switch in endothelial cells. Antioxidants 8: 315.

Lewerenz, J., Hewett, S.J., Huang, Y., Lambros, M., Gout, P.W., Kalivas, P.W., Massie, A., Smolders, I., Methner, A., Pergande, M. and Smith, S.B., 2013. The cystine/glutamate antiporter system xc− in health and disease: from molecular mechanisms to novel therapeutic opportunities. Antioxidants & Redox Signaling 18: 522-555.

Li, J.L. and Sunde, R.A., 2016. Selenoprotein transcript level and enzyme activity as biomarkers for selenium status and selenium requirements of chickens (*Gallus gallus*). PloS One 11: e0152392.

Li, R., Fan, X., Zhang, T., Song, H., Bian, X., Nai, R., Li, J. and Zhang, J., 2018. Expression of selenium-independent glutathione peroxidase 5 (GPx5) in the epididymis of Small Tail Han sheep. Asian-Australasian Journal of Animal Sciences 31: 1591-1597.

Li, S., Gao, F., Huang, J., Wu, Y., Wu, S. and Lei, X.G., 2018. Regulation and function of avian selenogenome. Biochimica et Biophysica Acta (BBA)-General Subjects 1862: 2473-2479.

Likittrakulwong, W., Poolprasert, P. and Roytrakul, S., 2019. Morphological trait, molecular genetic evidence and proteomic determination of different chickens (*Gallus gallus*) Breeds. Journal of Applied Biology & Biotechnology 7: 065-070.

Lin, S.L., Wang, C.W., Tan, S.R., Liang, Y., Yao, H.D., Zhang, Z.W. and Xu, S.W., 2014. Selenium deficiency inhibits the conversion of thyroidal thyroxine (T 4) to triiodothyronine (T 3) in chicken thyroids. Biological Trace Element Research 161: 263-271.

Liu, C.P., Fu, J., Xu, F.P., Wang, X.S. and Li, S., 2015. The role of heat shock proteins in oxidative stress damage induced by Se deficiency in chicken livers. Biometals 28: 163-173.

Liu, N., Ding, K., Wang, J.Q., Jia, S.C., Wang, J.P. and Xu, T.S., 2017. Detoxification, metabolism, and glutathione pathway activity of aflatoxin B1 by dietary lactic acid bacteria in broiler chickens. Journal of Animal Science 95: 4399-4406.

Liu, N., Wang, J., Liu, Z., Wang, Y. and Wang, J., 2018. Effect of supplemental yeast cell walls on growth performance, gut mucosal glutathione pathway, proteolytic enzymes and transporters in growing broiler chickens. Journal of Animal Science 96: 1330-1337.

Liu, N., Wang, J.Q., Liu, Z.Y., Chen, Y.K. and Wang, J.P., 2018a. Effect of cysteamine hydrochloride supplementation on the growth performance, enterotoxic status, and glutathione turnover of broilers fed aflatoxin B1 contaminated diets. Poultry Science 97: 3594-3600.

Liu, T., Ma, Q., Zhao, L., Jia, R., Zhang, J., Ji, C. and Wang, X., 2016b. Protective effects of sporoderm-broken spores of *Ganderma lucidum* on growth performance, antioxidant capacity and immune function of broiler chickens exposed to low level of aflatoxin B1. Toxins 8: 278.

Liu, Z., Qu, Y., Wang, J. and Wu, R., 2016a. Selenium deficiency attenuates chicken duodenal mucosal immunity via activation of the NF-κb signaling pathway. Biological Trace Element Research 172: 465-473.

López-Grueso, M.J., González-Ojeda, R., Requejo-Aguilar, R., McDonagh, B., Fuentes-Almagro, C.A., Muntané, J., Bárcena, J.A. and Padilla, C.A., 2019. Thioredoxin and glutaredoxin regulate metabolism through different multiplex thiol switches. Redox Biology 21: 101049.

Luan, Y., Zhao, J., Yao, H., Zhao, X., Fan, R., Zhao, W., Zhang, Z. and Xu, S., 2016. Selenium deficiency influences the mRNA expression of selenoproteins and cytokines in chicken erythrocytes. Biological Trace Element Research 171: 427-436.

Lubos, E., Loscalzo, J. and Handy, D.E., 2011. Glutathione peroxidase-1 in health and disease: from molecular mechanisms to therapeutic opportunities. Antioxidants & Redox Signaling 15: 1957-1997.

Lundström, J. and Holmgren, A., 1990. Protein disulfide-isomerase is a substrate for thioredoxin reductase and has thioredoxin-like activity. Journal of Biological Chemistry 265: 9114-9120.

Ma, D., Shan, A., Chen, Z., Du, J., Song, K., Li, J. and Xu, Q., 2005. Effect of *Ligustrum lucidum* and *Schisandra chinensis* on the egg production, antioxidant status and immunity of laying hens during heat stress. Archives of Animal Nutrition 59: 439-447.

Ma, D., Shan, A., Li, J., Zhao, Y. and Guo, X., 2009. Influence of an aqueous extract of *Ligustrum lucidum* and an ethanol extract of *Schisandra chinensis* on parameters of antioxidative metabolism and spleen lymphocyte proliferation of broilers. Archives of Animal Nutrition 63: 66-74.

Ma, Y., Zheng, Y.X., Dong, X.Y. and Zou, X.T., 2018a. Effect of mercury chloride on oxidative stress and nuclear factor erythroid 2-related factor 2 signalling molecule in liver and kidney of laying hens. Journal of Animal Physiology and Animal Nutrition 102: 1199-1209.

Ma, Y., Zhu, M., Miao, L., Zhang, X., Dong, X. and Zou, X., 2018. Mercuric chloride induced ovarian oxidative stress by suppressing Nrf2-Keap1 signal pathway and its downstream genes in laying hens. Biological Trace Element Research 185: 185-196.

Maddipati, K.R. and L.J. Marnett, L.J., 1987. Characterization of the major hydroperoxide-reducing activity of human plasma. Purification and properties of a selenium-dependent glutathione peroxidase. The Journal of Biological Chemistry 262: 17398-17403.

Mahmoud, K.Z. and Edens, F.W., 2003. Influence of selenium sources on age-related and mild heat stress-related changes of blood and liver glutathione redox cycle in broiler chickens (*Gallus domesticus*). Comparative Biochemistry and Physiology Part B: Biochemistry and Molecular Biology 136: 921-934.

Mahmoud, K.Z. and Hijazi, A.A., 2007. Effect of vitamin A and/or E on plasma enzymatic antioxidant systems and total antioxidant capacity of broiler chickens challenged with carbon tetrachloride. Journal of Animal Physiology and Animal Nutrition 91: 333-340.

Mann, K. and Mann, M., 2008. The chicken egg yolk plasma and granule proteomes. Proteomics 8: 178-191.

Marchionatti, A.M., Perez, A.V., de Barboza, G.E.D., Pereira, B.M. and de Talamoni, N.G.T., 2008. Mitochondrial dysfunction is responsible for the intestinal calcium absorption inhibition induced by menadione. Biochimica et Biophysica Acta (BBA)-General Subjects 1780: 101-107.

Mehmeti, I., Lortz, S., Avezov, E., Jörns, A. and Lenzen, S., 2017. ER-resident antioxidative GPx7 and GPx8 enzyme isoforms protect insulin-secreting INS-1E β-cells against lipotoxicity by improving the ER antioxidative capacity. Free Radical Biology and Medicine 112: 121-130.

Meister, A., 1992. On the antioxidant effects of ascorbic acid and glutathione. Biochemical Pharmacology 44: 1905-1915.

Meister, A. and Anderson, M.E., 1983. Glutathione. Annual Review of Biochemistry 52: 711-760.

Meyer, Y., Buchanan, B.B., Vignols, F. and Reichheld, J.P., 2009. Thioredoxins and glutaredoxins: unifying elements in redox biology. Annual Review of Genetics 43: 335-367.

Mézes, M., Salyi, G., Banhidi, G.Y. and Szeberenyi, S., 1992. Effect of acute salinomycin-tiamulin toxicity on the lipid peroxide and antioxidant status of broiler chicken. Acta Veterinaria Hungarica 40: 251-257.

Mézes, M. and Salyi, G., 1994. Effect of acute selenium toxicosis on the lipid peroxide status and the glutathione system of broiler chickens. Acta Veterinaria Hungarica 42: 459-463.

Milinković-Tur, S., Stojević, Z., Piršljin, J., Zdelar-Tuk, M., Poljičak-Milas, N., Ljubić, B. and Gradinski-Vrbanac, B., 2007. Effects of fasting and refeeding on the antioxidant system in cockerels and pullets. Acta Veterinaria Hungarica 55: 181-189.

Mills, G.C., 1957. Hemoglobin catabolism. 1. Glutathione peroxidase, and erythrocyte enzyme which protects hemoglobin from oxidative breakdown. The Journal of Biological Chemistry 229: 189-197.

Miyazaki, S., 1991. Effect of chemicals on glutathione peroxidase of chick liver. Research in Veterinary Science 51: 120-122.

Miyazaki, S. and Motoi, Y., 1996. Purification and characterisation of chicken liver monomeric glutathione peroxidase. British Poultry Science 37: 651-660.

Miyazaki, S. and Motoi, Y., 1992. Tissue distribution of monomeric glutathione peroxidase in broiler chicks. Research in Veterinary Science 53: 47-51.

Mörk, H., Lex, B., Scheurlen, M., Dreher, I., Schütze, N., Köhrle, J. and Jakob, F., 1998. Expression pattern of gastrointestinal selenoproteins--targets for selenium supplementation. Nutrition and Cancer 32: 64-70.

Mugesh, G. and Singh, H.B., 2000. Synthetic organoselenium compounds as antioxidants: glutathione peroxidase activity. Chemical Society Reviews 29: 347-357.

Mutryn, M.F., Brannick, E.M., Fu, W., Lee, W.R. and Abasht, B., 2015. Characterization of a novel chicken muscle disorder through differential gene expression and pathway analysis using RNA-sequencing. BMC Genomics 16: 399.

Nakade, M., Pelyhe, C., Kövesi, B., Balogh, K., Kovács, B., Szabó-Fodor, J., Zándoki, E., Mézes, M. and Erdélyi, M., 2018. Short-term effects of T-2 toxin or deoxynivalenol on glutathione status and expression of its regulatory genes in chicken. Acta Veterinaria Hungarica 66: 28-39.

Nam, S.Y., Baek, I.J., Lee, B.J., In, C.H., Jung, E.Y., Yon, J.M., Ahn, B., Kang, J.K., Yu, W.J. and Yun, Y.W., 2003. Effects of 17beta-estradiol and tamoxifen on the selenoprotein phospholipid hydroperoxide glutathione peroxidase (PHGPx) mRNA expression in male reproductive organs of rats. The Journal of Reproduction and Development 49: 389-396.

Nayernia, K., Diaconu, M., Aumuller, G., Wennemuth, G., Schwandt, I., Kleene, K., Kuehn, H. and Engel, W., 2004. Phospholipid hydroperoxide glutathione peroxidase: expression pattern during testicular development in mouse and evolutionary conservation in spermatozoa. Molecular Reproduction and Development 67: 458-464.

Naziroglu, M., Sahin, K., Simsek, H., Aydilek, N. and Ertas, O.N., 2000. The effects of food withdrawal and darkening on lipid peroxidation of laying hens in high ambient temperatures. DTW. Deutsche tierarztliche Wochenschrift 107: 199-202.

Nguyen, V.D., Saaranen, M.J., Karala, A.R., Lappi, A.K., Wang, L., Raykhel, I.B., Alanen, H.I., Salo, K.E., Wang, C.C. and Ruddock, L.W., 2011. Two endoplasmic reticulum PDI peroxidases increase the efficiency of the use of peroxide during disulfide bond formation. Journal of Molecular Biology 406: 503-515.

Nikolaienko, R., Bovo, E. and Zima, A.V., 2018. Redox dependent modifications of ryanodine receptor: basic mechanisms and implications in heart diseases. Frontiers in Physiology 9: 1775.

Noblanc, A., Kocer, A., Chabory, E., Vernet, P., Saez, F., Cadet, R., Conrad, M. and Drevet, J.R., 2011. Glutathione peroxidases at work on epididymal spermatozoa: an example of the dual effect of reactive oxygen species on mammalian male fertilizing ability. Journal of Andrology 32: 641-650.

Noblanc, A., Peltier, M., Damon-Soubeyrand, C., Kerchkove, N., Chabory, E., Vernet, P., Saez, F., Cadet, R., Janny, L., Pons-Rejraji, H. and Conrad, M., 2012. Epididymis response partly compensates for spermatozoa oxidative defects in snGPx4 and GPx5 double mutant mice. PloS One 7: e38565.

Ojano-Dirain, C., Iqbal, M., Wing, T., Cooper, M. and Bottje, W., 2005. Glutathione and respiratory chain complex activity in duodenal mitochondria of broilers with low and high feed efficiency. Poultry science 84: 782-788.

Okamura, N., Iwaki, Y., Hiramoto, S., Tamba, M., Bannai, S., Sugita, Y., Syntin, P., Dacheux, F. and Dacheux, J.L., 1997. Molecular cloning and characterization of the epididymis-specific glutathione peroxidase-like protein secreted in the porcine epididymal fluid. Biochimica et Biophysica Acta (BBA)-General Subjects 1336: 99-109.

Omaye, S.T. and Tappel, A.L., 1974. Effect of dietary selenium on glutathione peroxidase in the chick. The Journal of Nutrition 104: 747-753.

Omaye, S.T. and Tappel, A.L., 1974. Glutathione peroxidase, glutathione reductase, and thiobarbituric acid-reactive products in muscles of chickens and mice with genetic muscular dystrophy. Life Sciences 15: 137-145.

Ouyang, Y., Peng, Y., Li, J., Holmgren, A. and Lu, J., 2018. Modulation of thiol-dependent redox system by metal ions via thioredoxin and glutaredoxin systems. Metallomics 10: 218-228.

Oztürk-Urek, R., Bozkaya, L.A. and Tarhan, L., 2001. The effects of some antioxidant vitamin-and trace element-supplemented diets on activities of SOD, CAT, GSH-Px and LPO levels in chicken tissues. Cell Biochemistry and Function 19: 125-132.

Palamanda, J.R. and Kehrer, J.P. 1993. Involvement of vitamin E and protein thiols in the inhibition of microsomal lipid peroxidation by glutathione. Lipids 278: 427-431.

Panda, A.K., Ramarao, S.V., Raju, M.V.L.N. and Chatterjee, R.N., 2008. Effect of dietary supplementation with vitamins E and C on production performance, immune responses and antioxidant status of White Leghorn layers under tropical summer conditions. British Poultry Science 49: 592-599.

Pannala, V.R., Bazil, J.N., Camara, A.K.S. and Dash, R.K., 2014. A mechanistic mathematical model for the catalytic action of glutathione peroxidase. Free Radical Research 48: 487-502.

Partyka, A., Łukaszewicz, E. and Niżański, W., 2012. Effect of cryopreservation on sperm parameters, lipid peroxidation and antioxidant enzymes activity in fowl semen. Theriogenology 77: 1497-1504.

Partyka, A., Lukaszewicz, E. and Niżański, W., 2012. Lipid peroxidation and antioxidant enzymes activity in avian semen. Animal Reproduction Science 134: 184-190.

Payne, R.L. and Southern, L.L., 2005. Changes in glutathione peroxidase and tissue selenium concentrations of broilers after consuming a diet adequate in selenium. Poultry Science 84: 1268-1276.

Pelyhe, C., Kövesi, B., Zándoki, E., Kovács, B., Erdélyi, M., Kulcsár, S., Mézes, M. and Balogh, K., 2018. Multi-trichothecene mycotoxin exposure activates glutathione-redox system in broiler chicken. Toxicon 153: 53-57.

Peng, D., Belkhiri, A., Hu, T., Chaturvedi, R., Asim, M., Wilson, K.T., Zaika, A. and El-Rifai, W., 2012. Glutathione peroxidase 7 protects against oxidative DNA damage in oesophageal cells. Gut 61: 1250-1260.

Peng, X., Cui, H., He, Y., Cui, W., Fang, J., Zuo, Z., Deng, J., Pan, K., Zhou, Y. and Lai, W., 2012a. Excess dietary sodium selenite alters apoptotic population and oxidative stress markers of spleens in broilers. Biological Trace Element Research 145: 47-51.

Peng, Y.Z., Wang, Y.W., Ning, D. and Guo, Y.M., 2013. Changes of haematic parameters, redox status and mitochondrial complex activity in the heart and liver of broilers fed with different density diets under low ambient temperature. Avian Pathology 42: 327-334.

Perai, A.H., Kermanshahi, H., Moghaddam, H.N. and Zarban, A., 2015. Effects of chromium and chromium+ vitamin C combination on metabolic, oxidative, and fear responses of broilers transported under summer conditions. International Journal of Biometeorology 59: 453-462.

Petrovic, V., Marcincak, S., Popelka, P., Simkova, J., Martonova, M., Buleca, J., Marcincakova, D., Tuckova, M., Molnar, L. and Kovac, G., 2012. The effect of supplementation of clove and agrimony or clove and lemon balm on growth performance, antioxidant status and selected indices of lipid profile of broiler chickens. Journal of Animal Physiology and Animal Nutrition 96: 970-977.

Placha, I., Borutova, R., Gresakova, L., Petrovic, V., Faix, S. and Leng, L., 2009. Effects of excessive selenium supplementation to diet contaminated with deoxynivalenol on blood phagocytic activity and antioxidative status of broilers. Journal of Animal Physiology and Animal Nutrition 93: 695-702.

Placha, I., Takacova, J., Ryzner, M., Cobanova, K., Laukova, A., Strompfova, V., Venglovska, K. and Faix, S., 2014. Effect of thyme essential oil and selenium on intestine integrity and antioxidant status of broilers. British Poultry Science 55: 105-114.

Qin, T., Yin, Y., Yu, Q. and Yang, Q., 2015. Bursopentin (BP5) protects dendritic cells from lipopolysaccharide-induced oxidative stress for immunosuppression. PLoS One, 10: e0117477.

Rajashree, K., Muthukumar, T. and Karthikeyan, N., 2014. Comparative study of the effects of organic selenium on hen performance and productivity of broiler breeders. British Poultry Science 55: 367-374.

Rao, S.R., Prakash, B., Rajkumar, U., Raju, M.V.L.N., Srilatha, T. and Reddy, E.P.K., 2018. Effect of supplementing germinated sprouts of pulses on performance, carcass variables, immune and oxidative stress indicators in broiler chickens reared during tropical summer season. Tropical Animal Health and Production 50: 1147-1154.

Ramming, T., Hansen, H.G., Nagata, K., Ellgaard, L. and Appenzeller-Herzog, C., 2014. GPx8 peroxidase prevents leakage of H2O2 from the endoplasmic reticulum. Free Radical Biology and Medicine 70: 106-116.

Ramming, T., Okumura, M., Kanemura, S., Baday, S., Birk, J., Moes, S., Spiess, M., Jenö, P., Bernèche, S., Inaba, K. and Appenzeller-Herzog, C., 2015. A PDI-catalyzed thiol-disulfide switch regulates the production of hydrogen peroxide by human ero1. Free Radical Biology and Medicine 83: 361-372.

Ramnath, V. and Rekha, P.S., 2009. Brahma Rasayana enhances *in vivo* antioxidant status in cold-stressed chickens (*Gallus gallus domesticus*). Indian Journal of Pharmacology 41: 115-119.

Rao, S.V.R., Prakash, B., Kumari, K., Raju, M.V.L.N. and Panda, A.K., 2013. Effect of supplementing different concentrations of organic trace minerals on performance, antioxidant activity, and bone mineralization in Vanaraja chickens developed for free range farming. Tropical Animal Health and Production 45: 1447-1451.

Rao, S.R., Raju, M.V.L.N., Panda, A.K., Poonam, N.S., Murthy, O.K. and Sunder, G.S., 2012. Effect of dietary supplementation of organic chromium on performance, carcass traits, oxidative parameters, and immune responses in commercial broiler chickens. Biological Trace Element Research 147: 135-141.

Rejraji, H., Vernet, P. and Drevet, J.R., 2002. GPX5 is present in the mouse caput and cauda epididymidis lumen at three different locations. Molecular Reproduction and Development 63: 96-103.

Ren, X., Zou, L., Zhang, X., Branco, V., Wang, J., Carvalho, C., Holmgren, A. and Lu, J., 2017. Redox signaling mediated by thioredoxin and glutathione systems in the central nervous system. Antioxidants & Redox Signaling 27: 989-1010.

Ribas, V., García-Ruiz, C. and Fernández-Checa, J.C., 2014. Glutathione and mitochondria. Frontiers in Pharmacology 5: 151.

Rotruck, J.T., Pope, A.L., Ganther, H.E., Swanson, A.B., Hafeman, D.G. and Hoekstra, W.G., 1973. Selenium: biochemical role as a component of glutathione peroxidase. Science 179: 588-590.

Roveri, A., Maiorino, M., Nisii, C. and Ursini, F., 1994. Purification and characterization of phospholipid hydroperoxide glutathione peroxidase from rat testis mitochondrial membranes. Biochimica et Biophysica Acta 1208: 211-221.

Sahin, K., Orhan, C., Tuzcu, M., Sahin, N., Hayirli, A., Bilgili, S. and Kucuk, O., 2016. Lycopene activates antioxidant enzymes and nuclear transcription factor systems in heat-stressed broilers. Poultry Science 95: 1088-1095.

Salyi, G., Mezes, M. and Banhidi, G., 1990. Changes in the lipid peroxide status of broiler chickens in acute monensin poisoning. Acta Veterinaria Hungarica 38: 263-270.

Savaskan, N.E., Borchert, A., Bräuer, A.U. and Kuhn, H., 2007. Role for glutathione peroxidase-4 in brain development and neuronal apoptosis: specific induction of enzyme expression in reactive astrocytes following brain injury. Free Radical Biology and Medicine 43: 191-201.

Schwaab, V., Lareyre, J.J., Vernet, P., Pons, E. and Faure, J., 1998. Characterization, regulation of the expression and putative roles of two glutathione peroxidase. Journal of Reproduction and Fertility Supplement 53: 157-162.

Sen, C.K. and Packer, L., 2000. Thiol homeostasis and supplements in physical exercise. The American Journal of Clinical Nutrition 72: 553S-669S.

Sharma, A., Yuen, D., Huet, O., Pickering, R., Stefanovic, N., Bernatchez, P. and de Haan, J.B., 2016. Lack of glutathione peroxidase-1 facilitates a pro-inflammatory and activated vascular endothelium. Vascular Pharmacology 79: 32-42.

Shema, R., Kulicke, R., Cowley, G.S., Stein, R., Root, D.E. and Heiman, M., 2015. Synthetic lethal screening in the mammalian central nervous system identifies Gpx6 as a modulator of Huntington's disease. Proceedings of the National Academy of Sciences 112: 268-272.

Shen, X., Yi, D., Ni, X., Zeng, D., Jing, B., Lei, M., Bian, Z., Zeng, Y., Li, T. and Xin, J., 2014. Effects of Lactobacillus plantarum on production performance, immune characteristics, antioxidant status, and intestinal microflora of bursin-immunized broilers. Canadian Journal of Microbiology 60:193-202.

Singh, A., Rangasamy, T., Thimmulappa, R.K., Lee, H., Osburn, W.O., Brigelius-Flohé, R., Kensler, T.W., Yamamoto, M. and Biswal, S., 2006. Glutathione peroxidase 2, the major cigarette smoke-inducible isoform of GPX in lungs, is regulated by Nrf2. American Journal of Respiratory Cell and Molecular Biology 35: 639-650.

Song, G., Nesil, T., Cao, J., Yang, Z., Chang, S.L. and Li, M.D., 2016. Nicotine mediates expression of genes related to antioxidant capacity and oxidative stress response in HIV-1 transgenic rat brain. Journal of Neurovirology 22: 114-124.

Starrs, A.P., Orgeig, S., Daniels, C.B., Davies, M. and Lopatko, O.V., 2001. Antioxidant enzymes in the developing lungs of egg-laying and metamorphosing vertebrates. Journal of Experimental Biology 204: 3973-3981.

Storey, K.B., 1996. Metabolic adaptations supporting anoxia tolerance in reptiles: recent advances. Comparative Biochemistry and Physiology Part B: Biochemistry and Molecular Biology 113: 23-35.

Subbaiah, K.C.V., Raniprameela, D., Visweswari, G., Rajendra, W. and Lokanatha, V., 2011. Perturbations in the antioxidant metabolism during Newcastle disease virus (NDV) infection in chicken. Naturwissenschaften 98: 1019-1026.

Sun, B., Chen, C., Wang, W., Ma, J., Xie, Q., Gao, Y., Chen, F., Zhang, X. and Bi, Y., 2015. Effects of lycopene supplementation in both maternal and offspring diets on growth performance, antioxidant capacity and biochemical parameters in chicks. Journal of Animal Physiology and Animal Nutrition 99: 42-49.

Sun, X., Jiao, H., Zhao, J., Wang, X. and Lin, H., 2017. Unexpected effect of urate on hydrogen peroxide-induced oxidative damage in embryonic chicken cardiac cells. Free Radical Research 51: 693-707.

Sunde, R.A. and Hadley, K.B., 2010. Phospholipid hydroperoxide glutathione peroxidase (Gpx4) is highly regulated in male turkey poults and can be used to determine dietary selenium requirements. Experimental Biology and Medicine 235: 23-31.

Sunde, R.A., Sunde, G.R., Sunde, C.M., Sunde, M.L. and Evenson, J.K., 2015. Cloning, sequencing, and expression of selenoprotein transcripts in the turkey (*Meleagris gallopavo*). PLoS One 10: e0129801.

Sunde, R.A., 1993. Intracellular glutathione peroxidases – structure, regulation, and function. In: Burk, R.F. (ed.) Selenium in biology and human health. Springer-Verlag, New York, NY, USA, pp.45-77.

Surai, P.F., 1999. Tissue-specific changes in the activities of antioxidant enzymes during the development of the chicken embryo. British Poultry Science 40: 397-405.

Surai, P.F., 2000. Effect of selenium and vitamin E content of the maternal diet on the antioxidant system of the yolk and the developing chick. British Poultry Science 41: 235-243.

Surai, P.F., 2002a. Natural antioxidants in avian nutrition and reproduction. Nottingham University Press, Nottingham.

Surai, P.F., 2002b. Selenium in poultry nutrition: a new look at an old element. 1. Antioxidant properties, deficiency and toxicity. Worlds Poultry Science Journal 58: 333-347.

Surai, P.F., 2002c. Selenium in poultry nutrition: a new look at an old element. 2. Reproduction, egg and meat quality and practical applications. Worlds Poultry Science Journal 58: 431-450.

Surai, P.F., 2006. selenium in nutrition and health. Nottingham University Press, Nottingham, UK.

Surai, P.F., 2018. Selenium in poultry nutrition and health. Wageningen Academic Publishers: Wageningen, the Netherlands.

Surai P.F., Brillard J-P., Speake B.K., Blesbois E., Seigneurin F. and Sparks N.H.C. 2000. Phospholipid fatty acid composition, vitamin E content and susceptibility to lipid peroxidation of duck spermatozoa. Theriogenology 53: 1025-1039.

Surai, P.F. and Fisinin, V.I., 2015. Antioxidant-prooxidant balance in the intestine: applications in chick placement and pig weaning. Journal of Veterinary Science & Medicine 3, 1: 16.

Surai, P.F., Kochish, I.I., Fisinin, V.I. and Kidd, M.T., 2019. Antioxidant defence systems and oxidative stress in poultry biology: an update. Antioxidants 8, 7: 235.

Surai, P.F., Kuklenko, T.V., Ionov, I.A., Noble, R.C. and Sparks, N.H., 2000a. Effect of vitamin A on the antioxidant system of the chick during early postnatal development. British Poultry Science 41: 454-458.

Surai, P.F., Speake, B.K., Noble, R.C. and Sparks, N.H., 1999. Tissue-specific antioxidant profiles and susceptibility to lipid peroxidation of the newly hatched chick. Biological Trace Element Research 68: 63-78.

Surai, P.F., Noble, R.C. and Speake, B.K., 1999a. Relationship between vitamin E content and susceptibility to lipid peroxidation in tissues of the newly hatched chick. British Poultry Science 40: 406-410.

Surai, P.F., Blesbois, E., Grasseau, I., Ghalah, T., Brillard, J-P., Wishart, G., Cerolini, S. and Sparks, N.H.C., 1998a. Fatty acid composition, glutathione peroxidase and superoxide dismutase activity and total antioxidant activity of avian semen. Comparative Biochemistry and Physiology 120B: 527-533.

Surai, P.F., Cerolini, S., Wishart, G.J., Speake, B.K., Noble, R.C. and Sparks, N.H.C., 1998b. Lipid and antioxidant composition of chicken semen and its susceptibility to peroxidation. Poultry and Avian Biology Reviews 9: 11-23.

Surai, P.F., Kostjuk, I.A., Wishart, G., MacPherson, A., Speake, B., Noble, R.C., Ionov, I.A. and Kutz, E., 1998c. Effect of vitamin E and selenium of cockerel diets on glutathione peroxidase activity and lipid peroxidation susceptibility in sperm, testes and liver. Biological Trace Element Research 64: 119-132.

Takahashi, K., Avissar, N., Whitin, J. and Cohen, H., 1987. Purification and characterization of human plasma glutathione peroxidase: a selenoglycoprotein distinct from the known cellular enzyme. Archives of Biochemistry and Biophysics 256: 677-686.

Takebe, G., Yarimizu, J., Saito, Y., Hayashi, T., Nakamura, H., Yodoi, J., Nagasawa, S. and Takahashi, K., 2002. A comparative study on the hydroperoxide and thiol specificity of the glutathione peroxidase family and selenoprotein P. The Journal of Biological Chemistry 277: 41254-41258.

Tan, G.Y., Yang, L., Fu, Y.Q., Feng, J.H. and Zhang, M.H., 2010. Effects of different acute high ambient temperatures on function of hepatic mitochondrial respiration, antioxidative enzymes, and oxidative injury in broiler chickens. Poultry Science 89: 115-122.

Tanaka, C., Coling, D.E., Manohar, S., Chen, G.D., Hu, B.H., Salvi, R. and Henderson, D., 2012. Expression pattern of oxidative stress and antioxidant defense-related genes in the aging Fischer 344/NHsd rat cochlea. Neurobiology of Aging 33: 1842-e1-14.

Tang, J., Huang, X., Wang, L., Li, Q., Xu, J., Jia, G., Liu, G., Chen, X., Shang, H. and Zhao, H., 2017. Supranutritional dietary selenium depressed expression of selenoprotein genes in three immune organs of broilers. Animal Science Journal 88: 331-338.

Tappel, M.E., Chaudiere, J. and Tappel, A.L., 1982. Glutathione peroxidase activities of animal tissues. Comparative Biochemistry and Physiology Part B: Comparative Biochemistry 73: 945-949.

Taylor, A., Robson, A., Houghton, B.C., Jepson, C.A., Ford, W.C.L. and Frayne, J., 2013. Epididymal specific, selenium-independent GPX5 protects cells from oxidative stress-induced lipid peroxidation and DNA mutation. Human Reproduction 28: 2332-2342.

Taylor, R.M. and Sunde, R.A., 2016. Selenoprotein transcript level and enzyme activity as biomarkers for selenium status and selenium requirements in the turkey (*Meleagris gallopavo*). PLoS One 11: e0151665.

Thompson, K.H., Godin, D.V. and Lee, M., 1992. Tissue antioxidant status in streptozotocin-induced diabetes in rats. Effects of dietary manganese deficiency. Biological Trace Element Research 35: 213-224.

Toppo, S., Vanin, S., Bosello, V. and Tosatto, S.C., 2008. Evolutionary and structural insights into the multifaceted glutathione peroxidase (Gpx) superfamily. Antioxidants & Redox Signaling 10: 1501-1514.

Tu, B.P. and Weissman, J.S., 2004. Oxidative protein folding in eukaryotes: mechanisms and consequences. The Journal of Cell Biology 164: 341-346.

Ursini, F., Heim, S., Kiess, M., Maiorino, M., Roveri, A., Wissing, J. and Flohé, L., 1999. Dual function of the selenoprotein PHGPx during sperm maturation. Science 285: 1393-1396.

Ursini, F., Maiorino, M. and Gregolin, C., 1985. The selenoenzyme phospholipid hydroperoxide glutathione peroxidase. Biochimica et Biophysica Acta 839: 62-70.

Ushioda, R. and Nagata, K., 2019. Redox-mediated regulatory mechanisms of endoplasmic reticulum homeostasis. Cold Spring Harbor Perspectives in Biology 11: a033910.

Utomo, A., Jiang, X., Furuta, S., Yun, J., Levin, D.S., Wang, Y.C.J., Desai, K.V., Green, J.E., Chen, P.L. and Lee, W.H., 2004. Identification of a novel putative non-selenocysteine containing phospholipid hydroperoxide glutathione peroxidase (NPGPx) essential for alleviating oxidative stress generated from polyunsaturated fatty acids in breast cancer cells. Journal of Biological Chemistry 279: 43522-43529.

Venditti, P., Daniele, C.M., Balestrieri, M. and Meo, S.D., 1999. Protection against oxidative stress in liver of four different vertebrates. Journal of Experimental Zoology 284: 610-616.

Vernet, P., Faure, J., Dufaure, J.P. and Drevet, J.R., 1997. Tissue and developmental distribution, dependence upon testicular factors and attachment to spermatozoa of GPX5, a murine epididymis-specific glutathione peroxidase. Molecular Reproduction and Development 47: 87-98.

Vernet, P., Rigaudiere, N., Ghyselinck, N., Dufaure, J.P. and Drevet, J.R., 1996. *In vitro* expression of a mouse tissue specific glutathione-peroxidase-like protein lacking the selenocysteine can protect stably transfected mammalian cells against oxidative damage. Biochemistry and Cell Biology 74: 125-131.

Vernet, P., Rock, E., Mazur, A., Rayssiguier, Y., Dufaure, J.P. and Drevet, J.R., 1999. Selenium-independent epididymis-restricted glutathione peroxidase 5 protein (GPX5) can back up failing Se-dependent GPXs in mice subjected to selenium deficiency. Molecular Reproduction and Development 54: 362-370.

Voljč, M., Frankič, T., Levart, A., Nemec, M. and Salobir, J., 2011. Evaluation of different vitamin E recommendations and bioactivity of α-tocopherol isomers in broiler nutrition by measuring oxidative stress *in vivo* and the oxidative stability of meat. Poultry Science 90: 1478-1488.

Wang, F., Shu, G., Peng, X., Fang, J., Chen, K., Cui, H., Chen, Z., Zuo, Z., Deng, J., Geng, Y. and Lai, W., 2013. Protective effects of sodium selenite against aflatoxin B1-induced oxidative stress and apoptosis in broiler spleen. International Journal of Environmental Research and Public Health 10: 2834-2844.

Wang, X.F., Zhu, X.D., Li, Y.J., Liu, Y., Li, J.L., Gao, F., Zhou, G.H. and Zhang, L., 2015. Effect of dietary creatine monohydrate supplementation on muscle lipid peroxidation and antioxidant capacity of transported broilers in summer. Poultry science 94: 2797-2804.

Wang, Z.G., Pan, X.J., Zhang, W.Q., Peng, Z.Q., Zhao, R.Q. and Zhou, G.H., 2010. Methionine and selenium yeast supplementation of the maternal diets affects antioxidant activity of breeding eggs. Poultry Science 89: 931-937.

Wang, L., Zhang, L., Niu, Y., Sitia, R. and Wang, C.C., 2014. Glutathione peroxidase 7 utilizes hydrogen peroxide generated by Ero1α to promote oxidative protein folding. Antioxidants & Redox Signaling 20: 545-556.

Wang, Y., Yin, X., Yin, D., Lei, Z., Mahmood, T. and Yuan, J., 2019. Antioxidant response and bioavailability of methionine hydroxy analog relative to DL-methionine in broiler chickens. Animal Nutrition 5: 241-247.

Wei, F.X., Hu, X.F., Sa, R.N., Liu, F.Z., Li, S.Y. and Sun, Q.Y., 2014. Antioxidant capacity and meat quality of broilers exposed to different ambient humidity and ammonia concentrations. Genetics and Molecular Research 13: 3117-3127.

Wei, P.C., Hsieh, Y.H., Su, M.I., Jiang, X., Hsu, P.H., Lo, W.T., Weng, J.Y., Jeng, Y.M., Wang, J.M., Chen, P.L. and Chang, Y.C., 2012. Loss of the oxidative stress sensor NPGPx compromises GRP78 chaperone activity and induces systemic disease. Molecular Cell 48: 747-759.

Wu, Q.J., Wang, Y.Q. and Qi, Y.X., 2016. The protective effect of procyanidin against LPS-induced acute gut injury by the regulations of oxidative state. SpringerPlus 5: 1645.

Wu, X., Zhang, L., Miao, Y., Yang, J., Wang, X., Wang, C.C., Feng, J. and Wang, L., 2019. Homocysteine causes vascular endothelial dysfunction by disrupting endoplasmic reticulum redox homeostasis. Redox Biology 20: 46-59.

Xiao, Z., La Fontaine, S., Bush, A.I. and Wedd, A.G., 2019. Molecular mechanisms of glutaredoxin enzymes: versatile hubs for thiol-disulfide exchange between protein thiols and glutathione. Journal of Molecular Biology 431: 158-177.

Xie, W., Ma, W., Liu, P. and Zhou, F., 2019. Overview of thioredoxin system and targeted therapies for acute leukemia. Mitochondrion 47: 38-46.

Xu, D., Li, W., Huang, Y., He, J. and Tian, Y., 2014. The effect of selenium and polysaccharide of *Atractylodes macrocephala* Koidz.(PAMK) on immune response in chicken spleen under heat stress. Biological Trace Element Research 160: 232-237.

Xu, D. and Tian, Y., 2015. Selenium and polysaccharides of *Atractylodes macrocephala* koidz play different roles in improving the immune response induced by heat stress in chickens. Biological Trace Element Research 168: 235-241.

Xu, S.W., Yao, H.D., Zhang, J., Zhang, Z.W., Wang, J.T., Zhang, J.L. and Jiang, Z.H., 2013. The oxidative damage and disbalance of calcium homeostasis in brain of chicken induced by selenium deficiency. Biological Trace Element Research 151: 225-233.

Xu, S.Z., Lee, S.H., Lillehoj, H.S. and Bravo, D., 2015. Dietary sodium selenite affects host intestinal and systemic immune response and disease susceptibility to necrotic enteritis in commercial broilers. British Poultry Science 56: 103-112.

Yang, L., Tan, G.Y., Fu, Y.Q., Feng, J.H. and Zhang, M.H., 2010. Effects of acute heat stress and subsequent stress removal on function of hepatic mitochondrial respiration, ROS production and lipid peroxidation in broiler chickens. Comparative Biochemistry and Physiology Part C: Toxicology & Pharmacology 151: 204-208.

Yao, H., Zhao, W., Zhao, X., Fan, R., Khoso, P.A., Zhang, Z., Liu, W. and Xu, S., 2014. Selenium deficiency mainly influences the gene expressions of antioxidative selenoproteins in chicken muscles. Biological Trace Element Research 161: 318-327.

Yoboue, E.D., Rimessi, A., Anelli, T., Pinton, P. and Sitia, R., 2017. Regulation of calcium fluxes by GPX8, a type-II transmembrane peroxidase enriched at the mitochondria-associated endoplasmic reticulum membrane. Antioxidants & Redox Signaling 27: 583-595.

Young, A., Gill, R. and Mailloux, R.J., 2019. Protein S-glutathionylation: the linchpin for the transmission of regulatory information on redox buffering capacity in mitochondria. Chemico-Biological Interactions 299: 151-162.

Yu, J., Yao, H., Gao, X., Zhang, Z., Wang, J.F. and Xu, S.W., 2015. The role of nitric oxide and oxidative stress in intestinal damage induced by selenium deficiency in chickens. Biological Trace Element Research 163: 144-153.

Zenteno-Savin, T., Leger, J.S. and Ponganis, P.J., 2010. Hypoxemic and ischemic tolerance in emperor penguins. Comparative Biochemistry and Physiology Part C: Toxicology & Pharmacology 152: 18-23.

Zhang, P., Li, T., Wu, X., Nice, E.C., Huang, C. and Zhang, Y., in press. Oxidative stress and diabetes: antioxidative strategies. Frontiers of Medicine. https://doi.org/10.1007/s11684-019-0729-1

Zhang, C., Chen, K., Zhao, X. and Geng, Z., 2018. Protective effects of resveratrol against high ambient temperature-induced spleen dysplasia in broilers through modulating splenic redox status and apoptosis. Journal of the Science of Food and Agriculture 98: 5409-5417.

Zhang, G.F., Yang, Z.B., Wang, Y., Yang, W.R., Jiang, S.Z. and Gai, G.S., 2009. Effects of ginger root (*Zingiber officinale*) processed to different particle sizes on growth performance, antioxidant status, and serum metabolites of broiler chickens. Poultry Science 88: 2159-2166.

Zhang, G.G., Yang, Z.B., Wang, Y. and Yang, W.R., 2013. Effects of *Astragalus membranaceus* root processed to different particle sizes on growth performance, antioxidant status, and serum metabolites of broiler chickens1. Poultry Science 92: 178-183.

Zhang, J.L., Xu, B., Huang, X.D., Gao, Y.H., Chen, Y. and Shan, A.S., 2016. Selenium deficiency affects the mRNA expression of inflammatory factors and selenoprotein genes in the kidneys of broiler chicks. Biological Trace Element Research 171: 201-207.

Zhang, L., Hu, T.J., Liu, H.L. and Shuai, X.H., 2011. Inhibitory effect of *Sargassum* polysaccharide on oxidative stress induced by infectious bursa disease virus in chicken bursal lymphocytes. International Journal of Biological Macromolecules 49: 607-615.

Zhang, Z.W., Wang, Q.H., Zhang, J.L., Li, S., Wang, X.L. and Xu, S.W., 2012. Effects of oxidative stress on immunosuppression induced by selenium deficiency in chickens. Biological Trace element Research 149: 352-361.

Zhao, F.Q., Zhang, Z.W., Qu, J.P., Yao, H.D., Li, M., Li, S. and Xu, S.W., 2014. Cold stress induces antioxidants and Hsps in chicken immune organs. Cell Stress and Chaperones 19: 635-648.

Zhao, F.Q., Zhang, Z.W., Wang, C., Zhang, B., Yao, H.D., Li, S. and Xu, S.W., 2013. The role of heat shock proteins in inflammatory injury induced by cold stress in chicken hearts. Cell Stress and Chaperones 18: 773-783.

Zhao, X., Yao, H., Fan, R., Zhang, Z. and Xu, S., 2014. Selenium deficiency influences nitric oxide and selenoproteins in pancreas of chickens. Biological Trace Element Research 161: 341-349.

Zheng, X.C., Wu, Q.J., Song, Z.H., Zhang, H., Zhang, J.F., Zhang, L.L., Zhang, T.Y., Wang, C. and Wang, T., 2016. Effects of oridonin on growth performance and oxidative stress in broilers challenged with lipopolysaccharide. Poultry Science 95: 2281-2289.

Zhou, M., Zeng, D., Ni, X., Tu, T., Yin, Z., Pan, K. and Jing, B., 2016. Effects of *Bacillus licheniformis* on the growth performance and expression of lipid metabolism-related genes in broiler chickens challenged with *Clostridium perfringens*-induced necrotic enteritis. Lipids in Health and Disease 15: 48.

Zoidis, E., Demiris, N., Kominakis, A. and Pappas, A.C., 2014. Meta-analysis of selenium accumulation and expression of antioxidant enzymes in chicken tissues. Animal 8: 542-554.

Zoidis, E., Pappas, A.C., Georgiou, C.A., Komaitis, E. and Feggeros, K., 2010. Selenium affects the expression of GPx4 and catalase in the liver of chicken. Comparative Biochemistry and Physiology Part B: Biochemistry and Molecular Biology 155: 294-300.

Chapter 8
Sirtuins in avian biology

A new broom sweeps clean

8.1 Introduction

The redox balance in the cell/whole body is believed to be responsible for a regulation of an array of various physiological/biochemical processes, including cell signalling, gene expression and homeostasis maintenance (Chen *et al.*, 2014; Corsello *et al.*, 2018). In recent years, signalling role of RONS has received much attention (Moloney and Cotter, 2018; Santoro *et al.*, 2018) and vital functions of various transcription factors (Cuadrado *et al.*, 2018; Francois *et al.*, 2020) and vitagenes (Surai, 2020; Surai and Fisinin, 2016a,b; Surai *et al.*, 2019) have been described. Sirtuins belong to the vitagene family and their fundamental roles related to regulation of different pathways including regulation of transcription factors and antioxidant enzymes have receive substantial attention for the last 10 years. In 2017 sirtuin expression in chicken tissues has been characterised (Ren *et al.*, 2017), however, data on sirtuin expression and activity in various avian species are extremely limited and related to a few publications in the last 5 years.

It is the aim of this chapter to present an updated assessment of protective functions of sirtuins as an integral part of the vitagene family. This information can potentially be applied in avian biology and poultry health maintenance in commercial stress condition.

8.2 Protective functions of sirtuins

Sirtuins (SIRTs) are a highly conserved family of NAD+-dependent enzymes including seven members, SIRT1–SIRT7, which are ubiquitously distributed from eubacteria to mammals. They are located in different subcellular compartments: SIRT1, SIRT6 and SIRT7 are considered to be nuclear, SIRT2 is cytosolic, and SIRT3, SIRT4 and SIRT5 are located in mitochondria (Lee *et al.*, 2019). It seems likely that various stresses can affect SIRT location. Sirtuins possess deacetylases (all sirtuins), deacylase (SIRTs 1, 2, 4, 6), mono-ADP-ribosyltransferase (SIRT4, SIRT6), desuccinylase (SIRT5; SIRT7), demalonylase (SIRT5), deglutarylase (SIRT5), lipoamidase (SIRT4), decrotonylase (SIRT3), demyristoylase (SIRT6) and other activities (Dali-Youcef *et al.*, 2007; Zhao *et al.*, 2019a; Table 8.1).

The aforementioned activities of SIRTs make them important participants in regulation of post-translational modifications (PTMs) of various proteins including folding of

Peter F. Surai **Vitagenes in avian biology and poultry health**
DOI 10.3920/978-90-8686-906-0_8, © Wageningen Academic Publishers 2020

Table 8.1. Main sirtuins in mammals and birds (adapted from Foolad *et al*, 2019; Ghirotto *et al.*, 2019; Lee *et al.*, 2019; Song *et al.*, 2018; Tatone *et al.*, 2018).

Name	Location	Enzymatic activity	Target molecules[1]	Functions
SIRT1	nucleus cytosol	Deacetylase Deacylase	PTP1B, UCP2, PPARα, PPARγ, PPARγ2, p53, FOXO1, FOXO3, PGC-1α, NF-κB, CRTC2, SREBP, AP1, HIF-1α, AMPK, histones	Metabolism; mitochondrial biogenesis; cellular stress and senescence; chromatin regulation; cell cycle control and differentiation; DNA repair and genomic stability; apoptosis
SIRT2	cytosol nucleus	Deacetylase Demyristoylase ADP-ribosylase Deacylase	NF-κβ, PPARγ FOXO3, p53, histones	Metabolism; cell cycle regulation; cell differentiation and apoptosis; chromatin condensation/mitosis; DNA repair.
SIRT3	mitochondria	Deacetylase Decrotonylase Deacylase	NF-κβ, FOXO3, PGC-1α, p53, AMPK, CREB, PDH, LCAD, subunits of the ETC and ATP synthase, IDH, GDH, MnSOD, histones	Metabolism and energy production; mitochondrial biogenesis and function; AO activity and oxidative stress; metabolic homeostasis; chromatin silencing; DNA repair; apoptosis
SIRT4	mitochondria	ADP-ribosyl-transferase Lipoamidase Deacetylase Deacylase	PGC-1a, AMPK, adenine translocator, IDE, GDH, MCD1	Metabolism homeostasis (fatty acid oxidation); apoptosis
SIRT5	mitochondria	Desuccinylase Deacylase Deacetylase Demalonylase Deglutarylase	IDH2, G6P, CPS1, histones	AO defence and oxidative stress; metabolism homeostasis (fatty acid oxidation)
SIRT6	nucleus	ADP-ribosyl-transferase Deacetylase Deacylase	HIF-1α, Akt, FOXO1, IGF-1, NF-κB, p53, PARP1, GLUT1, LDH, PGK1, PFK-1, histones	Metabolism, homeostasis; inflammation; mitochondrial respiration repression; ribosome biosynthesis, genome stability, telomere maintenance; DNA repair; apoptosis
SIRT7	nucleus	Deacetylase Deacylase	TR4/TAK1, Cd36, PPARγ, p53, RNA Pol, histones	Metabolism; stress resistance; ribosome biogenesis; apoptosis

[1] Akt = serine/threonine kinase; AMPK = adenosine monophosphate kinase; AP1 = transcription factor; Cd36 = scavenger receptor; CPS1 = carbamoyl phosphate synthetase 1; CREB = cAMP responsive element binding protein; CRTC2 = CREB regulated transcription coactivator 2; ETC = electron transport chain; FOXO = forkhead box O; GDH = glutamate dehydrogenase; GLUT1 = glucose transporter1; HIF-1α = hypoxia-inducible factor 1α; IDE = insulin degrading enzyme; IDH = isocitrate dehydrogenase; IGF-1 = insulin-like growth factor 1; LDH = lactate dehydrogenase; LCAD = long-chain acyl-CoA dehydrogenase; MCD1 = mitotic chromosome determinant; NF-κB = transcription factor; p53 = tumour protein; PARP1 = poly-ADP-ribose polymerase; PDH = pyruvate dehydrogenase; PFK-1 = phosphofructokinase-1; PPAR = peroxisome proliferator-activated receptor; PGC-1α = peroxisome proliferator-activated receptor-γ coactivator; PGK1 = phosphoglycerate kinase 1; PTP1B = protein tyrosine phosphatase 1B; RNA Pol = RNA polymerase; SREBP = sterol regulatory element binding protein; TR4/TAK1 = nuclear receptor; UCP2 = uncoupling protein 2.

proteins structure, destination, activity and function. It is shown that more than 400 discrete types of modifications can occur and, more than 90,000 individual PTMs have been identified (Lothrop *et al.*, 2013). Indeed, PTMs are ubiquitous chemical modifications at specific amino acid residues occurring on nearly all proteins including phosphorylation, glycosylation, ubiquitination, nitrosylation, methylation, acetylation, lipidation, and proteolysis (Liu *et al.*, 2020b) and playing important roles in cellular signalling and stress adaptation. Interestingly, many domains within proteins are shown to be modified on multiple amino acid sidechains by diverse enzymes leading to a creation of a myriad of protein species (Lothrop *et al.*, 2013). Indeed, PTMs of transcription factors and antioxidant enzymes are important players in adaptive homeostasis maintenance and stress adaptation. Therefore, SIRTs regulate the biological functions of various molecules post-translationally by removing acetyl groups from protein substrates ranging from histones to transcription factors and are involved in regulation of redox balance in the cell (Singh *et al.*, 2018). In fact, main cytosolic and nuclear targets of SIRT1 are shown to include histones, p53, DNA damage proteins, FOXO1, -3 and -4, HSF1, PPARγ, PPARα, UCP2, NF-κB and HIF1α (Nogueiras *et al.*, 2012). Furthermore, SIRTs orchestrate cellular stress response and maintain genome integrity and protein stability (Radak *et al.*, 2013). In general, a number of biological processes, including cell growth and differentiations, apoptosis, chromatin condensation, energy transduction and glucose homeostasis are regulated via SIRTs expression (Dali-Youcef *et al.*, 2007). Furthermore, DNA repair and apoptosis (Lagunas-Rangel *et al.*, 2019), muscle and fat differentiation, neurogenesis, mitochondrial biogenesis, glucose and insulin homeostasis, hormone secretion, cell stress responses and circadian rhythms are proven to be regulated by SIRTs (Hubbard and Sinclair, 2014; Morris, 2013). Therefore, sirtuins are deeply involved in various stress-related pathways within the complex signalling network responsible for regulation of cell plasticity as specific sensors providing the crosstalk between environment and genome and are responsible for restoration of adaptive homeostasis under stress conditions (Lin *et al.*, 2018; Singh *et al.*, 2018).

Research assessing SIRT expression in poultry is sparse, but accumulating evidence indicates that SIRTs are highly conserved among organisms (Hickey *et al.*, 2012). Similar to mammals, stress can increase SIRT expression in birds, e.g. there was an upregulation of SIRT1 in the chicken hypothalamus, liver, and muscle in response to 48 h fasting (Fang *et al.*, 2014). On the other hand, HS was shown to downregulate SIRT1 in the chicken liver, while dietary supplementation of epigallocatechin gallate ameliorated detrimental effects of HS on SIRT1 expression (Xue *et al.*, 2017). The expression and regulation of sirtuin family members in chicken liver have been characterised (Ren *et al.*, 2017). In particular, it was shown that chicken SIRTs share the same conserved functional SIR2 domains. The chicken sirtuins are located in various cellular compartments, including the nucleus (cSIRT3 and cSIRT5), cytoplasm (cSIRT2 and cSIRT4), and in both, the cytoplasm and nucleus (cSIRT1, cSIRT6 and cSIRT7). All the sirtuins except cSIRT7 were characterised by a deacetylase activity. It was predicted that chicken sirtuins play roles in central intermediary metabolism (cSIRT1, cSIRT2, cSIRT5 and cSIRT6) and in amino acid biosynthesis (cSIRT3). Although cSIRT7 does not possess enzymatic properties, cSIRT4 has been suggested to participate in transcription regulation, with potential regulatory functions. In 30-week-old laying hens SIRTs were found to be expressed in the heart, liver, pectoralis, kidney, spleen, abdominal fat, duodenum, glandular stomach, pancreas, and lungs. An age-related regulation of gene expression (with increasing with sexual maturity) of *cSIRT1*, *cSIRT2*, *cSIRT4*, *cSIRT6*, and *cSIRT7* was observed in the chicken liver (Ren *et al.*, 2017). Recently, 24 target genes of SIRT1 in chicken embryonic liver have been identified. These genes are responsible for activation or inhibition of lipolysis and gluconeogenesis in embryos (Cogburn *et al.*, 2018). Interestingly, heat stress was shown to significantly downregulate SIRT1 in chicken muscles (Wan *et al.*, 2018). In the liver of the growing ducks SIRT1 expression was positively corelated with FOXO1 but significantly negatively correlated with FCR (Jin *et al.*, 2019). In skeletal muscles of growing broiler chickens expression of SIRT1 and SIRT3 was shown to be increased due to low dose oleuropein diet supplementation (Shimao *et al.*, 2019). Because of their roles in cellular stress responses, sirtuins would be expected to be important players in adaptive responses of poultry to stress and this topic awaits further investigation. It seems likely that major vitagenes in poultry, encoding elements of thioredoxin and glutathione systems are regulated via Nrf2 system and relationships between SIRTs and Nrf2 will be addressed in the next sections.

8.3 Sirtuins and oxidative stress

As an important part of the vitagene family SIRTs are deeply involved in antioxidant defences, redox balance and adaptive homeostasis maintenance.

8.3.1 Effect of oxidative stress on sirtuins

According to Santos *et al.* (2016) oxidative stress can affect the activity of sirtuins at different levels:

- inducing or inhibiting the expression of SIRT genes;
- posttranslational oxidative modifications of SIRTs;
- altering SIRT-protein interactions;
- changing NAD levels.

SIRT1 is the most well-studied member of the mammalian sirtuin family. It is shown to be localised predominantly in the nucleus, but recent multiple studies confirmed its presence also in the cytosol. Depending on the dosage and exposure time of the stressor, oxidative stress can increase or decrease SIRT1 expression and/or protein levels (Kwon and Ott, 2008). Low and/or moderate stress can induce SIRT1 expression and activity as adaptive response to changes in environment. It was suggested that SIRTs could initially counteract low levels of oxidative stress as part of the antioxidant stress response. In mice, the level of SIRT1 in the heart was significantly (4.3-fold) upregulated in response to oxidative stress caused by paraquat injection (Alcendor *et al.*, 2007). Similarly, moderate long-term exercise was shown to promote tissue adaptations, increasing muscle, liver and heart SIRT1 protein content and activity and increasing PGC-1α protein expression (Bayod *et al.*, 2012). Oxidative stress can also induce increased SIRT2 levels in differentiated PC12 cells (Nie *et al.*, 2014). Similarly, moderate increase of intracellular ROS stimulates SIRT3 gene expression and increases SIRT3 protein levels. For example, treatment of cells with 2,3-dimethoxy-1,4-napthoquinone, a reagent increasing mitochondria ROS level, was shown to increase SIRT3 mRNA and protein levels and simultaneously increased SOD2 activity in a time-course-dependent manner (Chen *et al.*, 2011). Importantly, SIRT4 expression was shown to be up-regulated by different anti-proliferative and senescence inducing stressors, including UVB and ionising radiation, due to inhibition of its negative regulator, microRNA miR-15b (Lang and Piekorz, 2018). It is important to mention that in many cases SIRT1 activation leads to increased expression of a range of transcription factors, including Nrf2, NF-κB, FOXO3A, p53, PGC-1α and others.

However, high oxidative stress usually decreases SIRT1 concentration and expression in various tissues. For example, in human lung fibroblasts, H_2O_2 has been found to supress activities of SIRT1, and accelerating senescence and aging (Furukawa *et al.*, 2007). Particulate matter-imposed oxidative stress in the lung tissue of rats was shown to decrease SIRT1 expression and antioxidant treatment (N-acetyl-L-cysteine) was found to ameliorate oxidative stress and improve SIRT1 expression (Yang *et al.*, 2018). In cultured skin keratinocytes, H_2O_2-induced oxidative stress was shown to downregulate SIRT1 dose and time dependently (Cao *et al.*, 2009). In fact, treatment with resveratrol (the SIRT1 activator) was found to prevent H_2O_2 induced cell death, decrease cell proliferation, and suppress senescence via modulation of p53 and JNK and activating the AMPK-FOXO3. In contrast, the SIRT1 inhibitors (sirtinol and nicotinamide) were discovered to enhance H_2O_2-induced cell death (Cao *et al.*, 2009, Ido *et al.*, 2015). In an experiment with embryonic stem cells, it was shown that H_2O_2-treatment was associated with a significant modulation in a number of SIRT1-associated genes involved in metabolism, apoptosis, ion transport, cell motility, and G protein signalling (Oberdoerffer *et al.*, 2008). In fact, oxidative and metabolic stress were shown to diminish SIRT1 activity in the hepatic cell line HepG2 (Shao *et al.*, 2019). Sirtuins inhibition in berberine treated human hepatoma cells was associated with decline in NAD(+)/NADH ratio, ATP generation, enhanced ROS production and decreased mitochondrial membrane potential (Shukla *et al.*, 2016). Oxidative and metabolic stress were shown to diminish SIRT1 activity in the hepatic cell line HepG2. High-glucose stimulation was shown to increase ROS levels in mouse podocytes and mitochondria associated with time-dependent decreases in the expression of SIRT1, PGC-1α, and its downstream genes NRF1 and mitochondrial transcription factor A (Zhang *et al.*, 2019). Interestingly, stress-inducing agents could promote the association of SIRT1 with SENP1, a SUMO specific protease that desumoylates proteins such as HDAC1, inactivating SIRT1, while cells depleted of SENP1 were shown to be more resistant to stress-induced apoptosis than control cells (Yang *et al.*, 2007). Indeed, SIRT1 response to stress is very complex and many various mechanisms are involved (balancing and counterbalancing measures) creating an adaptation opportunity for stress exposed cells/tissues. Similar results were reported with other sirtuins as well. In fact, the expression of SIRT2 was shown to be downregulated in annulus fibrosus cells due to oxidative stress imposed by tert-Butyl hydroperoxide treatment (Xu *et al.*, 2019). In human hepatoma HepG2 cells exposed to palmitic acid the level of SIRT3 expression in both nuclear and mitochondrial fractions in the cells was decreased with the increase in PA concentrations (Sharma *et al.*, 2019). Treatment of human cells (RPE cell line) with 100 μM H_2O_2 was shown to significantly decrease SIRT3 expression (Du *et al.*, 2019). Furthermore, a high fat high sucrose diet in mice was found to inhibit endogenous SIRT1 activity in mouse liver (Shao *et al.*, 2019). Resveratrol was found to decrease oxidative stress and ameliorate changes in antioxidant defences. In fact, SIRT1 protein expression levels were shown to be reduced during the process of aging, chronic diseases, and diverse extracellular stimuli (Kwon and Ott, 2008). Indeed, high or persistent oxidative stress in rats was shown to decrease nigral SIRT3 levels which further increases vulnerability to oxidative stress (Diaz-Ruiz *et al.*, 2020).

It seems likely that SIRT1 and SIRT6 are deeply involved in stress responses (D'Onofrio *et al.*, 2016). Indeed, initially oxidative stress response is associated with increased SIRT1 expression and protein levels and there is a closely cooperation with the positive cofactor AROS (to induce SIRT1 activity). However, SIRT1 also associated with the inhibitory co-factors DBC1 and SENP1 to decrease SIRT1 activity (Kwon and Ott, 2008) when stress situation is resolved, or stress is too high to handle it. Furthermore, oxidative stress can cause post-translation modifications of SIRTs. For example, an important post-translational modification of SIRT1, namely sumoylation at Lys 734, increasing SIRT1 deacetylase activity was identified. However, mutation of SIRT1 at Lys 734 or desumoylation by SENP1 (a nuclear desumoylase) was indicated to reduce its deacetylase activity (Yang *et al.*, 2007).

8.3.2 Activation/inhibition of sirtuins

Sirtuin 1

Sirtuin 1 (SIRT1) is the most well-studied member of the mammalian sirtuin family and many studies support SIRT1 activation as a protective strategy against oxidative stress imposed by various stresses (Singh *et al.*, 2018). SIRT1 was found to reduce oxidative stress by regulating antioxidant enzymes (e.g. SOD and catalase) and deacetylating its substrates (Cheng *et al.*, 2014), including FoxOs (Hori *et al.*, 2013), Nrf2 (Ding *et al.*, 2016) in response to oxidative stress. It seems likely that the SIRT1 signalling pathway may play a vital role in mitochondrial biogenesis. Indeed, downregulation of SIRT1 decreased PGC-1α expression and mitochondrial biogenesis machinery, increased Complex I dysfunction, protein oxidation and caspase-3 expression leading to neuronal cell damage in the hippocampus of rats (Chuang *et al.*, 2019). Oxidative stress markers were studied in HT22 cells and HT22 cells after exposure to TBA for 6 h with SIRT1 silencing (transfected with SIRT1 siRNA) or high expression (preconditioned with agonists SRT1720). It was shown that levels of TBA-induced oxidative stress were aggravated when SIRT1 was silenced (increasing the production of ROS and decreasing the activities of SOD and GSH) but alleviated when SIRT1 was enhanced (Ma *et al.*, 2018). SIRT1 silencing (small interfering RNA) in bone marrow-derived macrophage cultures was shown to enhance inflammatory cytokine programs, whereas SIRT1 activation was associated with decreased proinflammatory response in a SIRT1-dependent manner (Nakamura *et al.*, 2017).

It seems likely that SIRT1 effects on oxidative stress are condition-dependent, since SIRT1 inhibition was shown to increase cellular resistance to oxidative stress in mouse embryonic fibroblasts (Chua *et al.*, 2005), neurons (Li *et al.*, 2008; Sansone *et al.*, 2013), in human carcinoma cell lines and in the human embryonic kidney cell line HEK293 (Carnevale *et al.*, 2017). Interestingly, inhibition of miR-383 was shown to increase SIRT1 expression and to suppress oxidative stress and improving endothelial function (Hu *et al.*, 2020), associated with activation of another vitagene, namely HO-1 (Gou *et al.*, 2018). Treatment of A549 cells with sodium hydrosulphide (NaHS), a donor of H_2S, was shown to upregulate SIRT1 expression and attenuated cigarette smoke extract-induced oxidative stress, mitochondrial dysfunction, cellular senescence and apoptosis (Guan *et al.*, 2019). Further evidence related to protective effect of SIRT against oxidative stress came from various model systems. For example, hypercholesterolemia was shown to reduce significantly liver SIRT1 inducing hepatic oxidative stress in apolipoprotein E-deficient mice. Interestingly, in this model system melatonin was shown to upregulate SIRT1 and decrease oxidative stress (Bonomini *et al.*, 2018). Increased SIRT1 activity was found to protect against diabetes-induced oxidative stress and podocyte injury and effectively mitigated the progression of kidney disease in diabetic OVE26 mice (Hong *et al.*, 2018).

Since, SIRT1 has a wide range of targets that may be preferentially activated or inhibited in different conditions (Simmons *et al.*, 2015) a response to SIRT1 overexpression is condition dependent. For example, moderate overexpression of SIRT1 in mouse cardiac muscle was shown to protect against oxidative stress by increasing the expression of antioxidant defences, through FOXO-dependent mechanisms, while high level (12.5-fold) of SIRT1 overexpression was associated with increased oxidative stress, apoptosis and pathological changes in the heart (Alcendor *et al.*, 2007). Besides SIRT1, other sirtuins also participate in the cellular response to oxidative stress. Indeed, SIRT1 is considered to be a crucial player in the prevention of oxidative damage via a variety of mechanisms (Singh *et al.*, 2018). Recently, SIRT1-mediated inhibition on nucleolar stress response was suggested to represent a novel mechanism by which SIRT1 can modulate intracellular redox status (p53 accumulation) independent of lysine deacetylation (Bi *et al.*, 2019).

Sirtuin 2

Sirtuin 2 (SIRT2) is found ubiquitously throughout the body, localises in the cytoplasm, and the nucleus and have important roles in antioxidant- and redox-mediated cellular homeostasis maintenance in various stress conditions (Singh *et al.*, 2018). In particular, it was shown that SIRT2 can modulate mitochondrial biogenesis and control the production of ROS levels through PGC-1α (Krishnan *et al.*, 2012). Effects of SIRT2 manipulation on the antioxidant defence mechanisms are quite variable and it was suggested that through the deacetylation of specific substrates SIRT2 can play different roles in various conditions (Lee *et al.*, 2019). It seems likely that SIRT2 may affect mitochondrial biogenesis via deacetylation of PGC-1α and upregulation of antioxidant enzymes as a result of deacetylation of FOXO3a (Yu *et al.*, 2019). The loss of SIRT2 was found to increase oxidative stress, decrease ATP levels, and alter mitochondrial morphology at the cellular and tissue (i.e. brain) level. Furthermore, the autophagic/mitophagy processes were indicated to be dysregulated in SIRT2-deficient neurons and mouse embryonic fibroblasts (Liu *et al.*, 2017). Similarly, knockdown of SIRT2 aggravates the apoptosis and mitophagy caused by oxidative stress (Xu *et al.*, 2019). However, in an earlier study it was shown that SIRT2 silencing was associated with reduced ROS levels and decrease early stage apoptosis in the H_2O_2-treated cells (Nie *et al.*, 2014). Similarly, inhibition of SIRT2 by AGK2 was indicated to rescue differentiated PC12 cells from H_2O_2 induced toxicity (Nie *et al.*, 2014). Indeed, SIRT2 inactivation was shown to have protective roles in liver diseases. In fact, SIRT2 deficiency was found to attenuate the APAP-induced liver toxicity as a result of downregulation of c-Jun NH2-terminal kinase (JNK) activation (Lee *et al.*, 2019). The SIRT2 deficiency made mice less susceptible to APAP-induced hepatotoxicity. Indeed, SIRT2 inhibition was shown to confer neuroprotection by downregulation of FOXO3a and MAPK signalling pathways in ischemic stroke (She *et al.*, 2018). Similarly, SIRT2 overexpression was found to exacerbate H_2O_2-induced death of NIH3T3 cells (Wang *et al.*, 2007). In contrast, SIRT2 expression was shown to restrain oxidative stress and senescence of nucleus pulposus cells through inhibition of the p53/p21 pathway (Yang *et al.*, 2019). Similarly, in SH-SY5Y cells, elevated SIRT2 was shown to protect cells from rotenone or diquat induced cell death, while enzymatic inhibition of SIRT2 was associated with enhanced cell death (Singh *et al.*, 2017). The authors showed that SIRT2 protective effect was mediated, in part, through elevated SOD2 expression. Indeed, SIRT2 overexpression was associated with increased production of SOD1/2. Resveratrol was indicated to reduce the ROS levels in TNF-α induced HUVECs, and SIRT2 silencing significantly reversed the effect (Yu *et al.*, 2019). Therefore, SIRT2 plays a critical role in the modulation of the oxidative stress response (Singh *et al.*, 2018).

Sirtuin 3

In the SIRT family, the mitochondrial sirtuin, Sirtuin 3 (SIRT3) , plays a central role in stress resistance by binding to and regulating several enzymes involved in the cellular response to stressful conditions (Bause and Haigis, 2013). Indeed, SIRT3 was shown to regulate acetylation on multiple proteins, at multiple sites, across several metabolic pathways being a global regulator of mitochondrial protein acetylation with the capability of coordinating cellular responses to nutrient status, energy homeostasis and stress (Rardin *et al.*, 2013). It should be noted that more than 30% of all mitochondrial proteins are shown to be acetylated on at least one lysine residue (Anderson and Hirschey, 2012). Knockdown of SIRT3 was found to induce mitochondrial oxidative stress/damage, hyperacetylation of SOD2, and suppression of reendothelialisation capacity, while restoration of SIRT3 expression was found to eliminate mitochondrial oxidative stress via enhancing SOD2 deacetylation of EPCs in hypertension (He *et al.*, 2019). The authors suggested that SIRT3 could physically interact with SOD2, enhancing SOD2 activity by deacetylation of K68 and eliminating excess of mitochondrial ROS. In colon cancer cells SIRT3 silencing was shown to result in a reduced mitochondrial biogenesis and disfunction as evidenced by decreased expression of PGC-1α and mitochondrial transcription factor A and lower levels of OXPHOS complexes. In addition, SIRT3 knockdown was associated with diminished COX activity and oxygen consumption (Torrens-Mas *et al.*, 2019). SIRT3 silencing was shown to significantly reverse the ROS levels reduced by resveratrol in TNF-a induced HUVECs (Yu *et al.*, 2019). Downregulation of SIRT3 in human hepatoma HepG2 cells (by palmitic acid) was associated with increased oxidative stress, impairment of mitochondrial function, and cell apoptosis (Sharma *et al.*, 2019). SIRT3 silencing was shown to induce mitochondrial stress in tongue cancer cells by targeting mitochondrial fission and the JNK-Fis1 biological axis. Indeed, loss of SIRT3 promoted cell death associated with mitochondrial oxidative stress and apoptosis (Zhou *et al.*, 2019).

There are also several reports showing similar effects of SIRT3 deficiency on oxidative stress markers *in vivo*. For example, in SIRT3 deficient mice the activities of two major AO enzymes (MnSOD and Catalase) in kidney were markedly down-regulated while the expression of Caspase 3 was markedly increased and the ROS level was increased (Zhang *et al.*, 2018b). Sirtuin 3 deficiency was shown to exacerbates CCl4-induced oxidative stress and hepatic injury in mice (Li *et al.*, 2019c). In a rat kidney model decreased expression of SIRT3 protein was found to correlate with oxidative stress markers including increased MDA and reduced SOD activity (Zhou *et al.*, 2018). In murine models of pulmonary fibrosis and human lung fibroblasts, reduced SIRT3 expression in response to transforming growth factor beta 1 (TGFβ1) was shown to be associated with acetylation (inactivation) of major oxidative stress response regulators, such as SOD2 and isocitrate dehydrogenase 2 (Sosulski *et al.*, 2017). Sirtuin 3 overexpression was shown to attenuate oxidative stress parameters (as indicated by increased expression and activities of MnSOD and CAT) in rat vein grafts (Lu *et al.*, 2020). Activation of Sirtuin 3 and maintenance of mitochondrial integrity by *N*-Acetylcysteine was shown to have protective effects against bisphenol A-induced kidney and liver toxicity in rats (Peerapanyasut *et al.*, 2019). SIRT3 overexpression in cardiac microvascular endothelial cells was shown to attenuate mitochondrial reactive oxygen species generation (Wei *et al.*, 2017). SIRT3 overexpression in renal tubular epithelial cells was shown to antagonise high glucose-induced oxidative stress and apoptosis by controlling ROS levels ROS-sensitive Akt/FoxO signalling pathway (Jiao *et al.*, 2016). The aforementioned data clearly indicate that SIRT3 has an important role in protecting cells from oxidative damage and genotoxic stress being involved in the regulation of oxidative stress and adaptive homeostasis maintenance (Singh *et al.*, 2018).

Sirtuin 4

Sirtuin 4 (SIRT4) is a mitochondrial sirtuin which has been implicated in a range of human diseases including hyperinsulinemia and diabetes, liver disease, cancer, neurodegeneration, heart disease, aging, and pathogenic infections. (Betsinger and Cristea, 2019). SIRT4 has been shown to be involved in the regulation of ROS production in mitochondria (Singh *et al.*, 2018) and in regulation of mitochondrial quality control and mitophagy (Lang *et al.*, 2017). Increased expression of SIRT4 was associated with reduced O_2 consumption and decreased mitochondrial membrane potential and an increased generation of mitochondrial reactive oxygen species (mtROS). The author suggested that SIRT4 is stress triggered factor that causes mitochondrial dysfunction and impaired mitochondrial quality control through decreased mitophagy (Lang and Piekorz, 2018). In experiments involved Sirt4 knockout mice, Sirt4 was shown to inhibit the binding of MnSOD to SIRT3 and increased MnSOD acetylation levels leading to reduce its activity and increased ROS accumulation in the heart tissue upon angiotensin II stimulation (Luo *et al.*, 2017). Indeed, SIRT4 could affect antioxidant defences and decrease the oxidative stress (Yu *et al.*, 2019). On one hand, there was no effect of resveratrol on SIRT4 expression or activity. On the other hand, the inhibitory effect of resveratrol on TNF-α induced ROS production was reversed by siRNA targeting to SIRT1, SIRT2, SIRT3, SIRT4, and SIRT5 (Yu *et al.*, 2019) indicating the participation of SIRT4 in the antioxidant defences. Taking into account its multifaceted roles within the mitochondria, understanding how SIRT4 is regulated in response to fluctuating cellular conditions including redox status and oxidative stress, remains a major area of recent investigation (Betsinger and Cristea, 2019).

Sirtuin 5

Sirtuin 5 (SIRT5) is localised in the mitochondria and participates in regulation of oxidative stress as well as having important roles in cellular metabolism, detoxification, energy production, and mediation of the apoptosis pathway (Singh *et al.*, 2018). In cultured cells and mouse hearts under energy stress condition SIRT5 deficiency was shown to affect mitochondrial structure and functions. This was evidenced by suppression of mitochondrial NADH oxidation and inhibition of ATP synthase activity and decreased mitochondrial ATP production associated with increased AMP/ATP ratio, and induction of AMP-activated protein kinase (Zhang *et al.*, 2019a).

Overexpression of SIRT5 lowered the level of oxidative stress, countered the toxicity of H_2O_2 and markedly protected SH-EP neuroblastoma cells from stress-induced apoptosis (Liang *et al.*, 2017). SIRT5 was shown to attenuate cisplatin-induced oxidative stress, apoptosis and mitochondrial injury in human kidney HK-2 cells as a result of the regulation of Nrf2/HO-1 and Bcl-2 (Li *et al.*, 2019b). In this study SIRT5 overexpression was associated with maintenance of the mitochondrial density, mitochondrial membrane potentials and ameliorated intracellular ROS production during Cisplatin-induced oxidative stress. SIRT5 was shown to increase Nrf2, HO-1, and Bcl-2, it decreased Bax protein expression while inhibition of SIRT5 by siRNA showed the opposite effect on these proteins. It seems likely that protective antioxidant effect of SIRT5 was also associated with Nrf2-induced catalase activity, since catalase inhibitor 3-AT abolished the cytoprotective effect of SIRT5 (Li *et al.*, 2019b). Importantly, SIRT5 is considered to control mitochondrial function: SIRT5 overexpression increased ATP synthesis and oxygen consumption in HepG2 cells without affecting mitochondrial biogenesis (Buler *et al.*, 2014). In cell culture experiments it was shown that pyruvate kinase M2 over expressing cells are characterised by increased sensitivity to oxidative stress, while co-expression of SIRT5 can desensitise these cells to ROS (Xiangyun *et al.*, 2017). The author showed that PKM2 is succinylated at lysine 498 (K498) and succinylation was shown to increase its activity. Therefore, SIRT5 was found to bind to, desuccinylate and inhibit PKM2 activity. Interestingly, SIRT5 was found to desuccinylate and deglutarylate isocitrate dehydrogenase 2 (IDH2) and glucose-6-phosphate dehydrogenase (G6PD), respectively, leading to increased NADPH production (Zhou *et al.*, 2016). At the same time, knockdown or knockout of SIRT5 was found to lead to high levels of cellular ROS. Therefore, the authors concluded that SIRT5 inactivation can cause the inhibition of IDH2 and G6PD leading to decreasing NADPH production, lowering GSH, impairing the ability to scavenge ROS, and increasing cellular susceptibility to oxidative stress (Zhou *et al.*, 2016). Overall, it appears that SIRT5 plays important roles in creating a response of cells to oxidative stress leading to stress adaptation (Singh *et al.*, 2018).

Sirtuin 6

Sirtuin 6 (SIRT6) is located within the nucleus of the cell and has been considered as an important metabolic sensor connecting environmental signals to metabolic homeostasis and stress responses in animals (Singh *et al.*, 2018). The SIRT6-deficient (KO) mice were shown to develop severe liver injury characterised by a dramatic increase of oxidative stress and inflammation, whereas SIRT6 activation in the transgenic mice were found to be protected from alcohol-related liver disease via prevention of hepatic oxidative stress and inflammation (Kim *et al.*, 2019). Reducing levels of SIRT6 associated with activated FoxO3α phosphorylation as a result of usage of microRNA-351-5p was found to aggravate intestinal ischaemia/reperfusion injury in rodents through imposing oxidative stress (Hu *et al.*, 2018). SIRT6 upregulation was shown to protect against hepatic I/R injury via the maintenance of oxidative homeostasis and mitochondrial function leading to inhibition of the inflammatory responses and MAPK signalling and attenuating apoptosis and autophagy related hepatocyte death (Zhang *et al.*, 2018c). In mouse models of sepsis SIRT6 overexpression was shown to decrease apoptosis, increase the Nrf2 expression and enhanced levels of Nrf2-dependent AO enzymes NQO1 and HO1. As expected, down regulation of Nrf2 was found to abolish the protective effects of SIRT6 overexpression, including inhibiting anti-inflammatory and antioxidative genes expressions (Qin *et al.*, 2019). Furthermore, it was hypothesised that SIRT6 serves as a mediator of hormetic response, promoting longevity by stimulating DNA repair under stressful conditions (Mao *et al.*, 2011). Oxidative stress induced by excess of adiposity was related to a downregulation of hepatic SIRT6 expression in obese individuals and SIRT6 expression in the rat liver was found to be inversely related to plasmatic oxidative status (Carreira *et al.*, 2018). Importantly, SIRT6 has been considered as a stress response protein possessing cytoprotective activity that ameliorating oxidative stress and apoptosis under various cellular stress models including palmitate-induced pancreatic beta-cell dysfunction (Xiong *et al.*, 2016), cisplatin-induced acute kidney injury (Li *et al.*, 2018) and hepatic ischemia/reperfusion injury (Zhang *et al.*, 2018). Molecular associations of SIRT6 with Nrf2, NF-κB, FOXO3, PARP1, TNF-α and COX-2 (Singh *et al.*, 2018) clearly indicate a vital role of this sirtuin in oxidative stress response and stress adaptation.

Sirtuin 7

Sirtuin 7 (SIRT7) is the latest characterised sirtuin which is localised in nucleoli having multiple functions and participating in regulation of cell survival under different conditions of stress, lipid metabolism and protein synthesis (Singh *et al.*, 2018). SIRT7 was suggested to have protective and prosurvival functions and its expression to be essential element of strategy for cell survival under oxidative or genotoxic stress (Kim and Kim, 2013). Indeed, SIRT7 deficiency in mice was shown to be associated with mitochondrial dysfunction (Ryu *et al.*, 2014) and loss of stress resistance in mouse cardiomyocytes treated with H_2O_2 and adriamycin (Vakhrusheva *et al.*, 2008). Similarly, SIRT7 depletion was shown to cause mitochondrial disfunction with dramatically elevated ROS levels in mouse oocytes, while the forced expression of exogenous SIRT7 was reported to have a protective effect in maternal obesity-associated oxidative stress and meiotic defects in oocytes (Gao *et al.*, 2018). Furthermore, SIRT7 was shown to regulate a range of processes related to genome stability including transcriptional regulation, DNA replication and the DNA damage response. Indeed, SIRT7 depletion was confirmed to be associated with the impaired repair of DNA double-strand breaks leading to an increased mutation rate, sensitivity to different DNA damaging agents and abnormal rates of apoptosis (Vazquez *et al.*, 2017). It seems likely that SIRT7 can suppress LPS-induced inflammation and apoptosis via the NF-κB signalling pathway. In fact, it was shown that SIRT7 was significantly downregulated in LPS-treated dairy cow mammary epithelial cells (DCMECs) from breast tissues. Furthermore, SIRT7 knockdown significantly increased the LPS-stimulated production of inflammatory mediators, including ROS, IL-1β and IL-6, increased the phosphorylation of NF-κB p65 and promoted the translocation of NF-κB p65 to the cell nucleus. In contrast, SIRT7 overexpression had the opposite effects (Chen *et al.*, 2019). The aforementioned findings suggest that SIRT7 is an important regulator of cellular response and survival/adaptation under various stress conditions.

8.4 Nutritional regulation of sirtuins

Protective effects of SIRT upregulation by different nutritional and environmental factors were shown in many recent publications. It is well-known that resveratrol is a very potent sirtuin activator. In cerebral ischemia-reperfusion-induced injury mice model, it was found that resveratrol (30 mg/kg; i.p.) postconditioning (administered 5 mins before reperfusion) abrogated the IR-induced oxidative stress (TBARS, SOD, GSH). Importantly, Sirtinol, a SIRT1/2 selective inhibitor, was shown to significantly reverse the effect of resveratrol (Grewal *et al.*, 2019). Resveratrol was found to ameliorate endothelial dysfunction in diabetic and obese mice via activation of SIRT1 and peroxisome proliferator-activated receptor δ (PPARδ; Cheang *et al.*, 2019). Furthermore, hydrogen sulphide (NaHS) was shown to increase the expression of SIRT1 and SOD2 in neonatal mouse cardiomyocyte and a SIRT1 inhibitor (Ex 527) attenuated the cytoprotective effects of NaHS (Liu *et al.*, 2017).The protective effects of Apocynin, one of the main bioactive found in the roots of Canadian hemp (*Apocynum cannabinum*), against UV damages were shown in retinal pigment epithelium cells (RPEs) and retinal ganglion cells (RGCs). Furthermore, intravitreal application of Apocynin also improved retinal dysfunction caused by light damage. The protective effects of Apocynin were shown to be related to the induction of SIRT1 leading to decrease of oxidative stress and DNA damage and knocking down SIRT1 was confirmed to antagonise the protective effect of Apocynin against UV damages in both RPEs and RGCs (Liu *et al.*, 2020a). The low-concentration preconditioning of berberine (Chinese traditional medicine) was shown to have a mitohormetic effect against cellular senescence triggered by oxidative stress in some age-related diseases through the regulation of SIRT1 (Zhu *et al.*, 2017). Dietary silymarin caused the improvement of antioxidant defences in the liver of diabetic rats by promoting SIRT1, GPx and catalase activity (Kheiripour *et al.*, 2018). It was shown that an antioxidant compound N-acetylcysteine ameliorated cisplatin-induced renal senescence and renal interstitial fibrosis in mice through SIRT1 activation and p53 deacetylation (Li *et al.*, 2019a). Celastrol, a pentacyclic triterpenoid isolated from the root extracts of *Tripterygium wilfordii,* was shown to protect human retinal pigment epithelial cells against H_2O_2-induced oxidative stress, autophagy, and apoptosis via SIRT3 signal pathway. Indeed, the terpenoid elevated the messenger RNA (mRNA) and protein expression of SIRT3 while SIRT3 inhibition reversed the effects of celastrol on oxidative stress (Du *et al.*, 2019). Melatonin was shown to protect against myocardial ischemia-reperfusion oxidative stress and injury in mice by elevating SIRT3 expression and Mn-SOD activity (Feng *et al.*, 2018). It seems likely that nutritional modulation of SIRTs could be considered as an important direction in the development of the anti-stress strategy in poultry production.

8.5 Sirtuins and transcription factors

Transcription factors regulation takes place at several levels, including their synthesis, stability, cytoplasm nuclear traffic, DNA binding and nuclear transactivation (Marinho *et al.*, 2014). Sirtuins are shown to affect a great range of targets including specific transcription factors (Nrf2, NF-κB, FOXO, p53, PGC-1α), DNA damage response proteins and others; Chang and Guarente, 2014; Li, 2013; Singh *et al.*, 2018; Raynes *et al.*, 2013) which are important players in creating adequate adaptive response to various stresses. Transcription factors regulation takes place at several levels, including their synthesis, stability, cytoplasm-nuclear traffic, DNA binding and nuclear transactivation (Marinho *et al.*, 2014). Indeed, by affecting their various targets SIRTs participate in the regulation of multiple redox-sensitive cellular functions, including oxidative stress adaptation and stress resistance, DNA repair, mitochondrial biogenesis, inflammation, proteostasis maintenance, glucose, and lipid metabolism. In fact, SIRTs are believed to orchestrate cellular stress response and maintain genome integrity and protein stability (Radak *et al.*, 2013). Therefore, SIRTs are crucial players in stress adaptation and prevention of oxidative stress related damage via a variety of mechanisms, including regulation of variety of transcription factors. By helping stress adaptation SIRTS orchestrate cell survival under oxidative stress conditions.

8.5.1 Nrf2

Sirtuin-Nrf2 interactions are rather complex and condition-dependent. For example, SIRT1 can deacetylate/activate Nrf2, master regulator of enzymatic and non-enzymatic antioxidant defences (Ding *et al.*, 2016). It is known that acetylation of Nrf2 promotes DNA binding and target gene transcription, while deacetylation of Nrf2 disengages it from the ARE leading to transcriptional termination. Therefore, SIRT1 is considered to be a negative regulator of Nrf2 and consequently downregulates this adaptive response to oxidative stress (Kawai *et al.*, 2011). However, SIRT1 can promote nuclear accumulation and stimulate the activation of Nrf2 antioxidative pathway (Huang *et al.*, 2013). It seems likely that there is cross-talk between SIRT1 and the Keap1/Nrf2/ARE pathway. In fact, SIRT1 can enhance the activity of the Keap1/Nrf2/ARE pathway as a result of decreasing Keap1 expression and promoting ARE-binding ability in in rat glomerular mesangial cells (Huang *et al.*, 2017). On the other hand, Nrf2 inhibition in nucleus can down-regulate SIRT3 expression (Oh *et al.*, 2019). Indeed, SIRT1 can activate the Nrf2/Keap1 pathway to decrease ROS generation in mouse and human (Kulkarni *et al.*, 2014). In a recent paper relationship between SIRT1 and Nrf2 was further explored in a mouse model of doxorubicin-induced liver injury (Zhao *et al.*, 2019). On the one hand, up-regulation of miRNA-128-3p was shown to decrease SIRT1 expression and reduced the protein level of Nrf2 and increased Keap1 protein level. On the other hand, down-regulation of miRNA-128-3p in the rat liver was associated with increased SIRT1/Nrf2 signal and decreased Keap1 protein level. Furthermore, Nrf2 was found to increase the protein levels of SIRT3, NQO1 and HO-1, while inhibition of SIRT/Nrf2 pathway inhibited the protein levels of SIRT3, NQO1 and HO-1 in Dox-induced liver injury. In great contrast, the activation of SIRT1/Nrf2 pathway was shown to enhance the protein levels of SIRT3, NQO1 and HO-1 in Dox-induced liver damage (Zhao *et al.*, 2019).

Interestingly, SIRT6 can affect Nrf2 signalling by inhibiting Nrf2 repressor, Bach1 (Ka *et al.*, 2017) or deacetylating and reducing the ubiquitination of Nrf2 (Zhang *et al.*, 2017a,b). This leads to activation of Nrf2/ARE antioxidant signalling with promotion of the transactivation of Nrf2-regulated antioxidant genes (Pan *et al.*, 2016). SIRT6 was found to protect mouse brain from oxidative stress and cerebral ischemia/reperfusion injury through Nrf2 activation (Zhang *et al.*, 2017b). Furthermore, SIRT6 was shown to positively regulate Nrf2 expression and activated Nrf2-regulated anti-inflammatory and antioxidative enzymes associated with effective mitigation of LPS-induced human umbilical vein endothelial cells inflammatory responses (Qin *et al.*, 2019). In fact, SIRT6 can protect the brain from cerebral I/R injury by suppressing oxidative stress via Nrf2 activation (Zhang *et al.*, 2017b) or protect vascular endothelial cells from angiotensin II-induced apoptosis and oxidative stress by enhancing the activation of Nrf2/ARE redox signalling (Yang *et al.*, 2019b). SIRT6 was shown to protect retinal ganglion cells from hydrogen peroxide-induced damage and oxidative stress by promoting the activation of Nrf2/ARE signalling via inhibition of Bach1 (Yu *et al.*, 2019). SIRT6 overexpression in astrocytes was shown to induce Nrf2-mediated Hmoxl and Srxn1 gene expression and it seems likely that AO enzyme activation occurred at a posttranscriptional level (Harlan *et al.*, 2019). Interestingly, in Cr(VI)-

transformed cells binding of Nrf2 to antioxidant response element (ARE) of SIRT3 gene promoter was dramatically increased (Clementino *et al.*, 2019).

There is a range of publications addressing effect of nutritional and stress-factors on SIRTs and Nrf2 expression and activities. For example, resveratrol was shown to stimulate SIRT1 expression simultaneously with an increased expression of Nrf2 in a diabetic heart leading to alleviation of myocardial oxidative stress (Xu *et al.*, 2019a). Furthermore, it was shown that fasting increased Nrf2 target gene expression through Nrf2- and SIRT1-dependent mechanisms. For example, in intact mouse liver, fasting was found to induce Nrf2 target gene expression by at least 1.5 to 5-fold. Moreover, fasting-induced Nrf2 target gene expression was decreased in liver and hepatocytes of SIRT1 liver-specific null mice and Nrf2-null mice (Kulkarni *et al.*, 2014). Reduced SIRT1 activity and Nrf2 protein in the nuclear extract of left ventricular tissue due to trauma haemorrhage (T-H) were observed in rats. However, resveratrol (an activator of SIRT1) treatment restored the SIRT1 and Nrf2 levels, whereas the rats treated with sirtinol (sirtuin inhibitor) demonstrated levels of Nrf2 similar to those observed in rats subjected to T-H but not treated with resveratrol (Jian *et al.*, 2014). Furthermore, protective effect of pyrroloquinoline quinine on ultraviolet A irradiation-induced human dermal fibroblast senescence *in vitro* was shown to be associated with the anti-apoptotic SIRT1/Nrf2/HO-1 pathway (Zhang *et al.*, 2015). Interestingly, the anti-apoptotic effects of trimetazidine, an anti-ischemic metabolic agent, was found to result from decrease of pro-inflammatory cytokines, partly due to normalising the SIRT1/AMP-activated protein kinase (AMPK)/Nrf2/HO-1 and SIRT1/PPARα pathways in macrophages (Chen *et al.*, 2016). As one can expect, natural compounds can affect simultaneously various elements of cell redox signalling systems. For example, dioscin, a natural steroid saponin, found in various herbs was shown to directly bound to SIRT1, Keap1 and NF-κBp65 by hydrogen bonding and/or hydrophobic interactions (Zhang *et al.*, 2017a). Furthermore, in *in vivo* model systems of nephrotoxicity in rats and mice it was shown that dioscin up-regulated SIRT1 levels, altered levels of Nrf2 dependent enzymes (HO-1, glutathione-cysteine ligase subunits GCLC, GCLM) and Keap1. This, together with increased nuclear translocation of Nrf2 lead to decreased oxidative stress. Furthermore, dioscin was found to affect levels of AP-1, COX-2, HMGB1, IκB-α, IL-1β, IL-6 and TNF-α and decreased the ratio of acetylated NF-κB and normal NF-κB, to suppress inflammation (Zhang *et al.*, 2017c). Similarly, exposure to environmental chemicals was shown to perturb redox balance and cell signalling in various *in vitro* and *in vivo* systems. For example, benzyl butyl phthalate (BBP), a widespread endocrine disruptor, was shown to disturb antioxidant defences/redox balance in mesenchymal stem cells. In fact, expression of SIRT1, SIRT2 and SIRT7 were shown to be decreased in various time after cell treatment with BBP. Furthermore, SIRT1 targets and adipogenesis regulators (FOXO1 and β-catenin) were hyperacetylated at day 8 after BBP treatment and Nrf2 was also significantly decreased (Zhang and Choudhury, 2017). Melatonin was found to stimulate the SIRT1/Nrf2 signalling pathway to counteract lipopolysaccharide (LPS)-induced oxidative stress to rescue postnatal rat brain (Shah *et al.*, 2017). Nuclear factor of activated T cells (NFAT), a multifunctional cytokine family, was found to protect astrocytes against oxygen-glucose serum deprivation/restoration damage via the SIRT1/Nrf2 pathway

(Xia *et al.*, 2017). Interestingly, SIRT2 was found to regulate cellular iron homeostasis via deacetylation of transcription factor Nrf2 (Yang *et al.*, 2017). Indeed, iron as a major promotor of free radical production and oxidative stress, should be kept under strict control and SIRT2 seems to be one of those important factors regulating this process and by doing so help maintain cell homeostasis in various stress conditions. A summary of most recent findings confirming regulatory functions of SIRTs on Nrf2 expression and activity are shown in Table 8.2.

As can be seen from the above presented data SIRTs can affect Nrf2 expression and activity.

Table 8.2. Effect of Sirtuins on nuclear factor erythroid-2 related factor 2 (Nrf2) expression and activity.

Sirtuins	Effect	References
SIRT1	Mediating the effects of paeonol, a single phenolic compound extracted from the root bark of Cortex Moutan, on the activation of Nrf2/ARE pathway in high glucose-treated glomerular mesangial cells and streptozotocin-induced diabetic mice	Zhang *et al.*, 2018a
SIRT1	Down-regulation of sirtuin 1 by short hairpin RNA reversed the protective effects of chlorogenic acid on paraquat-induced alterations in Nrf2 expression in A549 cells	Kong *et al.*, 2019
SIRT1	Activation of SIRT1 by SRT2104 enhanced renal antioxidant activity via rescuing P53-induced inactivation of Nrf2	Ma *et al.*, 2019
SIRT1	Decreased levels of SIRT1 protein and activity in the kidneys of the Nrf2 KO mice compared with the WT mice	Ma *et al.*, 2019
SIRT3	SIRT3 overexpression inhibited the Ca oxalate-induced cell apoptosis in human proximal tubular cell line HK-2 cells, which was reversed by the NRF2 knockdown. SIRT3 overexpression alleviated the glyoxylate administration-induced renal damage, renal apoptosis in the kidneys from the stone model mice associated with activation of the NRF2/HO-1 pathway	Xi *et al.*, 2019
SIRT5	Upregulation of Nrf2 and increased protein expression in the human kidney HK-2 cell line	Li *et al.*, 2019b
SIRT5	Suppression of cisplatin-induced DNA damage in a ROS-dependent manner via regulation of the Nrf2/HO-1 pathway in ovarian cancer tissues	Sun *et al.*, 2019a
SIRT6	Nrf2 activation in LPS-induced human umbilical vein endothelial cells	Qin *et al.*, 2019
SIRT6	Enhancing the activation of Nrf2/ARE redox signalling in vascular endothelial cells	Yang *et al.*, 2019b
SIRT6	Activation of Nrf2/ARE signalling via inhibition of Bach1 in ganglion cells treated with H_2O_2	Yu *et al.*, 2019
SIRT6	Induction of Nrf2-mediated Hmox1 and Srxn1 gene expression in astrocytes	Harlan *et al.*, 2019
SIRT6	Inhibition of miR-4532 protected human lens epithelial cells from UV radiation-induced oxidative injury via activation of SIRT6-Nrf2 pathway	Sun *et al.*, 2019
SIRT6	SIRT6 overexpression in endothelial cells reversed a decrease in Nrf2 expression due to minute cholesterol crystal treatment. Endothelium-specific SIRT6 depletion suppressed Nrf2 expression in hyperlipidaemic mice	Jin *et al.*, 2020

8.5.2 NF-κB

SIRT1 can deacetylate the p65 subunit of NF-κB and inhibit NF-κB pro-inflammatory signalling (Du *et al.*, 2018; Guo *et al.*, 2014). In fact, SIRT1 can form a complex with the RelA/p65 subunit of NF-κB, deacetylating Lys310. This leads to a decreased NF-κB activity associated with downregulation of survival genes and increased apoptosis (Yeung *et al.*, 2004).

In high glucose-incubated tubular epithelial cells, downregulation of SIRT1 was found to increase the acetylation and activity of NF-κB, with simultaneous increasing the expression of Keap1 and inhibiting the Nrf2/ARE pathway (Zhou *et al.*, 2015). Similarly, diabetes-induced downregulation of SIRT1 leads to activation of NF-κB signalling, which promotes the activation of downstream pro-inflammatory factors and inhibits antioxidative stress Nrf2/ARE pathway (Wang *et al.*, 2019). In primary microglia levels of NF-κB p65 and proinflammatory cytokines (TNF-α and IL-12) decreased significantly after a sirtuin agonist (SRT1720) treatment (Lu *et al.*, 2020). In gout patients, resveratrol was shown to upregulate SIRT1 associated with downregulation of NF-κB expression and decrease gouty inflammation (Yang *et al.*, 2019c). Furthermore, resveratrol was indicated to protect against TNF-α-induced injury in human umbilical endothelial cells through promoting SIRT1-induced repression of NF-κB and p38 MAPK (Pan *et al.*, 2016). Melatonin, another compound with antioxidant activities, was shown to increase the protein levels of SIRT1, and reduced the levels of acetylated-Lys310 in the p65 subunit of NF-κB in SH-SY5Y cells treated with H_2O_2 (Nopparat *et al.*, 2017).

It seems likely that crosstalk between Sirtuins, NF-κB and FOXO is responsible for inflammation resolution during immune response. Indeed, there is an antagonistic regulation between the SIRT1 and NF-κB signalling pathways: SIRT1 inhibits NF-κB activation and NF-κB signalling and inflammatory response can suppress the SIRT1 activity (Kauppinen *et al.*, 2013). Furthermore, FOXO1 is suggested plays a cooperative role in inflammatory signalling through NF-κB. For example, several genes such as the IL-1β promoter are found to contain both FOXO1 and NF-κB response elements and, therefore, both FOXO1 and NF-κB are needed to induce IL-1β transcription. Furthermore, FOXO1 was suggested to serve as a coactivator of NF-κB in the nucleus to amplify NF-κB signalling and regulating inflammation (Wang *et al.*, 2014). Interestingly, FOXO1 expression by dendritic cells is shown to be an important event for optimal induction of an adaptive immune response to bacterial challenge (Song *et al.*, 2018a).

Other sirtuins seem to have similar effects on NF-κB expression and activity. For example, mitochondrial Sirt3 was shown to play an important role in providing antioxidant protection against oxidative stress in diverse cell types. For example, in cardiomyocytes, SIRT3 was shown to prevent/alleviate oxidative stress and decrease oxidative stress-related apoptosis by activating the NF-κB pathway (Chen *et al.*, 2013). It seems likely that SIRT7 can suppress LPS-induced inflammation and apoptosis via the NF-κB signalling pathway. In fact, it was shown that SIRT7 was significantly downregulated in LPS-treated dairy cow mammary epithelial cells (DCMECs) from breast tissues. Furthermore, SIRT7 knockdown significantly increased the LPS-stimulated production of inflammatory mediators, including ROS, IL-1β and IL-6, increased the phosphorylation of NF-κB p65 and promoted the translocation of NF-κB p65 to the cell nucleus. In contrast, SIRT7 overexpression had the opposite effects (Chen *et al.*, 2019). In fact, SIRT7 regulates the nuclear export of NF-κB p65 by deacetylating Ras-related nuclear antigen (Sobuz *et al.*, 2019). A summary of recent findings confirming regulatory functions of SIRTs on NF-κB expression and activity are shown in Table 8.3.

As can be seen from the above presented data SIRTs can affect NF-κB expression and activity.

8.5.3 FOXO

Evidence is quickly accumulated to show a functional link between mammalian SIRTs and FOXO family of transcription factors during various stress conditions. It was shown that SIRT1 can deacetylate FOXO family transcription factors such as FOXO1, FOXO3a and FOXO4 (Brunet *et al.*, 2004; Kitamura *et al.*, 2005; Paik *et al.*, 2007; Rached *et al.*, 2010) and repress their ability to activate target gene transcription. disturbing regulation of mitochondrial gene expression and modulating of ROS levels (Ferber *et al.*, 2012). Indeed, in mammalian cells, SIRT1 was shown to control the cellular response to stress by regulating the FOXO transcription factors (Brunet *et al.*, 2004). Interestingly, nuclear localisation of FoxO1 is controlled by SIRT1 deacetylase (Hallows *et al.*, 2008). Sirtuin activation or hydrogen peroxide treatment were shown

Table 8.3. Effect of sirtuins (SIRT) on nuclear factor kappa beta (NF-κB) expression and activity.

Sirtuins	Effect	References
SIRT1	Diabetes-induced downregulation of SIRT1 leads to activation of NF-κB signalling, which promotes the activation of downstream pro-inflammatory factors and inhibits antioxidative stress Nrf2/ARE pathway	Wang *et al.*, 2019
SIRT1	Upregulation of SIRT1 lead to downregulation of NF-κB expression and decrease inflammation in gout patients	Yang *et al.*, 2019a
SIRT1	The level of NF-κB p65 in the nucleus is negatively regulated by SIRT1 activity in LPS-treated RAW 264.7 and CCl_4-treated mice	Lee *et al.*, 2019a
SIRT1	Prevention of nuclear translocation of NF-κB (p65) in human aortic smooth muscle cells	Caon *et al.*, 2020
SIRT1	Agomelatine enhanced SIRT-1 levels to negatively regulate the transcription and activation of p-NF-κB /p65	Savran *et al.*, 2020
SIRT7	SIRT7 knockdown significantly increased the LPS-stimulated production of inflammatory mediators, including ROS, IL-1β and IL-6, increased the phosphorylation of NF-κBp65 and promoted the translocation of NF-κBp-p65 to the cell nucleus	Chen *et al.*, 2019a
SIRT7	SIRT7 regulates the nuclear export of NF-κB p65 by deacetylating Ras-related nuclear antigen	Sobuz *et al.*, 2019

to override the phosphorylation-dependent nuclear exclusion of FOXO1 caused by growth factors leading to nuclear translocation of FOXO1 in hepatocytes with increased expression of target genes. Interestingly, under physiological conditions FOXO1 is shown to be readily diffusible within the nucleus but under oxidative stress or following treatment with the prototypical Sirtuin agonist resveratrol FOXO1 becomes restricted within a nuclear subdomain (Frescas *et al.*, 2005). Moderate overexpression of SIRT1 (~3-fold of normal level) was shown to decrease oxidative stress and inhibit apoptosis of osteoblasts via the FOXO1 and β-catenin signalling pathway (Yao *et al.*, 2018). Interestingly, FOXO1 transcription factor was shown to promote osteoblast proliferation and to maintain redox balance and control bone formation (Rached *et al.*, 2010). In particular, SIRT1 and FOXO3 are found to form a complex in cells in response to oxidative stress leading to increase of FOXO3-dependent induction of cell cycle arrest and resistance to oxidative stress, while inhibited FOXO3's ability to induce cell death. Therefore, SIRT1 is responsible for tipping FOXO-dependent responses away from apoptosis and toward stress resistance (Brunet *et al.*, 2004). Later it was confirmed that the SIRT1, FOXO, and β-catenin proteins may form a complex in the nucleus to regulate gene expression in nematodes and other species (Parker *et al.*, 2012). Neuroprotective effects of SIRT1 activation have been reported in several mouse models of Huntington's disease and FOXO3a acetylation is considered to be involved in SIRT1 protection (Jiang *et al.*, 2011). Furthermore, SIRT3 was shown to increase the FoxO3a-induced antioxidant genes, including MnSOD and catalase (Peserico *et al.*, 2013; Rangarajan *et al.*, 2015). In fact, FOXO3a was found to protects quiescent cells from oxidative stress by directly increasing their quantities of MnSOD messenger RNA and protein (Kops *et al.*, 2002). In addition, in mammals FOXO3a was shown to regulate the resistance of cells to stress by inducing DNA repair (Tran *et al.*, 2002). SIRT3 was shown to protect aged human mesenchymal stem cells against oxidative stress by positively regulating antioxidant enzymes (MnSOD and CAT) via increasing the expression of FOXO3a in the nucleus (Zhang *et al.*, 2018). It seems likely that there is an autofeedback mechanism where FOXOs can regulate SIRT1 protein expression. For example, FOXO1 depletion by siRNA was shown to decreases SIRT1 expression in human embryonic kidney cells and vascular smooth muscle cells. Furthermore, in FOXO1 knockdown mouse embryos cytoplasmic expression of SIRT1 decreased, while in FOXO3 and FOXO4 knockdown embryos SIRT1 cytoplasmic expression was translocated to nucleus (Kuscu *et al.*, 2019). Nuclear levels of FOXO3a were decreased as a result of TNFα stimulation of rheumatoid arthritis synovial fibroblasts (RASFs), while forced expression of FOXO3a reversed the inductive effects of TNFα on cysteine rich protein 61 (CYR-61). Furthermore, simvastatin was found to inhibit the nuclear export, phosphorylation, and acetylation of FOXO3a. Interestingly, forced expression of SIRT1 in RASFs caused decreased levels of CYR-61 and deacetylation of FOXO3a. Simvastatin treatment was associated with upregulation of the expression of SIRT1 and SIRT1/FOXO3a binding was enhanced in RASFs (Kok *et al.*, 2013). Chronic hyperglycaemia in mice was shown to accelerate aging process through a novel SIRT1 and p300 regulated pathway. Furthermore, high glucose was responsible for induced reduction in SIRT1 leading to increased oxidative stress mediated through FOXO1 which was prevented by SIRT1 activation (Mortuza *et al.*, 2013). In pancreatic cancer cells, capsaicin was shown to activate JNK and FOXO1, leading to the acetylation of

FOXO1 through SIRT1 and CREB-binding protein (Pramanik *et al.*, 2014). In human hepatoma cells berberine, a quaternary isoquinoline alkaloid, derived from the root and stem bark of numerous clinically important medicinal plants of Berberidaceae family (100 μM) was shown to significantly ($P<0.05$) inhibit sirtuins at transcriptional and translational levels. Furthermore, combination of sirtuin inhibitor (nicotinamide) with berberine was found to potentiate sirtuins inhibition and concomitant increase in acetylation and transcriptional activation of FOXOs and p53 (Shukla *et al.*, 2016). In old mice melatonin administration was able to reduce oxidative stress and inflammation in pancreas associated with increased expression of SIRT1, Pdx1 and FOXO3a (Tresguerres *et al.*, 2013). It was shown that amyloid beta toxicity may be attenuated through SIRT1 and FOXO3a antioxidant dependent pathways (Lin *et al.*, 2015). Coenzyme Q_{10} is shown to protect hepatocytes from ischemia reperfusion-induced oxidative stress via regulation of Nrf-2/FOXO3/SIRT1 signalling pathway. In fact, CoQ_{10} restored redox balance via marked activation of Nrf2 protein as well as up-regulation of both SIRT1 and FOXO3 genes (Mahmoud *et al.*, 2019). Pretreatment with icariin, a major active ingredient in traditional Chinese medicines, was found to protect cardiomyocytes from I/R-induced oxidative stress through activation of SIRT1 /FOXO1 signalling (Wu *et al.*, 2018).

8.5.4 p53

p53, a transcriptional factor highly regulated by post-transcriptional modifications and called the 'guardian of the genome' showing protective role in preventing mutations. The activity of p53 is shown to be mediated by its acetylation status. In fact, deacetylation can reduce p53 activity (Fujita, 2019). In fact, p53 is a well-characterised substrate of SIRT1 playing an important regulatory role in modulation of redox signalling (Maillet and Pervaiz, 2012) being critical in cell cycle checkpoint regulation, apoptosis, and tumour suppression. In fact, irreparable DNA damages by RONS are shown to cause the stabilisation and activation of p53 leading to the expression of pro-apoptotic proteins such as BAX and PUMA, with following apoptosis induction. Interestingly, the deacetylation of p53 by SIRT1 inhibits p53's pro-apoptotic activity (Hori *et al.*, 2013). The authors showed that the antimycin A-induced increase in ROS levels and apoptosis in C2C12 mouse myoblasts was enhanced by SIRT1 inhibitors nicotinamide and splitomicin, whereas it was suppressed by a SIRT1 activator, resveratrol, and a SIRT1 cofactor, NAD+. Interestingly, the same experiment with p53-knockdown cells or p53-deficient cells showed that splitomicin and resveratrol modulated apoptosis by p53-dependent and p53-independent pathways (Hori *et al.*, 2013). In non-stressed conditions, the transcription factor p53 is shown to decrease intracellular ROS and increase activities of AO enzymes, including SOD2 and GPX1. At the same time, downregulation of p53 is associated with an increase in intracellular ROS levels and DNA oxidation levels (Liu *et al.*, 2008; Holley *et al.*, 2010). The expression of p53 in glioma cells was demonstrated to be regulated by SIRT1 (Chen *et al.*, 2019a). The level of SIRT1 protein was shown to decrease while the level of protein p53 increased during HUVEC senescence and these changes could be partially recovered when cells were co-incubated with flavonoids (Guo *et al.*, 2018). The overexpression of SIRT1 was found to impair the p53/p21 pathway, thereby preventing elderly adipose tissue-derived mesenchymal stem cells from entering senescence and restoring the beige differentiation ability (Khanh *et al.*, 2018). SIRT2 expression was shown to suppress oxidative stress and senescence of nucleus pulposus cells as a result of inhibition of the p53/p21 pathway (Yang *et al.*, 2019a). Activation of SIRT1 was shown to attenuate high glucose-induced neuronal apoptosis by deacetylating p53 (Shi *et al.*, 2018). SIRT3 deficiency in endothelial cells displayed striking increases in acetylation of p53 (Zeng and Chenm, 2019). In breast cancer cells up-regulation of miRNA-211-5p was associated with a significant decline in the acetylation status of p53 (Yarahmadi *et al.*, 2019). In lung cancer cells, SIRT3 overexpression resulted in decreased mutant p53 stability as a result of increased proteasomal ubiquitination-mediated degradation of the protein leading to the reduced its expression and half-life (Tang *et al.*, 2020).

8.5.5 HSF1

SIRT1-driven deacetylation of heat shock factor 1 (HSF1), a transcriptional regulator of heat shock protein synthesis (specifically HSP70 and HSP90), enhances its binding to the promotor thus serving to enhance the HSR by increasing HS-induced chaperone expression in HeLa and HEK293 cells (Westerheide *et al.*, 2009). Interestingly, the C. elegans SIRT1 homolog SIR2.1 was shown to be linked to the HSR (Raynes *et al.*, 2012). The authors also indicated that caloric restriction synergistically with HS induced HSP70 gene expression in the worm, and the effect was shown to be dependent on Sir2.1 (Raynes *et al.*, 2012). Recently, two natural regulators of SIRTs were described and used for elucidation of mechanisms of interaction between SIRTs and HSF/HSP. In fact, active regulator of SIRT1 (AROS) is a nuclear protein, which directly regulates SIRT1 function. In particular, AROS is shown to enhance SIRT1-mediated deacetylation of p53 both *in vitro* and *in vivo*, and it inhibited p53-mediated transcriptional activity. AROS activity was abrogated by the SIRT1 inhibitors splitomicin and nicotinamide and by SIRT1 small interfering RNA (siRNA; Kim *et al.*, 2007). It was shown that DBC1 (deleted in breast cancer 1) acts as a native inhibitor of SIRT1 in human cells. It directly interacts with SIRT1 and inhibits SIRT1 activity *in vitro* and *in vivo* (Kim *et al.*, 2008). Therefore, in contrast to AROS, DBC1 decreases the deacetylase function of SIRT1. In fact, the AROS/DBC1-SIRT1 interactions are considered as important ways in which the cell can fine tune stress-induced SIRT1 activity (Raynes *et al.*, 2013a). Heat shock (HS) was shown to increase HSF1 acetylation, while the overexpression of SIRT1 inhibited HS-induced HSF1 acetylation. In fact, upon a 2-hour HS, there was a 4-fold increase of SIRT1 recruitment to the HSP70 promoter. It was shown that heat shock resulted in an increase in the cellular NAD+/NADH ratio and, therefore, induction of the HSR correlates with an increase in the cellular NAD+ /NADH ratio and with an increase in recruitment of SIRT1 to the HSP70 promoter. Similarly, SIRT1 inhibition (by overexpression of DBC1 overexpression) was shown to enhance HSF1 acetylation under both non stress and stress conditions, while SIRT1 activation (by overexpression of AROS) was found to reduce HS-induced HSF1 acetylation to a similar degree as SIRT1 itself (Raynes *et al.*, 2013). It seems likely that various cells have the ability to respond to heat stress and possibly to other stresses quickly though HSP70 deacetylation, followed by a slower, more traditional transcriptional response (Xu *et al.*, 2019b).

8.5.6 Sirtuins and other transcription factors

It is known that SIRT1 can regulate redox status and antioxidant defences through activation of peroxisome proliferator-activated receptor coactivator 1-α (PGC-1α), an upstream regulator of mitochondrial biogenesis and function. In fact, in mitochondria, PGC-1α and SIRT1 are shown to be directly involved in regulation of mitochondrial biogenesis mediating the crosstalk between nuclear and mitochondrial genome (Aquilano *et al.*, 2010). Furthermore, PGC-1α stimulates mitochondrial electron transport activity and induce suppression of ROS production through the induction of several key ROS-detoxifying enzymes. In fact, PGC-1α null mice was shown to be characterised by increased sensitivity to the neurodegenerative effects of oxidative stress (Kong *et al.*, 2010). Furthermore, PGC-1α was confirmed to be involved in mitochondrial biogenesis and energy management as a transcription coactivator of many genes, which are vital for cell survival. PGC-1α also regulated synthesis of mitochondrial protein to scavenge ROS (Chen *et al.*, 2011). On the other hand, PGC-1α was shown to stimulate SIRT3 expression in C2C12 cells at the mRNA and protein levels and PGC-1α knockdown by shRNA was indicated to reduce SIRT3 mRNA level. Interestingly, induction of gene expression of AO enzymes SOD2 and GPx1 by PGC-1α is SIRT-dependent and was shown to be impaired by SIRT3 knockdown in C2C12 myotubes. Furthermore, knockdown of SIRT3 was reported to diminish the effect of PGC-1α on cellular ROS level and mitochondrial biogenesis (Kong *et al.*, 2010). The authors suggested that PGC-1α induces SIRT3 expression and in turn, SIRT3 stimulates PGC-1α gene expression, thus forming a positive-feedback loop. Fibroblast growth factor-21 was shown to enhance mitochondrial functions and increases the activity of PGC-1α in human dopaminergic neurons via SIRT1-depenfent pathway (Mäkelä *et al.*, 2014). It was suggested that SIRT1 activation could lead to alleviation of mitochondrial oxidative stress via PGC-1α signalling. In fact, SIRT1 expression and activity were found to be enhanced in the rat hippocampus following status epilepticus (SE). At the same time, SIRT1 inhibition was shown to block the SE-associated increase in PGC-1α and mitochondrial antioxidant enzymes, including SOD2 and uncoupling protein 2 (UCP2). Additionally, the activation of SIRT1 was associated with enhanced mitochondrial electron transport chain complex I activity and increased ATP content (Wang *et al.*, 2015). In experiment with transgenic mice, apolipoprotein E was shown to regulate mitochondrial function through the PGC-1α-sirtuin 3 pathway (Yin *et al.*, 2019).

As can be seen from the above presented data SIRTs are important modulators of various transcription factors. Furthermore, in biological systems there is a network where sirtuins interact with each other providing an important protective power in stress conditions (Figure 8.1).

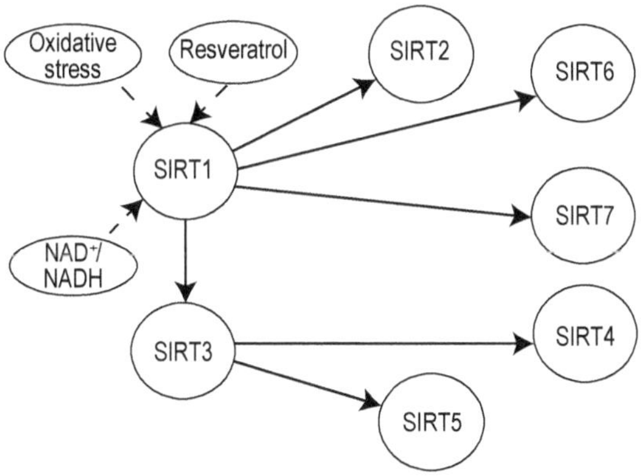

Figure 8.1. Sirtuin (SIRT) interactions in biological systems (adapted from Singh *et al.*, 2018; Wu *et al.*, 2014). There is a range of SIRT1 effectors including redox status (NAD+/NADH), oxidative stress (ROS) and specific nutrients (resveratrol). Interactions between different sirtuins make them effective regulators of adaptive homeostasis.

8.6 Conclusions

Stress adaptation is an important subject related to avian biology and poultry production. It is practically impossible to avoid commercially relevant stresses for growing chickens and producing laying hens. Therefore, the development of the vitagene concept was an important step in improving our understanding cell/body strategy to deal with stress-related overproduction of RONS and oxidative stress. The central role of the sirtuin family in stress adaptation as an important part of the vitagene network has been greatly appreciated. Indeed, sirtuins are:

- ubiquitously distributed from eubacteria to mammals in various tissues;
- located in different cell organelles;
- interacting with and regulate activity of a great number of protective target proteins, including antioxidant enzymes (SOD, catalase, GPx, GR, etc.) and transcription factors (Nrf2, NF-κB, HSF1, FOXO, PPARγ, PPARα, HIF1α, p53, etc.), histones and DNA damage proteins (Nogueiras *et al.*, 2012);
- involved in regulation of key genes and molecules participated in maintenance of redox homeostasis; orchestrate cellular response/adaptation to oxidative stress by maintenance of genome integrity and protein stability;
- controlled by posttranslational modifications and by the availability of NAD^+ (a stress-sensor).

It is important to emphasise that on the one hand, SIRTs interact with each other to provide an effective stress response. In fact, SIRT1 can interact directly with SIRTs 2, 6, and 7 as well as SIRT3, which further can interact with SIRTs 4 and 5 (Singh *et al.*, 2018). It was suggested that SIRT3 works co-ordinately with SIRT5 to regulate the acylation of mitochondrial proteins at distinct physiological states and subsequently affects multiple metabolic pathways in response to various stresses (Wu *et al.*, 2014). Cells with low SIRT1 levels were shown to have an ability to maintain their stress resistance and survival by increasing SIRT3 expression (Carnevale *et al.*, 2017). There are also other interactions between various sirtuins (Fang *et al.*, 2017; Kumari *et al.*, 2018; Kanwal *et al.*, 2019; Zhang *et al.*, 2019b). On the other hand, within the vitagene network SIRTs interact with other vitagenes, including direct and indirect activation of SOD (see above), deacetylation of HSP70 (Sun *et al.*, 2019b), HO-1 (via Nrf2 activation, Qin *et al.*, 2019), thioredoxin and glutathione systems (via Nrf2 activation or other mechanisms, Chang *et al.*, 2016; Tseng *et al.* 2014).

Therefore, the development of vitagene-regulating nutritional supplements, including those regulating SIRTs expression and activity, is on the agenda of many various companies worldwide (Surai *et al.*, 2018, 2019). A successful example could be a complex mixture of nutrients including carnitine, betaine, vitamins, minerals, organic acids, etc. which is commercially available in poultry (Grigorieva *et al.*, 2017; Shatskikh *et al.*, 2015). Indeed, prevention of detrimental consequences of stresses and improved performance in broilers (Grigorieva *et al.*, 2017), broiler breeders and layers (Shatskikh *et al.*, 2015) by using a vitagene-activation may, based on references in this chapter, optimise the antioxidant defence system. It seems likely that the vitagene concept of stress management could be used for maintenance/improvement of eggshell gland health, gut health and liver health of modern poultry under commercially relevant stress conditions (for review see Surai *et al.*, 2018). However, data on sirtuin expression and activities in various chicken tissues as well as in other avian species are extremely limited. Therefore, roles of sirtuin modulation by nutritional means in stress adaptation/protection in avian species await further investigation.

References

Alcendor, R.R., Gao, S., Zhai, P., Zablocki, D., Holle, E., Yu, X., Tian, B., Wagner, T., Vatner, S.F. and Sadoshima, J., 2007. Sirt1 regulates aging and resistance to oxidative stress in the heart. Circulation Research 100: 1512-1521.

Anderson, K.A. and Hirschey, M.D., 2012. Mitochondrial protein acetylation regulates metabolism. Essays in Biochemistry 52: 23-35.

Aquilano, K., Vigilanza, P., Baldelli, S., Pagliei, B., Rotilio, G. and Ciriolo, M.R., 2010. Peroxisome proliferator-activated receptor gamma co-activator 1alpha (PGC-1alpha) and sirtuin 1 (SIRT1) reside in mitochondria: possible direct function in mitochondrial biogenesis. Journal of Biological Chemistry 285: 21590-21599.

Bause, A.S. and Haigis, M.C., 2013. SIRT3 regulation of mitochondrial oxidative stress. Experimental Gerontology 48: 634-639.

Bayod, S., Del Valle, J., Lalanza, J.F., Sanchez-Roige, S., de Luxan-Delgado, B., Coto-Montes, A., Canudas, A.M., Camins, A., Escorihuela, R.M. and Pallas, M., 2012. Long-term physical exercise induces changes in sirtuin 1 pathway and oxidative parameters in adult rat tissues. Experimental Gerontology 47: 925-935.

Betsinger, C.N. and Cristea, I.M., 2019. Mitochondrial function, metabolic regulation, and human disease viewed through the prism of sirtuin 4 (SIRT4) functions. Journal of Proteome Research 18: 1929-1938.

Bi, X., Ye, Q., Li, D., Peng, Q., Wang, Z., Wu, X., Zhang, Y., Zhang, Q. and Jiang, F., 2019. Inhibition of nucleolar stress response by Sirt1: a potential mechanism of acetylation-independent regulation of p53 accumulation. Aging Cell 18: e12900.

Bonomini, F., Favero, G., Rodella, L.F., Moghadasian, M.H. and Rezzani, R., 2018. Melatonin modulation of sirtuin-1 attenuates liver injury in a hypercholesterolemic mouse model. BioMed Research International 2018: 7968452.

Brunet, A., Sweeney, L.B., Sturgill, J.F., Chua, K.F., Greer, P.L., Lin, Y., Tran, H., Ross, S.E., Mostoslavsky, R., Cohen, H.Y. and Hu, L.S., 2004. Stress-dependent regulation of FOXO transcription factors by the SIRT1 deacetylase. Science 303: 2011-2015.

Buler, M., Aatsinki, S.M., Izzi, V., Uusimaa, J. and Hakkola, J., 2014. SIRT5 is under the control of PGC-1α and AMPK and is involved in regulation of mitochondrial energy metabolism. The FASEB Journal 28: 3225-3237.

Cao, C., Lu, S., Kivlin, R., Wallin, B., Card, E., Bagdasarian, A., Tamakloe, T., Wang, W.J., Song, X., Chu, W.M. and Kouttab, N., 2009. SIRT1 confers protection against UVB-and H_2O_2-induced cell death via modulation of p53 and JNK in cultured skin keratinocytes. Journal of Cellular and Molecular Medicine 13: 3632-3643.

Caon, I., Bartolini, B., Moretto, P., Parnigoni, A., Caravà, E., Vitale, D.L., Alaniz, L., Viola, M., Karousou, E., De Luca, G. and Hascall, V.C., 2020. Sirtuin 1 reduces hyaluronan synthase 2 expression by inhibiting nuclear translocation of NF-kB and expression of the long non-coding RNA HAS2-AS1. Journal of Biological Chemistry 295: 3485-3496.

Carnevale, I., Pellegrini, L., D'Aquila, P., Saladini, S., Lococo, E., Polletta, L., Vernucci, E., Foglio, E., Coppola, S., Sansone, L. and Passarino, G., 2017. SIRT1-SIRT3 axis regulates cellular response to oxidative stress and etoposide. Journal of Cellular Physiology 232: 1835-1844.

Carreira, M.C., Izquierdo, A.G., Amil, M., Casanueva, F.F. and Crujeiras, A.B., 2018. Oxidative stress induced by excess of adiposity is related to a downregulation of hepatic SIRT6 expression in obese individuals. Oxidative Medicine and Cellular Longevity 2018: 6256052.

Chang, H.C. and Guarente, L., 2014. SIRT1 and other sirtuins in metabolism. Trends in Endocrinology & Metabolism 25: 138-145.

Chang, J.W., Kim, H., Baek, C.H., Lee, R.B., Yang, W.S. and Lee, S.K., 2016. Up-regulation of SIRT1 reduces endoplasmic reticulum stress and renal fibrosis. Nephron 133: 116-128.

Cheang, W.S., Wong, W.T., Wang, L., Cheng, C.K., Lau, C.W., Ma, R.C.W., Xu, A., Wang, N., Huang, Y. and Tian, X.Y., 2019. Resveratrol ameliorates endothelial dysfunction in diabetic and obese mice through sirtuin 1 and peroxisome proliferator-activated receptor δ. Pharmacological Research 139: 384-394.

Chen, C..J, Fu, Y.C., Yu, W. and Wang, W., 2013. SIRT3 protects cardiomyocytes from oxidative stress-mediated cell death by activating NF-κB. Biochemical and Biophysical Research Communications 430: 798-803.

Chen, H., Lin, R., Zhang, Z., Wei, Q., Zhong, Z., Huang, J. and Xu, Y., 2019. Sirtuin 1 knockdown inhibits glioma cell proliferation and potentiates temozolomide toxicity via facilitation of reactive oxygen species generation. Oncology Letters 17: 5343-5350.

Chen, J., Lai, J., Yang, L., Ruan, G., Chaugai, S., Ning, Q., Chen, C. and Wang, D.W., 2016. Trimetazidine prevents macrophage-mediated septic myocardial dysfunction via activation of the histone deacetylase sirtuin 1. British Journal of Pharmacology 173: 545-561.

Chen, K.L., Li, L., Li, C.M., Wang, Y.R., Yang, F.X., Kuang, M.Q. and Wang, G.L., 2019. SIRT7 Regulates lipopolysaccharide-induced inflammatory injury by suppressing the NF-κB signaling pathway. Oxidative Medicine and Cellular Longevity 2019: 3187972.

Chen, X., Li, S. and Liu, L., 2014. Engineering redox balance through cofactor systems. Trends in Biotechnology 32: 337-343.

Chen, Y., Zhang, J., Lin, Y., Lei, Q., Guan, K.L., Zhao, S. and Xiong, Y., 2011. Tumour suppressor SIRT3 deacetylates and activates manganese superoxide dismutase to scavenge ROS. EMBO Reports 12: 534-541.

Cheng, Y., Takeuchi, H., Sonobe, Y., Jin, S., Wang, Y., Horiuchi, H., Parajuli, B., Kawanokuchi, J., Mizuno, T. and Suzumura, A., 2014. Sirtuin 1 attenuates oxidative stress via upregulation of superoxide dismutase 2 and catalase in astrocytes. Journal of Neuroimmunology 269: 38-43.

Chua, K.F., Mostoslavsky, R., Lombard, D.B., Pang, W.W., Saito, S.I., Franco, S., Kaushal, D., Cheng, H.L., Fischer, M.R., Stokes, N. and Murphy, M.M., 2005. Mammalian SIRT1 limits replicative life span in response to chronic genotoxic stress. Cell Metabolism 2: 67-76.

Chuang, Y.C., Chen, S.D., Jou, S.B., Lin, T.K., Chen, S.F., Chen, N.C. and Hsu, C.Y., 2019. Sirtuin 1 regulates mitochondrial biogenesis and provides an endogenous neuroprotective mechanism against seizure-induced neuronal cell death in the hippocampus following status epilepticus. International Journal of Molecular Sciences 20: 3588.

Clementino, M., Kim, D. and Zhang, Z., 2019. Constitutive activation of NAD-dependent sirtuin 3 plays an important role in tumorigenesis of chromium(VI)-transformed cells. Toxicological Sciences 169: 224-234.

Cogburn, L.A., Trakooljul, N., Chen, C., Huang, H., Wu, C.H., Carré, W., Wang, X. and White, H.B., 2018. Transcriptional profiling of liver during the critical embryo-to-hatchling transition period in the chicken (*Gallus gallus*). BMC Genomics 19: 695.

Corsello, T., Komaravelli, N. and Casola, A., 2018. Role of hydrogen sulfide in NRF2-and sirtuin-dependent maintenance of cellular redox balance. Antioxidants 7: 129.

Cuadrado, A., Manda, G., Hassan, A., Alcaraz, M.J., Barbas, C., Daiber, A., Ghezzi, P., León, R., López, M.G., Oliva, B. and Pajares, M., 2018. Transcription factor NRF2 as a therapeutic target for chronic diseases: a systems medicine approach. Pharmacological Reviews 70: 348-383.

D'Onofrio, N., Servillo, L., Giovane, A., Casale, R., Vitiello, M., Marfella, R., Paolisso, G. and Balestrieri, M.L., 2016. Ergothioneine oxidation in the protection against high-glucose induced endothelial senescence: involvement of SIRT1 and SIRT6. Free Radical Biology and Medicine 96: 211-222.

Dali-Youcef, N., Lagouge, M., Froelich, S., Koehl, C., Schoonjans, K. and Auwerx, J., 2007. Sirtuins: the 'magnificent seven', function, metabolism and longevity. Annals of Medicine 39: 335-345.

Diaz-Ruiz, C., Villar-Cheda, B., Dominguez-Meijide, A., Garrido-Gil, P., Guerra, M.J. and Labandeira-Garcia, J.L., 2020. Aging-related overactivity of the angiotensin/at1 axis decreases sirtuin 3 levels in the substantia nigra, which induces vulnerability to oxidative stress and neurodegeneration. The Journals of Gerontology: Series A 75: 416-424.

Ding, Y.W., Zhao, G.J., Li, X.L., Hong, G.L., Li, M.F., Qiu, Q.M., Wu, B. and Lu, Z.Q., 2016. SIRT1 exerts protective effects against paraquat-induced injury in mouse type II alveolar epithelial cells by deacetylating NRF2 *in vitro*. International Journal of Molecular Medicine 37: 1049-1058.

Du, Y.G., Zhang, K.N., Gao, Z.L., Dai, F., Wu, X.X. and Chai, K.F., 2018. Tangshen formula improves inflammation in renal tissue of diabetic nephropathy through SIRT1/NF-κB pathway. Experimental and Therapeutic Medicine 15: 2156-2164.

Du, Z., Zhang, W., Wang, S., Zhang, J., He, J., Wang, Y., Dong, Y. and Huo, M., 2019. Celastrol protects human retinal pigment epithelial cells against hydrogen peroxide mediated oxidative stress, autophagy, and apoptosis through sirtuin 3 signal pathway. Journal of Cellular Biochemistry 120: 10413-10420.

Fang, J., Ianni, A., Smolka, C., Vakhrusheva, O., Nolte, H., Krüger, M., Wietelmann, A., Simonet, N.G., Adrian-Segarra, J.M., Vaquero, A. and Braun, T., 2017. Sirt7 promotes adipogenesis in the mouse by inhibiting autocatalytic activation of Sirt1. Proceedings of the National Academy of Sciences 114: E8352-E8361.

Fang, X.L., Zhu, X.T., Chen, S.F., Zhang, Z.Q., Zeng, Q.J., Deng, L., Peng, J.L., Yu, J.J., Wang, L.N., Wang, S.B. and Gao, P., 2014. Differential gene expression pattern in hypothalamus of chickens during fasting-induced metabolic reprogramming: functions of glucose and lipid metabolism in the feed intake of chickens. Poultry Science 93: 2841-2854.

Feng, J., Chen, X., Liu, R., Cao, C., Zhang, W., Zhao, Y. and Nie, S., 2018. Melatonin protects against myocardial ischemia-reperfusion injury by elevating Sirtuin3 expression and manganese superoxide dismutase activity. Free Radical Research 52: 840-849.

Ferber, E.C., Peck, B., Delpuech, O., Bell, G.P., East, P. and Schulze, A., 2012. FOXO3a regulates reactive oxygen metabolism by inhibiting mitochondrial gene expression. Cell Death and Differentiation 19: 968-979.

Foolad, F., Khodagholi, F. and Javan, M., 2019. Sirtuins in multiple sclerosis: the crossroad of neurodegeneration, autoimmunity and metabolism. Multiple Sclerosis and Related Disorders 34: 47-58.

Francois, M., Donovan, P. and Fontaine, F., 2020. Modulating transcription factor activity: interfering with protein-protein interaction networks. Seminars in Cell & Developmental Biology 99: 12-19.

Frescas, D., Valenti, L. and Accili, D., 2005. Nuclear trapping of the forkhead transcription factor FoxO1 via Sirt-dependent deacetylation promotes expression of glucogenetic genes. Journal of Biological Chemistry 280: 20589-20595.

Fujita, K., 2019. p53 isoforms in cellular senescence- and ageing-associated biological and physiological functions. International Journal of Molecular Sciences 20: 6023.

Furukawa, A., Tada-Oikawa, S., Kawanishi, S. and Oikawa, S., 2007. H_2O_2 accelerates cellular senescence by accumulation of acetylated p53 via decrease in the function of SIRT1 by NAD+ depletion. Cellular Physiology and Biochemistry 20: 45-54.

Gao, M., Li, X., He, Y., Han, L., Qiu, D., Ling, L., Liu, H., Liu, J. and Gu, L., 2018. SIRT7 functions in redox homeostasis and cytoskeletal organization during oocyte maturation. The FASEB Journal 32: 6228-6238.

Ghirotto, B., Terra, F.F., Câmara, N.O.S. and Basso, P.J., 2019. Sirtuins in B lymphocytes metabolism and function. World Journal of Experimental Medicine 9: 1-13.

Gou, L., Zhao, L., Song, W., Wang, L., Liu, J., Zhang, H., Huang, Y., Lau, C.W., Yao, X., Tian, X.Y. and Wong, W.T., 2018. Inhibition of miR-92a suppresses oxidative stress and improves endothelial function by upregulating heme oxygenase-1 in db/db mice. Antioxidants & Redox Signaling 28: 358-370.

Grewal, A.K., Singh, N. and Singh, T.G., 2019. Effects of resveratrol postconditioning on cerebral ischemia in mice: role of the sirtuin-1 pathway. Canadian Journal of Physiology and Pharmacology 97: 1094-1101.

Grigorieva, M.A., Velichko, O.A., Shabaldin, S.V., Fisinin, V.I. and Surai, P.F., 2017. Vitagene regulation as a new strategy to fight stresses in poultry production. Agricultural Biology (Sel'skokhozyaistvennaya biologiya) 52: 716-730.

Guan, R., Cai, Z., Wang, J., Ding, M., Li, Z., Xu, J., Li, Y., Li, J., Yao, H., Liu, W. and Qian, J., 2019. Hydrogen sulfide attenuates mitochondrial dysfunction-induced cellular senescence and apoptosis in alveolar epithelial cells by upregulating sirtuin 1. Aging 11: 11844-11864.

Guo, R., Liu, B., Wang, K., Zhou, S., Li, W. and Xu, Y., 2014. Resveratrol ameliorates diabetic vascular inflammation and macrophage infiltration in db/db mice by inhibiting the NF-κB pathway. Diabetes and Vascular Disease Research 11: 92-102.

Guo, Y., Xing, L., Qian, C., Ding, Z. and Jin, B., 2018. Involvement of flavonoids from the leaves of *Carya cathayensis* Sarg. in sirtuin 1 expression in HUVEC senescence. Evidence-based Complementary and Alternative Medicine 2018: 8246560.

Hallows, W.C., Albaugh, B.C. and Denu, J.M., 2008. Where in the cell is SIRT3? Functional localization of an NAD^+-dependent protein deacetylase. Biochemical Journal 411: e11-e13.

Harlan, B.A., Pehar, M., Killoy, K.M. and Vargas, M.R., 2019. Enhanced SIRT6 activity abrogates the neurotoxic phenotype of astrocytes expressing ALS-linked mutant SOD1. The FASEB Journal 33: 7084-7091.

He, J., Liu, X., Su, C., Wu, F., Sun, J., Zhang, J., Yang, X., Zhang, C., Zhou, Z., Zhang, X. and Lin, X., 2019. Inhibition of mitochondrial oxidative damage improves reendothelialization capacity of endothelial progenitor cells via SIRT3 (Sirtuin 3)-enhanced SOD2 (superoxide dismutase 2) deacetylation in hypertension. Arteriosclerosis, Thrombosis, and Vascular Biology 39: 1682-1698.

Hickey, A.J., Juellig, M., Aitken, J., Loomes, K., Hauber, M.E. and Phillips, A.R., 2012. Birds and longevity: does flight driven aerobicity provide an oxidative sink? Ageing Research Reviews 11: 242-253.

Holley, A.K., Dhar, S.K. and St Clair, D.K., 2010. Manganese superoxide dismutase vs. p53: regulation of mitochondrial ROS. Mitochondrion 10: 649-661.

Hong, Q., Zhang, L., Das, B., Li, Z., Liu, B., Cai, G., Chen, X., Chuang, P.Y., He, J.C. and Lee, K., 2018. Increased podocyte Sirtuin-1 function attenuates diabetic kidney injury. Kidney International 93: 1330-1343.

Hori, Y.S., Kuno, A., Hosoda, R. and Horio, Y., 2013. Regulation of FOXOs and p53 by SIRT1 modulators under oxidative stress. PLoS One 8: e73875.

Hu, B., Gong, Z. and Bi, Z., 2020. Inhibition of miR-383 suppresses oxidative stress and improves endothelial function by increasing sirtuin 1. Brazilian Journal of Medical and Biological Research 53: e8616.

Hu, Y., Tao, X., Han, X., Xu, L., Yin, L., Sun, H., Qi, Y., Xu, Y. and Peng, J., 2018. MicroRNA-351-5p aggravates intestinal ischaemia/reperfusion injury through the targeting of MAPK13 and Sirtuin-6. British Journal of Pharmacology 175: 3594-3609.

Huang, K., Gao, X. and Wei, W., 2017. The crosstalk between Sirt1 and Keap1/Nrf2/ARE anti-oxidative pathway forms a positive feedback loop to inhibit FN and TGF-β1 expressions in rat glomerular mesangial cells. Experimental Cell Research 361: 63-72.

Huang, K., Huang, J., Xie, X., Wang, S., Chen, C., Shen, X., Liu, P. and Huang, H., 2013. Sirt1 resists advanced glycation end products-induced expressions of fibronectin and TGF-β1 by activating the Nrf2/ARE pathway in glomerular mesangial cells. Free Radical Biology and Medicine 65: 528-540.

Hubbard, B.P. and Sinclair, D.A., 2014. Small molecule SIRT1 activators for the treatment of aging and age-related diseases. Trends in Pharmacological Sciences 35: 146-154.

Ido, Y., Duranton, A., Lan, F., Weikel, K.A., Breton, L. and Ruderman, N.B., 2015. Resveratrol prevents oxidative stress-induced senescence and proliferative dysfunction by activating the AMPK-FOXO3 cascade in cultured primary human keratinocytes. PloS One 10: e0115341.

Jian, B., Yang, S., Chaudry, I.H. and Raju, R., 2014. Resveratrol restores sirtuin 1 (SIRT1) activity and pyruvate dehydrogenase kinase 1 (PDK1) expression after hemorrhagic injury in a rat model. Molecular Medicine 20: 10-16.

Jiang, M., Wang, J., Fu, J., Du, L., Jeong, H., West, T., Xiang, L., Peng, Q., Hou, Z., Cai, H., Seredenina, T., Arbez, N., Zhu, S., Sommers, K., Qian, J., Zhang, J., Mori, S., Yang, X.W., Tamashiro, K.L., Aja, S., Moran, T.H., Luthi-Carter, R., Martin, B., Maudsley, S., Mattson, M.P., Cichewicz, R.H., Ross, C.A., Holtzman, D.M., Krainc, D. and Duan, W., 2011. Neuroprotective role of Sirt1 in mammalian models of Huntington's disease through activation of multiple Sirt1 targets. Nature Medicine 18: 153-158.

Jiao, X., Li, Y., Zhang, T., Liu, M. and Chi, Y., 2016. Role of Sirtuin3 in high glucose-induced apoptosis in renal tubular epithelial cells. Biochemical and Biophysical Research Communications 480: 387-393.

Jin, S., Yang, L., Fan, X., Wu, M., Xu, Y., Chen, X., Lin, Z. and Geng, Z., 2019. Effect of divergence in residual feed intake on expression of lipid metabolism-related genes in the liver of meat-type ducks. Journal of Animal Science 97: 3947-3957.

Jin, Z., Xiao, Y., Yao, F., Wang, B., Zheng, Z., Gao, H., Lv, X., Chen, L., He, Y., Wang, W. and Lin, R., 2020. SIRT6 inhibits cholesterol crystal-induced vascular endothelial dysfunction via Nrf2 activation. Experimental Cell Research 387: 111744.

Ka, S.O., Bang, I.H., Bae, E.J. and Park, B.H., 2017. Hepatocyte-specific sirtuin 6 deletion predisposes to nonalcoholic steatohepatitis by up-regulation of Bach1, an Nrf2 repressor. The FASEB Journal 31: 3999-4010.

Kanwal, A., Pillai, V.B., Samant, S., Gupta, M. and Gupta, M.P., 2019. The nuclear and mitochondrial sirtuins, Sirt6 and Sirt3, regulate each other's activity and protect the heart from developing obesity-mediated diabetic cardiomyopathy. The FASEB Journal 33: 10872-10888.

Kauppinen, A., Suuronen, T., Ojala, J., Kaarniranta, K. and Salminen, A., 2013. Antagonistic crosstalk between NF-κB and SIRT1 in the regulation of inflammation and metabolic disorders. Cellular Signalling 25: 1939-1948.

Kawai, Y., Garduño, L., Theodore, M., Yang, J. and Arinze, I.J., 2011. Acetylation-deacetylation of the transcription factor Nrf2 (nuclear factor erythroid 2-related factor 2) regulates its transcriptional activity and nucleocytoplasmic localization. Journal of Biological Chemistry 286: 7629-7640.

Khanh, V.C., Zulkifli, A.F., Tokunaga, C., Yamashita, T., Hiramatsu, Y. and Ohneda, O., 2018. Aging impairs beige adipocyte differentiation of mesenchymal stem cells via the reduced expression of Sirtuin 1. Biochemical and Biophysical Research Communications 500: 682-690.

Kheiripour, N., Karimi, J., Khodadadi, I., Tavilani, H., Goodarzi, M.T. and Hashemnia, M., 2018. Silymarin prevents lipid accumulation in the liver of rats with type 2 diabetes via sirtuin1 and SREBP-1c. Journal of Basic and Clinical Physiology and Pharmacology 29: 301-308.

Kim, E.J., Kho, J.H., Kang, M.R. and Um S.J., 2007. Active regulator of SIRT1 cooperates with SIRT1 and facilitates suppression of p53 activity. Molecular Cell 28: 277-290.

Kim, H.G., Huang, M., Xin, Y., Zhang, Y., Zhang, X., Wang, G., Liu, S., Wan, J., Ahmadi, A.R., Sun, Z. and Liangpunsakul, S., 2019. The epigenetic regulator SIRT6 protects the liver from alcohol-induced tissue injury by reducing oxidative stress in mice. Journal of Hepatology 71: 960-969.

Kim, J.E., Chen, J. and Lou, Z., 2008. DBC1 is a negative regulator of SIRT1. Nature 451: 583-586.

Kim, W. and Kim, J.E., 2013. SIRT7 an emerging sirtuin: deciphering newer roles. Journal of Physiology and Pharmacology 64: 531-534.

Kitamura, Y.I., Kitamura, T., Kruse, J.P., Raum, J.C,, Stein, R., Gu, W. and Accili, D., 2005. FoxO1 protects against pancreatic beta cell failure through NeuroD and MafA induction. Cell Metabolism 2: 153-163.

Kok, S.H., Lin, L.D., Hou, K.L., Hong, C.Y., Chang, C.C., Hsiao, M., Wang, J.H., Lai, E.H. and Lin, S.K., 2013. Simvastatin inhibits cysteine-rich protein 61 expression in rheumatoid arthritis synovial fibroblasts through the regulation of sirtuin-1/FoxO3a signaling. Arthritis and Rheumatism 65: 639-649.

Kong, D., Ding, Y., Liu, J., Liu, R., Zhang, J., Zhou, Q., Long, Z., Peng, J., Li, L., Bai, H. and Hai, C., 2019. Chlorogenic acid prevents paraquat-induced apoptosis via Sirt1-mediated regulation of redox and mitochondrial function. Free Radical Research 53: 680-693.

Kong, X., Wang, R., Xue, Y., Liu, X., Zhang, H., Chen, Y., Fang, F. and Chang, Y., 2010. Sirtuin 3, a new target of PGC-1alpha, plays an important role in the suppression of ROS and mitochondrial biogenesis. PloS One 5: e11707.

Kops, G.J., Dansen, T.B., Polderman, P.E., Saarloos, I., Wirtz, K.W., Coffer, P.J., Huang, T.T., Bos, J.L., Medema, R.H. and Burgering, B.M., 2002. Forkhead transcription factor FOXO3a protects quiescent cells from oxidative stress. Nature 419: 316-321.

Krishnan, J., Danzer, C., Simka, T., Ukropec, J., Walter, K.M., Kumpf, S., Mirtschink, P., Ukropcova, B., Gasperikova, D., Pedrazzini, T. and Krek, W., 2012. Dietary obesity-associated Hif1α activation in adipocytes restricts fatty acid oxidation and energy expenditure via suppression of the Sirt2-NAD+ system. Genes & Development 26: 259-270.

Kulkarni, S.R., Donepudi, A.C., Xu, J., Wei, W., Cheng, Q.C., Driscoll, M.V., Johnson, D.A., Johnson, J.A., Li, X. and Slitt, A.L., 2014. Fasting induces nuclear factor E2-related factor 2 and ATP-binding Cassette transporters via protein kinase A and Sirtuin-1 in mouse and human. Antioxidants & Redox Signaling 20: 15-30.

Kumari, P., Popescu, D., Yue, S., Bober, E., Ianni, A. and Braun, T., 2018. SIRT7 inhibits SIRT1-mediated activation of SUV39H1. Cell Cycle 17: 1403-1412.

Kuscu, N., Gungor-Ordueri, N.E., Sozen, B., Adiguzel, D. and Celik-Ozenci, C., 2019. FoxO transcription factors 1 regulate mouse preimplantation embryo development. Journal of Assisted Reproduction and Genetics 36: 2121-2133.

Kwon, H.S. and Ott, M., 2008. The ups and downs of SIRT1. Trends in Biochemical Sciences 33: 517-525.

Lagunas-Rangel, F.A., 2019. Current role of mammalian sirtuins in DNA repair. DNA Repair 80: 85-92.

Lang, A. and Piekorz, R.P., 2018. Novel role of the SIRT4-OPA1 axis in mitochondrial quality control. Cell Stress 2: 1-3.

Lang, A., Anand, R., Altinoluk-Hambüchen, S., Ezzahoini, H., Stefanski, A., Iram, A., Bergmann, L., Urbach, J., Böhler, P., Hänsel, J. and Franke, M., 2017. SIRT4 interacts with OPA1 and regulates mitochondrial quality control and mitophagy. Aging 9: 2163-2189.

Lee, S.H., Lee, J.H., Lee, H.Y. and Min, K.J., 2019. Sirtuin signaling in cellular senescence and aging. BMB Reports 52: 24-34.

Lee, S.J., Kim, S.J., Lee, H.S. and Kwon, O.S., 2019a. PKCδ mediates NF-κB inflammatory response and downregulates SIRT1 expression in liver fibrosis. International Journal of Molecular Sciences 20: 4607.

Li, X., 2013. SIRT1 and energy metabolism. Acta Biochimica et Biophysica Sinica 45: 51-60.

Li, C., Xie, N., Li, Y., Liu, C., Hou, F.F. and Wang, J., 2019a. N-acetylcysteine ameliorates cisplatin-induced renal senescence and renal interstitial fibrosis through sirtuin1 activation and p53 deacetylation. Free Radical Biology and Medicine 130: 512-527.

Li, W., Yang, Y., Li, Y., Zhao, Y. and Jiang, H., 2019b. Sirt5 attenuates cisplatin-induced acute kidney injury through regulation of Nrf2/HO-1 and Bcl-2. BioMed Research International 2019: 4745132.

Li, X., Song, S., Xu, M., Hua, Y., Ding, Y., Shan, X., Meng, G. and Wang, Y., 2019c. Sirtuin3 deficiency exacerbates carbon tetrachloride-induced hepatic injury in mice. Journal of Biochemical and Molecular Toxicology 33: e22249.

Li, Y., Xu, W., McBurney, M.W. and Longo, V.D., 2008. SirT1 inhibition reduces IGF-I/IRS-2/Ras/ERK1/2 signaling and protects neurons. Cell Metabolism 8: 38-48.

Li, Z., Xu, K., Zhang, N., Amador, G., Wang, Y., Zhao, S., Li, L., Qiu, Y. and Wang, Z., 2018. Overexpressed SIRT6 attenuates cisplatin-induced acute kidney injury by inhibiting ERK1/2 signaling. Kidney International 93: 881-892.

Liang, F., Wang, X., Ow, S.H., Chen, W. and Ong, W.C., 2017. Sirtuin 5 is anti-apoptotic and anti-oxidative in cultured SH-EP neuroblastoma cells. Neurotoxicity Research 31: 63-76.

Lin, C.L., Huang, W.N., Li, H.H., Huang, C.N., Hsieh, S., Lai, C. and Lu, F.J., 2015. Hydrogen-rich water attenuates amyloid β-induced cytotoxicity through upregulation of Sirt1-FoxO3a by stimulation of AMP-activated protein kinase in SK-N-MC cells. Chemico-Biological Interactions 240: 12-21.

Lin, S., Xing, H., Zang, T., Ruan, X., Wo, L. and He, M., 2018. Sirtuins in mitochondrial stress: indispensable helpers behind the scenes. Ageing Research Reviews 44: 22-32.

Liu, A.J., Li, B., Yang, M., Liu, Y., Liu, Y.L. and Su, J.W., 2017. Sirtuin 1 mediates hydrogen sulfide-induced cytoprotection effects in neonatal mouse cardiomyocytes. Chinese Medical Journal 130: 2346-2353.

Liu, B., Chen, Y. and St Clair, D.K., 2008. ROS and p53: a versatile partnership. Free Radical Biology and Medicine 44: 1529-1535.

Liu, F., Lin, C., Hong, J., Cai, C., Zhang, W., Zhang, J. and Guo, L., 2020a. Apocynin protects retina cells from ultraviolet radiation damage via inducing sirtuin 1. Journal of Drug Targeting 28: 330-338.

Liu, G., Park, S.H., Imbesi, M., Nathan, W.J., Zou, X., Zhu, Y., Jiang, H., Parisiadou, L. and Gius, D., 2017a. Loss of NAD-dependent protein deacetylase sirtuin 2 alters mitochondrial protein acetylation and dysregulates mitophagy. Antioxidants & Redox Signaling 26: 849-863.

Liu, X., Liu, R., Bai, Y., Jiang, H., Fu, X. and Ma, S., 2020b. Post-translational modifications of protein in response to ionizing radiation. Cell Biochemistry and Function 38: 283-289.

Lothrop, A.P., Torres, M.P. and Fuchs, S.M., 2013. Deciphering post-translational modification codes. FEBS Letters 587: 1247-1257.

Lu, H., Sun, L., Chen, W., Zhou, Y., Liu, K., Chen, J., Zhang, Z., Zhang, C. and Tian, H., 2020. Sirtuin 3 therapy attenuates aging expression, oxidative stress parameters, and neointimal hyperplasia formation in vein grafts. Annals of Vascular Surgery 64: 303-317.

Luo, H., Chiang, H.H., Louw, M., Susanto, A. and Chen, D., 2017. Nutrient sensing and the oxidative stress response. Trends in Endocrinology & Metabolism 28: 449-460.

Ma, F., Wu, J., Jiang, Z., Huang, W., Jia, Y., Sun, W. and Wu, H., 2019. P53/NRF2 mediates SIRT1's protective effect on diabetic nephropathy. Biochimica et Biophysica Acta (BBA)-Molecular Cell Research 1866: 1272-1281.

Ma, J., Song, D., Zhang, Y., Chen, L., Zhang, S., Jia, J., Chen, T., Guo, C., Tian, L., Gao, A. and Niu, P., 2018. SIRT1 exhibits antioxidative effects in HT22 cells induced by tert-butyl alcohol. Environmental Toxicology 33: 142-148.

Mahmoud, A.R., Ali, F.E.M., Abd-Elhamid, T.H. and Hassanein, E.H.M., 2019. Coenzyme Q(10) protects hepatocytes from ischemia reperfusion-induced apoptosis and oxidative stress via regulation of Bax/Bcl-2/PUMA and Nrf-2/FOXO-3/Sirt-1 signaling pathways. Tissue & Cel 60:1-13.

Maillet, A. and Pervaiz, S., 2012. Redox regulation of p53, redox effectors regulated by p53: a subtle balance. Antioxidants & Redox Signaling 16: 1285-1294.

Mäkelä, J., Tselykh, T.V., Maiorana, F., Eriksson, O., Do, H.T., Mudò, G., Korhonen, L.T., Belluardo, N. and Lindholm, D., 2014. Fibroblast growth factor-21 enhances mitochondrial functions and increases the activity of PGC-1α in human dopaminergic neurons via Sirtuin-1. Springerplus 3: 2.

Mao, Z., Hine, C., Tian, X., Van Meter, M., Au, M., Vaidya, A., Seluanov, A. and Gorbunova, V., 2011. SIRT6 promotes DNA repair under stress by activating PARP1. Science 332: 1443-1446.

Marinho, H.S., Real, C., Cyrne, L., Soares, H. and Antunes, F., 2014. Hydrogen peroxide sensing, signaling and regulation of transcription factors. Redox Biology 2: 535-562.

Moloney, J.N. and Cotter, T.G., 2018, August. ROS signalling in the biology of cancer. Seminars in Cell & Developmental Biology 80: 50-64.

Morris, B.J., 2013. Seven sirtuins for seven deadly diseases of aging. Free Radical Biology and Medicine 56: 133-171.

Mortuza, R., Chen, S., Feng, B., Sen, S. and Chakrabarti, S., 2013. High glucose induced alteration of SIRTs in endothelial cells causes rapid aging in a p300 and FOXO regulated pathway. PLoS One 8: e54514.

Nakamura, K., Kageyama, S., Ke, B., Fujii, T., Sosa, R.A., Reed, E.F., Datta, N., Zarrinpar, A., Busuttil, R.W. and Kupiec-Weglinski, J.W., 2017. Sirtuin 1 attenuates inflammation and hepatocellular damage in liver transplant ischemia/reperfusion: from mouse to human. Liver Transplantation 23: 1282-1293.

Nie, H., Hong, Y., Lu, X., Zhang, J., Chen, H., Li, Y., Ma, Y. and Ying, W., 2014. SIRT2 mediates oxidative stress-induced apoptosis of differentiated PC12 cells. Neuroreport 25: 838-842.

Nogueiras, R., Habegger, K.M., Chaudhary, N., Finan, B., Banks, A.S., Dietrich, M.O., Horvath, T.L., Sinclair, D.A., Pfluger, P.T. and Tschöp, M.H., 2012. Sirtuin 1 and sirtuin 3: physiological modulators of metabolism. Physiological Reviews 92: 1479-1514.

Nopparat, C., Sinjanakhom, P. and Govitrapong, P., 2017. Melatonin reverses H_2O_2-induced senescence in SH-SY5Y cells by enhancing autophagy via sirtuin 1 deacetylation of the RelA/p65 subunit of NF-κB. Journal of Pineal Research 63: e12407.

Oberdoerffer, P., Michan, S., McVay, M., Mostoslavsky, R., Vann, J., Park, S.K., Hartlerode, A., Stegmuller, J., Hafner, A., Loerch, P. and Wright, S.M., 2008. SIRT1 redistribution on chromatin promotes genomic stability but alters gene expression during aging. Cell 135: 907-918.

Oh, J.Y., Choi, G.E., Lee, H.J., Jung, Y.H., Chae, C.W., Kim, J.S., Lee, C.K. and Han, H.J., 2019. 17β-estradiol protects mesenchymal stem cells against high glucose-induced mitochondrial oxidants production via Nrf2/Sirt3/ MnSOD signaling. Free Radical Biology and Medicine 130: 328-342.

Paik, J.H., Kollipara, R., Chu, G., Ji, H., Xiao, Y., Ding, Z., Miao, L., Tothova, Z., Horner, J.W., Carrasco, D.R., Jiang, S., Gilliland, D.G., Chin, L., Wong, W.H,. Castrillon, D.H. and DePinho, R.A., 2007. FoxOs are lineage-restricted redundant tumor suppressors and regulate endothelial cell homeostasis. Cell 128: 309-323.

Pan, W., Yu, H., Huang, S. and Zhu, P., 2016. Resveratrol protects against TNF-α-induced injury in human umbilical endothelial cells through promoting sirtuin-1-induced repression of NF-KB and p38 MAPK. PloS One, 11: e0147034.

Parker, J.A., Vazquez-Manrique, R.P., Tourette, C., Farina, F., Offner, N., Mukhopadhyay, A., Orfila, A.M., Darbois, A., Menet, S., Tissenbaum, H.A. and Neri, C., 2012. Integration of β-catenin, sirtuin, and FOXO signaling protects from mutant huntingtin toxicity. The Journal of Neuroscience 32: 12630-12640.

Peerapanyasut, W., Kobroob, A., Palee, S., Chattipakorn, N. and Wongmekiat, O., 2019. Activation of sirtuin 3 and maintenance of mitochondrial integrity by N-acetylcysteine protects against bisphenol A-induced kidney and liver toxicity in rats. International Journal of Molecular Sciences 20: 267.

Peserico, A., Chiacchiera, F., Grossi, V., Matrone, A., Latorre, D., Simonatto, M., Fusella, A., Ryall, J.G., Finley, L.W., Haigis, M.C., Villani, G., Puri, P.L., Sartorelli, V. and Simone, C., 2013. A novel AMPK-dependent FoxO3A-SIRT3 intramitochondrial complex sensing glucose levels. Cellular and Molecular Life Sciences 70: 2015-2029.

Pramanik, K.C., Fofaria, N.M., Gupta, P. and Srivastava, S.K., 2014. CBP-mediated FOXO-1 acetylation inhibits pancreatic tumor growth by targeting SirT. Molecular Cancer Therapeutics 13: 687-698.

Qin, Y., Cao, L. and Hu, L., 2019. Sirtuin 6 mitigated LPS-induced human umbilical vein endothelial cells inflammatory responses through modulating nuclear factor erythroid 2-related factor 2. Journal of Cellular Biochemistry 120: 1305-11317.

Rached, M.T., Kode, A., Xu, L., Yoshikawa, Y., Paik, J.H., Depinho, R.A. and Kousteni, S., 2010. FoxO1 is a positive regulator of bone formation by favoring protein synthesis and resistance to oxidative stress in osteoblasts. Cell Metabolism 11: 147-160.

Radak, Z., Koltai, E., Taylor, A.W., Higuchi, M., Kumagai, S., Ohno, H., Goto, S. and Boldogh, I., 2013. Redox-regulating sirtuins in aging, caloric restriction, and exercise. Free Radical Biology and Medicine 58: 87-97.

Rangarajan, P., Karthikeyan, A., Lu, J., Ling, E.A. and Dheen, S.T., 2015. Sirtuin 3 regulates Foxo3a-mediated antioxidant pathway in microglia. Neuroscience 311: 398-414.

Rardin, M.J., Newman, J.C., Held, J.M., Cusack, M.P., Sorensen, D.J., Li, B., Schilling, B., Mooney, S.D., Kahn, C.R., Verdin, E. and Gibson, B.W., 2013. Label-free quantitative proteomics of the lysine acetylome in mitochondria identifies substrates of SIRT3 in metabolic pathways. Proceedings of the National Academy of Sciences 110: 6601-6606.

Raynes, R., Brunquell, J. and Westerheide, S. D., 2013a. Stress inducibility of SIRT1 and its role in cytoprotection and cancer. Genes & Cancer 4: 172-182.

Raynes, R., Leckey, B.D. Jr., Nguyen, K. and Westerheide, S.D., 2012. Heat shock and caloric restriction have a synergistic effect on the heat shock response in a sir2.1-dependent manner in Caenorhabditis elegans. Journal of Biological Chemistry 287: 29045-29053.

Raynes, R., Pombier, K.M., Nguyen, K., Brunquell, J., Mendez, J.E. and Westerheide, S.D., 2013. The SIRT1 modulators AROS and DBC1 regulate HSF1 activity and the heat shock response. PLoS One 8: e54364.

Ren, J., Xu, N., Ma, Z., Li, Y., Li, C., Wang, Y., Tian, Y., Liu, X. and Kang, X., 2017. Characteristics of expression and regulation of sirtuins in chicken (*Gallus gallus*). Genome 60: 431-440.

Ryu, D., Jo, Y.S., Sasso, G.L., Stein, S., Zhang, H., Perino, A., Lee, J.U., Zeviani, M., Romand, R., Hottiger, M.O. and Schoonjans, K., 2014. A SIRT7-dependent acetylation switch of GABPβ1 controls mitochondrial function. Cell Metabolism 20: 856-869.

Sansone, L., Reali, V., Pellegrini, L., Villanova, L., Aventaggiato, M., Marfe, G., Rosa, R., Nebbioso, M., Tafani, M., Fini, M. and Russo, M.A., 2013. SIRT1 silencing confers neuroprotection through IGF-1 pathway activation. Journal of Cellular Physiology 228: 1754-1761.

Santoro, M.M., 2018, Fashioning blood vessels by ROS signalling and metabolism. Seminars in Cell & Developmental Biology 80: 35-42.

Santos Costa, L., Escande, C. and Denicola, A., 2016. Potential modulation of sirtuins by oxidative stress. Oxidative Medicine and Cellular Longevity 2016: 9831825.

Savran, M., Aslankoc, R., Ozmen, O., Erzurumlu, Y., Savas, H.B., Temel, E.N., Kosar, P.A. and Boztepe, S., 2020. Agomelatine could prevent brain and cerebellum injury against LPS-induced neuroinflammation in rats. Cytokine 127: 154957.

Shah, S.A., Khan, M., Jo, M.H., Jo, M.G., Amin, F.U. and Kim, M.O., 2017. Melatonin stimulates the SIRT1/Nrf2 signaling pathway counteracting lipopolysaccharide (LPS)-induced oxidative stress to rescue postnatal rat brain. CNS Neuroscience & Therapeutics 23: 33-44.

Shao, D., Yao, C., Kim, M.H., Fry, J., Cohen, R.A., Costello, C.E., Matsui, R., Seta, F., McComb, M.E. and Bachschmid, M.M., 2019. Improved mass spectrometry-based activity assay reveals oxidative and metabolic stress as sirtuin-1 regulators. Redox Biology 22: 101150.

Sharma, G., Parihar, A., Parihar, P. and Parihar, M.S., 2019. Downregulation of sirtuin 3 by palmitic acid increases the oxidative stress, impairment of mitochondrial function, and apoptosis in liver cells. Journal of Biochemical and Molecular Toxicology 33: e22337.

Shatskikh, E., Latypova, E., Fisinin, V., Denev, S. and Surai, P., 2015. Molecular mechanisms and new strategies to fight stresses in egg-producing birds. Agricultural Science and Technology 7: 3-10.

She, D.T., Wong, L.J., Baik, S.H. and Arumugam, T.V., 2018. SIRT2 inhibition confers neuroprotection by downregulation of FOXO3a and MAPK signaling pathways in ischemic stroke. Molecular Neurobiology 55: 9188-9203.

Shi, X., Pi, L., Zhou, S., Li, X., Min, F., Wang, S., Liu, Z. and Wu, J., 2018. Activation of sirtuin 1 attenuates high glucose-induced neuronal apoptosis by deacetylating p53. Frontiers in Endocrinology 9: 274.

Shimao, R., Muroi, H., Furukawa, K., Toyomizu, M. and Kikusato, M., 2019. Effects of low-dose oleuropein diet supplementation on the oxidative status of skeletal muscles and plasma hormonal concentration of growing broiler chickens. British Poultry Science 60: 784-789.

Shukla, S., Sharma, A., Pandey, V.K., Raisuddin, S. and Kakkar, P., 2016. Concurrent acetylation of FoxO1/3a and p53 due to sirtuins inhibition elicit Bim/PUMA mediated mitochondrial dysfunction and apoptosis in berberine-treated HepG2 cells. Toxicology and Applied Pharmacology 291: 70-83.

Simmons, G.E., Pruitt, W.M. and Pruitt, K., 2015. Diverse roles of SIRT1 in cancer biology and lipid metabolism. International Journal of Molecular Sciences 16: 950-965.

Singh, C.K., Chhabra, G., Ndiaye, M.A., Garcia-Peterson, L.M., Mack, N.J. and Ahmad, N., 2018. The role of sirtuins in antioxidant and redox signaling. Antioxidants & Redox Signaling 28: 643-661.

Singh, P., Hanson, P.S. and Morris, C.M., 2017. Sirtuin-2 protects neural cells from oxidative stress and is elevated in neurodegeneration. Parkinson's Disease 2017: 2643587.

Sobuz, S.U., Sato, Y., Yoshizawa, T., Karim, F., Ono, K., Sawa, T., Miyamoto, Y., Oka, M. and Yamagata, K., 2019. SIRT7 regulates the nuclear export of NF-κB p65 by deacetylating Ran. Biochimica et Biophysica Acta (BBA)-Molecular Cell Research 1866: 1355-1367.

Song, J., Yang, B., Jia, X., Li, M., Tan, W., Ma, S., Shi, X. and Feng, L., 2018. Distinctive roles of sirtuins on diabetes, protective or detrimental? Frontiers in Endocrinology 9: 724.

Song, L., Dong, G., Guo, L. and Graves, D.T., 2018a. The function of dendritic cells in modulating the host response. Molecular Oral Microbiology 33: 13-21.

Sosulski, M.L., Gongora, R., Feghali-Bostwick, C., Lasky, J.A. and Sanchez, C.G., 2017. Sirtuin 3 deregulation promotes pulmonary fibrosis. The Journals of Gerontology: Series A 72: 595-602.

Sun, F., Jiang, X., Wang, X., Bao, Y., Feng, G., Liu, H., Kou, X., Zhu, Q., Jiang, L. and Yang, Y., 2019b. Vincristine ablation of Sirt2 induces cell apoptosis and mitophagy via Hsp70 acetylation in MDA-MB-231 cells. Biochemical Pharmacology 162: 142-153.

Sun, G.L., Huang, D., Li, K.R. and Jiang, Q., 2019. MicroRNA-4532 inhibition protects human lens epithelial cells from ultra-violet-induced oxidative injury via activating SIRT6-Nrf2 signaling. Biochemical and Biophysical Research Communications 514: 777-784.

Sun, X., Wang, S., Gai, J., Guan, J., Li, J., Li, Y., Zhao, J., Zhao, C., Fu, L. and Li, Q., 2019a. SIRT5 promotes cisplatin resistance in ovarian cancer by suppressing DNA damage in a ROS-dependent manner via regulation of the Nrf2/HO-1 pathway. Frontiers in Oncology 9: 754.

Surai, P.F., 2020. Antioxidants in poultry nutrition and reproduction: an update. Antioxidants 9, 2: 105.

Surai, P.F. and Fisinin, V.I., 2016a. Vitagenes in poultry production. Part 3. Vitagene concept development. World's Poultry Science Journal 72: 793-804.

Surai, P.F. and Fisinin, V.I., 2016b. Antioxidant system regulation: from vitamins to vitagenes. In: Watson, R.R. and De Meester, F. (eds) Handbook of cholesterol. Wageningen Academic Publishers, Wageningen, the Netherlands, pp. 451-481.

Surai, P.F., Kochish, I.I., Fisinin, V.I. and Kidd, M.T., 2019a. Antioxidant defence systems and oxidative stress in poultry biology: an update. Antioxidants 8, 7: 235.

Surai, P.F., Kochish, I.I., Fisinin, V.I., Grozina, A.A. and Shatskikh, E.V., 2018. Molecular mechanisms of gut health support in poultry: role of microbiota. Agricultural Technologies: Moscow, Russia.

Tang, X., Li, Y., Liu, L., Guo, R., Zhang, P., Zhang, Y., Zhang, Y., Zhao, J., Su, J., Sun, L. and Liu, Y., 2020. Sirtuin 3 induces apoptosis and necroptosis by regulating mutant p53 expression in small-cell lung cancer. Oncology Reports 43(2): 591-600.

Tatone, C., Di Emidio, G., Barbonetti, A., Carta, G., Luciano, A.M., Falone, S. and Amicarelli, F., 2018. Sirtuins in gamete biology and reproductive physiology: emerging roles and therapeutic potential in female and male infertility. Human Reproduction Update 24: 267-289.

Torrens-Mas, M., Hernández-López, R., Pons, D.G., Roca, P., Oliver, J. and Sastre-Serra, J., 2019. Sirtuin 3 silencing impairs mitochondrial biogenesis and metabolism in colon cancer cells. American Journal of Physiology-Cell Physiology 317: C398-C404.

Tran, H., Brunet, A., Grenier, J.M., Datta, S.R., Fornace, A.J. Jr., DiStefano, P.S., Chiang, L.W. and Greenberg, M.E., 2002. DNA repair pathway stimulated by the forkhead transcription factor FOXO3a through the Gadd45 protein. Science 296: 530-534.

Tresguerres, J.A., Cuesta, S., Kireev, R.A., Garcia, C., Acuña-Castroviejo, D. and Vara, E., 2013. Beneficial effect of melatonin treatment on age-related insulin resistance and on the development of type 2 diabetes. Hormone Molecular Biology and Clinical Investigation 16: 47-54.

Tseng, A.H.H., Wu, L.H., Shieh, S.S. and Wang, D.L., 2014. SIRT3 interactions with FOXO3 acetylation, phosphorylation and ubiquitinylation mediate endothelial cell responses to hypoxia. Biochemical Journal 464: 157-168.

Vakhrusheva, O., Smolka, C., Gajawada, P., Kostin, S., Boettger, T., Kubin, T., Braun, T. and Bober, E., 2008. Sirt7 increases stress resistance of cardiomyocytes and prevents apoptosis and inflammatory cardiomyopathy in mice. Circulation Research 102: 703-710.

Vazquez, B.N., Thackray, J.K. and Serrano, L., 2017. Sirtuins and DNA damage repair: SIRT7 comes to play. Nucleus 8: 107-115.

Wan, X., Ahmad, H., Zhang, L., Wang, Z. and Wang, T., 2018. Dietary enzymatically treated *Artemisia annua* L. improves meat quality, antioxidant capacity and energy status of breast muscle in heat-stressed broilers. Journal of the Science of Food and Agriculture 98: 3715-3721.

Wang, F., Nguyen, M., Qin, F.X.F. and Tong, Q., 2007. SIRT2 deacetylates FOXO3a in response to oxidative stress and caloric restriction. Aging Cell 6: 505-514.

Wang, F., Shang, Y., Zhang, R., Gao, X. and Zeng, Q., 2019. A SIRT1 agonist reduces cognitive decline in type 2 diabetic rats through antioxidative and anti-inflammatory mechanisms. Molecular Medicine Reports 19: 1040-1048.

Wang, S.J., Zhao, X.H., Chen, W., Bo, N., Wang, X.J., Chi, Z.F. and Wu, W., 2015. Sirtuin 1 activation enhances the PGC-1α/mitochondrial antioxidant system pathway in status epilepticus. Molecular Medicine Reports 11: 521-526.

Wang, Y., Zhou, Y. and Graves, D.T., 2014. FOXO transcription factors: their clinical significance and regulation. BioMed Research International 2014: 925350.

Wei, T., Huang, G., Gao, J., Huang, C., Sun, M., Wu, J., Bu, J. and Shen, W., 2017. Sirtuin 3 deficiency accelerates hypertensive cardiac remodeling by impairing angiogenesis. Journal of the American Heart Association 6: e006114.

Westerheide, S.D., Anckar, J., Stevens, S.M., Sistonen, L. and Morimoto, R.I., 2009. Stress-inducible regulation of heat shock factor 1 by the deacetylase SIRT1. Science 323: 1063-1066.

Wu, B., Feng, J.Y., Yu, L.M., Wang, Y.C., Chen, Y.Q., Wei, Y., Han, J.S., Feng, X., Zhang, Y., Di, S.Y., Ma, Z.Q., Fan, C.X. and Ha, X.Q., 2018. Icariin protects cardiomyocytes against ischaemia/reperfusion injury by attenuating sirtuin 1-dependent mitochondrial oxidative damage. British Journal of Pharmacology 175: 4137-4153.

Wu, Y.T., Wu, S.B. and Wei, Y.H., 2014. Roles of sirtuins in the regulation of antioxidant defense and bioenergetic function of mitochondria under oxidative stress. Free Radical Research 48: 1070-1084.

Xi, J., Jing, J., Zhang, Y., Liang, C., Hao, Z., Zhang, L. and Chen, Y., 2019. SIRT3 inhibited the formation of calcium oxalate-induced kidney stones through regulating NRF2/HO-1 signaling pathway. Journal of Cellular Biochemistry 120: 8259-8271.

Xia, X., Qu, B., Li, Y.M., Yang, L.B., Fan, K.X., Zheng, H., Huang, H.D., Gu, J.W., Kuang, Y.Q. and Ma, Y., 2017. NFAT5 protects astrocytes against oxygen-glucose-serum deprivation/restoration damage via the SIRT1/Nrf2 pathway. Journal of Molecular Neuroscience 61: 96-104.

Xiangyun, Y. and Xiaomin, N., 2017. Desuccinylation of pyruvate kinase M2 by SIRT5 contributes to antioxidant response and tumor growth. Oncotarget 8: 6984-6993.

Xiong, X., Sun, X., Wang, Q., Qian, X., Zhang, Y., Pan, X. and Dong, X.C., 2016. SIRT6 protects against palmitate-induced pancreatic β-cell dysfunction and apoptosis. The Journal of Endocrinology 231: 159-165.

Xu, G., Zhao, X., Fu, J. and Wang, X., 2019a. Resveratrol increase myocardial Nrf2 expression in type 2 diabetic rats and alleviate myocardial ischemia/reperfusion injury (MIRI). Annals of Palliative Medicine 8: 565-575.

Xu, L., Nitika Hasin, N., Cuskelly, D.D., Wolfgeher, D., Doyle, S., Moynagh, P., Perrett, S., Jones, G.W. and Truman, A.W., 2019b. Rapid deacetylation of yeast Hsp70 mediates the cellular response to heat stress. Scientific Reports 9: 16260.

Xu, W.N., Yang, R.Z., Zheng, H.L., Yu, W., Zheng, X.F., Li, B., Jiang, S.D. and Jiang, L.S., 2019. PGC-1α acts as an mediator of Sirtuin2 to protect annulus fibrosus from apoptosis induced by oxidative stress through restraining mitophagy. International Journal of Biological Macromolecules 136: 1007-1017.

Xue, B., Song, J., Liu, L., Luo, J., Tian, G. and Yang, Y., 2017. Effect of epigallocatechin gallate on growth performance and antioxidant capacity in heat-stressed broilers. Archives of Animal Nutrition 71: 362-372.

Yang, L., Duan, Z., Liu, X. and Yuan, Y., 2018. N-acetyl-l-cysteine ameliorates the PM2. 5-induced oxidative stress by regulating SIRT-1 in rats. Environmental Toxicology and Pharmacology 57: 70-75.

Yang, M., Peng, Y., Liu, W., Zhou, M., Meng, Q. and Yuan, C., 2019. Sirtuin 2 expression suppresses oxidative stress and senescence of nucleus pulposus cells through inhibition of the p53/p21 pathway. Biochemical and Biophysical Research Communications 513:616-622.

Yang, M., Peng, Y., Liu, W., Zhou, M., Meng, Q. and Yuan, C., 2019a. Sirtuin 2 expression suppresses oxidative stress and senescence of nucleus pulposus cells through inhibition of the p53/p21 pathway. Biochemical and Biophysical Research Communications 513: 616-622.

Yang, Q.B., He, Y.L., Zhong, X.W., Xie, W.G. and Zhou, J.G., 2019a. Resveratrol ameliorates gouty inflammation via upregulation of sirtuin 1 to promote autophagy in gout patients. Inflammopharmacology 27: 47-56.

Yang, X., Park, S.H., Chang, H.C., Shapiro, J.S., Vassilopoulos, A., Sawicki, K.T., Chen, C., Shang, M., Burridge, P.W., Epting, C.L., Wilsbacher, L.D., Jenkitkasemwong, S., Knutson, M., Gius, D. and Ardehali, H., 2017. Sirtuin 2 regulates cellular iron homeostasis via deacetylation of transcription factor NRF2. The Journal of Clinical Investigation 127: 1505-1516.

Yang, Y., Fu, W., Chen, J., Olashaw, N., Zhang, X., Nicosia, S.V., Bhalla, K. and Bai, W., 2007. SIRT1 sumoylation regulates its deacetylase activity and cellular response to genotoxic stress. Nature cell Biology 9: 1253-1262.

Yang, Y., Tian, T., Wang, Y., Li, Z., Xing, K. and Tian, G., 2019b. SIRT6 protects vascular endothelial cells from angiotensin II-induced apoptosis and oxidative stress by promoting the activation of Nrf2/ARE signaling. European Journal of Pharmacology 859: 172516.

Yao, H., Yao, Z., Zhang, S., Zhang, W. and Zhou, W., 2018. Upregulation of SIRT1 inhibits H_2O_2-induced osteoblast apoptosis via FoxO1/β-catenin pathway. Molecular Medicine Reports 17: 6681-6690.

Yarahmadi, S., Abdolvahabi, Z., Hesari, Z., Tavakoli-Yaraki, M., Yousefi, Z., Seiri, P., Hosseinkhani, S. and Nourbakhsh, M., 2019. Inhibition of sirtuin 1 deacetylase by miR-211-5p provides a mechanism for the induction of cell death in breast cancer cells. Gene 711: 143939.

Yeung, F., Hoberg, J.E., Ramsey, C.S., Keller, M.D., Jones, D.R., Frye, R.A. and Mayo, M.W., 2004. Modulation of NF-κB-dependent transcription and cell survival by the SIRT1 deacetylase. The EMBO Journal 23: 2369-2380.

Yin, J., Nielsen, M., Carcione, T., Li, S. and Shi, J., 2019. Apolipoprotein E regulates mitochondrial function through the PGC-1α-sirtuin 3 pathway. Aging 11: 11148-11156.

Yu, H., Pan, W., Huang, H., Chen, J., Sun, B., Yang, L. and Zhu, P., 2019. Screening analysis of sirtuins family expression on anti-inflammation of resveratrol in endothelial cells. Medical science monitor: international medical journal of experimental and clinical research 25: 4137-4148.

Zeng, H. and, Chen, J.X., 2019. Sirtuin 3, endothelial metabolic reprogramming, and heart failure with preserved ejection fraction. Journal of Cardiovascular Pharmacology 74: 315-323.

Zhang, J. and Choudhury, M., 2017. The plasticizer BBP selectively inhibits epigenetic regulator sirtuin during differentiation of C3H10T1/2 stem cell line. Toxicology In Vitro 39: 75-83.

Zhang, C., Wen, C., Lin, J. and Shen, G., 2015. Protective effect of pyrroloquinoline quinine on ultraviolet A irradiation-induced human dermal fibroblast senescence *in vitro* proceeds via the anti-apoptotic sirtuin 1/nuclear factor-derived erythroid 2-related factor 2/heme oxygenase 1 pathway. Molecular Medicine Reports 12: 4382-4388.

Zhang, C., Zhai, Z., Tang, M., Cheng, Z., Li, T., Wang, H. and Zhu, W.G., 2017a. Quantitative proteome-based systematic identification of SIRT7 substrates. Proteomics 17: 1600395.

Zhang, D.Y., Zhang, C.F., Fu, B.C., Sun, L., Wang, X.Q., Chen, W., Liu, W., Liu, K.Y., Du, G.Q., Ma, C.Y., Jiang, S.L., Li, R.K. and Tian, H., 2018. Sirtuin3 protects aged human mesenchymal stem cells against oxidative stress and enhances efficacy of cell therapy for ischaemic heart diseases. Journal of Cellular and Molecular Medicine 22: 5504-5517.

Zhang, L., Chen, Z., Gong, W., Zou, Y., Xu, F., Chen, L. and Huang, H., 2018a. Paeonol ameliorates diabetic renal fibrosis through promoting the activation of the Nrf2/ARE pathway via up-regulating Sirt1. Frontiers in Pharmacology 9: 512.

Zhang, M., Wu, J., Sun, R., Tao, X., Wang, X., Kang, Q., Wang, H., Zhang, L., Liu, P., Zhang, J. and Xia, Y., 2019. SIRT5 deficiency suppresses mitochondrial ATP production and promotes AMPK activation in response to energy stress. PloS One 14: e0211796.

Zhang, Q., Liu, X., Zhang, J., Yang, J. and Bu, P., 2018b. Sirtuin 3 deficiency aggravates contrast-induced acute kidney injury. Journal of Translational Medicine 16: 313.

Zhang, S., Jiang, S., Wang, H., Di, W., Deng, C., Jin, Z., Yi, W., Xiao, X., Nie, Y. and Yang, Y., 2018c. SIRT6 protects against hepatic ischemia/reperfusion injury by inhibiting apoptosis and autophagy related cell death. Free Radical Biology and Medicine 115: 18-30.

Zhang, T., Chi, Y., Ren, Y., Du, C., Shi, Y. and Li, Y., 2019. Resveratrol reduces oxidative stress and apoptosis in podocytes via Sir2-related enzymes, sirtuins1 (SIRT1)/peroxisome proliferator-activated receptor γ co-activator 1α (PGC-1α) axis. Medical Science Monitor 25: 1220-1231.

Zhang, W., Wei, R., Zhang, L., Tan, Y. and Qian, C., 2017b. Sirtuin 6 protects the brain from cerebral ischemia/reperfusion injury through NRF2 activation. Neuroscience 366: 95-104.

Zhang, Y., Tao, X., Yin, L., Xu, L., Xu, Y., Qi, Y., Han, X., Song, S., Zhao, Y., Lin, Y., Liu, K. and Peng, J., 2017c. Protective effects of dioscin against cisplatin-induced nephrotoxicity via the microRNA-34a/sirtuin 1 signalling pathway. British Journal of Pharmacology 174: 2512-2527.

Zhao, L., Cao, J., Hu, K., He, X., Yun, D., Tong, T. and Han, L., 2019a. Sirtuins and their biological relevance in aging and age-related diseases. Aging and Disease 11(4): 927-945.

Zhao, X., Jin, Y., Li, L., Xu, L., Tang, Z., Qi, Y., Yin, L. and Peng, J., 2019. MicroRNA-128-3p aggravates doxorubicin-induced liver injury by promoting oxidative stress via targeting Sirtuin-1. Pharmacological Research 146: 104276.

Zhou, J., Shi, M., Li, M., Cheng, L., Yang, J. and Huang, X., 2019. Sirtuin 3 inhibition induces mitochondrial stress in tongue cancer by targeting mitochondrial fission and the JNK-Fis1 biological axis. Cell Stress Chaperones 24: 369-383.

Zhou, L., Wang, F., Sun, R., Chen, X., Zhang, M., Xu, Q., Wang, Y., Wang, S., Xiong, Y., Guan, K.L. and Yang, P., 2016. SIRT5 promotes IDH2 desuccinylation and G6PD deglutarylation to enhance cellular antioxidant defense. EMBO Reports 17: 811-822.

Zhou, L., Xu, D.Y., Sha, W.G., Shen, L., Lu, G.Y., Yin, X. and Wang, M.J., 2015. High glucose induces renal tubular epithelial injury via Sirt1/NF-kappaB/microR-29/Keap1 signal pathway. Journal of Translational Medicine 13: 352.

Zhou, Q., Lv, D., Xia, Y., Zhao, Z. and Zou, H., 2018. Decreased expression of sirtuin 3 protein correlates with early stage chronic renal allograft dysfunction in a rat kidney model. Experimental and Therapeutic Medicine 15: 3725-3732.

Zhu, X., Yue, H., Guo, X., Yang, J., Liu, J., Liu, J., Wang, R. and Zhu, W., 2017. The preconditioning of berberine suppresses hydrogen peroxide-induced premature senescence via regulation of sirtuin 1. Oxidative Medicine and Cellular Longevity 2017: 2391820.

Part III.
Nutritional modulation of vitagenes

Diets cures more than the lancet

Chapter 9
Carnitine

Delays are dangerous

9.1 Introduction

L-carnitine (LC), a naturally occurring and widely distributed in nature compound was discovered in 1905 by Russian scientists Gulewitsch and Krimberg (1905) who named the substance from the Latin word for flesh ('carnus'). For the last 50 years this nutrient has received a substantial attention from medical sciences and poultry and pig nutritionists. Main dietary sources of carnitine in poultry/animal nutrition are animal-derived feed ingredients while grains and their by-products are quite poor in carnitine (Borum, 1983). Because of the cereal grains and their by-products represent the major component of poultry and pig diets and endogenous carnitine synthesis depends on many factors, it could well be that chickens and pigs may face carnitine deficiency in some situations such as stress and high performance. For example, chicken embryos have a limited capacity to synthesise LC during incubation (Casillas and Newburgh, 1969) and freshly laid eggs from hens fed diets of plant origin possess low concentrations of LC (Chiodi *et al.*, 1994). There is also a metabolic need for supplemental carnitine in young pigs (Heo *et al.*, 2000a). Furthermore, LC in the maternal diet of breeders affected carcase traits of their progeny, decreasing carcase fat and increased breast meat in progeny fed on high nutrient density diets (Kidd *et al.*, 2005). Therefore, several early studies on pigs, fish, foal, quail and broiler chickens demonstrate a growth improvement and other beneficial effects by feeding extra dietary LC (Szilágyi, 1998). In a recent review it was concluded that in poultry production, LC has a multi-functional purpose, which includes growth promotion, strengthening the immune system and antioxidant effects (Golzar Adabi, *et al.*, 2011).

Therefore, the aim of this chapter is a critical review of recent data related to antioxidant action of carnitine *in vitro* and *in vivo* as a possible mechanism of its protective action against various stresses in animal and poultry production. Vitagene-regulating and immunomodulating properties of carnitine will also be analysed in details.

9.2 Absorption and metabolism of carnitine

Carnitine absorption and metabolism have been reviewed extensively elsewhere (Rebouche, 2004; Evans and Fornazini, 2003) and can be summarised as follows. Dietary LC is absorbed by active and passive transfer across enterocyte membranes. Bioavailability of dietary LC is about 54-87%, depending on the amount of L-carnitine in the diet. However, bioavailability of LC dietary supplements is substantially lower,

DOI 10.3920/978-90-8686-906-0_9, © Wageningen Academic Publishers 2020

comprising about 14-18% of dose (Pochini *et al.*, 2013). It seems likely that there are species-specific differences in carnitine assimilation from the diet. For example, in supplemented pigs (500 mg/kg diet), LC absorption and degradation in the intestinal tract was estimated at 30-40% and 60-70% of LC respectively (Heo *et al.*, 2000b). In more recent publication it has been shown that young pigs have a high capacity to absorb carnitine from the diet and plasma and tissue carnitine concentrations in young pigs can be markedly increased by supplementation of carnitine (Fischer *et al.*, 2009). Indeed, absorption rate of the supplemented carnitine in the small intestine was greater than 95% for the lower doses (25, 50, 100 mg/kg) and greater than 90% for the higher doses (200, 500, 1000 mg/kg). Furthermore, dietary supplementation of carnitine caused a dose-dependent increase of free carnitine, acetyl carnitine (ALC) and total carnitine concentrations in plasma, liver, kidney, heart and skeletal muscle. In fact, at the highest dose of 1000 mg/kg, plasma and tissue total carnitine concentrations were 3- to 6-fold higher than in the unsupplemented control group (Fischer *et al.*, 2009). In pigs, LC salts have a similar bioavailability to that of free LC while LC esters have a lower one (Eder *et al.*, 2005).

Free LC, absorbed from the diet or synthesised in the liver and kidney is then taken up from the blood stream by other tissues through active carrier-mediated transport systems which maintain high tissue/plasma concentration ratios (Pochini *et al.*, 2013). Delivery of carnitine into cells, its distribution within the body and intracellular homeostasis, including LC reabsorption in the kidney, are controlled by organic cation transporters (OCTN; Ringseis and Eder, 2009). In particular, active sodium-dependent high affinity OCTN2 transporter is responsible for LC transport in the kidney and other tissues (Tamai *et al.*, 1998) and OCTN2 is regulated by PPARα indicating that lipid also may affect LC transport into tissues (Wen *et al.*, 2010). Recent comparative analysis indicates that the role of PPAR*α* as a regulator of carnitine homeostasis is well conserved across different species, including rat, mouse, pig, cattle, chicken, and human (Ringseis *et al.*, 2012). There are species-specific differences in OCTN2 expression. In particular, in humans, OCTN2 is expressed strongly in kidney but only weakly in other tissues (Tamai *et al.*, 1998), whereas in rats OCTN2 is highly expressed in kidney and testis and to a lesser extent in liver (Spaniol *et al.*, 2001). It is interesting to note that non-proliferating species are able to cover their increased demand for carnitine during fasting by an increased LC synthesis and uptake into cells. Indeed, fasting increases the activity of gamma-butyrobetaine dioxygenase (BBD) in liver and kidney and up-regulates the expression of OCTN2 in various tissues of pigs, probably mediated by PPARα activation (Ringseis *et al.*, 2009). It was shown that treatment with a PPARα agonist causes an upregulation of OCTN2 in liver, muscle and enterocytes from small intestine of pigs with following increases carnitine concentrations in liver and muscle probably by enhancing carnitine uptake into cells (Ringseis *et al.*, 2008a). Chicken enterocytes maintain a steady-state LC gradient of 5 to 1 and 90% of the transported LC remains in a readily diffusive form. In chicken brush-border membrane LC transport is Na+-, membrane voltage- and pH-dependent and has a high affinity for LC (km 26-31 μM) (Duran *et al.*, 2002). Therefore, the aforementioned transporter has properties similar to those of OCTN2.

LC and its short-chain esters do not bind to plasma proteins and excess carnitine is excreted via the kidneys (Rebouche, 2004). LC is eliminated from the body mainly via urinary excretion. Dietary sources of LC for human are mainly meat products, while plant foods contain considerably lower carnitine levels. For example, carnitine content in ground beef is shown to be 582 mg/100 g, while chicken, macaroni and corn flakes contain only 24, 0.13 and 0.013 mg carnitine per 100 g (Krahenbuhl, 2000). Carnitine contents in grains (wheat, barley, corn, etc.) is about 5-7 mg/kg, while in soya meal and sunflower meal carnitine concentrations are 15 and 5 mg/kg respectively (Arslan, 2006). Since carnitine is present primarily in foods of animal origin, farm animal and poultry commercial diets formulated mainly from plant ingredients could be deficient in this element. In general, chicken diet is shown to contain 17.8-22.9 mg carnitine/kg (Mast *et al.*, 2000).

In animals and man, LC is synthesised mainly in the liver with following transport to other tissues. For example, in pigs liver and kidney are shown to be the only tissues with a considerable activity of γ-butyrobetaine dioxygenase (BBD), the last enzyme of carnitine synthesis (Fischer *et al.*, 2009a). Therefore, in pigs, like in humans, these tissues might be regarded as the major sites of carnitine synthesis. It is interesting to note that about 98% of the carnitine body pool is localised in tissues that utilise fatty acids as their primary dietary fuel namely skeletal and heart muscle. It is well-known that synthesis of LC requires the essential amino acids lysine and methionine as well as such micronutrients as iron, ascorbic acid, vitamin B6 and niacin. Therefore, an imbalanced diet and various stresses can create a need for external LC supplementation (Walter, 2000). It should be taken into account that carnitine synthesis is a reasonably slow process and does not readily keep up with fast changes of the metabolic requirements in stress conditions (Bohles, 2000). In fact, only about 25% of carnitine is shown to come from its *de novo* synthesis while about 75% of carnitine is coming from the diet (Lohninger *et al.*, 2005). There are also species-specific differences in rate of carnitine synthesis with pigs having a lower rate of carnitine synthesis in tissues than humans (Fischer *et al.*, 2009a). In fact, carnitine biosynthesis in pigs fed diets without LC supplementation were estimated at 6.71 and 10.63 μmol/kg/day in low protein and high protein groups respectively (Heo *et al.*, 2000b). Kinetics of carnitine palmitoyltransferase-I, a major carnitine-dependent regulatory enzyme of lipid metabolism, required for the transport of long-chain fatty acids across the inner mitochondria membrane, were shown to be altered by dietary variables and suggest a metabolic need for supplemental carnitine in young pigs (Heo *et al.*, 2000a).

In mammalian cells and body fluids, carnitine is known to be present either as free carnitine, short-chain acyl carnitine, or long-chain acylcarnitine. In particular, acylcarnitines are shown to account for about 20% of total carnitine in serum and 10-15% in muscle and liver (Ferrari *et al.*, 1992). For example, acyl and free carnitine concentrations in human serum are, respectively, 12.8±7.4 and 67±21.8 μmol/l (Rubio *et al.*, 1998) and circulating carnitine accounts for only about 0.5% of body carnitine (Stanley, 2004). Mammalian tissues contain relatively high amounts of LC, ranging between low μM to low mM, with the highest concentrations in heart and skeletal muscles. Indeed, the carnitine concentrations are ≈1 mM in rat skeletal muscle, ≈3 mM in human muscle and may be up to 15 mM in ruminant muscle (Bremer, 1983). For example, the concentration of LC in fresh semitendinosus muscle from broiler chicken, pig, beef cattle, deer, horse and goat muscle were 0.69, 1.09, 1.86-3.57, 4.57, 4.95 and 11.36 μmol/g wet weight, respectively (Shimada *et al.*, 2004). In general, carnitine concentrations in tissues of pigs are generally markedly lower than those reported for humans (Fischer *et al.*, 2009a,b). For example, free carnitine levels in pig plasma at day 28 are shown to be 18.4 μmol/l and increased up to 37.7 μmol/l after 20 days LC dietary supplementation at 400 mg/day (Lösel and Rehfeldt, 2013). Plasma carnitine concentration is comparable with those of other antioxidants, including vitamin E (20-30 μM) and ascorbic acid (26.1-84.6 μM (Hu, 2011). In piglets at birth carnitine concentrations in liver, kidney, heart and muscles were (μmol/g) 200±22, 123±27, 316±22 and 284-333, respectively, while in plasma carnitine concentration (μmol/l) was 21.8±3.6 (Fischer *et al.*, 2009a). The authors showed that concentrations of total carnitine in plasma, liver and kidney were highest at birth and thereafter declined until an age of 4 weeks. In contrast, carnitine concentrations in heart and skeletal muscle rose from birth until an age of 3-4 weeks and thereafter declined. In 10-week old pigs total carnitine concentrations (nmol/g) in liver, kidney, skeletal muscle and heart were 33.7±1.6; 110±3; 628±25; 329±12, respectively, while in plasms carnitine concentration was 5.91±0.4 μmol/l (Ringseis *et al.*, 2009). A progressive increase in carnitine levels in the liver is seen from birth to 24 h in fed piglets (from 0.14 up to 0.33 mM), the level at 24 h being equal to that of a 3 week-old animal and double that of a 24h-old fasted animal (Bieber *et al.*, 1973). Heart, kidney, and skeletal muscle also contained appreciable levels (0.05-0.2 mM) of carnitine at birth. It is interesting to note that LC concentration in red muscle were significantly higher than those in white muscle suggesting that LC concentration in muscle is related to oxygen metabolism and to myofiber types (Shimada *et al.*, 2004). Indeed, carnitine concentrations in skeletal muscle of pigs, being around 600 nmol/g (0.6 mM), are comparable with those of rat muscle but they are markedly lower than those of human muscle. Furthermore, concentrations of carnitine in other tissues such as liver, kidney or brain of pigs are also three- to five-fold lower than those of the respective human tissues suggesting that pigs have generally a lower carnitine status than humans (Fischer *et al.*, 2009a). A pig weighing 100 kg, for example, has a LC pool of about 24 g with about 85% of this is present in muscle (20.4 g), about 8% in the gastrointestinal tract (1.92 g), 3.5% in the liver (0.84 g) and only 0.3% in the blood (0.07 g; Harmeyer, 2002). Piglets of LC treated sows (125 mg LC/day during pregnancy and 250 mg LC/day during lactation) had higher concentrations of LC in plasma and carcass at birth and on days 10 and 20 of age than control piglets (Birkenfeld *et al.*, 2006). It is interesting to note that that LC supplementation improves the growth performance in light piglets of primiparous sows (Birkenfeld *et al.*, 2005).

There is an intense exchange between the plasma and tissue carnitine pools. In fact, in tissues, carnitine can be acylated, transported back into plasma as acylcarnitine and eventually be excreted via the urine (Morand *et al.*, 2014). Acetyl-L-carnitine (ALC), the short-chain ester of carnitine, is endogenously produced within mitochondria and peroxisomes and is involved in the transport of acetyl-moieties across the membranes of these organelles. It was shown that an increase in the plasma carnitine pool, without changing LC concentration in muscles, may be sufficient to influence certain metabolic pathways in skeletal muscle (Morand *et al.*, 2014). This is a very important finding explaining protective effects of LC in various stress conditions, including recovery after intense exercise (Volec *et al.*, 2002). Indeed, carnitine-supplemented exercise-stressed animals could achieve high physical performance with comparatively low metabolic perturbance (Morand *et al.*, 2014).

Concentrations of LC and ALC change under altered dietary conditions and ALC is the most extensively investigated derivative of the carnitine formulations (Calabrese *et al.*, 2009). It seems likely that there are species-specific differences in carnitine assimilation from the diet. For example, in contrast to reports in humans, ALC and propionyl-L-carnitine (PLC) showed no significant bioavailability as parent substance in mice (Morand *et al.*, 2014). Indeed, both acylcarnitines were shown to be hydrolysed before reaching the systemic circulation. Similarly, in pigs most orally administered acetylcarnitine is hydrolysed before reaching the systemic circulation (Eder *et al.*, 2005). Comparison of data from various studies with mice, rats, pigs, cows, and laying hens and from human studies shows that carnitine homeostasis is well conserved across different species (Ringseis *et al.*, 2012).

First attempts to determine carnitine in chicken egg showed its level comprising <3 μg/g dry tissue (approximately 0.678 ug/g wet tissue or 4.2 nmol/g) (Fraenkel, 1953) and it was suggested that carnitine is synthesised in sizeable quantities in the growing chick embryo. Similar carnitine concentrations were reported by Chiodi *et al.* (1994). Indeed, in the egg yolk LC concentration was shown to be 4.64 nmol/g, while ALC was 3 times lower comprising 1.32 nmol/g. Furthermore, LC concentration in the egg albumin was extremely low – 0.029 nmol/g, while ALC was about three-fold higher (0.102 nmol/g) (Chiodi *et al.*, 1994). It is interesting to note that at day 2 of embryonic development the carnitine values found in the chick embryos were represented by 12% free carnitine, 29% ALC and 59% long chain acylcarnitines. There is an opportunity to manipulate LC level in eggs. Indeed, LC dietary supplementation (125 mg/kg diet) was associated with a significant (approximately by 30%; from about 13 nM/g wet yolk to about 17 nM/g wet yolk. e.g. from 0.012 up to 0.017 mM) increase carnitine concentration in egg yolk (Zhai *et al.*, 2008). Supplementation of the diets of young broiler breeder hens with 25 mg/kg of LC has been shown to increase LC concentrations in the yolk sacs and livers of 18-d broiler embryos and to subsequently influence yolk sac fatty acid β-oxidation (Peebles *et al.*, 2007). Carnitine in eggs and embryonic tissues could affect chicken embryonic development and have some epigenetic effects such as a decrease in abdominal fat in the progeny (Peebles *et al.*, 2007) and decreased carcass fat and increased breast meat in those progeny fed high nutrient density diets (Kidd *et al.*, 2005). Indeed, such a manipulation of LC level in egg represents an important model for further studies of health-promoting properties of carnitine in poultry production. In fact, free carnitine concentration in embryonic heart, brain and liver was found to be in a range of 2.98-5.84 μmol/g dry wt. and did not change between days 10 and 21 of the development (Casillas and Newburgh, 1969). The authors also showed that ALC was not detected in any organ until the 17th day of incubation and represented about 20% of the total carnitine in each organ on the day of hatch, while concentrations of long-chain acylcarnitine generally represented from 5 to 10% of the total carnitine in each case. In the embryonic liver total carnitine concentration was about 0.35 μmol/g dry wt at days 11 and 18 and decreased down to about 0.2 μmol/g dry wt at time of hatching and increased posthatch up to 0.75 μmol/g dry wt at day 180 (Rinaudo *et al.*, 1991). In terms of carnitine concentration on wet tissue basis it would represent about 50 nmol/g wet tissue at day 11, increasing up to 230 at day 18, decreasing down to 104 at hatch and increasing again up to 240 nmol/g wet tissues. It is interesting to underline that in muscles total carnitine concentration increased from about 0.6 up to 4 μmol/g dry wt (from about 120 up to 900 nmol/g wet tissue). In the heart total carnitine concentration at time of thatching and at day 180 posthatch was about 90 and 235 nmol/g wet tissue, respectively, while in the brain it was about 66 and 57 nmol/g wet tissue (Rinaudo *et al.*, 1991). In another study, carnitine concentration (nmol/mg non-collagenous protein) in chicken embryonic heart increased from about 1 up to 1.5 between days 7 and 17 with the following sharp decrease down to about 0.5 at time of hatching (Kargas *et al.*, 1985).

As LC is not regarded as an essential nutrient, no values for dietary reference intake or recommended daily allowance have been set. Main carnitine function in the body include (Vaz and Wanders, 2002): (1) transport of activated long-chain fatty acids from the cytosol to the mitochondrial matrix, where β-oxidation takes place; (2) transfer of the products of peroxisomal β-oxidation, including acetyl-CoA, to the mitochondria for oxidation to CO_2 and H_2O in the Krebs cycle; (3) modulation of the acyl-CoA/CoA ratio; (4) storage of energy as acetyl-carnitine; (5) modulation of toxic effects of poorly metabolised acyl groups by excreting them as carnitine esters. Other established functions of LC include the preservation of membrane integrity and mitochondria functions as well as apoptosis inhibition (Karlic and Lohninger, 2004). Indeed, recent evidence suggests that LC and ALC supplementation is associated with a modulation of the antioxidant systems responsible for attenuation of oxidative stress, and mitochondrial dysfunction associated to various pathological conditions (Ribas *et al.*, 2014). The molecular mechanisms accounting for the positive effect of LC on many physiological parameters in farm animals and poultry are not yet fully understood, but many biological functions (Figure 9.1) and protective effects of LC reported in literature have been postulated to be related to its antioxidant action (Surai, 2015a,b,c).

9.3 Antioxidant action of carnitine

The cytoprotective effects of carnitine are believed to be associated with a decrease in oxidative stress. There are several possible mechanisms of antioxidant action of carnitine. Firstly, carnitine can directly scavenge free radicals. Secondly, and more importantly, LC can decrease free radical formation by inhibiting specific enzymes responsible for free radical production, or by maintaining the integrity of mitochondria and especially electron-transport chain of mitochondria in stress conditions. Thirdly, carnitine can chelate transition metals (Fe^{2+} and Cu^{+}), preventing

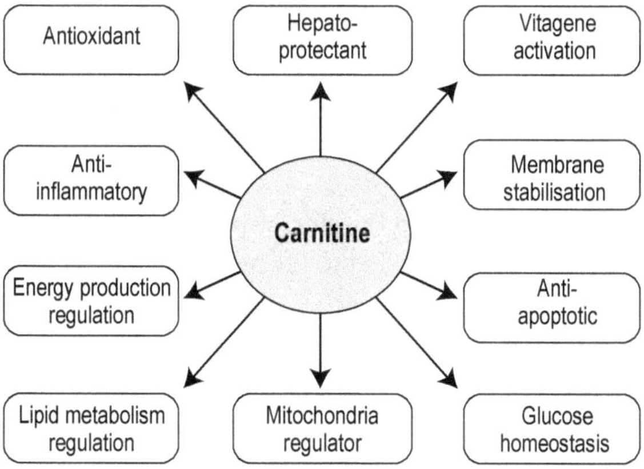

Figure 9.1. Some important function of carnitine in biological systems (adapted from Surai, 2015a,b,c).

their participation in ROS formation via Fenton reaction. Fourthly, carnitine can have a sparing effect on major membrane antioxidant, namely vitamin E. Fifthly, LC can participate in the maintenance of optimal redox status and redox signalling in the cell by activating a range of antioxidant enzymes and non-enzymatic antioxidants, mainly via transcription factors, including Nrf2 and NF-κB. Finally, carnitine can activate an array of vitagenes, responsible for the synthesis of protective molecules, including HSP, thioredoxin (Trx), SIRTs, SOD, etc., and providing additional protection in stress conditions (Surai, 2015a,b,c).

9.3.1 Direct free radical scavenging

When considering relevance *in vitro* studies of antioxidant properties of various compounds, it is necessary to take into account if effective antioxidant concentrations are achievable in blood and target tissues. Furthermore, the location of the compound within the cell (cytosol vs membranes) should also be considered. In fact, the early (1990s) studies in this area did not provide a convincing evidence on direct AO properties of carnitine. However, later publications (2006-2011) presented data indicating a possibility of the direct AO action of carnitine and its derivatives in biological tissues (for review and refences see Surai, 2005b). In could also well be that LC is an important element of the antioxidant defences in the gut (Surai and Fisinin, 2015; Surai, 2018), since LC concentration there could be quite high. Indeed, the antioxidant properties of carnitine in the gut is difficult to overestimate and clearly there is a need for more research in this area.

9.3.2 Chelating properties of carnitine

It is necessary to take into account that iron and copper are powerful promoters of free radical reactions and therefore their availability in 'catalytic' forms is carefully regulated *in vivo* (Halliwell, 2012). Indeed, reactions of Fe^{2+} and Cu^{+} with H_2O_2 are a source of the most powerful hydroxyl radical (*OH) in the Fenton reaction:

$$H_2O_2 + Fe^{2+}/Cu^{+} \longrightarrow {}^{*}OH + OH^{-} + Fe^{3+}/Cu^{2+}$$

Transition metal ions also accelerate the decomposition of lipid hydroperoxides into cytotoxic products such as aldehydes, alkoxyl radicals and peroxyl radicals:

$$LOOH + Fe^{2+} \longrightarrow LO^{*} + Fe^{3+} + OH^{-}$$

$$LOOH + Fe^{3+} \longrightarrow LOO^{*} + Fe^{2+} + H^{+}$$

Therefore, organisms have evolved to keep transition metal ions safely sequestered in storage or transport proteins. In this way, the metal-binding proteins prevent formation of hydroxyl radical by preventing them from participation in radical reactions. In addition, meatal chelating is another process involved in prevention of Fe and Cu from participation in ROS formation. Therefore, metal chelating properties of carnitine (Surai, 2015b) could contribute to its antioxidant action. Indeed, Reznick *et al.* (1992) showed that propionyl L-carnitine (PLA) at quite high concentration (75 mM) suppressed hydroxyl radical production in the Fenton system, probably by chelating the iron required for the generation of hydroxyl radicals. The author suggested that decrease in hydroxyl radical generation was due to chelating ability of PLC. However, in the same system, deferoxamine, a well-known chelating agent, was effective at 0.1 mM concentration indicating that propionyl carnitine is mild iron chelator. Furthermore, in the systems based on iron-induced ascorbate oxidation only very modest chelating ability of PLC was seen at 10 mM concentration and no chelating by LC even at 100 mM was observed. In the same system deferoxamine was much more effective at 0.1 mM (Reznick *et al.*, 1992). It is interesting to note that 14 years later quite strong chelating ability of LC was shown by using spectrophotometric assay with ferrozine which can quantitatively form complexes with Fe^{2+} (Gulcin, 2006). Therefore, in the presence of chelating agents, the complex formation is disrupted, resulting in a decrease in the red colour of the complex. In particular, it was shown that the metal chelating effects of LC was concentration-dependent between 5 to 30 μg/ml (approximately 0.03-0.19 mM). It is very important to note that the chelating ability of LC was comparable with that of EDTA, since LC exhibited 98.9% chelation of ferrous ion at 30 mg/ml (0.19 mM) concentration, while the percentages of metal chelating capacity of 30 μg/ml (approx. 0.1 mM) EDTA was found to be 80.7%. The chelating effect of LC is suggested to connect with the complex formation between the hydroxyl and carboxylate groups of LC molecule and metal ion (Gulcin, 2006). Indeed, the discrepancy of chelating ability of LC obtained by Gulcin (2006) and Reznick *et al.*, (1992) can be explained by different methodological approaches but, clearly, this issue needs further investigation using modern analytical techniques applied to various biological systems. It sems likely that dietary carnitine can have chelating properties in the chicken gut providing additional protection against RONS.

9.3.3 Protective effects of carnitine on mitochondria

Mitochondria are the primary cellular consumers of oxygen and contain numerous redox enzymes capable of transferring single electrons to oxygen, generating the ROS superoxide (O_2^-). It is well appreciated that mitochondrial enzymes known to generate ROS include the tricarboxylic acid cycle enzymes aconitase and α-ketoglutarate dehydrogenase; the electron-transport chain (ETC) complexes I, II and III; pyruvate dehydrogenase and glycerol-3-phosphate dehydrogenase; dihydroorotate dehydrogenase; the monoamine oxidases A and B; and cytochrome *b*5 reductase (Lin *et al.*, 2006). Furthermore, mitochondrial insults, including oxidative damage itself, can cause an imbalance between ROS production and removal, resulting in net ROS production. For example, ROS, which is an inevitable by-product of oxidative phosphorylation, induce protein modifications, lipid peroxidation and mitochondrial DNA damage, which ultimately results in mitochondrial dysfunction (Sekine and Ichijo, 2015). In fact, mitochondria are considered to be the main source of ROS in biological systems (Surai, 2018).

One of the mechanisms responsible for the decrease in oxidative stress is the protective effect of carnitine on mitochondrial structure and function. Indeed, our recent review on protective effects of carnitine on mitochondria (Surai, 2015a,b) indicates that in the case of oxidative stress carnitine protects/repairs mitochondria by triggering pro-survival cell signalling. Indeed, LC and its derivatives can affect expression of a range genes responsible for a synthesis of various proteins in the cell. For example, in rat liver mitochondria a significant age-related change occurred in 31 proteins involved in several metabolisms and ALC supplementation (1.5% solution *ad libitum*) altered the levels of 26 proteins. In particular, ALC reversed the age-related alterations of 10 mitochondrial proteins relative to mitochondrial cristae morphology, to the oxidative phosphorylation and antioxidant systems, to urea cycle, to purine biosynthesis (Musicco *et al.*, 2011). Indeed, in addition to stimulation of β-oxidation by increasing the mRNA expression of carnitine palmitoyltransferase 1A and preventing free fatty acid induced oxidative stress (Jun *et al*, 2011), carnitine can also prevent and/or ameliorate oxidative mitochondrial dysfunction. In the aged rat heart *in vivo* administration of ALC provides acetyl groups for protein acetylation and affects the amount of mitochondrial proteins (Kerner *et al.*, 2015). Therefore, LC can protect against mitochondrial dysfunctions associated with oxidative stress caused by a series of conditions such as aging, ischemia reperfusion, inflammation, degenerative diseases and drug toxicity, *in vivo* or *in vitro.* Indeed, carnitine can be considered as a mitochondria-specific antioxidant, responsible for mitochondria integrity maintenance and regulation of ROS production and ROS signalling. Furthermore, Nrf2 is considered to be an important player in supporting the structural and functional integrity of the mitochondria, and this role is particularly crucial under conditions of stress (Dinkova-Kostova and Abramov, 2015). Indeed, there is a crosstalk between mitochondria and the Nrf2 system, including the interaction between Nrf2-mediated antioxidant mechanisms and the regulation of mitochondrial integrity (Itoh *et al.*, 2015), associated with removal of damaged mitochondria and the biogenesis of healthy mitochondria. It could well be that carnitine can participate in this crosstalk.

9.3.4 Inhibition of free-radical generating enzymes by carnitine

Our recent reviews (Surai, 2015a,b) clearly showed that LC can inhibit main ROS-producing enzymes, namely xanthine oxidase (XO) and NADPH oxidase and in this way contributing to improved antioxidant defences. Interestingly, LC was also shown to decrease activity of myeloperoxidase, another RONS-producing enzyme (Li *et al.*, 2019).

9.3.5 *In vitro* antioxidant effects of carnitine

Clear evidence of protective effect of carnitine against oxidative stress caused by various chemicals came from *in vitro* studies with cell culture or isolated cells or organelles. This includes human dermal fibroblasts, cerebellar granular cell culture, LDL, human hepatocytes, cultured porcine oocytes, neuroblastoma cells, and neurones of newly born rats. LC was also able to decrease DNA damage caused by toxicants in various experimental systems. Effective LC concentrations showing antioxidant protective effects were within the physiological range of LC concentrations varying from 9 μM up to 1-10 mM (for review and references see Surai, 2015a,b). Indeed, the aforementioned data confirmed antioxidant action of carnitine in physiologically relevant concentrations in various *in vitro* systems.

9.3.6 Antioxidant effects of carnitine against oxidative stress *in vivo*: protection against toxicants

Protective effect of LC and its derivatives on the antioxidant systems of the body are shown in various models of oxidative stress/toxicity caused by CCl_4, cisplantin, ethanol, 3-Nitropropionic acid, valproate, diethylnitrosamine, doxorubicin, adriamycin, tamoxifen, indomethacin, acetaminophen, Cd, aflatoxin, thioacetamide, methotrexate. LC and its derivatives were effective in decreasing oxidative stress caused by neurotoxic agents such as glutamate, quinolinic acid or 3-nitropropionic acid, rotenone, dexamethasone, aminoglycosides, scopolamine, 1-methyl-4-phenyl-1,2,3,6-tetrahydropyridine, silver nano-particles (for review and references see Surai, 2015a,b).

Carnitine was also effective in prevention of oxidative stress in the L-buthionine-sulfoximine-induced cataract model, selenite-induced cataractogenesis, LPS-treated rats, high dietary copper in laying hens, high fat diet in rats, fructose-fed rats, experimental glaucoma, acetic acid-induced colitis, atherosclerotic rats. Furthermore, carnitine showed clear antioxidant protective activities in caerulein-induced acute pancreatitis, adjuvant arthritis and in hyperthyroid rats as well as in a mouse model of non-alcoholic steatohepatitis, in mice with familial amyotrophic lateral sclerosis, streptozotocin-induced diabetes in rats, Fanconi syndrome, chemically induced model of maple syrup urine disease and unilateral ureteral obstruction-induced oxidative stress (for review and references see Surai, 2015a,b).

9.3.7 Antioxidant effects of carnitine against oxidative stress *in vivo*: established clinical models

The possibility of modulating with natural antioxidants the progression of various diseases with ROS species-related pathogenesis has been widely discussed in the literature. It has been established that nutritional antioxidants play an important role in decreasing detrimental consequences of the oxidative stress-related pathologies. Therefore, several lines of evidence from animal experiments and clinical studies exist indicating that LC supplementation is effective in preventing oxidative stress under various pathological conditions. In particular, there is a strong evidence to confirm antioxidant-related protective effects of carnitine in various clinical models, including hypoxia, ischemia-reperfusion, ionising radiation, spontaneously hypertensive rats, exercise renal failure with drug-induced nephrotoxicity and ageing model (Surai *et al.*, 2015a,b; Figure 9.2).

9.3.8 Antioxidant effects of carnitine against oxidative stress *in vivo*: patients with various diseases

Promising antioxidant activities of carnitine have been found following supplementation in patients with cardiovascular diseases such as peripheral arterial disease, chronic heart failure, or stable angina, coronary artery disease patients, ischemic cardiomyopathy, renal disease with haemodialysis, multiple sclerosis, mild cognitive impairment and mild Alzheimer's disease, age related macular degeneration, disorders of propionate metabolism, after major surgery, alcoholic steatohepatitis, phenylketonuria and other patients (for review and references see Surai, 2015a,b,c). In a recent meta-analysis 74 reports testing carnitine and its derivatives were considered, including 18 trials related to kidney disease with success ratio 0.58. Furthermore, trails with other diseases were more successful with success ratio to be 1, including

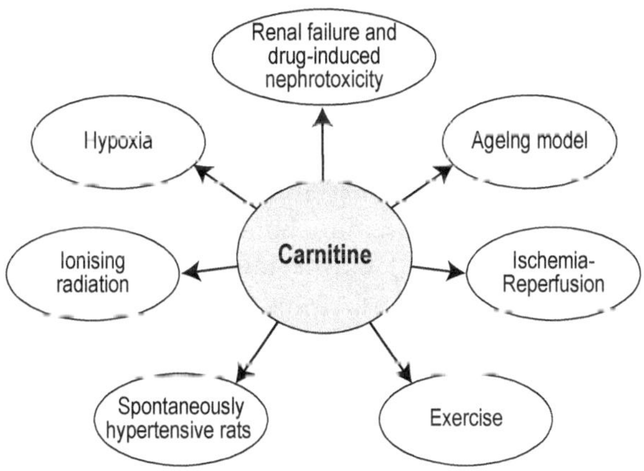

Figure 9.2. Antioxidant effects of carnitine against oxidative stress *in vivo* in established clinical models (adapted from Surai, 2015a,b).

13 reports on diabetes trials, 9 on heart and vessel disease, 6 on liver disease, 9 on neurological diseases and 6 on genetic diseases (Pagano *et al.*, 2014). It is also interesting to mention that carnitine supplementation can be beneficial in healthy subjects as well. For example, single dose administration of LC (2.0 g, per os) was shown to improve antioxidant activities in healthy subjects. Indeed, there was a gradual increase in plasma concentrations of SOD, GPx, CAT and total antioxidative capacity (T-AOC) in the first 3.5 h following LC administration. Furthermore, a positive correlation was found between LC concentration and the antioxidant index of SOD (r=0.992, $P<0.01$), GPx (r=0.932, $P<0.01$), catalase (r=0.972, $P<0.01$) or T-AOC (r=0.934, $P<0.01$; Cao *et al.*, 2011). The beneficial effects of carnitine derivatives on progressive motility, vitality and sperm DNA fragmentation were demonstrated in men with idiopathic oligoasthenozoospermia (Micic *et al.*, 2019). The nutraceutical value of carnitine and its use in dietary supplements has been recently reviewed (Durazzo *et al.*, 2020).

9.4 Carnitine and Nrf2 regulation

The recent evidences indicate that main protective effects of LC and ALC were associated with preservation or increased activity of antioxidant enzymes and GSH in various tissues affected by stress conditions. The mechanisms involved in the regulation of antioxidant enzymes by LC *in vivo* have not been precisely determined yet. However, it seems likely that transcription factors, including Nrf2, are involved in this regulation. First, it was shown *in vitro* that treatment of astrocytes with ALC (30-100 μM) induces HO-1 in a dose- and time-dependent manner and that this effect was associated with up-regulation of HSP60 as well as high expression of the redox-sensitive transcription factor Nrf2 in the nuclear fraction of treated cells (Calabrese *et al.*, 2006). Adipocytes cultured in the presence of ALC (0.1, 1 and 10 μM) for 24 h were characterised by increased expression of Nrf2 (Shen *et al.*, 2008). However, the treatments with lipoic acid (LA) or ALC alone at the same concentrations showed little effect on mitochondrial function and Nrf2 expression. Nrf2 regulated augmented antioxidant response on administration of ALC was shown to be a crucial factor in ameliorating hypoxia-induced neurodegeneration and memory impairment (Barhwal *et al.*, 2009). Indeed, a decrease in free radical generation, lipid peroxidation and protein oxidation was also observed along with a concomitant increase in Trx and GSH levels on administration of ALC during exposure to hypobaric hypoxia. It was also demonstrated that administration of ALC to hypoxic rats effectively protected hippocampal neurons from mitochondrial dysfunction, excitotoxicity, and neurodegeneration (Hota *et al.*, 2012). Furthermore, ALC caused increased expression of Nrf2 in the nuclear fraction of rats with a concomitant decrease in expression of the protein in the cytosolic fraction. In addition, ALC administration resulted in PPAR-γ coactivator-1α and nuclear respiratory factor-1-induced mitochondrial biogenesis, the expression of which was regulated by an extracellular-related kinase-nuclear factor erythroid 2-related factor 2 (ERK-Nrf2)-mediated mechanism. Indeed, ALC administered hypoxic rats showed increased DNA-binding ability of Nrf2 along with upregulation of Nrf1. The authors provided evidence for Nrf2-mediated regulation of mitochondrial biogenesis through Nrf1 following ALC supplementation (Hota *et al.*, 2012). Administration of LC to L-NAME-induced hypertensive rats prevented decrease in Nrf2 expression in the renal cortex (Zambrano *et al.*, 2013). Indeed, it was shown that LC can also significantly protect ischemia-reperfusion injury due to the overexpression of HO-1 induced by activated Keapl-Nrf2-ARE signalling pathway (Li *et al.*, 2012). Furthermore, ALC administration to human lens epithelial cells treated with homocysteine, restored (increased) the levels of antioxidant proteins, including SOD, GPx, Catalase, Nrf2, Keap1 and GSH (Yang *et al.*, 2015). In high glucose stimulated retinal ganglion cells (RGCs), LC treatment was shown to be associated with an increased level of Nrf2 (Figure 9.3), HO-1, γ-GCS and with a reduced expression of Keap1 protein (Cao *et al.*, 2015).

L-carnitine was shown to protect HL7702 cells against H_2O_2-induced cell damage through Akt-mediated activation of Nrf2 signalling pathway (Li *et al.*, 2016). Furthermore, LC was reported to attenuate cardiac function loss by inhibiting ROS production by activating p38 MAPK/Nrf2 signalling in hearts exposed to irradiation (Fan *et al.*, 2017). The protective effects of L-carnitine on myocardial ischaemia-reperfusion injury in patients with rheumatic valvular heart disease undergoing surgery was shown to be associated with the suppression of NF-κB pathway and the activation of Nrf2 pathway (Li *et al.*, 2019). In FHM cells, silencing Nrf2 (Nrf2-siRNApretreatment) was found to weaken the protective effect AO effects of L-carnitine (Zhang *et al.*, 2019) confirming a regulatory role of LC on Nrf2 expression and activity. Furthermore, in fish under oxidative stress imposed by oxidised fish oil consumption LC was shown to have protective effect on Nrf2: ameliorated increased expression of Keap1 and prevented stress-related reduction in expression of Nrf2 and HO-1 (Zhang *et al.*, 2019). Inclusion of LC (5 mM) into the incubation medium for human hepatoma HepG2 cells supplemented with fructose was shown to reduce ROS production and simultaneously increased protein content of SOD2 and Nrf2 (Montesano *et al.*, 2020). Similarly, LC was shown to increase the activity of the Nrf2/

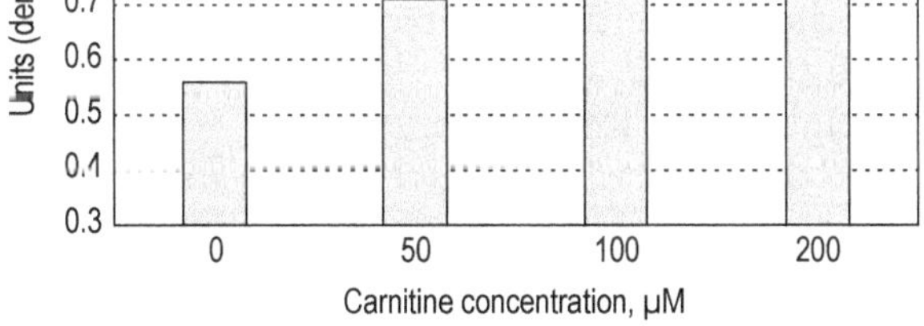

Figure 9.3. Effect of carnitine on nuclear factor erythroid-2 related factor 2 (Nrf2) expression in the retinal ganglion cells (adapted from Cao *et al.*, 2015).

HO-1 signalling pathway in H9c2 cells. while inhibition of the Nrf2/HO-1 attenuated the protective effect of LC on the cells (Zhao *et al.,* 2020).

9.5 Carnitine and NF-κB regulation

Initially, it was shown that ALC selectively induces the expression of metabotropic glutamate receptor 2 by acting as a donor of acetyl groups, hyperacetylating p65/RelA and thus changing the activity of the NF-κB family of transcription factors (Chiechio *et al.*, 2006). Indeed, the *in vitro* pro-neurogenic effects of ALC appear to be mediated by affecting the NF-κB pathway and in particular by p65 acetylation, and subsequent NF-κB-mediated upregulation of metabotropic glutamate receptor 2 (mGlu2) expression (Cuccurazzu *et al.*, 2013). *In vivo*, PLC treatment for 15 days after injury resulted in a reduction of relative rat aortic intimal volume, and a reduction of NF-κB (Orlandi *et al.*, 2007). The authors also showed that the PLC-induced attenuation of NF-κB activity in intimal cells was due to the increase of IκB-α bioavailability, as the result of a parallel induction of IκB-α synthesis and reduction of phosphorylation and degradation. LC (8.3-13.1 mM) was found to significantly inhibit LPS-induced transactivation of NF-κB in LPS-stimulated macrophage cells (Koc *et al.*, 2011). It has been shown that the molecular regulation of antioxidant enzymes through an inhibition of the renin-angiotensin system and a modulation of the NF-κB/IκB system are major mechanisms of the protective antioxidant effect of carnitine (Miguel-Carrasco *et al.*, 2010). A decrease in the expression of transcription factors Nrf2 and PPARα, together with an increase in NF-κB expression, was observed in the renal cortex of L-NAME induced hypertensive rats compared with control rats (0.3-, 0.8-, and 13-fold, respectively). The simultaneous administration of LC was reported to attenuate these alterations (Zambrano *et al.*, 2013, 2013a, 2014). It has been also shown that LC reduces NF-κB transactivational activity and then the production of TNFα, ICAM1, and MCP-1 in carboplatin-treated renal tubular cells (Sue *et al.*, 2014). In advanced nephrotoxicity rat model TNF-α and NF-κB protein expression were shown to be increased after contrast media administration and CAR supplementation (200 mg/kg) reduced both TNF-α and NF-κB expressions in the kidney (Kunak *et al.*, 2016). An elevation in the myocardial expression of pro-inflammatory cytokines, together with an increase in the mRNA expression of NF-κB, was observed in rats treated with an anti-cancer drug sunitinib and co-administration of LC (400 mg LC/kg/day) was shown to inhibit all those alterations (Blanca *et al.*, 2016). In a mouse model with cancer cachexia the inhibitory effect of LC on the increased expression level of NF-κB p65 in the peripheral mouse blood mononuclear cells was shown to be markedly weakened by GW9662, a selective inhibitor of PPAR-γ (Jiang *et al.*, 2016). In the 19 d chicken embryo LC injection into the egg was shown to prevent NF-κB activation in the heart associated with perfluorooctanoic acid-induced developmental cardiotoxicity (Zhao *et al.*, 2017). ALC is shown to attenuate As-induced oxidative stress and hippocampal mitochondrial dysfunction and decreases NF-κB expression (Keshavarz-Bahaghighat *et al.*, 2018). In the myocardium of patients after unclamping the aorta L-carnitine added to the crystalloid cardioplegic solution increased level/activity/expression of SOD and CAT and Nrf2 while decreased levels of MDA, protein carbonyl content and suppressed the activation of NF-κB (Li *et al.*, 2019). Interestingly, NF-κB is shown to affect LC metabolism. For example, it seems likely that the bovine organic cation transporter 2 (OCTN2) gene and carnitine uptake are regulated by NF-κB. Indeed, cell treatment with TNFα was shown to activate NF-κB and increase the mRNA and protein concentration of OCTN2 associated with stimulated the uptake of carnitine in MDBK cells. In contrast, TNFα and the NF-κB inhibitor (BAY 11-7085) was able to completely block the effect of TNFα on OCTN2 mRNA and protein concentration and uptake of carnitine (Zhou *et al.*, 2015). The protective effects of LC on myocardial ischaemia-reperfusion injury in patients undergoing surgery was shown to be associated with the suppression of NF-κB pathway and the activation of Nrf2 pathway (Li *et al.*, 2019). In rats LC supplementation was shown to ameliorate doxorubicin-induced increase in NF-κB expression in the cardiac muscles (Aziz *et al.*, 2020). Therefore, inhibitory effects of carnitine on the NF-κB activated by various stress factors could be an important protective mechanism of the antioxidant defences in the body.

9.6 Effect of carnitine on vitagene network

Considering molecular mechanisms of antioxidant protective action of carnitine, it is necessary to consider its possible involvement in vitagene regulation. Indeed, it has been suggested that carnitine can affect signalling pathways that result in activation of vitagene network encoding survival proteins and affecting redox-sensitive intracellular pathways (Calabrese *et al.*, 2008).

9.6.1 Superoxide dismutase

Protective effects of LC on antioxidant defences, including SOD and GPx activities were shown in various stress-models *in vitro* and *in vivo*. In particular, it was shown that increased Lipoic acid (LA) or acetyl-L-carnitine (ALC) resulted in increased total antioxidant capacity and SOD and GPx activities and decreased levels of MDA in serum and liver of birds (Jia *et al.*, 2014). Notably, birds fed diets containing 50 mg/kg of LA and 50 mg/kg of ALC had higher serum and liver SOD activities than those fed diets containing 100 mg/kg of LA or ALC alone. In laying hens reared in a hot and humid climate L-threonine supplementation at 0.2% maximised the SOD activity in both serum and liver (Azzam *et al.*, 2012). Serum SOD increased linearly and quadratically in laying hens receiving excess dietary tryptophan (0·4 g/kg) (Dong *et al.*, 2012). Broilers given a diet containing 5.9 g/kg methionine had enhanced serum SOD activity and decreased hepatic MDA content at day 7 (Chen *et al.*, 2013). For example, co-administration of ALC and $NaAsO_2$ was shown to significantly suppress the oxidative damage induced by $NaAsO_2$, including restoration/preservation of SOD and CAT activities (Sepand *et al.*, 2016). Nephroprotective effect of carnitine against glycerol and contrast-induced kidney injury in rats was associated with a restoration of altered antioxidant defences, including SOD activity (Kunak *et al.*, 2016). In renal tubular cells LC was indicated to restore SOD and CAT activity altered/reduced by leptin treatment to those levels found in untreated cells (Blanca *et al.*, 2016). Carnitine treatment was shown to significantly increase antioxidant enzyme (CAT, SOD and GPx) expression/levels in mouse testis (Roy *et al.*, 2017). L-carnitine was found to normalise chronic sleep deprivation induced reduction in the hippocampus ratio of GSH/GSSG and activity of CAT, GPx, and SOD in rats (Alzoubi *et al.*, 2017). In rats raised on a diet supplemented with lead acetate LC was shown to have ameliorative effects by maintaining and protecting the level of the of antioxidant enzymes SOD, CAT and glutathione peroxidase in the blood (El-Sherbini *et al.*, 2017). LC supplementation significantly modulated the adverse effects of copper access on seminiferous tubules damage, testes function, spermatogenesis, and sperm quality in rats by preserving antioxidant defences, including SOD activity, in stress conditions (Khushboo *et al.*, 2018). ALC was reported to attenuate arsenic-induced liver injury by abrogation of mitochondrial dysfunction, antioxidant enzymes (SOD and CAT) declining, inflammation, and apoptosis in rat liver (Bodaghi-Namileh *et al.*, 2018). Interestingly, combined but not single administration of L-carnitine and vitamin C was shown to ameliorate cisplatin-induced gastric mucosa damage (Adefisayo *et al.*, 2018) and nephrotoxicity in rats (Alabi *et al.*, 2018) associated with prevention of toxicity-induced antioxidant enzymes activity (SOD, CAT, GPx) reduction. LC was shown to prevent SOD, CAT and GPx activities decline in the cardiac muscle due to aspartame-induced oxidative stress in Wistar albino rats (Al-Eisa *et al.*, 2018). LC increased activities of SOD and CAT while decreased levels of MDA and protein carbonyl content in the myocardium of patients after unclamping the aorta (Li *et al.*, 2019). Administration of LC was found to attenuate the Doxorubicin (DOX)-evoked reduction in SOD activity in the heat tissues of rats (Aziz *et al.*, 2020). Furthermore, LC supplementation was shown to significantly increase erythrocyte SOD, CAT and GPx activities, and decreased MDA level in high stocking density-stressed laying hens (Çetin and Güçlü, 2020). Therefore, the aforementioned results clearly showed that under various stress conditions LC or ALC can modulate antioxidant defences by preserving/restoring SOD activity.

9.6.2 Heme oxygenase 1 and other heat shock proteins

First, it was shown *in vitro* that ALC (30-100 μM) induces vitagene HO-1 in astrocytes in a dose- and time-dependent manner. This effect was associated with up-regulation of another vitagene HSP60 (Calabrese *et al.*, 2005). Similarly, ALC (150 mg/kg b.w orally for 4 months) induced vitagenes HO-1, HSP70 and SOD-2 in senescent rats. This protective effect of ALC was associated with other changes: upregulation of GSH levels, prevention of age-related changes in mitochondrial respiratory chain complex expression and decrease in protein carbonyls and HNE formation (Calabrese *et al.*, 2006a). In an *in vitro* study with human endothelial cells in culture carnitine and its acyl derivatives (at 0.5-2 mM) were shown to increase gene and protein expression of HO-1 (Calò *et al.*, 2006). Similarly, in humans and in an animal model it was shown that carnitine-mediated improvement in response to erythropoietin involves induction of HO-1 (Calò *et al.*, 2008). ALC 100 μM was also effective in primary cortical neuronal cultures: significantly attenuating amyloid-beta peptide 1-42-induced cytotoxicity, protein oxidation, lipid peroxidation, and apoptosis in a dose-dependent manner by upregulation of HSPs (Abdul *et al.*, 2006). It seems likely that ALC exerted protective effects against oxidative stress in part by upregulating the levels of GSH and HSPs. Indeed, LC treatment can increase level of HO-1 in the retinal ganglion cells (Cao *et al.*, 2015a). It was shown that LC (50 μM, 100 μM and 200 μM) had protective effects on high glucose-induced oxidative stress in the retinal ganglion cells (RGCs). Indeed, in high glucose stimulated RGCs, LC treatment was associated with an increased level of Nrf2, HO-1 and γ-glutamyl cysteine synthetase (Cao *et al.*, 2015a). Furthermore, ALC administration to human lens epithelial cells treated with homocysteine had a protective effect indicative by restored (increased) levels of antioxidant proteins, including SOD, GPx, Catalase, Nrf2, Keap1 and GSH (Yang *et al.*, 2015). Furthermore, LC was demonstrated to protect HL7702 cells against H_2O_2-induced oxidative stress and cell damage through Akt-mediated activation of Nrf2/HO-1 signalling pathway (Li *et al.*, 2016). Therefore, LC and its derivatives can perform their antioxidant function via activating HO-1. Indeed, HO induction occurs together with the induction of other heat shock proteins during various stressful conditions. Particularly, manipulation of endogenous cellular defence mechanisms, via the heat shock response, through nutritional antioxidants, including carnitine, may represent an innovative approach to therapeutic intervention in diseases (Calabrese *et al.*, 2004) and protection against stresses. Indeed, by maintaining or recovering the activity of vitagenes, it is possible to improve adaptive ability of animals/poultry to withstand various stresses.

9.6.3 Sirtuins

It seems likely that carnitine can affect sirtuins, another vitagene playing an important role in cell adaptation to various stresses. In fact, both oxidative stress and mitochondrial damage are associated with reduced levels of renal sirtuin 3. Therefore, as expected, treatment with ALC restored SIRT3 expression and activity, improved renal function, and decreased tubular injury in mice (Morigi *et al.*, 2015). It has been shown that ALC and sirtuins together affect mitochondria acetylation/deacetylation and thereby have the potential to regulate the cellular redox state, energy homeostasis and cell adaptation to stress (Marcovina *et al.*, 2013). It was shown that LC alleviated epithelial-mesenchymal transition-associated renal fibrosis caused by perfluorooctanesulphonat through a SIRT1- and PPARγ-dependent mechanism (Chou *et al.*, 2017). From the aforementioned data it is clear that carnitine can be considered as an important regulator of the vitagene network.

9.7 Sparing effects of carnitine on vitamin E

It is well-known that vitamin E is main chain-breaking antioxidant in biological membranes having a unique role in the antioxidant systems (Surai, 2002; Surai *et al.*, 2019). In particular, vitamin E recycling mechanisms are considered to be the most important part of vitamin E efficacy in antioxidant defences. Indeed, when all essential elements of vitamin E recycling are present together with other antioxidant mechanisms, even low vitamin E levels in membranes, for example, in brain, can be sufficient to effectively protect the tissue against lipid peroxidation (Surai, 2002;

Surai *et al.*, 2019). It seems likely that as a part of the antioxidant systems carnitine can have a sparing effect on vitamin E absorption and assimilation. For example, dietary LC (150 mg/kg diet) increased the rates and amounts of lymphatic absorption of α-tocopherol and fat in ovariectomised rats (Zou *et al.*, 2005) and enhanced liver α-tocopherol in aging ovariectomised rats (Clark *et al.*, 2007). Similarly, carnitine dietary supplementation decreased lipid peroxidation and promotes increased concentrations of retinol and α-tocopherol in free-living women (Sachan *et al.*, 2005). Furthermore, administration of LC (1.5 g/l with drinking water) to rats intoxicated with ethanol significantly decreased lipids and proteins oxidation in the serum and liver and the level of vitamin E was increased by about 20% in the liver and blood serum in comparison to the ethanol group (Augustyniak and Skrzydlewska, 2009). In the irradiated rats treated with LC 1.5 mg/kg b.w, i.p. concentrations of vitamins E were higher than in those rats that were only exposed to 2.45-GHz radiation (Türker *et al.*, 2011). Furthermore, metabolomics analysis shows that α-tocopherol deficiency in rats was associated with a compensatory increase in carnitine content in the liver (Moazzami *et al.*, 2011). A combination of dietary LC and Vitamin E was shown to effectively overcome OA (≤1.0 mg/kg) induced immunosuppression in chickens (Bhatti *et al.*, 2018). In a tilmicosin-induced cardiotoxicity model in rats a combination of LC and vitamin E was shown to be significantly more effective than individual antioxidants in prevention of a toxicity-induced decrease in SOD and CAT activity and GSH concentration in the cardiac tissues (Aboubakr *et al.*, 2020). Therefore, molecular mechanisms of carnitine-vitamin E interactions need further investigation, but the effect of such interactions on the total antioxidant systems of the body could be quite significant.

9.8 Carnitine as a part of antioxidant mixtures

Based on the concept of integrated antioxidant systems in the body it is obvious that dietary supplementation of synergistic mixtures of various antioxidants could have higher protective effects in comparison with individual antioxidants, including carnitine. Indeed, it is the case in biological systems. For example, LC and vitamin E in combination are shown to be effective in ameliorating ochratoxin A-altered haematological and serum biochemical parameters in White Leghorn cockerels (Abidin *et al.*, 2013). A combination of LC and vitamin C was shown to decrease the risk of ischemia-induced necrosis in damaged tissues in rats (Arslan *et al.*, 2005). Similarly, supplementation of vitamin E, vitamin C, and LC in combination can attenuate the oxidative stress associated with intermittent hypobaric hypoxia in rats (Devi *et al.*, 2007). There are a number of studies showing antioxidant protective effects of carnitine in combination with another mitochondria-related antioxidant, namely lipoic acid (Abdul and Butterfield, 2007; Hagen *et al.*, 2002, 2002a; Kumaran *et al.*, 2005; Liu *et al.*, 2002; McMackin *et al.*, 2007; Savitha and Panneerselvam, 2007; Savitha *et al.*, 2005; Tamilselvan *et al.*, 2007). Similarly, a more complex antioxidant mixture, containing CoQ, LC, α-tocopherol and selenium was effective in decreasing DNA damage in the liver of fumonisin B1-treated rats (Atroshi *et al.*, 1999). Furthermore, a synergistic combination of ALC, folate and vitamin E provided a protection against oxidative stress resulting from exposure of human neuroblastoma cells to amyloid-beta (Dhitavat *et al.*, 2005). In this system, vitamin E prevents *de novo* membrane oxidative damage, folate maintains levels of the endogenous antioxidant GSH and ALC prevents A-beta-induced mitochondrial damage and ATP depletion providing superior protection to that derived from each agent alone (Dhitavat *et al.*, 2005). Supplementation of pregnant and lactating sow diet with carnitine-containing bioactive substances (a blend of flax seed, rapeseed, linden inflorescence, taurine, LC and tocopherol acetate) improved maturation of the small intestinal epithelium in their offspring during the early postnatal period (Strzałkowski *et al.*, 2007). Recently, it has been demonstrated that ALC, L-α-lipoic acid and silymarin had similar antioxidant effects in cisplatin-induced myocardial injury (El-Awady *et al.*, 2011). It would be advisable to assess antioxidant effects of a combination of carnitine and silymarin taking into account that both are considered to be hepatoprotectors and both are characterised by antioxidant properties (Surai, 2015d). In fact, it is proven that the therapeutic effect of silymarin combined with LC on non-alcoholic fatty liver disease in patients was higher than in silymarin used alone (Liu *et al.*, 2009). From the data presented above it is clear that the development of carnitine-containing antioxidant mixtures could be considered as an important step in stress prevention and treatment of stressed livestock animals.

9.9 Specific protective effects of carnitine in poultry production

Based on the aforementioned data it is clear that protective effects of carnitine and its derivatives are most pronounced in various stress conditions. Indeed, by decreasing negative consequences of stresses carnitine can improve productive and reproductive performance and general health of growing chickens, parent stock, and commercial layers (Arslan, 2006; Golzar Adabi *et al.*, 2011; Ringseis *et al.*, 2018).

9.9.1 Immunity

There is a substantial body of evidence indicating that LC dietary supplementation has immunomodulating effects on humoral and cell mediated immunity in chickens. For example, dietary LC supplementation (100 mg/kg diet) appeared to be beneficial for chickens in enhancing specific humoral responses on vaccination indicative by prevention of apoptotic death of B lymphocytes and enhanced IgG production in chickens, after both the primary and the secondary immunisation (Mast *et al.*, 2000). In fact, a long-lasting increased IgG response due to dietary LC supplementation may be of major practical importance in the enhancement of protective immunity on vaccination. Similarly, enhanced specific antibody response to bovine serum albumin in pigeons due to LC supplementation (1 g/l drinking water) was observed (Janssens *et al.*, 2000). In fact, both BSA-specific IgG and IgM responses were enhanced by about 10% by LC supplementation. The effects of supplementing Leghorn-type chickens with dietary LC (1 g/kg diet) after hatching for 4 weeks were assessed in a 12-week study (Deng *et al.*, 2006).

It was concluded that a short-term supply of dietary LC to a conventional commercial feed after hatching enhanced subsequent humoral immunity in Leghorn-type chickens. Indeed, an increased relative thymus weight and an enhanced serum primary antibody response to a mitogen in LC-fed birds were detected. Furthermore, LC in the diet of broiler chickens (100 mg/kg diet) can enhance or advance the acute phase protein response (Buyse *et al.*, 2007). Indeed, after a LPS challenge of male broiler chickens the elevations in circulating hemopexin and alpha-1 acid glycoprotein levels were more pronounced in the LC supplemented chickens than in control birds. In ascites-susceptible broilers serum IgG content was improved by LC supplementation (75-100 mg/kg diet) (Geng *et al.*, 2007). Adding LC (300 mg/kg diet) into the chicken diet had a significant effect on Newcastle disease antibody titre at day 32 (Parsaeimehr *et al.*, 2014). It seems likely that a combination of LC and methionine can also improve humoral immunity. Indeed, highest levels of IgG and WBC were found in birds fed 130% NRC methionine + 150 mg/kg LC (Reza *et al.*, 2015).

Supplementation of LC (200 mg/kg) to broiler chickens reared at high altitude increased plasma nitric oxide and immune responsiveness, which manifested in an increased toe-web thickness index measured at 24 h following the injection of phytohemagglutinin, an *in vivo* indicator of cell-mediated immune responses (Khajali and Khajali, 2014). Furthermore, supplemental LC (100-400 mg/kg diet) enhanced the humoral and cell mediated immune responses in Japanese quail as evidenced with better antibody titres against Newcastle disease virus and greater wing web swelling in response to PHA-L injection, respectively (Abdel-Fattah *et al.*, 2014). Indeed, carnitine can have beneficial impact on chicken immunity participating in preventing infection in commercial poultry production. Immunomodulating properties of carnitine were also shown in farm and laboratory animals. For example, in pigs, white blood cell and lymphocyte concentrations were increased by LC dietary supplementation (250 mg/kg diet) for 10-weeks (Chen *et al.*, 2008).

There is also evidence that LC can improve innate immunity by modulating macrophage and neutrophil activities. For example, treatment with LC (300 mg/kg b.w) significantly improved neutrophil functions, delayed-type hypersensitivity responses and the concentrations of immunoglobulins A and G in aged rats (Thangasamy *et al.*, 2008). LC is also capable of restoring the age-related changes of neutrophil functions. Indeed, the neutrophils of aged rats exhibited an increase in superoxide anion production and decline in phagocytosis and chemotaxis when compared with that in young rat neutrophils. Superoxide anion production in aged rats was significantly decreased by LC treatment (50 mg/kg b.w for 30 days) which was accompanied with a significant enhancement of chemotactic and phagocytic activities which were restored to control levels (Izgut-Uysal *et al.*, 2003, 2004). It has been shown that LC restored lymphocyte proliferative response and the lytic activity of macrophages in aged rats (Elliott *et al.*, 1990; Jirillo *et al.*, 1986; Thomas *et al.*, 1999). In cultured mouse hybridoma cells LC is reported to stimulate growth and antibody production (Berchiche *et al.*, 1994), while in leukemic cells isovaleryl carnitine improved phagocytosis and cell killing activity (Ferrara *et al.*, 2005).

Carnitine was shown to regulate immune response in various inflammation-related diseases in animal models and in humans. For example, in rodents, treatment with LC (50-100 mg/kg body weight) markedly suppressed the LPS-induced cytokine production, improving their survival during cachexia and septic shock (Ruggiero *et al.*, 1993; Winter *et al.*, 1995). Moreover LC (200 mg/kg b.w, i.p.) was reported to improve immune responses in mice exposed to low frequency high intensity magnetic field (Arafa *et al.*, 2003). *In vivo*, protection from trinitrobenzene sulphonic acid colitis was observed in LC-treated mice and was attributed to the abrogation of both innate and adaptive immune responses (Fortin *et al.*, 2009). Indeed, LC has been shown to reduce CD4+ and CD8+ T cell numbers and IL-2 production in splenocytes isolated from LC treated mice (Athanassakis *et al.*, 2001) and reduce TNF-α production in *Staphylococcus aureus*-stimulated human polymorphonuclear cells (Fattorossi *et al.*, 1993).

Decreased serum TNFα levels have been reported after LC supplementation in surgical patients and AIDS patients (Delogu *et al.*, 1993; De Simone *et al.*, 1993). LC administration was shown to ameliorate effects of LPS on cellular and humoral immunity in testis through reduction of IL-2 and by buffering the oxidative stress-induced damage (Abd-Allah *et al.*, 2009). It seems likely that an anti-apoptotic action of carnitine is of great importance in its immunomodulating properties. For example, it was shown that LC inhibited apoptosis of white blood cells (Sener *et al.*, 2006) and CD4+ and CD8+ cells (De Simone *et al.*, 1993; Moretti *et al.*, 2002). Similarly, supplementation with a carnitine-containing formula (alpha-tocopherol, alpha-lipoic acid, coenzyme Q (10), carnitine, and selenomethionine) to healthy individuals was shown to modulate the process of apoptosis under *in vivo* conditions (Mosca *et al.*, 2002).

Interestingly, LC and its derivatives have been shown to reduce apoptosis through the mitochondrial pathway (Furuno *et al.*, 2001; Ishii *et al.*, 2000) and this appears to be linked with downregulating the transduction of the pro-apoptotic Fas signal and suppressing the generation of ceramide, an important endogenous mediator of apoptosis (Di Marzio *et al.*, 1997). This anti-apoptotic effect of carnitine has been observed in different cells and organelles, including neurons (Ye *et al.*, 2014), cardiomyocytes (Mao *et al.*, 2014), hepatocytes (Revoltella *et al.*, 1994), bone marrow cells (Abd-Allah *et al.*, 2005), neuroblastoma cell line (Bavari *et al.*, 2016), retinal ganglion cells (Cao *et al.*, 2014), renal tubular cells (Sue *et al.*, 2014), embryonic neural stem cells (Liu *et al.*, 2014), spinal cord and mitochondria (Zhang *et al.*, 2015). Therefore, LC-mediated cytoprotection and immunomodulating properties are due, in part, to inhibition of the mitochondrial apoptotic pathway (Wang *et al.*, 2007).

Gut immunity plays an important role in protection against various pathogens (Surai and Fisinin, 2015). It seems likely that LC may re-establish equilibrium between pro-inflammatory and anti-inflammatory cytokines, reducing the former and/or increasing the latter. This action is extremely important in the gut, since the interplay between both innate and adaptive immune responses is crucial to perpetuate inflammation in the gut in various stress conditions. Indeed, LC can suppress DC and macrophage co-stimulatory molecule expression dose-dependently [Fortin *et al.*, 2009). Therefore, there is a therapeutic potential of LC in treating the acute and chronic aspects of intestinal inflammation. It was shown that carnitine deficiency resulted in the hyperactivation of CD4+ T cells and enhanced production of the classical Th1 cytokine, IFN-γ (Fortin *et al.*, 2009) and leads to increased apoptosis of enterocytes, villous atrophy, inflammation and gut injury (Sonne *et al.*, 2012). Similarly, mice deficient in the carnitine transporter, OCTN2, develop spontaneous atrophy of intestinal epithelial cells and colonic inflammation (Shekhawat *et al.*, 2007). In contrast, LC treatment significantly inhibited both APC and CD4+ T cell function, as assessed by the expression of classical activation markers, proliferation and cytokine production (Fortin *et al.*, 2009). Indeed, LC has a protective effect on the intestinal mucosa by preventing ROS production (Surai and Fisinin, 2015) However, the role of LC on immunological functions in various stress conditions still remains to be explored and the precise mechanisms of immunomodulating action of LC remain elusive. There are a range of potential mechanisms, which may be related to this effect. It is well recognised that sophisticated antioxidant defences directly and indirectly protect the host against the damaging effects of cytokines and oxidants. In particular, indirect protection is afforded by antioxidants, which reduce activation of NF-κB, thereby preventing up-regulation of cytokine production by oxidants. On the other hand, cytokines increase both oxidant production and antioxidant defences, thus minimising damage to the host.

Antioxidants prevent oxidative stress-induced damage to immune cells. It is necessary to take into account that cellular integrity is very important for receiving and responding to the messages needed to coordinate an immune response. The immune system generates ROS as part of its defence function and these ROS are an important weapon to kill pathogens. However, chronic overproduction of ROS can cause damage to immune cells and compromise their function (Wu and Meydani, 1998). In fact immune cells are rich in PUFAs which are very susceptible to free radical attack. It is well recognised that many immunological functions are membrane dependent. These are antigen recognition, receptor expression, secretion of antibodies and cytokines, lymphocyte transformation, and contact cell lysis (Wu and Meydani, 1998). In particular, the receptors are important for antigen recognition and the secretion of various chemical mediators such as interferon, tumour necrosis factor, prostaglandins and interleukins. Therefore, lipid peroxidation can change membrane structure and properties (e.g. fluidity, permeability, flexibility, etc.) which would affect immune cell functions. In contrast, antioxidants are able to prevent those damaging effects of ROS and maintain immune function.

If immune system is considered as 'an army' fighting against invaders (microorganisms, viruses, etc.) then one would expect them to have something like mobile phones to receive and send signals to each other. Remarkably enough, major immune cells (macrophages, neutrophils, T- and B-lymphocytes) have on their surface something like 'mobile phones' called receptors. Those receptors are extremely sensitive to communicating molecules, but they are also sensitive to free radicals and can be easily damaged (Surai, 2002, 2006, 2018). In such a situation without proper communication all those huge armies of immune cells would become useless. They also can start fighting each other as well and eventually destroying immunocompetence and causing autoimmune reactions. If immune cells are considered as 'soldiers' using chemical weapon to kill enemy than special ammunition protecting them from their own weapon would be crucial for effective battle. In the case of immune cells such ammunition is represented by natural antioxidants. Indeed, if not properly protected, macrophage functions could be compromised including initial overproduction of free radicals with consecutive damages to specific enzymatic systems resulting in decreasing efficiency of oxidative burst and apoptosis. Based on the presented model it is clear that antioxidant defence is a crucial factor of immune defence in the body. Indeed, the crucial role of various receptors in immunocompetence processes and receptor sensitivity to ROS need further investigation, since exact biological/biochemical mechanisms by which oxidants/antioxidants regulate immunity are still ill-defined. In particular, research data have shown that TCR-induced ROS generation may be an important regulator of T cell signal transduction and gene expression (Williams and Kwon, 2004).

Data on the redox dependence of signal transduction in T cells are quickly growing. Some data suggest an underlying regulatory role for ROS in controlling the susceptibility of T cells to apoptosis (Hildeman, 2004) and innate immunity efficacy (Kohchi *et al.*, 2009). In particular, recently it has been shown that mitochondria-dependent signalling controls innate and adaptive immune responses (Weinberg *et al.*,2015). Clearly, oxidative stress leads to accrual of damaged/misfolded proteins, lipids and causes inflammation (Surai, 2018; Surai *et al.*, 2009). It is important to mention that individual cells and multicellular organisms have developed intricate effective mechanisms to utilise ROS and RNS to modulate homeostasis and respond to threats. Therefore, ROS and RNS are active participants in innate and acquired immune responses. Antioxidant nutrients such as carnitine may protect against oxidant-mediated inflammation and tissue damage by virtue of their ability to upregulate the antioxidant defences and optimise redox signalling, including activation of Nrf2 and prevention of the activation of NF-κB. In fact, NF-κB is required for maximal transcription of many inflammatory cytokines and adhesion molecules (Hughes, 1999). Therefore, LC may have great anti-inflammatory properties via downregulation of TNF-α and inhibition of NF-κB. Thus, maintaining adequate antioxidant status may provide a useful approach in attenuating the cellular injury and dysfunction observed in some inflammatory disorders (Conner and Grisham, 1996).

It is necessary to underline that, non-toxic concentrations of ROS and RNS play an important role in regulating the expression of genes involved in the inflammatory response and in modulating apoptosis (Jourd'heuil *et al.*, 1997). At the same time, an immune response requires extensive communication between a wide range of cell types (Klasing, 1998) and special cell receptors are of great importance in this communication. Therefore, protective effect of antioxidants, including carnitine, in prevention of membrane and receptor damages due to peroxidation could provide an important way of enhancing the immune system. In addition to the aforementioned mechanisms of immunomodulating properties of carnitine evidence from both animal and human studies suggests that, at pharmacological doses, LC may mimic some of the actions of glucocorticoids, including their well-known immunomodulatory effect. Indeed, LC can activate glucocorticoid receptor-α (GRα) and, through this mechanism, regulate glucocorticoid-responsive genes, potentially sharing some of the biological and therapeutic properties of glucocorticoids (Alesci *et al.*, 2003). In fact, LC reduced the binding capacity of GRα, induced its nuclear translocation, and stimulated its transcriptional activity. Moreover, LC suppressed TNFα and IL-12 release from human monocytes in glucocorticoid-like fashion (Alesci *et al.*, 2004). It was suggested that LC is a 'nutritional modulator' of the GR, by acting as an agonist-like compound (Manoli *et al.*, 2004).

While the above data suggest immunomodulating and anti-inflammatory roles for LC, some early studies have been reported contradictory results, in part reflecting the complexity of the immune response and great variation between experimental conditions (De Simone *et al.*, 1994; Kouttab *et al.*, 1993). Therefore, LC is shown to enhance immunocompetence of birds by improving humoral and cell-mediated immunity. Furthermore, LC can decrease negative consequences of post-vaccination stress and increase vaccination success and clearly effects of carnitine on innate and acquired immunity in avian species awaits further investigation.

9.9.2 Ascites

Ascites syndrome (pulmonary hypertension syndrome, PHS) is a serious metabolic disease causing important economic losses to the poultry meat industry. It seems likely that interactions of genetic, physiological, environmental, and management factors are responsible for this syndrome (Baghbanzadeh and Decuypere, 2008). Furthermore, it is proven that the elevated ROS production and compromised antioxidant defences are involved in the development of ascites (Arab *et al.*, 2006; Bottje and Wideman, 1995; Enkvetchakul *et al.*, 1993). Therefore, protective effects of nutritional antioxidants are of great importance (Bottje *et al.*, 1997; Villar-Patino *et al.*, 2002). In this respect, based on results showing positive effects of LC dietary supplementation on growing chickens at various temperature regimes, Buyse *et al.* (2001) suggested that LC is a potential agent for reducing the incidence of metabolic diseases in broiler chickens. Indeed, LC (75 or 150 mg/kg) or LC+CoQ10 dietary supplementation increased SOD activity and reduce ascites mortality of broilers (Geng *et al.*, 2004). Similarly, supplemental LC (100 mg/kg diet) reduced plasma MDA, increased SOD, inhibited remodelling and postponed the occurrence of PHS for 1 week in cold-exposed broilers (Tan *et al.*, 2008). Indeed, in broilers reared under low temperature environment dietary LC supplementation (100 mg/kg) reduced organ index, enhanced antioxidative capacity of the heart (SOD and GPx), and enhanced liver enzymes activity involved in tricarboxylic acid cycle, and reduced serum glucose and triglyceride (Wang *et al.*, 2013).

Dietary LC (50-150 mg/kg diet) improves pulmonary hypertensive response in broiler chickens subjected to hypobaric hypoxia and reduces ascites mortality in broiler chickens by increased NO production, reduced MDA concentration, and reduced right ventricular hypertrophy (Yousefi *et al.*, 2013). Supplementation of LC (200 mg/kg diet) had also beneficial effects on preventing lipid peroxidation and pulmonary hypertension in broiler chickens raised at high altitude (2100 m above sea level; Khajali and Khajali, 2014). Furthermore, dietary LC supplementation (100 mg/kg diet) of reduced-protein diets had beneficial effects in preventing pulmonary arterial hypertension mortality mainly through enhancing blood NO concentration (Sharifi *et al.*, 2015). Clearly, LC supplementation can be used to decrease detrimental consequences of ascites.

9.9.3 Other commercially relevant stresses

The reduction of heat production in exercising pigeons after LC supplementation (Janssens *et al.*, 1998) could be very relevant for protective effect of carnitine in heat stressed animals/birds. For example, LC supplementation with drinking water significantly prevented deterioration of some egg quality characteristics (relative albumen weight and height) of layers under high environmental temperature (Celik *et al.*, 2004). It was also shown that dietary supplemental LC (50 mg/kg diet) or LC + ascorbic acid had positive effects on body weight gain and carcass weight under high temperature conditions (Celik and Oztürkcan, 2003). LC (1 g/kg diet) or its combination with vitamin E (200 mg/kg diet) ameliorated ochratoxin A-induced alterations in haematological and serum biochemical parameters (Abidin *et al.*, 2013). LC dietary supplementation (400 mg/kg diet) also has a protective effect on lipid peroxidation and drop in performance of laying hens fed high cupper diet (Güçlü *et al.*, 2011). Carnitine can also help with stresses associated with chicken placement and their first days of life. For example, Nouboukpo *et al.* (2009) supplemented LC in drinking water (30-60 mg/l) to broiler chickens and observed improved growth rate for the first 7 days of rearing. Indeed, recent data indicate that ALC supplementation at low levels (50 or 100 mg/kg) improved antioxidative ability (increased total antioxidant capacity and SOD and GPx activities and decreased levels of MDA in serum and liver of birds), energy metabolism, and lipid metabolism in broilers (Jia *et al.*, 2014). There were also synergistic effects of the combined supplementation of ALC and another antioxidant, namely lipoic acid, indicative by serum and liver SOD activities and serum glucose and TG levels (Jia *et al.*, 2014). Carnitine can also help with aggravated stresses in poultry. For example, LC supplementation (50-500 mg/kg diet) of a practical layer diet of old (65-week-old) laying hens kept in cages for 8 weeks improved egg white quality indicative by increased Haugh units (Rabie *et al.*, 1997).

9.10 Conclusions

Carnitine is a newcomer into the antioxidant family. Antioxidant properties of carnitine have been demonstrated *in vitro* and *in vivo* using various model systems as well as clinical observations in patients with various diseases. In the aforementioned *in vitro* and *in vivo* studies, the antioxidant properties of LC and its derivatives are demonstrated by:

- restoration of the endogenous AO enzymes (SOD, CAT, GPx, GR and GST) and non-enzymatic antioxidants (vitamins E and C) in the liver and other tissues of stressed animals;
- increased intracellular concentration of GSH in liver and other tissues;
- decreased lipid and protein oxidation, detected as reduced MDA/TBARS and carbonyl content;
- decreased DNA fragmentation/damage and apoptosis;
- reduced secretion of ALT, AST, ALP, γ-GT from the liver into the plasma due to hepatic injuries caused by ROS;
- restored Nrf2 and HO-1 activities;
- reduced NF-κB expression and concentration of pro-inflammatory cytokines, including tumour necrosis factor;
- vitagene activation and increased synthesis of HSPs, thioredoxins and sirtuins.

From the data presented above, it could be concluded that carnitine is an important element of the antioxidant defence systems and modulator of vitagene expression. It is well-known that carnitine is synthesised in the body and also obtained from the feed. It seems likely that under stress condition carnitine synthesis can be compromised. Furthermore, because of limited amount of feed-derived carnitine provided by the diet, carnitine inadequacy could be the case in commercial poultry. Therefore, carnitine, in combination with other nutrients, supplied with water or feed, can be used in management of commercially-relevant stresses in industrial egg and meat production.

References

Abd-Allah, A.R., Al-Majed, A.A., Al-Yahya, A.A., Fouda, S.I. and Al-Shabana, O.A., 2005. L-carnitine halts apoptosis and myelosuppression induced by carboplatin in rat bone marrow cell cultures (BMC). Archives of Toxicology 79: 406-413.

Abd-Allah, A.R., Helal, G.K., Al-Yahya, A.A., Aleisa, A.M., Al-Rejaie, S.S. and Al-Bakheet, S.A., 2009. Pro-inflammatory and oxidative stress pathways which compromise sperm motility and survival may be altered by L-carnitine. Oxidative Medicine and Cellular Longevity 2: 73-81.

Abdel-Fattah, S.A., El-Daly, E.F. and Ali, N.G., 2014. Growth performance, immune response, serum metabolites and digestive enzyme activities of Japanese quail fed supplemental L-carnitine. Global Vet 12: 277-286.

Abdul, H.M. and Butterfield, D.A., 2007. Involvement of PI3K/PKG/ERK1/2 signaling pathways in cortical neurons to trigger protection by cotreatment of acetyl-L-carnitine and α-lipoic acid against HNE-mediated oxidative stress and neurotoxicity: Implications for Alzheimer's disease. Free Radical Biology and Medicine 42: 371-384.

Abdul, H.M., Calabrese, V., Calvani, M. and Butterfield, D.A., 2006. Acetyl-L-carnitine-induced up-regulation of heat shock proteins protects cortical neurons against amyloid-beta peptide 1-42-mediated oxidative stress and neurotoxicity: Implications for Alzheimer's disease. Journal of Neuroscience Research 84: 398-408.

Abidin, Z., Khan, M.Z., Khatoon, A., Saleemi, M.K., Khan, A. and Javed, I., 2013. Ameliorative effects of L-carnitine and vitamin E (α-tocopherol) on haematological and serum biochemical parameters in White Leghorn cockerels given ochratoxin A contaminated feed. British Poultry Science 54: 471-477.

Aboubakr, M., Elsayd, F., Soliman, A., Fadl, S.E., El-Shafey, A. and Abdelhiee, E.Y., 2020. L-carnitine and vitamin E ameliorate cardiotoxicity induced by tilmicosin in rats. Environmental Science and Pollution Research International 27: 23026-23034. https://doi.org/10.1007/s11356-020-08919-6

Adefisayo, M.A., Adeyemi, W.J. and Alabi, Q.K., 2018. Combined but not single administration of vitamin C and L-carnitine ameliorates cisplatin-induced gastric mucosa damage in male rats. Canadian Journal of Physiology and Pharmacology 96: 830-838.

Alabi, Q.K., Akomolafe, R.O., Olukiran, O.S., Nafiu, A.O., Adefisayo, M.A., Owotomo, O.I., Omole, J.G. and Olamilosoye, K.P., 2018. Combined administration of L-carnitine and ascorbic acid ameliorates cisplatin-induced nephrotoxicity in rats. Journal of the American College of Nutrition 37: 387-398.

Al-Eisa, R.A., Al-Salmi, F.A., Hamza, R.Z. and El-Shenawy, N.S., 2018. Role of L-carnitine in protection against the cardiac oxidative stress induced by aspartame in Wistar albino rats. PloS one 13: e0204913.

Alesci, S., De Martino, M.U., Kino, T. and Ilias, I., 2004. L-carnitine is a modulator of the glucocorticoid receptor alpha. Annals of the New York Academy of Sciences 1024: 147-152.

Alesci, S., De Martino, M.U., Mirani, M., Benvenga, S., Trimarchi, F., Kino, T. and Chrousos, G.P., 2003. L-carnitine: a nutritional modulator of glucocorticoid receptor functions. The FASEB Journal 17: 1553-1555.

Alzoubi, K.H., Rababa'h, A.M., Owaisi, A. and Khabour, O.F., 2017. L-carnitine prevents memory impairment induced by chronic REM-sleep deprivation. Brain Research Bulletin 131: 176-182.

Arab, H.A., Jamshidi, R., Rassouli, A., Shams, G. and Hassanzadeh, M.H., 2006. Generation of hydroxyl radicals during ascites experimentally induced in broilers. British Poultry Science 47: 216-222.

Arafa, H.M., Abd-Allah, A.R., El-Mahdy, M.A., Ramadan, L.A. and Hamada, F.M., 2003. Immunomodulatory effects of L-carnitine and q10 in mouse spleen exposed to low-frequency high-intensity magnetic field. Toxicology 187: 171-181.

Arslan, C., 2006. L-carnitine and its use as a feed additive in poultry feeding a review. Revue de Médecine Vétérinaire 157: 134-142.

Arslan, E., Basterzi, Y., Aksoy, A., Majka, C., Unal, S., Sari, A. and Demirkan, F., 2005. The additive effects of carnitine and ascorbic acid on distally burned dorsal skin flap in rats. Medical Science Monitor 11: BR176-BR180.

Athanassakis, I., Mouratidou, M., Sakka, P., Evangeliou, A., Spilioti, M. and Vassiliadis, S., 2001. L-carnitine modifies the humoral immune response in mice after *in vitro* or *in vivo* treatment. International Immunopharmacology 1: 1813-1822.

Atroshi, F., Rizzo, A., Biese, I., Veijalainen, P., Saloniemi, H., Sankari, S. and Andersson, K., 1999. Fumonisin B1-induced DNA damage in rat liver and spleen: effects of pretreatment with coenzyme Q10, L-carnitine, α-tocopherol and selenium. Pharmacological Research 40: 459-467.

Augustyniak, A. and Skrzydlewska, E., 2009. L-carnitine in the lipid and protein protection against ethanol-induced oxidative stress. Alcohol 43: 217-223.

Aziz, M.M., Abd El Fattah, M.A., Ahmed, K.A. and Sayed, H.M., 2020. Protective effects of olmesartan and L-carnitine on doxorubicin-induced cardiotoxicity in rats. Canadian Journal of Physiology and Pharmacology 98: 183-193.

Azzam, M.M., Dong, X.Y., Xie, P. and Zou, X.T., 2012. Influence of L-threonine supplementation on goblet cell numbers, histological structure and antioxidant enzyme activities of laying hens reared in a hot and humid climate. British Poultry Science 53: 640-645.

Baghbanzadeh, A. and Decuypere, E., 2008. Ascites syndrome in broilers: physiological and nutritional perspectives. Avian Pathology 37: 117-126.

Barhwal, K., Hota, S.K., Jain, V., Prasad, D., Singh, S.B. and Ilavazhagan, G., 2009. Acetyl-L-carnitine (ALCAR) prevents hypobaric hypoxia-induced spatial memory impairment through extracellular related kinase-mediated nuclear factor erythroid 2-related factor 2 phosphorylation. Neuroscience 161: 501-514.

Bavari, M., Tabandeh, M.R., Najafzadeh Varzi, H. and Bahramzadeh, S., 2016. Neuroprotective, antiapoptotic and antioxidant effects of L-carnitine against caffeine-induced neurotoxicity in SH-SY5Y neuroblastoma cell line. Drug and Chemical Toxicology 39: 157-166.

Berchiche, L., Legrand, C., Capiaumont, J., Belleville, F. and Nabet, P., 1994. Effect of L-carnitine and acylcarnitine derivatives on the proliferation and monoclonal antibody production of mouse hybridoma cells in culture. Journal of Biotechnology 34: 175-183.

Bhatti, S.A., Khan, M.Z., Hassan, Z.U., Saleemi, M.K., Khatoon, A., Abidin, Z.U. and Hameed, M.R., 2018. Dietary L-carnitine and vitamin-E; a strategy to combat ochratoxin-A induced immunosuppression. Toxicon 153: 62-71.

Bieber, L.L., Markwell, M.A.K., Blair, M. and Helmrath, T.A., 1973. Studies on the development of carnitine palmitoyltransferase and fatty acid oxidation in liver mitochondria of neonatal pigs. Biochimica et Biophysica Acta (BBA)-Lipids and Lipid Metabolism 326: 145-154.

Birkenfeld, C., Doberenz, J., Kluge, H. and Eder, K., 2006. Effect of L-carnitine supplementation of sows on L-carnitine status, body composition and concentrations of lipids in liver and plasma of their piglets at birth and during the suckling period. Animal Feed Science and Technology 129: 23-38.

Birkenfeld, C., Ramanau, A., Kluge, H., Spilke, J. and Eder, K., 2005. Effect of dietary l-carnitine supplementation on growth performance of piglets from control sows or sows treated with l-carnitine during pregnancy and lactation. Journal of Animal Physiology and Animal Nutrition 89: 277-283.

Blanca, A.J., Ruiz-Armenta, M.V., Zambrano, S., Miguel-Carrasco, J.L., Arias, J.L., Arévalo, M., Mate, A., Aramburu, O. and Vázquez, C.M., 2016. Inflammatory and fibrotic processes are involved in the cardiotoxic effect of sunitinib: protective role of L-carnitine. Toxicology Letters 241: 9-18.

Blanca, A.J., Ruiz-Armenta, M.V., Zambrano, S., Salsoso, R., Miguel-Carrasco, J.L., Fortuño, A., Revilla, E., Mate, A. and Vázquez, C.M., 2016. Leptin induces oxidative stress through activation of NADPH oxidase in renal tubular cells: antioxidant effect of L-carnitine. Journal of Cellular Biochemistry 117: 2281-2288.

Bodaghi-Namileh, V., Sepand, M.R., Omidi, A., Aghsami, M., Seyednejad, S.A., Kasirzadeh, S. and Sabzevari, O., 2018. Acetyl-L-carnitine attenuates arsenic-induced liver injury by abrogation of mitochondrial dysfunction, inflammation, and apoptosis in rats. Environmental toxicology and Pharmacology 58: 11-20.

Bohles, H., 2000. Basic concept of L-carnitine supplementation. Annals of Nutrition and Metabolism 44: 77-78.

Borum, P.R., 1983. Carnitine. Annual Reviews of Nutrition 3: 233-259.

Bottje, W.G., Erf, G.F., Bersi, T.K., Wang, S., Barnes, D. and Beers, K.W., 1997. Effect of dietary dl-alpha-tocopherol on tissue alpha-and gamma-tocopherol and pulmonary hypertension syndrome (ascites) in broilers. Poultry Science 76: 1506-1512.

Bottje, W.G. and Wideman, R.F., 1995. Potential role of free radicals in the pathogenesis of pulmonary hypertension syndrome. Poultry and Avian Biology Reviews 6: 211-231.

Bremer, J., 1983. Carnitine – metabolism and functions. Physiological Reviews 63: 1420-1480.

Buyse, J., Janssens, G.P.J. and Decuypere, E., 2001. The effects of dietary L-carnitine supplementation on the performance, organ weights and circulating hormone and metabolite concentrations of broiler chickens reared under a normal or low temperature schedule. British Poultry Science 42: 230-241.

Buyse, J., Swennen, Q., Niewold, T.A., Klasing, K.C., Janssens, G.P., Baumgartner, M. and Goddeeris, B.M., 2007. Dietary L-carnitine supplementation enhances the lipopolysaccharide-induced acute phase protein response in broiler chickens. Veterinary Immunology and Immunopathology 118: 154-159.

Calabrese, V., Colombrita, C., Sultana, R., Scapagnini, G., Calvani, M., Butterfield, D.A. and Stella, A.G., 2006. Redox modulation of heat shock protein expression by acetylcarnitine in aging brain: relationship to antioxidant status and mitochondrial function. Antioxidants & Redox Signaling 8: 404-416.

Calabrese, V., Cornelius, C., Dinkova-Kostova, A.T. and Calabrese, E.J., 2009. Vitagenes, cellular stress response, and acetylcarnitine: relevance to hormesis. Biofactors 35: 146-160.

Calabrese, V., Cornelius, C., Mancuso, C., Pennisi, G., Calafato, S., Bellia, F., Bates, T.E., Stella, A.M.G., Schapira, T., Kostova, A.T.D. and Rizzarelli, E., 2008. Cellular stress response: a novel target for chemoprevention and nutritional neuroprotection in aging, neurodegenerative disorders and longevity. Neurochemical Research 33: 2444-2471.

Calabrese, V., Stella, A.M.G., Calvani, M. and Butterfield, D.A., 2006a. Acetylcarnitine and cellular stress response: roles in nutritional redox homeostasis and regulation of longevity genes. The Journal of Nutritional Biochemistry 17: 73-88.

Calabrese, V., Ravagna, A., Colombrita, C., Scapagnini, G., Guagliano, E., Calvani, M., Butterfield, D.A. and Giuffrida Stella, A.M., 2005. Acetylcarnitine induces heme oxygenase in rat astrocytes and protects against oxidative stress: involvement of the transcription factor Nrf2. Journal of Neuroscience Research 79: 509-521.

Calabrese, V., Stella, A.G., Butterfield, D.A. and Scapagnini, G., 2004. Redox regulation in neurodegeneration and longevity: role of the heme oxygenase and HSP70 systems in brain stress tolerance. Antioxidants & Redox Signaling 6: 895-913.

Calò, L.A., Davis, P.A., Pagnin, E., Bertipaglia, L., Naso, A., Piccoli, A., Corradini, R., Spinello, M., Savica, V. and Libera, L.D., 2008. Carnitine mediated improved response to erythropoietin involves induction of haem oxygenase-1: studies in humans and in an animal model. Nephrology Dialysis Transplantation 23: 890-895.

Calò, L.A., Pagnin, E., Davis, P.A., Semplicini, A., Nicolai, R., Calvani, M. and Pessina, A.C., 2006. Antioxidant effect of L-carnitine and its short chain esters: relevance for the protection from oxidative stress related cardiovascular damage. International Journal of Cardiology 107: 54-60.

Cao, Y., Li, X., Shi, P., Wang, L.X. and Sui, Z.G., 2014. Effects of L-carnitine on high glucose-induced oxidative stress in retinal ganglion cells. Pharmacology 94: 123-130.

Cao, Y., Li, X., Wang, C.J., Li, P., Yang, B., Wang, C.B. and Wang, L.X., 2015. Role of NF-E2-related factor 2 in neuroprotective effect of L-carnitine against high glucose-induced oxidative stress in the retinal ganglion cells. Biomedicine & Pharmacotherapy 69: 345-348.

Cao, Y., Qu, H.J., Li, P., Wang, C.B., Wang, L.X. and Han, Z.W., 2011. Single dose administration of L-carnitine improves antioxidant activities in healthy subjects. The Tohoku Journal of Experimental Medicine 224: 209-213.

Casillas, E.R. and Newburgh, R.W., 1969. Carnitine and derivatives in embryonic chick tissue. Biochimica et Biophysica Acta (BBA)-General Subjects 184: 566-577.

Celik, L. and Öztürkcan, O., 2003. Effects of dietary supplemental L-carnitine and ascorbic acid on performance, carcass composition and plasma L-carnitine concentration of broiler chicks reared under different temperature. Archives of Animal Nutrition 57: 27-38.

Celik, L.B., Tekeli, A. and Öztürkcan, O., 2004. Effects of supplemental l-carnitine in drinking water on performance and egg quality of laying hens exposed to a high ambient temperature. Journal of Animal Physiology and Animal Nutrition 88: 229-233.

Çetin, E. and Güçlü, B.K., 2020. Effect of dietary L-carnitine supplementation and energy level on oxidant/antioxidant balance in laying hens subjected to high stocking density Journal of Animal Physiology and Animal Nutrition 104: 136-143.

Chen, Y.J., Kim, I.H., Cho, J.H., Yoo, J.S., Wang, Q., Wang, Y. and Huang, Y., 2008. Evaluation of dietary L-carnitine or garlic powder on growth performance, dry matter and nitrogen digestibilities, blood profiles and meat quality in finishing pigs. Animal Feed Science and Technology 141: 141-152.

Chen, Y.P., Chen, X., Zhang, H. and Zhou, Y.M., 2013. Effects of dietary concentrations of methionine on growth performance and oxidative status of broiler chickens with different hatching weight. British Poultry Science 54: 531-537.

Chiechio, S., Copani, A., De Petris, L., Morales, M.E.P., Nicoletti, F. and Gereau IV, R.W., 2006. Transcriptional regulation of metabotropic glutamate receptor 2/3 expression by the NF-κB pathway in primary dorsal root ganglia neurons: a possible mechanism for the analgesic effect of L-acetylcarnitine. Molecular Pain 2: 1744-8069.

Chiodi, P., Ciani, B., Kentroti, S., Maccari, F., Vernadakis, A., Angelucci, L. and Ramacci, M.T., 1994. Carnitine and derivatives in the central nervous system of chick embryo. The International Journal of Biochemistry 26: 711-720.

Chou, H.C., Wen, L.L., Chang, C.C., Lin, C.Y., Jin, L. and Juan, S.H., 2017. From the cover: L-carnitine via PPARγ- and Sirt1-dependent mechanisms attenuates epithelial-mesenchymal transition and renal fibrosis caused by perfluorooctanesulfonate. Toxicological Sciences 160: 217-229.

Clark, R.M., Balakrishnan, A., Waters, D., Aggarwal, D., Owen, K.Q. and Koo, S.I., 2007. L-carnitine increases liver α-tocopherol and lowers liver and plasma triglycerides in aging ovariectomized rats. The Journal of Nutritional Biochemistry 18: 623-628.

Conner, E.M. and Grisham, M.B., 1996. Inflammation, free radicals, and antioxidants. Nutrition 12: 274-277.

Cuccurazzu, B., Bortolotto, V., Valente, M.M., Ubezio, F., Koverech, A., Canonico, P.L. and Grilli, M., 2013. Upregulation of mGlu2 receptors via NF-κ B p65 acetylation is involved in the proneurogenic and antidepressant effects of acetyl-L-carnitine. Neuropsychopharmacology 38: 2220-2230.

De, C.S., Famularo, G., Tzantzoglou, S., Trinchieri, V., Moretti, S. and Sorice, F., 1994. Carnitine depletion in peripheral blood mononuclear cells from patients with AIDS: effect of oral L-carnitine. AIDS 8: 655-660.

De Simone, C., Tzantzoglou, S., Famularo, G., Moretti, S., Paoletti, F., Vullo, V. and Delia, S., 1993. High dose L-carnitine improves immunologic and metabolic parameters in AIDS patients. Immunopharmacology and Immunotoxicology 15: 1-12.

Delogu, G., De Simone, C., Famularo, G., Fegiz, A., Paoletti, F. and Jirillo, E., 1993. Anaesthetics modulate tumour necrosis factor α: effects of L-carnitine supplementation in surgical patients. Preliminary results. Mediators of Inflammation 2: S33-S36.

Deng, K., Wong, C.W. and Nolan, J.V., 2006. Long-term effects of early-life dietary L-carnitine on lymphoid organs and immune responses in Leghorn-type chickens. Journal of Animal Physiology and Animal Nutrition 90: 81-86.

Devi, S.A., Vani, R., Subramanyam, M.V.V., Reddy, S.S. and Jeevaratnam, K., 2007. Intermittent hypobaric hypoxia-induced oxidative stress in rat erythrocytes: protective effects of vitamin E, vitamin C, and carnitine. Cell Biochemistry and Function 25: 221-231.

Dhitavat, S., Ortiz, D., Rogers, E., Rivera, E. and Shea, T.B., 2005. Folate, vitamin E, and acetyl-L-carnitine provide synergistic protection against oxidative stress resulting from exposure of human neuroblastoma cells to amyloid-beta. Brain Research 1061: 114-117.

Di, L.M., Alesse, E., Roncaioli, P., Muzi, P., Moretti, S., Marcellini, S., Amicosante, G., De, C.S. and Cifone, M.G., 1997. Influence of L-carnitine on CD95 cross-lining-induced apoptosis and ceramide generation in human cell lines: correlation with its effects on purified acidic and neutral sphingomyelinases *in vitro*. Proceedings of the Association of American Physicians 109: 154-163.

Dinkova-Kostova, A.T. and Abramov, A.Y., 2015. The emerging role of Nrf2 in mitochondrial function. Free Radical Biology and Medicine 88: 179-188.

Dong, X.Y., Azzam, M.M.M., Rao, W., Yu, D.Y. and Zou, X.T., 2012. Evaluating the impact of excess dietary tryptophan on laying performance and immune function of laying hens reared under hot and humid summer conditions. British Poultry Science 53: 491-496.

Duran, J.M., Peral, M.J. and Calonge, M.L., 2002. Functional characterization of intestinal L-carnitine transport. Journal of Membrane Biology, 185^ 65-74.

Durazzo, A., Lucarini, M., Nazhand, A., Souto, S.B., Silva, A.M., Severino, P., Souto, E.B., and Santini, A., 2020. The nutraceutical value of carnitine and its use in dietary supplements. Molecules 25: E2127.

Eder, K., 2005. Effects of L-carnitine supplementation in sows. Monatshefte für Chemie 136: 535-1544.

El-Awady, E.S.E., Moustafa, Y.M., Abo-Elmatty, D.M. and Radwan, A., 2011. Cisplatin-induced cardiotoxicity: Mechanisms and cardioprotective strategies. European Journal of Pharmacology 650: 335-341.

Elliott, G.R., Lauwen, A.P.M. and Bonta, I.L., 1990. The effect of acute feeding of carnitine, acetyl carnitine and propionyl carnitine on basal and A23187-stimulated eicosanoid release from rat carrageenan-elicited peritoneal macrophages. British Journal of Nutrition 64: 497-503.

El-Sherbini, E.S., El-Sayed, G., El Shotory, R., Gheith, N., Abou-Alsoud, M., Harakeh, S.M. and Karrouf, G.I., 2017. Ameliorative effects of L-carnitine on rats raised on a diet supplemented with lead acetate. Saudi Journal of Biological Sciences 24: 1410-1417.

Enkvetchakul, B., Bottje, W., Anthony, N., Moore, R. and Huff, W., 1993. Compromised antioxidant status associated with ascites in broilers. Poultry Science 72: 2272-2280.

Evans, A.M. and Fornasini, G., 2003. Pharmacokinetics of L-carnitine. Clinical Pharmacokinetics 42: 941-967.

Fan, Z., Han, Y., Ye, Y., Liu, C. and Cai, H., 2017. L-carnitine preserves cardiac function by activating p38 MAPK/Nrf2 signalling in hearts exposed to irradiation. European journal of Pharmacology 804: 7-12.

Fattorossi, A., Biselli, R., Casciaro, A., Tzantzoglou, S. and de Simone, C., 1993. Regulation of normal human polyrnorphonuclear leucocytes by carnitine. Mediators of Inflammation 2: S37-S41.

Ferrara, F., Bertelli, A. and Falchi, M., 2005. Evaluation of carnitine, acetylcarnitine and isovalerylcarnitine on immune function and apoptosis. Drugs Under Experimental and Clinical Research 31: 109-114.

Ferrari, R., Di Mauro, S. and Shervood G., 1992. L-carnitine and its role in medicine: from function to therapy, Academic Press. London, UK.

Fischer, M., Keller, J., Hirche, F., Kluge, H., Ringseis, R. and Eder, K., 2009. Activities of γ-butyrobetaine dioxygenase and concentrations of carnitine in tissues of pigs. Comparative Biochemistry and Physiology Part A: Molecular & Integrative Physiology 153: 324-331.

Fischer, M., Varady, J., Hirche, F., Kluge, H. and Eder, K., 2009. Supplementation of L-carnitine in pigs: absorption of carnitine and effect on plasma and tissue carnitine concentrations. Archives of Animal Nutrition 63: 1-15.

Fortin, G., Yurchenko, K., Collette, C., Rubio, M., Villani, A.C., Bitton, A., Sarfati, M. and Franchimont, D., 2009. L-carnitine, a diet component and organic cation transporter OCTN ligand, displays immunosuppressive properties and abrogates intestinal inflammation. Clinical & Experimental Immunology 156: 161-171.

Fraenkel, G., 1953. Studies on the distribution of vitamin BT (carnitine). The Biological Bulletin 104: 359-371.

Furuno, T., Kanno, T., Arita, K., Asami, M., Utsumi, T., Doi, Y., Inoue, M. and Utsumi, K., 2001. Roles of long chain fatty acids and carnitine in mitochondrial membrane permeability transition. Biochemical Pharmacology 62: 1037-1046.

Geng, A., Guo, Y. and Yuan, J., 2004. Effects of dietary L-carnitine and coenzyme Q10 supplementation on performance and ascites mortality of broilers. Archives of Animal Nutrition 58: 473-482.

Geng, A., Li, B. and Guo, Y., 2007. Effects of dietary L-carnitine and coenzyme Q10 at different supplemental ages on growth performance and some immune response in ascites-susceptible broilers. Archives of Animal Nutrition 61: 50-60.

Adabi, S.G., Cooper, R.G., Ceylan, N. and Corduk, M., 2011. L-carnitine and its functional effects in poultry nutrition. World's Poultry Science Journal 67: 277-296.

Güçlü, B.K., Kara, K., Çakır, L., Çetin, E. and Kanbur, M., 2011. Carnitine supplementation modulates high dietary copper-induced oxidative toxicity and reduced performance in laying hens. Biological Trace Element Research 144: 725-735.

Gülcin, İ., 2006. Antioxidant and antiradical activities of L-carnitine. Life Sciences 78: 803-811.

Gulewitsch, W.L. and Krimberg, R., 1905. Zur Kenntnis der Extraktivstoffe der Muskeln. II. Mitteilung. Über das Carnitin. Hoppe-Seyler′ s Zeitschrift für physiologische Chemie 45: 326-330.

Hagen, T.M., Liu, J., Lykkesfeldt, J., Wehr, C.M., Ingersoll, R.T., Vinarsky, V., Bartholomew, J.C. and Ames, B.N., 2002. Feeding acetyl-L-carnitine and lipoic acid to old rats significantly improves metabolic function while decreasing oxidative stress. Proceedings of the National Academy of Sciences 99: 1870-1875.

Hagen, T.M., Moreau, R., Suh, J.H. and Visioli, F., 2002. Mitochondrial decay in the aging rat heart: evidence for improvement by dietary supplementation with acetyl-L-carnitine and/or lipoic acid. Annals of the New York Academy of Sciences 959: 491-507.

Halliwell, B., 2012. Free radicals and antioxidants: updating a personal view. Nutrition Reviews 70: 257-265.

Heo, K., Lin, X.I., Odle, J. and Han, I.K., 2000. Kinetics of carnitine palmitoyltransferase-I are altered by dietary variables and suggest a metabolic need for supplemental carnitine in young pigs. The Journal of Nutrition 130: 2467-2470.

Heo, K.N., Odle, J. and Han, I.K., 2000. Effects of dietary L-carnitine and protein level on plasma carnitine, energy and carnitine balance, and carnitine biosynthesis of 20 kg pigs. Asian-Australasian Journal of Animal Sciences 13: 1568-1575.

Harmeyer, J., 2002. The physiological role of L-carnitine. Lohman Information 27: 15-21.

Hildeman, D.A., 2004. Regulation of T-cell apoptosis by reactive oxygen species. Free Radical Biology and Medicine 36: 1496-1504.

Hota, K.B., Hota, S.K., Chaurasia, O.P. and Singh, S.B., 2012. Acetyl-L-carnitine-mediated neuroprotection during hypoxia is attributed to ERK1/2-Nrf2-regulated mitochondrial biosynthesis. Hippocampus 22: 723-736.

Hu, M.L., 2011. Dietary polyphenols as antioxidants and anticancer agents: more questions than answers. Chang Gung Medical Journal 34: 449-460.

Hughes, D.A., 1999. Effects of dietary antioxidants on the immune function of middle-aged adults. Proceedings of the Nutrition Society 58: 79-84.

Ishii, T., Shimpo, Y., Matsuoka, Y. and Kinoshita, K., 2000. Anti-apoptotic effect of acetyl-L-carnitine and L-carnitine in primary cultured neurons. The Japanese Journal of Pharmacology 83: 119-124.

Itoh, K., Ye, P., Matsumiya, T., Tanji, K. and Ozaki, T., 2015. Emerging functional cross-talk between the Keap1-Nrf2 system and mitochondria. Journal of Clinical Biochemistry and Nutrition 56: 91-97.

İzgüt-Uysal, V.N., Ağaç, A., Karadoğan, İ. and Derin, N., 2003. Effects of L-carnitine on neutrophil functions in aged rats. Mechanisms of Ageing and Development 124: 341-347.

Izgüt-Uysal, V.N., Agaç, A., Karadogan, I. and Derin, N., 2004. Peritoneal macrophages function modulation by L-carnitine in aging rats. Aging Clinical and Experimental Research 16: 337-341.

Janssens, G.P., Buyse, J., Seynaeve, M., Decuypere, E. and De Wilde, R., 1998. The reduction of heat production in exercising pigeons after L-carnitine supplementation. Poultry Science 77: 578-584.

Janssens, G.P.J., Mast, J., Goddeeris, B.M., Cox, E., Hesta, M. and De Wilde, R.O.M., 2000. Enhanced specific antibody response to bovine serum albumin in pigeons due to L-carnitine supplementation. British Poultry Science 41: 448-453.

Jia, R., Bao, Y.H., Zhang, Y., Ji, C., Zhao, L.H., Zhang, J.Y., Gao, C.Q. and Ma, Q.G., 2014. Effects of dietary α-lipoic acid, acetyl-L-carnitine, and sex on antioxidative ability, energy, and lipid metabolism in broilers. Poultry Science 93: 2809-2817.

Jiang, F., Zhang, Z., Zhang, Y., Wu, J., Yu, L. and Liu, S., 2016. L-carnitine ameliorates the liver inflammatory response by regulating carnitine palmitoyltransferase I-dependent PPARγ signaling. Molecular Medicine Reports 13: 1320-1328.

Jirillo, E., Altamura, M., Munno, I., Pellegrino, N.M., Sabato, R., Fabio, S.D. and Simone, C.D., 1991. Effects of acetyl-L-carnitine oral administration on lymphocyte antibacterial activity and TNF-α levels in patients with active pulmonary tuberculosis. A randomized double blind versus placebo study. Immunopharmacology and immunotoxicology 13: 135-146.

Jourd'heuil, D., Morise, Z., Conner, E.M. and Grisham, M.B., 1997. Oxidants, transcription factors, and intestinal inflammation. Journal of Clinical Gastroenterology 25: S61-S72.

Jun, D.W., Cho, W.K., Jun, J.H., Kwon, H.J., Jang, K.S., Kim, H.J., Jeon, H.J., Lee, K.N., Lee, H.L., Lee, O.Y. and Yoon, B.C., 2011. Prevention of free fatty acid-induced hepatic lipotoxicity by carnitine via reversal of mitochondrial dysfunction. Liver International 31: 1315-1324.

Kargas, S.A., Bruyere Jr, H.J., Gilbert, E.F. and Shug, A.L., 1985. Changes in carnitine levels in the embryonic chick heart during development. Comparative Biochemistry and physiology. B, Comparative Biochemistry 82: 525-527.

Karlic, H. and Lohninger, A., 2004. Supplementation of L-carnitine in athletes: does it make sense? Nutrition 20: 709-715.

Kerner, J., Yohannes, E., Lee, K., Virmani, A., Koverech, A., Cavazza, C., Chance, M.R. and Hoppel, C., 2015. Acetyl-L-carnitine increases mitochondrial protein acetylation in the aged rat heart. Mechanisms of Ageing and Development 145: 39-50.

Keshavarz-Bahaghighat, H., Sepand, M.R., Ghahremani, M.H., Aghsami, M., Sanadgol, N., Omidi, A., Bodaghi-Namileh, V. and Sabzevari, O., 2018. Acetyl-L-carnitine attenuates arsenic-induced oxidative stress and hippocampal mitochondrial dysfunction. Biological Trace Element Research 184: 422-435.

Khajali, F. and Khajali, Z., 2014. L-carnitine supplementation decreases lipid peroxidation and improves cardiopulmonary function in broiler chickens reared at high altitude. Acta Veterinaria Hungarica 62: 489-499.

Khushboo, M., Murthy, M.K., Devi, M.S., Sanjeev, S., Ibrahim, K.S., Kumar, N.S., Roy, V.K. and Gurusubramanian, G., 2018. Testicular toxicity and sperm quality following copper exposure in Wistar albino rats: ameliorative potentials of L-carnitine. Environmental Science and Pollution Research 25: 1837-1862.

Kidd, M.T., McDaniel, C.D., Peebles, E.D., Barber, S.J., Corzo, A., Branton, S.L. and Woodworth, J.C., 2005. Breeder hen dietary L-carnitine affects progeny carcase traits. British Poultry Science 46: 97-103.

Klasing, K.C., 1998. Nutritional modulation of resistance to infectious diseases. Poultry Science 77: 1119-1125.

Koc, A., Ozkan, T., Karabay, A.Z., Sunguroglu, A. and Aktan, F., 2011. Effect of L-carnitine on the synthesis of nitric oxide in RAW 264· 7 murine macrophage cell line. Cell Biochemistry and function 29: 679-685.

Kohchi, C., Inagawa, H., Nishizawa, T. and Soma, G.I., 2009. ROS and innate immunity. Anticancer Research 29: 817-821.

Kouttab, N.M. and De Simone, C., 1993. Modulation of cytokine production by carnitine. Mediators of Inflammation 2: S25-S28.

Krahenbuhl, S., 2000. L-carnitine and vegetarianism. Annals of Nutrition & Metabolism 44: 81-82.

Kumaran, S., Panneerselvam, K.S., Shila, S., Sivarajan, K. and Panneerselvam, C., 2005. Age-associated deficit of mitochondrial oxidative phosphorylation in skeletal muscle: role of carnitine and lipoic acid. Molecular and Cellular Biochemistry 280: 83-89.

Kunak, C.S., Ugan, R.A., Cadirci, E., Karakus, E., Polat, B., Un, H., Halici, Z., Saritemur, M., Atmaca, H.T. and Karaman, A., 2016. Nephroprotective potential of carnitine against glycerol and contrast-induced kidney injury in rats through modulation of oxidative stress, proinflammatory cytokines, and apoptosis. The British Journal of Radiology 89: 20140724.

Li, J., Zhang, Y., Luan, H., Chen, X., Han, Y. and Wang, C., 2016. L-carnitine protects human hepatocytes from oxidative stress-induced toxicity through Akt-mediated activation of Nrf2 signaling pathway. Canadian Journal of Physiology and Pharmacology 94: 517-525.

Li, M., Xu, S., Geng, Y., Sun, L., Wang, R., Yan, Y., Wang, H., Li, Y., Yi, Q., Zhang, Y. and Hao, J., 2019. The protective effects of L-carnitine on myocardial ischaemia-reperfusion injury in patients with rheumatic valvular heart disease undergoing CPB surgery are associated with the suppression of NF-κB pathway and the activation of Nrf2 pathway. Clinical and Experimental Pharmacology and Physiology 46: 1001-1012.

Li, K., Gao, J.G., Sun, Y.B. and Hou, S.C., 2012. The antioxidation of L-carnitine on renal ischemia-reperfusion injury and its underlying mechanism. Journal of Modern Urology 3: 12.

Li, J., Zhang, Y., Luan, H., Chen, X., Han, Y. and Wang, C., 2016. L-carnitine protects human hepatocytes from oxidative stress-induced toxicity through Akt-mediated activation of Nrf2 signaling pathway. Canadian Journal of Physiology and Pharmacology 94: 517-525.

Lin, M.T. and Beal, M.F., 2006. Mitochondrial dysfunction and oxidative stress in neurodegenerative diseases. Nature 443: 787-795.

Liu, Z., Shen, H. and Zhang, J., 2009. Clinical effect of silymarin combined with levocarnitine on non-alcoholic fatty liver disease. China Modern Medicine 7: p.R575.

Liu, F., Rainosek, S.W., Sadovova, N., Fogle, C.M., Patterson, T.A., Hanig, J.P., Paule, M.G., Slikker Jr, W. and Wang, C., 2014. Protective effect of acetyl-L-carnitine on propofol-induced toxicity in embryonic neural stem cells. Neurotoxicology 42: 49-57.

Liu, J., Killilea, D.W. and Ames, B.N., 2002. Age-associated mitochondrial oxidative decay: improvement of carnitine acetyltransferase substrate-binding affinity and activity in brain by feeding old rats acetyl-L-carnitine and/or R-α-lipoic acid. Proceedings of the National Academy of Sciences 99: 1876-1881.

Lohninger, A., Pittner, G. and Pittner, F., 2005. L-carnitine: new aspects of a known compound – a brief survey. Monatshefte für Chemie 136: 1255-1268.

Lösel, D. and Rehfeldt, C., 2013. Effects of L-carnitine supplementation to suckling piglets on carcass and meat quality at market age. Animal 7: 1191-1198.

Manoli, I., De Martino, M.U., Kino, T. and Alesci, S., 2004. Modulatory effects of L-carnitine on glucocorticoid receptor activity. Annals of the New York Academy of Sciences 1033: 147-157.

Mao, C.Y., Lu, H.B., Kong, N., Li, J.Y., Liu, M., Yang, C.Y. and Yang, P., 2014. Levocarnitine protects H9c2 rat cardiomyocytes from H_2O_2-induced mitochondrial dysfunction and apoptosis. International Journal of Medical Sciences 11: 1107-1115.

Marcovina, S.M., Sirtori, C., Peracino, A., Gheorghiade, M., Borum, P., Remuzzi, G. and Ardehali, H., 2013. Translating the basic knowledge of mitochondrial functions to metabolic therapy: role of L-carnitine. Translational Research 161: 73-84.

Mast, J., Buyse, J. and Goddeeris, B.M., 2000. Dietary L-carnitine supplementation increases antigen-specific immunoglobulin G production in broiler chickens. British Journal of Nutrition 83: 161-166.

McMackin, C.J., Widlansky, M.E., Hamburg, N.M., Huang, A.L., Weller, S., Holbrook, M., Gokce, N., Hagen, T.M., Keaney Jr, J.F. and Vita, J.A., 2007. Effect of combined treatment with α-lipoic acid and acetyl-l-carnitine on vascular function and blood pressure in patients with coronary artery disease. The Journal of Clinical Hypertension 9: 249-255.

Micic, S., Lalic, N., Djordjevic, D., Bojanic, N., Bogavac-Stanojevic, N., Busetto, G.M., Virmani, A. and Agarwal, A., 2019. Double-blind, randomised, placebo-controlled trial on the effect of L-carnitine and L-acetylcarnitine on sperm parameters in men with idiopathic oligoasthenozoospermia. Andrologia 51: e13267.

Miguel-Carrasco, J.L., Monserrat, M.T., Mate, A. and Vázquez, C.M., 2010. Comparative effects of captopril and L-carnitine on blood pressure and antioxidant enzyme gene expression in the heart of spontaneously hypertensive rats. European Journal of Pharmacology 632: 65-72.

Moazzami, A.A., Andersson, R. and Kamal-Eldin, A., 2011. Changes in the metabolic profile of rat liver after α-tocopherol deficiency as revealed by metabolomics analysis. NMR in Biomedicine 24: 499-505.

Montesano, A., Senesi, P., Vacante, F., Mollica, G., Benedini, S., Mariotti, M., Luzi, L. and Terruzzi, I., 2020. L-carnitine counteracts *in vitro* fructose-induced hepatic steatosis through targeting oxidative stress markers. Journal of Endocrinological Investigation 43: 493-503.

Morand, R., Bouitbir, J., Felser, A., Hench, J., Handschin, C., Frank, S. and Krähenbühl, S., 2014. Effect of carnitine, acetyl-, and propionylcarnitine supplementation on the body carnitine pool, skeletal muscle composition, and physical performance in mice. European Journal of Nutrition 53: 1313-1325.

Moretti, S., Famularo, G., Marcellini, S., Boschini, A., Santini, G., Trinchieri, V., Lucci, L., Alesse, E. and De Simone, C., 2002. L-carnitine reduces lymphocyte apoptosis and oxidant stress in HIV-1-infected subjects treated with zidovudine and didanosine. Antioxidants and Redox Signaling 4: 391-403.

Morigi, M., Perico, L., Rota, C., Longaretti, L., Conti, S., Rottoli, D., Novelli, R., Remuzzi, G. and Benigni, A., 2015. Sirtuin 3-dependent mitochondrial dynamic improvements protect against acute kidney injury. The Journal of Clinical Investigation 125: 715-726.

Mosca, L., Marcellini, S., Perluigi, M., Mastroiacovo, P., Moretti, S., Famularo, G., Peluso, I., Santini, G. and De Simone, C., 2002. Modulation of apoptosis and improved redox metabolism with the use of a new antioxidant formula. Biochemical Pharmacology 63: 1305-1314.

Musicco, C., Capelli, V., Pesce, V., Timperio, A.M., Calvani, M., Mosconi, L., Cantatore, P. and Gadaleta, M.N., 2011. Rat liver mitochondrial proteome: changes associated with aging and acetyl-L-carnitine treatment. Journal of Proteomics 74: 2536-2547.

Nouboukpo, K.E., Tona, K., Agbonon, A., Gbeassor, M., Buyse, J. and Decuypere, E., 2010. Effects of L-carnitine supplementation in drinking water on layer-type chick juvenile performance. Archiv für Geflügelkunde 74: 116-120.

Orlandi, A., Francesconi, A., Marcellini, M., Di Lascio, A. and Spagnoli, L.G., 2007. Propionyl-L-carnitine reduces proliferation and potentiates Bax-related apoptosis of aortic intimal smooth muscle cells by modulating nuclear factor-κB activity. Journal of Biological Chemistry 282: 4932-4942.

Pagano, G., Aiello Talamanca, A., Castello, G., Cordero, M.D., d'Ischia, M., Gadaleta, M.N., Pallardó, F.V., Petrović, S., Tiano, L. and Zatterale, A., 2014. Current experience in testing mitochondrial nutrients in disorders featuring oxidative stress and mitochondrial dysfunction: rational design of chemoprevention trials. International Journal of Molecular Sciences 15: 20169-20208.

Parsaeimehr, K., Afrouziyeh, M. and Hoseinzadeh, S., 2014. The effects of L-carnitine and different levels of animal fat on performance, carcass characteristics, some blood parameters and immune response in broiler chicks. Iranian Journal of Applied Animal Science 4: 561-566.

Peebles, E.D., Kidd, M.T., McDaniel, C.D., Tanksley, J.P., Parker, H.M., Corzo, A. and Woodworth, J.C., 2007. Effects of breeder hen age and dietary L-carnitine on progeny embryogenesis. British Poultry Science 48: 299-307.

Pochini, L., Scalise, M., Galluccio, M. and Indiveri, C., 2013. OCTN cation transporters in health and disease: role as drug targets and assay development. Journal of Biomolecular Screening 18: 851-867.

Rabie, M.H., SziláGyi, M. and Gippert, T., 1997. Effects of dietary L-carnitine on the performance and egg quality of laying hens from 65-73 weeks of age. British Journal of Nutrition 78: 615-623.

Rebouche, C.J., 2004. Kinetics, pharmacokinetics, and regulation of l-carnitine and acetyl-l-carnitine metabolism. Annals of the New York Academy of Sciences 1033: 30-41.

Revoltella, R.P., Dal Canto, B., Caracciolo, L. and D'Urso, C.M., 1994. L-carnitine and some of its analogs delay the onset of apoptotic cell death initiated in murine C2. 8 hepatocytic cells after hepatocyte growth factor deprivation. Biochimica et Biophysica Acta (BBA)-Molecular Cell Research 1224: 333-341.

Reza, Y.M., Saeid, S.S. and Javad, N., 2015. The effects of different levels of L-carnitine and methionine on immune response and some blood parameters in female broiler. Research Opinions in Animal and Veterinary Sciences 5: 94-98.

Reznick, A.Z., Kagan, V.E., Ramsey, R., Tsuchiya, M., Khwaja, S., Serbinova, E.A. and Packer, L., 1992. Antiradical effects in L-propionyl carnitine protection of the heart against ischemia-reperfusion injury: the possible role of iron chelation. Archives of Biochemistry and Biophysics 296: 394-401.

Ribas, G.S., Vargas, C.R. and Wajner, M., 2014. L-carnitine supplementation as a potential antioxidant therapy for inherited neurometabolic disorders. Gene 533: 469-476.

Rinaudo, M.T., Curto, M., Bruno, R., Piccinini, M. and Marino, C., 1991. Acid soluble, short chain esterified and free carnitine in the liver, heart, muscle and brain of pre and post hatched chicks. International Journal of Biochemistry 23: 59-65.

Ringseis, R. and Eder, K., 2009. Influence of pharmacological PPARα activators on carnitine homeostasis in proliferating and non-proliferating species. Pharmacological Research 60: 179-184.

Ringseis, R., Keller, J. and Eder, K., 2018. Basic mechanisms of the regulation of L-carnitine status in monogastrics and efficacy of L-carnitine as a feed additive in pigs and poultry. Journal of Animal Physiology and Animal Nutrition 102: 1686-1719.

Ringseis, R., Luci, S., Spielmann, J., Kluge, H., Fischer, M., Geissler, S., Wen, G., Hirche, F. and Eder, K., 2008. Clofibrate treatment up-regulates novel organic cation transporter (OCTN)-2 in tissues of pigs as a model of non-proliferating species. European Journal of Pharmacology 583: 11-17.

Ringseis, R., Lüdi, S., Hirche, F. and Eder, K., 2008a. Treatment with pharmacological peroxisome proliferator-activated receptor α agonist clofibrate increases intestinal carnitine absorption in rats. Pharmacological Research 58: 58-64.

Ringseis, R., Wege, N., Wen, G., Rauer, C., Hirche, F., Kluge, H. and Eder, K., 2009. Carnitine synthesis and uptake into cells are stimulated by fasting in pigs as a model of nonproliferating species. The Journal of Nutritional Biochemistry 20: 840-847.

Ringseis, R., Wen, G. and Eder, K., 2012. Regulation of genes involved in carnitine homeostasis by PPARα across different species (rat, mouse, pig, cattle, chicken, and human). PPAR Research 2012: 868317.

Roy, V.K., Verma, R. and Krishna, A., 2017. Carnitine-mediated antioxidant enzyme activity and Bcl2 expression involves peroxisome proliferator-activated receptor-γ coactivator-1α in mouse testis. Reproduction, Fertility and Development 29: 1057-1063.

Rubio, J.C., de Bustos, F., Molina, J.A., Jiménez-Jiménez, F.J., Benito-León, J., Martín, M.A., Campos, Y., Ortí-Pareja, M., Cabrera-Valdivia, F. and Arenas, J., 1998. Cerebrospinal fluid carnitine levels in patients with Alzheimer's disease. Journal of the Neurological Sciences 155: 192-195.

Ruggiero, V., D'Urso, C.M., Albertoni, C., Campo, S., Foresta, P. and Martelli, E.A., 1993. LPS-induced serum TNF production and lethality in mice: effect of L-carnitine and some acyl-derivatives. Mediators of Inflammation 2: S43-S50.

Sachan, D.S., Hongu, N. and Johnsen, M., 2005. Decreasing oxidative stress with choline and carnitine in women. Journal of the American College of Nutrition 24: 172-176.

Savitha, S. and Panneerselvam, C., 2007. Mitigation of age-dependent oxidative damage to DNA in rat heart by carnitine and lipoic acid. Mechanisms of Ageing and Development 128: 206-212.

Savitha, S., Sivarajan, K., Haripriya, D., Kokilavani, V. and Panneerselvam, C., 2005. Efficacy of levo carnitine and alpha lipoic acid in ameliorating the decline in mitochondrial enzymes during aging. Clinical Nutrition 24: 794-800.

Sekine, S. and Ichijo, H., 2015. Mitochondrial proteolysis: its emerging roles in stress responses. Biochimica et Biophysica Acta (BBA)-General Subjects 1850: 274-280.

Şener, G., Ekşioğlu-Demiralp, E., Cetiner, M., Ercan, F., Şirvancı, S., Gedik, N. and Yeğen, B.C., 2006. L-carnitine ameliorates methotrexate-induced oxidative organ injury and inhibits leukocyte death. Cell Biology and Toxicology 22: 47-60.

Sepand, M.R., Razavi-Azarkhiavi, K., Omidi, A., Zirak, M.R., Sabzevari, S., Kazemi, A.R. and Sabzevari, O., 2016. Effect of acetyl-L-carnitine on antioxidant status, lipid peroxidation, and oxidative damage of arsenic in rat. Biological Trace Element Research 171: 107-115.

Sharifi, M., Hassanpour, H. and Khajali, F., 2015. Dietary L-carnitine supplement counteracts pulmonary hypertensive response in broiler chickens fed reduced-protein diets and subjected to cool condition and hypobaric hypoxia. The Journal of Poultry Science 52: 206-212.

Shekhawat, P.S., Srinivas, S.R., Matern, D., Bennett, M.J., Boriack, R., George, V., Xu, H., Prasad, P.D., Roon, P. and Ganapathy, V., 2007. Spontaneous development of intestinal and colonic atrophy and inflammation in the carnitine-deficient jvs (OCTN2−/−) mice. Molecular Genetics and Metabolism 92: 315-324.

Shen, W., Liu, K., Tian, C., Yang, L., Li, X., Ren, J., Packer, L., Cotman, C.W. and Liu, J., 2008. R-α-Lipoic acid and acetyl-L-carnitine complementarily promote mitochondrial biogenesis in murine 3T3-L1 adipocytes. Diabetologia 51: 165-174.

Shimada, K., Sakuma, Y., Wakamatsu, J., Fukushima, M., Sekikawa, M., Kuchida, K. and Mikami, M., 2004. Species and muscle differences in L-carnitine levels in skeletal muscles based on a new simple assay. Meat Science 68: 357-362.

Sonne, S., Shekhawat, P.S., Matern, D., Ganapathy, V. and Ignatowicz, L., 2012. Carnitine deficiency in OCTN2−/− newborn mice leads to a severe gut and immune phenotype with widespread atrophy, apoptosis and a pro-inflammatory response. PLoS One 7: e47729.

Spaniol, M., Brooks, H., Auer, L., Zimmermann, A., Solioz, M., Stieger, B. and Krähenbühl, S., 2001. Development and characterization of an animal model of carnitine deficiency. European Journal of Biochemistry 268: 1876-1887.

Stanley, C.A., 2004. Carnitine deficiency disorders in children. Annals of the New York Academy of Sciences 1033: 42-51.

Strzałkowski, A.K., Godlewski, M.M., Hallay, N., Kulasek, G., Gajewski, Z. and Zabielski, R., 2007. The effect of supplementing sow with bioactive substances on neonatal small intestinal epithelium. Journal of Physiology and Pharmacology 58: 115-122.

Sue, Y.M., Chou, H.C., Chang, C.C., Yang, N.J., Chou, Y. and Juan, S.H., 2014. L-carnitine protects against carboplatin-mediated renal injury: AMPK-and PPARα-dependent inactivation of NFAT3. PloS One 9: e104079.

Surai, P.F., 2002. Natural antioxidants in avian nutrition and reproduction. Nottingham University Press, Nottingham, UK.

Surai, P.F., 2006, Selenium in nutrition and health. Nottingham University Press, Nottingham, UK.

Surai, P.F., 2015a. Antioxidant action of carnitine: molecular mechanisms and practical applications. EC Veterinary Science 2.1: 66-84.

Surai, P.F., 2015b. Carnitine enigma: from antioxidant action to vitagene regulation. Part 1. Absorption, metabolism and antioxidant activities. Journal of Veterinary Science & Medicine 3, 2: 14.

Surai P.F., 2015c. Carnitine enigma: from antioxidant action to vitagene regulation. Part 2. Transcription factors and practical applications. Journal of Veterinary Science & Medicine 3, 2: 17.

Surai, P.F., 2015d. Silymarin as a natural antioxidant: an overview of the current evidence and perspectives. Antioxidants 4: 204-247.

Surai, P.F., 2018. Selenium in poultry nutrition and health. Wageningen Academic Publishers, Wageningen, the Netherlands.

Surai, P.F. and Fisinin, V.I., 2015. Antioxidant-prooxidant balance in the intestine: applications in chick placement and pig Weaning. Journal of Veterinary Science & Medicine 3, 1: 16.

Surai, P.F., Kochish, I.I., Fisinin, V.I. and Kidd, M.T., 2019. Antioxidant defence systems and oxidative stress in poultry biology: an update. Antioxidants 8, 7: 235.

Szilagyi, M., 1998. L-carnitine as essential methylated compound in animal metabolism. an overview. Acta Biologica Hungarica 49: 209-218.

Tamai, I., Ohashi, R., Nezu, J.I., Yabuuchi, H., Oku, A., Shimane, M., Sai, Y. and Tsuji, A., 1998. Molecular and functional identification of sodium ion-dependent, high affinity human carnitine transporter OCTN2. Journal of Biological Chemistry 273: 20378-20382.

Tamilselvan, J., Jayaraman, G., Sivarajan, K. and Panneerselvam, C., 2007. Age-dependent upregulation of p53 and cytochrome c release and susceptibility to apoptosis in skeletal muscle fiber of aged rats: role of carnitine and lipoic acid. Free Radical Biology and Medicine 43: 1656-1669.

Tan, X., Hu, S.H. and Wang, X.L., 2008. The effect of dietary l-carnitine supplementation on pulmonary hypertension syndrome mortality in broilers exposed to low temperatures. Journal of Animal Physiology and Animal Nutrition 92: 203-210.

Thangasamy, T., Subathra, M., Sittadjody, S., Jeyakumar, P., Joyee, A.G., Mendoza, E. and Chinnakkanu, P., 2008. Role of L-carnitine in the modulation of immune response in aged rats. Clinica Chimica Acta 389: 19-24.

Thomas, S., Fischer, F.P., Mettang, T., Pauli-Magnus, C., Weber, J. and Kuhlmann, U., 1999. Effects of L-carnitine on leukocyte function and viability in hemodialysis patients: a double-blind randomized trial. American Journal of Kidney Diseases 34: 678-687.

Türker, Y., Nazıroğlu, M., Gümral, N., Çelik, Ö., Saygın, M., Çömlekçi, S. and Flores-Arce, M., 2011. Selenium and L-carnitine reduce oxidative stress in the heart of rat induced by 2.45-GHz radiation from wireless devices. Biological Trace Element Research 143: 1640-1650.

Vaz, F.M. and Wanders, R.J., 2002. Carnitine biosynthesis in mammals. Biochemical Journal 361: 417-429.

Villar-Patiño, G., Díaz-Cruz, A., Ávila-González, E., Guinzberg, R., Pablos, J.L. and Piña, E., 2002. Effects of dietary supplementation with vitamin C or vitamin E on cardiac lipid peroxidation and growth performance in broilers at risk of developing ascites syndrome. American Journal of Veterinary Research 63: 673-676.

Volek, J.S., Kraemer, W.J., Rubin, M.R., Gómez, A.L., Ratamess, N.A. and Gaynor, P., 2002. L-carnitine L-tartrate supplementation favorably affects markers of recovery from exercise stress. American Journal of Physiology-Endocrinology and Metabolism 282: E474-E482.

Walter, P., 2000. Introduction. Annals of Nutrition & Metabolism 44: 77.

Wang, C., Sadovova, N., Ali, H.K., Duhart, H.M., Fu, X., Zou, X., Patterson, T.A., Binienda, Z.K., Virmani, A., Paule, M.G. and Slikker Jr, W., 2007. L-carnitine protects neurons from 1-methyl-4-phenylpyridinium-induced neuronal apoptosis in rat forebrain culture. Neuroscience 144: 46-55.

Wang, Y.W., Ning, D., Peng, Y.Z. and Guo, Y.M., 2013. Effects of dietary L-carnitine supplementation on growth performance, organ weight, biochemical parameters and ascites susceptibility in broilers reared under low-temperature environment. Asian-Australasian Journal of Animal Sciences 26: 233-240.

Weinberg, S.E., Sena, L.A. and Chandel, N.S., 2015. Mitochondria in the regulation of innate and adaptive immunity. Immunity 42: 406-417.

Wen, G., Ringseis, R. and Eder, K., 2010. Mouse OCTN2 is directly regulated by peroxisome proliferator-activated receptor α (PPARα) via a PPRE located in the first intron. Biochemical Pharmacology 79: 768-776.

Williams, M.S. and Kwon, J., 2004. T cell receptor stimulation, reactive oxygen species, and cell signaling. Free Radical Biology and Medicine 37: 1144-1151.

Winter, B.K., Fiskum, G. and Gallo, L.L., 1995. Effects of L-carnitine on serum triglyceride and cytokine levels in rat models of cachexia and septic shock. British Journal of Cancer 72: 1173-1179.

Wu, D.O. and Meydani, S.N., 1998. Antioxidants and immune function. In: Papas, A.M. (ed) Antioxidant status, diet, nutrition, and health. CRC Press, Boca Raton, CA, USA, pp: 371-400.

Yang, S.P., Yang, X.Z. and Cao, G.P., 2015. Acetyl-L-carnitine prevents homocysteine-induced suppression of Nrf2/Keap1 mediated antioxidation in human lens epithelial cells. Molecular Medicine Reports 12: 1145-1150.

Ye, J., Han, Y., Chen, X., Xie, J., Liu, X., Qiao, S. and Wang, C., 2014. L-carnitine attenuates H_2O_2-induced neuron apoptosis via inhibition of endoplasmic reticulum stress. Neurochemistry International 78: 86-95.

Yousefi, A., Khajali, F., Hassanpour, H. and Khajali, Z., 2012. Dietary L-carnitine improves pulmonary hypertensive response in broiler chickens subjected to hypobaric hypoxia. The Journal of Poultry Science 50: 143-149.

Zambrano, S., Blanca, A.J., Ruiz-Armenta, M.V., Miguel-Carrasco, J.L., Arévalo, M., Vázquez, M.J., Mate, A. and Vázquez, C.M., 2013. L-carnitine protects against arterial hypertension-related cardiac fibrosis through modulation of PPAR-γ expression. Biochemical Pharmacology 85: 937-944.

Zambrano, S., Blanca, A.J., Ruiz-Armenta, M.V., Miguel-Carrasco, J.L., Arévalo, M., Mate, A. and Vázquez, C.M., 2014. L-carnitine attenuates the development of kidney fibrosis in hypertensive rats by upregulating PPAR-γ. American Journal of Hypertension 27: 460-470.

Zambrano, S., Blanca, A.J., Ruiz-Armenta, M.V., Miguel-Carrasco, J.L., Revilla, E., Santa-María, C., Mate, A. and Vázquez, C.M., 2013. The renoprotective effect of L-carnitine in hypertensive rats is mediated by modulation of oxidative stress-related gene expression. European Journal of Nutrition 52: 1649-1659.

Zhai, W., Neuman, S.L., Latour, M.A. and Hester, P.Y., 2008. The effect of male and female supplementation of L-carnitine on reproductive traits of white leghorns. Poultry Science 87: 1171-1181.

Zhang, D.M., Guo, Z.X., Zhao, Y.L., Wang, Q.J., Gao, Y.S., Yu, T., Chen, Y.K., Chen, X.M. and Wang, G.Q., 2019. L-carnitine regulated Nrf2/Keap1 activation *in vitro* and *in vivo* and protected oxidized fish oil-induced inflammation response by inhibiting the NF-κB signaling pathway in Rhynchocypris lagowski Dybowski. Fish & Shellfish Immunology 93: 1100-1110.

Zhang, Z.Y., Fan, Z.K., Cao, Y., Jia, Z.Q., Li, G., Zhi, X.D., Yu, D.S. and Lv, G., 2015. Acetyl-L-carnitine ameliorates mitochondrial damage and apoptosis following spinal cord injury in rats. Neuroscience Letters 604: 18-23.

Zhao, M., Jiang, Q., Wang, W., Geng, M., Wang, M., Han, Y. and Wang, C., 2017. The roles of reactive oxygen species and nitric oxide in perfluorooctanoic acid-Induced developmental cardiotoxicity and L-carnitine mediated protection. International Journal of Molecular Sciences 18: 1229.

Zhao, T., Chen, S., Wang, B. and Cai, D., 2020. L-carnitine reduces myocardial oxidative stress and alleviates myocardial ischemia-reperfusion injury by activating nuclear transcription-related factor 2 (Nrf2)/heme oxygenase-1 (HO-1) signaling pathway. Medical Science Monitor 26: e923251.

Zhou, X., Ringseis, R., Wen, G. and Eder, K., 2015. The pro-inflammatory cytokine tumor necrosis factor α stimulates expression of the carnitine transporter OCTN2 (novel organic cation transporter 2) and carnitine uptake via nuclear factor-κB in Madin-Darby bovine kidney cells. Journal of Dairy Science 98: 3840-3848.

Zou, W., Noh, S.K., Owen, K.Q. and Koo, S.I., 2005. Dietary L-carnitine enhances the lymphatic absorption of fat and α-tocopherol in ovariectomized rats. The Journal of Nutrition 135: 753-756.

Chapter 10
Taurine

What is worth doing at all is worth doing well

10.1 Introduction

Taurine (Tau) was discovered in 1927 and there were several milestones in Tau research related to poultry production. Initially, research in 1950s was the first step in understanding Tau role in poultry and it was related to studies with chicken embryo and discovery that in chicken Tau can be synthesised from sulphur (Martin *et al.*, 1966). Furthermore, it was established that chicken can obtain Tau from both internal synthesis and from the feed. However, only animal by-products were shown to be comparatively rich sources of Tau, while in grains (main poultry feed ingredient) its level is negligible. Interestingly, it was discovered that Tau was an unidentified factor of the fish meal in poultry diets stimulating chicken growth (Machlin *et al.*, 1955; Monson, 1969). The second milestone in Tau research (1990th) was associated with a range of publications related to attempts to improve chicken performance by inclusion of Tau in the diet. The results were quite variable, inconclusive and in most cases inclusion of various levels of Tau into the balanced chicken diet did not affect their performance (Blair *et al.*, 1991). Since then, Tau in poultry production has not been a priority of poultry research. The third milestone of the Tau poultry research (2000s) was associated with several publications confirming that inclusion of Tau into the chicken diet was mostly without significant effects on chicken growth and development (Yuan and Wang, 2010). Finally, the fourth milestone in the Tau research in poultry (2010-2018), is related to use of different stress models to investigate possible protective effects of Tau.

On the one hand, it is well accepted that commercial poultry production is associated with a range of stresses, including environmental, technological, nutritional and internal/biological stressors (Surai and Fisinin, 2016a,b). It was clearly shown that during stress adaptation there are changes in chicken metabolism and the requirements for many nutrients change. On the other hand, oxidative stress was proven to be a molecular mechanism of detrimental consequences of commercially relevant stresses (Surai, 2016, 2018) and the vitagene concept of stress adaptation was developed (Surai and Fisinin, 2016c,d). Importantly, recent data on the protective effects of Tau in various stress models are related to transcription factors (Ma *et al.*, 2016; Sun *et al.*, 2018) and vitagenes modulation (Belal *et al.*, 2018; Lu *et al.*, 2019). Furthermore, protective effects of Tau in different human diseases (Chen *et al.*, 2016; De Luca *et al.*, 2015; Inam-U-Llah *et al.*, 2018; Miyazaki and Matsuzaki, 2014) are also of great importance. Therefore, it is time to review recent research evidence to see examples

DOI 10.3920/978-90-8686-906-0_10, © Wageningen Academic Publishers 2020

if Tau can be part of the anti-stress strategy in modern poultry production. In fact, the modern high productive birds differ substantially from the chickens used 20-30 years ago being more susceptible to stresses (Soleimani *et al.*, 2011; Xu *et al.*, 2018). Furthermore, changes in poultry diets with an almost complete removal of animal by-products makes the situation of meeting Tau requirement in stressful commercial conditions even more challenging. However, until now there was no single review devoted to the role of Tau in poultry production.

Therefore, the aim of this paper is to review research data related to Tau in poultry nutrition with a specific emphasis on its antioxidant properties, protective role in stress conditions and vitagene-modulating action.

10.2 Taurine sources

A Tau story stated 190 years ago when a new compound was isolated from ox bile by Tiedemann and Gmelin (1827) which is known today as Tau (2-aminoethanesulfonic acid). Chemically, Tau does not contain a carboxyl group and its amino group is at the β-position, but it is commonly called a non-protein sulphur amino acid, since it is not incorporated into proteins but is widely distributed in animal tissues. (Seidel *et al.*, 2018). Farm animals and poultry obtain Tau from two main sources, namely the diet and from endogenous synthesis. The main natural dietary sources of Tau are feed ingredients of animal origin (Huxtable, 1996) and terrestrial plants including grains (major feed ingredient for farm animals and poultry) contain only traces of Tau (Kataoka and Ohnishi, 1986). Feed processing and heat treatment could be a source of Tau level variability in different feed ingredients, since the amount of Tau that remained in a feed ingredient after cooking is dependent upon the method of feed preparation (Spitze *et al.*, 2003).

Tau synthesis *in vivo* is shown to be species-specific. For example, rodents are characterised by a high synthetic ability, while humans have an intermediate capacity to synthesise Tau. At the same time cats are shown to lack the ability to synthesise this amino acid as a result of a limited activity of key enzyme (Edgar *et al.*, 1998). It is shown that Tau synthesis in animals takes place mainly in the liver from cysteine catalysed by two enzymes namely cysteine dioxygenase (CDO) and cysteine sulphinate decarboxylase (CSD) leading to formation of cysteine sulphinate and hypotaurine, which is further oxidised to Tau. In addition, other tissues including brain, lungs, skeletal muscle and adipose tissue have an ability to synthesise Tau (for review see Lambert *et al.* 2015).

10.3 Taurine absorption and metabolism

Dietary Tau is well absorbed from the gastrointestinal tract and characterised by a comparatively high (about 90%) bioavailability (Sved *et al.*, 2007). It seems likely that carrier mediated active transport in the brush border membrane is responsible for Tau delivery to enterocytes with transport to the liver via portal vein and further release into circulation. Generally speaking, Tau absorption is considered to be quite a fast process (Schaffer and Kim, 2018). For example, in humans, plasma Tau concentrations are shown to reach a peak of 0.53 mM just 1.5 h after oral administration of Tau and plasma elimination half-life of Tau is about 1.0 h (Ghandforoush-Sattari *et al.*, 2010). It seems likely that Tau absorption depends on the dosage and after a single oral low or high dose administration in rats, the maximal plasma Tau concentrations were found at 15 and 45 min, respectively (Nielsen *et al.*, 2017).

At the next stage, Tau can enter cells via two different transporter systems. The first one is the high affinity and low transport capacity, Na^+–Cl^-–Tau symport transporter (*TauT*). The second one is the low affinity but high transport capacity proton-coupled amino acid transporter (PAT1; see Ribeiro *et al.*, 2018). It seems likely that, under physiological conditions, TauT-mediated uptake is the major absorptive mechanism at low Tau concentrations, whereas at higher Tau concentrations PAT1-mediated uptake is the main one (Anderson *et al.*, 2009). It is known that *TauT* expression depends on Tau status of the cells: a high Tau concentration downregulates *TauT*, while low Tau concentration could upregulate *TauT*. Furthermore, Tau is reabsorbed into circulation through the renal tubules in the kidney. Since Tau is not incorporated in proteins, concentration of this amino acids in many animal tissues could be comparatively high.

10.4 Biological roles of taurine

Tau involves in a variety of physiological functions related to homeostasis of skeletal muscle, the retina, the central nervous and cardiovascular systems (Schaffer and Kim, 2018). Major biological functions of Tau are summarised in Table 10.1 and Figure 10.1. Indeed, Tau participates in regulation of major biochemical pathways in the body and modulates basic processes, including enzyme activity, receptor regulation, osmotic pressure, cation homeostasis, cell development and cell signalling (Schaffer *et al.*, 2010).

Taurine also involved in AO protection, energy metabolism, gene expression, ER stress, cell quality control, Ca^{2+}-homeostasis as well as neuromodulation and osmoregulation are among major Tau biological functions (Schaffer and Kim, 2018).

Table 10.1. Regulatory roles of taurine in biological functions.

Function	Recent references
Antioxidant	Cheong *et al.*, 2013; Grove and Carpowicz, 2017; Patel *et al.*, 2016
Membrane stabilisation	Schaffer *et al.*, 1995; Zhang *et al.*, 2014
Mitochondrial integrity maintenance	Hansen *et al.*, 2010; Jong *et al.*, 2012; Schaffer and Kim, 2018; Seidel *et al.*, 2018
Vitagene activation	Belal *et al.*, 2018; Kang and Kim, 2013; Lu *et al.*, 2019; Surai *et al.*, 2017
Bile acid conjugation	Bouckenooghe *et al.*, 2006; Ciaula *et al*, 2017; Johnson *et al.*, 1991; Kevresan *et al.*, 2006
Ca homeostasis	Henry and MacCormack, 2018; Jakaria *et al.*, 2019; Scheffer and Kim, 2018
FA oxidation	Chen *et al.*, 2009; Fukuda *et al.*, 2011
Energy metabolism	Schaffer *et al.*, 2016; Wen *et al.*, 2019
Osmoregulation	Cuisinier *et al.*, 2002; Schaffer and Kim, 2018; Warskulat *et al.*, 2007
Thermoregulation	Frosini *et al.*, 2003, 2000; Page *et al.*, 2019
Detoxification	Devi *et al.*, 2009; Qvartskhava *et al.*, 2019; Vanitha *et al.*, 2016
Neuroprotection	Jakaria *et al.*, 2019; Kilb and Fukuda, 2017; Wu *et al.*, 2005; Zhu *et al.*, 2016
Anti-inflammatory	Kim and Cha *et al.*, 2014; Marcinkiewicz and Kontny, 2014; Seidel *et al.*, 2018
Immunomodulation	Marcinkiewicz, 2010; Schuller-Levis and Park, 2004; Walczewska *et al.*, 2015

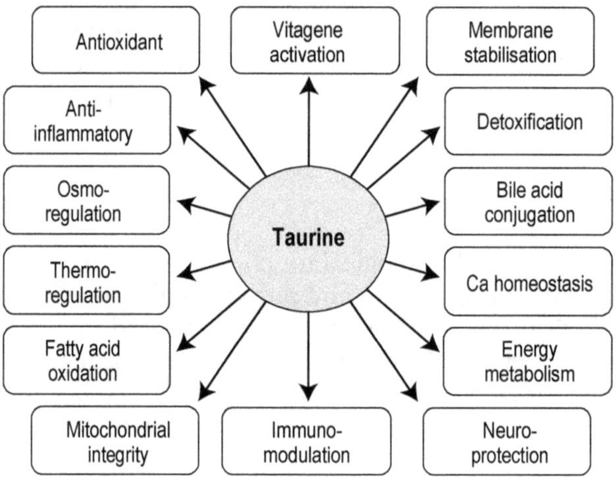

Figure 10.1. Major biological functions of taurine.

10.5 Antioxidant properties of taurine

It is well established that Tau could contribute to the antioxidant defences in different ways (Figure 10.2), including:

- Direct free radical scavenging (Cheong *et al.*, 2013; Grove and Carpowicz, 2017; Patel *et al.*, 2016).

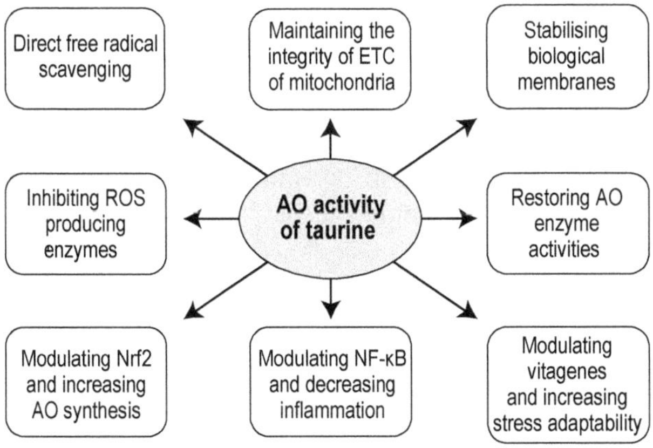

Figure 10.2. Antioxidant (AO) action of taurine.

- Maintaining the integrity of electron-transport chain of mitochondria in stress conditions (Hansen *et al.*, 2010; Jong *et al.*, 2012; Schaffer and Kim, 2018; Seidel *et al.*, 2018).
- Stabilising biological membranes (Schaffer *et al.*, 1995; Zhang *et al.*, 2014).
- Inhibiting specific enzymes responsible for free radical production (Che *et al.*, 2018; Das and Sil, 2012; Feng *et al.*, 2017; Hou *et al.*, 2018).
- Modulating transcription factors, including Nrf2 and increasing synthesis of AO enzymes (Lou *et al.*, 2018; Lu *et al.*, 2019; Sun *et al.*, 2018) or NF-κB and decreasing inflammation (Dai *et al.*, 2015; Ma *et al.*, 2016; Wu *et al.*, 2017; Zhao *et al.*, 2018).
- Regulating vitagenes and improved adaptation to stress via increasing synthesis of SOD, HSP, thioredoxin, TR and SIRTs (Belal *et al.*, 2018; Kang and Kim, 2013; Lu *et al.*, 2019; Surai *et al.*, 2017).

10.5.1 Direct free radical scavenging

The beginning of the Tau-related antioxidant story was not very promising. For example, in 1983 it was found that pre-treatment with Tau (3% aqueous solution, p.o., for 3 days) facilitated lipid peroxide formation in the rat liver associated with carbon tetrachloride (CCl_4: 2 ml/kg, b.w., i.p.) administration (Nakashima *et al.*, 1982). In addition, in 1990th it was shown that Tau reacted poorly with superoxide, hydrogen peroxide, and hydroxyl radicals (Aruoma *et al.* 1988). In particular, it was found that 1.5 mM final concentrations of Tau did not decrease the rate of reduction of 100 μM cytochrome C by the hypoxanthine/xanthine oxidase system. Furthermore, Tau at 3 mM final concentrations had no effect on the rate of reduction of 100 μM suggesting that Tau reactions with superoxide radical (O_2^{*-}) if any, proceed with rate constants less than 10^3 $M^{-1}s^{-1}$. Similarly, Tau scavenging of OH* generated by pulse radiolysis was shown to be very poor (k_2 2.42×10^6 m^{-1} s^{-1}; Aruoma *et al.* 1988). In general, direct AO activity of Tau has been the subject of great controversy. For example, in some studies Tau failed to scavenge the superoxide radical generated by a xanthine-xanthine oxidase system (Hanna *et al.*, 2004). Therefore, the data on direct measurements of scavenging activity of Tau in physiologically relevant concentrations (1.5-3 mM) showed that it is a weak scavenger for various radicals such as hydrogen peroxide, oxygen superoxide, or peroxynitrite (Aruoma *et al.*, 1988; Mehta and Dawson, 2001; Shi *et al.*, 1997). This was confirmed later, when Tau was shown to have very weak radical scavenging activities with the IC50 values of Tau against DPPH radical, alkyl radical, and hydroxyl radical scavenging activity being above 4 mg/ml (>30 mM; Du *et al.* 2010).

However, when Tau was used in higher concentrations (15-60 mM) free-radical scavenging activity was clearly observed. For example, ability of Tau to scavenge free radicals *in vitro* was studied in 2 different model systems, including measuring a decrease in luminol luminescence generated by superoxide from hypoxanthine/xanthine oxidase or due to peroxide formation from glucose/glucose oxidase systems (Kilic *et al.*, 1999). It was shown that Tau (30 mM) in both systems was responsible for the decrease of luminescence to zero, confirming a free-radical (superoxide and H_2O_2) scavenging activity of this compound. Spin trapping also confirmed superoxide-scavenging activity of taurine (Kilic *et al.*, 1999). Later it was shown that Tau in concentrations >15 mM, acted as a good *in vitro* scavenger of ROS (peroxyl radical and superoxide anion) and RNS (nitric oxide and peroxynitrite; Oliveira *et al.*, 2010). However, Tau did not react with 1 mM H_2O_2 at concentrations tested. This was confirmed later by Grove and Carpowicz (2017). The total reactive antioxidant potential of Tau against peroxyl radicals, produced by thermal decomposition of the free radical generator, AAPH, was also studied. In this model system decomposition of AAPH yields molecular nitrogen and two carbon radicals (R*), which rapidly react with oxygen to form peroxyl radicals (ROO*). In the aforementioned test system Tau (60 mM) was shown to be effective in blocking this process completely. In addition, the authors also showed that AO activity of Tau was responsible for inhibition of *t*-BHP-induced damage to lipids in liver slices and to preservation of -SH groups, as well as protection of -SH pool from oxidation (Oliveira *et al.*, 2010). It is interesting to note that Tau prevented both the formation of nitrotyrosine adducts and the decrease in SOD activity caused by peroxynitrite (Oliveira *et al.*, 2010). Recently, using a spin-trapping electron method it has been shown that Tau has a dose-dependent DPPH radical scavenging activity and IC50 (the concentration inhibiting 50% of free radical generation) was shown to be 0.144 mg/ml (about 1.2 mM), while the IC 50 of hydroxyl, alkyl and superoxide radical scavenging activity was shown to be 0.241 mg/ml (~1.9 mM), 0.206 mg/ml (~1.7 mM) and >8 mM respectively (Cheong *et al.*, 2013). Therefore, it seems likely that depending on experimental conditions Tau can scavenge various free radicals, but the superoxide radical scavenging activity by Tau was shown to be weaker than for DPPH, hydroxyl or alkyl radicals. Indeed, the reactivity of Tau is minimal towards the DPPH and peroxyl radicals, comparatively weak towards the hydroxyl radical, and moderate towards the superoxide anion (Patel *et al.*, 2016). Recently it has been shown that Tau and hypotaurine can react with superoxide anion *in vitro* to form the novel molecule peroxytaurine. In fact, gravimetric and spectroscopy assays suggested a stoichiometry of two superoxide anions reacting with one hypotaurine or two Tau molecules (Grove and Carpowicz, 2017). Tau is also found to be a DNA protector, which at physiological concentrations provides antioxidant activity to prevent DNA damage (Cheong *et al.*, 2013; Messina and Dawson 2000). It seems likely that Tau concentrations >15 mM are not achievable in plasma and most tissues of humans and animals. However, in some tissues including heart and eye lenses, etc., direct Tau free-radical scavenging could have beneficial effects. Therefore, the aforementioned data clearly indicate that Tau is not a characteristic direct scavenger of ROS in biological systems.

An additional benefit of antioxidant activity, including free-radical scavenging, of Tau could be expected in phagocyte cells. For example, Tau was shown to react with hypochlorous acid to form taurine chloramine (TauCl, Auroma *et al.*, 1988) and this reaction could occur in activated neutrophils. In fact, administration of Tau or TauCl to activated macrophages *in vitro* was shown to modulate the inflammatory response by inhibiting the generation/expression of nitric oxide, prostaglandin E, tumour necrosis factor, and interleukin (Gupta, 2006). It seems likely that one of the most important functions of TauCl is related to restriction of excessive cytotoxicity of the neutrophil, preventing damage during inflammation (Gupta, 2006). In fact, TauCl was found to be released from activated neutrophils due to their apoptosis with following inhibition of the production/expression of inflammatory mediators such as, superoxide anion, nitric oxide, TNF-α, interleukins, and prostaglandins in inflammatory cells/tissues. Furthermore, TauCl could increase the expressions of antioxidant proteins, such as heme oxygenase 1, peroxiredoxin, thioredoxin, glutathione peroxidase, and catalase in macrophages (Kim and Cha, 2014). Therefore, TauCl released from activated neutrophils may potentially be responsible for the recovery processes of cells affected by inflammatory oxidative stresses. By neutralising hypochlorous acid, an important ROS generated by phagocyte myeloperoxidase (Marcinkiewicz and Kontny, 2014), Tau is converted to TauCl, which is less toxic than hypochlorous acid. In general, TauCl, by interfering with the production of several pro-inflammatory mediators and affecting the transcription factor NF-κB) (Marcinkiewicz and Kontny, 2014) can modulate the immune system.

10.5.2 Protective effects of taurine on mitochondria

It is well appreciated that mitochondria are the main cellular consumers of oxygen and primary source of free radicals in the cell. It is proven that mitochondrial enzymes producing ROS include (Lin and Beal, 2006):

- aconitase and α-ketoglutarate dehydrogenase, important the tricarboxylic acid (TCA) cycle enzymes;
- complexes I, II and III of the electron-transport chain (ETC);
- glycerol-3-phosphate dehydrogenase and pyruvate dehydrogenase;
- dihydroorotate dehydrogenase;
- the monoamine oxidases (MAO) A and B;
- cytochrome b5 reductase.

In general, ubiquinone reduction site of complex I, the outer quinone-binding site of the Q-cycle in complex III and glycerol 3-phosphate dehydrogenase are considered as major contributors to ROS production (Brand, 2010). In fact, updated data indicate that there are eleven sites of ROS (superoxide and/or hydrogen peroxide) formation in mammalian mitochondria (Brand, 2016; Wong *et al.*, 2017). Furthermore, various mitochondrial insults leading to oxidative stress, can substantially increase ROS production with following increase in protein modifications, lipid peroxidation and mitochondrial DNA damage associated with mitochondrial dysfunction (Sekine and Ichijo, 2015). For many years the detrimental effects of ROS have been a focus of research, but it is now clear that mitochondrially generated ROS have a light side being involved in regulation of intracellular signal transduction pathways (Forrester *et al.*, 2018; Larosa and Remacle, 2018), leading to cell adaptation to stress (Calabrese *et al.*, 2010). Tau localisation in mitochondria (Jong *et al.*, 2011, 2010; Lobo *et al.*, 2000a,b) with quite high (20-30 mM) mitochondrial concentration (Hansen *et al.*, 2010) are important findings for deeper understanding crucial roles of Tau in mitochondrial functions.

In order to elucidate possible mechanisms determining the antioxidant activity of Tau, specific Tau antagonist and Tau transport inhibitor, β-alanine, were used (Jong *et al.*, 2012). The authors showed that exposure of isolated cardiomyocytes to medium containing β-alanine caused a significant (by 45%) decrease in Tau content leading to an elevation in mitochondrial oxidative stress. In the same experiment activities of respiratory chain complexes I and III decreased by 50-65% indicating a reduction in electron transport activity. In fact, Tau deficiency caused impaired synthesis of key subunits of the electron transport chain (e.g. ND5 and ND6) leading to increased mitochondrial ROS concentrations. As expected, restoration of Tau concentration in mitochondria by co-administration of Tau with β-alanine was shown to prevent the aforementioned detrimental changes, including ROS production, in mitochondria (Jong *et al.*, 2012). The authors concluded that Tau protects the mitochondria against excessive ROS generation and oxidative stress being an important regulator of mitochondrial synthesis of proteins essential for the assembly of active respiratory chain complexes and electron transport chain activity. Furthermore, Tau has been suggested to be involved in maintenance of the correct translation and expression of the mitochondrial respiratory proteins. Indeed, Tau containing modified uridine has been discovered at wobble position in mitochondrial transfer RNA (tRNA) (Suzuki *et al.* 2002). Therefore, it has been suggested that taurine-modified tRNA to play a crucial role in the translation (through stabilising codon-anticodon interaction) of proteins participating in electron transport (Kirino *et al.* 2004). In particular, a mitochondrial protein MTO1 sharing evolutionary conservation with yeast mto1, has been found to be involved in tRNA modification and protein synthesis. Therefore, the deficiency of MTO1 led to a complete loss of Tau modification of mt-tRNA, which caused the embryonic lethality in mice at a very early developmental stage (Fakruddin *et al.*, 2018). Furthermore, the essential role of Tau modification in mitochondrial translation and in protein homeostasis in the cell was confirmed by showing impaired mitochondrial translation and respiratory activity associated with MTO1 deficiency. Indeed, MTO1-deficient cells were characterised by abnormal mitochondrial morphology and the mistargeted proteins were aggregated and misfolded in the cytoplasm leading to cytotoxic unfolded protein response (Fakruddin *et al.*, 2018). Therefore, Tau depletion could cause a decrease in Tau-modified tRNA, impaired electron transport capacity leading to further increase in ROS production. In such conditions, oxidative stress would contribute to further mitochondrial dysfunction. Therefore, Tau is likely to inhibit the ROS production in the cell via regulation of mitochondria integrity and functionality (Schaffer *et al.* 2009), while in Tau-deficient tissues mitochondrial ROS generation was reported to be enhanced (Jong *et al.*, 2015a,b, 2011). Indeed, defective translation of mitochondrial encoded proteins is responsible for ETC dysfunction and superoxide generation (Schaffer *et al.*, 2009, 2014). Indeed, Tau is a key determinant of oxidative phosphorylation and it plays an essential role in mitochondria. In particular, it was reported that mitochondrial ROS generation was enhanced in Tau-deficient heart (Jong *et al.*, 2015a,b, 2011) and Tau transporter knockout (taut−/−) mice were found to exhibit liver mitochondria dysfunction characterised by severe lower respiratory control ratio (Warskulat *et al.*, 2006). Furthermore, in transgenic mice knockout for the Tau transporter mitochondrial dysfunction in cardiomyocytes, characterised by reduced ATP production and elevated superoxide generation was observed (Jong *et al.*, 2015a). Therefore, Tau dietary supply/therapy can restore mitochondrial protein biosynthesis, improve mitochondrial function and reduce ROS formation (Rikimaru *et al.*, 2012; Schaffer *et al.*, 2014).

Furthermore, Tau can preserve a redox balance in the mitochondrial matrix by regulating GSH levels within this organelle (Hansen and Grunnet, 2013; Hansen *et al.*, 2015). This could have long lasting consequences in terms of health maintenance (Mezes *et al.*, 1997). In fact, Tau plays important roles as mitochondrial matrix buffer stabilising mitochondrial pH gradient (Hansen *et al.*, 2006) regulating the activity of specific enzymes such as pyruvate dehydrogenase and maintaining mitochondria homeostasis (Hansen *et al.*, 2015). In addition, Tau could stimulate mitochondrial Mn-SOD, regulate mitochondrial calcium homeostasis and preserve mitochondrial function (El Idrissi, 2008; Warskulat *et al.*, 2006). Therefore, Tau may block/decrease generation of ROS by inhibiting cytoplasmic calcium increases and preventing mitochondrial dysfunction (El Idrissi and Trenkner, 1999). Interestingly, cardiomyocytes cultured in Tau-deficient medium were shown to suffer from oxidative stress caused by a disruption of the mitochondria electron transport chain and increased superoxide production (Chang *et al.*, 2004; Parvez *et al.*, 2008). Furthermore, re-introduction of Tau in the medium restored the electron transport chain integrity leading to decrease in the superoxide production (Jong *et al.*, 2012). Indeed, the damaged electron transport chain is responsible for excessive production of ROS, which interfering with other important physiological events within the cell, including Ca^{2+} transport (Schaffer *et al.*, 2015). Therefore, from the one hand, Tau has a unique mitochondrial protective effect due to prevention of intracellular calcium overload (El Idrissi and Trenkner 1999). On the other hand, Tau was shown to decrease intracellular calcium levels following an excitotoxic insult (Wu *et al.*, 2005).

In stress conditions Tau could interact with mitochondria to maintain mitochondrial homeostasis resulting in reducing the leakage of ROS formed inside mitochondria (Hansen *et al.*, 2006). The severe collapse of brain mitochondrial membrane potential decreased mitochondrial dehydrogenases activity, mitochondrial swelling, and depleted mitochondrial ATP were observed in Mn-exposed mitochondria. At the same time, Tau administration at physiological concentrations (0.1-10 mM) was indicated to preserve mitochondrial ATP, decrease mitochondrial depolarisation and swelling, and increase mitochondrial dehydrogenases activity (Ahmadi *et al.*, 2018). Tau was shown to enhance antioxidant defences in a traumatic brain injury model through increased mitochondrial ETC activity (Niu *et al.*, 2018). Recently in an *in vitro* study with mouse muscle cells, it has been shown that the treatment with 100 μM MDA leads to increased oxidative stress, inhibition of mitochondrial biogenesis and the reduction of the cell survival rates. The pre-treatment with 0.1 μM Tau was shown to reduce MDA-induced death rate via inhibition of oxidative stress, restoration of mitochondrial membrane potential, ATP production, reduced protein adduct formation and increased expression of Nrf-2 and the factors of mitochondrial biogenesis (Cai *et al.*, 2018). The authors suggested that the cytoprotective effects of Tau may be due to an induction of mitochondrial biogenesis. It should be mentioned that oxidised proteins and amino acids are potential promoters of oxidative stress (Estevez and Xiong, 2019). Further evidence on protective role of Tau in mitochondria come from various studies related to toxic effects of various chemicals known to be promoters of the oxidative stress in mitochondria, including ozone, nitrogen dioxide, bleomycin, amiodarone, arsenic, iron, adriamycin and catecholamines. In most of cases Tau treatment was shown to decrease/prevent detrimental consequences related to increased ROS formation (Schaffer and Kim, 2018; Schaffer *et al.*, 2009). Taurine is also suggested to be able to limit oxidative stress by preventing damage to ROS-sensitive AO enzymes (Schaffer and Kim, 2018).

The antioxidant properties of Tau have been widely studied in different experimental models with an important conclusion that protective effects of Tau in many cases are related to regulation/preservation of mitochondrial functions in various stress conditions.

10.5.3 Membrane-stabilising activity of Tau

More than 40 years ago Huxtable and Bressler (1973) proposed that taurine acts as a membrane stabiliser. Later membrane-stabilising effects of Tau were confirmed in an *in vitro* study with isolated segments of frog retina under illumination (Pasantes-Morales and Cruz, 1984) and in lymphoblastoid cultures stressed by retinoids (Pasantes-Morales *et al.*, 1984). Therefore, Tau was suggested to be a physiological stabiliser of photoreceptor membranes (Pasantes-Morales and Cruz, 1985a). In particular, it was shown that Tau acts at the level of the cell membrane by binding to the membrane at or near the insulin receptor leading to its hypoglycaemic effects in rats (Kulakowski and Maturo, 1984). One possibility is that Tau effect on biological membranes is mediated by prevention of lipid peroxidation in membranes. For example, in the light-induced damage model Tau (25 mM) was found to attenuate the increase in lipid peroxidation leading to an entire protection of rod outer segment structure (Pasantes-Morales and Cruz, 1985b). Tau was also shown to have an inhibitory effect on protein phosphorylation by direct effects of kinase system (Lombardini, 1985). This was confirmed by Leibowitz *et al.* (1988) showing that Tau inhibited phosphate incorporation into rat retinal proteins. It has been proposed that a membrane stabilising effect for Tau could be mediated via modification of the cell membrane fluidity by raising the PE/PC ratio (Hamaguchi *et al.*, 1991). Furthermore, Schaffer *et al.* (1995) summarised 5 mechanisms of stabilising effects of Tau on biological membranes. First of all, a reduction of lipid peroxidation and oxidative injury to membranes could be an important element of cytoprotective action of Tau. Secondly, alteration of protein phosphorylation could affect membrane structure, activities of various proteins affecting important membrane-dependent functions, for-example, Ca transport. Thirdly, interaction of Tau with the neutral phospholipids could lead to alteration in various cellular processes including membrane calcium binding and function. Finally, inhibition of phospholipid N-methylation by Tau with changes in membrane composition and structure could be a driving force of protective roles of Tau (Schaffer *et al.* (1995). Indeed, the interactions of Tau with membranes could lead to its stabilisation. For example, the protection against toxic compounds such as streptozotocin could be mediated by membrane stabilisation and maintaining membrane permeability to water and ions (Timbrell *et al.*, 1995). Indeed, Tau could prevent detrimental changes in oxidative stress-induced membrane permeability, stabilise membrane and maintaining its important biological properties (Schuller-Levis and Park 2004; Timbrell *et al.* 1995).

Protective effects of Tau in mitochondria are especially important in stress conditions. The oxidative stress is known to cause oxidation of proteins, PUFAs and mtDNA in mitochondria, associated with morphologic and functional abnormalities in mitochondria and further increase of ROS production and aggravated oxidative stress. For example, it was shown that the neuroprotective effects of Tau against the toxicity of nickel depended on its roles in improving mitochondrial function (Xu *et al.*, 2015). In particular, a membrane stabilising effect of Tau in NiCl2 toxicity was evident as an inhibited LDH release into plasma. Moreover, Tau pre-treatment also ameliorated reduced ATP production, disrupted the mitochondrial membrane potential and decreased mtDNA content associated with nickel exposure (Xu *et al.*, 2015). Similar protective effect of Tau was observed in an iron-overload model. For example, Tau supplementation substantially (by 40%) increased liver Tau levels, maintained antioxidant defences and redox balance (GSH and antioxidant enzyme activities) and improved liver function, reduced apoptosis, ROS formation/lipid peroxidation and mitochondrial swelling and prevented the loss of the mitochondrial membrane potential (Zhang *et al.*, 2014). Treatment with Tau prior to KBrO3 administration greatly attenuated the activities of AO enzymes in rat hemolysate and it was suggested that membrane stabilisation could be responsible for such a protection (Ahmad and Mahmood, 2016).

Our research with poultry semen (Surai, 2002) and a range of recent publications indicate that spermatozoa could be an ideal model to study membrane-stabilising effects of various compounds. From the one hand, spermatozoa membranes are rich in PUFA and are very sensitive to various stresses (Surai *et al.*, 2003). On the other hand, there is a range of easily measurable physiological and biochemical parameters of spermatozoa reflecting their membrane integrity and stability. Furthermore, semen storage *in vitro* and cryopreservation are commonly used procedures in biology of reproduction of farm animals and poultry and adding test compounds into semen diluent can generate a great deal of results reflecting their possible membrane-stabilising properties. For example, protective effect of TAU on lipid peroxidation of the membranes was observed in rabbit (Alvarez and Storey, 1983) and canine (Michael *et al.*, 2009) spermatozoa. Interestingly, exposure of chimpanzee spermatozoa to osmotic stress was accompanied by an increase in cells with disintegrated plasma and outer acrosomal membranes and Tau was able to partially prevent those damages with a concomitant enhancement of cell viability (Ozasa and Gould, 1982). Tau (50 mM) was shown to have significant positive effect on quality of mithun semen during liquid storage *in vitro* at 5 °C and protective effects of Tau on sperm parameters were associated with enhancing the function of antioxidant enzymes, preventing efflux of cholesterol from cell membranes and decreased MDA production (Perumal *et al.*, 2013).

Tau added to the semen extender was also shown to reduce cryodamage to the buffalo (Shiva Shankar Reddy *et al.*, 2010), buffalo and cattle (Chhillar *et al.*, 2012; Kumar *et al.*, 2013), donkey (Bottrel *et al.*, 2018), dog (Martins-Bessa *et al.*, 2009), chicken (Partyka *et al.*, 2017), ram (Banday *et al.*, 2017; Sánchez-Partida *et al.*, 1997), goldfish (Kutluyer *et al.*, 2016), Sea bass (Martínez-Páramo *et al.*, 2013) spermatozoa.

Tau is shown to contribute to the antioxidant defence networks in different ways. First of all, by direct free radical scavenging. Secondly, by maintaining the integrity of electron-transport chain of mitochondria in stress conditions. Thirdly, by preventing free radical formation as a result of inhibition of specific enzymes responsible for free radical production, Fourthly, by maintaining optimal redox status of the cell via activating a range of antioxidant enzymes and non-enzymatic antioxidants, mainly via interacting with such transcription factors as Nrf2 and NF-κB. Finally, by activating vitagenes, responsible for the synthesis of protective molecules, including SOD, HSP, thioredoxin, sirtuins, etc., leading to additional protection in stress conditions and providing an effective adaptative response to stresses. From the data analysed above it is clear that Tau plays an important role as natural antioxidant. However, on one hand, direct antioxidant effect of Tau due to scavenging free radicals could be related only to few tissues with high (>15-20 mM) Tau concentrations, including heart and eye lenses. On the other hand, stabilising effects of Tau on mitochondria, a prime site of free radical formation, characterised by comparatively high Tau concentration, deserves more attention. In fact, by maintaining optimal Tau status of mitochondria it is possible to minimise free radical production. This mechanism of protective Tau action is of great importance in various stress conditions including commercial animal and poultry production. Furthermore, membrane-stabilising effects of Tau could also contribute to its protective action in stress conditions. In particular, adding Tau to semen extenders for various animal species was shown to have membrane-stabilising and antioxidant effects. It seems likely that indirect antioxidant activities of Tau associated with modulation of various transcription factors and upregulation of antioxidant enzymes are major molecular mechanisms of Tau protective actions and they will be considered in the second part of the review.

10.6 Taurine and transcription factors

10.6.1 Nrf2

Initially, cytoprotective actions of Tau was related to its ability to deal with toxic products produced by immune cells. Indeed, Tau can protects neutrophils from the toxicity of HOCl with formation of TauCl which is released by activated neutrophils. For example, HO-1 was shown to be induced by TauCl and taurine bromamide in J774.2 macrophages (Olszanecki and Marcinkiewicz, 2004) and TauCl was shown to protect various cells from oxidative stress-induced cell death by inducing antioxidant enzymes including HO-1, peroxiredoxin and thioredoxin (Sun Jang *et al.*, 2009). In fact, upregulation of HO-1 expression and elevation of HO activity induced by TauCl via the Nrf-2 pathway (Kim *et al.*, 2010) is considered to be an adaptive mechanism of cytoprotection by suppressing overproduction of NO and O_2^- from activated phagocytic cells. The authors demonstrated that exposure of murine macrophages up to 0.7 mM TauCl enhanced the expression of HO-1 mRNA, HO-1 protein and HO activity in a dose-dependent manner. TauCl administered *in vivo* was found to increase Nrf-2-regulated antioxidant enzyme (HO-1, Prx-1, and Trx-1) expression in murine peritoneal macrophages (Kang and Kim, 2013). TAU-Cl pretreatment significantly abrogated toxic effects of azathioprine in rat testes including alleviation of detrimental changes in Nrf2 and HO-1 expression (Schaalan *et al.*, 2018). In HT22 cells, TauCl was indicated to attenuate oxidative stress-related cellular damage induced by glutamate or H_2O_2 through ERK and p38, Nrf2, and HO-1 pathways (Cheong and Lee, 2017). Furthermore, TauCl potentiated the HO-1 expression via increased the Nrf2 translocation to the nucleus in mouse BV2 microglial cells (Lee *et al.*, 2017b). TauCl was shown to upregulate Nrf2-mediated HO-1 expression and subsequent production of CO leading to increase of the efferocytic activity of murine macrophages (Kim *et al.*, 2015).

In recent years various *in vitro* and *in vivo* stress-models have been used to elucidate effect of Tau on Nrf2-mediated cytoprotection. In particular, it was clearly demonstrated that xylose-taurine reduced H_2O_2-induced oxidative stress in cultured hepatocytes through the activation of Nrf2 signalling including NQO1 and HO-1 (Park *et al.*, 2017). Tau was also shown to protects mouse spermatocytes from ionising radiation-induced damage as a result of activation of Nrf2/HO-1 (Yang *et al.*, 2017). Tau increased the resistance against glutamate-induced oxidative stress in HT22 cells via the p38/Nrf2-dependent HO-1 expression (Lee and Cheong, 2017). Recent results of Sun *et al.* (2018) demonstrated that Tau can protect SK-N-SH cells against CORT-induced oxidative stress by activation of ERK and Nrf2 signalling pathway. In fact, CORT was shown to induce neuronal apoptotic cell death by increasing mitochondrial ROS production and intracellular ROS concentration, decreasing the $\Delta\Psi$m and the ATP level in SK-N-SH cells. Pretreatment with taurine was shown to block the aforementioned changes in antioxidant defence mechanisms and suppressed CORT-induced apoptotic cell death (Sun *et al.*, 2018). On the other hand, in the same study, inhibition of ERK or silencing Nrf2 were found to block the protective effects of Tau on cell viability in the CORT model of neuronal damage. Cytoprotective effect of Tau against H_2O_2-induced toxicity in osteoblast cells was shown to be associated with restoration of Nrf2 and the downstream antioxidants (HO-1, NQO1, GCLC) inhibited by the oxidative stress (Lou *et al.*, 2018). This protective effect was executed by activating ERK and the Wnt/β-catenin pathway in osteoblast cells.

Cytoprotective effects of Tau associated with Nrf2 activation was confirmed in various animal models studies. For example, Tau was demonstrated to ameliorate neuropathy and reduced oxidative stress (via regulating NF-κB and Nrf2/HO-1 signalling cascades) in diabetic rat brain. In fact, Tau supplementation partially reduced lipid peroxidation (serum MDA concentration) and partially alleviated neuroinflammation as reflected by suppressed NF-κB expression and enhanced Nrf2, HO-1, GLUT1,3 expressions in the diabetic rats (Agca *et al.*, 2014). Furthermore, Tau supplementation was shown to prevent the reduction of Nrf2 and inhibit ROS production and autophagy in the pancreas of As_2O_3-treated rats (Bai *et al.*, 2016). Similarly, Tau attenuated As_2O_3-induced autophagy in mouse cerebrum through a prevention of detrimental changes in Nrf2 pathway caused by the toxicant (Piao *et al.*, 2017). Dietary Tau supplementation (5 g/kg diet) of chronically heat-stressed chickens alleviated mitochondrial damages and significantly improved antioxidant defences as demonstrated by increasing the mRNA expressions of Nrf2, NAD(P)H quinone dehydrogenase 1 (NQO1) and heme oxygenase 1 (HMOX1) and decreasing lipid peroxidation (ROS and MDA) in breast muscles (Lu *et al.*, 2019).

Therefore, it is possible to conclude from the aforementioned data that Tau and TauCl, similar to other AO, acts as indirect antioxidants through their ability to induce Nrf2 transactivation. This leads to the additional synthesis of an array of antioxidant molecules responsible for providing effective protection in various stress conditions and decreasing oxidative stress. In general, recent experimental studies suggest that dietary Tau can activate antioxidant pathways such as Nrf2/HO-1.

10.6.2 NF-κB

Research devoted to the effects of Tau on NF-κB expression started in 2000's. For example, the combined treatment with Tau and niacin was shown to block lung injury and fibrosis by down-regulating bleomycin-induced activation and nuclear localisation of NF-κB in mice (Gurujeyalakshmi *et al.*, 2000). Further development of the research in this area was associated with usage of various *in vitro* model systems. First, it is known that Tau reacts with hypochlorous acid, an important ROS with high damaging potential, produced by the phagocyte myeloperoxidase. As a result, more stable but less toxic taurine chloramine (TauCl) possessing anti-inflammatory activities and serving as a modulator of the immune system is formed (Kim *et al.*, 2010). In fact, data from several laboratories clearly demonstrated that TauCl is a powerful regulator of inflammation capable to down-regulate the production of pro-inflammatory cytokines in both rodent and human leukocytes (Schuller-Levis and Park, 2003). For example, TauCl diminished the activity of NF-κB and, to a lesser extent, that of AP-1 transcription factor in fibroblast-like synoviocytes isolated from rheumatoid arthritis patients (Kontny *et al.*, 2000). Later it was shown that TauCl inhibited NF-κB activation without affecting AP-1 activity in murine peritoneal macrophages (Kim and Kim, 2005). TauCl decreased NF-κB migration into the nucleus of activated alveolar macrophages and caused a more sustained presence of IκB in the cytoplasm (Barua *et al.*, 2001). TauCl inhibits production of chemokines by activated rat alveolar macrophages through a reduction of nuclear NF-κB binding activity (Liu and Quinn, 2002). Met45 oxidation of IκB-α in Jurkat cells is believed to be a main molecular mechanism determining the TauCl-induced inhibition of NF-κB activation (Kanayama *et al.*, 2002). Therefore, Tau may help inhibiting excessive inflammatory reaction in neutrophils. Furthermore, production of inflammatory mediators by activated C6 cells (Liu *et al.*, 2003) or alveolar macrophages (Quinn *et al.*, 2003) was shown to be attenuated by TauCl attenuation of NF-κB activation. Results of *in vitro* studies with Jurkat cells (T-lymphocytes) and YJ cells (myeloid-committed eosinophils) indicated that TauBr can inhibit degradation of IκB-α and TNF-α-induced NF-κB activation (Tokunaga *et al.*, 2007). TauCl was shown to suppress excessive inflammatory response in human eosinophils *in vitro* (Martinez-Losa *et al.*, 2009). Ogino *et al.* (2009) suggested that Tau-Cl oxidises IκB and inhibits NF-κB activation through oxidant transfer to NH_4^+ leading to NH_2Cl formation. By reviewing existing evidence Kim and Cha (2009) concluded that in phagocytes TauCl can inhibit LPS-induced iNOS expression and NO production as a result of direct inhibition of Ras activation, ERK1/2 phosphorylation and NF-κB activation.

At the next stage of the research Tau was used in various *in vitro* systems. For example, in a model system based on Jurkat cells it was demonstrated that Tau treatment was associated with reduced NF-κB activation imposed by IL-2 treatment (Maher *et al.*, 2005). Interestingly a concentration of Tau as low as 10 nM had anti-inflammatory effects *in vitro* as showed by Tau inhibition of nitrite release from alveolar macrophages after LPS stimulation (Ward *et al.*, 2011). In human embryo endothelial cells exposed to methamphetamine Tau (10 mM) attenuated dysregulations of NF-κB, p-ERK as well as Bax (Shao *et al.*, 2012). Furthermore, in human intestinal epithelial Caco 2 cells *in vitro* Tau increased expression of genes encoding proteins that protect against inflammation (Gondo *et al.*, 2012). Similarly, in mouse mammary epithelial cell line after *Streptococcus uberis* challenge significantly increases in the expression of TLR-2, the activities of transcription factors NF-κB and NFAT, and related cytokines (TNF-α, IL-1β, IL-6, G-CSF, IL-2, KC, IL-15, FasL, MCP-1, and LIX) in culture supernatants were observed. Taurine administration (45 mM) was shown to downregulate all the aforementioned parameters and the activities of NF-κB and NFAT (Dai *et al.*, 2016). In phorbol 12-myristate 13-acetate and calcium ionophore A23187 (PMACI)-triggered human mast cell line (HMC-1 cells), Tau (1 and10 mM) showed a significant inhibition of the nuclear accumulation of NF-κB and phosphorylation of IκB-α (Nam *et al.*, 2017). Xylose-Taurine was shown to reduce the inflammatory responses in LPS-stimulated Raw264.7 macrophages reflected by inhibition of the activation of NF-κB expression and the phosphorylation of IκB-α (Lee *et al.*, 2017c). It seems likely that protective effect of Tau on the NF-κB activation depends on the concentration used and experimental design. For example, in IL-1β-stimulated fibroblast-like synoviocytes increased expression of NF-κB was not inhibited by 0.6 mM TauCl (Kim *et al.*, 2010b).

Finally, anti-inflammatory protective properties of Tau have been demonstrated in many models of oxidant-induced injury. In fact, dietary taurine (4%) reduced the apoptosis of photoreceptors in the rat retina associated with inhibition of AP-1 (c-fos/c-jun subunits) expression, up-regulation of NF-κB (p65) expression and decreasing caspase-1 expression (Yu *et al.*, 2007). Tau treatment was associated with increased phosphorylation of NF-κB proteins and decreased phosphorylation of p38 protein leading decreased apoptosis in the livers of NZB/W F1 mice that were fed a high-cholesterol diet (Hsu *et al.*, 2008). Oral administration of taurine (100 mg/kg body weight for 5 days) prior to $NaAsO_2$ intoxication was found to downregulate NF-κB expression in rat testes and decreased As-induced oxidative stress and apoptosis (Das *et al.*, 2009). Similar protective effect of Tau against As-induced NF-κB activation was also observed in cardiomyocytes (Ghosh *et al.*, 2009) and in kidney of As-intoxicated rats (Roy *et al.*, 2009). In *S. uberis*-induced mastitis in rats pretreatment with taurine (100 mg/kg BW daily) attenuated inflammatory changes and significantly decreased mRNA expression of NF-κB p65 and NF-κB DNA binding activity in mammary gland (Miao *et al.*, 2011). It was also shown that down-regulation of NF-κB and PARP expression in male rats by Tau is responsible for its anti-inflammatory mechanism during experimental ischemic stroke (Sun *et al.*, 2012). Taurine pretreatment (300 mM, 1 ml/kg BW) inhibited NF-κB activity and p38 MAPK expression and of Kupffer cells in rats with severe acute pancreatitis (Wei *et al.*, 2012). In human intestinal epithelial Caco-2 cells *in vitro* Tau increased expression of genes encoding proteins that protect against inflammation (Gondo *et al.*, 2012). In the streptozotocin-induced diabetic rats Tau partially alleviated neuroinflammation as indicted by suppressed NF-κB expression and enhanced Nrf2, HO-1, GLUT1,3 expressions (Agca *et al*, 2014). In cisplatin-induced cardiotoxicity in mice, Tau administration (150 mg/kg BW) significantly mitigated the ROS production, alleviated the overexpression of NF-κB, and inhibited the elevation of pro-inflammatory cytokines, adhesion molecules, and chemokines (Chowdhury *et al.*, 2016). In rats with alcoholic liver disease treated with LPS, Tau was found to down-regulate the gene expressions of NF-κB, TNF-α, CD14, TLR4, and reduce the levels of hepatic TNF-α, IL-1β and IL-6 (Wu *et al.*, 2017). Tau was shown to blocks paraquat and maneb-induced activation of NF-κB in brainstem of mice (Che *et al.*, 2018; Hou *et al.*, 2018).

The anti-inflammatory activity of Tau has been demonstrated in a number of *in vitro* and *in vivo* studies. It was proven that Tau and TauCl can down-regulate NF-κB expression and activity with following reduction in pro-inflammatory cytokine synthesis and healthy resolution of inflammation. Therefore, from the one hand, the modulation of signalling pathways (e.g. the Nrf2/Keap1/ARE and NF-κB pathways) by Tau could result in increased expression of genes encoding for a range of cytoprotective molecules, including AO enzymes and phase II detoxification enzymes. On the other hand, decreased expression of NF-κB-regulated genes would reduce production of pro-inflammatory cytokines helping controlling cell/tissue homeostasis.

10.7 Effect of taurine on vitagene expression

10.7.1 Superoxide dismutase

Inclusion of Tau into incubation medium of *in vitro* incubated astrocytes or neurons (Niu *et al.*, 2018), hepatocytes treated with $NaAsO_2$ prevented SOD activity decline due to toxic effect of As (Sinha *et al.*, 2007) or in morphine-treated C6 cells (Zhow *et al.*, 2011). Tau also prevented the SOD activity decline in the liver of CCl4 treated rats (Abdel-Moneim *et al.*, 2015) or in STZ-treated rats (Javed *et al.*, 2013), in cortex and hippocampus of STZ-treated rats (Reeta *et al.*, 2017), in brain of NaF-exposed rats (Adedara *et al.*, 2017b), in brain of cisplatin-treated rats (Owoeye *et al.*, 2018), in testes of Doxorubicin (DOX)-treated rats (Das *et al.*, 2012) and in the liver of DOX-exposed mice (Nagai *et al.*, 2016). In rats treated with DOX, taurine zinc dose-dependently increased liver SOD activity (Wang *et al.*, 2015), in lungs of hamsters treated by LPS (Bhavsar *et al.*, 2010, 2009) or in zebrafish liver of thioacetamide-treated rats (Hammes *et al.*, 2012). Tau was also shown to have protective effect against SOD inactivation in other stress conditions, including photochemical stress (fluorescent light) in rats (Yu *et al.*, 2007), in brain of γ-irradiation of rats (El-Maraghi *et al.*, 2018), low temperature stress in pufferfish (Cheng *et al.*, 2018) and disease/clinical states including diabetes (Mathew *et al.*, 2013, Wang *et al.*, 2013) and nephrectomy (Guan *et al.*, 2007). On the other hand, in rats KBrO3 administration significantly increased activity of SOD in hemolysate and plasma, while taurine was able to return the SOD activity to initial level (Ahmad and Mahmood, 2016).

10.7.2 HSP70

In primary hepatocytes *in vitro* Tau (10 mM) did not affect heat-induced HSP70 production under normo-osmotic conditions. However, during hyper-osmotic exposure, when the induction of HSP70 by heat was blocked, Tau was able to restore heat-induced HSP70 protein and mRNA expression to nearly control levels (Kurz *et al.*, 1998). It was shown that donor preconditioning with Tau could protect kidney grafts from oxidative stress and injury after experimental transplantation. In fact, Tau treatment (1.5 ml. i.v., 100 and 300 mM, 10 min before nephrectomy) of donor rats significantly decreased tubular damage and caspase 3 expression, while HSP72 expression increased by 17% and 77% of controls, respectively (Guan *et al.*, 2007). In the same study Tau treatment was responsible for a significant (by 95%) increase expression of SOD. Interestingly, Tau transporter knockout (TauTKO) mice exhibited cardiomyopathy and cardiac atrophy with significantly elevated HSP70 in both heart and skeletal muscle (Ito *et al.*, 2008). Tau treatment (2.5%) was shown to significantly increase endothelial HSP70 in rabbits fed on high cholesterol diet (Zulli *et al.*, 2009). Protective effects of Tau against hepatic abnormality in NZB/W F1 mice fed a hypercholesterolemic diet was associated with a significant decrease in HSP70 expression in the liver (Hsu *et al.*, 2010). It was shown that in white rabbits fed on an atherogenic diet Tau (2.5%) significantly increased myocardial HSP70 expression (by 42.7%) and HSF-1 (by 12%) were observed (Setyarani *et al.*, 2014). In the liver of sea bass Tau (0.2, 0.5 or 1% in the diet) caused statistically significant increase in levels of Hsp70 comparing to control fish fed on non-supplemented diet (Feidantsis *et al.*, 2014). In rats exposed to cold-restraint stress Tau zinc (100 and 200 mg/kg BW) upregulated HSP70 expression and decreased oxidative stress in the gastric mucosa (Yu *et al.*, 2015). In stress-induced hypertensive rats HSP70 expression in pituitary and adrenal was almost doubled in comparison to normal non-stressed animals. At the same time, inclusion into the diet of Tau inhibitor (β-alanine) significantly decreased HSH70 expression in the tissues, while Tau supplementation (200 mg/kg BW daily) was able to significantly increase HSP70 expression in the pituitary and adrenal tissues of the stressed rats (Lv *et al.*, 2017). In pufferfish (Takifugu obscurus) dietary Tau supplementation (550 and 850 mg/kg diet) was shown to increase HSP70 mRNA expression in the liver. Furthermore, under low temperature stress (13 °C) the HSP70 mRNA expression levels in Tau supplemented group were significantly increased, comparing with the control group, with the highest response being observed at 1740 mg/kg Tau additive group (Cheng *et al.*, 2018). At the same tine Mn-SOD expression was also increased due to Tau supplementation. Similarly, increased dietary supplementation with taurine (from 2.5 up to17.1 g/kg diet) significantly improved ability of fish to resist ammonia stress by increasing gene expression of HSP70, as well as CAT and GR, reducing ROS production and blood cell apoptosis (Tan *et al.*, 2018). In broilers exposed to heat stress Tau supplementation (1% with water) reduced HSP70 expression (at the gene and protein levels) in the liver and muscle tissues (Belal *et al.*, 2018), indicating that taurine supplementation could improve broiler thermotolerance. Indeed, in comparison to the previous study with stressed fish and laboratory animals, heat-stressed poultry responded differently in terms of HSP70 expression.

10.7.3 Heme oxygenase-1

Tau in various forms is shown to affects HO-1 activity in different model systems. First of all, two Tau derivatives, namely TauCl and TauBr, which are produced in inflammation sites by neutrophils and eosinophils, respectively, were shown to induce HO-1 dose-dependently (0.03-0.3 mM) in resting and LPS-stimulated J774.2 macrophages (Olszanecki and Marcinkiewicz, 2004). It was suggested that HO-1 activation is responsible for anti-inflammatory effects of TauCl/TauBr probably by inhibition of COX-2 activity and decrease of PGE(2) production (Olszanecki *et al.*, 2008) and decrease production of pro-inflammatory cytokines (Muz *et al.*, 2008). It was suggested that HO-1 and TauCl could independently regulate the production of proinflammatory cytokines with TauCl/TauBr providing an important link between the two antioxidant systems namely the cysteine pathway and the heme oxygenase system (Marcinkiewicz *et al.*, 2009). In the next study it was confirmed that TauCl can increase the expression of HO-1 and enhances HO-1 activity simultaneously with direct inhibition of Ras activation, ERK1/2 phosphorylation and NF-κB activation (Kim and Cha, 2009). This provides important elements of the protective mechanism of Tau cytoprotection against overproduction of O_2^- and NO by activated phagocytes. It was suggested that elevation of HO-1 and other vitagene (thioredoxin) activity and increased expression of transcription factors (Nrf2) induced by TauCl which can be produced in the activated neutrophils and released to surrounding cells, is an important protective mechanism against cytotoxicity caused by various inflammatory prooxidants, including H_2O_2 (Sun Jang *et al.*, 2009). Indeed, TauCl induced HO-1 expression via Nrf2 activation in murine macrophages (Kim *et al.*, 2010). In addition to HO-1 activation TauCl can improve redox status of the macrophages by increasing GSH level (Piao *et al.*, 2011) and be involved in the recovery processes of cells affected by inflammatory oxidative stresses (Kim and Cha, 2014). In addition, TauCl can help resolution of inflammation by enhancing the efferocytic activity of macrophages (Kim *et al.*, 2015). Protective role of TauCl in prevention of inflammatory damages to various tissues are associated with decreased protein expressions of PGE_2, COX-2, NO and iNOS in LPS-treated cells. At the same time there is an inhibition of pro-inflammatory cytokines (IL-6, IL-1β and TNF-α) production associated with decreased nuclear translocation and DNA-binding of NF-κB and activation of Nrf2 and expression and additional synthesis of HO-1 (Lee *et al.*, 2017). Furthermore, additional protective mechanisms of TauCl in oxidative stress-stimulated mouse hippocampal HT22 cells include ERK and p38 pathways (Cheong and Lee, 2017). It is interesting to note that induction of HO-1 expression and subsequent CO production caused by TauCl are shown to be essential for phagocytosis of fungi by macrophages (Kim *et al.*, 2018) indicating vital roles of TauCl in host defence against fungal infection. In fact, TauCl potentiates phagocytic activity of macrophages through upregulation of Nrf2 leading to increased HO-1 expression. In fact, abolishing HO-1 activity by pharmacologic means or knockdown techniques prevented TauCl-induced phagocytosis. On the other hand, the overexpression of HO-1 increased the phagocytic ability of macrophages (Kim *et al.*, 2018b). In addition to the aforementioned data related to TauCl and TauBr as important protective elements against oxidative-stress related damages, there is a growing body of information to show the same for Tau. For example, it was shown that rats with STZ-induced diabetes were characterised by significant increases of 4.9- and 6.6-fold in HO-1 mRNA expressions in the liver and kidneys in comparison to the non-diabetic rats. Tau supplementation in diabetic rats was shown to significantly suppress the elevated expression of HO-1 mRNA in kidneys (41.6%) and liver (25.2% suppression) compared to that of untreated diabetic rats (Yao *et al.*, 2009). Taurine was shown to attenuate oxidative stress and alleviates cardiac failure in type I diabetic rats and the protective effect was associated with vitagene activation namely increased level of HO-1 protein and increased SOD activity in myocardial tissue (Wang *et al.*, 2013). In different conditions, Tau enhanced the HO-1 expressions in the diabetic rats (Agca *et al.*, 2014). In doxorubicin-induced hepatotoxicity model Tau in the form of zinc salt (40, 80 or 160 mg/kg) dose-dependently increased rat liver HO-1 mRNA and protein expression (Wang *et al.*, 2015). In cultured rat astrocytes Tau (5 mM) was shown to prevent ammonia-induced upregulation of HO-1 and cell senescence (Oenarto *et al.*, 2016). Similar to the aforementioned data related to TauCl, Tau is also shown to increase the resistance against oxidative stress/damages induced by glutamate in HT22 cells via the p38/Nrf2-dependent HO-1 expression (Lee and Cheong, 2017b). Similarly, pre-treatment with xylose-taurine was shown to increase the expression of Nrf2, NQO1 and HO-1 in cultured hepatocytes (Park *et al.*, 2017). Furthermore, Nrf2 signalling pathway and increased expression of HO-1 were shown to mediate protective effects of Tau against corticosterone-induced cell death in human neuronal cells in culture (Sun *et al.*, 2018). As mentioned above,

Tau supplementation was associated with improved antioxidant defences on chicken muscles, including increased HO-1 expression (Lu *et al.*, 2019).

10.7.4 Thioredoxin system

In As_2O_3-treated rats with compromised antioxidant system in pancreas, including reduced Trx expression, Tau increased Trx expression returning it to the initial level (Bai *et al.*, 2016). Interestingly, Nrf-2 expression followed the same pattern and was recovered by Tau treatment confirming finding that Trx synthesis is under Nrf2 control. More work was done with TauCl. For example, in macrophages TauCl provided protection from death caused by H_2O_2 by decreasing production of ROS associated with increased expression of peroxyredoxin-1, thioredoxin-1 and HO-1, the antioxidants normally induced by the activation of Nrf2 (Sun Jang *et al.*, 2009). In fact, the authors showed that TauCl increased nuclear translocation of Nrf2 and binding to the anti-oxidant response element. Interestingly, TauCl can increase the expressions of antioxidant proteins, including Trx, Prdx, GPx and HO-1 in activated neutrophils (Kim and Cha *et al.*, 2013) and murine primary macrophages (Kang and Kim, 2013). Proteomic analysis of the liver of rates exposed to carbon tetrachloride and treated with a combination of antioxidants (taurine, epigallocatechin gallate and genistein) has been recently performed. It was found that combination therapy could protect liver against oxidative stress through up-regulating expression of a range of antioxidants, including Txn1 (Cao *et al.*, 2015, 2017). In young and middle-aged rats Tau (200 mg/kg BW ip) was shown to significantly increase TR activities in the liver (Yildrim *et al.*, 2007). The activity of TR was significantly reduced (2.2-folds) in rat hemolysates due to KBrO3 treatment and this change was abrogated by pre-treatment with taurine (100 mg/kg BW, daily; Ahmad and Mahmood, 2016). Similar protective effect of Tau on TR activity in KBrO3 toxicity was observed in the rat intestine (Ahmad *et al.*, 2015). In Cac-2 cells TAU (25-100 mM) was shown to increase the mRNA expression and activity of thioredoxin interacting protein (TXNIP; Gondo *et al.*, 2012), which is from one hand involved in Trx inhibition and from the other hand it participates in Trx translocation into the nucleus and has a range of other beneficial functions, including cellular redox regulation (Alhawiti *et al.*, 2017). Therefore, it was suggested that dietary taurine is an important modulator of redox homeostasis of endothelial cells via interacting with TXNIP (Vincent *et al.*, 2014).

10.7.5 Sirtuins

Protective effects of Tau on hepatic steatosis induced by thioacetamide (TAA) in zebrafish was also shown. Indeed, SIRT1 mRNA expression was 2.6-fold higher in the Tau-TAA group than in the TAA group (Hammes *et al.*, 2012). Tau was shown to recover SIRT-1 gene expressions in arsenite-treated SH-SY5Y cells (Chou *et al.*, 2013). It was shown that Tau (100 μM) significantly increased the expression of SIRT1 in the amyloid β 1-42 -treated SK-N-SH cells. Furthermore, the protection of taurine was prevented by SIRT1 siRNA (Sun *et al.*, 2014). In cafeteria diet-induced steatohepatitis in rats, Tau supplementation was shown to recover disease-associated decrease of SIRT-1 concentration in the liver (Abd Elwahab *et al.*, 2017).

The aforementioned data clearly indicate that Tau/TauCl plays key roles in regulating vitagenes and improving animal adaptation to stress. First of all, Tau was shown to upregulate HSP70, main protective mechanism maintaining cell proteostasis. Secondly, Tau increases expression of other HSPs including HO-1. Thirdly, Tau regulated Trx system, including increased expression and concentration of Trx, TR and TXNIP, important elements in maintaining optimal cellular redox balance. Fourthly, in different *in vitro* and *in vivo* systems Tau was shown to prevent inhibition of expression and activity of SOD, main AO enzyme of the first level of AO defence, responsible for detoxification of superoxide radical. Fifthly, there are several recent publications showing a regulatory role of Tau on the sirtuin functions.

10.8 Taurine metabolism in poultry

In 1950-1970s chicken was used as a model to prove that chicks can utilise dietary sulphate to synthesise Tau. For example, Lowe and Roberts (1955) showed that within 36 minutes after administration of sulphate-S^{35} to a chick embryo, labelled Tau was detected. Interestingly, sulphate-S^{35} was not found in other sulphur-containing amino acids, including cystine, cysteine or methionine in the chick embryo (Martin *et al.*, 1966). Therefore, chicks fed diets with serine or the serine-sparing compounds glycine and ethanolamine were shown to efficiently synthesise Tau from sulphate (Miraglia *et al.*, 1966; Tomichek *et al.*, 1972). The authors suggested that the synthesis of Tau from sulphate proceeds by the dehydration of serine, transfer of sulphate from *3'-phosphoadenosine-5'-phosphosulphate* to form cysteic acid, with following its decarboxylation. In one-day-old chicks, conversion of sulphate-sulphur into Tau was shown to take place in various tissues, including liver, heart, spleen and kidney (Miraglia *et al.*, 1966). In general, the highest Tau concentration was found in the heart being 2-fold higher than that in the liver, kidney or spleen. Brain tissue was shown to be most active in taking up Tau, but by day 7 the process has reached its limits when determined by the amount of Tau available in the yolk (Van Gelder and Bélanger, 1988). In fact, in chicken egg Tau was found exclusively in egg yolk (101-136 µg/g or about 1 mM) and not in egg albumin (Kim *et al.*, 2006) and therefore the chicken embryo is supplied Tau from the egg yolk and from the synthesis in the embryo.

About 50 mg Tau was shown to be synthesised by chicken embryo during development. Indeed, after injection of S^{35}-labeled sulphate into the egg, about 65% of the dose was recovered as Tau in the newly hatched chick (Machlin *et al.*, 1955). It was shown that the Tau content of the chicken embryo goes up rapidly through the 3rd day of incubation and reached a stable level on the 4th day. However, there was a redistribution of Tau between various embryonic tissues, since in the embryonic heart the concentration of Tau increased steadily from the 7th day through to hatching reaching its maximum concentration at hatching (Gonzales and Awapara, 1955). Between day 14th and hatching Tau concentration in the chicken brain slightly increased (from about 10-11 mM up to 12-14 mM), while in muscles it increased 2.5-fold from about 4 mM up 10 mM (Sato *et al.*, 2009). Interestingly, from day 19 of the embryonic development (pipping) to newly hatched chick, Tau concentration in the liver and heart doubled and in plasma in the same period Tau concentration increased more than 2.5-fold. Furthermore, in the chick at day 19 of the embryonic development Tau content in the brain comprised about 26.8% of the total Tau in the body, while in the brain of the newly hatched chick and adult chicken contained 20.6 and 6.4% of total body Tau (Levi and Morisi, 1971). It seems likely that increased Tau concentration in embryonic tissues is an adaptive mechanism to deal with overproduction of free radicals. Tau's importance for embryonic AO defence was confirmed when oxidative stress imposed by ethanol in the egg caused a 5-fold decrease in Tau level in the chicken embryonic (day 11) brain (from 3.45 mM down to 0.63 mM; Berlin *et al.*, 2010). Indeed, our previous research clearly showed that the hatching process is a high oxidative stress condition (Surai, 2002; Surai *et al.*, 1996). Therefore, increased concentration of Tau, possessing antioxidant activities (Lu *et al.*, 2019; Ma *et al.*, 2016; Zhang *et al.*, 2014), provides protection against oxidative stress and maintains chicken health at hatching time and in early postnatal development.

Tau is also synthesised from Cys in the liver of chicken (Jacobsen and Smith, 1968). Therefore, birds can use dietary sulphur amino acids or inorganic sulphur to synthesise Tau, but molecular mechanisms of regulation of such synthesis are presently unknown. Furthermore, a contribution to Tau status of chicken made by its synthesis and dietary sources needs further investigation. We can speculate that internal Tau synthesis could not be adequate to produce enough Tau to meet its requirement and additional dietary supply of this nutrient, especially in stress conditions, could be beneficial. It is interesting to note that Tau injection into the egg on the third day of incubation was associated with morphological and molecular changes in muscle development during chicken embryogenesis. In particular, in breast muscle increasing number of muscles cells were observed (Zielińska *et al.*, 2010). Furthermore, increased fibre diameter, the total number of nuclei and the proportion of proliferating cell nuclear antigen-positive cells were also found due to Tau injection. Furthermore, the gene expression of FGF-2 and PCNA was downregulated (Zielinska *et al.*, 2012).

It is well accepted that Tau is taken up by tissues through a Na^+/Cl^--dependent Tau transporter (TauT; Takahashi and Hatta, 2017). In fact, TauT is shown to be highly expressed in chicken embryo brains at days 4, 6, 9 and 13. TauT was also expressed in other embryonic tissues, including heart and eye at day 3 of the development (Kim *et al.*, 2006). Interestingly, the Tau synthetic enzyme CSD was expressed first (12 h) and only then TauT gene expression (48 h) took place in the developing chicken embryo (Kim *et al.*, 2006). Indeed, a portion of the dietary requirement for sulphate by the growing chick could be utilised in the synthesis of Tau, a process which was shown to be increased in parallel with chicken growth and the total Tau in the liver increased due to methionine or sulphate supplementation (Martin, 1972). An essential enzyme in Tau synthesis, called 3'-phosphoadenosine 5'-phosphosulphate transferase, was isolated and partially purified from chicken liver and methionine was shown to be directly involved in Tau synthesis (Miraglia and Martin, 1969).

During the first two weeks posthatch the plasma Tau concentration increased almost 3-fold (from 40 to 117 μM) and remained constant thereafter (Porter *et al.*, 1991). In chickens, a large portion of the Tau administered was shown to be incorporated into tissues and slowly excreted. For example, only 16% of the Tau-S^{35} injected was excreted 1 day after its administration and even after 9 days less than 40% of the Tau dose was excreted (Machlin and Pearson, 1957). Furthermore, there is a tissue-specificity in Tau incorporation. In fact, the heart was shown to incorporate 34 times more S^{35} in comparison to the breast muscle. Interestingly, the tissues of the gastrointestinal tract and the spleen were indicated to incorporate large amounts of labelled Tau. The muscles of the leg incorporated nine-fold more S^{35} than breast muscles. It is important to underline that after oral administration of labelled Tau, major part (80%) of the S^{35} recovered in the excreta as Tau, and only 7% as Tau conjugates, 4% as bile acids, 2% as sulphate and about 7% were represented by unidentified compounds (Machlin and Pearson, 1957). Tau is characterised by immunomodulating properties (Schuller-Levis and Park, 2004) and chick B cells were shown to possess a specialised Tau transport system dependent on age of birds (Porter and Martin, 1992). Furthermore, Tau participates in regulation of Ca^{2+} uptake and (Ca^{2+}+Mg^{2+})-ATPase activity in developing chick B-cells (Porter and Martin, 1993).

It was found that tissue Tau concentration can be regulated by dietary means. For example, in growing chickens fed on a diet without animal by-products (main source of feed-derived Tau) a slight increase in cardiac Tau concentration due to supplementation of Tau (1 mg/kg) was observed (Blair *et al.*, 1991). When chickens were fed on a diet containing animal by-products, supplementation of Tau with water significantly (2-5-fold) increased Tau level in plasma but did not affect its level in the heart (Ruiz-Feria and Wideman, 2001). Interestingly, in the same study the usage of β-alanine (an antagonist of Tau membrane transport) supplementation significantly reduced free Tau concentrations in cardiac tissues but not in the plasma. There was a linear increase in the Tau content of both the thigh muscle (from 1.66 g/kg up to 3.14 g/kg) and breast muscle (from 0.1 g/kg up to 0.4 g/kg) with increasing dietary Tau from 0 up to 8 g/kg. However, only high doses of Tau supplementation (2 g/kg and 8 g/kg) significantly increased Tau content in thigh and breast muscles. In the second experiment, when 22-day old chicks were supplemented with 4 g/kg Tau for 3 weeks, Tau concentration in the liver, thigh muscle and breast muscle increased by 1.56-, 1.89 and 1.89-times. It was also shown that a week-long supplementation of Tau (4 g/kg) was adequate to significantly increase its content in the breast muscle, but it took 3 weeks to significantly increase Tau concentration in the liver and thigh muscle. Importantly, Tau supplementation for a week could improve antioxidant defences as indicated by a significant (2-fold) decrease in serum lipid peroxidation (MDA; Huang *et al.*, 2014a,b). It is also possible to increase Tau content in the egg yolk by dietary means. For example, dietary Tau supplementation (2.5%) was responsible for a 58% increase in Tau concentration in the egg yolk within a week. This level did not increase further for the next 3 weeks of feeding (Lee *et al*, 2010).

10.9 Effects of dietary taurine on growing chickens

It seems likely that under physiological conditions, internal Tau synthesis and dietary provision of this amino acid are adequate to meet its requirement. Therefore, inclusion of Tau into the balanced chicken diet does not affect bird performance. For example, Tau supplementation of 0.025, 0.05 or 0.10% to the broiler diet did not affect growth of male broiler chicks (Blair *et al.*, 1991). Similarly, no differences in growth performance of broiler chickens or turkeys fed on diets supplemented with Tau were reported by Tufft and Jensen (1992). Twenty years later, the effect of Tau dietary supplementation on growth performance, meat quality, oxidative status and muscle Tau content in broilers was studied (Huang *et al.*, 2014a). For this purpose, one-day-old male Cobb chicks were fed on a diet supplemented with 0, 0.125, 0.50, 2.00 or 8.00 g/kg Tau from 1 to 42 d of age. In the study, Tau did not affect daily gain, FCR and meat quality parameters (Huang *et al.*, 2014a,b). Similarly, in an earlier study from the same department Tau (0.5 g/kg) did not affect chicken final body weight at 6-week dosage (Yuan and Wang, 2010). Furthermore, the supplementation of Tau in broiler diets (0.5%) did not affect FI, BWG and FCR (Alzawqari *et al.*, 2016). Interestingly, when growing chickens fed a diet containing animal by-products (a source of dietary Tau), additional Tau supplementation in drinking water (0.75-3%) caused growth inhibiting effect (Ruiz-Feria and Wideman, 2001). There are several early reports (Campbel and Classen, 1989; Divakaran, 2006; Martin and Patrick 1961; Monson 1969) indicating some improvement in growing birds' performance due to dietary Tau supplementation, but in general, those responses were related only to some periods of bird's development, were small and inconsistent. It seems likely that bird's genetics play an important role in response to dietary Tau supplementation. For example, White Leghorn chicks selected for high (HA) antibody response to SRBC responded with greater growth with a relatively low level of Tau, whereas a line selected for low (LA) antibody response to SRBC characterised by increased Tau level in the heart and showed improved growth with a relatively high level of Tau (Dunnington *et al.*, 1996). The authors suggested that the growth enhancement by Tau is a complex function of both the level of Tau and the genetics of the chicks.

Recently, an attempt to replace methionine by Tau (partly or completely) in the chicken diet was described: dietary methionine (2% in the diet) was replaced with Tau (25, 50, 75 and 100% replacement) and growth performance and blood index of AA+ broilers were assessed. (Lv *et al.*, 2017a). It was shown that inclusion of Tau instead of Met (25-100% replacement) was associated with significantly improved daily gain and FCR. However, there was no difference in those parameters between Tau supplemented groups. Bird mortality in the control group was quite high (9.24%) and 50 or 75% Met replacement by Tau was associated with a slight but significant decrease in mortality (down to 8.97%). In the same groups (50 and 75% Met replacement by Tau) IgA, IgM, GPx and total AOA in chicken serum were also significantly increased in comparison to the control group (Lv *et al.*, 2017).

Surprisingly, a negative effect of Tau supplementation on weight of the duodenum and jejunum and the length of jejunum in 6-week-old broilers was reported (Yuan and

Wang, 2010). In later study from the same laboratory, body weight, ADG, ADFI and FCR were not affected by Tau supplementation at 0.25, 0.50, 1.00, or 2.00 g/kg of diet the age of 7, 22, or 42 days. The authors reported that dietary Tau impaired intestinal mucosal development as indicated by decreased length and weight of the small intestine, villus width, surface area, and crypt depth in the duodenum and jejunum (Huang *et al.*, 2014b). It is difficult to explain how such negative consequences at the gut level due to Tau supplementation did not affect growth, development and FCR of experimental birds. Furthermore, these findings have not been validated or repeated in other studies from the same department or by different research groups. Clearly, this question deserves more attention and warrants further investigation.

It was shown that dietary Tau (0.5%) has the potential to decrease egg weight without affecting egg production, feed conversion, or body weight (Yamazaki and Takemasa, 1998). This could be an important finding for breeder nutrition in the second part of reproduction when egg size and mass is difficult to control. Furthermore, dietary Tau (0.1-0.75%) did not affect chicken growth and development for 6 weeks, however, Tau was shown to interact significantly with Met on daily gain and feed consumption (Lee *et al.*, 2004). The authors suggested that the growth response obtained by Tau supplements is a result of interactions with sulphur amino acids and modulation of broiler lipid metabolism. It was shown that dietary Tau of 0.15% in the chicken diet significantly increased the apparent metabolisable energy and crude fat digestibility, enhanced the activity of lipase in pancreas and small intestine (Zeng *et al.*, 2012). Furthermore, compared to the control group, the 0.15% Tau significantly decreased serum total cholesterol, triglycerides, free fatty acids, glucose and liver TG, FFA (Zeng *et al.*, 2012). Interestingly, supplementation of high fat diet with Tau (0.5%) was associated with improved antioxidant defences as indicated by decreased lipid peroxidation in the layer liver, however, performance was not affected by Tau supplementation (Shim *et al.*, 2010).

There are species and age-related differences in Tau synthesis, metabolism and requirement. In particular, it was shown that the Tau-supplemented diet had a beneficial effect on immune responses and performance in growing Japanese quail (Wang *et al.*, 2009). One-day-old quails were fed with maize-soybean meal-based diets containing 0, 0.01, or 0.05% Tau for 42 days. It was shown that dietary 0.05% Tau significantly improved feed efficiency and decreased feed intake. The same Tau-treated group also showed some improvements in immune indexes. In particular, in comparison to the control group, relative weights of the bursa of Fabricius and thymus were greater and the response to phytohemagglutinin measured as the foot web index was significantly increased by 0.05% Tau supplementation (Wang *et al.*, 2009). Interestingly, sixty-day Tau supplementation (1% in drinking water) to hyperlipidaemia- and atherosclerosis-prone Japanese quail fed high-cholesterol (1%) diet significantly reduced serum non-HDL cholesterol (2-fold) and dramatically (by 44%) decreased the serum triglyceride level (by 44%; Murakami *et al.*, 2010). Adding 0.5 g/kg Tau to quail diet significantly increased Tau in the serum (by 38%) and egg yolk (by 14%), significantly improved egg production (by 4%) and FCR (by 9%), reduced egg yolk cholesterol concentration, serum triglyceride level and did not affect egg quality (Wang *et al.*, 2010, 2010a). It is interesting that genetic selection for resistance to dietary-induced atherosclerosis in quail is associated with changes in Tau metabolism. For example, cholesterol fed atherosclerosis-resistant quail had significantly increased number of ileum microbiome genes associated with Tau and hypotaurine metabolism (Liu *et al.*, 2018).

10.10 Protective effects of taurine in stress conditions

Considering Tau as an important part of the antioxidant defence network/system, it is reasonable to suggest its protective effects under various stress conditions. Indeed, by decreasing oxidative stress, a driving force of detrimental consequences of various stresses in poultry production, Tau can help fighting heat stress, immunological challenge or stress associated with increased stocking density of poultry.

10.10.1 Heat stress

Heat stress (HS) is the most common environmental stressor in the poultry industry worldwide (Nawab *et al.*, 2018) and increased metabolic activity of modern poultry genotypes could be responsible for the reduction in heat resistance (Soleimani *et al.*, 2011). At the molecular level, oxidative stress is considered to be a driving force of the detrimental consequences of HS (Akbarian *et al.*, 2016; Farag and Alagawany, 2018; Habibian *et al.*, 2015). Therefore, natural antioxidants, including vitamin E (Rehman *et al.*, 2017; Surai *et al.*, 2019), selenium (Habibian *et al.*, 2015; Surai and Kochish, 2019), ascorbic acid (Chand *et al.*, 2014; Egbuniwe *et al.*, 2018), carnitine (Celik *et al.*, 2004), betaine (Ratriyanto and Mosenthin, 2018; Sayed and Downing, 2011), and phytochemicals (Akbarian *et al.*, 2016; Surai, 2015) are shown to be protective in heat stressed birds.

In an experiment conducted by Shim *et al.* (2006) broiler chicks from d22 to d42 were fed balanced diet and kept either at 22 °C (control), 34 °C (heat stress, HS) or 34 °C and additionally supplemented with 0.8% Tau (HST). As expected, HS significantly decreased BW and BWG, while Tau supplementation had a significant protective effect on BWG and prevented HS-induced reduction in the relative liver and gall bladder weights. Furthermore, dietary Tau supplementation resulted in the recovery, to control levels, of serum triglyceride concentrations, the amounts of total lipids, triglycerides, and cholesterol in the liver affected by the heat stress condition (Shim *et al.*, 2006). In a similar fashion, the body weight of the Tau-supplemented (0.1%) chickens housed in HS conditions (34 °C, weeks 3-5) was partly recovered being significantly higher (by 9.0%) than that of the HS group (Belal *et al.*, 2018). It seems likely that Tau can improve broiler thermotolerance as indicated by significant prevention of the HS-induced increase in expression of HSP70 in the chicken liver and muscles.

Effects of Tau supplementation on growth performance, jejunal histology, and appetite-related genes expressions of broilers under HS (32 °C) from 28 to 42 days have been recently assessed (He *et al.*, 2019). Tau supplementation (5 g/kg) did not prevent negative consequences of HS on chicken growth and development but significantly mitigated the impairment of jejunal morphology and prevented elevation of the concentrations of serum ghrelin in broilers subjected to HS. Furthermore, Tau increased mRNA expression of appetite-related genes in the intestine of heat-stressed birds (He *et al.*, 2019). Interestingly, metabolomic analysis identified the sulphur amino acid metabolic pathway, including Tau and hypotaurine metabolism, as a primary point affected by HS in chickens. In particular, hypotaurine concentration in plasma heat-stressed chickens decreased by 62% (Tomonaga *et al.*, 2018). Tau supplementation (5 g/kg for 2 weeks) significantly alleviated the increase in lightness and drip loss and the decrease in pH and sheer force of breast meat in chronic heat-stressed (32 °C from day 28 for 2 weeks) broilers (Lu *et al.*, 2019). Compared with the HS group, Tau supplementation significantly decreased oxidative stress as indicated by decreased levels of ROS and MDA due to upregulation of the antioxidant defence mechanisms. Indeed, Tau increased the mRNA expressions of Nrf2 and two important Nrf2-regulated AO enzymes namely NAD(P)H quinone dehydrogenase 1 and HO-1. In addition, Tau supplementation significantly alleviated mitochondrial damage in breast muscles caused by chronic heat exposure (Lu *et al.*, 2019).

10.10.2 Lipopolysaccharide treatment

Lipopolysaccharide (LPS) is an endotoxin released from the cell wall of pathogenic bacteria and LPS-induced acute intestinal disruption/injury is a well-established animal model for studying gastrointestinal physiology (Lai *et al.*, 2011; Gil-Cardoso *et al.*, 2019). Interestingly, increased ROS production and oxidative stress are major consequences of inflammatory response to LPS (Hussain *et al.*, 2017).

In early post-hatched chickens (3 day old) challenged with LPS, Tau dietary supplementation (3.6 mg/chick/d for 7 days) was shown to improve the antioxidant status in the duodenum, leading to amelioration of LPS-induced chicken duodenal inflammation and improvement of goblet cell function (Xiao *et al.*, 2018). The authors clearly showed that Tau supplementation of LPS-challenged chicks restored antioxidant status in duodenum, including amelioration of decreased major antioxidant system indexes including GSH and GSH/GSSG ratio, GPx and SOD activity and prevention of lipid peroxidation. This was associated with prevention of acute intestinal mucosal inflammation as indicated by significant down-regulation of the mRNA level of cellular adhesion molecule *ICAM-5* and *VCAM-1* and down-regulation of the increased proinflammatory IL-1β and TNF-α and up-regulate the decreased anti-inflammatory IL-10 levels. Furthermore, a substantial decrease (from 84% down to 12%) in diarrhoea associated with intestinal mucosal hyperemia was also observed. Furthermore, a restoration of goblet cells density and secretary functions in duodenum and improved mucosal renewal in duodenum of chickens under LPS challenge were also observed (Xiao *et al.*, 2018).

10.10.3 Increased stocking/housing density

High stocking density (SD) is known to be a stressful condition (Puron *et al.*, 1995), detrimentally affecting the welfare and gut health of broiler chicks and predisposing them to various gut disorders including necrotic enteritis (Tsiouris *et al.*, 2015). Furthermore, high stocking density is associated with oxidative stress as indicated by decreased GSH concentrations and GSH/GSSG ratios (Simitzis *et al.*, 2012), increased serum MDA level and reduced GPx activity (Simsek *et al.*, 2009) or detrimental changes in serum corticosterone, ovotransferrin, ceruloplasmin and brain heat shock protein (HSP 70) expression (Najafi *et al.*, 2015).

Effects of dietary Tau (0.1%) and increased SD on hepatic, oviduct and renal functions in green-shell laying hens were investigated. Layers were randomly divided into 3 groups: a free-range group, a caged group with low-density, and a caged group with high density. Each group was further divided into the control (C) and Tau-supplemented groups. After 15 days of supplementation, tissues were collected and analysed. It was shown that cage housing environment is associated with various stresses causing damages to the liver. In fact, liver integrity indexes (serum alanine aminotransferase and aspartate aminotransferase) in the caged hens were significantly elevated in comparison to the free-range group, detrimental changes in liver integrity were significantly ameliorated by dietary Tau (Ma *et al.*, 2014). Furthermore, cage environment also was associated with increased serum inducible nitric oxide synthase activity and dietary Tau was shown to significantly reduce this activity in the low-density group. Most importantly, NF-κB DNA-binding activity in the liver significantly increased due to HD housing conditions when compared with the other 2 housing patterns and it was significantly decreased by Tau. In addition, Tau was demonstrated to reduce the expression of pro-inflammatory TNF-α mRNA (a downstream effector molecule of the NF-κB pathway) in all 3 rearing conditions. This was associated with improved antioxidant defences as indicated by decreased MDA levels in serum and liver of Tau-treated birds and increased serum total AOA in the low-density group. Therefore, the aforementioned data clearly demonstrated protective effect of dietary Tau against hepatic damage in laying hens due to HD rearing conditions. In the second study with the same design protective effect of dietary Tau (0.1% for 15 days) on oviduct function in green-shell laying hens were studied (Dai *et al.*, 2015).

It was shown that layer mortality was decreased in Tau-supplemented groups independent of rearing systems. Tau also improved egg production in free range and low-density housing birds and had a tendency to increase it in HD housed birds. The authors suggested that dietary Tau supplementation can increase egg production via improved oviduct integrity and function, especially in the case of stress conditions created by HD housing. Indeed, HD chicken housing caused increased infiltration of erythrocytes and neutrophils into the oviduct and these changes were associated with upregulation of NF-κB (NF-κB DNA binding activity). These pathophysiological changes in the oviduct significantly attenuated by Tau supplementation. In general, NF-κB activation is a central event in stress-related tissue damage. Furthermore, Tau supplementation significantly reduced the expression of pro-inflammatory

TNF-α mRNA (a downstream effector molecule of the NF-κB pathway) in the free range and low-density groups and restored IL-10 expression in the oviduct, altered by HD housing, to the initial level (Dai *et al.*, 2015). The authors also showed that supplementation with Tau significantly increased SOD activity in the oviduct of free-range hens, but differences in SOD activity in birds with other housing systems did not reach significance.

The same experimental design was used in a later study by the same group related to kidney integrity (Ma *et al.*, 2016). In particular, it was shown that layers raised under HD conditions (from 12 up to 21 weeks) were characterised by increased morphological lesions of renal tubules (renal function impairment) compared with the free range and low-density housing groups. It was demonstrated that Tau supplementation decreased morphological renal damage, similar to the effect of rearing birds free range or in low-density housing. Interestingly, that indexes of renal function (serum creatinine, blood urea nitrogen concentrations and N-acetyl-β-D-glucosaminidase activity) in HD housing groups were elevated (reflecting renal function impairment) in comparison to those in the free range and low-density housing groups. Again, treatment with Tau was shown to prevent the elevation of these parameters. It seems likely that HD density housing caused oxidative stress associated with decreased antioxidant defences as indicated by decreased SOD activity and T-AOC levels and increased MDA (the common index of oxidative stress) among HD housed birds in comparison to those in free range and low-density housed groups. Interestingly, Tau supplementation decreased oxidative stress in kidney of HD housed birds as indicated by decreased MDA level and there was a trend in increasing SOD activity and T-AOC levels in kidney tissue of the same group (Ma *et al.*, 2016). The authors speculated that renal impairment in laying hens reared in HD environment resulted from an increase in inflammation. First, NF-κB, a constitutively expressed transcription factor regulating synthesis of pro-inflammatory cytokines and playing a central role in stress adaptation was studied. It was shown that Tau down regulated NF-κB DNA binding in the low-density housing group, but the difference in HD group was not significant. Interestingly, inhibition of NF-κB activity is an important function of Tau (Sun *et al.*, 2012). Compared with the free range and low-density housing groups, the mRNA expression of proinflammatory cytokines, namely TNF-α and IFN-γ, in the HD housing group increased significantly. Tau was shown to significantly decrease mRNA expression of TNF-α and IFN-γ, in the kidney of high-density group confirming that Tau supplementation can attenuate renal inflammation (Das and Sil, 2012). Therefore, the impairment of kidney function in the HD housing group may be associated with a combination of the inflammatory cascade and oxidative stress and Tau supplementation can ameliorate those detrimental changes (Ma *et al.*, 2016). Indeed, Tau treatment can down-regulate NF-κB and inhibit the expression of the NF-κB-driven inflammatory mediators (cytokines).

Taken together, the aforementioned data clearly showed protective effects of Tau against oxidative stress, inflammation and damages to tissue integrity in liver, oviduct and kidney due to stress conditions imposed by HD housing during birds rearing. The responses in the liver were more pronounced than in the oviduct or kidney, probably, reflecting Tau concentration alteration during feeding period with the liver to be much more responsive than other tissues. The limitations of the aforementioned studies are associated with a short (15 days) duration of Tau supplementation and it could well be that longer exposure to Tau could potentially show better protective effect in stress conditions of HD rearing. Tau application during rearing period (from 12 to 21 week) could also be beneficial. Furthermore, a range of Tau doses should also be tested to find an optimal one.

10.10.4 Pulmonary hypertension syndrome and sudden death syndrome in broilers

There are two important conditions in poultry production where oxidative stress is involved, namely pulmonary hypertension syndrome (PHS; Bottje and Wideman, 1995; Bottje *et al.*, 1995) and sudden death syndrome (SDS; Nain *et al.*, 2008; Olkowski *et al.*, 2008; Reid and Tervit, 1995). Therefore, nutritional antioxidants including vitamin C (Xiang *et al.*, 2002), vitamin E (Bottje *et al.*, 1995) and other antioxidants (Rajani *et al.*, 2011) were shown to decrease oxidative stress and reduce PSH incidence/severity. However, usage of the same vitamins to decrease SDS was not successful (Nain *et al.*, 2008). Interestingly after more than 35 yr of research the etiology of SDS is not fully understood (Leeson, 2007).

Potential protective role of Tau in the development of PHS could be related to its Ca^{2+}-regulating, membrane-stabilising and antioxidant functions. In particular, cardiac Tau release into the plasma was suggested to be a protective mechanism in response to the induction of pulmonary hypertension, hypoxemia, and right-side heart failure (Ruiz-Feria *et al.*, 1999). Various attempts to decrease PHS by supplementing diet with Tau were rather unsuccessful: neither the incidence of PHS nor specific predictors of PHS susceptibility were affected by Tau or its antagonist (β-alanine) supplementation (Ruiz-Feria and Wideman, 2001). However, it seems likely that Tau still has some protective effects on cardiac functions and may be important in the late stages of PHS. For example, additional stress in the form of low oxygen challenges together with depleted Tau reserves by β-alanine supplementation caused rapid deterioration of cardiac function associated with increased reductions in cardiac output, stroke volume, and mean systemic arterial pressure in comparison to control broilers (Ruiz-Feria and Wideman, 2001). In a recent study, a low dose of Tau supplementation (40 mg/kg) significantly decreased ascites mortality (18 vs 28%) in broiler chickens (Faraji *et al.*, 2019). Furthermore, Tau supplementation prevented detrimental effects of high doses of guanidinoacetic acid (2.25 g/kg) on weight gain and FCR. In addition, Tau also decreased abdominal fat proportion (1.15 vs 1.67%) by significant upregulating PPARα (Faraji *et al.*, 2019).

It was shown that total mortality and SDS were lower for broilers fed Tau-supplemented (0.05; 0.1 or 0.2%) diets (Campbell and Classen, 1989). Similarly, SDS-related chicken deaths were significantly reduced by supplementation of the feed with 0.5 g Tau/kg, this led to a reduction in SDS deaths as a percentage of total deaths at 3 and 6 wk of age (Blair *et al.*, 1991). However, in the study conducted in the same department it was found that dietary supplementation with β-alanine (Tau antagonist) decreased cardiac Tau concentration almost 10-fold without significant effect on SDS mortality rates (Jacob *et al.*, 1991). The authors interpreted the results as an evidence that Tau does not play a major role in the etiology of SDS.

Interestingly, Tau levels were found to be increased in breast muscles of birds affected by white striping (probably as an adaptive mechanism to hypoxic conditions) and the Tau metabolism was among the most perturbed pathways in the affected muscles (Boerboom *et al.*, 2018). Therefore, more detailed research on the effect of Tau dietary supplementation on the white stripping condition in growing birds await further investigation.

10.10.5 Prevention of oxidative stress caused by various toxicants

Data related to protective effect of natural antioxidants against toxic effect of various commonly used chemicals could be used to substantiate antioxidant beneficial properties of Tau. For example, Tau was shown to have protective effects against doxorubicin-induced cardiotoxicity in perfused chicken hearts (Hamaguchi *et al.*, 1989). Oral administration of Tau (200 mg/kg for seven days) partially ameliorated isoprenaline-induced myocardial damage (preventing necrotic changes) in chickens (Ohta *et al.*, 1986) and reduced oxidative stress in chicken heart as indicated by restoration of GSH level and inhibition of lipid peroxidation (Ohta *et al.*, 1988). In various model systems based on *in vitro* studies with cell cultures and *in vivo* studies with laboratory animals, protective antioxidant effects of Tau against other chemicals, causing oxidative stress, including CCl4 (Abdel-Moneim *et al.*, 2015), thiacetamide (TAA; Balkan *et al.*, 2001), fluoride (Adedara *et al.*, 2017), cisplatin (CP; Owoeye *et al.*, 2018) were proven. Therefore, Tau was shown to be protective against toxicant-imposed oxidative stress in a variety of *in vitro* and *in vivo* systems (Figure 10.3).

The results of the aforementioned studies showing protective effects of Tau can be summarised as follows. Tau was shown to restore antioxidant enzyme activities (SOD, GPx, GR, GST, Catalase) and non-enzymatic antioxidants (GSH, ascorbic acid) compromised by toxicant treatment. Furthermore, Tau was found to restore the redox status of the cells/tissues with protective effects of Tau observed *in vitro* in cell cultures and *in vivo* in the brain, testis, kidney, liver and other tissues. Furthermore, Tau was able to ameliorate changes in main indexes of lipid peroxidation (MDA and 4-hydroxyalkenals) and DNA oxidation (8-OHdG) compromised by toxicant treatment. In addition, Tau was shown to upregulate transcription factors, including Nrf2, responsible for adaptive synthesis of a range of antioxidant molecules. At the same time, in many cases Tau decreased expression of another transcription factor, namely NF-κB, responsible for synthesis of pro-inflammatory cytokines and inflammation in general. Therefore, it is clear at present that, molecular mechanisms of antioxidant protective actions of Tau are diverse, and restoration of expression of transcription factors and antioxidant enzymes activities in stress conditions, described above are of great importance.

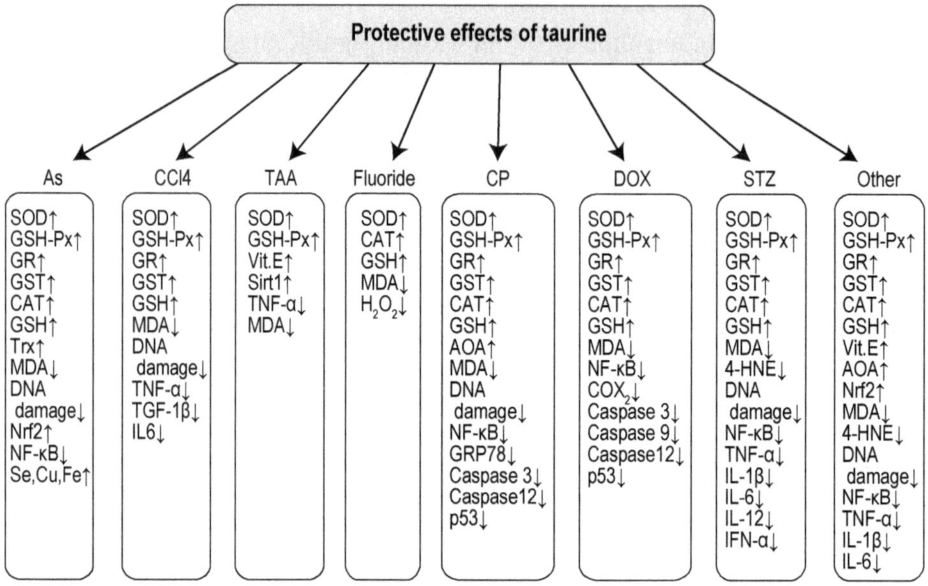

Figure 10.3. Protective effects of taurine against various toxicants.

10.10.6 *In vitro* semen manipulation

Our research with poultry semen (Surai *et al.*, 2003) and recent publications by others (Agarwal and Majzoub, 2017; Bui *et al.*, 2018; Losano *et al.*, 2018) indicate that spermatozoa could be an ideal model to study oxidative stress and protective effects of various compounds. It is well-known that spermatozoa membranes are rich in PUFA and therefore are very sensitive to various stresses (Surai, 2002). Furthermore, there is a range of easily measurable physiological and biochemical parameters of spermatozoa reflecting redox status, membrane integrity and stability. In addition, various *in vitro* manipulations including semen storage and cryopreservation are commonly used procedures in reproductive biology of farm animals and poultry. Therefore, by adding test compounds into semen diluent it is possible to assess their protective antioxidant and membrane-stabilising properties.

Effect of L-carnitine (LC), hypotaurine (HT), and Tau on the quality of frozen-thawed chicken semen were studied (Partyka *et al.*, 2017). Pooled semen samples were divided into seven aliquots, diluted with a special cryopreservation medium (control) or the same medium containing 1 mM LC, 5 mM LC, 1 mM HT, 10 mM HT, 1 mM Tau, and 10 mM Tau and subjected to cryopreservation. In comparison to the control medium, Tau inclusion into the diluent was associated with:

- increased percentage of spermatozoa without apoptosis and membrane reorganisation changes;
- elevated percentage of live spermatozoa without lipid peroxidation;
- increased percentage of spermatozoa with high mitochondrial potential;
- decreased DNA fragmentation in spermatozoa (Partyka *et al.*, 2017).

10.11 Taurine essentiality and requirement in poultry

Tau is considered to be an essential nutrient for several animal species, including the cat, certain dogs, the fox, some monkeys and the anteater (Schaffer *et al.*, 2014). Although in major mammalian and avian species Tau can be synthesised and under physiological conditions probably would cover major part of the requirement and the rest would be obtained from the diet. However, under stress conditions, when Tau requirement would go up, but feed consumption is usually decreased, the Tau synthesis would not be adequate, and thus dietary consumption of Tau is necessary to maintain its homeostasis (Kadam and Prabhasankar, 2010). Therefore, under stress conditions Tau could be an essential or semi-essential nutrient for many mammals and birds. This could explain variability in feeding trial results presented above: in most of cases there was not a significant effect of Tau supplementation on chicken growth and development or layer performance. Differences in challenges of experimental birds in various trials could explain variability of some results, related to the nutritional value of the supplemented Tau. Therefore, in various stress models Tau showed significant protective effects reflecting its important physiological roles and its participation in redox status maintenance, transcription factors and vitagene activation and antioxidant defences regulation. It is generally accepted that commercial poultry production is associated with a range of stressful situations, including technological, environmental (Surai and Fisinin, 2016a), nutritional and internal/biological (Surai and Fisinin, 2016b) stresses. Therefore, anti-stress protective actions of Tau under commercially relevant stress conditions deserve more attention and further investigation.

Analysis of major studies related to nutritional value of Tau in poultry nutrition indicated that Tau doses used in various studies greatly varied, but generally were in a range 0.05-0.5% of the diet with 0.1% to be most effective. However, there are several publications where an effective Tau dosage was substantially (10-20-fold) lower than 0.1%. For example, optimal Tau level for layers was found to be 80 mg/kg (0,008%; Li *et al.*, 2008), optimal egg yolk lipid composition was obtained in layers supplemented with 50 mg/kg (0.005%) Tau (Yin, 2006), while Tau at 100 mg/kg (0.01%) was shown to decrease percentage of broken and soft eggs in quail (Wang *et al.*, 2010). Furthermore, as mentioned earlier, Tau supplementation at 40 mg/kg significantly reduced ascites mortality (18 vs 28%) in broiler chickens (Faraji *et al.*, 2019). Therefore, there is a need to conduct more studies devoted to protective effects of low Tau dosage alone or in combination with other antioxidants. Since Tau has a comparatively high water solubility it can also be applied with water via drinking system.

Indeed, the aforementioned data related to protective effects of Tau in stressed poultry, medical Tau applications (Jakaria *et al.*, 2019; Schaffer and Kim, 2018; Siedel *et al.*, 2019) and research in other animals, including pigs (Huang and Peng, 2008; Tang *et al.*, 2018; Xu *et al.*, 2019) regarding Tau and its essentiality forms the basis for further research on Tau in poultry.

10.12 Conclusions

Traditionally, as a sulphur-containing, non-proteinaceous amino acid, Tau was considered to be non-essential for poultry, since it is synthesised in the body. Recent changes in breeder diet formulation associated with removal of animal-derived feed ingredients, main feed sources of Tau for poultry, make such a diet Tau-deficient. Furthermore, Tau synthesis can be compromised under various stress conditions. In fact, commercial poultry production is associated with a range of stresses and the development of anti-stress nutritional strategy is an important part of recent poultry research work. Therefore, antioxidant activities and regulating role of Tau in expression of various transcription factors and vitagenes in poultry under various stress conditions deserve more attention. In fact, under laboratory conditions with small numbers of birds and well-controlled environment and balanced diets, protective effects of dietary Tau most likely would not be observed. This was confirmed in a range of trials conducted 30-50 years ago. However, recent data related to various stress models, including HS, immunological challenge, increased stocking density and presence of toxicants in the diet, clearly showed protective effects of dietary Tau in poultry. It could well be that under stress conditions and with diets without animal protein Tau could become semi-essential amino acid for poultry and its supplementation with feed or water could have beneficial protective effects. This subject warrants further investigation.

References

Abd Elwahab, A.H.A., Ramadan, B.K., Schaalan, M.F. and Tolba, A.M., 2017. A novel role of SIRT1/FGF-21 in taurine protection against cafeteria diet-induced steatohepatitis in rats. Cellular Physiology and Biochemistry 43: 644-659.

Abdel-Moneim, A.M., Al-Kahtani, M.A., El-Kersh, M.A. and Al-Omair, M.A., 2015. Free radical-scavenging, anti-inflammatory/anti-fibrotic and hepatoprotective actions of taurine and silymarin against CCl4 induced rat liver damage. PLoS One 10: e0144509.

Adedara, I.A., Olabiyi, B.F., Ojuade, T.D., Idris, U.F., Onibiyo, E.M. and Farombi, E.O., 2017. Taurine reverses sodium fluoride-mediated increase in inflammation, caspase-3 activity, and oxidative damage along the brain-pituitary-gonadal axis in male rats. Canadian Journal of Physiology and Pharmacology 95: 1019-1029.

Agarwal, A. and Majzoub, A., 2017. Laboratory tests for oxidative stress. Indian Journal of Urology 33: 199-206.

Agca, C.A., Tuzcu, M., Hayirli, A. and Sahin, K., 2014. Taurine ameliorates neuropathy via regulating NF-κB and Nrf2/HO-1 signaling cascades in diabetic rats. Food and Chemical Toxicology 71: 116-121.

Ahmad, M.K., Khan, A.A., Ali, S.N. and Mahmood, R., 2015. Chemoprotective effect of taurine on potassium bromate-induced DNA damage, DNA-protein cross-linking and oxidative stress in rat intestine. PloS One 10: e0119137.

Ahmad, M.K. and Mahmood, R., 2016. Protective effect of taurine against potassium bromate-induced hemoglobin oxidation, oxidative stress, and impairment of antioxidant defense system in blood. Environmental Toxicology 31: 304-313.

Akbarian, A., Michiels, J., Degroote, J., Majdeddin, M., Golian, A. and De Smet, S., 2016. Association between heat stress and oxidative stress in poultry; mitochondrial dysfunction and dietary interventions with phytochemicals. Journal of Animal Science and Biotechnology 7: 37.

Alhawiti, N.M., Al Mahri, S., Azhar Aziz, M., Shafi Malik, S. and Mohammad, S., 2017. TXNIP in metabolic regulation: physiological role and therapeutic outlook. Current Drug Targets 18: 1095-1103.

Alvarez, J.G. and Storey, B.T., 1983. Taurine, hypotaurine, epinephrine and albumin inhibit lipid peroxidation in rabbit spermatozoa and protect against loss of motility. Biology of Reproduction 29: 548-555.

Alzawqari, M.H., Al-Baadani, H.H., Alhidary, I.B., Al-Owaimer, A.N. and Abudabos, A.M., 2016. Effect of taurine and bile acid supplementation and their interaction on performance, serum components, ileal viscosity and carcass characteristics of broiler chickens. South African Journal of Animal Science 46: 448-457.

Anderson, C.M., Howard, A., Walters, J.R., Ganapathy, V. and Thwaites, D.T., 2009. Taurine uptake across the human intestinal brush-border membrane is via two transporters: H+-coupled PAT1 (SLC36A1) and Na+-and Cl−-dependent TauT (SLC6A6). The Journal of Physiology 587: 731-744.

Aruoma, O.I., Halliwell, B., Hoey, B.M. and Butler, J., 1988. The antioxidant action of taurine, hypotaurine and their metabolic precursors. Biochemical Journal 256: 251-255.

Bai, J., Yao, X., Jiang, L., Qiu, T., Liu, S., Qi, B., Zheng, Y., Kong, Y., Yang, G., Chen, M. and Liu, X., 2016. Taurine protects against As2O3-induced autophagy in pancreas of rat offsprings through Nrf2/Trx pathway. Biochimie 123: 1-6.

Balkan, J., Doggru-Abbasoğlul, S., Kanbaglil, Ö., Cevikbas, U., Aykac-Toker, G. and Uysal, M., 2001. Taurine has a protective effect against thioacetamide-induced liver cirrhosis by decreasing oxidative stress. Human & Experimental Toxicology 20: 251-254.

Banday, M.N., Lone, F.A., Rasool, F., Rashid, M. and Shikari, A., 2017. Use of antioxidants reduce lipid peroxidation and improve quality of crossbred ram sperm during its cryopreservation. Cryobiology 74: 25-30.

Barua, M., Liu, Y. and Quinn, M.R., 2001. Taurine chloramine inhibits inducible nitric oxide synthase and TNF-α gene expression in activated alveolar macrophages: decreased NF-κB activation and IκB kinase activity. The Journal of Immunology 167: 2275-2281.

Belal, S.A., Kang, D.R., Cho, E.S.R., Park, G.H. and Shim, K.S., 2018. Taurine reduces heat stress by regulating the expression of heat shock proteins in broilers exposed to chronic heat. Brazilian Journal of Poultry Science 20: 479-486.

Berlin, K.N., Cameron, L.M., Gatt, M. and Miller Jr, R.R., 2010. Reduced de novo synthesis of 5-methyltetrahydrofolate and reduced taurine levels in ethanol-treated chick brains. Comparative Biochemistry and Physiology Part C: Toxicology & Pharmacology 152: 353-359.

Bhavsar, T.M., Cantor, J.O., Patel, S.N. and Lau-Cam, C.A., 2009. Attenuating effect of taurine on lipopolysaccharide-induced acute lung injury in hamsters. Pharmacological Research 60: 418-428.

Bhavsar, T.M., Patel, S.N. and Lau-Cam, C.A., 2010. Protective action of taurine, given as a pretreatment or as a posttreatment, against endotoxin-induced acute lung inflammation in hamsters. Journal of Biomedical Science 17: S19.

Blair, R., Jacob, J.P. and Gardiner, E.E., 1991. Lack of an effect of taurine supplementation on the incidence of sudden death syndrome in male broiler chicks. Poultry Science 70: 554-560.

Boerboom, G., Van Kempen, T., Navarro-Villa, A. and Pérez-Bonilla, A., 2018. Unravelling the cause of white striping in broilers using metabolomics. Poultry Science 97: 3977-3986.

Bottje, W., Enkvetchakul, B., Moore, R. and McNew, R., 1995. Effect of α-tocopherol on antioxidants, lipid peroxidation, and the incidence of pulmonary hypertension syndrome (ascites) in broilers. Poultry Science 74: 1356-1369.

Bottje, W.G. and Wideman, R.F. Jr., 1995. Potential role of free radicals in the pathogenesis of pulmonary hypertension syndrome. Poultry and Avian Biology Reviews 6: 211-231.

Bottrel, M., Acha, D., Ortiz, I., Hidalgo, M., Gósalvez, J., Camisão, J. and Dorado, J., 2018. Cryoprotective effect of glutamine, taurine, and proline on post-thaw semen quality and DNA integrity of donkey spermatozoa. Animal Reproduction Science 189: 128-135.

Bouckenooghe, T., Remacle, C. and Reusens, B., 2006. Is taurine a functional nutrient? Current Opinion in Clinical Nutrition & Metabolic Care 9: 728-733.

Brand, M.D., 2016. Mitochondrial generation of superoxide and hydrogen peroxide as the source of mitochondrial redox signaling. Free Radical Biology & Medicine 100: 14-31.

Bui, A.D., Sharma, R., Henkel, R. and Agarwal, A., 2018. Reactive oxygen species impact on sperm DNA and its role in male infertility. Andrologia 50: e13012.

Cai, J.G., Luo, L.M., Tang, H. and Zhou, L., 2018. Cytotoxicity of malondialdehyde and cytoprotective effects of taurine via oxidative stress and PGC-1α signal pathway in C2C12 Cells. Molecular Biology (Moscow) 52: 616-627.

Calabrese, V., Cornelius, C., Stella, A.M. and Calabrese, E.J., 2010. Cellular stress responses, mitostress and carnitine insufficiencies as critical determinants in aging and neurodegenerative disorders: role of hormesis and vitagenes. Neurochemical Research 35: 1880-1915.

Caletti, G., Herrmann, A.P., Pulcinelli, R.R., Steffens, L., Morás, A.M., Vianna, P., Chies, J.A.B., Moura, D.J., Barros, H.M.T. and Gomez, R., 2018. Taurine counteracts the neurotoxic effects of streptozotocin-induced diabetes in rats. Amino Acids 50: 95-104.

Campbell, GL. and Classen, HL., 1989. Effect of dietary taurine supplementation on sudden death syndrome in broiler chickens. Canadian Journal of Animal Science 69: 509-512.

Cao, W., Li, Y., Li, M., Zhang, X. and Liao, M., 2017. Txn1, Ctsd and Cdk4 are key proteins of combination therapy with taurine, epigallocatechin gallate and genistein against liver fibrosis in rats. Biomedicine & Pharmacotherapy 85: 611-619.

Cao, W., Zhou, Y., Li, Y., Zhang, X., He, M., Zang, N., Zhou, Y. and Liao, M., 2015. iTRAQ-based proteomic analysis of combination therapy with taurine, epigallocatechin gallate, and genistein on carbon tetrachloride-induced liver fibrosis in rats. Toxicology Letters 232: 233-245.

Celik, L.B., Tekeli, A. and Öztürkcan, O., 2004. Effects of supplemental l-carnitine in drinking water on performance and egg quality of laying hens exposed to a high ambient temperature. Journal of Animal Physiology and Animal Nutrition 88: 229-233.

Chand, N., Naz, S., Khan, A., Khan, S. and Khan, R.U., 2014. Performance traits and immune response of broiler chicks treated with zinc and ascorbic acid supplementation during cyclic heat stress. International Journal of Biometeorology 58: 2153-2157.

Chang, L., Xu, J.X., Zhao, J., Pang, Y.Z., Tang, C.S. and Qi, Y.F., 2004. Taurine antagonized oxidative stress injury induced by homocysteine in rat vascular smooth muscle cells. Acta Pharmacologica Sinica 25: 341-346.

Che, Y., Hou, L., Sun, F., Zhang, C., Liu, X., Piao, F., Zhang, D., Li, H. and Wang, Q., 2018. Taurine protects dopaminergic neurons in a mouse Parkinson's disease model through inhibition of microglial M1 polarization. Cell Death & Disease 9: 435.

Chen, W., Guo, J., Zhang, Y. and Zhang, J., 2016. The beneficial effects of taurine in preventing metabolic syndrome. Food & Function 7: 1849-1863.

Chen, X., Sebastian, B.M., Tang, H., McMullen, M.M., Axhemi, A., Jacobsen, D.W. and Nagy, L.E, 2009. Taurine supplementation prevents ethanol-induced decrease in serum adiponectin and reduces hepatic steatosis in rats. Hepatology 49: 1554-1562.

Cheng, C.H., Guo, Z.X. and Wang, A.L., 2018. The protective effects of taurine on oxidative stress, cytoplasmic free-Ca(2+) and apoptosis of pufferfish (*Takifugu obscurus*) under low temperature stress. Fish Shellfish Immunology 77: 457-464.

Cheong, S.H. and Lee, D.S., 2017. Taurine chloramine prevents neuronal HT22 cell damage through Nrf2-Related heme oxygenase-1. Advances in Experimental Medicine and Biology 975: 145-157.

Cheong, S.H., Moon, S.H., Lee, S.J., Kim, S.H. and Chang, K.J., 2013. Antioxidant and DNA protection effects of taurine by electron spin resonance spectroscopy. Advances in Experimental Medicine and Biology 776: 167-177.

Chhillar, S., Singh, V.K., Kumar, R. and Atreja, S.K., 2012. Effects of Taurine or Trehalose supplementation on functional competence of cryopreserved Karan Fries semen. Anim. Reproduction Science 135: 1-7.

Chou, C.T., Lin, W.F., Kong, Z.L., Chen, S.Y. and Hwang, D.F., 2013. Taurine prevented cell cycle arrest and restored neurotrophic gene expression in arsenite-treated SH-SY5Y cells. Amino Acids 45: 811-819.

Chowdhury, S., Sinha, K., Banerjee, S. and Sil, P.C., 2016. Taurine protects cisplatin induced cardiotoxicity by modulating inflammatory and endoplasmic reticulum stress responses. Biofactors 42: 647-664.

Di Ciaula, A., Garruti, G., Lunardi Baccetto, R., Molina-Molina, E., Bonfrate, L., Wang, D.Q. and Portincasa, P., 2018. Bile acid physiology. Annals of Hepatology 16: 4-14.

Cuisinier, C., De Welle, J.M., Verbeeck, R.K., Poortmans, J.R., Ward, R., Sturbois, X. and Francaux, M., 2002. Role of taurine in osmoregulation during endurance exercise. European Journal of Applied Physiology 87: 489-495.

Dai, B., Zhang, J., Liu, M., Lu, J., Zhang, Y., Xu, Y., Miao, J. and Yin, Y., 2016. The role of Ca2+ mediated signaling pathways on the effect of taurine against *Streptococcus uberis* infection. Veterinary Microbiology 192: 26-33.

Dai, B., Zhang, Y.S., Ma, Z.L., Zheng, L.H., Li, S.J., Dou, X.H., Gong, J.S. and Miao, J.F., 2015. Influence of dietary taurine and housing density on oviduct function in laying hens. Journal of Zhejiang University, Science B 16: 456-464.

Das, J., Ghosh, J., Manna, P. and Sil, P.C., 2012. Taurine protects rat testes against doxorubicin-induced oxidative stress as well as p53, Fas and caspase 12-mediated apoptosis. Amino acids 42: 1839-1855.

Das, J., Ghosh, J., Manna, P., Sinha, M. and Sil, P.C., 2009. Taurine protects rat testes against $NaAsO_2$-induced oxidative stress and apoptosis via mitochondrial dependent and independent pathways. Toxicology Letters 187: 201-210.

Das, J. and Sil, P.C., 2012. Taurine ameliorates alloxan-induced diabetic renal injury, oxidative stress-related signaling pathways and apoptosis in rats. Amino Acids 43: 1509-1523.

De Luca, A., Pierno, S. and Camerino, D.C., 2015. Taurine: the appeal of a safe amino acid for skeletal muscle disorders. Journal of Translational Medicine 13: 243.

Devi, S.L., Viswanathan, P. and Anuradha, C.V., 2009. Taurine enhances the metabolism and detoxification of ethanol and prevents hepatic fibrosis in rats treated with iron and alcohol. Environmental Toxicology and Pharmacology 27: 120-126.

Divakaran, S., 2006. Taurine: An amino acid rich in fish meal. In: Cruz Suárez, L.E., Ricque Marie, D., Nieto López, M.G., Tapia Salazar, M., Villarreal Cavazos,D., Puello Cruz, A.C. and García Ortega, A. (eds) Avances en nutricion acuicola VIII. VIII Symposium Internacional de Nutición Acuícola. Nov. 15-17. Universídad Autónoma de Neuvo León, Monterrey. Neuvo León, México, pp. 310-317.

Du, H., You, J.S., Zhao, X., Park, J.Y., Kim, S.H. and Chang, K.J., 2010. Antiobesity and hypolipidemic effects of lotus leaf hot water extract with taurine supplementation in rats fed a high fat diet. Journal of Biomedical Science 17: S42.

Dunnington, E.A., Siegel, P.B. and Stewart, K.K., 1996. Effects of dietary taurine on growth and *Escherichia coli* resistance in chickens. Poultry Science 75: 1330-1333.

Edgar, S.E., Kirk, C.A., Rogers, Q.R. and Morris, J.G., 1998. Taurine status in cats is not maintained by dietary cysteinesulfinic acid. The Journal of Nutrition 128: 751-757.

Egbuniwe, I.C., Ayo, J.O. and Ocheja, O.B., 2018. Betaine and ascorbic acid modulate indoor behavior and some performance indicators of broiler chickens in response to hot-dry season. Journal of Thermal Biology 76: 38-44.

El Idrissi, A., 2008. Taurine increases mitochondrial buffering of calcium: role in neuroprotection. Amino Acids 34: 321-328.

El Idrissi, A. and Trenkner, E., 1999. Growth factors and taurine protect against excitotoxicity by stabilizing calcium homeostasis and energy metabolism. Journal of Neuroscience 19: 9459-9468.

El-Maraghi, E.F., Abdel-Fattah, K.I., Soliman, S.M. and El-Sayed, W.M., 2018. Taurine provides a time-dependent amelioration of the brain damage induced by γ-irradiation in rats. Journal of Hazardous Materials 359: 40-46.

Estévez, M. and Xiong, Y., 2019. Intake of oxidized proteins and amino acids and causative oxidative stress and disease: recent scientific evidences and hypotheses. Journal of Food Science 84: 387-396.

Fakruddin, M., Wei, F.Y., Suzuki, T., Asano, K., Kaieda, T., Omori, A., Izumi, R., Fujimura, A., Kaitsuka, T., Miyata, K. and Araki, K., 2018. Defective mitochondrial tRNA taurine modification activates global proteostress and leads to mitochondrial disease. Cell Reports 22: 482-496.

Farag, M.R. and Alagawany, M., 2018. Physiological alterations of poultry to the high environmental temperature. Journal of Thermal Biology 76: 101-106.

Faraji, M., Karimi Dehkordi, S., Zamiani Moghadam, A.K., Ahmadipour, B. and Khajali, F., 2019. Combined effects of guanidinoacetic acid, coenzyme Q10 and taurine on growth performance, gene expression and ascites mortality in broiler chickens. Journal of Animal Physiology and Animal Nutrition 103: 162-169.

Feidantsis, K., Kaitetzidou, E., Mavrogiannis, N., Michaelidis, B., Kotzamanis, Y. and Antonopoulou, E., 2014. Effect of taurine-enriched diets on the Hsp expression, MAPK activation and the antioxidant defence of the European sea bass (*Dicentrarchus labrax*). Aquaculture Nutrition 20: 431-442.

Feng, Y., Sun, F., Gao, Y., Yang, J., Wu, G., Lin, S. and Hu, J., 2017. Taurine decreased uric acid levels in hyperuricemic rats and alleviated kidney injury. Biochemical and Biophysical Research Communications 489: 312-318.

Forrester, S.J., Kikuchi, D.S., Hernandes, M.S., Xu, Q. and Griendling, K.K., 2018. Reactive oxygen species in metabolic and inflammatory signaling. Circulation Research 122: 877-902.

Frosini, M., Sesti, C., Saponara, S., Donati, A., Palmi, M., Valoti, M., Machetti, F. and Sgaragli, G., 2000. Effects of taurine and some structurally related analogues on the central mechanism of thermoregulation: a structure-activity relationship study. Advances in Experimental Medicine and Biology 483: 273-282.

Frosini, M., Sesti, C., Saponara, S., Ricci, L., Valoti, M., Palmi, M., Machetti, F. and Sgaragli, G., 2003. A specific taurine recognition site in the rabbit brain is responsible for taurine effects on thermoregulation. British Journal of Pharmacology 139: 487-494.

Fukuda, N., Yoshitama, A., Sugita, S., Fujita, M. and Murakami, S., 2011. Dietary taurine reduces hepatic secretion of cholesteryl ester and enhances fatty acid oxidation in rats fed a high-cholesterol diet. Journal of Nutritional Science and Vitaminology 57: 144-149.

Ghandforoush-Sattari, M., Mashayekhi, S., Krishna, C.V., Thompson, J.P. and Routledge, P.A., 2010. Pharmacokinetics of oral taurine in healthy volunteers. Journal of Amino Acids, 2010: 3.

Ghosh, J., Das, J., Manna, P. and Sil, P.C., 2009. Taurine prevents arsenic-induced cardiac oxidative stress and apoptotic damage: Role of NF-κB, p38 and JNK MAPK pathway. Toxicology and Applied Pharmacology 240: 73-87.

Gil-Cardoso, K., Comitato, R., Ginés, I., Ardévol, A., Pinent, M., Virgili, F., Terra, X. and Blay, M., 2019. Protective effect of proanthocyanidins in a rat model of mild intestinal inflammation and impaired intestinal permeability induced by LPS. Molecular Nutrition & Food Research 63:1800720.

Gondo, Y., Satsu, H., Ishimoto, Y., Iwamoto, T. and Shimizu, M., 2012. Effect of taurine on mRNA expression of thioredoxin interacting protein in Caco-2 cells. Biochemical and Biophysical Research Communications 426: 433-437.

Gonzales, F. and Awapara, J., 1955. The taurine concentration of developing chick emryo. Experimental Cell Research 9: 353-355.

Grove, R.Q. and Karpowicz, S.J., 2017. Reaction of hypotaurine or taurine with superoxide produces the organic peroxysulfonic acid peroxytaurine. Free Radical Biology and Medicine 108: 575-584.

Guan, X., Dei-Anane, G., Liang, R., Gross, M.L., Nickkholgh, A., Kern, M., Ludwig, J., Zeier, M., Büchler, M.W., Schmidt, J. and Schemmer, P., 2008. Donor preconditioning with taurine protects kidney grafts from injury after experimental transplantation. Journal of Surgical Research 146: 127-134.

Gupta, R.C., 2006. Taurine analogues and taurine transport: therapeutic advantages. Advances in Experimental Medicine and Biology 583: 449-467.

Gurujeyalakshmi, G., Wang, Y. and Giri, S.N., 2000. Taurine and niacin block lung injury and fibrosis by down-regulating bleomycin-induced activation of transcription nuclear factor-κB in mice. Journal of Pharmacology and Experimental Therapeutics 293: 82-90.

Habibian, M., Sadeghi, G., Ghazi, S. and Moeini, M.M., 2015. Selenium as a feed supplement for heat-stressed poultry: a review. Biological Trace Element Research 165: 183-193.

Hamaguchi, T., Azuma, J. and Schaffer, S., 1991. Interaction of taurine with methionine: inhibition of myocardial phospholipid methyltransferase. Journal of Cardiovascular Pharmacology 18: 224-230.

Hamaguchi, T., Azuma, J., Harada, H., Takahashi, K., Kishimoto, S. and Schaffer, S.W., 1989. Protective effect of taurine against doxorubicin-induced cardiotoxicity in perfused chick hearts. Pharmacological Research 21: 729-734.

Hammes, T.O., Pedroso, G.L., Hartmann, C.R., Escobar, T.D.C., Fracasso, L.B., da Rosa, D.P., Marroni, N.P., Porawski, M. and da Silveira, T.R., 2012. The effect of taurine on hepatic steatosis induced by thioacetamide in zebrafish (Danio rerio). Digestive Diseases and Sciences 57: 675-682.

Hanna, J., Chahine, R., Aftimos, G., Nader, M., Mounayar, A., Esseily, F. and Chamat, S., 2004. Protective effect of taurine against free radicals damage in the rat myocardium. Experimental and Toxicologic Pathology 56: 189-194.

Hansen, S.H. and Grunnet, N., 2013. Taurine, glutathione and bioenergetics. Advances in Experimental Medicine and Biology 776: 3-12.

Hansen, S.H., Andersen, M.L., Birkedal, H., Cornett, C. and Wibrand, F., 2006. The important role of taurine in oxidative metabolism. Advances in Experimental Medicine and Biology 583: 129-135.

Hansen, S.H., Andersen, M.L., Cornett, C., Gradinaru, R. and Grunnet, N., 2010. A role for taurine in mitochondrial function. Journal of Biomedical Science 17: S23.

Hansen, S.H., Birkedal, H., Wibrand, F. and Grunnet, N., 2015. Taurine and regulation of mitochondrial metabolism. Advances in Experimental Medicine and Biology 803: 397-405.

He, X., Lu, Z., Ma, B., Zhang, L., Li, J., Jiang, Y., Zhou, G. and Gao, F., 2019. Effects of dietary taurine supplementation on growth performance, jejunal morphology, appetite-related hormones, and genes expression in broilers subjected to chronic heat stress. Poultry Science 98: 2719-2728.

Henry, E.F. and MacCormack, T.J., 2018. Taurine protects cardiac contractility in killifish, Fundulus heteroclitus, by enhancing sarcoplasmic reticular Ca2+ cycling. Journal of Comparative Physiology B 188: 89-99.

Hou, L., Che, Y., Sun, F. and Wang, Q., 2018. Taurine protects noradrenergic locus coeruleus neurons in a mouse Parkinson's disease model by inhibiting microglial M1 polarization. Amino Acids 50: 547-556.

Hsu, T.C., Chiang, S.Y., Wu, J.H., Tsai, C.C., Huang, C.Y., Chen, Y.C. and Tzang, B.S., 2008. Treatment with taurine attenuates hepatic apoptosis in NZB/W F1 mice fed with a high-cholesterol diet. Journal of Agricultural and Food Chemistry 56: 9685-9691.

Hsu, T.C., Chen, Y.C., Tsai, C.C., Wu, J.H., Li, S.L. and Tzang, B.S., 2010. Protective effects of taurine against hepatic abnormality in NZB/W F1 mice fed a hypercholesterolemic diet. Food Chemistry 119: 62-68.

Huang, C., Guo, Y. and Yuan, J., 2014a. Dietary taurine impairs intestinal growth and mucosal structure of broiler chickens by increasing toxic bile acid concentrations in the intestine. Poultry Science 93: 1475-1483.

Huang, C.X., Wang, B., Min, Z. and Yuan, J., 2014b. Dietary inclusion level and time effects of taurine on broiler performance, meat quality, oxidative status and muscle taurine content. British Poultry Science 55: 598-604.

Huang, R.S. and Peng, Z.L., 2008. Effect of diet type and dietary taurine supplementation on growth performance of weaning pigs. Cereal & Feed Industry 9: 44-45.

Hussain, T., Tan, B., Liu, G., Murtaza, G., Rahu, N., Saleem, M. and Yin, Y., 2017. Modulatory mechanism of polyphenols and Nrf2 signaling pathway in LPS challenged pregnancy disorders. Oxidative Medicine and Cellular Longevity 2017, 8254289.

Huxtable, R. and Bressler, R., 1973. Effect of taurine on a muscle intracellular membrane. Biochimica et Biophysica Acta (BBA)-Biomembranes 323: 573-583.

Huxtable, R.J., 1996. Taurine. Past, present, and future. Advances in Experimental Medicine and Biology 403: 641-650.

Inam-U-Llah, Piao, F., Aadil, R.M., Suleman, R., Li, K., Zhang, M., Wu, P., Shahbaz, M. and Ahmed, Z., 2018. Ameliorative effects of taurine against diabetes: a review. Amino Acids 50: 487-502.

Ito, T., Kimura, Y., Uozumi, Y., Takai, M., Muraoka, S., Matsuda, T., Ueki, K., Yoshiyama, M., Ikawa, M., Okabe, M. and Schaffer, S.W., 2008. Taurine depletion caused by knocking out the taurine transporter gene leads to cardiomyopathy with cardiac atrophy. Journal of Molecular and Cellular Cardiology 44: 927-937.

Jacob, J.P., Blair, R., Hart, L.E. and Gardiner, E.E., 1991. The effect of taurine transport antagonists on cardiac taurine concentration and the incidence of sudden death syndrome in male broiler chickens. Poultry Science 70: 561-567.

Jacobsen, J.G. and Smith, L.H., 1968. Biochemistry and physiology of taurine and taurine derivatives. Physiological Reviews 48: 424-511.

Jakaria, M., Azam, S., Haque, M.E., Jo, S.H., Uddin, M.S., Kim, I.S. and Choi, D.K., 2019. Taurine and its analogs in neurological disorders: Focus on therapeutic potential and molecular mechanisms. Redox Biology 24: 101223.

Javed, H., Khan, A., Vaibhav, K., Khan, M.M., Ahmad, A., Ahmad, M.E., Ahmad, A., Tabassum, R., Islam, F., Safhi, M.M. and Islam, F., 2013. Taurine ameliorates neurobehavioral, neurochemical and immunohistochemical changes in sporadic dementia of Alzheimer's type (SDAT) caused by intracerebroventricular streptozotocin in rats. Neurological Sciences 34: 2181-2192.

Johnson, M.R., Barnes, S., Kwakye, J.B. and Diasio, R.B., 1991. Purification and characterization of bile acid-CoA: amino acid N-acyltransferase from human liver. Journal of Biological Chemistry 266: 10227-10233.

Jong, C.J., Azuma, J. and Schaffer, S.W., 2011. Role of mitochondrial permeability transition in taurine deficiency-induced apoptosis. Experimental & Clinical Cardiology 16: 125-128.

Jong, C.J., Azuma, J. and Schaffer, S., 2012. Mechanism underlying the antioxidant activity of taurine: prevention of mitochondrial oxidant production. Amino Acids 42: 2223-2232.

Jong, C.J., Ito, T. and Schaffer, S.W., 2015. The ubiquitin-proteasome system and autophagy are defective in the taurine-deficient heart. Amino Acids 47: 2609-2622.

Jong, C. J., Ito, T., Azuma, J. and Schaffer, S., 2015. Taurine depletion decreases GRP78 expression and downregulates perk-dependent activation of the unfolded protein response. Advances in Experimental Medicine and Biology 803: 571-579.

Jong, C.J., Ito, T., Mozaffari, M., Azuma, J. and Schaffer, S., 2010. Effect of β-alanine treatment on mitochondrial taurine level and 5-taurinomethyluridine content. Journal of Biomedical Science, 17: S25.

Kadam, S.U. and Prabhasankar, P., 2010. Marine foods as functional ingredients in bakery and pasta products. Food Research International 43: 1975-1980.

Kanayama, A., Inoue, J.I., Sugita-Konishi, Y., Shimizu, M. and Miyamoto, Y., 2002. Oxidation of IκBα at methionine 45 is one cause of taurine chloramine-induced inhibition of NF-κB activation. Journal of Biological Chemistry 277: 24049-24056.

Kang, I.S. and Kim, C., 2013. Taurine chloramine administered *in vivo* increases NRF2-regulated antioxidant enzyme expression in murine peritoneal macrophages. Advances in Experimental Medicine and Biology 775: 259-267.

Kataoka, H. and Ohnishi, N., 1986. Occurrence of taurine in plants. Agricultural and Biological Chemistry 50: 1887 1888.

Kevresan, S., Kuhajda, K., Kandrac, J., Fawcett, J.P. and Mikov, M., 2006. Biosynthesis of bile acids in mammalian liver. European Journal of Drug Metabolism and Pharmacokinetics 31: 145-156.

Kilb, W. and Fukuda, A., 2017. Taurine as an essential neuromodulator during perinatal cortical development. Frontiers in Cellular Neuroscience 11: 328.

Kilic, F., Bhardwaj, R., Caulfeild, J. and Trevithick, J. R., 1999. Modelling cortical cataractogenesis 22: is *in vitro* reduction of damage in model diabetic rat cataract by taurine due to its antioxidant activity? Experimental Eye Research 69. 291-300.

Kim, C. and Cha, Y. N., 2009. Production of reactive oxygen and nitrogen species in phagocytes is regulated by taurine chloramine. Advances in Experimental Medicine and Biology 643: 463-472.

Kim, C. and Cha, Y.N., 2014. Taurine chloramine produced from taurine under inflammation provides anti-inflammatory and cytoprotective effects. Amino Acids 46: 89-100.

Kim, C., Jang, J.S., Cho, M.R., Agarawal, S.R. and Cha, Y.N., 2010. Taurine chloramine induces heme oxygenase-1 expression via Nrf2 activation in murine macrophages. International Immunopharmacology 10: 440-446.

Kim, J.W. and Kim, C., 2005. Inhibition of LPS-induced NO production by taurine chloramine in macrophages is mediated though Ras-ERK-NF-κB. Biochemical Pharmacology 70: 1352-1360.

Kim, K.S., Choi, H.M., Kim, C., Jeong, J.S., Yoo, M.C. and Yang, H.I., 2010. Effect of taurine chloramine on the production of matrix metalloproteinases (MMPs) in adiponectin-or IL-1β-stimulated fibroblast-like synoviocytes. Journal of Biomedical Science 17: S27.

Kim, S.H., Zhong, X., Kim, W., Kim, K., Suh, Y.G., Kim, C., Joe, Y., Chung, H.T., Cha, Y.N. and Surh, Y.J., 2018. Taurine chloramine potentiates phagocytic activity of peritoneal macrophages through up-regulation of dectin-1 mediated by heme oxygenase-1-derived carbon monoxide. The FASEB Journal 32: 2246-2257.

Kim, W., Kim, H.U., Lee, H.N., Kim, S.H., Kim, C., Cha, Y.N., Joe, Y., Chung, H.T., Jang, J., Kim, K. and Suh, Y.G., 2015. Taurine chloramine stimulates efferocytosis through upregulation of Nrf2-mediated heme oxygenase-1 expression in murine macrophages: possible involvement of carbon monoxide. Antioxidants & Redox Signaling 23: 163-177.

Kim, W., Kim, S.H., Jang, J.H., Kim, C., Kim, K., Suh, Y.G., Joe, Y., Chung, H.T., Cha, Y.N. and Surh, Y.J., 2018. Role of heme oxygenase-1 in potentiation of phagocytic activity of macrophages by taurine chloramine: Implications for the resolution of zymosan A-induced murine peritonitis. Cellular Immunology 327: 36-46.

Kim, C. and Cha, Y.N., 2014. Taurine chloramine produced from taurine under inflammation provides anti-inflammatory and cytoprotective effects. Amino Acids 46: 89-100.

Kim, H.W., Yoon, S.H., Park, T., Kim, B.K., Park, K.K. and Lee, D.H., 2006. Gene expressions of taurine transporter and taurine biosynthetic enzyme during mouse and chicken embryonic development. Advances in Experimental Medicine and Biology 583: 69-77.

Kirino, Y., Yasukawa, T., Ohta, S., Akira, S., Ishihara, K., Watanabe, K. and Suzuki, T., 2004. Codon-specific translational defect caused by a wobble modification deficiency in mutant tRNA from a human mitochondrial disease. Proceedings of the National Academy of Sciences 101: 15070-15075.

Kontny, E., Szczepańska, K., Kowalczewski, J., Kurowska, M., Janicka, I., Marcinkiewicz, J. and Maśliński, W., 2000. The mechanism of taurine chloramine inhibition of cytokine (interleukin-6, interleukin-8) production by rheumatoid arthritis fibroblast-like synoviocytes. Arthritis & Rheumatism 43: 2169-2177.

Kulakowski, E.C. and Maturo, J., 1984. Hypoglycemic properties of taurine: not mediated by enhanced insulin release. Biochemical Pharmacology 33: 2835-2838.

Kumar, R., Singh, V.K., Chhillar, S. and Atreja, S.K., 2013. Effect of supplementation of taurine or trehalose in extender on immunolocalization of tyrosine phosphoproteins in buffalo and cattle (Karan Fries) cryopreserved spermatozoa. Reproduction in Domestic Animals 48: 407-415.

Kurz, A.K., Schliess, F. and Häussinger, D., 1998. Osmotic regulation of the heat shock response in primary rat hepatocytes. Hepatology 28: 774-781.

Kutluyer, F., Öğretmen, F. and Inanan, B.E., 2016. Cryopreservation of goldfish (*Carassius auratus*) spermatozoa: effects of extender supplemented with taurine on sperm motility and DNA damage. CryoLetters 37: 41-46.

Lai, H.T., Nieuwland, M.G., Kemp, B., Aarnink, A.J. and Parmentier, H.K., 2011. Effects of repeated intratracheally administered lipopolysaccharide on primary and secondary specific antibody responses and on body weight gain of broilers. Poultry Science 90: 337-351.

Lambert, I.H., Kristensen, D.M., Holm, J.B. and Mortensen, O.H., 2015. Physiological role of taurine – from organism to organelle. Acta Physiologica 213: 191-212.

Larosa, V. and Remacle, C., 2018. Insights into the respiratory chain and oxidative stress. Bioscience Reports 38: BSR20171492.

Lee, D.S. and Cheong, S.H., 2017a. Taurine have neuroprotective activity against oxidative damage-induced HT22 cell death through heme oxygenase-1 pathway. Advances in Experimental Medicine and Biology 975: 159-171.

Lee, D.S., Kwon, K.H. and Cheong, S.H., 2017b. Taurine chloramine suppresses LPS-induced neuroinflammatory responses through Nrf2-mediated heme oxygenase-1 expression in mouse BV2 microglial cells. Advances in Experimental Medicine and Biology 975: 131-143.

Lee, W., Kang, N., Park, S.Y., Cheong, S.H., Chang, K.J., Kim, S.H., Um, J.H., Han, E.J., Kim, E.A., Jeon, Y.J. and Ahn, G., 2017c. Xylose-taurine reduced suppresses the inflammatory responses in lipopolysaccharide-stimulated Raw264. 7 macrophages. Advances in Experimental Medicine and Biology 975: 633-642.

Lee, D.N., Cheng, Y.H., Chuang, Y.S., Shive, J.L., Lian, Y.M., Wei, H.W. and Weng, C.F., 2004. Effects of dietary taurine supplementation on growth performance, serum constituents and antibody production of broilers. Asian-Australasian Journal of Animal Sciences 17: 109-115.

Lee, S.M., Lim, H.S., Namgung, N., Lee, B.H. and Paik, I.K., 2010. Taurine enrichment of eggs with feather meal and pyridoxine. Asian-Australasian Journal of Animal Sciences 23: 622-629.

Leeson, S., 2007. Metabolic challenges: past, present, and future. Journal of Applied Poultry Research 16: 121-125.

Levi, G. and Moris, G., 1971. Free amino acids and related compounds in chick brain during development. Brain Research 26: 131-140.

Li, X., Pan, S., Yuan, Y., Sun, Y., 2008. Effect of taurine on production performance in latter period layers [J]. China Poultry 5.

Liebowitz, S.M., Lombardini, J.B. and Allen, C.I., 1988. Effects of aminocycloalkanesulfonic acid analogs of taurine on ATP-dependent calcium ion uptake and protein phosphorylation. Biochemical Pharmacology 37: 1303-1309.

Lin, M.T. and Beal, M.F., 2006. Mitochondrial dysfunction and oxidative stress in neurodegenerative diseases. Nature 443: 787-795.

Liu, Y., Barua, M., Serban, V. and Quinn, M.R., 2003. Production of inflammatory mediators by activated C6 cells is attenuated by taurine chloramine inhibition of NF-kappaB activation. Advances in Experimental Medicine and Biology 526: 365-372.

Liu, Y. and Quinn, M.R., 2002. Chemokine production by rat alveolar macrophages is inhibited by taurine chloramine. Immunology Letters 80: 27-32.

Liu, S., Tun, H.M., Leung, F.C., Bennett, D.C., Zhang, H. and Cheng, K.M., 2018. Interaction of genotype and diet on small intestine microbiota of Japanese quail fed a cholesterol enriched diet. Scientific Reports 8: 2381.

Lobo, M.V., Alonso, F.J. and del Río, R.M., 2000. Immunohistochemical localization of taurine in the male reproductive organs of the rat. The Journal of Histochemistry and Cytochemistry 48. 313-320.

Lobo, M.V., Alonso, F.J. and Martin del Rio, R., 2000a. Immunocytochemical localization of taurine in different muscle cell types of the dog and rat. The Histochemical Journal 32: 53-61.

Lombardini, J.B., 1985. Inhibition by taurine of the phosphorylation of rat retinal membranes. Progress in Clinical and Biological Research 179: 383-393.

Losano, J.D.A., Angrimani, D.S.R., Ferreira Leite, R., Simões da Silva, B.D.C., Barnabe, V.H. and Nichi. M., 2018. Spermatic mitochondria: role in oxidative homeostasis, sperm function and possible tools for their assessment. Zygote 26: 251-260.

Lou, J., Han, D., Yu, H., Yu, G., Jin, M. and Kim, S.J., 2018. Cytoprotective effect of taurine against hydrogen peroxide-induced oxidative stress in UMR-106 cells through the Wnt/β-catenin signaling pathway. Biomolecules & Therapeutics 26: 584-590.

Lowe, I.P. and Roberts, E., 1955. Incorporation of radioactive sulfate sulfur into taurine and other substances in the chick embryo. Journal of Biological Chemistry 212: 477-483.

Lu, Z., He, X., Ma, B., Zhang, L., Li, J., Jiang, Y., Zhou, G. and Gao, F., 2019. Dietary taurine supplementation improves breast meat quality in chronic heat-stressed broilers via activating the Nrf2 pathway and protecting mitochondria from oxidative attack. Journal of the Science of Food and Agriculture 99: 1066-1072.

Lv, Q., Yang, Q., Cui, Y., Yang, J., Wu, G., Liu, M., Ning, Z., Cao, S., Dong, G. and Hu, J., 2017. Effects of taurine on ACE, ACE2 and HSP70 expression of hypothalamic-pituitary-adrenal axis in stress-induced hypertensive rats. Advances in Experimental Medicine and Biology 975: 871-886.

Lv, Q., Sun, L., Cui, Y., Yang, J., Yang, Q., Yu, X., Liu, M., Ning, Z., Hu, J., 2017a. Effects of replacement of methionine in diets with taurine on growth performance and blood index in broilers. Advances in Experimental Medicine and Biology 975: 989-1000.

Ma, Z., Zhang, J., Ma, H., Dai, B., Zheng, L., Miao, J. and Zhang, Y., 2014. The influence of dietary taurine and reduced housing density on hepatic functions in laying hens. Poultry Science 93: 1724-1736.

Ma, Z.L., Gao, Y., Ma, H.T., Zheng, L.H., Dai, B., Miao, J.F. and Zhang, Y.S., 2016. Effects of taurine and housing density on renal function in laying hens. Journal of Zhejiang University. Science. B 17: 952-964.

Machlin, L.J. and Pearson, P.B., 1957. Metabolism of taurine in the growing chicken. Archives of Biochemistry and. Biophysics 70: 35-42.

Machlin, L.J., Pearson, P.B. and Denton, C., 1955. The utilization of sulfate sulfur for the synthesis of taurine in the developing chick embryo. Journal of Biological Chemistry 212: 469-475.

Maher, S.G., Condron, C.E.M., Bouchier-Hayes, D.J. and Toomey, D.M., 2005. Taurine attenuates CD3/ interleukin-2-induced T cell apoptosis in an *in vitro* model of activation-induced cell death (AICD). Clinical & Experimental Immunology 139: 279-286.

Marcinkiewicz, J., Walczewska, M., Olszanecki, R., Bobek, M., Biedroń, R., Dulak, J., Józkowicz, A., Kontny, E. and Maślinski, W., 2009. Taurine haloamines and heme oxygenase-1 cooperate in the regulation of inflammation and attenuation of oxidative stress. Advances in Experimental Medicine and Biology 643: 439-450.

Marcinkiewicz, J. and Kontny, E., 2014. Taurine and inflammatory diseases. Amino Acids 46: 7-20.

Marcinkiewicz, J., 2010. Taurine bromamine (TauBr)-its role in immunity and new perspectives for clinical use. Journal of Biomedical Science 17: S3.

Marcinkiewicz, J. and Kontny, E., 2014. Taurine and inflammatory diseases. Amino Acids 46: 7-20.

Martin, W.G. and Patrick, H., 1961. The effect of taurine on the sulfate-S' retention by chicks. Poultry Science 40: 267-268

Martin, W.G., 1972. Sulfate metabolism and taurine synthesis in the chick. Poultry Science 51: 608-612.

Martin, W.G., Miraglia, R.J., Spaeth, D.G. and Patrick, H., 1966. Synthesis of taurine from sulfate by the chick. Proceedings of the Society for Experimental Biology and Medicine 122: 841-844.

Martinez-Losa, M., Cortijo, J., Piqueras, L., Sanz, M.J. and Morcillo, E.J., 2009. Taurine chloramine inhibits functional responses of human eosinophils *in vitro*. Clinical & Experimental Allergy 39: 537-546.

Martínez-Páramo, S., Diogo, P., Dinis, M.T., Soares, F., Sarasquete, C. and Cabrita., E., 2013. Effect of two sulfur-containing amino acids, taurine and hypotaurine in European sea bass (*Dicentrarchus labrax*) sperm cryopreservation. Cryobiology 66: 333-338.

Martins-Bessa, A., Rocha, A. and Mayenco-Aguirre, A., 2009. Effects of taurine and hypotaurine supplementation and ionophore concentrations on post-thaw acrosome reaction of dog spermatozoa. Theriogenology 71: 248-253.

Mathew, E., Barletta, M.A. and Lau-Cam, C.A., 2013. The effects of taurine and thiotaurine on oxidative stress in the aorta and heart of diabetic rats. Advances in Experimental Medicine and Biology 775: 345-369.

Mehta, T.R. and Dawson, R. Jr., 2001. Taurine is a weak scavenger of peroxynitrite and does not attenuate sodium nitroprusside toxicity to cells in culture. Amino Acids 20: 419-433.

Messina, S.A. and Dawson, R. Jr., 2000. Attenuation of oxidative damage to DNA by taurine and taurine analogs. Advances in Experimental Medicine and Biology 483: 355-367.

Mezes, M., Surai, P., Salyi, G., Speake, B.K., Gaal, T. and Maldjian, A., 1997. Nutritional metabolic diseases of poultry and disorders of the biological antioxidant defence system. Acta Veterinaria Hungarica 45: 349-360.

Miao, J., Zheng, L., Zhang, J., Ma, Z., Zhu, W. and Zou, S., 2011. The effect of taurine on the toll-like receptors/nuclear factor kappa B (TLRs/NF-κB) signaling pathway in *Streptococcus uberis*-induced mastitis in rats. International Immunopharmacology 11: 1740-1746.

Michael, A.J., Alexopoulos, C., Pontiki, E.A., Hadjipavlou-Litina, D.J., Saratsis, P., Ververidis, H.N. and Boscos, C.M., 2009. Effect of antioxidant supplementation in semen extenders on semen quality and reactive oxygen species of chilled canine spermatozoa. Animal Reproduction Science 112: 119-135.

Miraglia, R.J. and Martin, W.G., 1969. The synthesis of taurine from sulfate II. Chick liver phosphoadenosinephosphosulfate-transferase. Proceedings of the Society for Experimental Biology and Medicine 132: 640-644.

Miraglia, R.J., Martin, W.G., Spaeth, D.G. and Patrick, H., 1966. On the synthesis of taurine from sulfate by the chick. I. Influential dietary factors. Proceedings of the Society for Experimental Biology and Medicine 123: 725-730.

Miyazaki, T. and Matsuzaki, Y., 2014. Taurine and liver diseases: a focus on the heterogeneous protective properties of taurine. Amino Acids 46: 101-110.

Monson, W.J., 1969. Evidence that taurine may be one of the elusive unidentified factors. Poultry Science 48: 2069-2074.

Murakami, S., Sakurai, T., Tomoike, H., Sakono, M., Nasu, T. and Fukuda, N., 2010. Prevention of hypercholesterolemia and atherosclerosis in the hyperlipidemia- and atherosclerosis-prone Japanese (LAP) quail by taurine supplementation. Amino Acids 38: 271-278.

Muż, B., Kontny, E., Marcinkiewicz, J. and Maśliński, W., 2008. Heme oxygenase-1 participates in the anti-inflammatory activity of taurine chloramine. Amino Acids 35: 397-402.

Nagai, K., Fukuno, S., Oda, A. and Konishi, H., 2016. Protective effects of taurine on doxorubicin-induced acute hepatotoxicity through suppression of oxidative stress and apoptotic responses. Anti-cancer Drugs 27: 17-23.

Nain, S., Wojnarowicz, C., Laarveld, B. and Olkowski, A.A., 2008. Effects of dietary vitamin E and C supplementation on heart failure in fast growing commercial broiler chickens. British Poultry Science 49: 697-704.

Najafi, P., Zulkifli, I., Soleimani, A.F. and Kashiani, P., 2015. The effect of different degrees of feed restriction on heat shock protein 70, acute phase proteins, and other blood parameters in female broiler breeders. Poultry Science 94: 2322-2329.

Nakashima, T., Taniko, T. and Kuriyama, K., 1982. Therapeutic effect of taurine administration on carbon tetrachloride-induced hepatic injury. The Japanese Journal of Pharmacology 32: 583-589.

Nam, S.Y., Kim, H.M. and Jeong, H.J., 2017. The potential protective role of taurine against experimental allergic inflammation. Life Sciences 184: 18-24.

Nawab, A., Ibtisham, F., Li, G., Kieser, B., Wu, J., Liu, W., Zhao, Y., Nawab, Y., Li, K., Xiao, M. and An, L., 2018. Heat stress in poultry production: mitigation strategies to overcome the future challenges facing the global poultry industry. Journal of Thermal Biology 78: 131-139.

Nielsen, C.U., Bjerg, M., Ulaganathan, N. and Holm, R., 2017. Oral and intravenous pharmacokinetics of taurine in sprague-dawley rats: the influence of dose and the possible involvement of the proton-coupled amino acid transporter, PAT1, in oral taurine absorption. Physiological Reports 5: e13467.

Niu, X., Zheng, S., Liu, H. and Li, S., 2018. Protective effects of taurine against inflammation, apoptosis, and oxidative stress in brain injury. Molecular Medicine Reports 18: 4516-4522.

Oenarto, J., Karababa, A., Castoldi, M., Bidmon, H.J., Görg, B. and Häussinger, D., 2016. Ammonia-induced miRNA expression changes in cultured rat astrocytes. Scientific Reports 6: 18493,

Ogino, T., Than, T.A., Hosako, M., Ozaki, M., Omori, M. and Okada, S., 2009. Taurine chloramine: a possible oxidant reservoir. Advances in Experimental Medicine and Biology 643: 451-461.

Ohta, H., Azuma, J., Awata, N., Hamaguchi, T., Tanaka, Y., Sawamura, A., Kishimoto, S. and Sperelakis, N., 1988. Mechanism of the protective action of taurine against isoprenaline induced myocardial damage. Cardiovascular Research 22: 407-413.

Ohta, H., Azuma, J., Onishi, S., Awata, N., Takihara, K. and Kishimoto, S., 1986. Protective effect of taurine against isoprenaline-induced myocardial damage. Basic Research in Cardiology 81: 473-481.

Oliveira, M.W., Minotto, J.B., de Oliveira, M.R., Zanotto-Filho, A., Behr, G.A., Rocha, R.F., Moreira, J.C. and Klamt, F., 2010. Scavenging and antioxidant potential of physiological taurine concentrations against different reactive oxygen/nitrogen species. Pharmacological Reports 62: 185-193.

Olkowski, A.A., Wojnarowicz, C., Nain, S., Ling, B., Alcorn, J.M. and Laarveld, B., 2008. A study on pathogenesis of sudden death syndrome in broiler chickens. Research in Veterinary Science 85: 131-140.

Olszanecki, R., Kurnyta, M., Biedroń, R., Chorobik, P., Bereta, M. and Marcinkiewicz, J., 2008. The role of heme oxygenase-1 in down regulation of PGE 2 production by taurine chloramine and taurine bromamine in J774. 2 macrophages. Amino Acids 35: 359-364.

Olszanecki, R. and Marcinkiewicz, J., 2004. Taurine chloramine and taurine bromamine induce heme oxygenase-1 in resting and LPS-stimulated J774. 2 macrophages. Amino Acids 27: 29-35.

Owoeye, O., Adedara, I.A. and Farombi, E.O., 2018. Pretreatment with taurine prevented brain injury and exploratory behaviour associated with administration of anticancer drug cisplatin in rats. Biomedicine & Pharmacotherapy 102: 375-384.

Ozasa, H. and Gould, K.G., 1982. Protective effect of taurine from osmotic stress on chimpanzee spermatozoa. Archives of Andrology 9: 121-126.

Page, L.K., Jeffries, O. and Waldron, M., 2019. Acute taurine supplementation enhances thermoregulation and endurance cycling performance in the heat. European Journal of Sport Science 19: 1101-1109.

Park, S.Y., Ahn, C.B., Chang, K.J., Kim, S.H., Lee, W., Um, J.H., Han, E.J., Jeon, Y.J., Cheong, S.H. and Ahn, G., 2017. Hepatoprotective effects of xylose-taurine reduced against hydrogen peroxide-induced oxidative stress in cultured hepatocytes. Advances in Experimental Medicine and Biology 975: 621-631.

Partyka, A., Rodak, O., Bajzert, J., Kochan, J. and Niżański, W., 2017. The effect of L-carnitine, hypotaurine, and taurine supplementation on the quality of cryopreserved chicken semen. BioMed Research International 2017: 7279341.

Parvez, S., Tabassum, H., Banerjee, B.D. and Raisuddin, S., 2008. Taurine prevents tamoxifen-induced mitochondrial oxidative damage in mice. Basic & Clinical Pharmacology & Toxicology 102: 382-387.

Pasantes-Morales, H. and Cruz, C., 1984. Protective effect of taurine and zinc on peroxidation-induced damage in photoreceptor outer segments. Journal of Neuroscience Research 11: 303-311.

Pasantes-Morales, H. and Cruz, C., 1985a. Taurine: a physiological stabilizer of photoreceptor membranes. Progress in Clinical and Biological Research 179: 371-381.

Pasantes-Morales, H. and Cruz, C., 1985b. Taurine and hypotaurine inhibit light-induced lipid peroxidation and protect rod outer segment structure. Brain Research 330: 154-157.

Pasantes-Morales, H., Wright, C.E. and Gaull, G.E., 1984. Protective effect of taurine, zinc and tocopherol on retinol-induced damage in human lymphoblastoid cells. The Journal of Nutrition 114: 2256-2261.

Patel, S.N., Pandya, K., Clark, G.J., Parikh, M.C. and Lau-Cam, C.A., 2016. Comparison of taurine and pantoyltaurine as antioxidants *in vitro* and in the central nervous system of diabetic rats. Experimental and Toxicologic Pathology 68: 103-112.

Perumal, P., Vupru, K. and Rajkhowa, C., 2013. Effect of addition of taurine on the liquid storage (5 °C) of mithun (*Bos frontalis*) semen. Veterinary Medicine International 2013: 165348.

Piao, F., Zhang, Y., Yang, L., Zhang, C., Shao, J., Liu, X., Li, Y. and Li, S., 2017. Taurine attenuates as 2 O 3-induced autophagy in cerebrum of mouse through Nrf2 pathway. Advances in Experimental Medicine and Biology 975: 863-870.

Piao, S., Cha, Y.N. and Kim, C., 2011. Taurine chloramine protects RAW 264.7 macrophages against hydrogen peroxide-induced apoptosis by increasing antioxidants. Journal of Clinical Biochemistry and Nutrition 49: 50-56.

Porter, D.W. and Martin, W.G., 1992. Taurine uptake into chick B cells. Proceedings of the Society for Experimental Biology and Medicine 199: 243-248.

Porter, D.W. and Martin, W.G., 1993. Taurine regulation of Ca2+ uptake and (Ca (2+)+ Mg2+)-ATPase in developing chick B-cells. Comparative Biochemistry and Physiology. Comparative Physiology 106: 309-312.

Porter, D.W., Walker, S.A., Martin, W.G., Lee, P. and Kaczmarczyk, W., 1991. Taurine uptake in chicken leukocytes and erythrocytes. Comparative Biochemistry and Physiology. A, Comparative Physiology 98: 305-309.

Puron, D., Santamaria, R., Segura, J.C. and Alamilla, J.L., 1995. Broiler performance at different stocking densities. Journal of Applied Poultry Research 4: 55-60.

Quinn, M.R., Barua, M., Liu, Y. and Serban, V., 2003. Taurine chloramine inhibits production of inflammatory mediators and iNOS gene expression in alveolar macrophages; a tale of two pathways: part I, NF-kappaB signaling. Advances in Experimental Medicine and Biology 526: 341-348.

Qvartskhava, N., Jin, C.J., Buschmann, T., Albrecht, U., Bode, J.G., Monhasery, N., Oenarto, J., Bidmon, H.J., Görg, B. and Häussinger, D., 2019. Taurine transporter (TauT) deficiency impairs ammonia detoxification in mouse liver. Proceedings of the National Academy of Sciences 116: 6313-6318.

Rajani, J., Torshizi, M.K. and Rahimi, S., 2011. Control of ascites mortality and improved performance and meat shelf-life in broilers using feed adjuncts with presumed antioxidant activity. Animal Feed Science and Technology 170: 239-245.

Ratriyanto, A. and Mosenthin, R., 2018. Osmoregulatory function of betaine in alleviating heat stress in poultry. Journal of Animal Physiology and Animal Nutrition 102: 1634-1650.

Reeta, K.H., Singh, D. and Gupta, Y.K., 2017. Chronic treatment with taurine after intracerebroventricular streptozotocin injection improves cognitive dysfunction in rats by modulating oxidative stress, cholinergic functions and neuroinflammation. Neurochemistry International 108: 146-156.

Rehman, Z.U., Chand, N. and Khan, R.U., 2017. The effect of vitamin E, L-carnitine, and ginger on production traits, immune response, and antioxidant status in two broiler strains exposed to chronic heat stress. Environmental Science and Pollution Research 24: 26851-26857.

Reid, G.M. and Tervit, H., 1995. Sudden infant death syndrome and superoxide/nitric oxide. Medical Hypotheses 45: 395-397.

Ribeiro, R.A., Bonfleur, M.L., Batista, T.M., Borck, P.C. and Carneiro, E.M., 2018. Regulation of glucose and lipid metabolism by the pancreatic and extra-pancreatic actions of taurine. Amino Acids 50: 1511-1524.

Rikimaru, M., Ohsawa, Y., Wolf, A.M., Nishimaki, K., Ichimiya, H., Kamimura, N., Nishimatsu, S.I., Ohta, S. and Sunada, Y., 2012. Taurine ameliorates impaired the mitochondrial function and prevents stroke-like episodes in patients with MELAS. Internal Medicine 51: 3351-3357.

Roy, A., Manna, P. and Sil, P.C., 2009. Prophylactic role of taurine on arsenic mediated oxidative renal dysfunction via MAPKs/NF-κ B and mitochondria dependent pathways. Free Radical Research 43: 995-1007.

Ruiz-Feria, C.A., Beers, K.W., Kidd, M.T. and Wideman, R.F. Jr., 1999. Plasma taurine levels in broilers with pulmonary hypertension syndrome induced by unilateral pulmonary artery occlusion. Poultry Science 78: 1627-1633.

Ruiz-Feria, C.A. and Wideman, R.F. Jr., 2001. Taurine, cardiopulmonary hemodynamics, and pulmonary hypertension syndrome in broilers. Poultry Science 80: 1607-1618.

Sanchez-Partida, L.G., Setchell, B.P. and Maxwell, W.M.C., 1997. Epididymal compounds and antioxidants in diluents for the frozen storage of ram spermatozoa. Reproduction, Fertility and Development 9: 689-696.

Sato, M., Tomonaga, S., Denbow, D.M. and Furuse, M., 2009. Changes in free amino acids in the brain during embryonic development in layer and broiler chickens. Amino Acids 36: 303-308.

Sayed, M.A.M. and Downing, J., 2011. The effects of water replacement by oral rehydration fluids with or without betaine supplementation on performance, acid-base balance, and water retention of heat-stressed broiler chickens. Poultry Science 90: 157-167.

Schaalan, M.F., Ramadan, B.K. and Abd Elwahab, A.H., 2018. Ameliorative effect of taurine-chloramine in azathioprine-induced testicular damage; a deeper insight into the mechanism of protection. BMC Complementary and Alternative Medicine 18: 255.

Schaffer, S.W., Shimada-Takaura, K., Jong, C.J., Ito, T. and Takahashi, K., 2016. Impaired energy metabolism of the taurine-deficient heart. Amino Acids 48: 549-558.

Schaffer, S. and Kim, H.W., 2018. Effects and mechanisms of taurine as a therapeutic agent. Biomolecules & Therapeutics 26: 225-241.

Schaffer, S.W., Azuma, J. and Madura, J.D., 1995. Mechanisms underlying taurine-mediated alterations in membrane function. Amino Acids 8: 231-246.

Schaffer, S.W., Azuma, J. and Mozaffari, M., 2009. Role of antioxidant activity of taurine in diabetes. Canadian Journal of Physiology and Pharmacology 87: 91-99.

Schaffer, S.W., Azuma, J. and Madura, J.D., 1995. Mechanisms underlying taurine-mediated alterations in membrane function. Amino Acids 8: 231-246.

Schaffer, S.W., Ito, T. and Azuma, J., 2014. Clinical significance of taurine. Amino Acids 46: 1-5.

Schaffer, S.W., Jong, C.J., Ramila, K.C. and Azuma, J., 2010. Physiological roles of taurine in heart and muscle. Journal of Biomedical Science 17: S2.

Schaffer, S. W., Ramila, K. C., Jong, C. J., Shetewy, A., Shimada, K., Ito, T., Azuma, J. and Cioffi, E., 2015. Does taurine prolong lifespan by improving heart function? Advances in Experimental Medicine and Biology 803: 555-570.

Schuller-Levis, G.B. and Park, E., 2003. Taurine: new implications for an old amino acid. FEMS microbiology letters, 226:195-202.

Schuller-Levis, G.B. and Park, E., 2004. Taurine and its chloramine: modulators of immunity. Neurochemical Research 29: 117-126.

Seidel, U., Huebbe, P. and Rimbach, G., 2019. Taurine: a regulator of cellular redox homeostasis and skeletal muscle function. Molecular Nutrition & Food Research 63: 1800569.

Sekine, S. and Ichijo, H., 2015. Mitochondrial proteolysis: its emerging roles in stress responses. Biochimica et Biophysica Acta (BBA)-General Subjects 1850: 274-280.

Setyarani, M., Zinellu, A., Carru, C. and Zulli, A., 2014. High dietary taurine inhibits myocardial apoptosis during an atherogenic diet: association with increased myocardial HSP70 and HSF-1 but not caspase 3. European Journal of Nutrition 53: 929-937.

Shao, X., Hu, Z., Hu, C., Bu, Q., Yan, G., Deng, P., Lv, L., Wu, D., Deng, Y., Zhao, J. and Zhu, R., 2012. Taurine protects methamphetamine-induced developmental angiogenesis defect through antioxidant mechanism. Toxicology and Applied Pharmacology 260: 260-270.

Shi, X., Flynn, D.C., Porter, D.W., Leonard, S.S., Vallyathan, V. and Castranova, V., 1997. Efficacy of taurine based compounds as hydroxyl radical scavengers in silica induced peroxidation. Annals of Clinical & Laboratory Science 27: 365-374.

Shim, K.S., Hwang, K.T., Son, M.W. and Park, G.H., 2006. Lipid metabolism and peroxidation in broiler chicks under chronic heat stress. Asian-Australasian Journal of Animal Sciences 19: 1206-1211.

Shim, K.S., Park, G.H., Na, C.S., Ji, J.R. and Choe, H.S., 2010. Effects of taurine supplementation on performance, egg quality, blood parameter and liver lipid and lipid peroxidation levels of laying hens fed high fat diet. Korean Journal of Poultry Science 37: 337-345.

Reddy, N.S.S., Mohanarao, G.J. and Atreja, S.K., 2010. Effects of adding taurine and trehalose to a tris-based egg yolk extender on buffalo (*Bubalus bubalis*) sperm quality following cryopreservation. Animal Reproduction Science 119: 183-190.

Simitzis, P.E., Kalogeraki, E., Goliomytis, M., Charismiadou, M.A., Triantaphyllopoulos, K., Ayoutanti, A., Niforou, K., Hager-Theodorides, A.L. and Deligeorgis, S.G., 2012. Impact of stocking density on broiler growth performance, meat characteristics, behavioural components and indicators of physiological and oxidative stress. British Poultry Science 53: 721-730.

Simsek, U.G., Dalkilic, B., Ciftci, M. and Yuce, A., 2009. The influences of different stocking densities on some welfare indicators, lipid peroxidation (MDA) and antioxidant enzyme activities (GSH, GSH-Px, CAT) in broiler chickens. Journal of Animal and Veterinary Advances 8: 1568-1572.

Sinha, M., Manna, P. and Sil, P.C., 2007. Taurine, a conditionally essential amino acid, ameliorates arsenic-induced cytotoxicity in murine hepatocytes. Toxicology *in vitro* 21: 1419-1428.

Soleimani, A.F., Zulkifli, I., Omar, A.R. and Raha, A.R., 2011. Physiological responses of 3 chicken breeds to acute heat stress. Poultry Science 90: 1435-1440.

Spitze, A.R., Wong, D.L., Rogers, Q.R. and Fascetti, A.J., 2003. Taurine concentrations in animal feed ingredients; cooking influences taurine content. Journal of Animal Physiology and Animal Nutrition 87: 251-262.

Jang, J.S., Piao, S., Cha, Y.N. and Kim, C., 2009. Taurine chloramine activates Nrf2, increases HO-1 expression and protects cells from death caused by hydrogen peroxide. Journal of Clinical Biochemistry and Nutrition 45: 37-43.

Sun, M., Zhao, Y., Gu, Y. and Xu, C., 2012. Anti-inflammatory mechanism of taurine against ischemic stroke is related to down-regulation of PARP and NF-κB. Amino Acids 42: 1735-1747.

Sun, Q., Jia, N., Yang, J. and Chen, G., 2018. Nrf2 signaling pathway mediates the antioxidative effects of taurine against corticosterone-induced cell death in HUMAN SK-N-SH cells. Neurochemical Research 43: 276-286.

Sun, Q., Hu, H., Wang, W., Jin, H., Feng, G. and Jia, N., 2014. Taurine attenuates amyloid β 1-42-induced mitochondrial dysfunction by activating of SIRT1 in SK-N-SH cells. Biochemical and Biophysical Research Communications 447: 485-489.

Surai, P.F., 2002. Natural antioxidants in avian nutrition and reproduction. Nottingham University Press, Nottingham, UK.

Surai, P.F., 2015. Silymarin as a natural antioxidant: an overview of the current evidence and perspectives. Antioxidants 4: 204-247.

Surai, P.F., 2016. Antioxidant systems in poultry biology: superoxide dismutase. Journal of Animal Research and Nutrition 1: 8.

Surai, P.F., 2018. Selenium in poultry nutrition and health. Wageningen Academic Publishers, Wageningen, the Netherlands.

Surai, P.F., 2018a. Taurine and carnitine in poultry production: from vitagene activation to chicken health maintenance. Ptitsevodstvo.ua (Ukrainian Poultry Science) 1-2: 12-17.

Surai, P.F. and Fisinin, V.I., 2016a. Vitagenes in poultry production. Part 1. Technological and environmental stresses. World's Poultry Science Journal 72: 721-733.

Surai, P.F. and Fisinin, V.I., 2016b. Vitagenes in poultry production. Part 2. Nutritional and internal stresses. World's Poultry Science Journal 72: 761-772.

Surai, P.F. and Fisinin, V.I., 2016c. Vitagenes in poultry production. Part 3. Vitagene concept development. World's Poultry Science Journal 72: 793-804.

Surai, P.F. and Fisinin, V.I., 2016d. Natural antioxidants and stresses in poultry production: from vitamins to vitagenes. Proc. XXV World Poultry Congress, Invited Lecture Papers, Beijing, China, 5-9 September, 2016, pp. 116-121.

Surai, P.F. and Kochish, I.I., 2019. Nutritional modulation of the antioxidant capacities in poultry: the case of selenium. Poultry Science 98: 4231-4239.

Surai, P.F., Kochish, I.I. and Fisinin, V.I., 2017. Antioxidant systems in poultry biology: nutritional modulation of vitagenes. European Journal of Poultry Science 81: 1612-9199.

Surai, P.F., Kochish, I.I. and Kidd, M.T., 2020. Taurine in poultry nutrition. Animal Feed Science and Technology 260: 114339.

Surai, P.F., Kochish, I.I., Fisinin, V.I., Grozina, A.A. and Shatskikh, E.V., 2018. Molecular mechanisms of chicken gut health maintenance: role of microbiota. Agricultural Technologies, Moscow, Russia.

Surai, P.F., Kochish, I.I., Romanov, M.N. and Griffin, D.K., 2019. Nutritional modulation of the antioxidant capacities in poultry: the case of vitamin E. Poultry Science 98: 4030-4041.

Surai, P., Noble, R. and Speake, B., 1996. Tissue-specific differences in antioxidant distribution and susceptibility to lipid peroxidation during development of the chick embryo. Biochimica et Biophysica Acta 1304: 1-10.

Surai, P.F., Speake B.K. and Sparks, N.H.C., 2003. Comparative aspects of lipid peroxidation and antioxidant protection in avian semen. Chapter 15. In: De Vriese, S. and Christophe, A. (eds) Male fertility and lipid metabolism. AOCS Press, Urbano, IL, USA, pp.211-249.

Suzuki, T., Suzuki, T., Wada, T., Saigo, K. and Watanabe, K., 2002. Taurine as a constituent of mitochondrial tRNAs: new insights into the functions of taurine and human mitochondrial diseases. The EMBO Journal 21: 6581-6589.

Sved, D.W., Godsey, J.L., Ledyard, S.L., Mahoney, A.P., Stetson, P.L., Ho, S., Myers, N.R., Resnis, P. and Renwick, A.G., 2007. Absorption, tissue distribution, metabolism and elimination of taurine given orally to rats. Amino Acids 32: 459-466.

Takahashi, Y. and Hatta, H., 2017. Effects of taurine administration on exercise-induced fatigue and recovery. The Journal of Physical Fitness and Sports Medicine 6: 33-39.

Tan, X., Sun, Z., Zhu, X. and Ye, C., 2018. Dietary supplementation with taurine improves ability to resist ammonia stress in hybrid snakehead (*Channa maculatus*♀ × *Channa argus*♂). Aquaculture Research 49: 3400-3410.

Tang, Z., Liu, J., Sun, Z., Li, J., Sun, W., Mao, J. and Wang, Y., 2018. Protective effects of taurine on growth performance and intestinal epithelial barrier function in weaned piglets challenged without or with lipopolysaccharide. Animal Production Science 58: 2011-2020.

Tang, Z, Liu, J, Sun, Z, Li, J, Sun, W, Mao, J. and Wang, Y., 2018. Protective effects of taurine on growth performance and intestinal epithelial barrier function in weaned piglets challenged without or with lipopolysaccharide. Animal Production Science 58: 2011-2020.

Tiedemann, F. and Gmelin, L., 1827. Einige neue Bestandtheile der Galle des Ochsen. Annalen der Physik 85: 326-337.

Timbrell, J.A., Seabra, V. and Waterfield, C.J., 1995. The *in vivo* and *in vitro* protective properties of taurine. General Pharmacology: The Vascular System 26: 453-462.

Tokunaga, S., Kanayama, A. and Miyamoto, Y., 2007. Modification of IκBα by taurine bromamine inhibits tumor necrosis factor α-induced NF-κB activation. Inflammation Research 56: 479-486.

Tomichek, R.C., Sass, N.L. and Martin, W.G., 1972. Role of vitamins in taurine synthesis from sulfate by the chick. The Journal of Nutrition 102: 313-318.

Tomonaga, S., Okuyama, H., Tachibana, T. and Makino, R., 2018. Effects of high ambient temperature on plasma metabolomic profiles in chicks. Animal Science Journal 89: 448-455.

Tsiouris, V., Georgopoulou, I., Batzios, C., Pappaioannou, N., Ducatelle, R. and Fortomaris, P., 2015. High stocking density as a predisposing factor for necrotic enteritis in broiler chicks. Avian Pathology 44: 59-66.

Tufft, LS. and Jensen, LS., 1992. Influence of dietary taurine on performance and fat retention in broilers and turkey poults fed varying levels of fat. Poultry Science 71: 880-885.

Van Gelder, N.M. and Belanger, F., 1988. Embryonic exposure to high taurine: a possible nutritional contribution to Friedreich's ataxia. Journal of Neuroscience Research 20: 383-389.

Vanitha, M.K., Baskaran, K., Periyasamy, K., Selvaraj, S., Ilakkia, A., Saravanan, D., Venkateswari, R., Revathi Mani, B., Anandakumar, P. and Sakthisekaran, D., 2016. Modulatory effect of taurine on 7, 12-dimethylbenz (a) anthracene-induced alterations in detoxification enzyme system, membrane bound enzymes, glycoprotein profile and proliferative cell nuclear antigen in rat breast tissue. Journal of Biochemical and Molecular Toxicology 30: 414-423.

Vincent, A., Scheers, N., Gabrielsson, B. and Sandberg, A.S., 2014. Taurine may affect redox homeostasis via regulation of thioredoxin interacting protein in endothelial cells (830.29). The FASEB Journal 28: 830-29.

Walczewska, M., Ciszek-Lenda, M., Surmiak, M., Kozlowska, A., Jozefowski, S. and Marcinkiewicz, J., 2015. Impact of taurine on innate and adaptive immunity as the result of HOCl neutralization. Advances in Experimental Medicine and Biology 803:109-120.

Wang, G.G., Li, W., Lu, X.H., Zhao, X. and Xu, L., 2013. Taurine attenuates oxidative stress and alleviates cardiac failure in type I diabetic rats. Croatian Medical Journal 54: 171-179.

Wang, Y., Mei, X., Yuan, J., Lu, W., Li, B. and Xu, D., 2015. Taurine zinc solid dispersions attenuate doxorubicin-induced hepatotoxicity and cardiotoxicity in rats. Toxicology and Applied Pharmacology 289: 1-11.

Wang, F., Dong, X., Zhang, X., Tong, J.M., Zhang, Q., 2010a. Effects of taurine on egg production, immune responses and fat metabolism in laying quails. Journal of Food Science and Biotechnology 3: 381-384.

Wang, F.R., Dong, X.F., Tong, J.M., Zhang, X.M., Zhang, Q. and Wu, Y.Y., 2009. Effects of dietary taurine supplementation on growth performance and immune status in growing Japanese quail (*Coturnix coturnix japonica*). Poultry Science 88: 1394-1398.

Wang, F.R., Dong, X.F., Zhang, X.M., Tong, J.M., Xie, Z.G. and Zhang, Q., 2010. Effects of dietary taurine on egg production, egg quality and cholesterol levels in Japanese quail. Journal of the Science of Food and Agriculture 90: 2660-2663.

Ward, R.J., Lallemand, F., De Witte, P., Crichton, R.R., Piette, J., Tipton, K., Hemmings, K., Pitard, A., Page, M., Della Corte, L. and Taylor, D., 2011. Anti-inflammatory actions of a taurine analogue, ethane β-sultam, in phagocytic cells, *in vivo* and *in vitro*. Biochemical Pharmacology 81: 743-751.

Warskulat, U., Borsch, E., Reinehr, R., Heller-Stilb, B., Mönnighoff, I., Buchczyk, D., Donner, M., Flögel, U., Kappert, G., Soboll, S., Beer, S., Pfeffer, K., Marschall, H.U., Gabrielsen, M., Amiry-Moghaddam, M., Ottersen, O.P., Dienes, H.P. and Häussinger, D., 2006. Chronic liver disease is triggered by taurine transporter knockout in the mouse. FASEB Journal 20: 574-576.

Warskulat, U., Heller-Stilb, B., Oermann, E., Zilles, K., Haas, H., Lang, F. and Häussinger, D., 2007. Phenotype of the taurine transporter knockout mouse. Methods in Enzymology 428: 439-458.

Wei, S., Huang, Q., Li, J., Liu, Z., You, H., Chen, Y. and Gong, J., 2012. Taurine attenuates liver injury by downregulating phosphorylated p38 MAPK of Kupffer cells in rats with severe acute pancreatitis. Inflammation 35: 690-701.

Wen, C., Li, F., Zhang, L., Duan, Y., Guo, Q., Wang, W., He, S., Li, J. and Yin, Y., 2019. Taurine is involved in energy metabolism in muscles, adipose tissue, and the liver. Molecular Nutrition & Food Research 63: 1800536.

Wong, H.S., Dighe, P.A., Mezera, V., Monternier, P.A. and Brand, M.D., 2017. Production of superoxide and hydrogen peroxide from specific mitochondrial sites under different bioenergetic conditions. Journal of Biological Chemistry 292: 16804-16809.

Wu, G., Yang, Q., Yu, Y., Lin, S., Feng, Y., Lv, Q., Yang, J. and Hu, J., 2017. Taurine inhibits kupffer cells activation induced by lipopolysaccharide in alcoholic liver damaged rats. Advances in Experimental Medicine and Biology 975: 789-800.

Wu, H., Jin, Y., Wei, J., Jin, H., Sha, D. and Wu, J.Y., 2005. Mode of action of taurine as a neuroprotector. Brain Research 1038: 123-131.

Xiang, R.P., Sun, W.D., Wang, J.Y. and Wang, X.L., 2002. Effect of vitamin C on pulmonary hypertension and muscularisation of pulmonary arterioles in broilers. British Poultry Science 43: 705-712.

Xiao, M., Mi, Y., Liu, L., Lv, C., Zeng, W., Zhang, C. and Li, J., 2018. Taurine regulates mucosal barrier function to alleviate lipopolysaccharide-induced duodenal inflammation in chicken. Amino Acids 50: 1637-1646.

Xu, M., Che, L., Gao, K., Wang, L., Yang, X., Wen, X., Jiang, Z., Wu, D., 2019. Effects of dietary taurine supplementation to gilts during late gestation and lactation on offspring growth and oxidative stress. Animals 9: 220.

Xu, S., He, M., Zhong, M., Li, L., Lu, Y., Zhang, Y., Zhang, L., Yu, Z. and Zhou, Z., 2015. The neuroprotective effects of taurine against nickel by reducing oxidative stress and maintaining mitochondrial function in cortical neurons. Neuroscience Letters 590: 52-57.

Xu, Y., Lai, X., Li, Z., Zhang, X. and Luo, Q., 2018. Effect of chronic heat stress on some physiological and immunological parameters in different breed of broilers. Poultry Science 97: 4073-4082.

Yamazaki, M. and Takemasa, M., 1998. Effects of dietary taurine on egg weight. Poultry Science 77: 1024-1026.

Yang, W., Huang, J., Xiao, B., Liu, Y., Zhu, Y., Wang, F. and Sun, S., 2017. Taurine protects mouse spermatocytes from ionizing radiation-induced damage through activation of Nrf2/HO-1 signaling. Cellular Physiology and Biochemistry 44: 1629-1639.

Yao, H.T., Lin, P., Chang, Y.W., Chen, C.T., Chiang, M.T., Chang, L., Kuo, Y.C., Tsai, H.T. and Yeh, T.K., 2009. Effect of taurine supplementation on cytochrome P450 2E1 and oxidative stress in the liver and kidneys of rats with streptozotocin-induced diabetes. Food and Chemical Toxicology 47: 1703-1709.

Yildirim, Z., Kiliç, N., Ozer, C., Babul, A., Take, G. amd Erdogan D., 2007. Effects of taurine in cellular responses to oxidative stress in young and middle-aged rat liver. Annals of the New York Academy of Sciences 1100: 553-561.

Ying, Y.Y.Y., 2006. Effects of different taurine levels in dietary on lipid metabolism in layers. China Feed 15.

Yu, C., Mei, X.T., Zheng, Y.P. and Xu, D.H., 2015. Taurine zinc solid dispersions protect against cold-restraint stress-induced gastric ulceration by upregulating HSP70 and exerting an anxiolytic effect. European Journal of Pharmacology 762: 63-71.

Yu, X., Chen, K., Wei, N., Zhang, Q., Liu, J. and Mi, M., 2007. Dietary taurine reduces retinal damage produced by photochemical stress via antioxidant and anti-apoptotic mechanisms in Sprague-Dawley rats. British Journal of Nutrition 98: 711-719.

Yu, X., Xu, Z., Mi, M., Xu, H., Zhu, J., Wei, N., Chen, K., Zhang, Q., Zeng, K., Wang, J. and Chen, F., 2008. Dietary taurine supplementation ameliorates diabetic retinopathy via anti-excitotoxicity of glutamate in streptozotocin-induced Sprague-Dawley rats. Neurochemical Research 33: 500-507.

Yuan, J.M. and Wang, Z.H., 2010. Effect of taurine on intestinal morphology and utilisation of soy oil in chickens. British Poultry Science 51: 540-545.

Zeng, D.S., Gao, Z.H., Huang, X.L., Zhao, J.H., Huang, G.Q. and Duo, L., 2012. Effect of taurine on lipid metabolism of broilers. Journal of Applied Animal Research 40: 86-89.

Zhang, Z., Liu, D., Yi, B., Liao, Z., Tang, L., Yin, D. and He, M., 2014. Taurine supplementation reduces oxidative stress and protects the liver in an iron-overload murine model. Molecular Medicine Reports 10: 2255-2262.

Zhao, W., Jia, L., Yang, H.J., Xue, X., Xu, W.X., Cai, J.Q., Guo, R.J. and Cao, C.C., 2018. Taurine enhances the protective effect of Dexmedetomidine on sepsis-induced acute lung injury via balancing the immunological system. Biomedicine & Pharmacotherapy 103: 1362-1368.

Zhou, J., Li, Y., Yan, G., Bu, Q., Lv, L., Yang, Y., Zhao, J., Shao, X., Deng, Y., Zhu, R. and Zhao, Y., 2011. Protective role of taurine against morphine-induced neurotoxicity in C6 cells via inhibition of oxidative stress. Neurotoxicity Research 20: 334-342.

Zhu, X.Y., Ma, P.S., Wu, W., Zhou, R., Hao, Y.J., Niu, Y., Sun, T., Li, Y.X. and Yu, J.Q., 2016. Neuroprotective actions of taurine on hypoxic-ischemic brain damage in neonatal rats. Brain Research Bulletin 124: 295-305.

Zielinska, M., Sawosz, E., Grodzik, M., Balcerak, M., Wierzbicki, M., Skomial, J., Sawosz, F. and Chwalibog, A., 2012. Effect of taurine and gold nanoparticles on the morphological and molecular characteristics of muscle development during chicken embryogenesis. Archives of Animal Nutrition 66: 1-13.

Zielińska, M.K., Sawosz, E., Chwalibog, A., Ostaszewska, T., Kamaszewski, M., Grodzik, M. and Skomiał, J., 2010. Nano nutrition of chicken embryos. Effect of gold and taurine nanoparticles on muscle development. Journal of Animal and Feed Sciences 663: 127.

Zulli, A., Lau, E., Wijaya, B.P., Jin, X., Sutarga, K., Schwartz, G.D., Learmont, J., Wookey, P.J., Zinellu, A., Carru, C. and Hare, D.L., 2009. High dietary taurine reduces apoptosis and atherosclerosis in the left main coronary artery: association with reduced CCAAT/enhancer binding protein homologous protein and total plasma homocysteine but not lipidemia. Hypertension 53: 1017-1022.

Chapter 11
Silymarin

You never know what you can do till you try

11.1 Introduction

Silymarin (SM) is a C25 containing flavonoid mixture, extracted from the *Silybum marianum* (milk thistle) plant. Today's standardised (according to its silibinin, often called silybin, content) SM extract contains approximately 65 to 80% flavonolignans (silybin A and silybin B, isosilybin A, isosilybin B, silychristin and silydianin), with small amounts of flavonoids, and approximately 20% to 35% of fatty acids and polyphenolic compounds possessing a range of metabolic regulatory effects (Comelli *et al.*, 2007). Silybin was discovered as the first member of a new family of natural compounds called flavonolignans in 1959 (Biedermann *et al.*, 2014) and it is known as the predominant and primary active ingredient in SM (Hackett *et al.*, 2013; Křen *et al.*, 2013). That is why compounds containing milk thistle ingredients showing silybin content and silybin antioxidant, as well as other activities in various model systems, are used to explain the biological activity of SM. In particular, SM has been the gold standard drug to treat liver disorders of different etiologies and milk thistle extracts have been used as traditional herbal remedies ('liver tonics') for almost 2000 years. Therefore, SM is most well-known for its antioxidant and chemoprotective effects on the liver (Federico *et al.*, 2008; Loguercio *et al.*, 2007; Post-White *et al.*, 2007; Stiuso *et al.*, 2014) and it is often prescribed and self-prescribed as a complementary and alternative hepatoprotective medicine (Testino *et al.*, 2013). SM is being studied as a hepato-, neuro-, nephro- and cardio-protective ingredient due to its strong antioxidant and tissue regenerative properties There is a range of recent comprehensive reviews covering various routes and mechanisms of action of SM in animal models and human trials very often referring to its antioxidant properties (for review and references see Surai, 2015). However, it seems likely that direct antioxidant (AO) activity of polyphenols does not contribute directly to the antioxidant defence of the body (Surai, 2014) and recently a lot of attention has been paid to SM/silybin impact on the induction of cellular antioxidant defence via the modulation of various transcription factors, including Nrf2 and NF-κB and respective gene and protein expressions. In fact, potential molecular proliferative signalling targets for anti-cancer activity of silibinin include the receptor tyrosine kinase, STAT, androgen receptor and NF-κB pathways (Li *et al.*, 2010), however, anti-cancer activity of SM is beyond the scope of this chapter.

Therefore, this chapter focuses on recent studies on silymarin antioxidant effects in various *in vitro* and *in vivo* model systems in the context of its contribution to the antioxidant systems regulation, participation in cell signalling and vitagene regulation.

DOI 10.3920/978-90-8686-906-0_11,

A special attention will be paid to protective effects of SM in the gut. Important protective roles of SM in poultry nutrition and its hepatoprotective properties will also be characterised.

11.2 Absorption and metabolism of silibinin

SM and its main constituent silibinin sources, metabolism and bioavailability have already been studied extensively. It has been shown that after oral consumption silibinin is characterised by comparatively low availability, e.g. in rats it is only about 0.95% (Wu *et al.*, 2007). In fact, after the oral administration of the standardised milk thistle extract Legalon, flavonolignans were rapidly absorbed and eliminated (Zhu *et al.*, 2013) with a half-life for silibinin of 6 h (Barzaghi *et al.*, 1990; Flory *et al.*, 1980). The main biotransformation route of silybin and derivatives was identified to be conjugation (Křen *et al.*, 2013). It is interesting to note that silibinin, similar to other flavonoids is recognised by the body as a foreign matter and quickly metabolised via phase II enzymes. Indeed, oral consumption of silibinin was associated with a significant increase in both glutathione S-transferase (GST) and quinone reductase (QR) activities in liver, lung, stomach, skin and small bowel in a dose- and time-dependent manner (Zhao *et al.*, 1999). Silibinin in the systemic circulation was found mainly in conjugated form (Han *et al.*, 2004; D'Andrea *et al.*, 2005). In fact, after oral SM administration to healthy volunteers, only 10% (Weyhenmeyer *et al.*, 1992) to 17% (Calani *et al.*, 2012) of the total silibinin in the plasma was found in the free unconjugated form. Indeed, mono-, di-, and sulpho-glucuronides are shown to be formed, and 31 metabolites have been identified (Rickling *et al.*, 1995). Indeed, as mentioned above, silibinin in humans and rats shows fast elimination of both the free and conjugated forms with the mean elimination half-life being 6.32 h (Lorenz *et al.*, 1984). Therefore, similar to other flavonoids, after oral consumption silibinin, the main constituent of SM, is characterised by comparatively low availability, fast metabolism and its concentration in plasma is mainly in nano-molar range and only in some cases reaching micro-molar concentrations.

11.3 Antioxidant properties of silymarin

It should be noted that SM can contribute to the antioxidant defences in different ways. Firstly, by direct free radical scavenging. Secondly, by preventing free radical formation by inhibiting specific enzymes responsible for free radical production, or by maintaining the integrity of electron-transport chain of mitochondria in stress conditions. Thirdly, by participating in the maintenance of optimal redox status of the cell by activating a range of antioxidant enzymes and non-enzymatic antioxidants, mainly via transcription factors, including Nrf2 and NF-κB. Finally, by activating an array of vitagenes, responsible for the synthesis of protective molecules, including SOD, HSPs, elements of thioredoxin (Trx) and glutathione (GSH) systems, sirtuins and providing additional protection in stress conditions. In most studies pure silybin, as the main component of SM, was used, however, in some cases SM also showed antioxidant action *in vivo*.

11.3.1 Direct free radical scavenging

The effects of silibinin on the formation of ROS and eicosanoids by human platelets, white blood and endothelial cells were studied (Rickling *et al.*, 1995). Silibinin showed to be a strong scavenger of HOCl (IC_{50} 7 μM), but not of O_2^- (IC_{50}>200 μM) produced by human granulocytes. Similarly, production of O_2^- and NO in isolated rat Kupffer cells were inhibited by silibinin in a dose-dependent manner, with IC_{50} 80 μM (Dehmlow *et al.*, 1996a,b). Indeed, silybin has no superoxide anion scavenging capability but was able to significantly decrease (at 20 μM) hem-mediated low-density lipoprotein (LDL) oxidation and showed slight inhibition of hydroxyl radical formation (Varga *et al.*, 2006). The rate constants of silybin with OH radical (1.8×10^{10} dm^3/mol/s) is diffusion controlled, suggesting that silybin is a potent free radical scavenger (Fu *et al.*, 2009). Indeed, silibinin (2.5 μM) significantly decreased the concentration of H_2O_2 in Ab1-42-stressed neurons and prevented oxidative-related injuries (Yin *et al.*, 2011). Furthermore, silybin (10 μM and higher) is shown to have protective activity in ameliorating DNA damage in a model system (Fu *et al.*, 2009). Treatment *in vitro* with silibinin significantly inhibited spontaneous O_2^- and H_2O_2 release and TNF-α production by monocytes from pre-eclamptic women. The main effect of silibinin was obtained at 50 μM concentration (Cristofalo *et al.*, 2013). Thus, the authors concluded that silibinin exerts anti-oxidative and anti-inflammatory effects on monocytes from pre-eclamptic pregnant women by inhibiting the *in vitro* endogenous release of ROS and TNF-α production.

Antioxidant action of SM (equal to 62 μM of silybin) has been shown in various model systems. For example, IC_{50} for H_2O_2 was 38 μM, while IC_{50} for NO was 266 μM (Domitrović *et al.*, 2013). It was shown that SM inhibited 82.7% lipid peroxidation of linoleic acid emulsion; while BHA, BHT, α-tocopherol and Trolox indicated inhibition of 83.3, 82.1, 68.1 and 81.3% respectively on peroxidation of linoleic acid emulsion at the same concentration (Köksal *et al.*, 2009). In another *in vitro* study free radical scavenging activity and antioxidant properties of SM (>200 μM) were showed by four different assays (Asghar and Masood, 2008). It is important to mention that the free radical scavenging activity of pure individual compounds of the SM is reported to vary considerably, with silydianin and silychristin being 2-10-fold more active than the silibinin and on a mass basis, SM is shown to be about 8-fold more potent than silibinin as a free radical scavenger (Dvorak *et al.*, 2003). Indeed, SM and silybin (silibinin) are not the same compounds and their AO activities could differ substantially depending on their concentrations in the target tissues. Therefore, in the following sections a distinction will be made between AO activities of SM and its active constituent silybin.

It seems likely, that SM/silibinin AO effects could be on the cells further upstream of the ROS and TNFα release. In fact, SM has been shown to alter trafficking within cells (Blaising *et al.*, 2013; Crocenzi *et al.*, 2004) and affect the energy status of the cell (Colturato *et al.*, 2012), so the ROS scavenging effects, if any, could be secondary. Furthermore, many of the effects of SM/silibinin described *in vitro* occur at concentrations not currently achievable in humans or animals. Therefore, a direct scavenging ROS by silibinin in biological systems (except the gut) is not likely to substantially contribute to the antioxidant protection.

11.3.2 Protective effects of silybin on mitochondria, a main source of free radical production in the cell

Mitochondria are the primary cellular consumers of oxygen and contain numerous redox enzymes capable of transferring single electrons to oxygen, generating the ROS superoxide (O_2^-). It is well documented that mitochondrial enzymes known to generate ROS include the tricarboxylic acid (TCA) cycle enzymes aconitase and α-ketoglutarate dehydrogenase; the electron-transport chain (ETC) complexes I, II and III; pyruvate dehydrogenase and glycerol-3-phosphate dehydrogenase; dihydroorotate dehydrogenase; the monoamine oxidases (MAO) A and B; and cytochrome b5 reductase (Lin and Beal, 2006). Furthermore, mitochondrial insults, including oxidative damage itself, can cause an imbalance between ROS production and removal, resulting in net ROS production. For example, ROS can induce protein modifications, lipid peroxidation and mitochondrial DNA damage, which ultimately results in mitochondrial dysfunction (Sekine and Ichijo, 2015). Many studies have focused on the detrimental effects of ROS, but it is now clear that mitochondrially generated ROS are also involved in regulating intracellular signal transduction pathways that result in cell adaptation to stress (Surai *et al.*, 2019). One of the mechanisms responsible for the decrease in oxidative stress is the protective effect of SM/silibinin on mitochondrial structure and function. Indeed, SM protects mitochondria from pathological events by triggering pro-survival cell signalling. For example, silibinin supplementation is shown to optimise electron-transport chain, decreasing electron leakage and ROS formation and directly reducing activities of ROS-producing enzymes in the mitochondria. In rats subjected to ischemia/reperfusion (I/R), compared with the control group, a severe impairment of mitochondrial bioenergetics was observed. SM prevented the most significant changes that occurred in mitochondria during I/R (decreased ATP levels, membrane potential and state 3 respiration) and the associated cell dysfunction (Rolo *et al.*, 2015). Silibinin (100-500 μM), was evaluated for its protective effect against beta-adrenergic agonist isoproterenol-induced injury in cultured rat neonatal cardiac myocytes (Zhou *et al.*, 2006a). It was shown that silibinin addition was associated with increased SOD activity and upregulated mitochondrial membrane potential and with a prevention of mitochondrial dysfunction and cell injury (Rolo *et al.*, 2003). Silibinin, at a concentration as low as 10 μM, fully mitigated the rise in metabolic flow-driven ROS formation in perfused rat hepatocytes. In addition, studies on isolated liver mitochondria revealed that this low dose of silibinin depressed ROS production linked to the electron transfer chain activity (Detaille *et al.*, 2008). It has been shown that cold preservation and warm reperfusion of the rat liver were associated with increased lipid peroxidation and superoxide anion generation, as well as with decreased GSH, mitochondrial ATP content and respiratory control ratio (RCR). However, preservation conducted in presence of silibinin (100 μM) improved parameters affected by preservation and reperfusion. Indeed, silibinin promoted an increase of ATP and RCR by 39 and 16% respectively and decreased oxidative stress to values observed in livers never preserved nor perfused (Ligeret *et al.*, 2008). It has been suggested that the uncoupler-like activity of dehydrosilybin could be the basis of its ROS modulation effect in various experimental systems. In fact, dehydrosilybin uncoupled the respiration of isolated rat heart mitochondria with a very high potency in suppressing ROS formation in isolated rat heart mitochondria with IC_{50}=0.15 μM (Gabrielová *et al.*, 2010). It is interesting to note that silibinin in mitochondria was far more effective than its effect in a purely chemical system generating superoxide or in cells capable of oxidative burst, where the IC(50) for dehydrosilybin exceeds 50 μM. Changes in mitochondrial respiratory complexes in fatty hepatocytes were also attenuated by silibinin-vitamin E complex (15 mg vitamin E and 47 mg silybin) fed to rats with a major protective effect on Complex II subunit CII-30 (Grattagliano *et al.*, 2013). Similarly, silybin (0.4 g/kg) in complex with phospholipid (SILIPHOS) was effective in decreasing severe oxidative stress and preserving hepatic mitochondrial bioenergetics and mitochondrial proton leak and ATP reduction in non-alcoholic steatohepatitis induced by the methionine- and choline-deficient (MCD) diet (Serviddio *et al.*, 2010). Silymarin oil (10 ml/kg/BW) significantly increased levels of membrane fluidity and membrane potential of liver mitochondria (Zhu *et al.*, 2014). It has been suggested that protective mechanism of action of silibinin (50-200 μM) in intrastriatal MPP+-injected rats may be linked to maintenance of mitochondrial bioenergetics and integrity (Geed *et al.*, 2014). An *in vitro* study demonstrated that silibinin inhibits the activity of ROS-generating monoamine oxidase (MAO) that catalyses the oxidative deamination of monoamines (Mazzio *et al.*, 1998). The formation of leukotrienes via the 5-lipoxygenase pathway was indicated to be strongly inhibited by silibinin. In particular, in human granulocytes IC_{50}-values of 15 μM and 14.5 μM silibinin were detected for LTB4 and LTC4/D4/E4/F4 formation respectively. However, much higher silibinin concentrations (45-69 μM) were necessary to inhibit the cyclooxygenase pathway (Dehmlow *et al.*, 1996). Rats exposed to a carcinogen (1,2-di-methylhydrazine DMH) showed increased activities of phase I enzymes (cytochrome b5, cytochrome b5 reductase, cytochromeP450, cytochromeP450 reductase, cytochromP4502E1) in the liver and colonic mucosa as compared to control rats. Silibinin supplementation (50 mg/kg/BW) modulates the xenobiotic metabolising enzymes including decreasing activity of ROS producing cytochrome b5 reductase (Sangeetha *et al.*, 2012). Although energy generation in the mitochondrion is an essential and extremely important process for cell survival, excessive mitochondrial ROS production also has detrimental consequences for the cell and the whole body.

11.3.3 Inhibition of free-radical producing enzymes by silybin

Xanthine oxidase

Xanthine oxidase (XO) catalyses the terminal two steps of purine degradation, converting hypoxanthine to uric acid and is considered to be a critical source of both O^{2-} and H_2O_2 in inflammatory diseases (Doehner and Landmesser, 2011). In particular, XO in its oxidase form is considered to be a main source of oxidative stress and destructive free radicals in ischemia reperfusion injury associated with heart attacks, stroke and spinal cord injury, as well as being a destructive force in myocardial or renal hypoxia and infarctions (Varga *et al.*, 2004). It seems likely that silibinin can affect XO activity, and inhibition of XO by silibinin in the steady-state appears to be independent of the silibinin concentration. Furthermore, in the presence of 50 μM silibinin, the production of superoxide by XO was decreased by approximately 20% relative to the control at 100 μM xanthine (Pauff and Hille, 2009). When O^{2-} production was induced in xanthine/xanthine oxidase system, silybin inhibited uric acid formation with an IC_{50} of 32.2 μM (Varga *et al.*, 2004). Indeed, inhibition in uric acid formation occurred parallel with inhibition of cytochrome C reduction (e.g. with O^{2-} formation) suggesting that the observed effects are due to the inhibition of xanthine oxidase activity rather than the scavenging of O^{2-} radicals.

NADPH oxidase

ROS generation by NADPH oxidase enzyme complex plays a critical role in several physiological processes including host defence, posttranslational modification of proteins, cell differentiation and regulation of gene expression (Bedard and Krause, 2007). Silybin and its derivatives inhibited the NADPH oxidase activity in PMA-stimulated cell lysate in a concentration dependent manner. The most distinct inhibition was observed when cell lysate was pre-incubated with 5,7,4″-trimethylsilybin possessing high lipid solubility. Indeed, 50% inhibition of NADPH oxidase activity was achieved when using 5 μM of 5,7,4″-trimethylsilybin and 10 μM of silybin at final concentrations (Varga *et al.*, 2004). It is interesting to note that As-intoxicated rats showed a significant ($P<0.05$) up-regulation of myocardial NADPH (NOX) oxidase sub units such as NOX2 and NOX4 as well as Keap-1, and down-regulation of Nrf2 and heme oxygenase (HO-1) protein expressions. Pre-administration of silibinin (75 mg/kg/BW) remarkably recovered all those altered parameters to near normalcy in As-induced cardiotoxic rat (Muthumani and Prabu, 2014). Co-administration of silibinin (75 mg/kg/BW) along with As-resulted in a reversal of As-induced biochemical changes in the kidney, accompanied by a significant decrease in lipid peroxidation and an increase in the level of renal antioxidant defence system (Prabu and Muthumani, 2012). It also decreased the NADPH oxidase, iNOS and NF-κB over expression by As and upregulated the Nrf2 expression in the renal tissue. The effect of silybin in cultures of mouse podocytes and in the OVE26 mouse, a model of type 1 diabetes mellitus and diabetic nephropathy was investigated. Exposure of podocytes to high glucose increased intracellular superoxide production by 60%, NADPH oxidase activity by 90%, NOX4 expression by 100%, and the number of apoptotic cells by 150%, effects that were completely blocked by 10 μM silybin. These *in vitro* observations were confirmed by similar *in vivo* findings. The kidney cortex of vehicle-treated control OVE26 mice displayed greater NOX4 expression and twice as much superoxide production than cortex of silybin-treated mice (Khazimet *et al.*, 2013). Therefore, the *in vitro* data demonstrated that silybin decreased superoxide generation in cultured podocytes and the *in vivo* study showed a similar effect in the kidney cortex. This is not surprising considering that oxidative stress and enhanced generation of ROS is the underlying mechanism of injury to multiple cell types.

11.3.4 Antioxidant protective properties of SM (Silibinin) in prevention of toxic effects of various chemicals

The liver is a primary organ involved in biotransformation of food toxicants and drugs. Toxins absorbed from the intestinal tract first enter the liver resulting in a variety of liver disorders. Hepatic disorders are mainly caused by toxic chemicals (alcohol), xenobiotics (carbon tetrachloride, chlorinated hydrocarbons), anticancer (doxorubicin, cisplatin), immunosuppressant (cyclosporine), analgesic anti-inflammatory (paracetamol, thioacetami-de), anti-tubercular (isoniazid, rifampicin) drugs, radiations (gamma radiations), heavy metals (cadmium, arsenic), mycotoxins (aflatoxin, T-2 toxin, fumonisins, ochratoxins), galactosamine, lipopolysaccharides, etc. (Ingawale *et al.*, 2014). Increasing evidence supports the important role of oxidative stress as a key mechanism of hepatotoxicity. Therefore, the aforementioned toxicants were used in various experimental models to assess antioxidant protective properties of SM. These are especially important, since SM is primarily used in clinical practice to protect, support and repair the liver.

There is a growing body of evidence showing protective antioxidant effects of silymarin against toxic effects of arsenic, Carbone Tetrachloride, Thioacetamide, Cisplatin, Mn Toxicity, Benzo[a]pyrene, Doxorubicin, ethanol and other toxicants, including pyrogallol toxicity, bleomicyn-injected mice, acrolein toxicity in mice, diethylnitrosamine-induced oxidative stress in rat liver, sodium fluoride, and sodium nitrite toxicity (for review and references see Surai, 2015). It has been shown that major mycotoxins (aflatoxins, AF; ochratoxin A, OTA; T-2 toxin and fumonisins, FM) cause oxidative stress and apoptosis in the cell (Cimbalo *et al.*, 2020; Surai and Dvorska, 2005; Yan *et al.*, 2020). Indeed, mycotoxins generate ROS, which induce lipid and protein oxidation leading to changes in membrane integrity, cellular redox signalling, and in the antioxidant status of the cells. Silibinin is shown to be a potent protective compound against apoptosis and cytotoxicity caused by OTA (Essid *et al.*, 2012). In fact, silibinin in doses from 130 to 260 μM prevented chromatin condensation, caspase-3 activation, apoptotic DNA fragmentation and lipid peroxidation that were induced by OTA, H_2O_2 and ActD/TNF-α, respectively. Furthermore, silibinin (0.04-26.0 μM) had hepatoprotective effects against OTA or LPS-mediated TNF-alpha release and also reduced the cytotoxicity of both compounds (Al-Anati *et al.*, 2009). Pre-treatment of the rats with SM (100 mg/kg/BW) prior to AFB1 was found to show significant protection against AFB1-induced liver damage, as evidenced by a significant lowering of the activity of the serum enzymes glutamic-oxaloacetic transaminase (GOT) and glutamic-pyruvic transaminase (GPT) and enhanced hepatic reduced GSH status (Shyamal *et al.*, 2010). SM (100 mg/kg/BW) also ameliorated toxic liver damage caused by FB_1 in BALB/c mice by preventing elevation of the levels of caspase-8 and tumour necrosis factor-alpha mediators (Sozmen *et al.*, 2014).

Indeed, in animal experiments and *in vitro* studies SM/silibinin blocked the oxidative toxicities of As, CCl4, various mycotoxins, thioacetamide, cisplatin, Mn overload, benzo[a]pyrene, doxorubicin, ethanol and some other toxicants by decreasing an abnormally high production of ROS in living cells and restoring signal transduction and gene expression. In the aforementioned studies the free radical scavenging and antioxidant properties of SM and silibinin are demonstrated by: (1) restoration of the endogenous AO enzymes (SOD, CAT, GPx, GR and GST) and non-enzymatic antioxidants (vitamins E and C) in the liver and other tissues of stressed animals; (2) increased intracellular concentration of GSH in liver and other tissues; (3) decreased lipid and protein oxidation, detected as reduced MDA/TBARS and carbonyl content; (4) decreased DNA fragmentation/damage and apoptosis; (5) reduced secretion of ALT, AST, ALP from the liver into the plasma due to hepatic injuries caused by ROS; (6) restored Nrf2 and HO-1 activities; (7) reduced NF-κB expression and concentration of pro-inflammatory cytokines, including tumour necrosis factor (Surai, 2015).

11.3.5 Is effective silybin concentration *in vitro* achievable *in vivo*?

As mentioned above, because of their poor absorption and rapid elimination, plasma values of silybin will rarely exceed micromole concentrations. Therefore, the SM concentration used in most of the aforementioned studies are not achievable in the *in vivo* target tissues, where the silybin concentration is usually at least 10-100-fold lower. Indeed, the active free silibinin concentration in plasma after oral consumption of SM, depending on dose of supplementation, could be in the range 0.2-2.0 μM. For example, in healthy volunteers, after an oral administration of SM (equivalent to 120 mg silibinin), total (unconjugated + conjugated) silibinin concentration in plasma was 1.1-1.3 μg/ml (Kim *et al.*, 2003). Lower bioavailability of silibinin was shown in another study with higher dose of SM consumption (equivalent to 240 mg silibinin), where maximum silibinin concentration was about 0.7 μM (Lorenz *et al.*, 1984). Comparable silibinin concentrations (0.3-1.9 μM) were reported in healthy volunteers receiving about 240 mg silibinin (Hoh *et al.*, 2006). More recently, silybin A maximal concentration in human plasma was shown to be 0.2-0.6 μM and silybin B concentrations comprised about 30% of those shown for silybin A (Zhu *et al.*, 2013). The concentrations of free and conjugated silibinin in the target tissues of mice have been shown to reach their maximum levels within 1 h after 50 mg/kg/BW silibinin administration. The Cmax of free/conjugated silibinin in the tissues were 18.3/11.8; 8.9/5.8; 255/560; 2.9/8.9; 5.3/12.6; 12.0/21.9 μM for liver, lung, stomach, skin, prostate and pancreas, respectively (Zhao *et al.*, 1999). Indeed, the average concentration of most plant polyphenols in plasma rarely exceeds 1-2 μM in healthy subjects (Surai, 2014, 2015). Furthermore, the effect of SM in physiologically relevant concentrations would be relatively limited when taking into account concentrations of other antioxidants in plasma. For example, vitamin C (26.1 to 84.6 μM), vitamin E (20-30 μM), urate (150-200 μM), and albumin (several hundred μM) are found in plasma and polyphenol contribution to plasma TAC will certainly be less than 2% (Surai, 2014, 2015). A similar situation could be found in animals (chickens) where antioxidant concentrations in plasma are in the same range – vitamin C: 61-66 μM; vitamin E: 13-15 μM; and total antioxidant value in plasma could vary from 336.9 μM up to 740-830 μM and the general concentration of antioxidants in human plasma is about 1000 μM (Surai, 2014, 2015), therefore 20-50 μM additional antioxidant from dietary sources would be required to make a significant contribution to systemic antioxidant capacity (Halliwell *et al.*, 2005). While SM displays potent antioxidant activity *in vitro*, the bioactive forms of flavonoids *in vivo* are not those forms found in plants (i.e. flavonoid glycosides) due to their extensive biotransformation in the small intestine and hepatic metabolism upon absorption. Therefore, metabolic modifications occurring *in vivo* may substantially influence the antioxidant activity of dietary flavonoids. Silibinin, similar to other flavonoids, is recognised by the body as foreign matter and quickly metabolised via phase II enzymes. The aforementioned data clearly indicates that SM protects cells against the damaging effects of oxidative/nitrosative stress and inflammation. It also prevents unwanted cell proliferation in cancer cells (Fan *et al.*, 2014) as well as pathogen infection (Lim *et al.*, 2014).

11.4 Silymarin and Nrf2 regulation

An interesting approach to explain health-promoting properties of polyphenols could be to explore the modulation of signalling pathways, e.g. the Nrf2/Keap1/ARE and NF-κB pathways, resulting in increased expression of genes encoding for cytoprotective molecules, including AO enzymes and phase II detoxification enzymes. In addition, decreased expression of NF-κB-regulated genes would reduce production of pro-inflammatory cytokines. Indeed, it has been shown that some phytochemicals (sulforaphane, curcumin, EGCG, diallyl sulphide, resveratrol, etc.) at concentrations as low as 10 μM can activate Nrf2 signalling by inducing phosphorylation of Nrf2 via activation of upstream protein kinases and/or through direct interaction with Keap1 cysteine thiols (Surh *et al.*, 2008). It is interesting to note that anthocyanins activate the antioxidant transcription factor Nrf2, at a concentration as low as 1 μM (Zhang *et al.*, 2013). Silymarin's ability to upregulate the master antioxidant coordinator Nrf2 has also been investigated in various *in vitro* and *in vivo* model systems. Initially, the SM-containing supplement Protandim (ashwagandha, bacopa extract, green tea extract, SM, and curcumin) was shown to induce HO-1 in a cell culture primarily through Nrf2 activation (Velmurugan *et al.*, 2009). Protandim also induces Nrf2 nuclear localisation and antioxidant enzyme expression (Donovan *et al.*, 2012). However, it is difficult to distinguish the SM effect in the mixture. Furthermore, it was shown that SM (a water-soluble formulation Legalon-SIL containing a succinate-modified active compounds) at 50 and 100 μM significantly upregulated Nrf2 protein levels in CON1 cells after 48 h treatment (Mehrab-Mohseni *et al.*, 2015). SM (104 μM) protected human cells against the damage from the herbicide paraquat via triggering anti-oxidant related genes, including Nrf2, NQO1 and HO-1 (Podder *et al.*, 2012). In an *in vitro* experiment with human hepatic HepG2 cells, SM (100 μM) was shown to be protective against AAPH-induced oxidative stress and apoptosis and the antioxidant potential of SM was correlated with induction of antioxidant genes including HO-1, NQO-1, γ-GCLC and SOD via transcriptional activation of Nrf2 (Gokila Vani *et al.*, 2013).

In a dietary rat model of non-alcoholic steatohepatitis there was an increase in nuclear translocation of Nrf2 protein in SM-fed animals (Kim *et al.*, 2012) which could enhance the protective effect against oxidative stress. Numerically, there was also a trend in increasing Nrf2 endogenous expression protein in rat liver tissue damaged by CCl4 by pre-treatment with SM (100 mg/kg/BW (Wu *et al.*, 2013) as well as in an alcohol-pyrazole-fed rat model (50 mg/kg/BW (Choi *et al.*, 2013). It seems likely that the effect of SM on the Nrf2 activation in various oxidative-stress related animal models depends on the dose of SM used. For example, at 10 mg/kg/BW SM showed protective effects against dimerumic acid-induced changes in GSH and HO-1 levels, but did not change Nrf2 levels (Lee *et al.*, 2013). Indeed, SB-treatment (75 mg/kg/BW) ameliorated arsenic-induced nephrotoxicity by abrogation of oxidative stress, inflammation and apoptosis in rats (Prabu and Muthumani, 2012). It is interesting to note that beyond the preventive action of SB, in rats fed silibinin alone there was a significant increase in Nrf2 mRNA levels in the kidney. The same SB-treated rats were characterised by increased antioxidant defences as indicated by increased renal SOD and GPx activities, as well as enhanced levels of GSH, vitamins E and C. In fact, SB-induced activation of Nrf2 and HO-1 played a significant role in cellular defence against As-induced oxidative cardiac damage. In a recent study, co-treatment with SM (86 mg/kg/BW) significantly enhanced the antioxidant defence systems of the ethanol-consuming mice via the activation of Nrf2 and HO-1 expression in damaged livers (Cao *et al.*, 2015). Furthermore, SM (200 mg/kg/BW) significantly up-regulated HO-1 and Nrf-2 in the liver of AAPH-treated mice (Gokila Vani *et al.*, 2013). Recently Nrf2-activating effects of SM and its components have been reviewed (Vargas-Mendoza *et al.*, 2020) with a conclusion that further researches with in silico, *in vitro*, and *in vivo* models are necessary to elucidate molecular mechanisms of modulating effects of SM and/or its flavolignans on Nrf2.

Therefore, it is clear from the aforementioned data that SM/silibinin, similar to other polyphenols, acts as an indirect antioxidant through its ability to induce Nrf2 transactivation, which is responsible for the synthesis of an array of antioxidant molecules decreasing oxidative stress and providing effective protection in various stress conditions. This changes an attitude to SM, a compound that has been assumed to act typically as a free radical scavenger and traditional antioxidant. It is still not clear if physiologically relevant (up to 2 μM) silybin concentrations would upregulate Nrf2, but *in vivo* results obtained with various model systems clearly indicated an activation of Nrf2 by silibinin. In general, most of the experimental studies suggest that dietary SM/silibinin activates antioxidant pathways such as Nrf2/HO1 and downregulates NF-κB (see Section 11.5), and effect PPAR (Pferschy-Wenzig *et al.*, 2014; Prakash *et al.*, 2014), HIF-1 (Jung *et al.*, 2009; Lin *et al.*, 2013; Wang *et al.*, 2012) and STAT (Agarwal *et al.*, 2007; Hou *et al.*, 2010; Tyagi *et al.*, 2012) pathways.

11.5 Silymarin and NF-κB regulation

The anti-inflammatory activity of phenolic compounds has been demonstrated in a number of *in vitro* and *in vivo* studies and polyphenols may affect inflammation mainly as modulators of inflammatory redox signalling pathways (Gupta *et al.*, 2014). The polyphenolic compounds express anti-inflammatory activity by modulating the expression of pro-inflammatory genes such as cyclooxygenase, lipoxygenase, nitric oxide synthases and several important cytokines, mainly acting through nuclear factor-κB and mitogen-activated protein kinase signalling (Gupta *et al.*, 2014; Stevenson and Hurst, 2007). Due to the large number of studies that have demonstrated regulatory effects of SM/silybin on the expression of NF-κB in various *in vitro* and *in vivo* models there is insufficient space in this review to analyse all of them, so we will focus only on recent investigations addressing the issue. It is well established that various plant-derived polyphenols can suppress TNF-α activated, NF-κB-associated inflammatory pathways both *in vitro* and *in vivo*. These polyphenols include SM as well as curcumin, resveratrol, genistein, epigallocatechin gallate, flavopiridol, emodin, morin isoliquiritigenin, naringenin, ellagic acid, apigenin, kaempferol, catechins, myricetin, xanthohumol, fisetin, vitexin, escin, mangostin and others (Gupta *et al.*, 2014). SM has long been known to inhibit the activation of inflammatory NF-κB signalling (Ramasamy and Agarwal, 2008). Recent studies have confirmed that the protective anti-inflammatory effects of SM/silybin could be mediated by their inhibitory potential on the NF-κB, which is a key transcriptional factor for numerous genes involved in regulation of inflammation, immune system, cell differentiation, survival, apoptosis, etc. Here, we give some examples of participation of SM/silibinin in NF-κB pathway inhibition.

11.5.1 *In vitro* studies

Silybin is a potent inhibitor of NF-κB activation. Indeed, Manna *et al.* (2009) tested silybin in a number of *in vitro* human cell experimental systems and found it inhibited TNF-mediated NF-κB activation in a dose-dependent manner with maximum inhibition at 50 μM and it regulated NF-κB 100 times more effectively than aspirin. Furthermore, silibinin (80 μM) protected glial cells against peroxide-induced ROS formation, ATP depletion, and cell damage (Tsai *et al.*, 2010). Interestingly, the inhibition of peroxide-induced ROS by silybin was partially attenuated by inhibitors of NF-κB or protein kinase C (PKC), suggesting an involvement of NF-κB and PKC signalling pathways. Similarly, a significant decrease in NF-κB was observed, when PBMCs from pre-eclamptic patients were cultured with silibinin at 5 μM and 50 μM concentrations and stimulated with LPS (Giorgi *et al.*, 2012). Moreover, silibinin (50-200 μM) inhibited the nuclear translocation of nuclear factor NF-κB through inhibition of the phosphorylation of IκBα and suppressed NF-κB transcriptional activity in stimulated HMC-1 cells (Kim *et al.*, 2013). SM is also effective in inhibiting T cell activation and proliferation, by acting on pathways of NF-κB activation/translocation. In particular, in CD4+ splenocytes from C57/Bl6 mice, SM (50 μM) significantly inhibited CD4+ cells proliferation, inhibited IL-2 and IFN-gamma production and blocked nuclear translocation of transcription factor NF-κB. Moreover, SM inhibited p65/NF-κB phosphorylation in CD4+ T cell (Gharagozloo *et al.*, 2010). The significant upregulation of oxidative stress biomarkers including MDA, TNFα, IFN-γ and IL-1β genes was observed as well as NF-κB, COX-2 and iNOS proteins expression occurred upon heat stress in chicken hepatocytes. Furthermore, AO enzyme (SOD, CAT,

GR) activities decreased. SM (259 μM) was able to normalise the expression of all of these biomarkers in heat-induced chicken hepatocytes (Oskoueian *et al.*, 2014). Silibinin was clearly shown to inhibit the expression of NF-κB at transcriptional and translational levels (Kang *et al.*, 2019). Silibinin also downregulates the NF-κB pathway and NLRP1/NLRP3 inflammasomes in monocytes from pregnant women with preeclampsia (Matias *et al.*, 2019). However, there is a need for more studies with pure silybin at physiologically-relevant concentrations (0.2-2 μM) in order to confirm its modulating effect on the NF-κB pathway.

11.5.2 *In vivo* studies

It has been shown that silibinin (50 mg/kg/BW) had a protective effect against d-galactose-induced senescence due to promotion of cellular oxidoreductase activities and inhibition of NF-κB activation and ROS production (Wang *et al.*, 2011). Histopathological and immunohistochemical studies in the kidney of rats also shows that silibinin (75 mg/kg/BW) markedly reduced the toxicity of As and preserved the normal histological architecture of the renal tissue, inhibited the caspase-3 mediated tubular cell apoptosis and decreased the NADPH oxidase, iNOS and NF-κB overexpression by As (Prabu and Muthumani, 2012). In a model of experimental ischemic stroke in male rats, silibinin (100 mg/kg) significantly downregulated NF-κB in ischemic brain tissue after stroke (Wang *et al.*, 2012). In an experimental nonalcoholic steatohepatitis model, mice fed a methionine-choline deficient diet consistently had increased NF-κB p65 and p50 binding activity, while silibinin administration (20 mg/kg/BW/i.p.) significantly decreased the activity of both subunits (Salamone *et al.*, 2012). It was shown that CCl4 treatment of mice enhanced the NF-κB expression in the liver and induced hepatic fibrosis, while SM (200 mg/kg/BW) reduced the CCl4-induced hepatic NF-κB and improved CCl4-induced liver fibrosis (Li *et al.*, 2012). Protective effect of silibinin (150 mg/kg/BW/intragastric) in the rat model of cerebral ischemia was associated with decreased expression of the Bax, NF-κB protein and Bax, NF-κB mRNA in the brain (Liu *et al.*, 2014). SM is also shown to have NF-κB inhibiting effects. For example, mRNA and protein expressions of NF-κB were significantly up-regulated in d-galactosamine induced hepatotoxic rats and treatment with SM (25 mg/kg/BW) significantly down-regulated the expressions of these genes (Aristatile *et al.*, 2013). Furthermore, it has been shown that SM (250 mg/kg/BW) suppressed NF-κB signalling cascade in alcoholic liver fibrosis in guinea pigs (Abhilash *et al.*, 2013, 2014). In a mouse model of non-alcoholic steatohepatitis silibinin treatment was shown to alleviate the severity of hepatic inflammation and steatohepatitis due to the inhibition as a result of inhibition of NF-κB-signalling pathway (Lee *et al.*, 2018). Similarly, silybin was reported to alleviates hepatic steatosis and fibrosis in NASH mice by inhibiting oxidative stress and decreasing NF-κB signalling (Ou *et al.*, 2018). Protective role of Silibinin against myocardial ischemia/reperfusion injury-induced cardiac dysfunction was shown to be associated with decreased apoptosis, oxidative stress and inflammatory response through deactivation of NF-κB pathway (Chen *et al.*, 2020). Similarly, Silibinin was reported to attenuate renal fibrosis *in vitro* and *in vivo* via inhibition of NF-κB (Liu *et al.*, 2019).

Overall, SM/silibinin targets multiple signalling pathways, including NF-κB, towards inhibiting the secretion of pro-inflammatory cytokines. Together these findings suggest that the inhibitory effect of SM/silibinin on NF-κB signalling in a variety of cells, as well as in *in vivo* studies, could be an important mechanism of its anti-inflammation efficacy, and that targeting this signalling pathway by silibinin may have potential applications in the clinic and veterinary practice.

11.6 Effect of silymarin on vitagene expression

Emerging findings suggest a large number of potential mechanisms of action of SM (polyphenols) in preventing disease, which may be beyond their direct conventional antioxidant activities. It seems likely that SM, similar to other flavonoids, can affect the vitagene network. The recent relevant findings are reviewed below. First of all, SM/silybin affects HO-1 activity in different model systems. For example, As-intoxicated rats showed a significant up-regulation of myocardial NADPH (NOX) oxidase subunits such as NOX2 and NOX4 as well as Keap1 and down-regulation of Nrf2 and vitagene HO-1 protein expressions. Pre-administration of silibinin (75 mg/kg/BW) recovered all these altered parameters to near normalcy in As-induced cardiotoxic rat (Muthumani and Prabu, 2014). Similarly, in a model of liver injury caused by alcohol plus pyrazole, SM administration (50 mg/kg/BW) had a protective effect with a trend in restoring the decreased activity of HO-1 and Nrf2 (Choi *et al.*, 2013). SM (250 mg/kg/BW) possesses substantial protective effect against B(a)P-induced damages by increasing (restoring) HO-1 (vitagene) activity (Kiruthiga *et al.*, 2015). Similarly, *in vitro* SM (500 μM) reduced tBH-induced hepatocyte toxicity by activating HO-1 gene expression (Cerný *et al.*, 2009). Indeed, the enzyme HO-1 is an important regulatory molecule present in most mammalian cells. In fact, the main function of HO-1 is to break down the pro-oxidant molecule heme into three products; carbon monoxide (CO), biliverdin and free iron and actively participate in the antioxidant defence in the human/animal body (Venditti and Smith, 2014). In fact, HO-1 is a stress-inducible protein and can be induced by various oxidative and inflammatory signals. Therefore, HO-1 expression is regarded as an important adaptive cellular mechanism of AO defence in stress conditions (Ambegaokar *et al.*, 2014; Naito *et al.*, 2014). Treatment of primary human dermal fibroblasts with both SM and silybin (SB) was shown to reduce UVA-stimulated ROS generation, prevent GSH depletion, moderately increased HO-1 level and reduced HSP70 level (Rajnochová Svobodová *et al.*, 2018). Recently it has been shown that ameliorative effects of silymarin on HCl-induced acute lung injury in rats was associated with activation of the Nrf-2/HO-1 pathway (Ahmed *et al.*, 2019).

From the data presented above it is clear that SM/silibinin can upregulate HO-1 and improve antioxidant defences. Furthermore, SM/silibinin can affect other HSPs including HSP70. Indeed, in an *in vitro* system based on CHO-K1 cells treated with As, SM (5 μM) significantly decreased HSP70 expression previously elevated by As (Bongiovanni *et al.*, 2007). In another *in vitro* system based on heat-induced chicken hepatocytes, SM (259 μM) affected HSP70 expression significantly, preventing its alleviation by heat stress (Oskoueian *et al.*, 2014). A similar protective effect of SM (100 mg/kg/BW) on HSP70 was seen in rats given SM for 7 days prior to mesenteric ischemia-reperfusion (I-R) compared to I-R group (Demir *et al.*, 2014). It is interesting to note that silybin was identified as a novel HSP90 inhibitor (Zhao *et al.*, 2011). Similarly, SM (2 g/kg feed for 15 d after lambing) was shown to decrease expression of HSP70 gene in blood serum of post-parturient sanjabi ewes (Khamisabadi *et al.*, 2020). Therefore, silibinin can decrease HSP70 expression in stressed cells indicating improved AO defences and decrease stress by other means (e.g. Nrf2-related increased AO synthesis).

Thirdly, there are several recent publications showing a regulatory role of SM/ silibinin on the sirtuin functions. In fact, a protective effect of silibinin against chemotherapeutic reagent mitomycin C-induced cell death in A375-S2 cells was studied. It was shown that mitomycin C caused cell apoptosis and over-expression of p53, elevated translocation of p53 into the nucleus, and decreased SIRT1 expression; and these effects were alleviated by silibinin (100-150 μM) to a large extent (Jiang *et al.*, 2009). Silibinin protected cardiac myocytes against isoproterenol-induced injury through resuming mitochondrial function and regulating the expression of SIRT1 and Bcl-2 family members (Zhou *et al.*, 2006). Systemic administration of silibinin (50 mg/kg/BW) reversed streptozotocin-induced downregulation of SIRT1 expression in mice (Wang *et al.*, 2012). It is important to note that SIRT1 was activated by SM (500 μM) in UV-irradiated human malignant melanoma, A375-S2 cells (Li *et al.*, 2007). However, silibinin could have completely the opposite action in regulating SIRT1 expression in normal cardiac myocytes and breast cancer MCF-7 cells, where it downregulates SIRT1 and promotes apoptosis (Wang *et al.*, 2008). Similarly, silybin treatment downregulated SIRT1 in human lung adenocarcinoma cells (Wu *et al.*, 2014). Recent findings indicate that sirtuins-mediated signalling pathways are involved in metabolic reprogramming and upregulation of antioxidant defence systems (Wu *et al.*, 2014a). In mice with non-alcoholic fatty liver disease Silybin was shown to inhibit NLRP3 inflammasome assembly through the NAD^+/SIRT2 pathway (Zhang *et al.*, 2018). Furthermore, silibinin restored NAD^+ levels and induced the SIRT1/AMPK pathway in non-alcoholic fatty liver (Salomone *et al.*, 2017). Activation of sirtuin 3 by silybin was reported to attenuate mitochondrial dysfunction in cisplatin-induced acute kidney injury (Li *et al.*, 2017). Silymarin (20 mg/kg/day i.p, 4 weeks) was shown to increase SIRT1 levels in brain of aged rats (Sarubbo *et al.*, 2018). Silymarin administration (60 or 120 mg/kg) to diabetic rats was shown to protect against oxidative stress by promoting SIRT1, GPx and Catalase expression in the liver tissue (Kheiripour *et al.*, 2018). Therefore, SM-related activation or prevention of inhibition of sirtuins in stress conditions could be considered as an important adaptive mechanism responsible for maintaining redox-regulated homeostasis in the cell and in the whole body.

Finally, there is only a single publication devoted to the protective effect of SM on another vitagene called thioredoxins system. Silybin-phospholipid complex containing vitamin E (Realsil®, Indena, IBI-Lorenzini spa, Italy) was administered daily by gavage (15 mg vitamin E and 47 mg silybin complexed with phospholipids) to rats fed a choline-deprived or a high fat diet for 30 days and 60 days, respectively. Realsil was shown to be able to maintain Trx concentration in the rat liver fed on a high fat diet at day 60, while in the control group it dramatically decreased (Grattagliano *et al.*, 2013). Silymarin was shown to attenuate Paraquat-induced cytotoxicity in macrophage by regulating Trx/TXNIP complex (promoted the expression of Trx and antioxidant enzymes and suppress the TXNIP) and inhibiting NLRP3 inflammasome activation and apoptosis (Liu *et al.*, 2018). It is well accepted that the ability of SM to maintain Trx system in stress conditions is an important element of improving the adaptive ability of animals/humans via the vitagene network. Therefore, the vitagene network represents the major cellular pathway involved in the so called 'programmed cell life' as an opposite to apoptosis, providing an effective protection against oxidative stress and toxic products of ROS metabolism (Calabrese *et al.*, 2009, 2009a, 2010). In particular, nutritional antioxidants can be neuroprotective through the activation of hormetic pathways under the control of the Vitagene protein network (Calabrese *et al.*, 2014). Recently modulatory effect of SM on Nrf2-regulated regulated redox status and NF-κB-mediated inflammation in experimental gastric ulcer have been evaluated (Keshk *et al.*, 2017). As can be seen from Figure 11.1 SM partly prevented decrease in expression of Nrf2 and its target AO (SOD, GPx) and ameliorated increased NF-κB, TNFα, MDA and activity of myeloperoxidase (MPx).

The aforementioned results clearly indicate that, similar to other flavonoids SM, or its active ingredient silibinin, affect vitagene network decreasing detrimental consequences of stresses. Further research is needed to fully understand the molecular

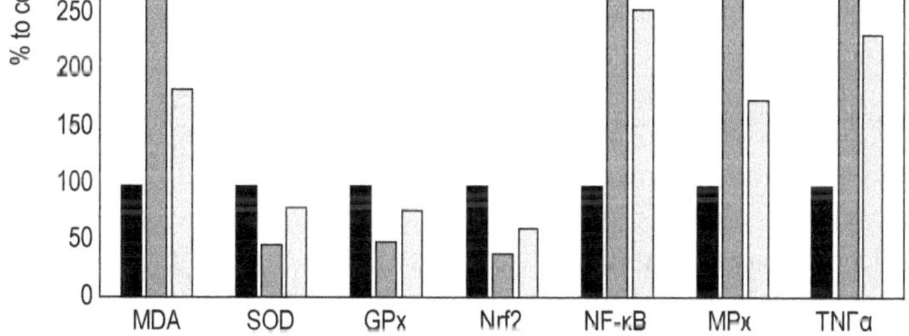

Figure 11.1. Protective effects of silymarin (SM) in experimental gastric ulcer (adapted from Keshk *et al.*, 2017).

mechanisms involved in the vitagene network activation by SM/silybin.

11.7 Protective effect of silymarin in the gut

It seems likely that the gut is the major place of antioxidant action of polyphenols (Surai and Fisinin, 2015; Surai *et al.*, 2003, 2004). Indeed, reduction of oxidative damage, modulation of colonic flora and variation in gene expression are involved in the modulation of intestinal function by polyphenols. For example, to study the molecular effects of wine polyphenols at the gene level, the microarray technology was used: rats were treated with 50 mg/kg/BW wine polyphenols for 14 days. Global expression analysis of 5,707 genes revealed an extensive down-regulation of genes involved in a wide range of physiological functions, such as metabolism, transport, signal transduction and intercellular signalling (Dolara *et al.*, 2005). It was shown that two major regulatory pathways were downregulated in the colon mucosa of polyphenols treated rats: inflammatory response and steroid metabolism. Since flavonoids are consumed in concentrations usually much higher than other antioxidant compounds, their protective effect during digestion is of great importance. For example, it has been shown that flavonoids not only prevented an accumulation of peroxidised lipids but can also switch prooxidant properties of heme-proteins to antioxidant ones (Kanner and Lapidot, 2001). Dietary polyphenols can also modulate *in vivo* oxidative damage in the gastrointestinal tract of rodents (Giovannelli *et al.*, 2000) supporting the hypothesis that dietary polyphenols might have both a protective and a therapeutic potential in oxidative damage-related pathologies. Indeed, the antioxidant-prooxidant balance (redox status) in various parts of the intestine would ultimately depend on the level of antioxidants and pro-oxidants provided by the diet and released by cells themselves, as well as on the level of absorption of both antioxidants and pro-oxidants (Surai and Fisinin, 2015; Surai *et al.*, 2003, 2004). In a model system mimicking stomach conditions, it was shown that both lipid peroxidation and co-oxidation of vitamin E and beta-carotene were inhibited at pH 3.0 by red wine polyphenols (Gorelik *et al.*, 2005). In the gut, an interaction between different antioxidants could bring about their maximum synergistic effects. In particular, it was found that the mixture of the green tea polyphenol, vitamin E and vitamin C could act synergistically to protect lipid peroxidation (Dai *et al.*, 2008). It has been found that consumption of partially oxidised food could increase lipid peroxidation in the stomach and the absorption of cytotoxic lipid peroxidation products into the body. The addition of red wine polyphenols to the meal may alter these outcomes (Gorelik *et al.*, 2008), therefore red wine polyphenols exert a beneficial effect by the novel function, absorption inhibition of the MDA (Gorelik *et al.*, 2008a). These findings explain the potentially harmful effects of oxidised fats found in foods/feeds (Surai *et al.*, 2003) and the important benefit of dietary polyphenols in the diet. Indeed, locating the biological site of action of polyphenols in the gastrointestinal tract may lead to a revision in our understanding of how phytochemicals, including silybin, work *in vivo* and may help to elucidate the mechanisms of the health-promoting action of silybin and other polyphenols (Kanner *et al.*, 2012).

Redox signalling in gut inflammation is complex and poorly understood. However, it is generally accepted that homeostatic control of the intestinal epithelial redox environment is central for nutrient digestion and absorption, stem cell proliferation, apical enterocyte apoptosis, and immune response (Circu and Aw, 2012). Indeed, polyphenols may play a role in intestinal mucosa integrity, inflammation and permeability (Martínez *et al.*, 2013; Surai, 2014; Surai *et al.*, 2004). For example, wine phenolics were able to prevent or delay the progression of intestinal diseases characterised by oxidative stress and inflammation, acting as both free radical scavengers and modulators of specific inflammation-related genes involved in cellular redox signalling (Biasi *et al.*, 2014). They exert their effects by modulating cell signalling pathways, mainly activated in response to oxidative and inflammatory stimuli, and Nrf2 and NF-κB are the principal downstream effectors (Biasi *et al.*, 2011). For example, pigs administered a grape seed and grape marc extract had a lower transactivation of NF-κB and Nrf2 and a lower expression of various target genes of these transcription factors in the duodenal mucosa (but not in the liver and plasma) than control pigs (Gessner *et al.*, 2013). Furthermore, the ratio of villus height: crypt depth and the gain:feed ratio was higher in the animals fed a gape extract than in control pigs. In mice, the dietary grape seed extract decreased proliferation and enhanced differentiation of epithelial cell and the changes in gut epithelium were associated with the suppression of NF-κB signalling (Yang *et al.*, 2014). Polyphenolic compounds could also affect gut integrity. For example, grape seed extract (0.1%) supplied to rats with drinking water for 21 days significantly increased the expression of gut junction protein occludin in the proximal colon and reduced faecal levels of the neutrophil protein calprotectin, compared with control (Goodrich *et al.*, 2012). Recently, it has been reported that a naturally occurring flavonoid procyanidin B2 could modulate cellular redox status and the antioxidant enzyme defence system in colonic cells protecting against oxidative stress and xenobiotics (Rodríguez-Ramiro *et al.*, 2011, 2011a).

It is possible to suggest that there is a biological reason for some antioxidants not to be absorbed completely and so provide antioxidant protection in lower parts of the intestine (Surai and Fisinin, 2015; Surai *et al.*, 2003, 2004). Comparatively low bioavailability and antioxidant potential of various flavonoids could be beneficial for humans/animals providing antioxidant protection in various parts of the digestive tract, including the large intestine where levels of other antioxidants would be quite low. The same is true for SM. For example, silibinin concentration in the gut could reach 800 μM (Köck *et al.*, 2013), a concentration exceeding the requirement for its direct antioxidant activity. Indeed, it is still unclear whether SM has any direct antioxidant effects *in vivo*, although it might be capable of exerting such effects within the gastrointestinal tract, where SM may come into direct contact with cells without having undergone absorption and metabolism (Surai *et al.*, 2004). It has been suggested that gastroprotective effects of SM might be related to the protection of gastric mucosal NO and non-protein sulfhydryl groups and the modulation of capsaicin-sensitive gastric sensory afferents (Shin *et al.*, 2013). The beneficial effects of SM, on a murine model of colitis was explained (Esmaily *et al.*, 2009). SM effectively reduced colonic damage, lipid peroxidation, neutrophil infiltration and the content of inflammatory cytokines alongside increasing total antioxidant capacity of colonic tissue. In trinitrobenzene sulfonic acid-induced colitis model, silibinin treatment significantly reduced several components of inflammatory colitis such as NF-κB activity, levels of IL-1β, TNFα, TBARS, protein carbonyl, myeloperoxidase activity, and an improvement in antioxidant capability of the colon tissue (Esmaily *et al.*, 2011). Furthermore, silibinin (50 mg/kg/BW) modulates gut microbial enzymes, colonic oxidative stress and Wnt/β-catenin signalling resulting in antiproliferative effect against 1,2-di-methylhydrazine-induced colon carcinogenesis in experimental rats (Sangeetha *et al.*, 2012).

It is also well-known that polyphenols are providing antioxidant protection in the lower intestine and can alter colonic flora (Moco *et al.*, 2012). Polyphenols, including silybin, are extensively metabolised by gut bacteria into a complex series of end-products that affect the functional ecology of symbiotic partners that can alter the host physiology (Moco *et al.*, 2012). It has been suggested that the consumption of a diet rich in plant foods with high dietary polyphenol content may enhance the gastrointestinal health of the host through microbiota modulation. Indeed, grape antioxidant dietary fibre intake stimulates proliferation of Lactobacillus and slightly affects the composition of Bifidobacterium species in the cecum of rats (Pozuelo *et al.*, 2012). It appears that dietary polyphenols may have the ability to modify the gut microbial balance, but this effect is indirect, i.e. it is mediated by biotransformation products, rather than the original plant compounds (Parkar *et al.*, 2013). It should be mentioned that polyphenol-microbiota interactions are complex and subject to large inter-individual variability, leading to different polyphenol-metabolising phenotypes or 'metabotypes' (Bolca *et al.*, 2013). Unfortunately, there is no data available on the effect of SM on gut microbiota, but we can suggest the same changes would occur as described above for other phytochemicals. Therefore, it seems likely that reduction of oxidative damage, decreased inflammation, modulation of colonic flora and variation in gene expression are involved in the modulation of intestinal function by polyphenols, including SM.

11.8 Silymarin in poultry

Based on multiple protective effects of SM shown above in various model systems one can expect its beneficial effect in commercial poultry production associated with stress conditions. Data on protective effects of SM and milk thistle seeds on poultry are shown in Table 11.1.

Data on SM effects on poultry are very limited and obtained mainly in the last 7 years. In normal physiological conditions SM can decrease oxidative stress imposed by commercial conditions of chicken housing (Ahmad *et al.*, 2020; Schiavone *et al.*, 2007). More pronounced protective effects of MTS or SM was observed in various stress models, including heat stress (Morovat *et al.*, 2016), high energy diet (Çeribaşı *et al.*, 2020), AF (Amiridumari *et al.*, 2013; Jahanian *et al.*, 2017) or ATA (Khatoon *et al.*, 2013) dietary contamination or CCL4 treatment (Baradaran *et al.*, 2019). In general, several important aspects of the beneficial properties of milk thistle on poultry growth performance in experimentally induced aflatoxicosis (Alhidary *et al.*, 2017; Zaker-

Table 11.1. Protective effects of silymarin (SM) on poultry.

No.	Experimental design	Effects of SM	References
1	180 male chicks (ROSS 508), basal diet (BD), BD +40 or 80 mg/kg SM, 60 days	Reduced lipid content of breast and thigh muscles, increased muscles resistance to oxidative stress.	Schiavone *et al.*, 2007
2	120 broiler chicks, BD, BD+ milk thistle seeds (MTS): 5, 10 or 15 g/kg, 42 days	MTS at 15 g/kg decreased the negative effects of natural summer stress in broilers: Serum MDA↓, antibody titer against ND↑, improved FCR for weeks 2, 3, 4 and 5.	Ahmad *et al.*, 2020
3	240 chicks (ROSS 308), BD and BD+100 mg/kg SM, 42 days, days 7-28, increased by 4 °C temperature for 4 hours	Improved performance, immunity and carcass characteristics and decreased the adverse effects of the heat stress	Morovat *et al.*, 2016
4	One 120 male 15-day-old Japanese quail chicks, 4 groups, BD, high energy diet (HD), Milk thistle seeds (MTS, 1%), HD+MTS for 35 days	HD imposed oxidative stress in testes (MDA↑, GPx↓) and disturbed spermatogenesis; almost all the disturbances were ameliorated by MTS.	Çeribaşı *et al.*, 2020
5	216 one-d-old Ross 308 male broiler chicks, 9 groups, 250-500 µg/kg AFB1, 0.5 and 1% MTS.	AFB1 disturbed metabolism of Ca, glucose and creatinine, increased AST activity (liver damage). MTS significantly reduced the effect of AFB1 on the above parameters	Amiridumari *et al.*, 2013
6	430 (7-day-old) Ross broiler chicks. BD, AF mix -0.5 or 2 mg/kg, SM 0.5 or 1 g/kg for 21 days.	SM improved performance of AF-challenged birds by suppressing ileal bacteria and enhancing absorptive surface area	Jahanian *et al.*, 2017
7	240 one-day-old birds, 12 groups. BD, OTA (1 or 2 mg/kg), SM (10 g/kg), Vit.E (200 mg/kg) for 42 days	SM and Vit. E alone or in combination ameliorated the immunotoxic effects induced by 1.0 mg OTA/kg, no protection against 2.0 mg OTA/kg.	Khatoon *et al.*, 2013
8	240-day-old broilers, 4 groups. BD, SM 100 mg/kg BW, CCl_4 1 ml/kg and SM+CCl4	CCl4 damaged liver and imposed oxidative stress by down-regulating CAT, GPx, and Mn-SOD. SM showed protective effects by ameliorating those changes	Baradaran *et al.*, 2019

Esteghamati *et al.*, in press) and SM hepatoprotective properties (Saeed *et al.*, 2017) and its usage as a protective agent against natural or chemical toxicities (Fanoudi *et al.*, 2020) have recently been reviewed. Protective effects of MTS or SM was related to prevention of oxidative stress and to maintenance of productive performance under stressful conditions. Based on the multiple protective actions of SM in various stress models *in vitro* and *in vivo* presented above, we can expect more research in this area related to a design of protective nutrient compositions to deal with commercially relevant stresses in egg and meat production.

11.9 Conclusions

Recent achievements in biochemistry and molecular biology, together with epidemiological data have changed our thinking about food. It has become increasingly clear that our diet plays a pivotal role in maintenance of human/animal health and a misbalanced diet can cause serious health-related problems. It seems likely that antioxidants are among the major regulators of many physiological processes and, therefore, a redox balance between antioxidants and prooxidants in the diet, gastro-intestinal tract, plasma and tissues is an important determinant of the state of health. This is also relevant to poultry nutrition. Since flavonoids are not well absorbed in the gut, their active concentration in the plasma and target tissues are comparatively low, but probably sufficient for Nrf2 activation and NF-κB suppression as well as vitagene activation. Indeed, it seems very likely that activation of the Keap1/Nrf2/ARE pathway and inhibition of NF-κB pathway, rather than direct free radical scavenging activity, may be the main mechanisms of the health benefits of phytochemicals (Calabrese *et al.*, 2012), including SM (Surai, 2015). Therefore, consumption of phytochemicals, including SM, could have a pre-conditioning effect on the antioxidant system of the body: they keep the antioxidant defence system alert. This could be related to mild pro-oxidant properties of many polyphenolics and help explaining the beneficial health-promoting effects for human of a diet rich in fruits and vegetables as important sources of the aforementioned chemicals (polyphenols and other phytochemicals) maintaining the body's ability to be highly adaptive to various stresses. Therefore, SM and its main component silibinin are part of the dietary phytochemical mixture responsible for regulation of the antioxidant defences in the gut and in the whole body. It could well be that some dietary constituents which are not well absorbed could have health-promoting properties by maintaining redox balance in the large intestine, where concentration of other antioxidants (vitamin E, carotenoids, ascorbate) could be low, but prooxidants (iron, oxidised PUFAs, etc.) and substrates of oxidation are still present (Surai, 2018; Surai and Fisinin, 2015; Surai *et al.*, 2003, 2004). There is also an interaction of polyphenolics, including SM, with gut microbiota and they can have prebiotic effects (Surai *et al.*, 2018). This protective effect in the large intestine could be responsible, for example, for bowl cancer prevention.

Therefore, there could be a biological reason for some nutrients not being well absorbed, but still being involved in antioxidant protection in the lower gut. Taking into account high concentrations of phytochemicals in the gut, it could well be that they play an essential part in maintaining an optimal antioxidant-prooxidant balance in the digestive tract responsible for additional health effects of phytochemicals (Surai, 2014), including SM (Surai, 2015). In poultry/animal nutrition and disease prevention strategy SM alone, or in combination with other hepato-active compounds (carnitine, betaine, vitamin B12, etc.), could have similar hepatoprotective effects as described in humans with similar mechanisms of protective action (Surai *et al.*, 2018).

In conclusion, there are many possible mechanisms by which SM can improve the antioxidant defence mechanisms in the body. They include direct and indirect SM actions (Figure 11.2).

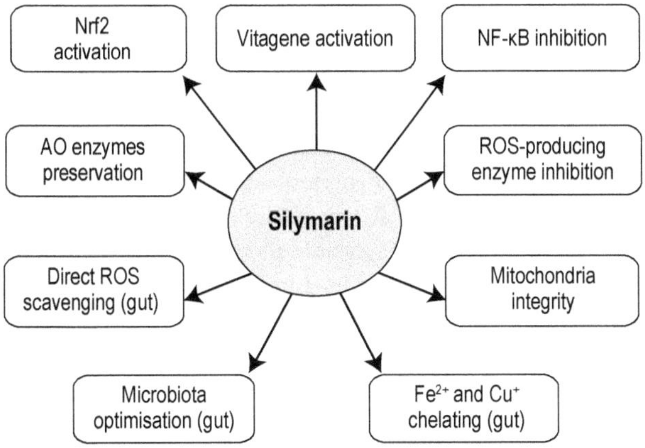

Figure 11.2. Protective actions of silymarin.

First of all, a direct scavenging free radicals and chelating free Fe and Cu are mainly effective in the gut. Secondly, preventing free radical formation by inhibiting specific ROS-producing enzymes, or improving the integrity of electron-transport chain of mitochondria in stress conditions as a result of SM consumption, is of great importance. Thirdly, maintaining an optimal redox balance in the cell by activating a range of antioxidant enzymes and non-enzymatic antioxidants, mainly via Nrf2 activation, is probably the main driving force of AO action of SM. Fourthly, decreasing inflammatory responses in the gut and other tissues by inhibiting NF-κB pathways is an emerging mechanism of SM protective effect in liver toxicity and diseases. Fifthly, activating vitagenes, responsible for synthesis of protective molecules, including HSP, Trx, SIRTs, etc., and providing additional protection in stress conditions deserves more attention in future research. Finally, effects on the microenvironment of the gut, including SM bacteria interactions, await future investigation.

Therefore, SM is shown to have unique protective actions in the body as an important part of the antioxidant defence mechanisms, vitagene and transcription factor modulator and an element of maintaining/creating adaptive homeostasis. Clearly, more research is needed to explain molecular mechanisms of protective effects of SM in maintaining avian/poultry/animal health under stress conditions of avian life in wild nature or commercial poultry production.

References

Abhilash, P.A., Harikrishnan, R. and Indira, M., 2013. Ascorbic acid is superior to silymarin in the recovery of ethanol-induced inflammatory reactions in hepatocytes of guinea pigs. Journal of Physiology and Biochemistry 69: 785-798.

Abhilash, P.A., Harikrishnan, R. and Indira, M., 2014. Ascorbic acid suppresses endotoxemia and NF-κB signaling cascade in alcoholic liver fibrosis in guinea pigs: A mechanistic approach. Toxicology and Applied Pharmacology 274: 215-224.

Agarwal, C., Tyagi, A., Kaur, M. and Agarwal, R., 2007. Silibinin inhibits constitutive activation of Stat3 and causes caspase activation and apoptotic death of human prostate carcinoma DU145 cells. Carcinogenesis 28: 1463-1470.

Ahmad, M., Chand, N., Khan, R.U., Ahmad, N., Khattak, I. and Naz, S., 2020. Dietary supplementation of milk thistle (*Silybum marianum*): growth performance, oxidative stress, and immune response in natural summer stressed broilers. Tropical Animal Health and Production 52: 711-715.

Ahmed, R.F., Moussa, R.A., Eldemerdash, R.S., Zakaria, M.M. and Abdel-Gaber, S.A., 2019. Ameliorative effects of silymarin on HCl-induced acute lung injury in rats; role of the Nrf-2/HO-1 pathway. Iranian Journal of Basic Medical Sciences 22: 1483-1492.

Al-Anati, L., Essid, E., Reinehr, R. and Petzinger, E., 2009. Silibinin protects OTA-mediated TNF-α release from perfused rat livers and isolated rat Kupffer cells. Molecular Nutrition & Food Research 53: 460-466.

Alhidary, I.A., Rehman, Z., Khan, R.U. and Tahir, M., 2017. Anti-aflatoxin activities of milk thistle (*Silybum marianum*) in broiler. World's Poultry Science Journal 73: 559-566.

Ambegaokar, S.S. and Kolson, D.L., 2014. Heme oxygenase-1 dysregulation in the brain: implications for HIV associated neurocognitive disorders. Current HIV Research 12: 174-188.

Amiridumari, H., Sarir, H., Afzali, N. and Fanimakki, O., 2013. Effects of milk thistle seed against aflatoxin B1 in broiler model. Journal of research in medical sciences: the official journal of Isfahan University of Medical Sciences 18: 786-790.

Baradaran, A., Samadi, F., Ramezanpour, S. S. and Yousefdoust, S., 2019. Hepatoprotective effects of silymarin on CCl_4-induced hepatic damage in broiler chickens model. Toxicology Reports 6: 788-794.

Aristatile, B., Al-Assaf, A.H. and Pugalendi, K.V., 2013. Carvacrol suppresses the expression of inflammatory marker genes in D-galactosamine-hepatotoxic rats. Asian Pacific Journal of Tropical Medicine 6: 205-211.

Asghar, Z. and Masood, Z., 2008. Evaluation of antioxidant properties of silymarin and its potential to inhibit peroxyl radicals *in vitro*. Pakistan Journal of Pharmaceutical Sciences 21: 249-254.

Barzaghi, N., Crema, F., Gatti, G., Pifferi, G. and Perucca, E., 1990. Pharmacokinetic studies on IdB 1016, a silybin-phosphatidylcholine complex, in healthy human subjects. European Journal of Drug Metabolism and Pharmacokinetics 15: 333-338.

Bedard, K. and Krause, K.H., 2007. The NOX family of ROS-generating NADPH oxidases: physiology and pathophysiology. Physiological Reviews 87: 245-313.

Biasi, F., Astegiano, M., Maina, M., Leonarduzzi, G. and Poli, G., 2011. Polyphenol supplementation as a complementary medicinal approach to treating inflammatory bowel disease. Current Medicinal Chemistry 18: 4851-4865.

Biasi, F., Deiana, M., Guina, T., Gamba, P., Leonarduzzi, G. and Poli, G., 2014. Wine consumption and intestinal redox homeostasis. Redox Biology 2: 795-802.

Biedermann, D., Vavříková, E., Cvak, L. and Křen, V., 2014. Chemistry of silybin. Natural Product Reports 31: 1138-1157.

Blaising, J., Lévy, P.L., Gondeau, C., Phelip, C., Varbanov, M., Teissier, E., Ruggiero, F., Polyak, S.J., Oberlies, N.H., Ivanovic, T. and Boulant, S., 2013. Silibinin inhibits hepatitis C virus entry into hepatocytes by hindering clathrin-dependent trafficking. Cellular Microbiology 15: 1866-1882.

Bolca, S., Van de Wiele, T. and Possemiers, S., 2013. Gut metabotypes govern health effects of dietary polyphenols. Current Opinion in Biotechnology 24: 220-225.

Bongiovanni, G.A., Soria, E.A. and Eynard, A.R., 2007. Effects of the plant flavonoids silymarin and quercetin on arsenite-induced oxidative stress in CHO-K1 cells. Food and Chemical Toxicology 45: 971-976.

Calabrese, V., Cornelius, C., Dinkova-Kostova, A.T. and Calabrese, E.J., 2009a. Vitagenes, cellular stress response, and acetylcarnitine: relevance to hormesis. Biofactors 35: 146-160.

Calabrese, V., Cornelius, C., Dinkova-Kostova, A.T., Iavicoli, I., Di Paola, R., Koverech, A., Cuzzocrea, S., Rizzarelli, E. and Calabrese, E.J., 2012. Cellular stress responses, hormetic phytochemicals and vitagenes in aging and longevity. Biochimica et Biophysica Acta (BBA)-Molecular Basis of Disease 1822: 753-783.

Calabrese, V., Cornelius, C., Mancuso, C., Barone, E., Calafato, S., Bates, T., Rizzarelli, E. and Kostova, A.T., 2009. Vitagenes, dietary antioxidants and neuroprotection in neurodegenerative diseases. Frontiers in Bioscience 14: 376-397.

Calabrese, V., Cornelius, C., Stella, A.M.G. and Calabrese, E.J., 2010. Cellular stress responses, mitostress and carnitine insufficiencies as critical determinants in aging and neurodegenerative disorders: role of hormesis and vitagenes. Neurochemical Research, 35: 1880-1915.

Calabrese, V., Scapagnini, G., Davinelli, S., Koverech, G., Koverech, A., De Pasquale, C., Salinaro, A.T., Scuto, M., Calabrese, E.J. and Genazzani, A.R., 2014. Sex hormonal regulation and hormesis in aging and longevity: role of vitagenes. Journal of Cell Communication and Signaling 8: 369-384.

Calani, L., Brighenti, F., Bruni, R. and Del Rio, D., 2012. Absorption and metabolism of milk thistle flavanolignans in humans. Phytomedicine 20: 40-46.

Çeribaşı, S., Türk, G., Özçelik, M., Doğan, G., Çeribaşı, A.O., Mutlu, S.İ., Erişir, Z., Güvenç, M., Güngören, G., Acısu, T.C., Akarsu, S.A., Kaya, Ş.Ö., Sönmez, M., Yüce, A., Çiftçi, M., Çambay, Z., Bağcı, E., Azman, M.A. and Şimşek, Ü.G., 2020. Negative effect of feeding with high energy diets on testes and metabolic blood parameters of male Japanese quails, and positive role of milk thistle seed. Theriogenology 144: 74-81.

Cao, Y.W., Jiang, Y., Zhang, D.Y., Wang, M., Chen, W.S., Su, H., Wang, Y.T. and Wan, J.B., 2015. Protective effects of Penthorum chinense Pursh against chronic ethanol-induced liver injury in mice. Journal of Ethnopharmacology 161: 92-98.

Černý, D., Canová, N.K., Martínek, J., Hořínek, A., Kmoníčková, E., Zídek, Z. and Farghali, H., 2009. Effects of resveratrol pretreatment on tert-butylhydroperoxide induced hepatocyte toxicity in immobilized perifused hepatocytes: involvement of inducible nitric oxide synthase and hemoxygenase-1. Nitric Oxide 20: 1-8.

Chen, Y.H., Lin, H., Wang, Q., Hou, J.W., Mao, Z.J. and Li, Y.G., 2020. Protective role of silibinin against myocardial ischemia/reperfusion injury-induced cardiac dysfunction. International Journal of Biological Sciences 16: 1972-1988.

Choi, M.K., Han, J.M., Kim, H.G., Lee, J.S., Lee, J.S., Wang, J.H., Son, S.W., Park, H.J. and Son, C.G., 2013. Aqueous extract of *Artemisia capillaris* exerts hepatoprotective action in alcohol-pyrazole-fed rat model. Journal of Ethnopharmacology 147: 662-670.

Cimbalo, A., Alonso-Garrido, M., Font, G. and Manyes, L., 2020. Toxicity of mycotoxins *in vivo* on vertebrate organisms: a review. Food and Chemical Toxicology 137: 111161.

Circu, M.L. and Aw, T.Y., 2012, September. Intestinal redox biology and oxidative stress. Seminars in Cell & Developmental Biology 23: 729-737.

Colturato, C.P., Constantin, R.P., Maeda Jr, A.S., Constantin, R.P., Yamamoto, N.S., Bracht, A., Ishii-Iwamoto, E.L. and Constantin, J., 2012. Metabolic effects of silibinin in the rat liver. Chemico-Biological Interactions 195: 119-132.

Comelli, M.C., Mengs, U., Schneider, C. and Prosdocimi, M., 2007. Toward the definition of the mechanism of action of silymarin: activities related to cellular protection from toxic damage induced by chemotherapy. Integrative Cancer Therapies 6: 120-129.

Cristofalo, R., Bannwart-Castro, C.F., Magalhaes, C.G., Borges, V.T.M., Peraçoli, J.C., Witkin, S.S. and Peracoli, M.T., 2013. Silibinin attenuates oxidative metabolism and cytokine production by monocytes from preeclamptic women. Free Radical Research 47: 268-275.

Crocenzi, F.A., Mottino, A.D. and Roma, M.G., 2004. Regulation of synthesis and trafficking of canalicular transporters and its alteration in acquired hepatocellular cholestasis. Experimental therapeutic strategies for its prevention. Current Medicinal Chemistry 11: 501-524.

D'Andrea, V., Pérez, L.M. and Pozzi, E.J.S., 2005. Inhibition of rat liver UDP-glucuronosyltransferase by silymarin and the metabolite silibinin-glucuronide. Life Sciences 77: 683-692.

Dai, F., Chen, W.F. and Zhou, B., 2008. Antioxidant synergism of green tea polyphenols with α-tocopherol and l-ascorbic acid in SDS micelles. Biochimie 90: 1499-1505.

Dehmlow, C., Erhard, J. and De Groot, H., 1996a. Inhibition of Kupffer cell functions as an explanation for the hepatoprotective properties of silibinin. Hepatology 23: 749-754.

Dehmlow, C., Murawski, N. and De Groot, H., 1996b. Scavenging of reactive oxygen species and inhibition of arachidonic acid metabolism by silibinin in human cells. Life sciences 58: 1591-1600.

Demir, M., Amanvermez, R., Polat, A.K., Karabıçak, I., Cınar, H., Kesicioğlu, T. and Polat, C., 2014. The effect of silymarin on mesenteric ischemia-reperfusion injury. Medical Principles and Practice 23: 140-144.

Detaille, D., Sanchez, C., Sanz, N., Lopez-Novoa, J.M., Leverve, X. and El-Mir, M.Y., 2008. Interrelation between the inhibition of glycolytic flux by silibinin and the lowering of mitochondrial ROS production in perfused rat hepatocytes. Life Sciences 82: 1070-1076.

Doehner, W. and Landmesser, U., 2011, Xanthine oxidase and uric acid in cardiovascular disease: clinical impact and therapeutic options. Seminars in Nephrology 31: 433-440.

Dolara, P., Luceri, C., De Filippo, C., Femia, A.P., Giovannelli, L., Caderni, G., Cecchini, C., Silvi, S., Orpianesi, C. and Cresci, A., 2005. Red wine polyphenols influence carcinogenesis, intestinal microflora, oxidative damage and gene expression profiles of colonic mucosa in F344 rats. Mutation Research 591: 237-246.

Domitrović, R., Jakovac, H., Marchesi, V.V. and Blažeković, B., 2013. Resolution of liver fibrosis by isoquinoline alkaloid berberine in CCl4-intoxicated mice is mediated by suppression of oxidative stress and upregulation of MMP-2 expression. Journal of Medicinal Food 16: 518-528.

Donovan, E.L., McCord, J.M., Reuland, D.J., Miller, B.F. and Hamilton, K.L., 2012. Phytochemical activation of Nrf2 protects human coronary artery endothelial cells against an oxidative challenge. Oxidative Medicine and Cellular Longevity 2012: 132931.

Dvořák, Z., Kosina, P., Walterová, D., Šimánek, V., Bachleda, P. and Ulrichová, J., 2003. Primary cultures of human hepatocytes as a tool in cytotoxicity studies: cell protection against model toxins by flavonolignans obtained from *Silybum marianum*. Toxicology Letters 137: 201-212.

Esmaily, H., Hosseini-Tabatabaei, A., Rahimian, R., Khorasani, R., Baeeri, M., Barazesh-Morgani, A., Yasa, N., Khademi, Y. and Abdollahi, M., 2009. On the benefits of silymarin in murine colitis by improving balance of destructive cytokines and reduction of toxic stress in the bowel cells. Central European Journal of Biology 4: 204-213.

Esmaily, H., Vaziri-Bami, A., Miroliaee, A.E., Baeeri, M. and Abdollahi, M., 2011. The correlation between NF-κB inhibition and disease activity by coadministration of silibinin and ursodeoxycholic acid in experimental colitis. Fundamental & Clinical Pharmacology 25: 723-733.

Essid, E., Dernawi, Y. and Petzinger, E., 2012. Apoptosis induction by OTA and TNF-α in cultured primary rat hepatocytes and prevention by silibinin. Toxins 4: 1139-1156.

Fan, L., Ma, Y., Liu, Y., Zheng, D. and Huang, G., 2014. Silymarin induces cell cycle arrest and apoptosis in ovarian cancer cells. European Journal of Pharmacology 743: 79-88.

Fanoudi, S., Alavi, M.S., Karimi, G. and Hosseinzadeh, H., 2020. Milk thistle (*Silybum marianum*) as an antidote or a protective agent against natural or chemical toxicities: a review. Drug and Chemical Toxicology 43: 240-254.

Federico, A., Niosi, M., Vecchio Blanco, C. D. and Loguercio, C., 2008. Emerging drugs for non-alcoholic fatty liver disease. Expert Opinion on Emerging Drugs 13: 145-158.

Flory, P.J., Krug, G., Lorenz, D. and Mennicke, W.H., 1980. Studies on elimination of silymarin in cholecystectomized patients. I. Biliary and renal elimination after a single oral dose. Planta Medica 38: 227-237.

Fu, H., Lin, M., Muroya, Y., Hata, K., Katsumura, Y., Yokoya, A., Shikazono, N. and Hatano, Y., 2009. Free radical scavenging reactions and antioxidant activities of silybin: mechanistic aspects and pulse radiolytic studies. Free Radical Research 43: 887-897.

Gabrielová, E., Jabůrek, M., Gažák, R., Vostálová, J., Ježek, J., Křen, V. and Modrianský, M., 2010. Dehydrosilybin attenuates the production of ROS in rat cardiomyocyte mitochondria with an uncoupler-like mechanism. Journal of Bioenergetics and Biomembranes 42: 499-509.

Geed, M., Garabadu, D., Ahmad, A. and Krishnamurthy, S., 2014. Silibinin pretreatment attenuates biochemical and behavioral changes induced by intrastriatal MPP+ injection in rats. Pharmacology Biochemistry and Behavior 117: 92-103.

Gessner, D.K., Fiesel, A., Most, E., Dinges, J., Wen, G., Ringseis, R. and Eder, K., 2013. Supplementation of a grape seed and grape marc meal extract decreases activities of the oxidative stress-responsive transcription factors NF-κB and Nrf2 in the duodenal mucosa of pigs. Acta Veterinaria Scandinavica 55: 18.

Gharagozloo, M., Velardi, E., Bruscoli, S., Agostini, M., Di Sante, M., Donato, V., Amirghofran, Z. and Riccardi, C., 2010. Silymarin suppress CD4+ T cell activation and proliferation: effects on NF-κB activity and IL-2 production. Pharmacological Research 61: 405-409.

Giorgi, V.S.I., Peracoli, M.T.S., Peracoli, J.C., Witkin, S.S. and Bannwart-Castro, C.F., 2012. Silibinin modulates the NF-κb pathway and pro-inflammatory cytokine production by mononuclear cells from preeclamptic women. Journal of Reproductive Immunology 95: 67-72.

Giovannelli, L., Testa, G., De Filippo, C., Cheynier, V., Clifford, M.N. and Dolara, P., 2000. Effect of complex polyphenols and tannins from red wine on DNA oxidative damage of rat colon mucosa *in vivo*. European Journal of Nutrition 39: 207-212.

Gokila Vani, M., Kumar, K.J., Liao, J.W., Chien, S.C., Mau, J.L., Chiang, S.S., Lin, C.C., Kuo, Y.H. and Wang, S.Y., 2013. Antcin C from *Antrodia cinnamomea* protects liver cells against free radical-induced oxidative stress and apoptosis *in vitro* and *in vivo* through Nrf2-dependent mechanism. Evidence-Based Complementary and Alternative Medicine 2013: 296082.

Goodrich, K.M., Fundaro, G., Griffin, L.E., Hulver, M.W., Ponder, M.A. and Neilson, A.P., 2012. Chronic administration of dietary grape seed extract increases colonic expression of gut tight junction protein occludin and reduces fecal calprotectin: a secondary analysis of healthy Wistar Furth rats. Nutrition Research 32: 787-794.

Gorelik, S., Lapidot, T., Shaham, I., Granit, R., Ligumsky, M., Kohen, R. and Kanner, J., 2005. Lipid peroxidation and coupled vitamin oxidation in simulated and human gastric fluid inhibited by dietary polyphenols: health implications. Journal of Agricultural and Food Chemistry 53: 3397-3402.

Gorelik, S., Ligumsky, M., Kohen, R. and Kanner, J., 2008. A novel function of red wine polyphenols in humans: prevention of absorption of cytotoxic lipid peroxidation products. The FASEB Journal 22: 41-46.

Gorelik, S., Ligumsky, M., Kohen, R. and Kanner, J., 2008. The stomach as a 'bioreactor': when red meat meets red wine. Journal of Agricultural and Food Chemistry 56: 5002-5007.

Grattagliano, I., Diogo, C.V., Mastrodonato, M., de Bari, O., Persichella, M., Wang, D.Q., Liquori, A., Ferri, D., Carratù, M.R., Oliveira, P.J. and Portincasa, P., 2013. A silybin-phospholipids complex counteracts rat fatty liver degeneration and mitochondrial oxidative changes. World Journal of Gastroenterology 19: 3007-3017.

Gupta, S.C., Tyagi, A.K., Deshmukh-Taskar, P., Hinojosa, M., Prasad, S. and Aggarwal, B.B., 2014. Downregulation of tumor necrosis factor and other proinflammatory biomarkers by polyphenols. Archives of Biochemistry and Biophysics 559: 91-99.

Hackett, E.S., Twedt, D.C. and Gustafson, D.L., 2013. Milk thistle and its derivative compounds: a review of opportunities for treatment of liver disease. Journal of Veterinary Internal Medicine 27: 10-16.

Halliwell, B., Rafter, J. and Jenner, A., 2005. Health promotion by flavonoids, tocopherols, tocotrienols, and other phenols: direct or indirect effects? Antioxidant or not? The American Journal of Clinical Nutrition 81: 268S-276S.

Han, Y.H., Lou, H.X., Ren, D.M., Sun, L.R., Ma, B. and Ji, M., 2004. Stereoselective metabolism of silybin diastereoisomers in the glucuronidation process. Journal of Pharmaceutical and Biomedical Analysis 34: 1071-1078.

Hoh, C., Boocock, D., Marczylo, T., Singh, R., Berry, D.P., Dennison, A.R., Hemingway, D., Miller, A., West, K., Euden, S. and Garcea, G., 2006. Pilot study of oral silibinin, a putative chemopreventive agent, in colorectal cancer patients: silibinin levels in plasma, colorectum, and liver and their pharmacodynamic consequences. Clinical Cancer Research 12: 2944-2950.

Hou, Y.C., Liou, K.T., Chern, C.M., Wang, Y.H., Liao, J.F., Chang, S., Chou, Y.H. and Shen, Y.C., 2010. Preventive effect of silymarin in cerebral ischemia-reperfusion-induced brain injury in rats possibly through impairing NF-κB and STAT-1 activation. Phytomedicine 17: 963-973.

Ingawale, D.K., Mandlik, S.K. and Naik, S.R., 2014. Models of hepatotoxicity and the underlying cellular, biochemical and immunological mechanism (s): a critical discussion. Environmental Toxicology and Pharmacology 37: 118-133.

Jahanian, E., Mahdavi, A. H., Asgary, S. and Jahanian, R., 2017. Effects of dietary inclusion of silymarin on performance, intestinal morphology and ileal bacterial count in aflatoxin-challenged broiler chicks. Journal of Animal Physiology and Animal Nutrition 101: e43-e54.

Jiang, Y.Y., Wang, H.J., Wang, J., Tashiro, S.I., Onodera, S. and Ikejima, T., 2009. The protective effect of silibinin against mitomycin C-induced intrinsic apoptosis in human melanoma A375-S2 cells. Journal of Pharmacological Sciences 111: 137-146.

Jung, H.J., Park, J.W., Lee, J.S., Lee, S.R., Jang, B.C., Suh, S.I., Suh, M.H. and Baek, W.K., 2009. Silibinin inhibits expression of HIF-1α through suppression of protein translation in prostate cancer cells. Biochemical and Biophysical Research Communications 390: 71-76.

Kanner, J., Gorelik, S., Roman, S. and Kohen, R., 2012. Protection by polyphenols of postprandial human plasma and low-density lipoprotein modification: the stomach as a bioreactor. Journal of Agricultural and Food Chemistry 60: 8790-8796.

Kanner, J. and Lapidot, T., 2001. The stomach as a bioreactor: dietary lipid peroxidation in the gastric fluid and the effects of plant-derived antioxidants. Free Radical Biology and Medicine 31: 1388-1395.

Kang, D. Y., Sp, N., Do Park, K., Lee, H. K., Song, K. D. and Yang, Y. M., 2019. Silibinin inhibits *in vitro* ketosis by regulating HMGCS2 and NF-kB: elucidation of signaling molecule relationship under ketotic conditions. *In vitro* cellular & developmental biology. Animal 55: 368-375.

Keshk, W.A., Zahran, S.M., Katary, M.A. and Ali, D.A.E., 2017. Modulatory effect of silymarin on nuclear factor-erythroid-2-related factor 2 regulated redox status, nuclear factor-κB mediated inflammation and apoptosis in experimental gastric ulcer. Chemico-Biological Interactions 273: 266-272.

Khatoon, A., Zargham Khan, M., Khan, A., Saleemi, M. K. and Javed, I., 2013. Amelioration of ochratoxin A-induced immunotoxic effects by silymarin and vitamin E in white leghorn cockerels. Journal of Immunotoxicology 10: 25-31.

Khamisabadi, H., 2020. Effects of silymarin on milk production, liver enzymes, oxidative status and HSP70 gene expression in postparturient Sanjabi ewes. Cellular and Molecular Biology 66: 76-81.

Khazim, K., Gorin, Y., Cavaglieri, R.C., Abboud, H.E. and Fanti, P., 2013. The antioxidant silybin prevents high glucose-induced oxidative stress and podocyte injury *in vitro* and *in vivo*. American Journal of Physiology-Renal Physiology 305: F691-F700.

Kheiripour, N., Karimi, J., Khodadadi, I., Tavilani, H., Goodarzi, M.T. and Hashemnia, M., 2018. Silymarin prevents lipid accumulation in the liver of rats with type 2 diabetes via sirtuin1 and SREBP-1c. Journal of Basic and Clinical Physiology and Pharmacology 29: 301-308.

Kim, B.R., Seo, H.S., Ku, J.M., Kim, G.J., Jeon, C.Y., Park, J.H., Jang, B.H., Park, S.J., Shin, Y.C. and Ko, S.G., 2013. Silibinin inhibits the production of pro-inflammatory cytokines through inhibition of NF-κB signaling pathway in HMC-1 human mast cells. Inflammation Research 62: 941-950.

Kim, M., Yang, S.G., Kim, J.M., Lee, J.W., Kim, Y.S. and Lee, J.I., 2012. Silymarin suppresses hepatic stellate cell activation in a dietary rat model of non-alcoholic steatohepatitis: analysis of isolated hepatic stellate cells. International Journal of Molecular Medicine 30: 473-479.

Kim, N.C., Graf, T.N., Sparacino, C.M., Wani, M.C. and Wall, M.E., 2003. Complete isolation and characterization of silybins and isosilybins from milk thistle (*Silybum marianum*). Organic & Biomolecular Chemistry 1: 1684-1689.

Kiruthiga, P.V., Karthikeyan, K., Archunan, G., Pandian, S.K. and Devi, K.P., 2015. Silymarin prevents benzo (a) pyrene-induced toxicity in Wistar rats by modulating xenobiotic-metabolizing enzymes. Toxicology and Industrial Health 31: 523-541.

Köck, K., Xie, Y., Hawke, R.L., Oberlies, N.H. and Brouwer, K.L., 2013. Interaction of silymarin flavonolignans with organic anion-transporting polypeptides. Drug Metabolism and Disposition 41: 958-965.

Köksal, E., Gülçin, İ., Beyza, S., Sarikaya, O. and Bursal, E., 2009. *In vitro* antioxidant activity of silymarin. Journal of Enzyme Inhibition and Medicinal Chemistry 24: 395-405.

Kren, V., Marhol, P., Purchartova, K., Gabrielova, E. and Modriansky, M., 2013. Biotransformation of silybin and its congeners. Current Drug Metabolism 14: 1009-1021.

Lee, B.H., Hsu, W.H., Hsu, Y.W. and Pan, T.M., 2013. Dimerumic acid attenuates receptor for advanced glycation end products signal to inhibit inflammation and diabetes mediated by Nrf2 activation and promotes methylglyoxal metabolism into d-lactic acid. Free Radical Biology and Medicine 60: 7-16.

Lee, S.J., Nam, M.J., Lee, D.E., Park, J.W., Kang, B.S., Lee, D.S., Lee, H.S. and Kwon, O.S., 2018. Silibinin ameliorates O-GlcN acylation and inflammation in a mouse model of nonalcoholic steatohepatitis. International Journal of Molecular Sciences 19: 2165.

Li, C.C., Hsiang, C.Y., Wu, S.L. and Ho, T.Y., 2012. Identification of novel mechanisms of silymarin on the carbon tetrachloride-induced liver fibrosis in mice by nuclear factor-κB bioluminescent imaging-guided transcriptomic analysis. Food and Chemical Toxicology 50: 1568-1575.

Li, L., Zeng, J., Gao, Y. and He, D., 2010. Targeting silibinin in the antiproliferative pathway. Expert Opinion on Investigational Drugs 19: 243-255.

Li, L.H., Wu, L.J., Tashiro, S.I., Onodera, S., Uchiumi, F. and Ikejima, T., 2007. Activation of the SIRT1 pathway and modulation of the cell cycle were involved in silymarin's protection against UV-induced A375-S2 cell apoptosis. Journal of Asian Natural Products Research 9: 245-252.

Li, Y., Ye, Z., Lai, W., Rao, J., Huang, W., Zhang, X., Yao, Z. and Lou, T., 2017. Activation of sirtuin 3 by silybin attenuates mitochondrial dysfunction in cisplatin-induced acute kidney injury. Frontiers in Pharmacology 8: 178.

Ligeret, H., Brault, A., Vallerand, D., Haddad, Y. and Haddad, P.S., 2008. Antioxidant and mitochondrial protective effects of silibinin in cold preservation-warm reperfusion liver injury. Journal of Ethnopharmacology 115: 507-514.

Lim, R., Morwood, C.J., Barker, G. and Lappas, M., 2014. Effect of silibinin in reducing inflammatory pathways in *in vitro* and *in vivo* models of infection-induced preterm birth. PLoS One 9: e92505.

Lin, C.H., Li, C.H., Liao, P.L., Tse, L.S., Huang, W.K., Cheng, H.W. and Cheng, Y.W., 2013. Silibinin inhibits VEGF secretion and age-related macular degeneration in a hypoxia-dependent manner through the PI-3 kinase/Akt/mTOR pathway. British Journal of Pharmacology 168: 920-931.

Lin, M.T. and Beal, M.F., 2006. Mitochondrial dysfunction and oxidative stress in neurodegenerative diseases. Nature 443: 787-795.

Liu, B.N., Han, B.X. and Liu, F., 2014. Neuroprotective effect of pAkt and HIF-1α on ischemia rats. Asian Pacific Journal of Tropical Medicine 7: 221-225.

Liu, Z., Sun, M., Wang, Y., Zhang, L., Zhao, H. and Zhao, M., 2018. Silymarin attenuated paraquat-induced cytotoxicity in macrophage by regulating Trx/TXNIP complex, inhibiting NLRP3 inflammasome activation and apoptosis. Toxicology in Vitro 46: 265-272.

Liu, K., Zhou, S., Liu, J., Wang, Y., Zhu, F. and Liu, M., 2019. Silibinin attenuates high-fat diet-induced renal fibrosis of diabetic nephropathy. Drug design, development and therapy, 13, 3117-3126.

Loguercio, C., Federico, A., Trappoliere, M., Tuccillo, C., De Sio, I., Di Leva, A., Niosi, M., D'Auria, M.V., Capasso, R. and Blanco, C.D.V., 2007. The effect of a silybin-vitamin e-phospholipid complex on nonalcoholic fatty liver disease: a pilot study. Digestive Diseases and Sciences 52: 2387-2395.

Lorenz, D., Lücker, P.W., Mennicke, W.H. and Wetzelsberger, N., 1984. Pharmacokinetic studies with silymarin in human serum and bile. Methods and Findings in Experimental and Clinical Pharmacology 6: 655-661.

Manna, S.K., Mukhopadhyay, A., Van, N.T. and Aggarwal, B.B., 1999. Silymarin suppresses TNF-induced activation of NF-κB, c-Jun N-terminal kinase, and apoptosis. The Journal of Immunology 163: 6800-6809.

Martínez, J.A., Etxeberría, U., Galar, A. and Milagro, F.I., 2013. Role of dietary polyphenols and inflammatory processes on disease progression mediated by the gut microbiota. Rejuvenation Research 16: 435-437.

Mateen, S., Raina, K. and Agarwal, R., 2013. Chemopreventive and anti-cancer efficacy of silibinin against growth and progression of lung cancer. Nutrition and Cancer 65: 3-11.

Matias, M. L., Gomes, V. J., Romao-Veiga, M., Ribeiro, V. R., Nunes, P. R., Romagnoli, G. G., Peracoli, J. C. and Peracoli, M., 2019. Silibinin Downregulates the NF-κB Pathway and NLRP1/NLRP3 Inflammasomes in Monocytes from Pregnant Women with Preeclampsia. Molecules 24: 1548.

Mazzio, E.A., Harris, N. and Soliman, K.F., 1998. Food constituents attenuate monoamine oxidase activity and peroxide levels in C6 astrocyte cells. Planta Medica 64: 603-606.

Mehrab-Mohseni, M., Sendi, H., Steuerwald, N., Ghosh, S., Schrum, L.W. and Bonkovsky, H.L., 2011. Legalon-SIL downregulates HCV core and NS5A in human hepatocytes expressing full-length HCV. World Journal of Gastroenterology 17: 1694-1700.

Moco, S., Martin, F.P.J. and Rezzi, S., 2012. Metabolomics view on gut microbiome modulation by polyphenol-rich foods. Journal of Proteome Research 11: 4781-4790.

Morovat, M., Chamani, M., Zarei, A. and Sadeghi, A. A., 2016. Dietary but not in ovo feeding of *Silybum marianum* extract resulted in an improvement in performance, immunity and carcass characteristics and decreased the adverse effects of high temperatures in broilers. British Poultry Science 57: 105-113.

Muthumani, M. and Prabu, S.M., 2014. Silibinin potentially attenuates arsenic-induced oxidative stress mediated cardiotoxicity and dyslipidemia in rats. Cardiovascular Toxicology 14: 83-97.

Muthumani, M. and Prabu, S.M., 2012. Silibinin potentially protects arsenic-induced oxidative hepatic dysfunction in rats. Toxicology Mechanisms and Methods 22: 277-288.

Naito, Y., Takagi, T. and Higashimura, Y., 2014. Heme oxygenase-1 and anti-inflammatory M2 macrophages. Archives of Biochemistry and Biophysics 564: 83-88.

Oskoueian, E., Abdullah, N., Idrus, Z., Ebrahimi, M., Goh, Y.M., Shakeri, M. and Oskoueian, A., 2014. Palm kernel cake extract exerts hepatoprotective activity in heat-induced oxidative stress in chicken hepatocytes. BMC Complementary and Alternative Medicine 14: 368.

Ou, Q., Weng, Y., Wang, S., Zhao, Y., Zhang, F., Zhou, J. and Wu, X., 2018. Silybin alleviates hepatic steatosis and fibrosis in NASH mice by inhibiting oxidative stress and involvement with the Nf-κB pathway. Digestive Diseases and Sciences 63: 3398-3408.

Parkar, S.G., Trower, T.M. and Stevenson, D.E., 2013. Fecal microbial metabolism of polyphenols and its effects on human gut microbiota. Anaerobe 23: 12-19.

Pauff, J.M. and Hille, R., 2009. Inhibition studies of bovine xanthine oxidase by luteolin, silibinin, quercetin, and curcumin. Journal of Natural Products 72:725-731.

Pferschy-Wenzig, E.M., Atanasov, A.G., Malainer, C., Noha, S.M., Kunert, O., Schuster, D., Heiss, E.H., Oberlies, N.H., Wagner, H., Bauer, R. and Dirsch, V.M., 2014. Identification of isosilybin a from milk thistle seeds as an agonist of peroxisome proliferator-activated receptor gamma. Journal of Natural Products 77: 842-847.

Podder, B., Kim, Y.S., Zerin, T. and Song, H.Y., 2012. Antioxidant effect of silymarin on paraquat-induced human lung adenocarcinoma A549 cell line. Food and Chemical Toxicology 50: 3206-3214.

Post-White, J., Ladas, E.J. and Kelly, K.M., 2007. Advances in the use of milk thistle (*Silybum marianum*). Integrative Cancer Therapies 6: 104-109.

Pozuelo, M.J., Agis-Torres, A., Hervert-Hernández, D., Elvira López-Oliva, M., Muñoz-Martínez, E., Rotger, R. and Goni, I., 2012. Grape antioxidant dietary fiber stimulates *Lactobacillus* growth in rat cecum. Journal of Food Science 77: H59-H62.

Prabu, S.M. and Muthumani, M., 2012. Silibinin ameliorates arsenic induced nephrotoxicity by abrogation of oxidative stress, inflammation and apoptosis in rats. Molecular Biology Reports 39: 11201-11216.

Prakash, P., Singh, V., Jain, M., Rana, M., Khanna, V., Barthwal, M.K. and Dikshit, M., 2014. Silymarin ameliorates fructose induced insulin resistance syndrome by reducing de novo hepatic lipogenesis in the rat. European Journal of Pharmacology 727: 15-28.

Rajnochová Svobodová, A., Gabrielová, E., Michaelides, L., Kosina, P., Ryšavá, A., Ulrichová, J., Zálešák, B. and Vostálová, J., 2018. UVA-photoprotective potential of silymarin and silybin. Archives of Dermatological Research 310: 413-424.

Ramasamy, K. and Agarwal, R., 2008. Multitargeted therapy of cancer by silymarin. Cancer Letters 269: 352-362.

Rickling, B., Hans, B., Kramarczyk, R., Krumbiegel, G. and Weyhenmeyer, R., 1995. Two high-performance liquid chromatographic assays for the determination of free and total silibinin diastereomers in plasma using column switching with electrochemical detection and reversed-phase chromatography with ultraviolet detection. Journal of Chromatography B: Biomedical Sciences and Applications 670: 267-277.

Rodríguez-Ramiro, I., Martín, M.Á., Ramos, S., Bravo, L. and Goya, L., 2011a. Comparative effects of dietary flavanols on antioxidant defences and their response to oxidant-induced stress on Caco2 cells. European Journal of Nutrition 50: 313-322.

Rodríguez-Ramiro, I., Ramos, S., Bravo, L., Goya, L. and Martín, M.Á., 2011. Procyanidin B2 and a cocoa polyphenolic extract inhibit acrylamide-induced apoptosis in human Caco-2 cells by preventing oxidative stress and activation of JNK pathway. The Journal of Nutritional Biochemistry 22: 1186-1194.

Rolo, A.P., Oliveira, P.J., Moreno, A.J. and Palmeira, C.M., 2003. Protection against post-ischemic mitochondrial injury in rat liver by silymarin or TUDC. Hepatology Research 26: 217-224.

Saeed, M., Babazadeh, D., Arif, M., Arain, M.A., Bhutto, Z.A., Shar, A.H., Kakar, M.U., Manzoor, R. and Chao, S., 2017. Silymarin: a potent hepatoprotective agent in poultry industry. World's Poultry Science Journal 73: 483-492.

Salamone, F., Galvano, F., Marino, A., Paternostro, C., Tibullo, D., Bucchieri, F., Mangiameli, A., Parola, M., Bugianesi, E. and Volti, G.L., 2012. Silibinin improves hepatic and myocardial injury in mice with nonalcoholic steatohepatitis. Digestive and Liver Disease 44: 334-342.

Salomone, F., Barbagallo, I., Godos, J., Lembo, V., Currenti, W., Cinà, D., Avola, R., D'Orazio, N., Morisco, F., Galvano, F. and Li Volti, G., 2017. Silibinin restores NAD^+ levels and induces the SIRT1/AMPK pathway in non-alcoholic fatty liver. Nutrients 9: 1086.

Sangeetha, N., Viswanathan, P., Balasubramanian, T. and Nalini, N., 2012. Colon cancer chemopreventive efficacy of silibinin through perturbation of xenobiotic metabolizing enzymes in experimental rats. European Journal of Pharmacology 674: 430-438.

Sarubbo, F., Ramis, M. R., Kienzer, C., Aparicio, S., Esteban, S., Miralles, A. and Moranta, D., 2018. Chronic silymarin, quercetin and naringenin treatments increase monoamines synthesis and hippocampal Sirt1 levels improving cognition in aged rats. Journal of Neuroimmune Pharmacology 13: 24-38.

Schiavone, A., Righi, F., Quarantelli, A., Bruni, R., Serventi, P. and Fusari, A., 2007. Use of *Silybum marianum* fruit extract in broiler chicken nutrition: influence on performance and meat quality. Journal of Animal Physiology and Animal Nutrition 91: 256-262.

Sekine, S. and Ichijo, H., 2015. Mitochondrial proteolysis: its emerging roles in stress responses. Biochimica et Biophysica Acta (BBA)-General Subjects 1850: 274-280.

Serviddio, G., Bellanti, F., Giudetti, A.M., Gnoni, G.V., Petrella, A., Tamborra, R., Romano, A.D., Rollo, T., Vendemiale, G. and Altomare, E., 2010. A silybin-phospholipid complex prevents mitochondrial dysfunction in a rodent model of nonalcoholic steatohepatitis. Journal of Pharmacology and Experimental Therapeutics 332: 922-932.

Shin, J.H., Lee, C.W., Oh, S.J., Yun, J., Lee, K., Park, S.K., Kim, H.M., Han, S.B., Kim, Y., Kim, H.C. and Kang, J.S., 2013. Protective effect of silymarin against ethanol-induced gastritis in rats: role of sulfhydryls, nitric oxide and gastric sensory afferents. Food and Chemical Toxicology 55: 353-357.

Shyamal, S., Latha, P.G., Suja, S.R., Shine, V.J., Anuja, G.I., Sini, S., Pradeep, S., Shikha, P. and Rajasekharan, S., 2010. Hepatoprotective effect of three herbal extracts on aflatoxin B1-intoxicated rat liver. Singapore Medical Journal 51: 326-331.

Sozmen, M., Devrim, A.K., Tunca, R., Bayezit, M., Dag, S. and Essiz, D., 2014. Protective effects of silymarin on fumonisin B1-induced hepatotoxicity in mice. Journal of Veterinary Science 15: 51-60.

Stevenson, D.E. and Hurst, R.D., 2007. Polyphenolic phytochemicals – just antioxidants or much more? Cellular and Molecular Life Sciences 64: 2900-2916.

Stiuso, P., Scognamiglio, I., Murolo, M., Ferranti, P., De Simone, C., Rizzo, M.R., Tuccillo, C., Caraglia, M., Loguercio, C. and Federico, A., 2014. Serum oxidative stress markers and lipidomic profile to detect NASH patients responsive to an antioxidant treatment: a pilot study. Oxidative Medicine and Cellular Longevity, 2014: 169216.

Surai, K.P., Surai, P.F., Speake, B.K. and Sparks, N.H., 2003. Antioxidant-prooxidant balance in the intestine: Food for thought. 1. Prooxidants. Nutritional Genomics & Functional Foods 1: 51-70.

Surai, K.P., Surai, P.F., Speake, B.K. and Sparks, N.H., 2004. Antioxidant-prooxidant balance in the intestine: food for thought 2. antioxidants. Current Topics in Nutraceutical Research 2: 27-46.

Surai, P. F., 2014. Polyphenol compounds in the chicken/animal diet: from the past to the future. Journal of Animal Physiology and Animal Nutrition 98: 19-31.

Surai, P.F., 2015. Silymarin as a natural antioxidant: an overview of the current evidence and perspectives. Antioxidants 4: 204-247.

Surai, P.F. and Dvorska, J.E., 2005. Effects of mycotoxins on antioxidant status and immunity. In: Diaz, D.E. (ed.) The mycotoxin blue book. Nottingham University Press, Nottingham, UK, pp. 93-137.

Surai, P.F. and Fisinin, V.I., 2015. Antioxidant-Prooxidant balance in the intestine: applications in chick placement and pig weaning. Journal of Veterinary Science & Medicine 3: 16.

Surai, P.F., Kochish, I.I., Fisinin, V.I. and Kidd, M.T., 2019. Antioxidant defence systems and oxidative stress in poultry biology: an update. Antioxidants 8: 235.

Surh, Y.J., 2008. NF-kappa B and Nrf2 as potential chemopreventive targets of some anti-inflammatory and antioxidative phytonutrients with anti-inflammatory and antioxidative activities. Asia Pacific Journal of Clinical Nutrition, 17: 269-272.

Testino, G., Leone, S., Ansaldi, F. and Borro, P., 2013. Silymarin and S-adenosyl-L-methionine (SAMe): two promising pharmacological agents in case of chronic alcoholic hepathopathy. A review and a point of view. Minerva Gastroenterologica e Dietologica 59: 341-356.

Tsai, M.J., Liao, J.F., Lin, D.Y., Huang, M.C., Liou, D.Y., Yang, H.C., Lee, H.J., Chen, Y.T., Chi, C.W., Huang, W.C. and Cheng, H., 2010. Silymarin protects spinal cord and cortical cells against oxidative stress and lipopolysaccharide stimulation. Neurochemistry International 57: 867-875.

Tyagi, A., Agarwal, C., Dwyer-Nield, L.D., Singh, R.P., Malkinson, A.M. and Agarwal, R., 2012. Silibinin modulates TNF-α and IFN-γ mediated signaling to regulate COX2 and iNOS expression in tumorigenic mouse lung epithelial LM2 cells. Molecular Carcinogenesis 51: 832-842.

Varga, Z., Seres, I., Nagy, E., Ujhelyi, L., Balla, G., Balla, J. and Antus, S., 2006. Structure prerequisite for antioxidant activity of silybin in different biochemical systems *in vitro*. Phytomedicine 13: 85-93.

Varga, Z., Ujhelyi, L., Kiss, A., Balla, J., Czompa, A. and Antus, S., 2004. Effect of silybin on phorbol myristate actetate-induced protein kinase C translocation, NADPH oxidase activity and apoptosis in human neutrophils. Phytomedicine 11: 206-212.

Velmurugan, K., Alam, J., McCord, J.M. and Pugazhenthi, S., 2009. Synergistic induction of heme oxygenase-1 by the components of the antioxidant supplement Protandim. Free Radical Biology and Medicine 46: 430-440.

Venditti, C.C. and Smith, G.N., 2014. Involvement of the heme oxygenase system in the development of preeclampsia and as a possible therapeutic target. Women's Health 10: 623-643.

Vargas-Mendoza, N., Morales-González, Á., Morales-Martínez, M., Soriano-Ursúa, M. A., Delgado-Olivares, L., Sandoval-Gallegos, E. M., Madrigal-Bujaidar, E., Álvarez-González, I., Madrigal-Santillán, E. and Morales-Gonzalez, J. A., 2020. Flavolignans from silymarin as Nrf2 bioactivators and their therapeutic applications. Biomedicines 8: E122.

Wang, C., Wang, Z., Zhang, X., Zhang, X., Dong, L., Xing, Y., Li, Y., Liu, Z., Chen, L., Qiao, H. and Wang, L., 2012. Protection by silibinin against experimental ischemic stroke: up-regulated pAkt, pmTOR, HIF-1α and Bcl-2, down-regulated Bax, NF-κB expression. Neuroscience Letters 529: 45-50.

Wang, H.J., Tashiro, S.I., Onodera, S. and Ikejima, T., 2008. Inhibition of insulin-like growth factor 1 receptor signaling enhanced silibinin-induced activation of death receptor and mitochondrial apoptotic pathways in human breast cancer MCF-7 cells. Journal of Pharmacological Sciences 107: 260-269.

Wang, Q., Liu, M., Liu, W.W., Hao, W.B., Tashiro, S.I., Onodera, S. and Ikejima, T., 2012. *In vivo* recovery effect of silibinin treatment on streptozotocin-induced diabetic mice is associated with the modulations of Sirt-1 expression and autophagy in pancreatic β-cell. Journal of Asian Natural Products Research 14: 413-423.

Wang, Q., Zou, L., Liu, W., Hao, W., Tashiro, S.I., Onodera, S. and Ikejima, T., 2011. Inhibiting NF-κB activation and ROS production are involved in the mechanism of silibinin's protection against D-galactose-induced senescence. Pharmacology Biochemistry and Behavior 98: 140-149.

Weyhenmeyer, R., Mascher, H. and Birkmayer, J., 1992. Study on dose-linearity of the pharmacokinetics of silibinin diastereomers using a new stereospecific assay. International Journal of Clinical Pharmacology, Therapy, and Toxicology 30: 134-138.

Wu, J.W., Lin, L.C., Hung, S.C., Chi, C.W. and Tsai, T.H., 2007. Analysis of silibinin in rat plasma and bile for hepatobiliary excretion and oral bioavailability application. Journal of Pharmaceutical and Biomedical Analysis 45: 635-641.

Wu, S., Yue, Y., Tian, H., Li, Z., Li, X., He, W. and Ding, H., 2013. Carthamus red from *Carthamus tinctorius* L. exerts antioxidant and hepatoprotective effect against CCl4-induced liver damage in rats via the Nrf2 pathway. Journal of Ethnopharmacology 148: 570-578.

Wu, Y.T., Wu, S.B. and Wei, Y.H., 2014a. Metabolic reprogramming of human cells in response to oxidative stress: implications in the pathophysiology and therapy of mitochondrial diseases. Current Pharmaceutical Design 20: 5510-5526.

Wu, Y.T., Wu, S.B. and Wei, Y.H., 2014. Roles of sirtuins in the regulation of antioxidant defense and bioenergetic function of mitochondria under oxidative stress. Free Radical Research 48: 1070-1084.

Yang, G., Wang, H., Kang, Y. and Zhu, M.J., 2014. Grape seed extract improves epithelial structure and suppresses inflammation in ileum of IL-10-deficient mice. Food & Function 5: 2558-2563.

Yang, C., Song, G. and Lim, W., 2020. Effects of mycotoxin-contaminated feed on farm animals. Journal of Hazardous Materials 389: 122087.

Yin, F., Liu, J., Ji, X., Wang, Y., Zidichouski, J. and Zhang, J., 2011. Silibinin: a novel inhibitor of Aβ aggregation. Neurochemistry International 58: 399-403.

Zaker-Esteghamati, H., Seidavi, A.R. and Bouyeh, M., in press. A review on the effect of *Silybum marianum* and its derivatives on broilers under healthy and aflatoxicosis conditions: part 1: performance, carcass and meat characteristics, and intestinal microflora. World's Poultry Science Journal. https://doi.org/10.1080/00439339.2020.1740068

Zhang, S., Yang, Y., Liang, Z., Duan, W., Yang, J., Yan, J., Wang, N., Feng, W., Ding, M., Nie, Y. and Jin, Z., 2013. Silybin-mediated inhibition of Notch signaling exerts antitumor activity in human hepatocellular carcinoma cells. PloS One 8: e83699.

Zhang, B., Xu, D., She, L., Wang, Z., Yang, N., Sun, R., Zhang, Y., Yan, C., Wei, Q., Aa, J., Liu, B., Wang, G. and Xie, Y., 2018. Silybin inhibits NLRP3 inflammasome assembly through the NAD^+/SIRT2 pathway in mice with nonalcoholic fatty liver disease. FASEB Journal 32: 757-767.

Zhao, H., Brandt, G.E., Galam, L., Matts, R.L. and Blagg, B.S., 2011. Identification and initial SAR of silybin: an Hsp90 inhibitor. Bioorganic & Medicinal Chemistry Letters 21: 2659-2664.

Zhao, J. and Agarwal, R., 1999. Tissue distribution of silibinin, the major active constituent of silymarin, in mice and its association with enhancement of phase II enzymes: implications in cancer chemoprevention. Carcinogenesis 20: 2101-2108.

Zhou, B., Wu, L.J., Li, L.H., Tashiro, S.I., Onodera, S., Uchiumi, F. and Ikejima, T., 2006. Silibinin protects against isoproterenol-induced rat cardiac myocyte injury through mitochondrial pathway after up-regulation of SIRT1. Journal of Pharmacological Sciences 102: 387-395.

Zhou, B., Wu, L.J., Tashiro, S.I., Onodera, S., Uchiumi, F. and Ikejima, T., 2006a. Silibinin protects rat cardiac myocyte from isoproterenol-induced DNA damage independent on regulation of cell cycle. Biological and Pharmaceutical Bulletin 29: 1900-1905.

Zhu, H.J., Brinda, B.J., Chavin, K.D., Bernstein, H.J., Patrick, K.S. and Markowitz, J.S., 2013. An assessment of pharmacokinetics and antioxidant activity of free silymarin flavonolignans in healthy volunteers: a dose escalation study. Drug Metabolism and Disposition 41: 1679-1685.

Zhu, S.Y., Dong, Y., Tu, J., Zhou, Y., Zhou, X.H. and Xu, B., 2014. *Silybum marianum* oil attenuates oxidative stress and ameliorates mitochondrial dysfunction in mice treated with D-galactose. Pharmacognosy Magazine 10: S92-S99.

Chapter 12
Natural antioxidants as vitagene modulators

Do not boast until you see the enemy dead

12.1 Introduction

Since vitagenes play an important role in stress management and adaptation, search for effective modulators of the vitagene network is an essential part of the strategy to deal with commercially relevant stresses in poultry/animal production. In previous chapters vitagene-modulating properties of carnitine, taurine and silymarin have been described.

This chapter will be devoted to vitagene-modulating properties of various nutrients including vitamins A, D, E, C, betaine, selenium, polyphenols singly or in various nutritional combinations.

12.2 Vitamin A

The importance of vitamin A in avian physiology is related to retinoic acid, a hormonal form of vitamin A, controlling gene transcription and retinaldehyde participating in vision and redox signalling (Hoyos *et al.*, 2012). Vitamin A is well studied in avian species and biochemical and physiological consequences of its deficiency (Sommer and West, 1996) and excess (Veltmann and Jensen, 1986) are well characterised. Vitamin A is an important constituent of commercial premixes for poultry feed supplying 5,000-15,000 IU/kg diet. At the commercial level of vitamin A supplementation basic requirements in this vitamin are met, however, vitamin A interactions with other nutrients, especially in stress conditions need further investigation.

Generally speaking, vitamin A is not a classical antioxidant in biological systems (Sies *et al.*, 1992) but it can show either antioxidant or prooxidant activities, depending on its concentration and experimental conditions. In fact, vitamin A can affect antioxidant defences indirectly, in particular, by regulating vitagenes. However, little is known on the effects of various doses of vitamin A on vitagene network and research in this area was mainly done *in vitro* with cell cultures or *in vivo* with laboratory animals. For example, in the liver of thyroidectomised rats retinoic acid was shown to increase the stimulation of HO-1 activity by thyroid hormone (Smith *et al.*, 1991). Similarly protective roles of vitamin A in liver damage induced by bile duct ligation in rats was shown: hepatic SOD activity was restored and the expression of HO-1 was increased (Wang *et al.*, 2014). Vitamin A deficiency in rats was shown to up-regulate hepatic and spleen heme oxygenase-1 gene expression while reducing the liver HO-1

DOI 10.3920/978-90-8686-906-0_12, © Wageningen Academic Publishers 2020

specific activity compared with the control (da Cunha *et al.*, 2014). Vitamin A oral supplementation (450 retinol equivalents/day) was found to induce oxidative stress and suppresses IL-10 and HSP70 in skeletal muscle of trained rats (Petiz *et al.*, 2017) as well as inhibited the increase in HSP70 expression in the liver acquired with exercise alone (Petiz *et al.*, 2017a).

A moderate vitamin A supplementation (4,000 IU/kg diet) in rats caused increased hepatic SOD in comparison to non-supplemented animals (Cha *et al.*, 2016). Furthermore, vitamin A deficiency was shown to decrease CuZn-SOD activities in rat aorta (Gatica *et al.*, 2005). More recently Pipaliya and Vaghaaiya (2012) demonstrated that in stressed rats (isoproterenol injection) vitamin A can restore heart SOD activity to normal level. There is also considerable evidence that vitamin A can cause oxidative stress and decrease SOD activity. Indeed, vitamin A supplementation (1000-9,000 IU/kg/day) to rat for 28 days induced oxidative stress and decreased SOD activity in lungs. The oxidative stress was associated with increased lipid peroxidation, protein carbonylation, and oxidation of protein thiol groups (Pasquali *et al.*, 2009). In ovariectomised rats vitamin A supplementation (1,500 IU/kg/day) reduced hippocampal SOD activity (Behr *et al.*, 2012). Also, vitamin A was shown to affect antioxidant defences, including SOD activity *in vitro* in Sertoli cells (Zanotto-Filho *et al.*, 2008).

It has been shown that retinol and/or its metabolites can transcriptionally upregulate thioredoxin gene. Indeed, vitamin A was shown to stimulate (8-10-fold) the expression of the thioredoxin gene in monkey tracheobronchial (TBE) epithelial cells (An and Wu, 1992). It was shown that retinoic acid (1 μM) can increase TR1 expression in cultured cells (the bovine mammary cell line MAC-T; Bruzelius *et al.*, 2010). Regulatory role of vitamin A in thioredoxin gene expression in human airway epithelial cells was also described by Chang *et al.* (2002). Recently it has been observed that thioredoxin interacting protein (TXNIP), which inhibits antioxidant activity of thioredoxin (Trx), was downregulated by all-trans retinoic acid (Shimizu *et al.*, 2018). In a model system based on bovine mammary epithelial cells, vitamin A promotes antioxidant functions by regulating the synthesis of GPx and TrxR and by reducing concentrations of IL-1 and NO *in vitro* by modulating Nrf2 and NF-κB signalling pathways (Shi *et al.*, 2018).

Together, these results suggest that vitamin A could affect the vitagene network including HSPs, SOD and thioredoxin system. However, more research work is needed to understand vitamin A interactions with other nutrients in avian species as a way of vitagene modulation.

12.3 Vitamin D

Vitamin D plays a special role in avian nutrition being involved in regulation of Ca metabolism during bone and eggshell formation. Vitamin D is also known to be associated with oxidative stress reduction, immunomodulation and anti-inflammatory process in many different tissues (Bikle, 2014). Since commercial poultry production is based on technologies preventing direct contact of birds with sun light, optimal vitamin D dietary supplementation is key for high productive and reproductive performance of breeders, layers and growing chickens. For the last 20 years doses of vitamin D supplementation of commercial poultry feed increased substantially and are in the range of 3,000-5,000 IU/kg diet. Vitamin D is not considered to be a direct antioxidant (Tagliaferri *et al.*, 2019) but its participation in maintaining redox balance and cell signalling is of great importance in avian nutrition. It has been suggested that health-promoting properties of vitamin D are facilitated through the ability of this vitamin to increase the expression of both Nrf2 and the anti-ageing protein Klotho, which are also major regulators of Ca^{2+} and redox signalling (Berridge, 2015).

In experiments with laboratory animals it was proven that vitamin D can affect the vitagene network taking part in regulation of cellular oxidative stress responses and exerting cytoprotective effects. In fact, in rat kidney vitamin D was shown to be considered as an important inducer of HSP70 (Kim *et al.*, 2005) and HSP70 can interact with the vitamin D receptor (VDR; Lutz *et al.*, 2001). Furthermore, cytoprotective effects of the analogue of 1,25-dihydroxyergocalciferol (paricalcitol) in spontaneously hypertensive rats were also linked to the HSP70 induction (Garcia *et al.*, 2014). Another HSP, namely HO-1, was shown to be down-regulated in the liver of vitamin D deficient mouse (Zhu *et al.*, 2015). In obese rats vitamin D deficiency increased the expression of the hepatic mRNA levels of HO-1 (Roth *et al.*, 2012). In rat model of reperfusion pre-treatment of animals with vitamin D3 caused a significant increase in HO-1 expression in leucocytes (Shih *et al.*, 2011). Furthermore, a substantial (1.9-fold) increase in expression of HO-1 was observed in the intestine of vitamin D-deficient rats due to a single dose of 1,25-D3 (Kutuzova and DeLuca, 2007). Vitamin D was shown to protect endothelial progenitor cells from AngII-induced vascular injury via the activation of the PPAR-γ/HO-1 signalling pathway (Xu *et al.*, 2019). Vitamin D was also indicated to improves intestinal barrier function in cirrhosis rats by upregulating HO-1 expression (Wang *et al.*, 2019). Furthermore, vitamin D deficiency was shown to aggravate hepatic oxidative stress and inflammation during chronic alcohol-induced liver injury by further increasing alcohol-induced upregulation of hepatic inducible nitric oxide synthase (*inos*), two NADPH oxidase subunits *p47phox* and *gp91phox*, and HO-1 (Hu *et al.*, 2020). In the same study vitamin D deficiency was shown to attenuate alcohol-induced upregulation of hepatic antioxidant enzyme genes, including superoxide dismutase (*sod*) 1 and GPx (*gshpx*). In contrast, vitamin D supplementation (500 or 1000 mg/kg) lead to normalisation of serum HO-1 concentration with a significant decrease compared with the acetaminophen-treated rats (Abo El-Magd *et al.*, 2020). In addition, vitamin D deficiency was reported to significantly elevated alcohol-induced upregulation of hepatic proinflammatory cytokines and chemokines. Taken together, these results suggest that vitamin D deficiency aggravates hepatic oxidative stress and inflammation during chronic alcohol-induced liver injury.

The modulating effects of vitamin D3 on the vitagene network can also be observed at transcriptional regulation of thioredoxin-interacting protein (TXNIP, Chen and DeLuca, 1994) and there is a significant positive correlation between vitamin D and

TXNIP (Omar *et al.*, 2018). This protein is involved in regulation of the cellular redox state and stress-induced cellular apoptosis (Pang *et al.*, 2009).

Vitamin D is suggested to be involved in SOD expression and activity regulation. For example, oral administration of vitamin D with calcium to diabetic rats was shown to significantly increase the activities of SOD, GPx and catalase compared with the untreated group (Alatawi *et al.*, 2018). Chronic immobilisation stress (CIS) in rats was shown to be associated with a significant decrease in tissue/cardiac reserves of GSH, SOD activity as well as increase in MDA, changes which were ameliorated by treatment with vitamin D (Hussien *et al.*, 2019). Regular consumption of vitamin D was reported to increase the SOD, GPx, CAT activity and reduce the MDA plasma levels in HD patients (Malekmakan *et al.*, 2020). Administration of vitamin D to diabetic rats resulted in a decrease of aortic MDA levels an increase of aortic SOD activity (Said, in press). Vitamin D is believed to counteract the activity of NADPH oxidase (a source of ROS) and improve antioxidant defence network by enhancing the activity of antioxidative enzymes such as SOD, while inhibiting NF-κB signalling (Kim *et al.*, 2020). Vitamin D3 pretreatment was shown to decreased IL-6 and MDA levels, whereas the activities of SOD, CAT, and total thiol content in the hippocampus tissue of LKPS-treated rats were significantly increased (Mokhtari-Zaer *et al.*,2020)

Furthermore, in endothelial cells vitamin D was shown to up-regulate the sirtuin-1 (SIRT-1) and prevented the SIRT-1 down-regulation induced by H_2O_2 (Polidoro *et al.*, 2013). It seems likely that, 1,25-D3 signalling via vitamin D receptor (VDR) and the regulation of the SIRT1 protein function are closely related (An *et al.*, 2010; Sabir *et al.*, 2017) and await further investigation. In fact, acetylation of VDR is believed to comprise a negative feedback loop that attenuates 1,25D-VDR signalling and SIRT1-catalysed deacetylation of VDR can amplify VDR signalling and 1,25D actions (Sabir *et al.*, 2017). Vitamin D deficiency in patients with coronary heart disease is reported to increase the inflammatory response by upregulating the expression of p65 protein and reducing SIRT1 protein expression in heart tissue (Liu *et al.*, 2018). When C2C12 myotubes were treated with high glucose (HG, 25 mM) and 1,25$(OH)_2D_3$ (0-50 nM), it was shown that 1,25$(OH)_2D_3$ supplementation upregulated insulin-independent SIRT1 expression (Manna *et al.*, 2018). Muscle SIRT1 mRNA expression was shown to be significantly inhibited by 68.9% in vitamin D-insufficient rats and muscle SIRT1 activity was significantly decreased (by 13.4%) in obese rats with vitamin D insufficiency, compared to vitamin D adequate obese animals (Chang *et al.*, 2019). Overall, the aforementioned studies reveal that vitamin D can modulate the vitagene network, including HSP, SOD, thioredoxin system and sirtuins.

12.4 Vitamin E

Vitamin E is known to be a major biological antioxidant and its role in avian biology has been extensively reviewed (Surai, 1999, 2002; Surai *et al.*, 2016, 2019). Commercial vitamin E supplementation of poultry diets varies from 20 up to 100 mg/kg diet and in stress conditions the level of vitamin E supplementation can be increased. For example in heat stressed layers high vitamin E doses (up to 250-500 mg/kg diet) showed protective effects illustrated by decreasing negative consequences of the stress on egg production and quality (Abd El-Hack *et al.*, 2017; Bollengier-Lee *et al.*, 1998, 1999). Furthermore, vitamin E participation in cell signalling regulation and modulation of gene transcription (Azzi *et al.*, 2018) and other non-antioxidant properties of α-tocopherol and its metabolites (Galli *et al.*, 2017) deserve more attention. It seems likely that in addition to direct antioxidant action and non-antioxidant signalling mechanisms vitamin E can affect adaptation to stress by modulation of the vitagene network.

Effect of vitamin E on HSPs depends on stress model used and generally speaking in many cases vitamin E decreases stress manifestation as exhibited by prevention of HSP70 upregulation. For example, in isolation-stressed quail, vitamin E or ascorbic acid were shown to prevent an increase in HSP70 expression in the brain and heart (Soleimani *et al.*, 2012). Indeed, increased vitamin E and ascorbic acid dietary supplementation can modulate expression of HSP70 providing additional cytoprotective effect in maintaining optimal protein refolding and reducing negative consequences of various stresses, including heat stress. By contrast, in most cases effects of vitamin E on HO-1 expression is shown to be different from the aforementioned effects on HSP70. In fact, in an *in vitro* experiment, α-tocopherol induced the expression of hypoxia-inducible factor-1 and its target genes, including vascular endothelial growth factor and HO-1 (Zhang *et al.*, 2004). Recently it has been shown that in renal proximal tubule cells vitamin E activated the HO-1 promoter through the extracellular signal-regulated kinase and protein kinase A (Reed *et al.*, 2015). According to Bellezza *et al.* (2012) in prostate cancer cells vitamin E increased nuclear translocation and ARE binding activity of Nrf2, resulting in up-regulation of downstream genes including HO-1, while decreasing NF-κB nuclear translocation. However, interactions between vitamin E and HO-1 depend on the level of stress imposed and experimental model and vitamin E form used. For example, in adult rats palm tocotrienols were shown to prevent the increases in hepatic HO-1 induced by phenylhydrazine (Kamisah *et al.*, 2014).

Vitamin E can also affect SOD activity. In early publications no marked changes of SOD activities were observed in rat liver due to vitamin E deficiency (Masugi and Nakamura, 1976). However, later it was shown that vitamin E deficiency caused decreased mitochondrial SOD activity in liver and brain (De and Darad, 1988). In line with previous observations, vitamin E supplementation (125 or 250 mg/kg diet) was associated with increased SOD activity in muscles of heat-stressed chickens (Ghazi Harsini *et al.*, 2012), while there was no effect of vitamin E on SOD activity in chickens kept in thermoneutral conditions. An elevated vitamin E level (60 vs 30 mg/kg) significantly increased SOD activity in serum of laying hens (Zduńczyk *et al.*, 2013). Additionally, vitamin E (250 mg/kg diet) restored erythrocyte SOD activity to the physiological level after it was reduced by dietary Cu (300 mg/kg diet) in chickens (Cinar *et al.*, 2014). It seems likely that the effect of vitamin E on SOD activity depends on many factors and in stress conditions such an effect is stronger than in normal physiological conditions. For example, vitamin E supplementation at 100 mg/kg of unstressed chickens did not affect hepatic SOD activity (Guo *et al.*, 2003). Indeed, vitamin E increased SOD activity in rat kidney with cisplatin-induced oxidative stress (Dillioglugil *et al.*, 2005) and in the serum and corpus cavernosum of the hypertensive rats (Ushiyama *et al.*, 2008).

Vitamin E can also affect sirtuin expression and activities. For example, long-term vitamin E-deficiency significantly decreased the expression of SIRT mRNA in mouse brain compared to short-term deficiency (Fukui *et al.*, 2014). According to Wu *et al.* (2006) a diet high in saturated fat and sucrose decreased SIRT2alpha levels in the rat hippocampus and cerebral cortex, while vitamin E supplementation normalised SIRT2 levels. A mild fluid percussion injury (FPI) in rats increased protein oxidation and reduced levels of SOD and SIRT2 in the brain, while vitamin E dietary supplementation counteracted all the observed detrimental effects of FPI (Wu *et al.*, 2010). The interactions between vitamin E and SIRTs could be related to the antioxidant function of vitamin E, but it can also interact with SIRT1 via its non-antioxidant properties (Brigelius-Flohé, 2009). Recently it has been shown that in rats stressed by D-galactosamine and LPS, dietary vitamin E (400 mg/kg, once daily for 2 weeks) significantly increased the expression of Trx1 and TrxR1 in liver (Hashem *et al.*, 2016). From the aforementioned data the general conclusion would appear to be that vitamin E is an important modulator of the vitagene network, including HSPs, SOD, SIRTs and thioredoxin system.

12.5 Ascorbic acid

Ascorbic acid (AA) is the most important water-soluble antioxidant provided with feed and synthesised within the animal/chicken body (Chakraborthy *et al.*, 2014). Because of adequate AA synthesis in avian kidney, vitamin C is not included into commercial vitamin or vitamin-mineral premixes for poultry. However, AA supplementation with diet or drinking water could help deal with various stresses, including heat stress (HS). For example, AA (150 mg/kg diet) was shown to significantly decrease negative effects of HS on chicken growth, feed intake, protein digestibility and FCR (Attia *et al.*, 2009).

Numerous studies have demonstrated that AA can also affect antioxidant defences and stress adaptation via vitagene modulation. For example, HS-induced HSP70 increased expression in chicken heart was two-fold lower in AA-fed birds in comparison with the control chickens. Furthermore, plasma corticosterone (a stress marker) and heart HSP70 levels were positively correlated, while plasma AA and heart HSP70 were negatively correlated (Mahmoud *et al.*, 2004) and the AA-supplemented chickens showed the decreased HSP70 responses to heat stress (42 °C) in heart and liver (Mahmoud *et al.*, 2003). Additionally, increased dietary AA or vitamin E supplementation in heat-stressed quail was associated with linear decrease in HSP70 expression in ovary and brain. However, under thermo-neutral conditions HSP70 was not affected by the tested vitamins (Sahin *et al.*, 2009). By contrast, effects of AA on HO-1 expression and activity is different from that shown for HSP70. Indeed, the HO-1 mRNA and protein level in various rat tissues, including kidney, liver, and lung were significantly induced by AA treatment (100 mg/kg b.w.) under normal and HS conditions (Zhao *et al.*, 2014).

The effect of AA on SOD in chickens depends on conditions, including stress intensity. In fact, it was shown that in normal physiological conditions vitamin E + selenium supplementation increased serum SOD activity in chickens while AA (100 mg/kg diet) did not affect it (Tras *et al.*, 2000). However, AA is shown to be an effective regulator of SOD *in vitro* and *in vivo* in various stress conditions. For example, AA could decrease mitochondrial superoxide production to physiological levels, decrease oxidative stress and prevent growth defects observed in SOD-depleted cells (Tamari *et al.*, 2013). There are no data available on the effects of AA on thioredoxin system or sirtuins in the body. Thus, AA can modulate the vitagene network by affecting HSPs and SOD activities and effects of AA on the thioredoxin system and sirtuins in avian species await investigation.

12.6 Selenium

Selenium is an important trace element and its biological activity is mediated via 25 selenoproteins in human and birds. Most of feed ingredients worldwide contain comparatively low Se levels to meet Se requirement of poultry and farm animals. Selenium is included into commercial diets for poultry via mineral or vitamin-mineral premixes in various forms at 0.1-0.3 mg/kg diet (Surai, 2018; Surai and Fisinin, 2014; Surai and Kochish, 2019). Increased Se-protein expression in response to various stresses is a common adaptive phenomenon observed in various model systems *in vitro* and *in vivo* (Surai, 2006, 2018).

Effects of Se on vitagene products (HSP, TR and SOD) was shown in various stress conditions, caused by Se deficiency or excess, HS or heavy metal toxicity. For example, it has been observed that mRNA levels and protein levels of HSP70 were significantly increased in spleen and thymus (Khoso *et al.*, 2016) and erythrocytes (Zhao *et al.*, 2016) of chickens in response to stress imposed by the Se-deficiency. However, Se supplementation was indicated to decrease the mRNA levels of HSP70 in the spleen of heat-stressed chickens (Xu *et al.*, 2014a). Of note is the fact that effects of Se on HSPs depend on the Se form used. For example, selenium yeast significantly attenuated HSP70 expression in enteropathogenic Escherichia coli-challenged or in heat-stressed male broiler chickens (Mahmoud and Edens, 2005). Indeed, Se-yeast-fed chickens were less affected by mild HS (indicated by decreased induction of HSP70) than sodium selenite-fed chickens (Mahmoud and Edens, 2003). Of interest, HSP70 concentrations were significantly upregulated in the heat-stress turkey embryo developing in eggs from non-Se-supplemented birds, while in the Se-supplemented groups the HSP70 concentrations in embryos were not affected by HS (Rivera *et al.*, 2005).

Nutritional stress caused by low-Se diet in chickens was associated with a significant decrease in the activities of SOD and GPx, and an increased MDA content in thymus, spleen, bursa of Fabricius and serum (Zhang *et al*, 2012). Interestingly, not only

Se deficiency (0.03 mg Se per kg of diet) but also Se excess (3 mg/kg) in chickens significantly lowered SOD and CAT activities in liver and serum (Xu *et al.*, 2014). It seems likely that SOD in adult birds is also dependent on Se status. For example, laying hens fed a Se-supplemented diet (0.3 mg/kg diet) were characterised by higher SOD and GPx activity and lower MDA content in plasma compared with those birds fed a control non-supplemented diet containing Se at 0.03 mg/kg (Jing *et al.*, 2015). It has been shown that Se efficacy is dependent on its form in the dietary supplements with organic selenium being more effective than sodium selenite (Surai, 2006, 2018; Surai and Fisinin, 2014). Indeed, GPx and SOD activities in serum were significantly higher in Se-Yeast-fed chickens in comparison to sodium selenite-fed birds (Chen *et al.*, 2014). In line with these findings, dietary Se-Yeast significantly increased the activities of CAT and SOD but reduced the MDA content in chicken breast muscle at 42 days of age (Ahmad *et al.*, 2012). Dietary SeMet (0.15 mg/kg) was observed to increase hepatic SOD activity in broilers in comparison to birds fed on sodium selenite-supplemented diet (Wang *et al.*, 2011). Similarly, SOD activities in chicken breast muscle were significantly elevated due to dietary Se-Met supplementation (0.075-0.225 mg/kg diet; Jiang *et al.*, 2009).

Selenium can also affect thioredoxin system in birds. Of note is the fact that TrxR is a Se-dependent stress-inducible enzyme and in Se deficiency its expression in animal tissues could be increased or decreased depending on experimental conditions (Surai, 2006, 2018). For example, Se deficiency in chickens was associated with reduced TrxR activity in adipose tissue (Liang *et al.*, 2014), decreased TrxR2 expression in thyroids (Lin *et al.*, 2014) and reduced activity of TrxR1 (by 50%), TrxR2 (by 83%) and TrxR3 (by 36%) in pancreas (Zhao *et al.*, 2014a). Furthermore, a low Se diet (0.028 mg/kg) or high Se diet (3 mg/kg) significantly decreased the TrxR expression in chicken kidney. In particular, the Se-deficient diet downregulated the mRNA expression of TrxR3 (Xu *et al.*, 2016). Selenium dietary supplementation (0.4 mg/kg diet) was shown to increase TrxR activity in duodenal mucosa, liver and kidney in chickens (Placha *et al.*, 2014). Compared with sodium selenite, Se-Yeast or SeMet significantly increased the activity and expression of Trx1 in the liver and kidney of broiler breeders and their offspring (Yuan *et al.*, 2012). Adding sodium selenite (100 nM) to LNCaP cells caused an increase of TrxR1 expression and a decrease of HSP70 expression (Roveri *et al.*, 2008). Therefore, the available evidence indicate that Se is involved in vitagene network regulation in birds: it affects HSPs, SOD and TrxR expression and concentrations in different tissues.

12.7 Betaine

Betaine is the trimethyl derivative of glycine and it is synthesised in liver and kidney of birds as well as obtained from diet. There is no established requirement for betaine, and it is not included into mainstream commercial premixes for poultry. It has been shown that betaine is an important nutrient regulating cell osmotic balance (Kidd *et al.*, 1997). There is a growing body of evidence showing beneficial effects of betaine on animal performance and carcass quality (Eklund *et al.*, 2005). Indeed, dietary betaine supplementation during physiologically challenging conditions (high metabolic demand of rapid growth, heat stress, disease, etc.) is an important nutritional tool to deal with stresses (Ratriyanto and Mosenthin, 2018; Saeed *et al.*, 2017).

Evidence from studies with laboratory animals is accumulating to show vitagene-modulating effects of betaine. For example, betaine (250 mg/kg/day for 2 weeks) was shown to prevent lipid peroxidation (TBARS) reversing the Cd-induced depletion of antioxidant defences including total antioxidant potential, GSH concentration and GPx, SOD and catalase activities in rat kidney (Hagar *et al.*, 2014). Furthermore, betaine supplementation (20 g/kg diet) was observed to increase SOD and GPx activity in mouse liver (Wang *et al.*, 2013a) and rat cerebellum (180 mg/mouse/day; Alirezaei *et al.*, 2011). Pretreatment of acetaminophen-exposed mice with 250, 500, and 1000 mg/kg betaine was shown to ameliorate hepatotoxicity (as evidenced by serum ALT and AST activity and histopathology finding). Betaine also reversed mitochondrial complex II activity, GSH level, SOD, GPx and CAT, glutamate cysteine ligase catalytic protein expression, and mitochondrial membrane potential simultaneously with suppressed mitochondrial lipid peroxidation and ROS formation (Khodayar *et al.*, 2018).

Betaine was reported to modulate oxidative stress and antioxidant defence mechanisms as evidenced by increasing hepatic GSH content and SOD, GPx, CAT and paraoxonase activities in a mouse model of methionine-choline deficiency-induced fatty liver disease (Veskovic *et al.*, 2019). The effects of dietary betaine supplementation (250, 500 or 1000 mg/kg) on growth performance, carcass characteristics, muscle amino acid contents, meat quality and antioxidant capacity in Cherry Valley ducks were evaluated. It was shown that with increasing betaine levels, the activities of SOD and CAT in breast muscles were increased (Chen *et al.*, 2019). Improved antioxidant defences, including SOD, GPx activities and GSH content in breast muscle were reported in broilers exposed to HS due to betaine (1 g/kg) dietary supplementation (Wen *et al.*, 2019). Betaine was found to protect against heat exposure-induced oxidative stress and apoptosis in bovine mammary epithelial cells via regulation of ROS production by upregulating SOD and CAT activities (Li *et al.*, 2019). Protective effects of betaine on diabetes induced disruption of the male mice blood-testis barrier was associated with improved antioxidant defences including SOD, CAT activities and GSH concentrations (Jiang *et al.*, 2019). Betaine supplementation in cisplatin-treated rats was associated with increased SOD, GPx and CAT activities and GSH concentration and decreased lipid peroxidation in the liver (Hagar *et al.*, 2019). Dietary betaine was shown to increase SOD, GPx activities and decreased MDA in blood of formalin receiving mice (Hassanpour *et al.*, 2020). Betaine was also found to ameliorate a lead-induced decrease in expression and activity of SOD, GPx and CAT in the testes of mouse (Elsheikh *et al.*, 2020). Considered together, the aforementioned studies demonstrate that betaine could affect the main elements of the vitagene network and improve cell adaptability to various stresses. Vitagene regulating properties of betaine in birds await investigations.

When regulatory effects of betaine on muscle differentiation and mitochondrial biogenesis in C2C12 Cells was studied, it was found that betaine significantly increased the expression of SIRT-1 mRNA and protein in the myotubes (Ma *et al.*, 2019). Upregulation of Nrf2-related cytoprotective genes expression by acetaminophen-induced acute hepatotoxicity in mice was shown and the protective role of betaine was found. Since betaine supplementation was responsible for decreasing Nrf2 and HO-1 expression, it seems likely that its protective effect was Nrf-2 and HO-1 independent and probably occurred via modifying cysteine supply as a precursor of GSH in the transsulphuration pathway in the liver (Khodayar *et al.*, 2020).

In lactating Holstein cows subjected to heat stress betaine supplementation (114 mg/kg of BW per d) was shown to improve thermotolerance of animals as measured by reduced respiration rate, reduced water intake, and elevated HSP70 expression in peripheral leukocytes. Furthermore, in the *in vitro* portion of this work, primary bovine mammary epithelial cells also showed an increased expression of HSP70 when treated with 25 mM betaine *in vitro* (Hall *et al.*, 2016). Similarly, in bovine mammary and uterine cells exposed to thermal shock *in vitro*, HSP70 expression was increased in betaine-treated cells at 8 h under heat stress and viability was improved at 8 h under thermoneutral conditions and at 16 h in heat stress (Xiao *et al.*, 2017). Therefore, the aforementioned data clearly showed that betaine can regulate the vitagene network, including SOD, HSP70, SIRT and various elements of glutathione system.

12.8 Polyphenols/flavonoids

In recent years polyphenols/phytochemicals have become a centre of research attentions reflecting their well-established health-promoting properties. Among many possible molecular mechanisms of their action antioxidant properties of polyphenols/flavonoids attracted a remarkable level of interest and became a fashionable topic of research (Hostetler *et al.*, 2017). However, recent reviews devoted to the antioxidant properties of polyphenols/flavonoids revealed that direct antioxidant properties of these compounds are unlikely to be the major mechanism of their action (Croft *et al.*, 2016; Surai, 2014). In fact, the main problem for these compounds as nutritional supplements is related to their poor bioavailability and very low concentrations (if any) of active polyphenols in target tissues. It seems likely that antioxidant effects of polyphenols/ phytochemicals are mediated via activation of various transcription factors, signalling pathways and vitagenes. In particular, silymarin (SM), a plant extract containing polyphenols (e.g. silibinin), can be used as a model compound to show effects of polyphenols on the vitagene network. In fact, molecular mechanisms of SM have been recently extensively reviewed (Surai, 2015). From data presented in the review it could be concluded that SM can potentiate the vitagene network responsible for decreasing detrimental consequences of stresses. Firstly, in various model systems SM was shown to improve antioxidant defences by upregulating HO-1. Secondly, SM consumption was observed to be associated with decrease HSP70 expression in stressed cells reflecting improved AO defences and decrease stress by other pathways/means (e.g. Nrf2-related increased AO synthesis). Finally, SM-related activation or prevention of inhibition of sirtuins in stress conditions could be an essential adaptive mechanism responsible for maintaining the redox-regulated homeostasis in the cell and in the whole body. Earlier, similar conclusions were drawn as a result of analysis of polyphenol effects on the vitagene network by Calabrese *et al.* (2012). It could well be that hormetic mechanisms of preconditioning by polyphenols which can impose modest stress and prepare cell/body to deal with subsequent stresses (Calabrese *et al.*, 2017) are important fundamentals in the vitagene network modulation. Further research is needed to fully understand the molecular mechanisms of the vitagene network modulation by polyphenols, including SM, in avian species.

The aforementioned data clearly indicate that vitamins A, E, C, selenium, carnitine, betaine and silymarin can affect antioxidant defences via modulating the vitagene network, including activity and/or expression of HSPs, SOD, thioredoxin system and sirtuins. In general, in stress conditions the protective effects of the aforementioned nutrients are most pronounced, while in normal physiological conditions additional supplementation of the studied antioxidant compounds could be without any significant effect on the oxidative stress markers. In particular, HSP70 expression is usually decreased due to the antioxidant dietary supplementation reflecting decreasing requirement for synthesis of this protective protein because of additional protection provided by other antioxidant mechanisms. Indeed, HSP70 is an important marker of the oxidative stress. On the other hand, improvement of HO-1, SOD, thioredoxin and TrxR expression and/or activity due to dietary supplementation of the aforementioned nutrients in stress conditions can be considered as an important mechanism of improved adaptation to stress. Unfortunately, the data on sirtuins in chicken tissues are very limited and this subject awaits further investigation.

12.9 Synergistic combinations of antioxidants

Based on the concept of the integrated antioxidant systems in the body it can be expected that dietary supplementation of synergistic mixtures of various antioxidant nutrients could have higher protective effects in comparison with individual antioxidants. Indeed, a mixture of antioxidants would be most effective if they have different/complimentary mechanisms of antioxidant action. For example, vitamin E and selenium in combination can give some extra positive effects due to their synergetic action at different levels of free radical production and lipid peroxidation (Surai, 2003). Indeed, in the case of selenium and vitamin E 'one plus one equals more than two'. Furthermore, dietary supplementation of a combination of carnitine and vitamin E is proven to be effective in ameliorating ochratoxin A compromised haematological and serum biochemical parameters in White Leghorn cockerels (Abidin *et al.*, 2013). A synergistic protective effect of carnitine and vitamin E was also shown in laboratory and farm animals. For example, in a recent experiment conducted in China, piglets (weaned at 21 d of age) from 14 to 35 d of age, had *ad libitum* access to a basal diet. Furthermore, piglets in the experimental group were fed the same basal diet supplemented with 6.75 g/kg of a complex antioxidant blend containing vitamins E and C, lipoic acid, tea polyphenols and microbial antioxidants.

It was shown that weaning stress was associated with an oxidative stress at day 28 characterised by a dramatic (almost 2-fold) decrease in piglet blood SOD activity and a significantly increased (by 49%) H_2O_2 level in serum. Interestingly, the antioxidant blend was able to significantly prevent the SOD activity drop, normalised H_2O_2 in serum and help maintaining intestinal barrier function (Zhu *et al.*, 2012). However, additional supply of vitamins in low stress conditions could downregulate activities of some vitagene-regulated antioxidants. For example, when nursery piglets were treated with a mixture of vitamins E and D via drinking water (120 IU vitamin E and 7,200 IU vitamin D per 3.79 l (1 gallon) drinking water), SOD activity in erythrocytes was significantly decreased (362.86 vs 324.17 units/ml; Copeland *et al.*, 2017). From the data presented above it is possible to conclude that the development of protective antioxidant mixtures could be considered as an important step in stress prevention and treatment in commercial poultry and animal production. However, it is a challenging task to develop an effective synergistic antioxidant mixture.

12.10 Conclusions

The vitagene network in cell and the whole organism is responsible for adaptation to various stresses in avian species and in human and animals, in general. Indeed, the products of the vitagenes (HSPs, HO-1, SOD, elements of Trx- and GSH-systems, sirtuins, etc.) are proven to work cooperatively in detecting and managing various forms of stress and cell injuries with establishment of adaptive homeostasis. Furthermore, our analysis of recent publications clearly indicates that the vitagene network can be modulated by nutritional means (Figure 12.1).

In particular, vitamins A, D, E, C, as well as selenium, carnitine, betaine and silymarin are shown to affect vitagenes and improve adaptive ability of birds to various stresses. In fact, protective effects of the aforementioned nutrients on vitagenes are most

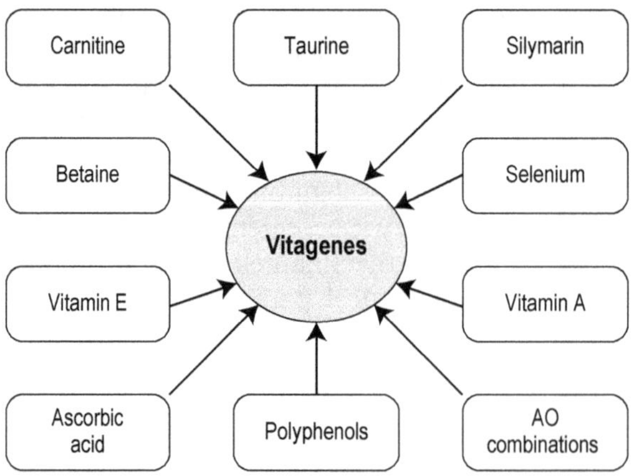

Figure 12.1. Nutritional modulation of vitagenes.

pronounced in stress conditions. Indeed, these results are the first step to go from the development of the vitagene concept in avian/poultry nutrition to designing commercial products which could help fight stresses in the commercial conditions of poultry production. However, further work is required to understand the molecular mechanisms of the interactions between vitagenes and various signalling systems and transcription factors in the cell to build an adequate adaptive response and minimise detrimental consequences of stresses in avian species. Interactions between gut microbiota, redox balance and the vitagene network in poultry/farm animals awaits further investigation.

References

Abd El-Hack, M.E., Mahrose, K., Arif, M., Chaudhry, M.T., Saadeldin, I.M., Saeed, M., Soomro, R.N., Abbasi, I.H. and Rehman, Z.U., 2017. Alleviating the environmental heat burden on laying hens by feeding on diets enriched with certain antioxidants (vitamin E and selenium) individually or combined. Environmental Science and Pollution Research 24: 10708-10717.

Abidin, Z., Khan, M.Z., Khatoon, A., Saleemi, M.K., Khan, A. and Javed, I., 2013. Ameliorative effects of L-carnitine and vitamin E (α-tocopherol) on haematological and serum biochemical parameters in White Leghorn cockerels given ochratoxin A contaminated feed. British Poultry Science 54: 471-477.

Abo El-Magd, N.F. and Eraky, S.M., 2020. The molecular mechanism underlining the preventive effect of vitamin D against hepatic and renal acute toxicity through the NrF2/ BACH1/ HO-1 pathway. Life Sciences 244: 117331.

Ahmad, H., Tian, J., Wang, J., Khan, M.A., Wang, Y., Zhang, L. and Wang, T., 2012. Effects of dietary sodium selenite and selenium yeast on antioxidant enzyme activities and oxidative stability of chicken breast meat. Journal of Agricultural and Food Chemistry 60: 7111-7120.

Alatawi, F.S., Faridi, U.A. and Alatawi, M.S., 2018. Effect of treatment with vitamin D plus calcium on oxidative stress in streptozotocin-induced diabetic rats. Saudi Pharmaceutical Journal 26: 1208-1213.

Alirezaei, M., Jelodar, G., Niknam, P., Ghayemi, Z. and Nazifi, S., 2011. Betaine prevents ethanol-induced oxidative stress and reduces total homocysteine in the rat cerebellum. Journal of Physiology and Biochemistry 67: 605-612.

An, B.S., Tavera-Mendoza, L.E., Dimitrov, V., Wang, X., Calderon, M.R., Wang, H.J. and White, J.H., 2010. Stimulation of Sirt1-regulated FoxO protein function by the ligand-bound vitamin D receptor. Molecular and Cellular Biology 30: 4890-4900.

An, G. and Wu, R., 1992. Thioredoxin gene expression is transcriptionally up-regulated by retinol in monkey conducting airway epithelial cells. Biochemical and Biophysical Research Communications 183: 170-175.

Attia, Y.A., Hassan, R.A. and Qota, E.M.A., 2009. Recovery from adverse effects of heat stress on slow-growing chicks in the tropics 1: effect of ascorbic acid and different levels of betaine. Tropical Animal Health and Production 41: 807-818.

Azzi, A., 2018. Many tocopherols, one vitamin E. Molecular Aspects of Medicine 61: 92-103.

Behr, G.A., Schnorr, C.E., Simões-Pires, A., Da Motta, L.L., Frey, B.N. and Moreira, J.C.F., 2012. Increased cerebral oxidative damage and decreased antioxidant defenses in ovariectomized and sham-operated rats supplemented with vitamin A. Cell Biology and Toxicology 28: 317-330.

Bellezza, I., Tucci, A., Galli, F., Grottelli, S., Mierla, A.L., Pilolli, F. and Minelli, A., 2012. Inhibition of NF-κB nuclear translocation via HO-1 activation underlies α-tocopheryl succinate toxicity. The Journal of Nutritional Biochemistry 23: 1583-1591.

Berridge, M.J., 2015. Vitamin D cell signalling in health and disease. Biochemical and Biophysical Research Communications 460: 53-71.

Bikle, D.D., 2014. Vitamin D metabolism, mechanism of action, and clinical applications. Chemistry & Biology 21: 319-329.

Bollengier-Lee, S., 1999. Optimal dietary concentration of vitamin E for alleviating the effect of heat stress on egg production in laying hens. British Poultry Science 40: 102-107.

Bollengier-Lee, S., Mitchell, M.A., Utomo, D.B., Williams, P.E.V. and Whitehead, C.C., 1998. Influence of high dietary vitamin E supplementation on egg production and plasma characteristics in hens subjected to heat stress. British Poultry Science 39: 106-112.

Brigelius-Flohé, R., 2009. Vitamin E: the shrew waiting to be tamed. Free Radical Biology and Medicine 46: 543-554.

Bruzelius, K., Sundler, R., Pagmantidis, V. and Åkesson, B., 2010. Regulation of selenoprotein mRNA expression by hormones and retinoic acid in bovine mammary cells. Journal of Trace Elements in Medicine and Biology 24: 251-256.

Calabrese, V., Cornelius, C., Dinkova-Kostova, A.T., Iavicoli, I., Di Paola, R., Koverech, A., Cuzzocrea, S., Rizzarelli, E. and Calabrese, E.J., 2012. Cellular stress responses, hormetic phytochemicals and vitagenes in aging and longevity. Biochimica et Biophysica Acta (BBA)-Molecular Basis of Disease 1822: 753-783.

Calabrese, V., Giordano, J., Crupi, R., Di Paola, R., Ruggieri, M., Bianchini, R., Ontario, M.L., Cuzzocrea, S. and Calabrese, E.J., 2017. Hormesis, cellular stress response and neuroinflammation in schizophrenia: early onset versus late onset state. Journal of Neuroscience Research 95: 1182-1193.

Cha, J.H., Yu, Q.M. and Seo, J.S., 2016. Vitamin A supplementation modifies the antioxidant system in rats. Nutrition Research and Practice 10: 26-32.

Chakraborthy, A., Ramani, P., Sherlin, H.J., Premkumar, P. and Natesan, A., 2014. Antioxidant and pro-oxidant activity of Vitamin C in oral environment. Indian Journal of Dental Research 25: 499-504.

Chang, E. and Kim, Y., 2019. Vitamin D ameliorates fat accumulation with AMPK/SIRT1 activity in C2C12 skeletal muscle cells. Nutrients 11: 2806.

Chang, W.H., Reddy, S.P.M., Di, Y.P.P., Yoneda, K., Harper, R. and Wu, R., 2002. Regulation of thioredoxin gene expression by vitamin A in human airway epithelial cells. American Journal of Respiratory Cell and Molecular Biology 26: 627-635.

Chen, G., Wu, J. and Li, C., 2014. Effect of different selenium sources on production performance and biochemical parameters of broilers. Journal of Animal Physiology and Animal Nutrition 98: 747-754.

Chen, K.S. and DeLuca, H.F., 1994. Isolation and characterization of a novel cDNA from HL-60 cells treated with 1, 25-dihydroxyvitamin D-3. Biochimica et Biophysica Acta (BBA)-Gene Structure and Expression 1219: 26-32.

Chen, R., Wen, C., Cheng, Y., Chen, Y., Zhuang, S. and Zhou, Y., 2019. Effects of dietary supplementation with betaine on muscle growth, muscle amino acid contents and meat quality in Cherry Valley ducks. Journal of Animal Physiology and Animal Nutrition 103: 1050-1059.

Cinar, M., Yildirim, E., Yigit, A.A., Yalcinkaya, I., Duru, O., Kisa, U. and Atmaca, N., 2014. Effects of dietary supplementation with vitamin C and vitamin E and their combination on growth performance, some biochemical parameters, and oxidative stress induced by copper toxicity in broilers. Biological Trace Element Research 158: 186-196.

Copeland, K.R., Scales, D.B., Hill, G.M., Link, J.E., Turner, K.C., Rozeboom, C.J. and Stuart, R.L., 2017. Effects of vitamins E and D on performance and antioxidant enzymes in nursery pigs. Journal of Animal Science 95: 192.

Croft, K.D., 2016. Dietary polyphenols: antioxidants or not? Archives of Biochemistry and Biophysics 595: 120-124.

da Cunha, M.S., Siqueira, E.M., Trindade, L.S. and Arruda, S.F., 2014. Vitamin A deficiency modulates iron metabolism via ineffective erythropoiesis. The Journal of Nutritional Biochemistry 25: 1035-1044.

De, A.K. and Darad, R., 1988. Physiological antioxidants and antioxidative enzymes in vitamin E-deficient rats. Toxicology Letters 44: 47-54.

Dillioglugil, M.O., Kir, H.M., Gulkac, M.D., Kanli, A.Ö., Ozdogan, H.K., Acar, O. and Dillioglugil, O., 2005. Protective effects of increasing vitamin E and a doses on cisplatin-induced oxidative damage to kidney tissue in rats. Urologia Internationalis 75: 340-344.

Eklund, M., Bauer, E., Wamatu, J. and Mosenthin, R., 2005. Potential nutritional and physiological functions of betaine in livestock. Nutrition Research Reviews 18: 31-48.

Elsheikh, N., Omer, N.A., Yi-Ru, W., Mei-Qian, K., Ilyas, A., Abdurahim, Y. and Wang, G.L., 2020. Protective effect of betaine against lead-induced testicular toxicity in male mice. Andrologia 52: e13600. https://doi.org/10.1111/and.13600

Fukui, K., Masuda, A., Hosono, A., Suwabe, R., Yamashita, K., Shinkai, T. and Urano, S., 2014. Changes in microtubule-related proteins and autophagy in long-term vitamin E-deficient mice. Free Radical Research 48: 649-658.

Galli, F., Azzi, A., Birringer, M., Cook-Mills, J.M., Eggersdorfer, M., Frank, J., Cruciani, G., Lorkowski, S. and Özer, N.K., 2017. Vitamin E: emerging aspects and new directions. Free Radical Biology and Medicine 102: 16-36.

García, I.M., Altamirano, L., Mazzei, L., Fornés, M., Cuello-Carrión, F.D., Ferder, L. and Manucha, W., 2014. Vitamin D receptor-modulated Hsp70/AT 1 expression may protect the kidneys of SHRs at the structural and functional levels. Cell Stress and Chaperones 19: 479-491.

Gatica, L., Alvarez, S., Gomez, N., Zago, M.P., Oteiza, P., Oliveros, L. and Gimenez, M.S., 2005. Vitamin A deficiency induces prooxidant environment and inflammation in rat aorta. Free Radical Research 39: 621-628.

Guo, Y., Zhang, G., Yuan, J. and Nie, W., 2003. Effects of source and level of magnesium and vitamin E on prevention of hepatic peroxidation and oxidative deterioration of broiler meat. Animal Feed Science and Technology 107: 143-150.

Hagar, H. and Al Malki, W., 2014. Betaine supplementation protects against renal injury induced by cadmium intoxication in rats: role of oxidative stress and caspase-3. Environmental Toxicology and Pharmacology 37: 803-811.

Hagar, H., Husain, S., Fadda, L.M., Attia, N.M., Attia, M. and Ali, H.M., 2019. Inhibition of NF-κB and the oxidative stress-dependent caspase-3 apoptotic pathway by betaine supplementation attenuates hepatic injury mediated by cisplatin in rats. Pharmacological Reports 71. 1025-1033.

Hall, L.W., Dunshea, F.R., Allen, J.D., Rungruang, S., Collier, J.L., Long, N.M. and Collier, R.J., 2016. Evaluation of dietary betaine in lactating Holstein cows subjected to heat stress. Journal of Dairy Science 99: 9745-9753.

Harsini, S.G., Habibiyan, M., Moeini, M.M. and Abdolmohammadi, A.R., 2012. Effects of dietary selenium, vitamin E, and their combination on growth, serum metabolites, and antioxidant defense system in skeletal muscle of broilers under heat stress. Biological Trace Element Research 148: 322-330.

Hashem, R.M., Hassanin, K.M., Rashed, L.A., Mahmoud, M.O. and Hassan, M.G., 2016. Effect of silibinin and vitamin E on the ASK1-p38 MAPK pathway in D-galactosamine/lipopolysaccharide induced hepatotoxicity. Experimental Biology and Medicine 241: 1250-1257.

Hassanpour, S., Rezaei, H. and Razavi, S.M., 2020. Anti-nociceptive and antioxidant activity of betaine on formalin and writhing tests induced pain in mice. Behavioural Brain Research 390: 112699.

Hostetler, G.L., Ralston, R.A. and Schwartz, S.J., 2017. Flavones: food sources, bioavailability, metabolism, and bioactivity. Advances in Nutrition 8: 423-435.

Hoyos, B., Acin-Perez, R., Fischman, D.A., Manfredi, G. and Hammerling, U., 2012. Hiding in plain sight: uncovering a new function of vitamin A in redox signaling. Biochimica et Biophysica Acta (BBA)-Molecular and Cell Biology of Lipids 1821: 241-247.

Hu, C.Q., Bo, Q.L., Chu, L.L., Hu, Y.D., Fu, L., Wang, G.X., Lu, Y., Liu, X.J., Wang, H. and Xu, D.X., 2020. Vitamin D deficiency aggravates hepatic oxidative stress and inflammation during chronic alcohol-induced liver injury in mice. Oxidative Medicine and Cellular Longevity 2020: 5715893.

Hussien, N.I., El-Wakeel, H.S., Souror, S.M. and Ahmed, I.A., 2019. Alleviation of cardiac mitochondrial dysfunction and oxidative stress underlies the protective effect of vitamin D in chronic stress-induced cardiac dysfunction in rats. General Physiology and Biophysics 38: 51-61.

Jiang, Y.P., Yang, J.M., Ye, R.J., Liu, N., Zhang, W.J., Ma, L., Zheng, P., Niu, J.G., Liu, P. and Yu, J.Q., 2019. Protective effects of betaine on diabetic induced disruption of the male mice blood-testis barrier by regulating oxidative stress-mediated p38 MAPK pathways. Biomedicine & Pharmacotherapy 120: 109474.

Jing, C.L., Dong, X.F., Wang, Z.M., Liu, S. and Tong, J.M., 2015. Comparative study of DL-selenomethionine vs sodium selenite and seleno-yeast on antioxidant activity and selenium status in laying hens. Poultry Science 94: 965-975.

Kamisah, Y., Lim, J.J., Lim, C.L. and Asmadi, A.Y., 2014. Inhibitory effects of palm tocotrienol-rich fraction supplementation on bilirubin-metabolizing enzymes in hyperbilirubinemic adult rats. PloS One 9: e89248.

Khodayar, M.J., Kalantari, H., Khorsandi, L., Rashno, M. and Zeidooni, L., 2018. Betaine protects mice against acetaminophen hepatotoxicity possibly via mitochondrial complex II and glutathione availability. Biomedicine & Pharmacotherapy 103: 1436-1445.

Khodayar, M.J., Kalantari, H., Khorsandi, L., Rashno, M. and Zeidooni, L., 2020. Upregulation of Nrf2-related cytoprotective genes expression by acetaminophen-induced acute hepatotoxicity in mice and the protective role of betaine. Human & Experimental Toxicology 39: 948-959.

Khoso, P.A., Liu, C., Khoso, M.H. and Li, S., 2016. Selenium deficiency activates heat shock protein expression in chicken spleen and thymus. Biological Trace Element Research 173: 492-500.

Kidd, M.T., Ferket, P.R. and Garlich, J.D., 1997. Nutritional and osmoregulatory functions of betaine. World's Poultry Science Journal 53: 125-139.

Kim, D.H., Meza, C.A., Clarke, H., Kim, J.S. and Hickner, R.C., 2020. Vitamin D and endothelial function. Nutrients 12: 575.

Kim, Y.O., Li, C., Sun, B.K., Kim, J.S., Lim, S.W., Choi, B.S., Kim, Y.S., Kim, J., Bang, B.K. and Yang, C.W., 2005. Preconditioning with 1, 25-dihydroxyvitamin D3 protects against subsequent ischemia-reperfusion injury in the rat kidney. Nephron Experimental Nephrology 100: e85-e94.

Kutuzova, G.D. and DeLuca, H.F., 2007. 1, 25-Dihydroxyvitamin D3 regulates genes responsible for detoxification in intestine. Toxicology and Applied Pharmacology 218: 37-44.

Li, C., Wang, Y., Li, L., Han, Z., Mao, S. and Wang, G., 2019. Betaine protects against heat exposure-induced oxidative stress and apoptosis in bovine mammary epithelial cells via regulation of ROS production. Cell Stress & Chaperones 24: 453-460.

Liang, Y., Lin, S.L., Wang, C.W., Yao, H.D., Zhang, Z.W. and Xu, S.W., 2014. Effect of selenium on selenoprotein expression in the adipose tissue of chickens. Biological Trace Element Research 160: 41-48.

Lin, S.L., Wang, C.W., Tan, S.R., Liang, Y., Yao, H.D., Zhang, Z.W. and Xu, S.W., 2014. Selenium deficiency inhibits the conversion of thyroidal thyroxine (T 4) to triiodothyronine (T 3) in chicken thyroids. Biological Trace Element Research 161: 263-271.

Liu, Y., Peng, W., Li, Y., Wang, B., Yu, J. and Xu, Z., 2018. Vitamin D deficiency harms patients with coronary heart disease by enhancing inflammation. Medical Science Monitor 24: 9376-9384.

Lutz, W., Kohno, K. and Kumar, R., 2001. The role of heat shock protein 70 in vitamin D receptor function. Biochemical and Biophysical Research Communications 282: 1211-1219.

Ma, J., Meng, X., Kang, S.Y., Zhang, J., Jung, H.W. and Park, Y.K., 2019. Regulatory effects of the fruit extract of *Lycium chinense* and its active compound, betaine, on muscle differentiation and mitochondrial biogenesis in C2C12 cells. Biomedicine & Pharmacotherapy 118: 109297.

Mahmoud, K.Z. and Edens, F.W., 2003. Influence of selenium sources on age-related and mild heat stress-related changes of blood and liver glutathione redox cycle in broiler chickens (*Gallus domesticus*). Comparative Biochemistry and Physiology Part B: Biochemistry and Molecular Biology 136: 921-934.

Mahmoud, K.Z. and Edens, F.W., 2005. Influence of organic selenium on hsp70 response of heat-stressed and enteropathogenic *Escherichia coli*-challenged broiler chickens (*Gallus gallus*). Comparative Biochemistry and Physiology Part C: Toxicology & Pharmacology 141: 69-75.

Mahmoud, K.Z., Edens, F.W., Eisen, E.J. and Havenstein, G.B., 2003. Effect of ascorbic acid and acute heat exposure on heat shock protein 70 expression by young white Leghorn chickens. Comparative Biochemistry and Physiology Part C: Toxicology & Pharmacology 136: 329-335.

Mahmoud, K.Z., Edens, F.W., Eisen, E.J. and Havenstein, G.B., 2004. Ascorbic acid decreases heat shock protein 70 and plasma corticosterone response in broilers (*Gallus gallus domesticus*) subjected to cyclic heat stress. Comparative Biochemistry and Physiology Part B: Biochemistry and Molecular Biology 137: 35-42.

Malekmakan, L., Karimi, Z., Mansourian, A., Pakfetrat, M., Roozbeh, J. and Rahimi Jaberi, K., 2020. Role of vitamin D in oxidative stress modulation in end-stage renal disease patients: a double-blind randomized clinical trial. Hemodialysis International 24: 367-373. https://doi.org/10.1111/hdi.12849

Manna, P., Achari, A.E. and Jain, S.K., 2018. 1,25$(OH)_2$-vitamin D_3 upregulates glucose uptake mediated by SIRT1/IRS1/GLUT4 signaling cascade in C2C12 myotubes. Molecular and Cellular Biochemistry 444: 103-108.

Masugi, F. and Nakamura, T., 1976. Effect of vitamin E deficiency on the level of superoxide dismutase, glutathione peroxidase, catalase and lipid peroxide in rat liver. International Journal for Vitamin and Nutrition Research 46: 187-191.

Mokhtari-Zaer, A., Hosseini, M., Salmani, H., Arab, Z. and Zareian, P., 2020. Vitamin D_3 attenuates lipopolysaccharide-induced cognitive impairment in rats by inhibiting inflammation and oxidative stress. Life Sciences 253: 117703.

Omar, D.F., Kamal, M.M., El-Hefnawy, M.H. and El-Mesallamy, H.O., 2018. Serum vitamin D and its upregulated protein, thioredoxin interacting protein, are associated with beta-cell dysfunction in adult patients with type 1 and type 2 diabetes. Canadian Journal of Diabetes 42: 588-594.

Pang, S.T., Hsieh, W.C., Chuang, C.K., Chao, C.H., Weng, W.H. and Juang, H.H., 2009. Thioredoxin-interacting protein: an oxidative stress-related gene is upregulated by glucose in human prostate carcinoma cells. Journal of Molecular Endocrinology 42: 205-214.

Pasquali, M.A., Gelain, D.P., Oliveira, M.R., Behr, G.A., Motta, L.L., Rocha, R.F., Klamt, F. and Moreira, J.C., 2009. Vitamin A supplementation induces oxidative stress and decreases the immunocontent of catalase and superoxide dismutase in rat lungs. Experimental Lung Research 35: 427-438.

Petiz, L.L., Girardi, C.S., Bortolin, R.C., Kunzler, A., Gasparotto, J., Rabelo, T.K., Matté, C., Moreira, J.C. and Gelain, D.P., 2017. Vitamin A oral supplementation induces oxidative stress and suppresses IL-10 and HSP70 in skeletal muscle of trained rats. Nutrients 9: 353.

Petiz, L.L., Kunzler, A., Bortolin, R.C., Gasparotto, J., Matté, C., Moreira, J. and Gelain, D.P., 2017a. Role of vitamin A oral supplementation on oxidative stress and inflammatory response in the liver of trained rats. Applied Physiology, Nutrition, and Metabolism 42: 1192-1200.

Pipaliya, H. and Vaghasiya, J., 2012. Cardio protective effect of vitamin A against isoproterenol-induced myocardial infarction. Journal of Nutritional Science and Vitaminology 58: 402-407.

Placha, I., Takacova, J., Ryzner, M., Cobanova, K., Laukova, A., Strompfova, V., Venglovska, K. and Faix, S., 2014. Effect of thyme essential oil and selenium on intestine integrity and antioxidant status of broilers. British Poultry Science 55: 105-114.

Polidoro, L., Properzi, G., Marampon, F., Gravina, G.L., Festuccia, C., Di Cesare, E., Scarsella, L., Ciccarelli, C., Zani, B.M. and Ferri, C., 2013. Vitamin D protects human endothelial cells from H2O2 oxidant injury through the Mek/Erk-Sirt1 axis activation. Journal of Cardiovascular Translational Research 6: 221-231.

Ratriyanto, A. and Mosenthin, R., 2018. Osmoregulatory function of betaine in alleviating heat stress in poultry. Journal of Animal Physiology and Animal Nutrition 102: 1634-1650.

Reed, D.K., Hall, S. and Arany, I., 2015. α-Tocopherol protects renal cells from nicotine-or oleic acid-provoked oxidative stress via inducing heme oxygenase-1. Journal of Physiology and Biochemistry 71: 1-7.

Rivera, R.E., Christensen, V.L., Edens, F.W. and Wineland, M.J., 2005. Influence of selenium on heat shock protein 70 expression in heat stressed turkey embryos (*Meleagris gallopavo*). Comparative Biochemistry and Physiology Part A: Molecular & Integrative Physiology 142: 427-432.

Roth, C.L., Elfers, C.T., Figlewicz, D.P., Melhorn, S.J., Morton, G.J., Hoofnagle, A., Yeh, M.M., Nelson, J.E. and Kowdley, K.V., 2012. Vitamin D deficiency in obese rats exacerbates nonalcoholic fatty liver disease and increases hepatic resistin and Toll-like receptor activation. Hepatology 55: 1103-1111.

Roveri, A., Vitale, M.P., Serain, E., Zaccarin, M., Mauri, P., Di Silvestre, D., De Palma, A., Gion, M., Toppo, S., Maiorino, M. and Ursini, F., 2008. Differential liquid phase proteomic analysis of the effect of selenium supplementation in LNCaP cells. Journal of Chromatography B 865: 63-73.

Sabir, M.S., Khan, Z., Hu, C., Galligan, M.A., Dussik, C.M., Mallick, S., Stone, A.D., Batie, S.F., Jacobs, E.T., Whitfield, G.K. and Haussler, M.R., 2017. SIRT1 enzymatically potentiates 1, 25-dihydroxyvitamin D3 signaling via vitamin D receptor deacetylation. The Journal of Steroid Biochemistry and Molecular Biology 172: 117-129.

Saeed, M., Babazadeh, D., Naveed, M., Arain, M.A., Hassan, F.U. and Chao, S., 2017. Reconsidering betaine as a natural anti-heat stress agent in poultry industry: a review. Tropical Animal Health and Production 49: 1329-1338.

Sahin, N., Tuzcu, M., Orhan, C., Onderci, M., Eroksuz, Y. and Sahin, K., 2009. The effects of vitamin C and E supplementation on heat shock protein 70 response of ovary and brain in heat-stressed quail. British Poultry Science 50: 259-265.

Said, M.A., in press. Vitamin D attenuates endothelial dysfunction in streptozotocin induced diabetic rats by reducing oxidative stress. Archives of Physiology and Biochemistry. https://doi.org/10.1080/13813455.2020.1741645

Shi, H.Y., Yan, S.M., Guo, Y.M., Zhang, B.Q., Guo, X.Y. and Shi, B.L., 2018. Vitamin A pretreatment protects NO-induced bovine mammary epithelial cells from oxidative stress by modulating Nrf2 and NF-κB signaling pathways. Journal of Animal Science 96: 1305-1316.

Shih, P.K., Chen, Y.C., Huang, Y.C., Chang, Y.T., Chen, J.X. and Cheng, C.M., 2011. Pretreatment of vitamin D3 ameliorates lung and muscle injury induced by reperfusion of bilateral femoral vessels in a rat model. Journal of Surgical Research 171: 323-328.

Shimizu, H., Tsubota, T., Kanki, K. and Shiota, G., 2018. All-trans retinoic acid ameliorates hepatic stellate cell activation via suppression of thioredoxin interacting protein expression. Journal of Cellular Physiology 233: 607-616.

Sies, H., Stahl, W. and Sundquist, A.R., 1992. Antioxidant functions of vitamins: vitamins E and C, beta-carotene, and other carotenoids. Annals of the New York Academy of Sciences 669: 7-20.

Smith, T.J. and Drummond, G.S., 1991. Retinoic acid can enhance the stimulation by thyroid hormone of heme oxygenase activity in the liver of thyroidectomized rats. Biochimica et Biophysica Acta (BBA)-General Subjects 1075: 119-122.

Soleimani, A.F., Zulkifli, I., Omar, A.R. and Raha, A.R., 2012. The relationship between adrenocortical function and Hsp70 expression in socially isolated Japanese quail. Comparative Biochemistry and Physiology Part A: Molecular & Integrative Physiology 161: 140-144.

Sommer, A., and West Jr., K.P., 1996. Vitamin A deficiency: health, survival, and vision. Oxford University Press, New York, NY, USA.

Surai, P.F., 1999. Vitamin E in avian reproduction. Poultry and Avian Biology Reviews 10: 1-60.

Surai, P.F., 2002. Natural antioxidants in avian nutrition and reproduction. Nottingham University Press, Nottingham, UK.

Surai, P.F., 2003. Selenium-vitamin E interactions: does 1+1 equal more than 2? In: Lyons, T.P. and Jacques, K.A. (eds) Nutritional biotechnology in the feed and food industries. Nottingham University Press, Nottingham, UK, pp.47-51.

Surai, P.F., 2006. Selenium in nutrition and health. Nottingham University Press, Nottingham, UK.

Surai, P.F., 2014. Polyphenol compounds in the chicken/animal diet: from the past to the future. Journal of Animal Physiology and Animal Nutrition 98: 19-31.

Surai, P.F., 2015. Silymarin as a natural antioxidant: an overview of the current evidence and perspectives. Antioxidants 4: 204-247.

Surai, P.F., 2018. Selenium in poultry nutrition and health. Wageningen Academic Publishers, Wageningen, the Netherlands.

Surai, P.F. and Fisinin, V.I., 2014. Selenium in poultry breeder nutrition: an update. Animal Feed Science and Technology 191: 1-15.

Surai, P.F. and Kochish, I.I., 2019. Nutritional modulation of the antioxidant capacities in poultry: the case of selenium. Poultry Science 98: 4231-4239.

Surai, P.F., Fisinin, V.I. and Karadas, F., 2016. Antioxidant systems in chick embryo development. Part 1. Vitamin E, carotenoids and selenium. Animal Nutrition 2: 1-11.

Surai, P.F., Kochish, I.I., Romanov, M.N. and Griffin, D.K., 2019. Nutritional modulation of the antioxidant capacities in poultry: the case of vitamin E. Poultry Science 98: 4030-4041.

Tagliaferri, S., Porri, D., De Giuseppe, R., Manuelli, M., Alessio, F. and Cena, H., 2019. The controversial role of vitamin D as an antioxidant: results from randomised controlled trials. Nutrition Research Reviews 32: 99-105.

Tamari, Y., Nawata, H., Inoue, E., Yoshimura, A., Yoshii, H., Kashino, G., Seki, M., Enomoto, T., Watanabe, M. and Tano, K., 2013. Protective roles of ascorbic acid in oxidative stress induced by depletion of superoxide dismutase in vertebrate cells. Free Radical Research 47: 1-7.

Tras, B., Inal, F., Bas, A.L., Altunok, V., Elmas, M. and Yazar, E., 2000. Effects of continuous supplementations of ascorbic acid, aspirin, vitamin E and selenium on some haematological parameters and serum superoxide dismutase level in broiler chickens. British Poultry Science 41: 664-666.

Ushiyama, M., Kuramochi, T., Yagi, S. and Katayama, S., 2008. Antioxidant treatment with α-tocopherol improves erectile function in hypertensive rats. Hypertension Research 31: 1007-1013.

Veltmann Jr, J.R. and Jensen, L.S., 1986. Vitamin A toxicosis in the chick and turkey poults. Poultry Science 65: 538-545.

Veskovic, M., Mladenovic, D., Milenkovic, M., Tosic, J., Borozan, S., Gopcevic, K., Labudovic-Borovic, M., Dragutinovic, V., Vucevic, D., Jorgacevic, B., Isakovic, A., Trajkovic, V. and Radosavljevic, T., 2019. Betaine modulates oxidative stress, inflammation, apoptosis, autophagy, and Akt/mTOR signaling in methionine-choline deficiency-induced fatty liver disease. European Journal of Pharmacology 848: 39-48.

Wang, G., Xiu, P., Li, F., Xin, C. and Li, K., 2014. Vitamin A supplementation alleviates extrahepatic cholestasis liver injury through Nrf2 activation. Oxidative Medicine and Cellular Longevity 2014: 273692.

Wang, L., Chen, L., Tan, Y., Wei, J., Chang, Y., Jin, T. and Zhu, H., 2013. Betaine supplement alleviates hepatic triglyceride accumulation of apolipoprotein E deficient mice via reducing methylation of peroxisomal proliferator-activated receptor alpha promoter. Lipids in Health and Disease 12: 34.

Wang, P.F., Yao, D.H., Hu, Y.Y. and Li, Y., 2019. Vitamin D improves intestinal barrier function in cirrhosis rats by upregulating heme oxygenase-1 expression. Biomolecules & Therapeutics 27: 222-230.

Wang, Y., Zhan, X., Zhang, X., Wu, R. and Yuan, D., 2011. Comparison of different forms of dietary selenium supplementation on growth performance, meat quality, selenium deposition, and antioxidant property in broilers. Biological Trace Element Research 143: 261-273.

Wen, C., Chen, Y., Leng, Z., Ding, L., Wang, T. and Zhou, Y., 2019. Dietary betaine improves meat quality and oxidative status of broilers under heat stress. Journal of the Science of Food and Agriculture 99: 620-623.

Wu, A., Ying, Z. and Gomez-Pinilla, F., 2006. Oxidative stress modulates Sir2α in rat hippocampus and cerebral cortex. European Journal of Neuroscience 23: 2573-2580.

Wu, A., Ying, Z. and Gomez-Pinilla, F., 2010. Vitamin E protects against oxidative damage and learning disability after mild traumatic brain injury in rats. Neurorehabilitation and Neural Repair 24: 290-298.

Xiao, Y., Rungruang, S., Hall, L.W., Collier, J.L., Dunshea, F.R. and Collier, R.J., 2017. Effects of niacin and betaine on bovine mammary and uterine cells exposed to thermal shock *in vitro*. Journal of Dairy Science 100: 4025-4037.

Xu, D., Li, W., Huang, Y., He, J. and Tian, Y., 2014a. The effect of selenium and polysaccharide of *Atractylodes macrocephala* Koidz.(PAMK) on immune response in chicken spleen under heat stress. Biological Trace Element Research 160: 232-237.

Xu, J.X., Cao, C.Y., Sun, Y.C., Wang, L.L., Li, N., Xu, S.W. and Li, J.L., 2014. Effects on liver hydrogen peroxide metabolism induced by dietary selenium deficiency or excess in chickens. Biological Trace Element Research 159: 174-182.

Xu, J.X., Zhang, C., Cao, C.Y., Zhu, S.Y., Li, H., Sun, Y.C. and Li, J.L., 2016. Dietary selenium status regulates the transcriptions of selenoproteome and activities of selenoenzymes in chicken kidney at low or super-nutritional levels. Biological Trace Element Research 170: 438-448.

Xu, W., Hu, X., Qi, X., Zhu, R., Li, C., Zhu, Y., Yin, S., Cheng, L. and Zhu, R., 2019. Vitamin D ameliorates angiotensin II-induced human endothelial progenitor cell injury via the PPAR-γ/HO-1 pathway. Journal of Vascular Research 56: 17-27.

Yuan, D., Zhan, X.A. and Wang, Y.X., 2012. Effect of selenium sources on the expression of cellular glutathione peroxidase and cytoplasmic thioredoxin reductase in the liver and kidney of broiler breeders and their offspring. Poultry Science 91: 936-942.

Zanotto-Filho, A., Gelain, D.P., Schröder, R., Souza, L.F., Pasquali, M.A., Klamt, F. and Moreira, J.C.F., 2009. The NFκB-mediated control of RS and JNK signaling in vitamin A-treated cells: duration of JNK-AP-1 pathway activation may determine cell death or proliferation. Biochemical Pharmacology 77: 1291-1301.

Zdunczyk, Z., Drazbo, A., Jankowski, J., Juskiewicz, J., Czech, A. and Antoszkiewicz, Z., 2013. The effect of different dietary levels of vitamin E and selenium on antioxidant status and immunological markers in serum of laying hens. Polish Journal of Veterinary Sciences 16: 333-339.

Zhang, B., Tanaka, J., Yang, L., Sakanaka, M., Hata, R., Maeda, N. and Mitsuda, N., 2004. Protective effect of vitamin E against focal brain ischemia and neuronal death through induction of target genes of hypoxia-inducible factor-1. Neuroscience 126: 433-440.

Zhang, Z.W., Wang, Q.H., Zhang, J.L., Li, S., Wang, X.L. and Xu, S.W., 2012. Effects of oxidative stress on immunosuppression induced by selenium deficiency in chickens. Biological Trace Element Research 149: 352-361.

Zhao, B., Fei, J., Chen, Y., Ying, Y.L., Ma, L., Song, X.Q., Huang, J., Chen, E.Z. and Mao, E.Q., 2014. Vitamin C treatment attenuates hemorrhagic shock related multi-organ injuries through the induction of heme oxygenase-1. BMC Complementary and Alternative Medicine 14: 442.

Zhao, J., Xing, H., Liu, C., Zhang, Z. and Xu, S., 2016. Effect of selenium deficiency on nitric oxide and heat shock proteins in chicken erythrocytes. Biological Trace Element Research 171: 208-213.

Zhao, X., Yao, H., Fan, R., Zhang, Z. and Xu, S., 2014. Selenium deficiency influences nitric oxide and selenoproteins in pancreas of chickens. Biological Trace Element Research 161: 341-349.

Zhu, L., Kong, M., Han, Y.P., Bai, L., Zhang, X., Chen, Y., Zheng, S., Yuan, H. and Duan, Z., 2015. Spontaneous liver fibrosis induced by long term dietary vitamin D deficiency in adult mice is related to chronic inflammation and enhanced apoptosis. Canadian Journal of Physiology and Pharmacology 93: 385-394.

Zhu, L.H., Zhao, K.L., Chen, X.L. and Xu, J.X., 2012. Impact of weaning and an antioxidant blend on intestinal barrier function and antioxidant status in pigs. Journal of Animal Science 90: 2581-2589.

Part VI.
Practical applications of the vitagene concept in commercial poultry production

You can't make an omelette without breaking eggs

Chapter 13
Performax concept development

Prevention is better than cure

13.1 Introduction

As can be seen from data presented in previous chapters, nutritional modulation of the antioxidant systems and vitagenes could be a basis for the development a new strategy for commercial stresses management in poultry production. Initially, increased doses of vitamin E have been used in commercial poultry production under various stress condition, including heat stress with various efficiency and variable success (Surai, 1999, 2002; Surai *et al.*, 2019a). Similarly, other nutrients possessing antioxidant activities, including selenium (Surai, 2018; Surai and Kochish, 2019), carotenoids (Surai, 2012a,b), polyphenols (Surai, 2014) have been also widely used (see previous chapters of this book) to deal with commercially-relevant stresses. However, efficacy of increased dietary supplementation of various antioxidants greatly varies.

The aim of this chapter is to present data related to the development and practical application of the first vitagene-modulating nutritional mixture as an important approach to deal with commercially-relevant stresses in egg and meat producing industries.

13.2 Usage of drinking system for vitagene-activating nutrient mixture delivery

At the stage of the development of vitagene-regulating nutrient mixture it was necessary to choose the best way of their delivery. Traditional usage of various feed additive has many advantages, however, in stress conditions this system has several limitations:

- Low flexibility. For example, in stress management there are situations when nutrients/drugs should be delivered to birds urgently. However, in commercial conditions, feed bins usually contain feed for few days and before that feed is used, technically impossible to add anything via feeding system.
- Difficulties with precise dosing. It is well-known that under stressful conditions feed consumption very often decreases. At the same time requirement in some nutrients, including antioxidants, can be increased. Therefore, in such conditions precise dosing via feed is almost impossible. However, in stress conditions poultry usually increase water consumption. Even if for some reason birds refuse the feed, they still drinking water. Therefore, a drinking system in a poultry house can be effectively used for the delivery vitagene-regulating nutrients.

DOI 10.3920/978-90-8686-906-0_13, © Wageningen Academic Publishers 2020

- Quick response system. By using drinking system in poultry hose it is possible to develop a quick response system. Let's consider a practical situation. A veterinarian or nutritionist/technologist discoveres a problem with 5-day-old birds in the poultry house – the birds look sleepy and there are 30,000 birds in the shed. It is clear that something must be done reasonably quickly to prevent further deterioration of the situation and possible mortality. However, there is a need for extra time for specialists to evaluate a situation, collect some samples of feed, blood, tissues, etc. and send them for analyses. It could take a day or several days before the picture is clear and diagnosis is available. That is exactly the time when vitagene upregulation by supplying some nutrients via drinking system could be a crucial factor 'to buy' some time for proper diagnostics. Vitagene upregulation can help birds to survive a day or few days before the necessary treatment is available. Depending on the cause of the problem, birds might need extra vaccination, antibiotics or other drugs and this would be supplied after the diagnosis is confirmed in the lab. Therefore, supporting birds in this case would help prevent extra losses. In this case vitagene upregulation seems to be an important approach in poultry management.
- Technological advances. Currently, drinking systems in poultry houses are widely used for vaccinations as well as for supplying some acidifiers, antibiotics, etc. The same drinking system can be used for urgent supply of some vitagene-activating nutrients. Various medicators/dosatrons (Figure 13.1) allow precise dosing of nutrients/preparations with drinking water. Indeed, by upregulating the vitagenes it is possible to help stressed birds deal with stresses and improve their stress adaptability.

It should be mentioned that old poultry sheds could lack dosatrons. In this case usually there is a water tank installed and it can be used for the same purpose. Therefore, there are situations in commercial poultry production where chickens need immediate attention and it is reasonable to use a drinking system to supply important vitagene-regulating nutrients in stress conditions.

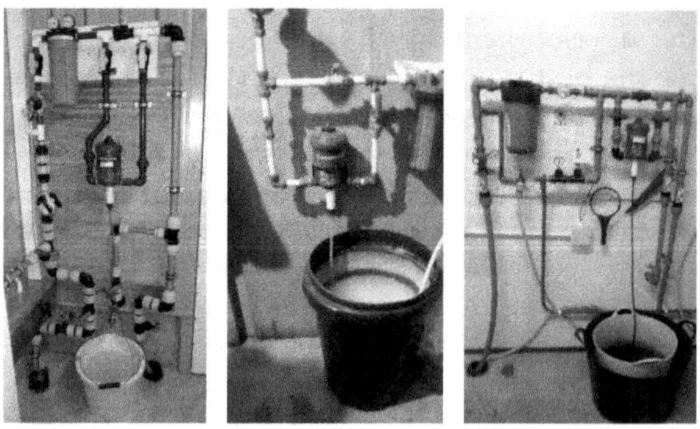

Figure 13.1. Dosatrons in poultry houses.

Our market analysis clearly showed that there has been a range of vitamin-mineral-amino acid mixtures produced for supplying to poultry via drinking system. Some of them are widely used in Eastern European countries. A closer look at the composition and history of the development of such mixtures indicate that they were developed 50-60 years ago when premix industry was still developing and there were cases of vitamin/mineral deficiency in commercially growing birds. Therefore, such nutritional mixture helped nutritionist to balance various diets when premixes were of low quality/stability. However, since then premix industry has made a great progress and today cases of vitamin and mineral/amino acid deficiencies in commercial conditions of many countries with the developed poultry industry are rather exception, than rule. One should also remember that additional supplementation of most vitamins (especially B-vitamins), minerals or amino acids beyond their requirements which are usually met by supplementing feed with commercial premixes at the stage of manufacturing, would not give any extras in terms of productive or reproductive performance and would increase cost of production of eggs or meat. It should be also mentioned that vitamin and mineral doses supplied with premixes are substantially higher than minimal physiological requirement in those nutrients. Therefore, usage of such preparations is just tribute to history and lack of knowledge in modern poultry nutritional standards.

In contrast to the aforementioned vitamin-mineral-amino acid mixtures, our new concept is based on the supplying a specific nutrient mixtures possessing vitagene-activating properties. This would increase expression of target genes (vitagenes) leading to improved ability of poultry to adapt to stress and increase stress resistance. In this case a unique composition of nutrients (a synergistic combination) is more important than concentrations of any individual compound in the preparation. Therefore, this mixture provides protective effects even if the diet is balanced in major nutrients. Indeed, as a result of additional synthesis of protective molecules (HSP70, HO-1, SOD, Trx-system, GSH-system and sirtuins) the body/tissue/cell can maintain a healthy redox balance providing optimal conditions for cell signalling and adaptation to stress, including environmental, technological, nutritional and internal stresses (Surai, 2019a,b; Surai *et al.*, 2019a). Therefore, by mobilising internal reserves to a greater extent, than without vitagene activation, the bird can withstand/resist increased stresses and overcome them with minimal loses.

However, closer look at various aspects of poultry nutrition and management showed several limitations of such approaches. First of all, technology of adding various nutrients to the feed to deal with stresses has low flexibility. Before the feed bins are empty (several tons of feed) it is practically impossible to change feed. Furthermore, under various stress conditions chickens usually are characterised by decreased feed consumption. However, there are situations in commercial poultry production where chickens need immediate attention. Therefore, it is reasonable to use a drinking system to supply important stress-fighting nutrients in stress conditions.

13.3 The development of multi-nutrient mixture for vitagene activation and increasing stress resistance of poultry

From the data presented in previous chapter it is clear that there is a range of vitagene-regulating nutrients and in many cases a combination of several nutrients were shown to be more effective that any single nutrient. Therefore, when developing vitagene-regulating nutrient mixture for improving chicken adaptability to stresses several points need to be taking into account.

- Vitagene-regulating nutrients. To upregulate vitagenes nutrient composition should include carnitine, betaine, taurine, ascorbic acid, vitamins A, D3, E, Se and possibly other nutrients in optimal proportions. In this case an optimal combination of nutrients seems to be more important that concentrations of any of the individual compound (Fisinin and Surai, 2011a,b; Surai, 2015a; Surai and Boroday, 2010; Surai and Fisinin, 2012a,b; Surai and Fotina, 2010a, 2011a, 2012a,b,c,d; Surai *et al.*, 2017, 2017a, 2018).
- Vitamin E recycling. To further upregulate antioxidant defences in stress conditions it is necessary to provide nutrients taking part in vitamin E recycling. Indeed, if the recycling of vitamin E from its oxidised form is effective, even low vitamin E concentration can provide optimal protection of biological membranes (Surai, 2002, 2006, 2018, Surai *et al.*, 2018, 2019a). Ascorbic acid, selenium (as a part of thioredoxin reductase), vitamins B1 and B2 are essential elements of vitamin E recycling.
- Osmolytes. Osmotic stress is considered to be an important part of intracellular changes observed in stress conditions. Betaine is considered to be an effective osmolyte providing protection against cellular dehydration (osmotic shock), regulating water balance and exerting a stabilising effect on tissue metabolism under various stress conditions (Ratriyanto and Mosenthin, 2018). This is especially important in the chicken gut. As mentioned earlier in this book, betaine also can upregulate some vitagenes. Betaine is also a source of methyl groups and showing a range of protective effects in the case of various stresses including heat stress, mycotoxin stress and others (Surai *et al.*, 2018).
- Liver and gut supporting nutrients. Since mycotoxins are considered to be main feed-derived stressors, and it is practically impossible to avoid mycotoxin feed contamination (Surai and Dvorska, 2005; Surai *et al.*, 2010), it is important to use nutrients supporting liver and gut, main sites of mycotoxin detoxification in the body (Surai *et al.*, 2018). In fact, major mycotoxins, including T-2 toxin, ochratoxins, aflatoxin, fumonisins, etc. are detoxified in the liver (Fisinin and Surai, 2012a,b,c,d) and DON is mainly detoxificated by gut microbiota (Fisinin and Surai, 2012e,f). Therefore, carnitine, betaine, vitamin E, Se, lysine and methionine are nutrients supporting the liver and gut of the chicken to help mycotoxin detoxification and minimise liver and gut damage (Surai *et al.*, 2018). It is important to mention that even when the diet contains some mycotoxin binders the problem is not solved, since it is impossible to bind all mycotoxin at the level of more that 50-60% of their presence in the diet. The rest of unbound mycotoxins have to be metabolised by the detoxification systems of the body. Therefore, vitagene activation (Figure13.2) is

considered as an important step in prevention/decrease detrimental consequences of mycotoxin contamination of the diet (Fisinin and Surai, 2012a,b,c,d,e,f).
- Immunomodulating nutrients, including carnitine, betaine, vitamins E and C, methionine, lysine, Se, Zn and Mn should also be an integral part of the anti-stress composition. In fact, oxidative stress can destroy practically all parts of the immunity. If we imagine that the immune system is an army fighting against invaders (microorganisms, viruses, etc.) then we would expect them to have communicating devices, something like mobile phones, to receive and send signals to each other. Remarkably enough, major immune cells (macrophages, neutrophils/heterophils, T- and B-lymphocytes) have on their surface 'mobile phones' called receptors. Those receptors are extremely sensitive to communicating molecules, but they are also sensitive to free radicals and can be easily damaged. In such a situation without proper communication all those huge armies of immune cells would become useless. They also can start fighting each other, as well as eventually destroying immunocompetence and causing autoimmune reactions. Indeed, communication between all types of immune cells (Surai, 2018) is a key for high immunocompetence.

If we imagine that immune cells are soldiers using chemical weapon to kill the enemy, then special gear protecting them from their own weapon would be a crucial for effective battle. In the case of immune cells, such gear is represented by natural antioxidants synthesised under vitagene governance, including HSP70, GSH, GPx, thioredoxin, thioredoxin reductase, sirtuins, etc. Indeed, if not properly protected, phagocyte functions could be compromised, including initial overproduction of

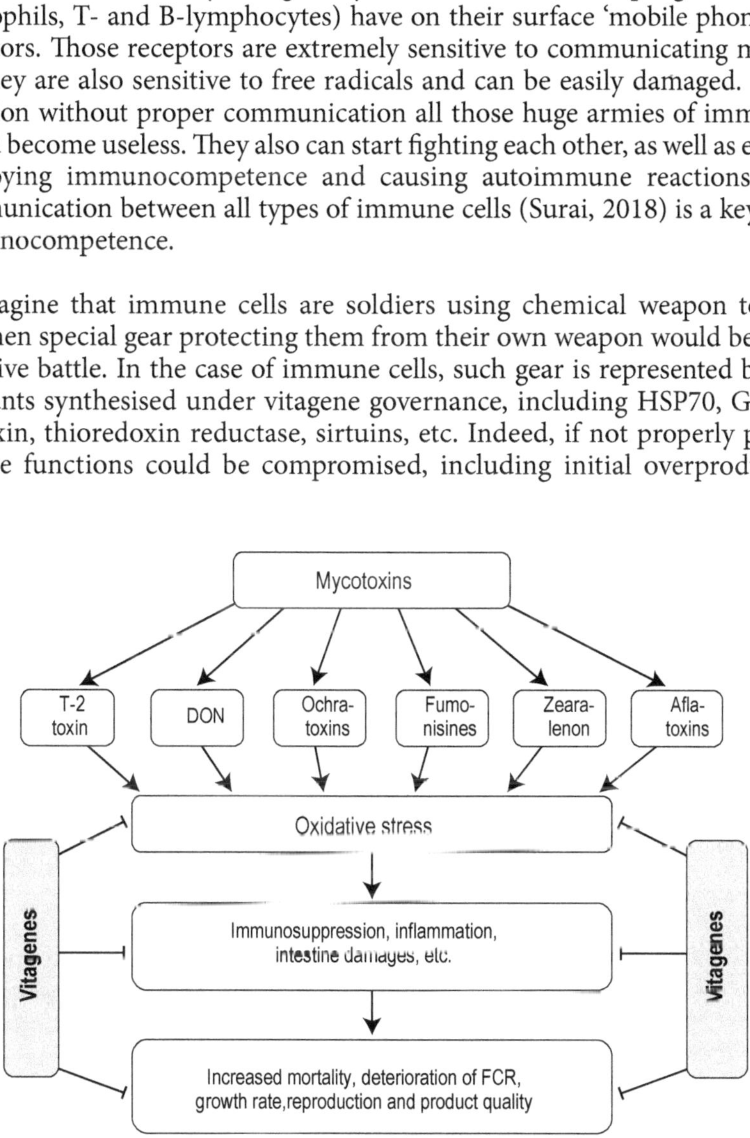

Figure 13.2. Mycotoxins, oxidative stress and vitagenes.

free radicals with consecutive damages to specific enzymatic systems resulting in decreasing efficiency of oxidative burst and apoptosis. Based on the presented model it is clear that antioxidant defence is a crucial factor of immune defence in the body and vitagenes play a central role in this defence. In fact, protective molecules, like HSP70, can 'cover up' receptors at time of stress helping immune-communication and maintaining immunocompetence in stress conditions (Fisinin and Surai, 2011, 2013a), including gut immunity (Fisinin and Surai, 2013b). This is the case in heat stress, mycotoxin stress and a range of other commercially relevant stresses (Surai *et al.*, 2018).

- Trace elements are of great importance in antioxidant defences (Surai, 2005). As mentioned above Mn is an integral part of the MnSOD, main mitochondrial AO enzyme. Furthermore, Zn is a part of another form of SOD, located in cytosol. Selenium is an integral part of at least 25 selenoproteins and more than half of them are involved in AO defence and redox balance maintenance (Surai, 2018). It is proven that Mg deficiency affect appetite and recently Mg is considered to be involved in activation of catalase (Estevez and Petrucci, 2019) and in coordination of antioxidant defences (Monteiro *et al.*, 2016). Therefore, a choice of minerals for antioxidant composition to fight stresses should include Mn, Zn, Se and Mg (Figure 13.3).
- Essential amino acids participate in various protective pathways in stress conditions. In particular, methionine (Met), a sulphur-containing amino acid, participates in the initiation and prolongation step of protein synthesis, transmethylation reactions, the synthesis of cysteine and cystine, being an important component of antioxidant systems (Mastrototaro *et al.*, 2016). Indeed, Met participates in synthesis of GSH and carnitine and activates methionine sulphoxide reductase A, to counteract oxidative stress (Martinez *et al.*, 2017). Another essential amino acid playing important role in carnitine synthesis and growth and development support is lysine (Surai, 2015). Therefore, a combine supplementation of methionine and lysine in stress conditions, when their requirements are increased, while

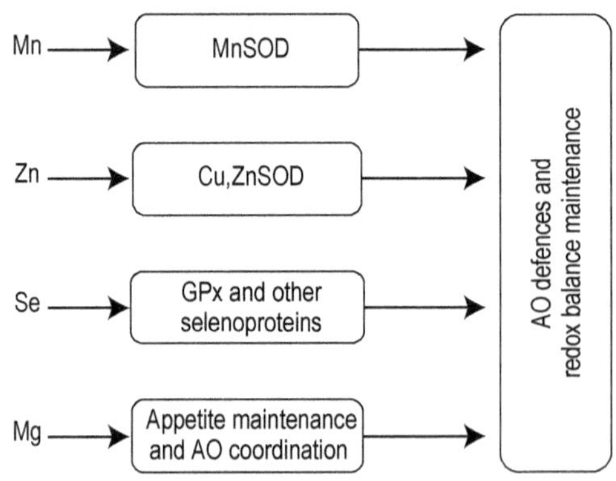

Figure 13.3. Minerals for antioxidant(AO) composition.

consumption could be compromised due to feed consumption reduction, could be an additional option during an effective antioxidant mixture development.

- Organic acids are shown to enhance poultry performance via regulation of microbiome and other mechanisms affecting the morphology and physiology of the gut and the immune system (Dittoe *et al.*, 2018). This is especially important in the case of feed grade antibiotic prohibition in many countries worldwide with developed poultry production. Therefore, adding an effective mixture of organic acids into the antioxidant composition would be also beneficial.
- Electrolytes. It is known that the bird's survivability during heat stress depends on the water consumption, which depends directly on bird's age and on electrolyte balance of the diet (Ahmad and Sarwar, 2006). In fact, under heat stress, the loss of carbon dioxide (CO_2) through respiration and bicarbonate ions through urine can disturb the acid-base balance. Therefore, electrolyte supplementation, either through drinking water or feed, was shown to help correcting blood acid-base balance having beneficial effects for broilers under different heat stress regimens (Ahmad and Sarwar, 2006; Mushtaq *et al.*, 2013). An optimal electrolyte balance and its correction in the body via water/feed supplementation were shown to have protective effects during transport stress (Schaefer *et al.*, 1997) and other stress situations (Olanrewaju *et al.*, 2007). Therefore, inclusion of electrolytes into an antioxidant composition for fighting stresses could help not only in heat stress conditions, but also in other stresses, including transportation stress and chicken placement.

The first commercial product based on the aforementioned concept of the development of the effective antioxidant-related, vitagene-regulating mixture was developed, successfully tested and now can be found on the market in different countries worldwide. Since such a product (Magic anti-stress mix/PerforMax) containing carnitine, betaine, vitamins, minerals, electrolytes, organic acids and amino acids was first introduced in Ukrainian and Russian markets (Fotina *et al.*, 2011; Shatskikh *et al.*, 2015, 2016; Velichko *et al.*, 2013), major research publications related to the results of the product tests in the university trials and in commercial poultry settings were published in a Russian/Ukrainian peer reviewed journal and they will be briefly reviewed below.

13.4 Effect of the vitagene-regulating anti-stress composition on rearing birds, layer and broiler breeders

The importance and efficacy of using the anti-stress composition (PerforMax/Magic Antistress Mix) for rearing birds and adult egg-type parent stock (Hy-Line) at one of the biggest egg producing farms in Russia (Borovskaya poultry farm, Tumen region) have been described recently and below there is a summary of major findings. The study was conducted with rearing birds as well as with adult laying hens and cockerels of the parent stock (Hy Line Brown, line SD). There were two poultry trials, two physiological trials and a big commercial trial. In total 53,400 birds were used with the major results were presented in the PhD thesis (Latipova, 2014a).

In the first phase of the experiment, the effect of Magic Antistress Mix (PerforMax) on the rearing birds was studied. There were a control and an experimental group with 2,000 females and 400 males in each group. The control group was fed in accordance with Hy Line recommendations with the diet formulated to meet all requirements in major nutrients and energy. The experimental group had an additional anti-stress supply via drinking water at the level of 100 g per 100L of drinking water during days 1-5, 9-13, 21-25, 27-31, 45-49, 63-67 and 75-79 which were related to stresses imposed by vaccinations, grading, and transfer to the breeding houses. The experiment lasted for 105 days (Shatskikh and Latipova, 2014). In general, growth and development of chickens were similar in both groups with slight increase in body weight (1,242 vs 1,200 g) at 15 weeks of age in the experimental female birds. Furthermore, usage of anti-stress composition was shown to improve birds uniformity ad age of 15 weeks (Shatskikh *et al.*, 2016; Figure 13.4).

Therefore, the experimental birds were closer to target body weight (1,230 g) than control birds. The usage of anti-stress composition positively affected testes development of 15-week old cockerels (2.47 vs 1.05 g, $P<0.05$) (Shatskikh and Latipova, 2014). It is interesting to note that in the experimental cockerels testes weight was also improved at 26 and 56 weeks of age (Shatskikh and Latipova, 2013). The liver of experimental birds was characterised by a significant increase in vitamin A content of the liver at various ages ($P<0.05$). There was an increase in Ca content of the bones of female birds at 15 weeks indicating better Ca reserves for future egg production. Results of balance experiments indicated that females and males of the experimental group were characterised by improved ($P<0.05$) usage of nitrogen, calcium and phosphorus from the diet. Indeed, usage of the anti-stress composition with water in period of chicken stress positively affected experimental birds (Shatskikh and Latipova, 2014).

The second phase of the experiment was a continuation of the first phase and lasted from day 106 until day 448 and similar to a previous experiment PerforMax (Magic Antistress Mix) was used with water at 100 g per 100 l of drinking water at days 106-107, 109-111, 148-157 and 238-246 associated with increased stresses of vaccinations, first egg laying, active egg production growth and peak of egg production (Latipova,

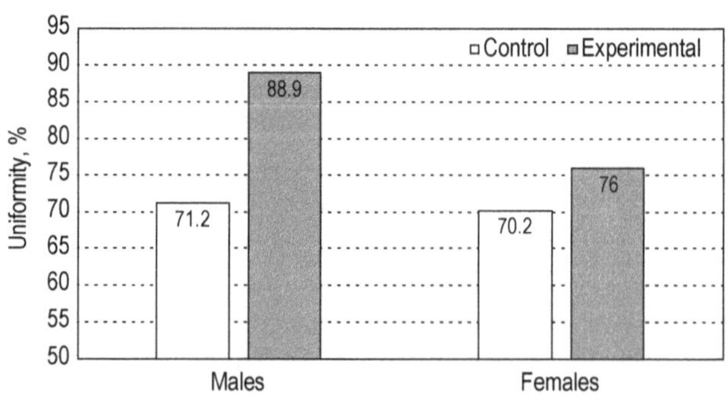

Figure 13.4. Uniformity of rearing birds at 15 weeks, % (adapted from Shatskikh *et al.*, 2016).

2014). The main results of egg production are shown in the Table 13.1. As can be seen from the presented data, usage of the anti-stress composition with drinking water at specific periods of the increased stress can improve breeder's performance. In particular there was an increase by 2% of the egg peak production and peak plato lasted about 50 days longer than that in the control birds. It is interesting to note that hen housed egg production in the control group (260.8 eggs) was higher than the target for the line (253.4 eggs) and in the experimental group it was increased by 6 eggs. Improved egg production was associated with increased weight of the oviduct in the experimental layers (Shackih and Latypova, 2013). It is also important to mention that FCR (feed per 10 eggs) was also improved by usage of the anti-stress composition and was better than the target for the line.

As a result, more than 12,000 additional eggs were obtained in the experimental group (Latipova and Shatskikh, 2014). It is interesting to note that eggshell strength at age 26, 36 and 56 weeks was improved in the experimental group by 2.8, 5.6 and 5.6% respectively (Figure 13.5). The most interesting finding was related to a significant increase carotenoid level in the egg yolk of the experimental birds. Since carotenoids were not supplied with the anti-stress composition this increase could be due to improved absorption of nutrients resulted from anti-stress composition usage. This can also explain improved FCR in the experimental birds. Vitamin A level in the egg yolk from the experimental layers was also increased probably reflecting its transfer from the anti-stress composition. Anti-stress composition usage was associated with improved fertility at 16, 40, 48 and 56 weeks by 2.5; 2.7; 2.8 and 3.7% respectively. In the same experimental group the hatch of condition chickens improved at 26, 32, 40, 48 and 56 weeks by 3.6; 2.1; 3.4; 4.9 and 4.3% respectively (Figure13.6 and Figure 13.7; Latipova, 2014).

Table 13.1. Egg production of Hy Line parent laying hens per 64 weeks (Latipova and Shatskikh, 2014).

Parameter[1]	Group	
	Control	Experimental
Total egg produced	504,958	517,200
Age of the first egg laid, days	108	106
Age of achieving 50-% laying, days	143	142
Peak of egg production, %	94.3	96.9
Age of achieving egg peak production, days	180	172
Plato of peak production, days	59	110
Egg production intensity, %	86.5	87.0
Eggs per hen housed (T: 253,4 eggs)	260.8	266.9
Eggs per hen day (T: 259,5 eggs)	263.6	268.1
Feed per 10 eggs, kg (T: 1,35 kg)	1.36	1.33

[1] T = target for Hy Line Brown.

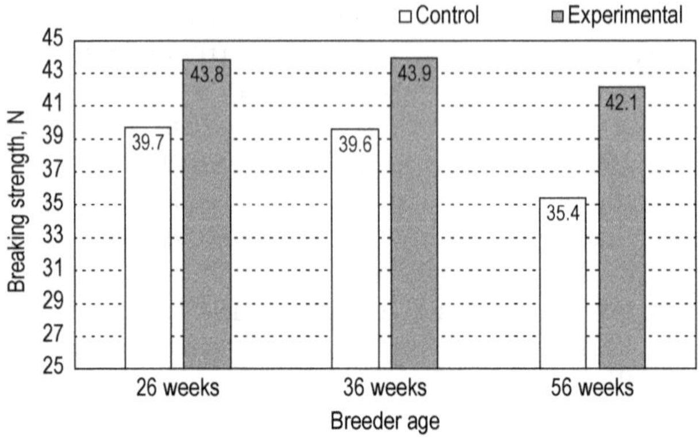

Figure 13.5. Egg shell breaking strength, N (adapted from Shatskikh *et al.*, 2016).

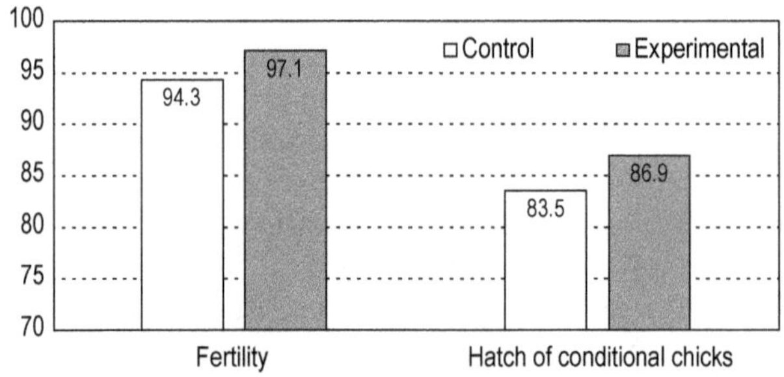

Figure 13.6. Reproduction of Hy Line breeders at 40 weeks, % (adapted from Shatskikh *et al.*, 2016).

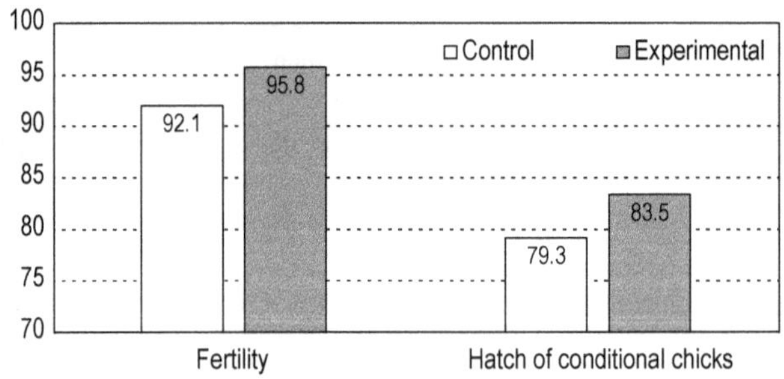

Figure 13.7. Reproduction of Hy Line breeders at 56 weeks, % (adapted from Shatskikh *et al.*, 2016).

Furthermore, vitamin A concentration in the liver of newly hatched chick from the experimental group was significantly (P<0.05) increased. Epigenetic effects could have important role in the poultry production, but there is a lack of scientific studies in this area. The aim of the present study was to evaluate a possible effect of supplying an anti-stress composition, containing vitagene-activating nutrients (carnitine, betaine, vitamins, etc.; PerforMax/Magic Antistress Mix) at 100 g/100 l drinking water to Hy Line breeders on their progeny chicks with specific emphasis to chick uniformity as an important determinant of rearing birds quality was evaluated (Shatskikh *et al.*, 2016). There were 1,938 layers and 176 cockerels in control and experimental groups and the experiment lasted from day 106 until day 448. The obtained data indicate that supplying the anti-stress composition to breeders was associated with a significant improvement of the uniformity (at day 28) in progeny chicks obtained from breeders of various ages: 26 weeks (81.3 vs 67.3%), 32 weeks (85.5 vs 76.8%), 40 weeks (83.2 vs 68.8%), 48 weeks (75.5 vs 68%) and 56 weeks (73.7 vs 62%) (Figure 13.8).

It is interesting to note that there was no difference in weight of day-old chicks between groups independent on breeder's age. Therefore, it seems likely that changes in egg composition due to supplying to breeders with important nutrients, including methyl donors (betaine, methionine, vitamin B12, etc.), could have epigenetic effects of the progeny chick. A positive effect of the usage of PerforMax/Magic Antistress Mix was confirmed in a big commercial trial and it was proven that the anti-stress composition supplied with water can improve hatching egg production efficacy and is an economically valuable tool to be used in the industry (Latipova *et al.*, 2014). Indeed, hatching egg production profitability increased from 19.7% in the control group up to 29.3% in the experimental group of breeders (Figure 13.9; Shatskikh *et al.*, 2016).

Positive effects of using vitagene-regulating nutrient mixture shown with layer breeders above can be transferred to the broiler breeders. Indeed, experimental data indicates that by decreasing commercially relevant stresses in broiler breeders it is possible to increase hen housed egg production, which was higher in experimental group than that in Ross standard for the line (Surai *et al.*, 2017b, 2018; Figure 13.10).

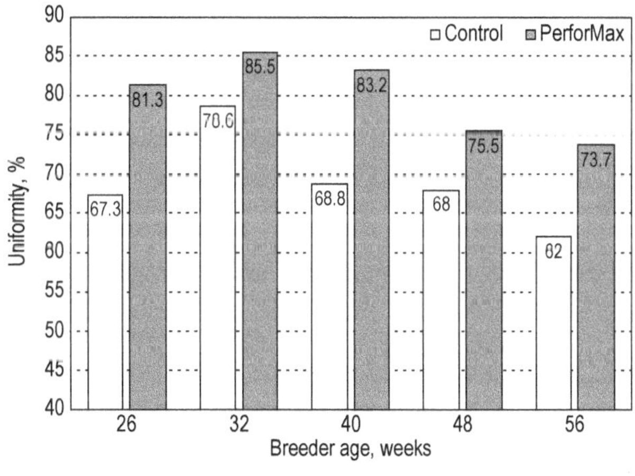

Figure 13.8. Day 28 uniformity of the chicks obtained from the eggs laid by hens of different ages, % (adapted from Shatskikh *et al.*, 2016).

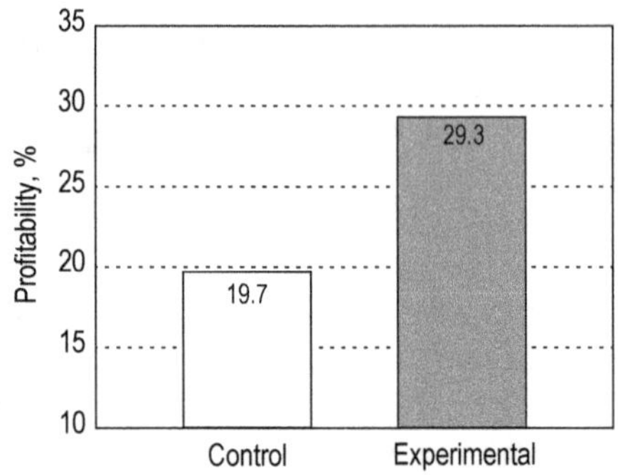

Figure 13.9. Hatching egg production profitability, %, (adapted from Shatskikh *et al.*, 2016).

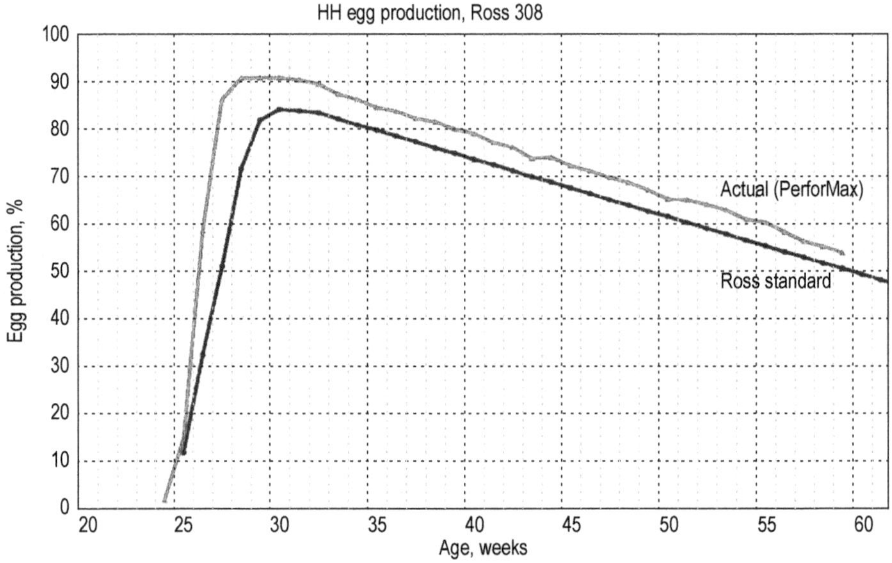

Figure 13.10. Effect of vitagene-regulating nutrient mixture on egg production in Ross breeders (adapted from Surai *et al.*, 2017b, 2018).

13.5 Effects of the vitagene-regulating anti-stress composition on broilers

The experiment was conducted at the poultry farm AO Prodo Tyumen Broiler (Tumen region, Russia) in a special poultry house designed and equipped for conducting experimental trials and measuring growth parameters weekly (Grigorieva *et al.*, 2017). Twelve thousand and four hundred newly hatched Arbor Acres chicks were divided into two equal groups with four replicates in each group and placed in a poultry house. Temperature, light and other parameters of chicken growth were maintained in accordance with recommendations of Arbor Acres accepted at the poultry farm. The diet was balances in all major nutrients in accordance with the same Arbor Acres recommendations accepted at the poultry farm. In additional to the balances died the experimental birds were supplemented with Magic Antistress Mix/PerforMax with drinking water at a dose of 1 g/l in stressful period of chicken growth in accordance with recommendations of the preparation developer (Feed-Food Ltd., UK). The preparation composition included vitagene-regulating mixture of vitamins, minerals, amino acids, carnitine, betaine, organic acids, etc. (Shatskikh *et al.*, 2016). Chickens were grown up to 35 days and information was collected weekly. Body weight and daily gain (g), mortality (%), daily feed consumption (g) and FCR were assessed in three growth periods (15-21 days, 22-28 days and 29-35 days) and for whole growth period (1-35 days). To compare broiler results from different flocks and different regions, the European Production Efficiency Factor (EPEF) is used. At the end of the experiment intestinal weight was assessed and internal organ examination was conducted to check nephritis, enteritis and intestinal atony. Efficacy of vaccination against infectious bursitis was assessed by ELISA in accordance with recommendations of test system producer (Bio-Tek Instruments, Inc., USA). Vaccination Index (VI) was calculated as a ratio of average antibody titre by ELISA to variation coefficient (Cv). During statistical analysis, average parameters (M) and the standard errors (±SEM) were determined. Significance of differences with Control was assessed by t-test by using ANOVA, differences were considered to be significant at $P<0.05$.

Chicken viability (Table 13.2) in both groups was within standards for the poultry farm. However, the mortality for 35 days in the control group (3.8%) was numerically higher in comparison to the experimental group. In commercial conditions such a difference could be significant for a meat producer. Usage of Magic Antistress Mix/PerforMax with drinking water was associated with decrease number of enteritis, nephritis and intestinal atony as well as pododermatitis (data not shown). For the first two weeks of the experimental period data in the control and experimental groups were not different (data not shown). Later (Table 13.2) there was an improvement in FCR in the experimental group in comparison to the control group resulting in decreased feed consumption per week.

There was no difference in body weight and daily gain in the control and experimental groups. Therefore, the main advantage of usage of vitagene activating composition for growing broilers was associated with improvement of FCR, which in the experimental group (1.50) was significantly better than in the control group (1.56). Taking into account all data of the chicken growth and development for 5 weeks EPEF in the control group was shown to be 356.8, while in the experimental group it was 372.8, being comparable with an efficacy of broiler production at the best farms in Russia and abroad. It is important to mention that in experimental chickens at day 35 intestinal weight was higher (by 5%) in comparison to the control chickens. Vaccination efficacy against infectious bursal disease assessed by ELISA methodology (BioTeck) at day 35 includes antibody titres, Cv expressed in % and a vaccination index calculated as a ratio of average antibody titre to Cv. Our data indicates that average antibody titre in 35 days old chickens was 4,500 in the control group and 5,258 in the experimental

Table 13.2. Effect of usage with drinking water of vitagene-stimulating additive on growth and development of Arbor Acres broilers (adapted from Grigorieva *et al.*, 2017).

Parameter			Control (n=6,200)	Experimental (n=6,200)[1]
15-21 days of age	Body weight	g	935.5±8.75	926.0±19.5
	Daily gain	g	68.6±2.4	68.3±1.6
	Mortality	%	0.40±0.04	0.30±0.05
	Feed consumption	g/chicken	101.4±1.8	99.5±1.9
	Feed conversion ratio		1.48±0.03	1.46±0.03*
22-28 days of age	Body weight	g	1,457.5±6.5	1,457.8±29.8
	Daily gain	g	74.6±0.6	76.1±3.0
	Mortality	%	0.50±0.05	0.3±0.08*
	Feed consumption	g/chicken	119.6±1.40	116.5±3.7
	Feed conversion ratio		1.61±0.02	1.54±0.04*
29-35 days of age	Body weight	g	2,019.0±7.50	2,007.5±27.5
	Daily gain	g	80.2±0.80	78.6±2.8
	Mortality	%	1.90±1.30	0.9±0.2
	Feed consumption	g/chicken	132.4±3.30	126.7±2.7
	Feed conversion ratio		1.65±0.04	1.62±0.04
1-35 days of age	Body weight	g	2,019.0±7.50	2,007.5±27.5
	Daily gain	g	56.5±0.20	56.2±0.8
	Chicken viability	%	96.2±1.20	97.7±0.5*
	Feed consumption	kg/chicken	3.15±0.07	3.02±0.02*
	Feed conversion ratio		1.56±0.04	1.50±0.02*
	EPEF[2]		356.8±10.8	372.8±9.6**

[1] *, ** means differences with Control are statistically significant at $P<0.05$ and $P<0.01$, respectively.

[2] European production efficiency factor.

group. It is interesting that Cv in the experimental group (24%) was lower than that in the control group (29%). As a result, the vaccination index in the experimental group was significantly higher than that in the control group (Figure 13.11).

As a result of FCR improvement due to water application of the anti-stress composition economics of meat production was improved (Table 13.2). Indeed, due to the PerforMax application an additional income per kg of meat production comprises 1.66 roubles. An earlier study conducted at the same poultry farm showed that supplying the anti-stress vitagene-activating composition with drinking water improved bird's performance. The biggest positive effect of such a supplementation was seen at the FCR: 1st week1.44 vs 1.55; 2nd week: 1.60 vs 1.66; 3rd week 1.70 vs 1.72; 4th week 1.83 vs 2.06 and for the whole growth period (39 days) 1.85 vs 2.04 and daily gain was improved by 1.85 g. The mortality was the same in both groups 3.0 vs 2.8%). By using the vitagene-regulating composition the cost of production of 1 kg of broiler body weight decreased by 13 cents (Velichko and Surai, 2014; Velichko *et al.*, 2013).

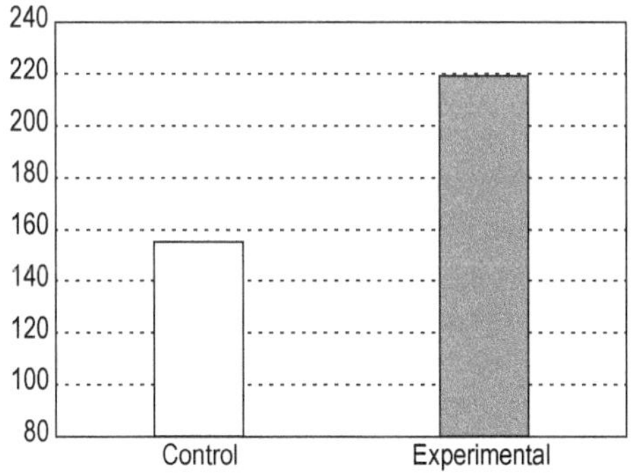

Figure 13.11. Vaccination index in the 35-day old chickens (adapted from Grigorieva *et al.*, 2017).

At small university trial the same vitagene-regulating additive supplied to broilers improved FCR (1.75 vs 1.88) and increased daily gain by 3 g (55.7 vs 52.7 g) (Fotina *et al.*, 2011). There was also a commercial trial with caged broilers comparing 2 control and 3 experimental (supplemented with the same vitagene-regulating mixture) sheds (Table 13.3).

The results of the commercial trial (Table 13.3) confirmed previous results showing beneficial effects of PerforMax supplementation on the broiler growth and development. Since molecular mechanisms of the stresses are similar in monogastric animals, the vitagene concept of fighting stresses was successfully applied for ducks (Surai *et al.*, 2012) and pigs as well (Surai and Litvinov, 2017; Surai and Melnichuk, 2012). In particular, using PerforMax (Magic Antistress Mix) for piglets to fight stresses at weaning (Gaponov *et al.*, 2011) and heat stress (Surai and Fotina, 2013a) is shown to be effective. Therefore, the technology of stress protection in poultry production, based on directed activation of vitagenes allows to increase adaptive ability of birds to commercially-relevant stresses. In particular, the technology can be effectively used for:

Table 13.3. Effect of PerforMax on broiler growth and development (adapted from Surai *et al.*, 2018).

Shed, chicken number	Days of growth	Body weight kg	Daily gain g	Feed conversion ratio	Survival %	EPEF[1]
Control, 22,605	41	2.302	55.2	1.88	92.2	275
Control, 22,585	40	2.271	55.8	1.92	93.4	276
PerforMax, 22,620	40	2.503	61.6	1.80	95.2	331
PerforMax, 22,603	41	2.594	62.3	1.85	94.8	324
PerforMax, 22,621	42	2.584	60.6	1.81	95.1	323
PerforMax/Control		+274 g	+6 g	-0.08	+2.2%	+50

[1] European production efficiency factor.

- Chicken support during and after their placement in the shed. This is a critical period of time when antioxidant defences and redox balance are crucial for effective development of the gut and immunity.
- Vaccination. Vitagene activation before and after vaccination is considered to be an effective means vaccination stress reduction and improvement vaccination efficacy.
- Nutritional stresses (for example, mycotoxin feed contamination). Support of the liver and gut, main sites of mycotoxin detoxification and metabolism? Are of great importance to decrease detrimental consequences of mycotoxin toxicity consequences.
- Heat or cold stress. Reduction of detrimental consequences of temperature stress could help prevent/decrease growth deterioration, immunosuppression and mortality.
- Technological stresses. Protection against technological stresses (e.g. problems with equipment, etc.) could help maintaining high productive and reproductive performance.
- Rearing bird transfer to breeder houses. Improved antioxidant defences and maintenance of the redox balance during this stressful period are of great importance to prevent feather pecking and stimulate egg production.
- Thinning. Vitagene activation during this stressful procedure could prevent detrimental consequences of the stress.
- Peak of egg production. This is the most demanding period of egg production and vitagene activation can help achieving higher peak of egg production and increasing its duration.
- Any other stressful period of bird's growth, development and production/ reproduction.

Protective effects of PerforMax in poultry production are summarised in the Figure 13.12. Indeed, activation of vitagenes by a specific nutrient mixture is shown to be an effective strategy to maintain good health and high productive and reproductive performance of birds in commercial stressful conditions.

13.6 Vitagene activation as an important strategy in stress prevention/ alleviation

Our research for the last 10 years allowed to transfer vitagene concept from medical science to poultry sciences (Fisinin and Surai, 2011a,b; Surai, 2011, 2015, 2016, 2018, 2019c,d, 2020a,b; Surai and Fisinin, 2012a,b, 2014, 2015, 2016a,b,c,d; Surai and Fotina, 2010a, 2011a, 2012a,b,c,d; Surai and Kochish, 2017, 2019; Surai *et al.*, 2017a,b, 2018, 2019a). Practical aspects of usage of the concept for effective protection against various stresses in commercial stressful conditions have been considered above. The vitagene concept is proved to be effective in broiler production (Fotina *et al.*, 2011, 2014; Velichko and Surai, 2014; Velichko *et al.*, 2013; Grigorieva *et al.*, 2017;), in nutrition and management of rearing birds, egg breeders and meat breeders (Latipova, 2014a,b; Latipova and Shatskikh, 2014, 2014a,b,c; Latipova *et al.*, 2014,

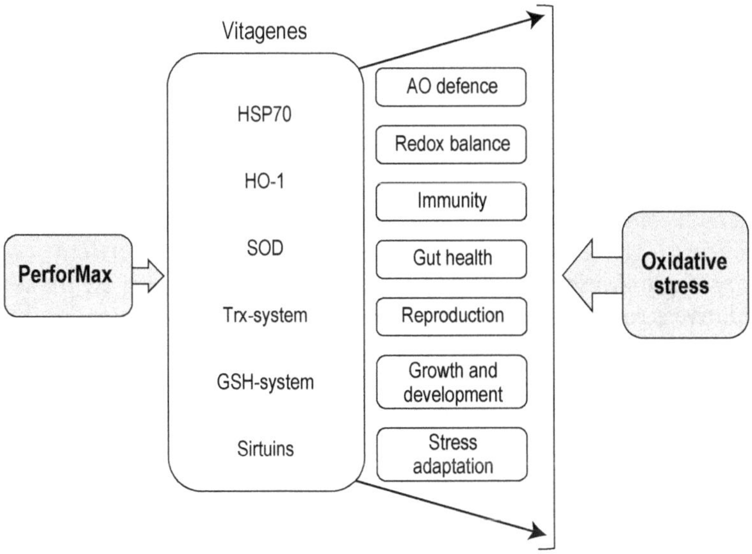

Figure 13.12. Protective effects of PerforMax in poultry production (Surai *et al.*, 2018).

2016; Shatskikh and Latipova, 2013, 2014, 2014a; Shatskikh *et al.*, 2015, 2015a, 2016, 2016a,b; Surai *et al.*, 2017a, 2018). The vitagene-regulating mixture was also effective for improvement of hatchability and chick quality (Fisinin and Surai, 2012g; Surai and Fisinin, 2013), at chicken placement (Surai and Fisinin, 2012h, 2015), for support of high immunocompetence (Fisinin and Surai, 2011, 2013a; 2013b; Surai and Fotina, 2012a,e, 2013b; Surai *et al.*, 2012), eggshell quality (Surai and Fotina, 2010b; 2011), to prevent/restrict feather pecking (Surai, 2012). Furthermore, vitagene activation is an important strategy to fight stresses associated with feed contamination by T-2 toxin (Fisinin and Surai, 2012c,d), DON (Fisinin and Surai, 2012e,f) and ochratoxins (Fisinin and Surai, 2012a,b), as well as showing protective action under heat stress in poultry (Surai and Fotina, 2013) and pigs (Surai, 2017a; Surai and Fotina, 2013a) as well under other stressful situations (Surai and Boroday, 2010; Surai and Fotina, 2010, 2011, 2012, 2012a,b,c). As mentioned above, this concept was also applicable in pig production (Gaponov *et al.*, 2012; Surai and Litvinov, 2017; Surai and Melnichuk, 2012) and is associated with maternal programming in poultry (Fisinin *et al.*, 2016). Therefore, it was proven that stress adaptation in poultry and farm animals takes place at the gene (vitagene) level and vitagene activation help the body to use its own reserves more efficiently for re-establishment of the homeostasis. Additional synthesis of the aforementioned protective molecules leads to re-establishment of the disturbed redox balance and cell signalling with following improvement of the stress resistance. Therefore, vitagene activation helps improve adaptive ability of poultry to various stress conditions. The vitagene interactions with various transcription factors, including Nrf2 и NF-κB deserves further research.

13.7 Conclusions

As can be seen from the data provided above vitagene activation could be considered as an effective means to deal with various stresses in poultry production. It is well accepted that commercial poultry production is associated with a range of stresses starting from laying eggs by breeders, their collection and incubation, continuing with chicken placement, vaccinations and growth and finalising at bird transportation to slaughtering. Therefore, the first vitagene-regulating nutrient composition (PerforMax/Magic Antistress Mix) was proven to be effective in those situations where oxidative stress takes place and vitagene activations was reported to improve bird's resistance to stresses and therefore maintain their productive and reproductive performance. Indeed, the aforementioned data are a good example of technology transfer from the laboratory research into the commercial chicken production. There are also indications that the vitagene concept and vitagene-regulating nutritional mixtures could be effective in fighting stresses in pig production as well as for health maintaining of other animals (calves and cattle, lambs and sheep, kids and goats, fur-animals, companion animals, zoo animals, fish, etc.) under various stress conditions (Surai *et al.*, 2018).

References

Ahmad, T. and Sarwar, M., 2006. Dietary electrolyte balance: implications in heat stressed broilers. World's Poultry Science Journal 62: 638-653.

Dittoe, D.K., Ricke, S.C. and Kiess, A.S., 2018. Organic acids and potential for modifying the avian gastrointestinal tract and reducing pathogens and disease. Frontiers in Veterinary Science 5: 216.

Estevez, M. and Petracci, M., 2019. Benefits of magnesium supplementation to broiler subjected to dietary and heat stress: improved redox status, breast quality and decreased myopathy incidence. Antioxidants 8: 456.

Fisinin, V.I. and Surai, P.F., 2011a. Effective protection against stresses in poultry production: from vitamins to vitagenes. Ptitsa i Pitseproducti (Poultry and Poultry Products, Moscow) 5: 23-26.

Fisinin, V.I. and Surai, P.F., 2011b. Effective protection against stresses in poultry production: from vitamins to vitagenes. Ptitsa i Pitseproducti (Poultry and Poultry Products, Moscow) 6: 10-13.

Fisinin, V.I. and Surai, P.F., 2011. Immunity in modern animal and poultry production: new discoveries and prospects. Zhivotnovodstvo Sjogodni (Today's Animal Production, Ukraine) 9: 40-47.

Fisinin, V.I. and Surai, P.F., 2012a. Mycotoxins and antioxidants: uncompromised fighting (ochratoxin A). Part 1., Kombikorma (Moscow) 3: 55-60.

Fisinin, V.I. and Surai, P.F., 2012b. Mycotoxins and antioxidants: uncompromised fighting (ochratoxin A). Part 2., Kombikorma (Moscow) 4: 59-60.

Fisinin, V.I. and Surai, P.F., 2012c. Mycotoxins and antioxidants: uncompromised fighting (T-2 toxin – metabolism and toxicity). Ptitsa i Pitseproducti (Poultry and Poultry Products, Moscow) 3: 38-41.

Fisinin, V.I. and Surai, P.F., 2012d. Mycotoxins and antioxidants: uncompromised fighting (T-2 toxin – mechanisms of toxicity and protection). Ptitsa i Pitseproducti (Poultry and Poultry Products, Moscow) 4: 36-39.

Fisinin, V.I. and Surai, P.F., 2012e. Properties and toxicity of deoxynivalenol (DON). Part1. Zhivotnovodstvo Rossiyi (Russian Animal Production, Moscow). Part 1. 5: 11-16.

Fisinin, V.I. and Surai, P.F., 2012f. Properties and toxicity of deoxynivalenol (DON). Part 2. Zhivotnovodstvo Rossiyi (Russian Animal Production, Moscow). 6: 3-5.

Fisinin, V.I. and Surai, P.F., 2012g. First days of chicken life: From stress prevention to effective adaptation. Ptitsevodstvo (Moscow) 2: 11-15.

Fisinin, V.I. and Surai, P.F., 2012h. Chicken early nutrition and muscle development. Ptitsevodstvo (Moscow) 3: 9-12.

Fisinin, V.I. and Surai, P.F., 2013a. Immunity in modern animal and poultry production: from theory to practice of immunomodulation. Ptitsevodstvo (Moscow) 5: 4-10.

Fisinin, V.I. and Surai, P.F., 2013b. Gut immunity in birds: facts and thoughts. Agricultural Biology (Moscow) 4: 1-25.

Fisinin, V.I., Shatskikh, E.V., Latipova, E.N. and Surai, P.F., 2016. Maternal effects in poultry production – from vitamins to vitagenes and epigenetics. Ptitsa i Pitseproducti (Poultry and Poultry Products, Moscow) 1: 29-33.

Fotina, A.A., Fotina, T.I. and Surai, P.F., 2011. Effect of antistress preparation 'Feed Food Magic Antistress Mix' on growing chickens. Annals of Sumy Agrarian University, Veterinary Medicine Series 2: 158-162.

Fotina, A., Fotina, T.I. and Surai, P., 2014. Effect of a water-soluble antistress composition on broiler chickens. Proceedings of the XIVth European Poultry Conference. Stavenger, Norway 23-27 June, 2014, p. 555.

Gaponov, I.V., Fotina, T.I. and Surai, P.F., 2012. Physiological and technological stresses at time of piglet weaning. Protective effect of the antistress preparation. Svinatstvo Ukraini (Pig Production of Ukraine, Kiev) 6: 6-9.

Grigorieva, M.A., Velichko, O.A., Shabaldin, S.V., Fisinin, V.I. and Surai, P.F., 2017. Vitagene regulation as a new strategy to fight stresses in poultry production. Agricultural Biology (Moscow) 52: 716-730.

Latipova, E.N., 2014a. Efficacy of usage of antistress preparations 'Vitaminocid' and 'Magic Antistress Mix' in poultry farming. PhD Thesis. Orenburg, Russia.

Latipova, E.N., 2014b. Reproductive parameters of layer breeders and progeny quality as a result of antistress preparation usage. Agrarniy Vestnik Urala (Agricultural Ural Journal) 120: 24-31.

Latipova, E.N., Shatskikh, E.V., Nesvet, E.G. Surai, P.F. and Zelenskaya, O.V., 2014. Usage of antistress preparations 'Vitamonocid' and 'Magic Antistres Mix' in poultry farming: research and practical advice. Ural Agricultural University, Ekaterinburg, Russia.

Latipova, E.N., Shatskikh, E.V. and Surai, P.F., 2016. Effect of an antistress dietary supplement on the reproductive performance of layer breeders. The Proceedings of the XXV Worlds Poultry Congress. 2016. Abstracts, S1-0137.

Latipova, E.N. and Shatskikh, E.V., 2014. 'Vitaminocid' and 'Magic Antistress Mix' in the diet of the egg breeders. Agrarniy Vestnik Urala (Agricultural Ural Journal) 119: 36-40.

Latipova, E.N. and Shatskikh, E.V., 2014a. Reproductive performance of egg breeders as a result of usage of 'Vitaminocid' and 'Magic Antistress Mix'. Glavniy Zootechnik (Chief Zootechnician, Moscow) 4: 34-40.

Latipova, E.N. and Shatskikh, E.V., 2014b. Reproductive parameters of layer breeders and progeny quality as affected by antistress preparation usage. Agrarniy Vestnik Urala (Agricultural Ural Journal) 120: 24-31.

Latipova, E.N. and Shatskikh, E.V., 2014c. Development of breeding birds and incubation egg quality as a result of usage antistress preparations. Bio 164: 4-8.

Martínez, Y., Li, X., Liu, G., Bin, P., Yan, W., Más, D., Valdivié, M., Hu, C.A.A., Ren, W. and Yin, Y., 2017. The role of methionine on metabolism, oxidative stress, and diseases. Amino Acids 49: 2091-2098.

Mastrototaro, L., Sponder, G., Saremi, B. and Aschenbach, J.R., 2016. Gastrointestinal methionine shuttle: priority handling of precious goods. IUBMB life 68: 924-934.
Monteiro, C.P., Matias, C.N., Bicho, M., Santa-Clara, H. and Laires, M.J., 2016. Coordination between antioxidant defences might be partially modulated by magnesium status. Magnesium Research 29: 161-168.
Mushtaq, M.M.H., Pasha, T.N., Mushtaq, T. and Parvin, R., 2013. Electrolytes, dietary electrolyte balance and salts in broilers: an updated review on growth performance, water intake and litter quality. World's Poultry Science Journal 69: 789-802.
Olanrewaju, H.A., Thaxton, J.P., Dozier III, W.A. and Branton, S.L., 2007. Electrolyte diets, stress, and acid-base balance in broiler chickens. Poultry Science 86: 1363-1371.
Ratriyanto, A. and Mosenthin, R., 2018. Osmoregulatory function of betaine in alleviating heat stress in poultry. Journal of Animal Physiology and Animal Nutrition 102: 1634-1650.
Schaefer, A.L., Jones, S.D.M. and Stanley, R.W., 1997. The use of electrolyte solutions for reducing transport stress. Journal of Animal Science 75: 258-265.
Shatskikh, E.V., Latipova, E.N., Nesvet, E.G. and Koburneev, I.V., 2016. Usage of anti-stress preparations in poultry production. Ural State University. Russia, Ekaterinburg, Russia.
Shatskikh, E., Latypova, E., Fisinin, V., Denev, S. and Surai, P., 2015. Molecular mechanisms and new strategies to fight stresses in egg-producing birds. Agricultural Science and Technology 7: 3-10.
Shatskikh, E.V. and Latipova, E.N. (2013). Effect of antistress preparations on the growth and development of egg type breeder birds. Agrarniy Vestnik Urala (Agricultural Ural Journal) 117: 33-37.
Shatskikh, E.V. and Latipova, E.N., 2014. Effect of antistress preparations on the development of rearing birds. Ptitsevodstvo (Moscow) 1: 22-27.
Shatskikh, E.V. and Latipova, E.N., 2014a. Development of internal organs in egg type rearing birds as a result of usage of antistress additives. Agrarniy Vestnik Urala (Agricultural Ural Journal) 122: 35-42.
Shatskikh, E.V, Latipova, E.N. and Surai, P.F., 2016a. Supplying an antistress composition with water to decrease negative consequences of commercially-relevant stresses in rearing birds. The Proceedings of the XXV Worlds Poultry Congress. 2016. Abstracts, S1-0140.
Shatskikh, E.V, Latipova, E.N., Fisinin, V.I. and Surai, P.F., 2016b. Epigenetic effects of an antioxidant composition in layer breeder diet. The Proceedings of the XXV Worlds Poultry Congress. 2016. Abstracts, S1-0142.
Shatskikh, E.V., Surai, P.F. and Latipova, E.N., 2015a. Morphological characteristics of blood of egg type birds as a result of usage of antistress additives. Agrarniy Vestnik Urala (Agricultural Ural Journal) 131: 44-48.
Surai, P.F. 1999. Vitamin E in avian reproduction. Poultry Avian Biology Reviws 10: 1-60.
Surai P.F. 2002. Natural antioxidants in avian nutrition and reproduction. Nottingham University Press, Nottingham, UK.
Surai, P.F., 2005. Minerals and antioxidants. In: Tucker, L.A. and Taylor-Pickard, J.A. (eds) Redefining mineral nutrition. Nottingham University Press, Nottingham, UK, pp. 147-177.
Surai, P.F., 2006. Selenium in nutrition and health. Nottingham University Press, Nottingham, UK.
Surai, P.F., 2011. Stresses in poultry production: from understanding mechanisms to the development of protection methods. Agrarnoye Resheniye (Russia) 4: 42-46.
Surai P F., 2012. New approaches in fighting stresses decrease cannibalism in chickens. Korma I Facti (Feed and Facts, Kiev) 2: 7-8.
Surai, P.F., 2012a. The antioxidant properties of canthaxanthin and its potential effects in the poultry eggs and on embryonic development of the chick. Part 1. World's Poultry Science Journal 68: 465-476.

Surai, P.F., 2012. The antioxidant properties of canthaxanthin and its potential effects in the poultry eggs and on embryonic development of the chick. Part 2. World's Poultry Science Journal 68: 717-726.

Surai, P.F., 2014. Polyphenol compounds in the chicken/animal diet: from the past to the future. Journal of Animal Physiology and Animal Nutrition 98: 19-31.

Surai, P.F., 2015a. Stresses in poultry production: from understanding mechanisms of development to the development of protection. Zhivotnovodstvo Rossii (Russian Animal Production, Moscow;), Special issue: 30-31.

Surai, P.F., 2016. Antioxidant systems in poultry biology: superoxide dismutase. Journal of Animal Nutrition 1: 8.

Surai, P.F., 2017a. What happens with pig under the heat stress and how to protect them from the stress. Korma I Facti (Feed and Facts, Kiev) 6-7: 35-37.

Surai, P.F., 2018. Selenium in poultry nutrition and health. Wageningen Academic Publishers, Wageningen, the Netherlands.

Surai, P.F., 2019a. Vitagenes in poultry production: stresses and antioxidants. Suchasne Ptakhivnitstvo (Modern Poultry Science, Ukraine) 9-10: 24-27.

Surai, P.F., 2019b. Vitagenes in poultry production: adaptation to commercially relevant stresses. Suchasne Ptakhivnitstvo (Modern Poultry Science, Ukraine) 7-8: 28-32.

Surai, P.F., 2019c. From healthy liver to general health and high productivity of poultry. Ptaknivnitstvo.ua 1-2: 4-7.

Surai, P.F., 2019d. Decreasing eggshell hairline cracks: from vitagenes to a ShellBone concept ShellBone. Ptaknivnitstvo.ua 5: 24-27.

Surai, P.F., 2020b. Antioxidants in poultry nutrition and reproduction: an update. Antioxidants 9(2): 105.

Surai, P.F., 2020a. Superoxide dismutase as a new entrant into the vitagene family in animals/poultry. EC Nutrition 15.3: 01-03.

Surai, P.F. and Boroday, V.P., 2010. Stresses in poultry production: from understanding of mechanisms of development to the development of methods of protection. Suchasne Ptakhivnitstvo (Modern Poultry Science, Ukraine) 7-8: 31-36.

Surai, P.F. and Dvorska, J.E., 2005. Effects of mycotoxins on antioxidant status and immunity. In: Diaz, D.E. (ed.) The mycotoxin blue book. Nottingham University Press, Nottingham, UK, pp. 93 137.

Surai, P.F. and Fisinin, V.I., 2012a. The modern antistress technologies in poultry: from antioxidants to vitagenes. Agricultural Biology (Moscow) 4: 3-13.

Surai, P.F. and Fisinin, V.I., 2012b. Modern methods for fighting stresses in poultry production: from antioxidants to sirtuins and vita-genes. Effectivne Ptakhivnitstvo (Effective Poultry Production, Ukraine) 8: 8-13.

Surai, P.F. and Fisinin, V.I., 2013. Natural antioxidants in hen's embryogenesis and antioxidant defence in postnatal development. Agricultural Biology (Moscow) 2: 3-17.

Surai, P.F. and Fisinin, V.I., 2014. Antioxidant systems of the body: from vitamin E to polyphenols and beyond. Proceedings of the 35th Western Nutrition Conference. September 24-25, 2014, Edmonton, Alberta, pp. 265-277.

Surai, P.F. and Fisinin, V.I., 2015. Antioxidant-prooxidant balance in the intestine: applications in chick placement and pig weaning. Journal of Veterinary Science & Medicine 3: 66-84.

Surai, P.F. and Fisinin, V.I., 2016a. Vitagenes in poultry production. Part 3. Vitagene concept development. World's Poultry Science Journal 72: 793-804.

Surai, P.F. and Fisinin, V.I., 2016b. Natural antioxidants and stresses in poultry production: from vitamins to vitagenes. Proc. XXV World Poultry Congress, Invited Lecture Papers, Beijing, China, 5-9 September, 2016, pp. 116-121.

Surai, P.F. and Fisinin, V.I., 2016c. Antioxidant system regulation: from vitamins to vitagenes. In: Watson, R.R. and De Meester, F. (eds) Handbook of cholesterol. Wageningen Academic Publishers, Wageningen, the Netherlands, pp. 451-481.

Surai, P.F., Fisinin, V.I., Shatskikh E.V and Latipova, E.N., 2017b. Modern methods of fighting stresses in poultry and pig production: vitagene concept in action. Sfera 2: 40-43.

Surai, P.F. and Fotina, T.I., 2010a. Once more about stresses. From changes in gene expression to usage with drinking water of antistress premix. Effectivne Ptakhivnitstvo (Effective Poultry Production, Ukraine) 8: 20-25.

Surai, P.F. and Fotina, T.I., 2010b. Eggshell quality in egg production. What is new in modern science? Korma I Facti (Feed and Facts, Ukraine) 4: 6-10.

Surai, P.F. and Fotina, T.I., 2011. Eggshell quality in egg production: introduction of innovative technologies into industry. Ptakhivnitstvo.ua (Modern Poultry Production, Ukraine) 1: 8-12.

Surai, P.F. and Fotina, T.I., 2011a. New approaches in fighting stresses in poultry production. From vitamins to vitagenes and sirtuins. Agrarnoye Resheniye (Russia) 9-10: 34-39.

Surai, P.F. and Fotina, T.I., 2012. Molecular mechanisms of immunosuppression. Is there light at the end of the tunnel? Part 1. Suchasna Veterinarna Medicina (Modern Veterinary Medicine, Ukraine) 6: 14-19.

Surai, P.F. and Fotina, T.I., 2012a. Vita-genes – next step after nutrigenomics: Lessons for poultry production. Ptakhivnitstvo (Kiev) 68: 429-435.

Surai, P.F. and Fotina, T.I., 2012b. Stresses in poultry production. Molecular mechanisms. Propozhizia (Ukraine) 9: 130-132.

Surai, P.F. and Fotina, T.I., 2012c. Stresses in poultry production. Methods of prevention. Propozhizia, Ukraine 10: 132-134.

Surai, P.F. and Fotina T.I., 2012d. In order to prevent stresses. Nashe Ptakhivnitstvo (Our Poultry Production, Ukraine) 3: 72-74.

Surai, P.F. and Fotina, T.I., 2012e. Molecular mechanisms of immunosuppression: is there light at the end of the tunnel? Part 1. Suchasna Veterinarna Madicina (Modern Veterinary Medicine, Ukraine) 6: 22-25.

Surai, P.F. and Fotina, T.I., 2013. Physiological mechanisms of the development of heat stress in poultry production. Tvarinnitstvo Syogodni (Today's Animal Production, Ukraine) 6: 54-60.

Surai, P.F. and Fotina, T.I., 2013a. Physiological mechanism and practical measures to decrease negative consequences of heat stress in pig production. Svinarstvo Ukraini (Pig Production of Ukraine, Kiev) 6: 12-15.

Surai, P.F. and Fotina, T.I., 2013b. Molecular mechanisms of immunosupression. Is there light at the end of the tunnel? Part 2. Suchasna Veterinarna Medicina (Modern Veterinary Medicine, Ukraine) 1: 26-27.

Surai P.F., Fotina A.A and Fotina T.I., 2012. Effect of Feed-Food Magic Antistress Mix on natural resistance of ducklings. Proceedings of Sumy National Agrarian University 7: 58-61.

Surai, P.F. and Kochish, I.I., 2017. Antioxidant systems and vitagenes in poultry biology: heat shock proteins. In: Asea, A.A.A. and Kaur, P. (ed.) Heat shock proteins in veterinary. Springer, Basel, Switzerland, pp. 123-177.

Surai, P.F. and Kochish, I.I., 2019. Nutritional modulation of the antioxidant capacities in poultry: the case of selenium. Poultry Science 98: 4231-4239.

Surai, PF and Kochish, I.I., 2018. Oxidative damage of biological molecules on animal metabolism and physiology. Proc. of the 2018 Animal Nutritional Conference of Canada. 2-3 May, 2018, Edmonton, Canada, pp. 234-251.

Surai, P.F., Kochish, I.I. and Fisinin, V.I., 2017a. Antioxidant systems in poultry biology: nutritional modulation of vitagenes. European Journal of Poultry Science 81: 1612-9199.
Surai, P.F., Kochish, I.I., Fisinin, V.I. and Kidd, M.T., 2019a. Antioxidant defence systems and oxidative stress in poultry biology: an update. Antioxidants 8, 7: 235.
Surai, P.F., Kochish, I.I., Fisinin, V.I., Grozina, A.A. and Shatskikh E.V., 2018. Molecular mechanisms of chicken gut health maintenance: role of microbiota. Agricultural Technologies, Moscow, Russia.
Surai, P.F. and Melnichuk, S.D., 2012. Mechanisms of protection from stresses in pig production: from vitamins to vitagenes. Svinarstvo Ukraini (Pig Production of Ukraine, Kiev) 2: 10-15.
Surai, P.F. and Litvinov, A., 2017. From vitamins to vitagenes. Modern method to fight stresses in pig production. Svinovodstvo (Pig Produuction, Russia) 3: 56-58.
Surai, P.F., Mezes, M., Fotina, T.I. and Denev, S.D., 2010. Mycotoxins in human diet: a hidden danger. In: De Meester, F., Zibadi, S. and Watson, R.R. (eds) Modern dietary fat intakes in disease promotion. Humana Press, Totowa, NJ, USA, pp. 275-303.
Velichko, O. and Surai, P.F., 2014. Effect of an antistress composition supplied with water on chick growth and development. Proceedings of the XIVth European Poultry Conference. Stavanger, Norway 23-27 June, 2014, p. 551.
Velichko, O.A., Shbaldin, S.V. and Surai, P.F., 2013. Practical aspects of usage of a vitagene concept in poultry production. Ptitsa I Ptitseproducti (Poultry and Poultry Products, Moscow) 4: 42-45.

Chapter 14
Shellbone concept development

Strike while the iron is hot

14.1 Introduction

World's egg production exceeds 80 million metric tons and eggshell quality determines the success of egg industry. Eggshell is shown to have an unusual combination of mechanical properties providing suitable conditions for the developing chick, which must be stiff and rigid but also brittle enough to be broken at hatching (Taylor *et al.*, 2016). However, when eggs are produced for human consumption, compromised eggshell quality causes considerable economic losses for the egg industry. For example, about 7-10% eggs are broken during laying in battery and furnished cages, plus breakage during egg collection (2-4%), transportation (1.2-2.7%), grading and packing (1.5-3.4%) is added (Mertens *et al.*, 2006) leading to huge losses in the egg industry. On average, losses from damaged eggshells account for >10% of total egg production (Fathi *et al.* 2010; Van Mourik *et al.* 2017). Since there is a trend to house layers longer than 80 weeks eggshell breakage could reach >30%. In fact, in hens around half-way (49 weeks) through the production period, both shell quality and bone strength were shown to deteriorate (Wistedt *et al.*, 2019). Since intact eggshell is an important barrier to various pathogens (Sa e Silva *et al.*, 2016), including Salmonella, broken eggshell is an important risk factor for human egg consumption. On the other hand, eggshell cracks in hatching eggs can substantially decrease hatchability, especially during egg storage before incubation (Nasri *et al.*, 2020). In particular hairline cracks in hatching eggs cause dramatic decrease in egg hatchability (Barnett *et al.*, 2004; Jabbar *et al.*, 2019) and create a lot of problems in egg producing industry (Surai, 2019d).

This chapter presents research background for the development of the vitagene modulating nutritional supplement to deal with eggshell quality deterioration and poultry leg problems in commercial poultry production.

14.2 Molecular mechanisms of egg shell quality deterioration and a choice of nutrients to design a feed supplement

Our recent research identified several important points related to understanding molecular mechanisms of eggshell quality deterioration in aging birds. They could be summarised as follows:

DOI 10.3920/978-90-8686-906-0_14, © Wageningen Academic Publishers 2020

- We believe that subclinical inflammation in the eggshell gland, associated with various commercial stresses, is a major cause of eggshell formation defects, including hairline cracks (Surai, 2019d).
- Decreased efficacy of vitamin D conversion into its active hormonal form (1,25D3) in ageing birds contributes substantially to eggshell quality deterioration (Surai, 2018a,b).
- Decreased synthesis of organic matrix components in ageing birds (Surai, 2018c) is an important contributor to decreased eggshell elasticity leading to increased percentage of crack eggs.

14.3 Taurine and shell gland

The ShellBone concept was developed to address the aforementioned problems. In particular, as it was shown in the previous chapter, taurine is considered to be an important nutrient regulating inflammation via affecting such transcription factors as NF-κB and Nrf2 (Figure 14.1).

14.4 Active vitamin D metabolites and eggshell formation

Furthermore, taurine also affects vitagenes leading to improved adaptability to stress and stress resistance (Surai, 2018d; Surai *et al.*, 2020). Furthermore, there is a strong evidence that vitamin D, especially its active hormonal form (1,25$(OH)_2$D3) showing strong anti-inflammatory properties. For example, 1,25$(OH)_2$D3 was found to be effective in attenuating ER stress and the NF-κB-driven inflammatory response in mammary epithelial (MCF-7) cells (Wen *et al.*, 2020). Furthermore, protective/ inhibitory effects of 1,25$(OH)_2$D3 on the inflammation associated with mastitis were

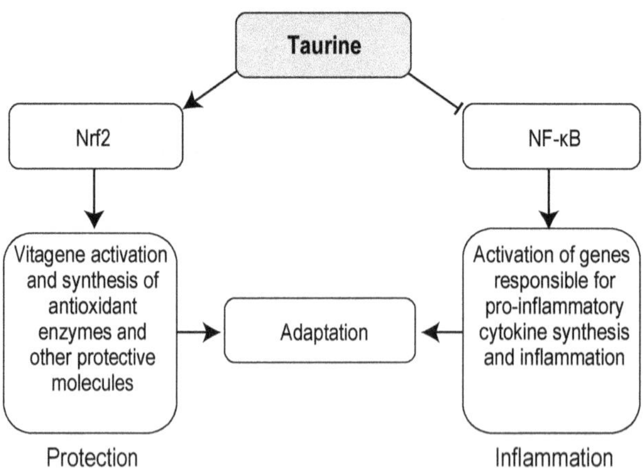

Figure 14.1. Taurine, inflammation, and stress adaptation (adapted from Surai *et al.*, 2020).

clearly shown (Lippolis *et al.*, 2011; Merriman *et al.*, 2017, 2018). Indeed, 1,25$(OH)_2$D3 was shown to exert its anti-inflammation effects through regulating the biosynthesis of pro-inflammatory molecules in the prostaglandin pathway or through NF-κB by affecting cytokine production and inflammatory response (Li *et al.*, 2019a; Lin and Li, 2016).

In laying hens, eggshell formation takes place daily in the hen's eggshell gland being one of the most rapid mineralisation processes known in nature. To produce the eggshell large amounts of $CaCO_3$ is required and the hen has to export about 2 g calcium comprising 10% of her total body calcium ach day (Nys and Le Roy, 2018). Eggshell calcium is derived from duodenal absorption and skeletal (medullary bone) stores. Importantly, duodenal calcium uptake was shown to be increased 6-fold during shell formation, but efficiency of Ca absorption is age-dependent and declines from 37 weeks of age in layers (Al-Batshan *et al.*, 1994).

Important roles of vitamin D in Ca absorption and metabolism are well established. Indeed, vitamin D affects expression and synthesis of Ca-binding proteins being an essential regulator of eggshell and bone formation and their integrity maintenance. An understanding of molecular mechanisms controlling 1,25$(OH)_2$D3 production is of great importance in explaining how the layer intestine adapts to fluctuations in calcium requirement and provision. In poultry, dietary vitamin D3 is shown to be absorbed in the small intestine and is transported via blood to the liver where it is converted into 25-hydroxycholecalciferol (25-OHD), the major circulating form of Vitamin D3. The next stage of Vitamin D3 metabolism includes transport of 25-OHD to the kidneys where it is converted into 1,25-dihydroxycholecalciferol (calcitriol, 1,25$(OH)_2$D3), the most biologically active, hormonal metabolite of the vitamin (Waldenstedt, 2006). Essential roles of 1,25$(OH)_2$D3 in eggshell formation is well established and during layer maturation (transfer from immature pullets to laying birds) concentration of 1,25$(OH)_2$D3 in plasma increased more than 5 fold (Bar, 2009). Furthermore, eggshell calcification is shown to be associated with a further 2-fold increase in the plasma level of 1,25$(OH)_2$D3 (Nys and Le Roy, 2018). However, with age, there is a progressive deterioration in the ability of the hen's liver to hydroxylate Vitamin D3 to 25-OH-D3 (Soares *et al.*,1976) and reduced kidney 25-OH-D3 hydroxylation resulted in inadequate production of 1,25$(OH)_2$D3 associated with decreased absorption of Ca (Stevens *et al.*, 1984) and decline in egg shell quality and deterioration in tibia strength with age.

Therefore, by using preformed 1.25(OH)2D3 it is possible to design a product of 'fast response' which can be used to address vitamin D inadequacy issues during egg production and broiler growth. In fact, stressful conditions, such as high bird density, heat stress, mycotoxicosis, enteritis, malabsorption syndromes, and certain immune disorders may impair absorption or liver hydroxylation of vitamin D3, which is one rationale for the use of vitamin D metabolites in layer/broiler feed. In fact, the expression of vitamin D receptor (VDR) was shown to decrease upon induction of ER stress and 1,25$(OH)_2$D3 was found to counter-regulates the inhibitory effect of ER stress on VDR expression in mammary epithelial cells (Wen *et al.*, 2020). There is a range of publications confirming an important role of supplying 1,25$(OH)_2$D3 on eggshell and bone formation. For example, when 1,25$(OH)_2$D3 was the only source of vitamin D in the layer diet increased feed consumption and egg production was observed, in comparison to non-supplemented diet (Frost *et al.*, 1990). The is a range of studies confirming that 1,25$(OH)_2$D3 supports egg production and eggshell mineralisation/quality and minimising eggshell breakage. For example, the effect of replacing dietary vitaminD3 by 1,25$(OH)_2$D3 in 32-week-old White Leghorn laying hens over 9 weeks on eggshell quality and egg production was studied (Tsang *et al.*, 1990a). The authors showed that after 4 weeks of experimental feeding, laying hens fed on a diet supplemented with 5 μg 1,25$(OH)_2$D3 /kg diet produced eggs with improved eggshell quality as indicated by significantly higher specific gravity and percentage shell than eggs from control hens fed on a diet supplemented with 27.5 μg D3/kg diet. Interestingly, eggs obtained from hens fed on a diet without vitamin D3 supplementation were characterised by very thin or soft shell, reflecting depletion of the birds' reserves of D3 or its metabolites within 4-week period (Tsang *et al.*, 1990a). Therefore, an optimal dose of 1,25$(OH)_2$D3 in the layer diets was associated with the best eggshell quality and tibial weights relative to the control group supplemented with vitamin D3 (Tsang *et al.*, 1990b). Interestingly, doubling the level of vitamin D3 supplementation in the control diet almost linearly increased the circulating concentration of 25-OH-D3 but did not change the concentration of 1,25$(OH)_2$D3, Ca, or egg specific gravity (SG), while replacing D3 by the optimal concentration of 1,25$(OH)_2$D3 improved egg SG (Tsang and Grunder, 1993). In Japanese quail dietary supplementation with 1.25D3 was associated with increased egg specific gravity, shell weight, percentage of shell, and egg breaking strength (Kaetzel and Soares, 1979). Bar *et al.* (1988) reported that the inclusion of 1,25$(OH)_2$D3 in the diet of aging hens decreased the rate of decline of shell weight and density in birds. In 56-week old laying hens, the addition of 1,25(OH)2D3 in the feed prevented a rapid decrease in egg production due to a Mycoplasma gallisepticum infection observed in hens fed the basal diet. (Carlos, and Edward, 1998). Dietary supplementation with 1,25$(OH)_2$D3 (as *Cestrum diurnum* leaf) helped maintain higher (80%) egg production in comparison to the control group supplemented with vitamin D3 (68%) or negative control without vitamin D supplementation (55%) at the end of production period (72 weeks; Chennaiah *et al.*, 2004). Feeding dried *Solanum glaucophyllum* plant material, as a source of 1,25$(OH)_2$D3, to laying hens has been shown to lead to thicker eggshells (Azcona *et al.*, 1982; Morris, 1977).

It seems likely that 1,25$(OH)_2$D3 is needed for normal chondrocyte metabolism (Boyan *et al.*,1999), therefore, affecting bone formation and metabolism. This is relevant not only to layer bone health maintenance (medullary bone is the major source of Ca for eggshell formation, providing about 40% of the Ca required) but also for commercially growing birds, where Tibial dyschondroplasia (TD) is a substantial problem. Indeed, TD, a bone deformity disease of rapidly growing birds, characterised by accumulation of non-mineral and avascular cartilage that causes chondrocyte death due to lack of blood supply, could cause great economic losses in broiler production. There is a growing body of evidence indicating that TD can be prevented or its incidence and severity reduced by using 1,25$(OH)_2$D3 (Edwards, 1990; Edwards *et al.*, 1992; Rennie and Whitehead, 1996; Rennie *et al.*, 1993) suggesting that the metabolic conversion of vitamin D3 to 1,25$(OH)_2$D3 is not sufficient under some stressful conditions to meet the requirement for this metabolite for physiological bone development.

Interestingly, 1,25$(OH)_2$D3 was shown to be effective not only in regulation of Ca absorption and metabolism, but also phosphorus utilisation in broilers can be improved by feeding 1,25(OH)2D3 (Edwards, 2002). Indeed, dietary supplement of 1,25$(OH)_2$D3 was shown to alleviate the incidence and severity of TD and rickets and increase bone mineralisation (Edwards *et al.*, 1992; Elliot and Edwards, 1997; Rennie *et al.*, 1993, 1995).

14.5 Manganese and eggshell quality

It has been shown that Mn plays an important role in maintaining structure and integrity of eggshell (Swiatkiewicz *et al.*, 2015; Xiao *et al.*, 2015). In particular, Mn dietary supplementation was shown to increase eggshell thickness and strength by improving its structure, including changes in the mammillary layer of the eggshell (Leach and Gross, 1983; Xiao *et al.*, 2014; Zhang *et al.*, 2017a,b).

Eggshell is considered to be bio-ceramics composed of six layers: the uncalcified inner and outer shell membranes, irregular cones corresponding to the mammillary knob layer, the palisade layer, thin vertical crystal layer and the outer layer, the cuticle (Nys and Le Roy, 2018). In the phase of shell formation (five hours after ovulation) the egg enters the uterus (shell gland) and rotates with the linear deposition of calcium carbonate (0.33 g/h in laying hens) with the mammillary and palisade layers are sequentially formation (Nys *et al.*, 2004). During about 20 hours, in the shell gland, the egg acquires the shell, consisting of 95% (by weight) calcium carbonate ($CaCO_3$) and about 3.5% is represented by an organic matrix consisting mainly of proteoglycans and glycoproteins in the calcified layers and fibrillar proteins with disulphide cross-links and collagen types I, V and X in the eggshell membranes (Nys *et al.*, 2004). It is proven that the eggshell ultrastructure (e.g. calcite crystal size, shape and orientation and other parameters) determine its biomechanical properties (Ahmed *et al.*, 2005; Fathi *et al.*, 2016; Radwan, 2015; Radwan *et al.*, 2010). There is a positive correlation between calcite crystal size inside the eggshell, their orientation and thickness of cones of mammillary layer (Dunn *et al.*, 2012). Indeed, the size, orientation and thickness of cones of mammillary knob layer determine major characteristics of commercial importance (e.g. strength) of the eggshell (Fernandez *et al.*, 2001; Nys *et al.*, 1999). Formation of cones of mammillary knob layer takes place at the beginning of the eggshell formation (from 5.5 up to 10 h to ovulation) associated with initiation of aggregation and growth in nucleation points (Nys *et al.*, 2004). It was shown that chicken eggshell contains two main sulphated polymers (proteoglycans), referred to as mammillan and ovoglycan which are involved in nucleation and growth of the eggshell calcite crystals. Based on existing evidences, the putative role of mammillan, a keratan sulphate proteoglycan, is in the nucleation of the first calcite crystals, while that of ovoglycan, a dermatan sulphate proteoglycan, is to regulate the growth and orientation of the later forming crystals of the chicken eggshell (Arias *et al.*, 2001).

Kinetics of $CaCO_3$ crystal formation and their morphologic structure are affectively modulated by organic matrix components found in the liquid of eggshell gland (Gautron *et al.*, 1997; Nys *et al.*, 1999). Furthermore, Mn supplementation with feed or water can improve eggshell ultrastructure due to modulation of sulphated proteoglycans in membranes and in the calcified eggshell (Xiao *et al.*, 2014; Zhang *et al.*, 2017a). In fact, dietary Mn supplementation is involved in the process of protein glycosylation and glycan metabolism and can improve the expression of genes encoding proteoglycans and glycoproteins in the eggshell gland, thus improving eggshell quality by increasing the mammillary-knob density during the initial deposition stage of shell formation (Zhang *et al.*, 2018). Since Mn is an integral part of SOD, main enzyme of the antioxidant defence network (Surai, 2016, 2020a), regulated at the vitagene level (Surai, 2020; Surai *et al.*, 2019b), therefore, an additional supplementation of Mn could be an effective way to increase antioxidant defences (Surai, 2005) in the eggshell gland leading to decreased inflammation (Surai, 2019d) and improvement of eggshell quality (Surai *et al.*, 2018).

14.6 Zinc and eggshell quality

During eggshell formation many different enzymes take part. For example, carbonic anhydrase (CA) is Zn-dependent enzyme responsible for reversible hydration of carbon dioxide and bicarbonate:

$$CO_2 + H_2O \longrightarrow HCO_3^- + H^+$$

Furthermore, HCO_3^- is a precursor of $CaCO_3$, main structural constituent of the eggshell. In fact, more than 70 years ago it was shown that CA affected eggshell quality and is the most important enzymes in calcite formation (Gutowska and Mitchell, 1945). CA was shown to contain 0.31-0.34% of Zn (Hove *et al.*, 1940). As expected, Zn deficiency is associated with decreased activity of a number of enzymes, including CA, in different animal species, including avian species (Ensminger *et al.*, 1990). Furthermore, partial or complete CA inhibition lead to decreased eggshell quality (Holm *et al.*, 2006; Nys *et al.*, 1999), reduced eggshell thickness (Holm *et al.*, 2006), and mass (Nys, 2001) and production of shell-less or soft-shelled eggs (Benesch *et al.*, 1944; Diamant and Schluns, 1964; Lundholm, 1990). On the other hand, increased CA activity in laying hens by estradiol was shown to increase eggshell quality at the second part of production period (Wistedt *et al.*, 2014). Interestingly, at 49 week of age, a dramatic reduction in duodenal CA activity, coinciding with decreased gland density in the shell gland was observed (Wistedt *et al.*, 2019). Similar age-dependent decrease in CA activity the shell gland of laying hens was observed (Snapir and Perek, 1070). Furthermore, increased Zn dietary supplementation for aged layers was shown to increase eggshell strength leading to decreased percentage of eggshell cracks and breakage (Zamani *et al.*, 2005) with simultaneous increased CA activity in the eggshell gland (Zhang *et al.*, 2017a). It seems likely that Zn can also improve eggshell quality as a result of direct regulation of eggshell formation by affecting growth stage of calcite crystals *in vitro* (Elzinga and Reeder, 2002). Interestingly, various combinations of Zn an Mn in the diet were shown to improve eggshell strength (Mabe *et al.*, 2003; Stefanello *et al.*, 2014; Swiatkiewicz and Koreleski, 2008).

14.7 Ascorbic acid and eggshell quality

About 80 years ago it was shown that poultry can synthesise ascorbic acid in their kidney (Grollman and Lehninger, 1957; Roy and Guha, 1958). However, under various stress conditions ascorbic acid synthesis can be compromised and dietary supplementation of this vitamin could have positive effects on productive and reproductive performance of poultry (McKee and Harrison, 1995). However, efficacy of such ascorbic acid dietary supplementation is condition-dependent and quite variable (Scott, 1975; Pardue and Thaxton, 1986). There is a range of publications claiming beneficial effects of ascorbic acid on egg shell strength (Balnave *et al.*, 1991; Bell and Marion, 1990; El-Boushy and van Albada, 1970; Orban *et al.*, 1993; Peebles and Brake, 1985; Perek and Kendler, 1963; Thornton and Moreng, 1958; Zapata and Gernat, 1995). However, molecular mechanisms of vitamin C effects on eggshell quality require further investigation. In particular, ascorbic acid participates in vitamin D metabolism in poultry (Weiser *et al.*, 1988a). It was shown that decreased 1,25$(OH)_2$D3 concentration in chicken serum was associated with decreased eggshell quality (Castaldo and Maurice, 1990) and dietary supplementation of 1,25$(OH)_2$D3 for layers at 71-weeks for 8 weeks decreased percentage of eggs with damaged/cracked eggshell, while additional usage of ascorbic acid in these conditions further improved eggshell quality (Weiser *et al.*, 1992). Therefore, participation of vitamin C in cholecalciferol hydroxylation (Weiser *et al.*, 1988a) and 1,25$(OH)_2$D3 synthesis (Farquharson *et al.*, 1998) may be a background of synergistic action of these two vitamins. Furthermore, ascorbic acid dietary supplementation was shown to increase egg production and hatchability (Balnave *et al.*, 1991; Bell and Marion, 1990; Orban *et al.*, 1993; Peebles and Brake, 1985; Zapata and Gernat, 1995), to decrease percentage of eggs with damaged eggshell (Weiser *et al.*, 1992) at the second part of the productive period. It seems likely that vitamin C is especially effective in poultry flocks with comparatively low egg production (Peebles and Brake, 1985). As mentioned above, vitamin C efficacy in poultry nutrition depends on stress strength and doses and combinations of other antioxidants included into the diet. For example, Whitehead *et al.* (1992) demonstrated that decreased feed consumption and egg production reduction by laying hens under heat stress (35 °C) were prevented by inclusion into the diet quite high (500 g/t) vitamin C levels. Interestingly, a combination of ascorbic acid and vitamin E was shown to significantly improve eggshell quality under heat stress.

It should be mentioned that vitamin C dietary supplementation also showed positive effects on layer bone strength (Chung *et al.*, 2005). At the same time, dietary vitamin C can decrease detrimental effects caused by various drugs (Ali *et al.*, 1984), heavy metal contamination (Hill, 1979; Toussant and Latshaw, 1994), and increase resistance to various poultry diseases (Amakye-Anim *et al.*, 2000). As mentioned above, 1,25$(OH)_2$D3 concentration in blood decreased in layers with eggshell problems (Castaldo and Maurice, 1990; Soares, 1984) and ascorbic acid dietary supplementation was shown to increase 1,25$(OH)_2$D3 concentration and improve bone calcification in birds (Weiser *et al.*, 1988). In general, vitamin C participates in bone formation by stimulating synthesis of hydroxyproline, an important amino acid in collagen synthesis. It should be taken into account that collagen fibrils comprise an important part of organic matrix of eggshell and bones (Weiser *et al.*, 1988a). Therefore, ascorbic acid plays an important role in the formation and development of bones as a co-factor in transcription and enzymatic reactions which regulate early differentiation of osteoblasts as well as biosynthesis of various molecules of organic matrix of bones, including collagen. In particular, ascorbic acid was shown to inhibit osteoclast activity and reduce their number, simultaneously increase differentiation of osteoblast leading to enhancement of bone development (Gabbay *et al.*, 2010).

14.8 Conclusions

Eggshell quality deterioration in ageing birds is a great problem in commercial poultry production. There is a range of publications indicating beneficial role of various nutrients on eggshell quality. For example, supplying ascorbic acid with feed or water in stress conditions was shown to have beneficial effects on eggshell and bone quality. It seems likely that a combination of ascorbic acid with other nutrients, including 1,25$(OH)_2$D3, taurine and minerals may show higher protective efficacy than its single usage. In general, it should be stated that for improvement of eggshell quality and solving problems with bone, including deformity and TD the following nutrients are of great importance:

- Taurine;
  - a major player in prevention of eggshell gland inflammation responsible for decreased eggshell quality, including hairline cracks;
  - activating vitagenes, responsible for stress adaptation and improved stress resistance;
- 1,25$(OH)_2$D3;
  - an active hormonal form of vitamin D, regulating Ca absorption and metabolism;
  - possesing anti-inflammatory properties and helping taurine in eggshell gland protection;
  - regulating vitagenes and helping stress adaptation;
- Mn;
  - an important co-factor of many enzymes, including those responsible for synthesis of organic matrix, determining eggshell composition and major properties, including strength and elasticity;
  - a cofactor of Mn-SOD, main antioxidant enzymes, regulated at vitagene level and helping stress adaptation;
- Zn;
  - a cofactor of CA, main enzymes of $CaCO_3$ formation in the egg shell;

  - an essential part of Cu,Zn-SOD, cytosolic form of the SOD, regulated at the vitagene level;
  - a vitagene activator, helping stress adaptation;
- Ascorbic acid;
  - natural antioxidant regulating antioxidant defences;
  - an essential element of organic matrix of the bone and eggshell formation;
  - a vitagene activator, helping stress resistance building.

Therefore, the ShellBone concept is based on usage of the combination of the aforementioned nutrients supplied with drinking water to improve eggshell quality and bone health in commercial poultry.

References

Ahmed, A.M.H., Rodríguez-Navarro, A.B., Vidal, M.L., Gautron, J., García-Ruiz, J.M. and Nys, A.Y., 2005. Changes in eggshell mechanical properties, crystallographic texture and in matrix proteins induced by moult in hens. British Poultry Science 46: 268-279.

Al-Batshan, H.A., Scheideler, S.E., Black, B.L., Garlich, J.D. and Anderson, K.E., 1994. Duodenal calcium uptake, femur ash, and eggshell quality decline with age and increase following molt. Poultry Science 73: 1590-1596.

Ali, B.H., Homeida, A.M. and Knifton, A., 1984. The effect of furazolidone on fertility of male chickens. Comparative Biochemistry and Physiology Part C: Comparative Pharmacology 78: 43-47.

Amakye-Anim, J., Lin, T.L., Hester, P.Y., Thiagarajan, D., Watkins, B.A. and Wu, C.C., 2000. Ascorbic acid supplementation improved antibody response to infectious bursal disease vaccination in chickens. Poultry Science 79: 680-688.

Arias, J.I., Jure, C., Wiff, J.P., Fernández, M.S., Fuenzalida, V. and Arias, J.L., 2001. Effect of sulfate content of biomacromolecules on the crystallization of calcium carbonate. MRS Online Proceedings Library Archive 711: 243-248.

Azcona, J.O., Bonino, M.F., Schang, S.J., Gallego, S.E. and Boland, R.L., 1982. Effect of calcinogenic substances from *Solanum malacoxylon* on eggshell thickness and performance of laying hen. Zootechnica International, February,12-13.

Balnave, D., Zhang, D. and Moreng, R.E., 1991. Use of ascorbic acid to prevent the decline in eggshell quality observed with saline drinking water. Poultry Science 70: 848-852.

Bar, A., 2009. Calcium transport in strongly calcifying laying birds: mechanisms and regulation. Comparative Biochemistry and Physiology Part A: Molecular & Integrative Physiology 152: 447-469.

Barnett, D.M., Kumpula, B.L., Petryk, R.L., Robinson, N.A., Renema, R.A. and Robinson, F.E., 2004. Hatchability and early chick growth potential of broiler breeder eggs with hairline cracks. Journal of Applied Poultry Research 13: 65-70.

Bell, D.E. and Marion, J.E., 1990. Vitamin C in laying hen diets. Poultry Science 69: 1900-1904.

Benesch, R., Barron, N.S. and Mawson, C.A., 1944. Carbonic anhydrase, sulphonamides and shell formation in the domestic fowl. Nature 153: 138-139.

Boyanabc, B.D., Sylvia, V.L., Dean, D.D., Pedrozo, H., Del Toro, F., Nemere, I., Posner, G.H. and Schwartz, Z., 1999. 1, 25-(OH) 2D3 modulates growth plate chondrocytes via membrane receptor-mediated protein kinase C by a mechanism that involves changes in phospholipid metabolism and the action of arachidonic acid and PGE2. Steroids 64: 129-136.

Carlos, A.B. and Edwards Jr, H.M., 1998. The effects of 1, 25-dihydroxycholecalciferol and phytase on the natural phytate phosphorus utilization by laying hens. Poultry Science 77: 850-858.

Castaldo, D.J. and Maurice, D.V., 1990. Plasma calcitriol in chickens producing strong and weak egg shells and the response of hens to exogenous calcitriol. International Journal for Vitamin and Nutrition Research 60: 320-323.

Chennaiah, S., Qadri, S.S.Y.H., Rao, S.R., Shyamsunder, G. and Raghuramulu, N., 2004. *Cestrum diurnum* leaf as a source of 1, 25 (OH) 2 Vitamin D3 improves egg shell thickness. The Journal of Steroid Biochemistry and Molecular Biology 89: 589-594.

Chung, M.K., Choi, J.H., Chung, Y.K. and Chee, K.M., 2005. Effects of dietary vitamins C and E on egg shell quality of broiler breeder hens exposed to heat stress. Asian-Australasian Journal of Animal Sciences 18: 545-551.

Diamants, T., and J. Schluns. 1964. Lokalisation und bedeutung der karboanhydrase im uterus von legehennen. Acta Histochemica 19: 296-302.

Dunn, I.C., Rodríguez-Navarro, A.B., Mcdade, K., Schmutz, M., Preisinger, R., Waddington, D., Wilson, P.W. and Bain, M.M., 2012. Genetic variation in eggshell crystal size and orientation is large and these traits are correlated with shell thickness and are associated with eggshell matrix protein markers. Animal genetics 43: 410-418.

Edwards Jr, H.M., 1990. Efficacy of several vitamin D compounds in the prevention of tibial dyschondroplasia in broiler chickens. The Journal of Nutrition 120: 1054-1061.

Edwards Jr, H.M., 2002. Studies on the efficacy of cholecalciferol and derivatives for stimulating phytate utilization in broilers. Poultry Science 81: 1026-1031.

Edwards, H.M., Elliot, M.A. and Sooncharernying, S., 1992. Effect of dietary calcium on tibial dyschondroplasia. Interaction with light, cholecalciferol, 1, 25-dihydroxycholecalciferol, protein, and synthetic zeolite. Poultry Science 71: 2041-2055.

El-Boushy, A.R. and Van Albada, M., 1970. The effect of vitamin C on egg shell quality under high environmental temperatures. Netherlands Journal of Agricultural Science 18: 62-71.

Elliot, M.A. and Edwards JNR, H.M., 1997. Effect of 1,25-dihydroxycholecalciferol, cholecalciferol and fluorescent lights on the development of tibial dyschondroplasia and rickets in broiler chickens. Poultry Science 76: 570-580.

Elzinga,E.J. and Reeder RJ. 2002. X-ray absorption spectroscopy study of Cu2+, and Zn2+, adsorption complexes at the calcite surface: Implications for site-specific metal incorporation preferences during calcite crystal growth. Geochimica et Cosmochimica Acta 66: 3943-3954.

Ensminger, M.E., Oldfield, J.E. and Heinemann, W.W., 1990. Feeds and nutrition. Second revised ed. The Ensminger Publishing Co. Clovis, CA, USA.

Farquharson, C., Berry, J.L., Mawer, E.B., Seawright, E. and Whitehead, C.C., 1998. Ascorbic acid-induced chondrocyte terminal differentiation: the role of the extracellular matrix and 1, 25-dihydroxyvitamin D. European Journal of Cell Biology 76: 110-118.

Fathi, M.M., El-Dlebshany, A.E., Bachine El-Deen, M., Radwan, L.M. and Rayan, G.N., 2016. Effect of long-term selection for egg production on eggshell quality of Japanese quail (*Coturnix japonica*). Poultry Science 95: 2570-2575.

Fathi, M.M., Afifi, Y.K. and El-Safty, A.S., 2010. Utlrastructural diversity of eggshell quality in some Egyptian local breeds of chicken. Egyptian Poultry Science 30: 813-827.

Fernandez, M.S., Moya, A., Lopez, L. and Arias, J.L., 2001. Secretion pattern, ultrastructural localization and function of extracellular matrix molecules involved in eggshell formation. Matrix Biology 19: 793-803.

Frost, T.J., Roland Sr, D.A. and Untawale, G.G., 1990. Influence of vitamin D3, 1α-Hydroxyvitamin D3, and 1, 25-dihydroxyvitamin D3 on eggshell quality, tibia strength, and various production parameters in commercial laying hens. Poultry Science 69: 2008-2016.

Gabbay, K.H., Bohren, K.M., Morello, R., Bertin, T., Liu, J. and Vogel, P., 2010. Ascorbate synthesis pathway: dual role of ascorbate in bone homeostasis. Journal of Biological Chemistry 285: 9510-19520.

Gautron, J., Hincke, M.T. and Nys, Y., 1997. Precursor matrix proteins in the uterine fluid change with stages of eggshell formation in hens. Connective Tissue Research 36: 195-210.

Grollman, A.P. and Lehninger, A.L., 1957. Enzymic synthesis of L-ascorbic acid in different animal species. Archives of Biochemistry and Biophysics 69: 458-467.

Gutowska, M.S. and Mitchell, C.A., 1945. Carbonic anhydrase in the calcification of the egg shell. Poultry Science 24: 159-167.

Hill, C.H., 1979. Studies on the ameliorating effect of ascorbic acid on mineral toxicities in the chick. The Journal of Nutrition 109: 84-90.

Holm, L., Blomqvist, A., Brandt, I., Brunström, B., Ridderstråle, Y. and Berg, C., 2006. Embryonic exposure to o, p'-DDT causes eggshell thinning and altered shell gland carbonic anhydrase expression in the domestic hen. Environmental Toxicology and Chemistry 25: 2787-2793.

Hove, E., Elvehjem, C.A. and Hart, E.B., 1940. The relation of zinc to carbonic anhydrase. Journal of Biological Chemistry 136: 425-434.

Jabbar, A., Hameed, A., Yousaf, A., Riaz, A. and Ditta, Y.A., 2019. The influence of hairline crack eggs on hatchery parameters and performance of chicks. World Veterinary Journal 9: 76-83.

Kaetzel Jr, D.M. and Soares Jr, J.H., 1979. Effects of cholecalciferol steroids on bone and egg shell calcification in Japanese quail. The Journal of Nutrition 109: 1601-1608.

Leach Jr, R.M. and Gross, J.R., 1983. The effect of manganese deficiency upon the ultrastructure of the eggshell. Poultry Science 62: 499-504.

Li, H., Li, W. and Wang, Q., 2019. 1, 25-dihydroxyvitamin D 3 suppresses lipopolysaccharide-induced interleukin-6 production through aryl hydrocarbon receptor/nuclear factor-κB signaling in oral epithelial cells. BMC Oral Health 19: 236.

Lin, Z. and Li, W., 2016. The roles of vitamin D and its analogs in inflammatory diseases. Current Topics in Medicinal Chemistry 16: 1242-1261.

Lippolis, J.D., Reinhardt, T.A., Sacco, R.A., Nonnecke, B.J. and Nelson, C.D., 2011. Treatment of an intramammary bacterial infection with 25-hydroxyvitamin D(3). PloS One 6: e25479.

Lundholm, C.E., 1990. The eggshell thinning action of acetazolamide; relation to the binding of Ca2+ and carbonic anhydrase activity of the shell gland homogenate. Comparative biochemistry and physiology. C. Comparative pharmacology 95: 85-89.

Mabe, I., Rapp, C., Bain, M.M. and Nys, Y., 2003. Supplementation of a corn-soybean meal diet with manganese, copper, and zinc from organic or inorganic sources improves eggshell quality in aged laying hens. Poultry Science 82: 1903-1913.

McKee, J.S. and Harrison, P.C., 1995. Effects of supplemental ascorbic acid on the performance of broiler chickens exposed to multiple concurrent stressors. Poultry Science 74: 1772-1785.

Merriman, K.E., Poindexter, M.B., Kweh, M.F., Santos, J.E. and Nelson, C.D., 2017. Intramammary 1, 25-dihydroxyvitamin D3 treatment increases expression of host-defense genes in mammary immune cells of lactating dairy cattle. The Journal of Steroid Biochemistry and Molecular Biology 173: 33-41.

Merriman, K.E., Powell, J.L., Santos, J.E. and Nelson, C.D., 2018. Intramammary 25-hydroxyvitamin D3 treatment modulates innate immune responses to endotoxin-induced mastitis. Journal of Dairy Science 101: 7593-7607.

Mertens, K., Bamelis, F., Kemps, B., Kamers, B., Verhoelst, E., De Ketelaere, B., Bain, M., Decuypere, E. and De Baerdemaeker, J., 2006. Monitoring of eggshell breakage and eggshell strength in different production chains of consumption eggs. Poultry Science 85: 1670-1677.

Morris, K.M.L., 1977. The effect on eggshell thickness of the inclusion of the calcinogenic plant *Solanum malacoxylon* in the diet of laying hens. The Veterinary Record 101: 502-504.

Nasri, H., Van den Brand, H., Najjar, T. and Bouzouaia, M., 2020. Egg storage and breeder age impact on egg quality and embryo development. Journal of Animal Physiology and Animal Nutrition 104: 257-268.

Nys, Y., 2001. Recent developments in layer nutrition for optimising shell quality. In: Proceedings of the 13th European Symposium on Poultry Nutrition. Blankenberge, Belgium, pp. 42-52.

Nys, Y. and Le Roy, N., 2018. Calcium homeostasis and eggshell biomineralization in female chicken. In: Feldman, D., Pike, J.W., Bouillon, R., Giovannucci, E., Goltzman, D. and Hewison, M. (eds) Vitamin D, 4th ed., Academic press, Cambridge, MA, USA, pp 361-382.

Nys, Y., Gautron, J., Garcia-Ruiz, J.M. and Hincke, M.T., 2004. Avian eggshell mineralization: biochemical and functional characterization of matrix proteins. Comptes Rendus Palevol 3: 549-562.

Nys, Y., Hincke, M.T., Arias, J.L., Garcia-Ruiz, J.M. and Solomon, S.E., 1999. Avian eggshell mineralization. Poultry and Avian Biology Reviews 10: 143-166.

Orban, J.I., Roland, D.A., Cummins, K. and Lovell, R.T., 1993. Influence of large doses of ascorbic acid on performance, plasma calcium, bone characteristics, and egg shell quality in broilers and laying hens. Poultry Science 72: 691-700.

Pardue, S.L. and Thaxton, J.P., 1986. Ascorbic acid in poultry: a review. World's Poultry Science Journal 42: 107-123.

Peebles, E.D. and Brake, J., 1985. Relationship of dietary ascorbic acid to broiler breeder performance. Poultry Science 64: 2041-2048.

Perek, M. and Kendler, J., 1963. Ascorbic acid as a dietary supplement for White Leghorn hens under conditions of climatic stress. British Poultry Science 4: 191-200.

Radwan, L.M., 2016. Eggshell quality: a comparison between Fayoumi, Gimieizah and Brown Hy-Line strains for mechanical properties and ultrastructure of their eggshells. Animal Production Science 56: 908-912.

Radwan, L.M., Fathi, M.M., Galal, A. and El-Dein, A.Z., 2010. Mechanical and ultrastructural properties of eggshell in two Egyptian native breeds of chicken. International Journal of Poultry Science 9: 77-81.

Rennie, J.S. and Whitehead, C.C., 1996. Effectiveness of dietary 25-and 1-hydroxycholecalciferol in combating tibial dyschondroplasia in broiler chickens. British Poultry Science 37: 413-421.

Rennie, J.S., Whitehead, C.C. and Thorp, B.H., 1993. The effect of dietary 1, 25-dihydroxycholecalciferol in preventing tibial dyschondroplasia in broilers fed on diets imbalanced in calcium and phosphorus. British Journal of Nutrition 69: 809-816.

Rennie, J.S., McCormack, H.A., Farquharson, C., Berry, J.L., Mawer, E.B. and Whitehead, C.C., 1995. Interaction between dietary 1, 25-dihydroxycholecalciferol and calcium and effects of management on the occurrence of tibial dyschondroplasia, leg abnormalities and performance in broiler chickens. British Poultry Science 36: 465-477.

Roy, R. and Guha, B., 1958. Species differences in regard to the biosynthesis of ascorbic acid. Nature 182: 319-320.

Sá e Silva, M., Susta, L., Moresco, K. and Swayne, D.E., 2016. Vaccination of chickens decreased Newcastle disease virus contamination in eggs. Avian Pathology 45: 38-45.

Scott, M.L., 1975. Environmental influences on ascorbic acid requirements in animals. Annals of the New York Academy of Sciences 258: 151-168.

Snapir, N. and Perek, M., 1970. Distribution of calcium, carbonic anhydrase and alkaline phosphatase activities in the uterus and isthmus of young and old white leghorn hens. Poultry Science 49: 1526-1531.

Soares, J.H. Jr., McLonghlin, C.M., Swerdel, M.R. and Bossard, E., 1976. Effect of hydroxy vitamin D metabolites on the mineralization of eggshell and bones. In: Proceedings of the Maryland Nutrition Conference, College Park, MD, USA, pp. 85-92.

Soares, J.H., Jr., 1984. Calcium metabolism and its control – a review. Poultry Science 63: 2075-2083.

Stefanello, C., Santos, T., Murakami, A., Martins, E. and Carneiro, T., 2014. Productive performance, eggshell quality, and eggshell ultrastructure of laying hens fed diets supplemented with organic trace minerals. Poultry Science 93:104-113.

Stevens, V.I., Blair, R. and Salmon, R.E., 1984. Influence of maternal vitamin D3 carry-over on kidney 25-hydroxyvitamin D3-l-hydroxylase activity of poults. Poultry Science 63: 765-774.

Surai, P.F., 2005. Minerals and antioxidants. In: Tucker, L.A. and Taylor-Pickard, J.A. (eds) Redefining mineral nutrition. Nottingham University Press, Nottingham, UK, pp. 147-177.

Surai, P.F., 2016. Antioxidant systems in poultry biology: superoxide dismutase. Journal of Animal Nutrition 1: 8.

Surai, P.F., 2018a. Egg shell and carcass quality in birds at the second part of production period. Vitamin D role. Ptakhivnitstvo.ua (Ukrainian Poultry Science, Kiev) 9: 6-11.

Surai, P.F., 2018b. Egg shell and carcass quality in birds at the second part of production period. Vitamin D role. Part 2. Ptakhivnitstvo.ua (Ukrainian Poultry Science, Kiev) 10: 16-17.

Surai, P.F., 2018c. Egg shell and carcass quality in birds at the second part of production period. Organic matrix role. Ptakhivnitstvo.ua (Ukrainian Poultry Science, Kiev) 11: 6-9.

Surai, P.F., 2018d. Taurine and carnitine in poultry production: from vitagene activation to chicken health maintenance. Ptakhivnitstvo.ua (Ukrainian Poultry Science, Kiev) 1-2: 12-17

Surai, P.F., 2019e. Decreasing eggshell hairline cracks: from vitagenes to a ShellBone concept ShellBone. Ptakhivnitstvo.ua (Ukrainian Poultry Science, Kiev) 5: 24-27.

Surai, P.F., 2020. Antioxidants in poultry nutrition and reproduction: an update. Antioxidants 9(2): 105.

Surai, P.F., 2020a. Superoxide dismutase as a new entrant into the vitagene family in animals/poultry. EC Nutrition 15.3: 01-03.

Surai, P.F., Kochish, I.I., Fisinin, V.I. and Kidd, M.T., 2019b. Antioxidant defence systems and oxidative stress in poultry biology: an update. Antioxidants 8(7): 235.

Surai P.F., Kochish I.I. and Kidd M.T., 2020. Taurine in poultry nutrition. Animal Feed Science and Technology 260: 114339.

Surai, P.F., Kochish, I.I., Fisinin, V.I., Grozina, A.A. and Shatskikh E.V., 2018. Molecular mechanisms of chicken gut health maintenance: role of microbiota. Agricultural Technologies, Moscow, Russia.

Świątkiewicz, S., Arczewska-Włosek, A., Krawczyk, J., Puchała, M. and Jozefiak, D., 2015. Dietary factors improving eggshell quality: an updated review with special emphasis on microelements and feed additives. World's Poultry Science Journal 71: 83-94.

Swiatkiewicz, S. and Koreleski, J., 2008. The effect of zinc and manganese source in the diet for laying hens on eggshell and bones quality. Veterinarni Medicina 53: 555-563.

Taylor, D., Walsh, M., Cullen, A. and O'Reilly, P., 2016. The fracture toughness of eggshell. Acta Biomaterialia 37: 21-27.

Thornton, P.A. and Moreng, R.E., 1958. The effect of ascorbic acid on egg quality factors. Poultry Science 37: 691-698.

Toussant, M.J. and Latshaw, J.D., 1994. Evidence of multiple metabolic routes in vanadium's effects on layers. Ascorbic acid differential effects on prepeak egg production parameters following prolonged vanadium feeding. Poultry Science 73: 1572-1580.

Tsang, C.P., Grunder, A.A. and Narbaitz, R., 1990a. Optimal dietary level of 1 alpha,25-dihydroxycholecalciferol for eggshell quality in laying hens. Poultry Science 69: 1702-1712.

Tsang, C.P.W., Grunder, A.A., Soares, J.H. and Narbaitz, R., 1990b. Effect of 1α, 25-dihydroxycholecalciferol on egg shell quality and egg production. British Poultry Science 31: 241-247.

Tsang, C.P.W. and Grunder, A.A., 1993. Effect of dietary contents of cholecalciferol, 1α, 25-dihydroxycholecalciferol and 24, 25-dihydroxycholecalciferol on blood concentrations of 25-hydroxycholecalciferol, 1α, 25-dihydroxycholecalciferol, total calcium and eggshell quality. British Poultry Science 34: 1021-1027.

Van Mourik, S., Alders, B.P.G.J., Helderman, F., Van de Ven, L.J.F. and Groot Koerkamp, P.W.G., 2017. Predicting hairline fractures in eggs of mature hens. Poultry Science, 96: 1956-1962.

Waldenstedt, L., 2006. Nutritional factors of importance for optimal leg health in broilers: a review. Animal Feed Science and Technology 126: 291-307.

Weiser, H., Schlachter, M. and Fenster, R., 1988a. The importance of vitamin C for the metabolism of vitamin D3 in poultry. In: Proceedings of the 18th World Poultry Congress, Nagoya, Japan. p. 831.

Weiser, H., Schlachter, M. and Bachmann, H., 1988. The importance of vitamin C for hydroxylation of vitamin D3 to 1,25(OH)2D3 and of 24R,25(OH)2D3 to a more active metabolite. In: Noeman, A.W., Schaefer, K., Grigolet, H.G. and Herrath, D. (eds) Vitamin D: molecular, cellular and clinical endocrinology. Walter de Gruyter, Berlin, Germany, pp. 644-653.

Weiser, H., Schlachter, M., Probst, H.P. and Korman, A.W., 1992. The relevance of ascorbic acid for bone metabolism. Proceedings of the 2nd Symposium on Ascorbic Acid in Domestic Animals, Kratause Ittingen, Switzerland, pp. 73-95.

Wen, G., Eder, K. and Ringseis, R., 2020. 1, 25-hydroxyvitamin D3 decreases endoplasmic reticulum stress-induced inflammatory response in mammary epithelial cells. Plos One 15: 0228945.

Whitehead, C.C., Mitchell M.A. and Njoku, P.C., 1992. Effects of ascorbic acid on egg yolk and shell precursor in heat-stressed laying hens. Proceedings of the 2nd symposium on ascorbic acid in domestic animals, Kratause Ittingen, Switzerland. pp. 262-269.

Wistedt, A., Ridderstråle, Y., Wall, H. and Holm, L., 2019. Age-related changes in the shell gland and duodenum in relation to shell quality and bone strength in commercial laying hen hybrids. Acta Veterinaria Scandinavica 61: 14.

Wistedt, A., Ridderstråle, Y., Wall, H. and Holm, L., 2014. Exogenous estradiol improves shell strength in laying hens at the end of the laying period. Acta Veterinaria Scandinavica 56: 34.

Xiao, J.F., Zhang, Y.N., Wu, S.G., Zhang, H.J., Yue, H.Y. and Qi, G.H., 2014. Manganese supplementation enhances the synthesis of glycosaminoglycan in eggshell membrane: a strategy to improve eggshell quality in laying hens. Poultry Science 93: 380-388.

Xiao, J.F., Wu, S.G., Zhang, H.J., Yue, H.Y., Wang, J., Ji, F. and Qi, G.H., 2015. Bioefficacy comparison of organic manganese with inorganic manganese for eggshell quality in Hy-Line Brown laying hens. Poultry Science 94: 1871-1878.

Zamani, A., Rahmani, H.R. and Pourreza, J., 2005. Supplementation of a corn-soybean meal diet with manganese and zinc improves eggshell quality in laying hens. Pakistan Journal of Biological Sciences 8: 1311-1317.

Zapata, L.F. and Gernat, A.G., 1995. The effect of four levels of ascorbic acid and two levels of calcium on eggshell quality of forced-molted White Leghorn hens. Poultry Science 74: 1049-1052.

Zhang, Y.N., Wang, J., Zhang, H.J., Wu, S.G. and Qi, G.H., 2017a. Effect of dietary supplementation of organic or inorganic manganese on eggshell quality, ultrastructure, and components in laying hens. Poultry Science 96: 2184-2193.

Zhang, Y.N., Zhang, H.J., Wang, J., Yue, H.Y., Qi, X.L., Wu, S.G. and Qi, G.H., 2017b. Effect of dietary supplementation of organic or inorganic zinc on carbonic anhydrase activity in eggshell formation and quality of aged laying hens. Poultry Science 96: 2176-2183.

Zhang, Y.N., Zhang, H.J., Wu, S.G., Wang, J. and Qi, G.H., 2018. Dietary manganese supplementation affects mammillary knobs of eggshell ultrastructure in laying hens. Poultry Science 97: 1253-1262.

Zhong, S., Fan, Y., Yan, Q., Fan, X., Wu, B., Han, Y., Zhang, Y., Chen, Y., Zhang, H. and Niu, J., 2017. The therapeutic effect of silymarin in the treatment of nonalcoholic fatty disease: a meta-analysis (PRISMA) of randomized control trials. Medicine 96: 9061.

Chapter 15
Vitatonic concept development

Desperate diseases must have desperate remedies

15.1 Introduction

The liver in birds is considered to be the main metabolic organ involved in a range of metabolic and homeostatic functions associated with the synthesis (amino-acids, proteins, lipids, coagulation factors, bile, immunoglobulins, etc.), metabolism (carbohydrate, protein, lipid, vitamin, etc.), storage (vitamins A, D3, B12, Fe and Cu), excretion, and detoxification processes (Figure15.1; Zaefarian *et al.*, 2019).

In poultry, lipid synthesis (lipogenesis) is shown to be very active being about 20 times greater than in adipose tissues. For example, in laying hens, major constituents of the egg yolk are synthesised in the liver and deposited in the egg. Since intensity of egg production at the peak can exceed 98%, the liver works under high metabolic stress and process of lipid synthesis and release can easily be disturbed leading to various metabolic disorders. There are some anatomical and physiological features predisposing poultry to fat deposition in the liver: due to the direct entry into the portal blood system, portomicrons have to pass through the liver before they reach the rest of the circulation (Zaefarian *et al.*, 2019). Furthermore, fat deposition in poultry

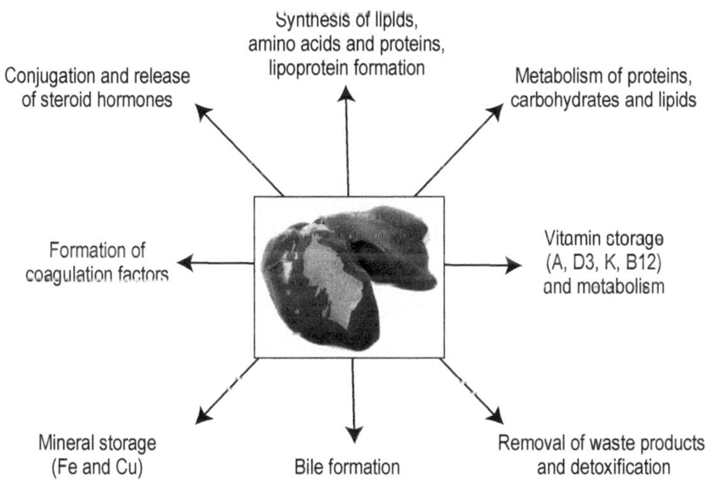

Figure 15.1. Main biological functions of the liver in poultry.

DOI 10.3920/978-90-8686-906-0_15, © Wageningen Academic Publishers 2020

liver depends on many nutritional and environmental factors, including (Fouad and El-Senousey, 2014; Zaefarian *et al.*, 2019):

- fat sources: PUFAs including omega-3 fatty acids usually decrease fat accumulation;
- feed restriction: re-feeding after feed restriction could increase fat accumulation in the liver;
- feed structure: fat synthesis and deposition was increased by feeding crumbled or pelleted diets to broilers;
- nutrient density: affects lipid accumulation in the liver;
- dietary fibre: affects microbiota and lipid accumulation in the liver;
- anti-nutritional factors: can disturb liver metabolism and increase lipid accumulation in the liver;
- various infectious agents and dietary toxins (aflatoxin, etc.): can disturb liver metabolism and increase lipid accumulation in the liver;
- biotin deficiency: predisposes birds to fatty liver;
- lack of exercise (cage housing): predisposes birds to fatty liver;
- positive energy balance: predisposes birds to fatty liver;
- hormonal (oestrogen) level: predisposes birds to fatty liver;
- heat stress: predisposes birds to fatty liver;
- other stresses: predispose birds to fatty liver.

In general, there is a range of pathological changes in the poultry liver due to the aforementioned factors, including fatty liver, hepatosis, necrosis, hepatitis and cholangitis (Zaefarian *et al.*, 2019), leading to decreased productive and reproductive performance of birds (Figure 15.2).

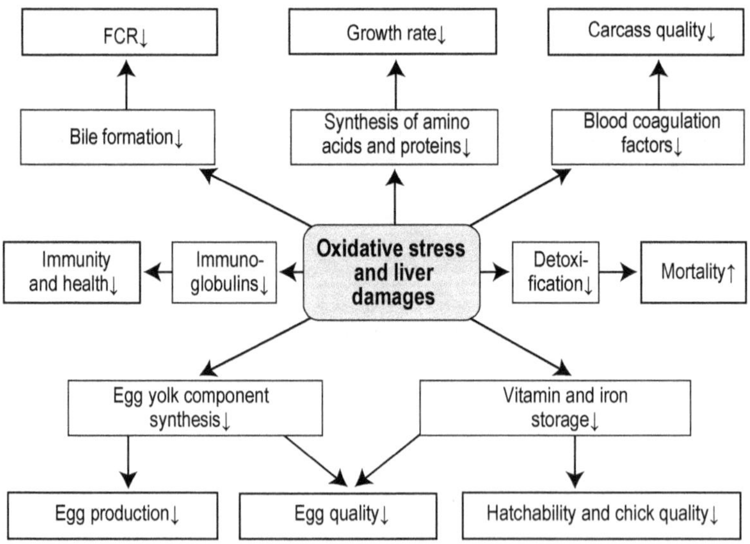

Figure 15.2. Effect of liver disturbance on productive and reproductive performance of poultry.

This chapter is devoted to the development vitagene-modulating nutritional supplementation strategy to deal with liver problems as well as with other health-related issues in poultry production.

15.2 Fatty-liver haemorrhagic syndrome

Fatty-liver haemorrhagic syndrome (FLHS) is a sporadic disease occurring in commercial caged, high productive, often overweight laying hens in the second half of the laying period leading to a fall in egg production in the affected flock (Ridell, 1991). The disease was first reported in 1956 by Couch who described it as fatty liver syndrome and a more precise terminology 'fatty liver haemorrhagic syndrome' differentiating it from fatty, non-haemorrhagic livers was introduced by Wolford and Polin (1972). The disease usually increases flock mortality (Butler, 1976) and dead hens are characterised by large blood clots in the liver and abdominal cavities and the liver is usually from a golden yellow to pale yellow or very pale brown in colour (Fowler, 1990; Ridell, 1991) with very high (40-70%) fat content (Ivy and Nesheim, 1973). However, after more than 60 years research in this field, the molecular mechanisms for the explanation of the hepatic necrosis and haemorrhage is still unknown. In fact, Squires and Leeson (1988) and Spurlock and Savage (1992) put forward a hypothesis indicating that the liver haemorrhages may be caused by an excessive lipid peroxidation of the unsaturated fatty acids accumulated within the hepatocytes, leading to oxidative damage to the cell and organelle membranes. This hypothesis was further developed with indication that oxidative stress, disruption of redox status, cell signalling and transcription factor (Nrf2 and NF-κB) are involved in the development of inflammation and FLHS (Surai *et al.*, 2018d, 2019c). Indeed, there is a growing evidence indicating that oxidative stress is a leading cause of detrimental consequences of fatty liver on metabolic changes in the body (Elvira-Torales *et al.*, 2019; Gillessen and Schmidt, 2020; Nagashimada and Ota, 2019; Ore and Akinloye, 2019; Surai *et al.*, 2018d; 2019c Xu *et al.*, 2019).

15.3 Vitagenes and fatty-liver haemorrhagic syndrome

Oxidative stress has been suggested as a conjoint pathological mechanism of FLHS, and it contributes to initiation and progression of liver injury, associated with lipid, protein and DNA damages and promotes several disorders of metabolic pathways and a disruption of antioxidant defence mechanisms and modulating pathways that control normal biological functions (Surai *et al.*, 2018). Therefore, the VitaTonic concept is based on usage of nutrient mixture upregulating vitagenes, decreasing oxidative stress in the liver and other tissues, showing hepatoprotective effect and maintaining liver health in commercial laying hens under commercially-relevant stress conditions. The first choice includes three hepatoprotective nutrients possessing vitagene-activating properties including carnitine, silymarin and betaine. In previous chapters detailed description of molecular mechanisms of protective effects of carnitine and silymarin have been described and in this chapter only information related to fatty liver prevention/treatment will be provided.

15.3.1 Carnitine

L-carnitine (LC) is shown to transport long-chain fatty acids across the inner mitochondrial membrane during β-oxidation helping liver deal with lipid excess and may play an important role in reducing fat deposition in modern strains of broiler chickens (Zaefarian *et al.*, 2019). It is well-known that LC is synthesised *in vivo* from lysine and methionine and obtained from animal-derived feed ingredients (Surai, 2015b,c,d). As mentioned in the Chapter 9, protective effects of carnitine and its derivatives are most pronounced in various stress conditions. Indeed, by decreasing negative consequences of stresses carnitine can improve productive and reproductive performance and general health of growing chickens, parent stock, and commercial layers (Arslan, 2006; Golzar Adabi *et al.*, 2011; Surai, 2018d). In a recent review of effects LC on poultry it was shown than under normal conditions in most cases there was no effect of this nutrient on productive and reproductive performance of growing broilers or layers (Ringseis *et al.*, 2018). However, effect of LC on lipid metabolism was confirmed in 7 from 25 trials reviewed: abdominal fat percentage was significantly reduced. Similarly, providing LC in drinking water was associated with a reduction of abdominal fat accumulation in Turkish native ducks and geese (Arslan *et al.*, 2003, 2004). It seems likely that protective effect of LC supplementation can be seen in various stress conditions when its requirement probably increased but internal synthesis reduced (Surai *et al.*, 2018). From the same 25 studies reviewed by Ringseis *et al.* (2018) in 3 studies LC reduced ascites mortality (Geng *et al.*, 2004; Wang *et al.*, 2013) and under high temperature conditions LC improved BW gain (Celik and Ozturkcan, 2003). The same conclusion can be drawn from 7 layer studies analysed by Ringseis *et al.* (2018): supplemental LC is able to prevent adverse effects of diet-induced oxidative stress on performance and lipid peroxidation in laying hens (Güçlü *et al.*, 2011). Protective effects of LC on laying hens was also observed at high ambient temperature (Celik *et al.*, 2004) as well as in layers of advanced (92 weeks) age (Kazemi-Fard *et al.*, 2015). whereas supplemental L-carnitine is largely ineffective under normal feeding conditions, that is absence of diet-induced stress. It was shown that dietary LC in the breeder diet was transferred to egg yolk and hatching rate was increased from 83% to 87% and from 82.4% to 85.3% in groups supplemented with 50 and 100 mg LC, respectively (Leibetseder, 1995). It seems likely that LC has some epigenetic effects on poultry breeders. For example, feeding female broiler breeders a diet supplemented with 25 mg/kg LC was shown to reduce the carcass fat content of their progeny (Kidd *et al.*, 2005).

As shown in the Chapter 9, protective effect of LC and its derivatives on the antioxidant systems of the body are shown in various models of oxidative stress/toxicity caused by CCl_4 cisplantin, ethanol, 3-Nitropropionic acid, valproate, diethylnitrosamine, doxorubicin, adriamycin, tamoxifen, indomethacin, acetaminophen, Cd, aflatoxin, thioacetamide, methotrexate, etc. LC and its derivatives were effective in decreasing oxidative stress caused by neurotoxic agents such as glutamate, quinolinic acid or

3-nitropropionic acid, rotenone, dexamethasone, aminoglycosides, scopolamine, 1-methyl-4-phenyl-1,2,3,6-tetrahydropyridine, silver nano-particles, etc. Carnitine was also effective in prevention of oxidative stress in the L-buthionine-sulfoximine-induced cataract model, selenite-induced cataractogenesis, LPS-treated rats, high dietary copper in laying hens, high fat diet in rats, fructose-fed rats, experimental glaucoma, acetic acid-induced colitis, atherosclerotic rats. Furthermore, carnitine showed clear antioxidant protective activities in caerulein-induced acute pancreatitis, adjuvant arthritis, and in hyperthyroid rats as well as in a mouse model of non-alcoholic steatohepatitis, in mice with familial amyotrophic lateral sclerosis, streptozotocin-induced diabetes in rats, ifosfamide-induced Fanconi syndrome, chemically induced model of maple syrup urine disease and unilateral ureteral obstruction-induced oxidative stress. etc. (for references see Chapter 9). Furthermore, carnitine was shown to be a more effective protectant in antioxidant mixtures than when it is used singly. Clearly, hepatoprotective properties of carnitine, together with its antioxidant and vitagene-regulating properties (Surai, 2018) make it an ideal component for the development of the vitagene-regulating nutrient mixture to maintain liver health in stress conditions of commercial egg and meat production. Immunomodulating properties of carnitine and protective effects of carnitine on immunity in stress condition, described in Chapter 9 are additional bonuses for inclusion of this nutrient in anti-stress vitagene-regulating mixture.

15.3.2 Silymarin

Detailed analysis of literature confirming an important role of silymarin in prevention/treatment of liver disorders have been presented in Chapter 11. Recent meta-analysis of studies devoted to usage of silymarin for prevention/treatment human non-alcoholic fatty liver confirmed an importance of the nutrient as a therapeutic element (Zhong *et al.*, 2017). In particular, antioxidant, anti-inflammatory (Lovelace *et al.*, 2015) and anti-apoptotic (Haddadi *et al.*, 2018) properties of silymarin are a background for its usage as a hepatoprotectant in poultry production (Saeed *et al.*, 2017a) and a combination of silymarin and carnitine was shown to be more effective than each of the nutrient used singly (Liu *et al.*, 2009; Mousa and Osman, 2016). The liver is a primary organ involved in biotransformation of food toxicants and drugs. Toxins absorbed from the intestinal tract first enter the liver resulting in tissue damages and a variety of liver disorders. Hepatic disorders are mainly caused by toxic chemicals (alcohol), xenobiotics (carbon tetrachloride, chlorinated hydrocarbons), anticancer (doxorubicin, cisplatin), immunosuppressant (cyclosporine), analgesic anti-inflammatory (paracetamol, thioacetamide), anti-tubercular (isoniazid, rifampicin) drugs, radiations (gamma radiations), heavy metals (cadmium, arsenic), mycotoxins (aflatoxin, T-2 toxin, fumonisins, ochratoxins), galactosamine, lipopolysaccharides, etc. and silymarin, as shown in the Chapter 11, was able to show substantial protective effect in the aforementioned model systems. Furthermore, hepatoprotective effects of silymarin was demonstrated in a range of various studies (Amiridumari *et al.*, 2013; Dumari *et al.*, 2014; Muhammad *et al.*, 2011; Tedesco *et al.*, 2004). Silymarin also decreased fat content of the chicken carcass and improved antioxidant defences in the chicken liver (Gengwu *et al.*, 2016).

Silymarin was also shown to decrease hepatotoxicity of aflatoxin B1, as evidenced by positive effects on feed consumption, daily gain and liver morphology of growing broilers (Fani *et al.*, 2013) and improved gut microbiota, helped maintain absorptive capacity of the small intestine (Jahanian *et al.*, 2017) and maintained spleen integrity (Kalorey *et al.*, 2005). Recently, hepatoprotective effects of silymarin have been shown in CCl4 induced toxicity in broilers (Baradaran *et al.*, 2019; Malayeri *et al.*, 2014). In particular, the CCl_4-challenged birds revealed considerable hepatic injures, however, liver damages were amended by the silymarin. Interestingly, a remarkable down-regulation in the expression of CAT, GPx, and Mn-SOD hepatic genes was detected in CCl_4-challenged birds, whereas silymarin prevented down-regulation of the aforementioned genes (Baradaran *et al.*, 2019). Similar hepatoprotective effect of silymarin was observed in CCl4-treated rats (Abdel-Moneim *et al.*, 2015; Al-Rasheed *et al.*, 2016), but a combination of silymarin+ taurine was more effective (Abdel-Moneim *et al.*, 2015). Silymarin-mediated protection against maneb- and paraquat-induced hepatotoxicity in rats was also shown (Ahmad *et al.*, 2013).

In general, silymarin functions in poultry are not restricted by hepatoprotection, it participates in regulation of many various physiological functions and biochemical pathways (Figure 15.3).

In particular, protective effects of SM on the chicken gut in various stress conditions described earlier, as well as immunoprotective actions in the case of aflatoxin B (Chand *et al.*, 2011) and ochratoxin (Khatoon *et al.*, 2013) in the chicken diet deserve more attention.

15.3.3 Betaine

Betaine is a naturally occurring trimethyl derivative of the amino acid glycine which can be found in relatively large quantities in aquatic invertebrates, sugar beets and their by-products (Metzler-Zebeli *et al.*, 2009). Biological functions of betaine are

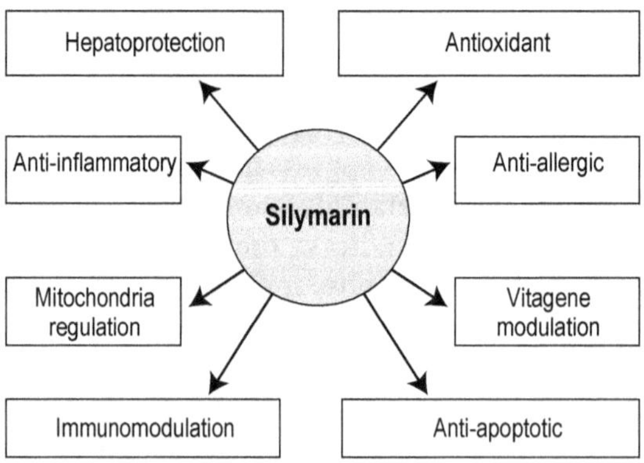

Figure 15.3. Biological functions of silymarin (adapted from Surai, 2015a).

based on its action as osmoprotectant assisting in cellular water homeostasis and as a methyl group donor via transmethylation. There is also a range of other betaine activities contributing to its protective effect on the chicken liver, gut and immunity (Figure15.4).

The significance of betaine in poultry nutrition was first shown in the early 1940s (Kidd *et al.*, 1997). Similar to other nutrients discussed above (e.g. taurine and carnitine), supplemental betaine may be beneficial for poultry during certain physiologically challenging conditions, including the high metabolic demand of rapid growth, various stresses (disease challenge, osmotic stress, heat stress, etc.; Kidd *et al.*, 2005, 1997). In fact, betaine is shown to be involved in the protein and lipid (energy) metabolism and it has been often referred to as a 'carcass modifier'. There is a substantial evidence that betaine can significantly reduce abdominal fat pad and improve breast meat yield in broiler chickens, tom turkeys, and meat ducks (for review see Metzler-Zebeli *et al.*, 2009). Indeed, the fat-lowering effect of betaine was proven in many different trials (Xing *et al.*, 2011; Zhan *et al.*, 2006). Betaine was shown to be effective in alleviating most of the disturbances in fatty liver disease in various animal models by reversing changes in fatty acid and lipid metabolism, relieving oxidative stress and restoring mitochondrial function (Day and Kempson, 2016). Furthermore, the mechanisms underlying hepatoprotective effects were also suggested to involve the stimulation of lipolysis and inhibition of lipogenesis (via gene expression and subsequent activity of lipolytic-/lipogenic-related proteins), stimulation of autocrine/endocrine IGF-1 release and insulin receptor signalling pathways, stimulation of growth hormone secretion, increased creatine synthesis and increases in protein synthesis (via intracellular hyper-hydration; Cholewa *et al.*, 2014). Betaine antioxidant action (Alirezaei *et al.*, 2012; Chen *et al.*, 2020; Li *et al.*, 2019b; Wen *et al.*, 2019), vitagene-regulating properties (Surai *et al.*, 2017, 2018, 2019) including HSP70 regulation (Dangi *et al.*, 2016), chaperone functions (Figueroa-Soto and Valenzuela-Soto, 2018; Roth *et al.*, 2012) and mitochondria-stabilising effects (Heidari *et al.*, 2018; Lee, 2015) could be of great importance for hepatoprotective actions of the nutrient. Protective

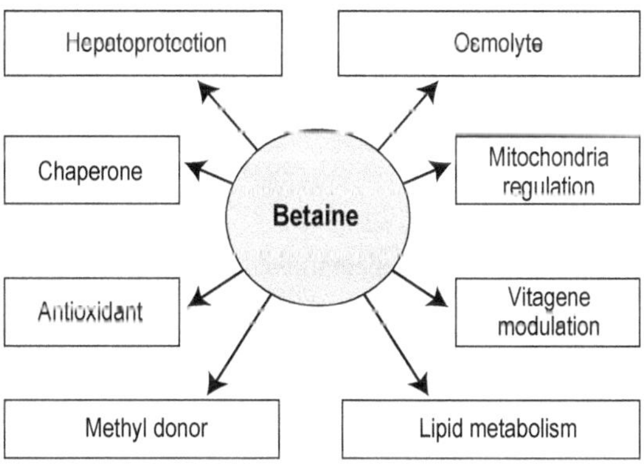

Figure 15.4. Biological functions of betaine.

effects of betaine under various stress conditions in poultry have been a topic of current research (Liu *et al.*, 2019; Ratriyanto and Mosenthin, 2018; Saeed *et al.*, 2017a; Wen *et al.*, 2019). Furthermore, gut protection under microbial stress (Fetterer *et al.*, 2003; Hamidi *et al.*, 2009), mycotoxin stress (Liew and Mohd-Redzwan, 2018; Surai and Fisinin, 2015) in chickens and immunomodulating betaine properties (Chand *et al.*, 2017; Kidd, 2004; Zhan, 2001) in poultry made this nutrient an important addition into the anti-stress composition to prevent various stresses and to maintain liver integrity and health.

15.4 Conclusions

As it was shown above, combining 3 hepatoprotectors having similar protective activities as well as each of them adding some extras (activating liver regeneration by silymarin, osmotic protection by betaine, decreasing blood lipids by carnitine, etc.) made the protective activity of the mixture more comprehensive and unique in comparison to individual use of the compounds (Figure 15.5).

It was also important to add into the developed composition a mixture of B-vitamins, regulating lipid metabolism in the liver (B1,B5,B6 and B12), Zn, Mn and Mg (see explanations in ShellBone discussions) and two organic acids to have extra protective effects on the gut. In general, anti-inflammatory properties of ShellBone and VitaTonic (Figure 15.6) related to the regulation of transcription factors (NF-κB and Nrf2) as well as vitagenes (HSP70, HO-1, SOD, thioredoxin- and glutathione systems, and sirtuins), provide an effective protection to the liver and shell gland as well as maintain gut health and high immunocompetence.

The vitagene concept successfully found its applications in poultry production and the development of 3 vitagene-regulating compositions described above confirmed its validity. Therefore, by modulating vitagenes via supplying special nutrient mixtures with drinking water it is possible to develop a 'system of quick response', which can help nutritionist/technologists/ veterinarian to deal with emergency situations in commercial poultry production associated with various environmental, technological,

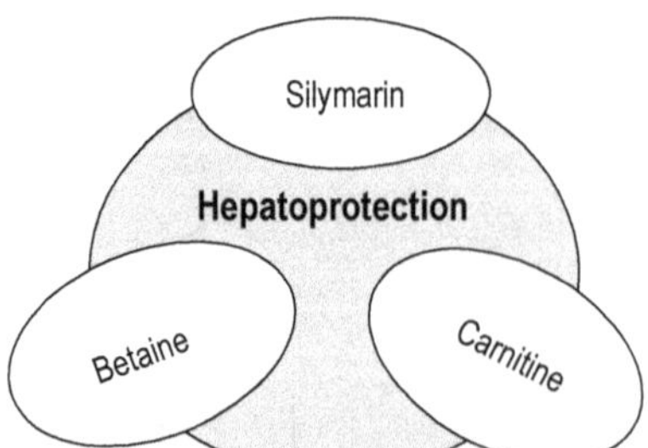

Figure 15.5. A synergistic mixture of hepatoprotectors (Surai, 2019).

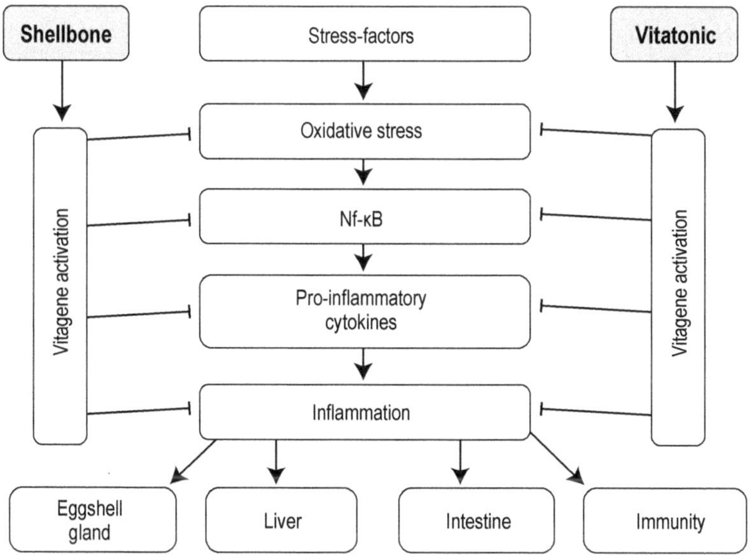

Figure 15.6. Protective effects of ShellBone and VitaTonic.

nutritional or internal stresses. In fact, vitagene activation is an important approach helping decrease negative consequences of various stresses. It should be mentioned, that the aim of the usage of the three aforementioned compositions is to help birds to mobilise their internal reserves to the maximum extent and usage such nutritional supplements do not affect general nutrition or nutrient balance. On the one hand, PerforMax is considered as a generic anti-stress supplement which can deal with most low to moderate stresses of the commercial egg and meat production. On the other hand, when stress consequences are more severe, including eggshell problems or fatty liver problems, ShellBone and VitaTonic can be used. These are additional 'instruments' in hands of nutritionists, technologists and veterinarians helping them with stress management.

References

Abdel-Moneim, A.M., Al-Kahtani, M.A., El-Kersh, M.A. and Al-Omair, M.A., 2015. Free radical-scavenging, anti-inflammatory/anti-fibrotic and hepatoprotective actions of taurine and silymarin against CCl4 induced rat liver damage. PLoS One 10: e0144509.

Ahmad, I., Shukla, S., Kumar, A., Singh, B.K., Kumar, V., Chauhan, A.K., Singh, D., Pandey, H.P. and Singh, C., 2013. Biochemical and molecular mechanisms of N acetyl cysteine and silymarin-mediated protection against maneb-and paraquat-induced hepatotoxicity in rats. Chemico-Biological Interactions 201: 9-18.

Alirezaei, M., Gheisari, H.R., Ranjbar, V.R. and Hajibemani, A., 2012. Betaine: a promising antioxidant agent for enhancement of broiler meat quality. British Poultry Science 53: 699-707.

Al-Rasheed, N., Faddah, L., Al-Rasheed, N., Bassiouni, Y.A., Hasan, I.H., Mahmoud, A.M., Mohamad, R.A. and Yacoub, H.I., 2016. Protective effects of silymarin, alone or in combination with chlorogenic acid and/or melatonin, against carbon tetrachloride-induced hepatotoxicity. Pharmacognosy Magazine 12: S337-S345.

Amiridumari, H., Sarir, H., Afzali, N. and Fanimakki, O., 2013. Effects of milk thistle seed against aflatoxin B1 in broiler model. Journal of Research in Medical Sciences 18: 786-790.

Arslan, C., 2006. L-carnitine and its use as a feed additive in poultry feeding a review. Revue de Médecine Vétérinaire 157: 134-142.

Baradaran, A., Samadi, F., Ramezanpour, S.S. and Yousefdoust, S., 2019. Hepatoprotective effects of silymarin on CCl4-induced hepatic damage in broiler chickens model. Toxicology Reports 6: 788-794.

Butler, E.J., 1976. Fatty liver diseases in the domestic fowl – a review. Avian Pathology 5: 1-14.

Celik, L. and Öztürkcan, O., 2003. Effects of dietary supplemental L-carnitine and ascorbic acid on performance, carcass composition and plasma L-carnitine concentration of broiler chicks reared under different temperature. Archives of Animal Nutrition 57: 27-38.

Celik, L.B., Tekeli, A. and Öztürkcan, O., 2004. Effects of supplemental l-carnitine in drinking water on performance and egg quality of laying hens exposed to a high ambient temperature. Journal of Animal Physiology and Animal Nutrition 88: 229-233.

Chand, N., Muhammad, D., Durrani, F.R., Qureshi, M.S. and Ullah, S.S., 2011. Protective effects of milk thistle (*Silybum marianum*) against aflatoxin B1 in broiler chicks. Asian-Australasian Journal of Animal Sciences 24: 1011-1018.

Chand, N., Naz, S., Maris, H., Khan, R.U., Khan, S. and Qureshi, M.S., 2017. Effect of betaine supplementation on the performance and immune response of heat stressed broilers. Pakistan Journal of Zoology, 49: 1857-1862,

Chen, R., Wen, C., Gu, Y., Wang, C., Chen, Y., Zhuang, S. and Zhou, Y., 2020. Dietary betaine supplementation improves meat quality of transported broilers through altering muscle anaerobic glycolysis and antioxidant capacity. Journal of the Science of Food and Agriculture 100: 2656-2663.

Cholewa, J.M., Guimaraes-Ferreira, L. and Zanchi, N.E., 2014. Effects of betaine on performance and body composition: a review of recent findings and potential mechanisms. Amino acids 46: 1785-1793.

Couch, J.R., 1956. Fatty livers in laying hens – a condition which may occur as a result of increased strain. Feedstuffs 28: 46-54.

Dangi, S.S., Dangi, S.K., Chouhan, V.S., Verma, M.R., Kumar, P., Singh, G. and Sarkar, M., 2016. Modulatory effect of betaine on expression dynamics of HSPs during heat stress acclimation in goat (*Capra hircus*). Gene 575: 543-550.

Day, C.R. and Kempson, S.A., 2016. Betaine chemistry, roles, and potential use in liver disease. Biochimica et Biophysica Acta (BBA)-General Subjects 1860: 1098-1106.

Dumari, M.A., Sarir, H., Makki, O.F. and Afzali, N. 2014. Effect of milk thistle (*Silybum marianum* l.) on biochemical parameters and immunity of broiler chicks fed aflatoxin b1 after three weeks. Iranian Journal of Toxicology 8: 1098-1103.

Elvira-Torales, L.I., García-Alonso, J. and Periago-Castón, M.J., 2019. Nutritional importance of carotenoids and their effect on liver health: A review. Antioxidants 8: 229.

Fani Makki, O., Afzali, N. and Omidi, A., 2013. Effect of different levels of silymarin (*Silybum marianum*) on growth rate, carcass variables and liver morphology of broiler chickens contaminated with aflatoxin B1. Poultry Science Journal 1: 105-116.

Fetterer, R.H., Augustine, P.C., Allen, P.C. and Barfield, R.C., 2003. The effect of dietary betaine on intestinal and plasma levels of betaine in uninfected and coccidia-infected broiler chicks. Parasitology Research 90: 343-348.

Figueroa-Soto, C.G. and Valenzuela-Soto, E.M., 2018. Glycine betaine rather than acting only as an osmolyte also plays a role as regulator in cellular metabolism. Biochimie 147: 89-97.

Fouad, A.M. and El-Senousey, H.K., 2014. Nutritional factors affecting abdominal fat deposition in poultry: a review. Asian-Australasian Journal of Animal Sciences 27: 1057-1068.

Fowler N.G., 1990. Nutritional disorders. In: Jordan, F.T.W. (ed.) Poultry diseases. W.B. Saunder, London, UK, pp. 297-319.

Geng, A., Guo, Y. and Yuan, J., 2004. Effects of dietary L-carnitine and coenzyme Q10 supplementation on performance and ascites mortality of broilers. Archives of Animal Nutrition 58: 473-482.

Gengwu, G., Ming, J., Fan, W., Wei, L., Juan, T. and Changgeng, Y., 2016. Effects of dietary silymarin on growth performance, enzymes of hepatic lipid metabolism and antioxidant ability in GIFT *Oreochromis niloticus*. Journal of Fisheries of China 40: 1309-1320.

Gillessen, A. and Schmidt, H. H., 2020. Silymarin as supportive treatment in liver diseases: a narrative review. Advances in Therapy 37: 1279-1301.

Golzar Adabi, S., Cooper, R.G., Ceylan, N. and Corduk, M., 2011. L-carnitine and its functional effects in poultry nutrition. World's Poultry Science Journal 67: 277-296.

Güçlü, B.K., Kara, K., Çakır, L., Çetin, E. and Kanbur, M., 2011. Carnitine supplementation modulates high dietary copper-induced oxidative toxicity and reduced performance in laying hens. Biological Trace Element Research 144: 725-735.

Haddadi, R., Nayebi, A.M. and Brooshghalan, S.E., 2018. Silymarin prevents apoptosis through inhibiting the Bax/caspase-3 expression and suppresses toll like receptor-4 pathway in the SNc of 6-OHDA intoxicated rats. Biomedicine & Pharmacotherapy 104: 127-136.

Hamidi, H., Pourreza, J. and Rahimi, H., 2009. Dietary betaine affect duodenal histology of broilers challenged with a mixed coccidial infection. Pakistan Journal of Biological Sciences 12: 291-295.

Heidari, R., Niknahad, H., Sadeghi, A., Mohammadi, H., Ghanbarinejad, V., Ommati, M.M., Hosseini, A., Azarpira, N., Khodaei, F., Farshad, O. and Rashidi, E., 2018. Betaine treatment protects liver through regulating mitochondrial function and counteracting oxidative stress in acute and chronic animal models of hepatic injury. Biomedicine & Pharmacotherapy 103: 75-86.

Ivy, C.A. and Nesheim, M.C., 1973. Factors influencing the liver fat content of laying hens. Poultry Science 52: 281-291.

Jahanian, E., Mahdavi, A.H., Asgary, S. and Jahanian, R., 2017. Effects of dietary inclusion of silymarin on performance, intestinal morphology and ileal bacterial count in aflatoxin-challenged broiler chicks. Journal of Animal Physiology and Animal Nutrition 101: e43-e54.

Kalorey, D.R., Kurkure, N.V., Ramgaonkar, J.S., Sakhare, P.S., Warke, S. and Nigot, N.K., 2005. Effect of Polyherbal Feed Supplement. Asian-Australasian Journal of Animal Sciences 18: 375-383.

Kazemi-Fard, M., Yousefi, S., Dirandeh, E. and Rezaei, M., 2015. Effect of different levels of L-carnitine on the productive performance, egg quality, blood parameters and egg yolk cholesterol in laying hens. Poultry Science Journal 3: 105-111.

Khatoon, A., Zargham Khan, M., Khan, A., Saleemi, M.K. and Javed, I., 2013. Amelioration of ochratoxin A-induced immunotoxic effects by silymarin and vitamin E in white Leghorn cockerels. Journal of Immunotoxicology 10: 25-31.

Kidd, M.T., 2004. Nutritional modulation of immune function in broilers. Poultry Science 83: 650-657.

Kidd, M.T., McDaniel, C.D., Peebles, E.D., Barber, S.J., Corzo, A., Branton, S.L. and Woodworth, J.C., 2005. Breeder hen dietary L-carnitine affects progeny carcase traits. British Poultry Science 46: 97-103.

Kidd, M.T., Ferket, P.R. and Garlich, J.D., 1997. Nutritional and osmoregulatory functions of betaine. World's Poultry Science Journal 53: 125-139.

Lee, I., 2015. Betaine is a positive regulator of mitochondrial respiration. Biochemical and Biophysical Research Communications 456: 621-625.

Leibetseder, J., 1995. Effects of L-carnitine in poultry. Archiv fur Tierernahrung 48: 97-108.

Li, C., Wang, Y., Li, L., Han, Z., Mao, S. and Wang, G., 2019. Betaine protects against heat exposure-induced oxidative stress and apoptosis in bovine mammary epithelial cells via regulation of ROS production. Cell Stress and Chaperones 24: 453-460.

Liew, W.P.P. and Mohd-Redzwan, S., 2018. Mycotoxin: its impact on gut health and microbiota. Frontiers in Cellular and Infection Microbiology 8: 60.

Liu, W., Yuan, Y., Sun, C., Balasubramanian, B., Zhao, Z. and An, L., 2019. Effects of dietary betaine on growth performance, digestive function, carcass traits, and meat quality in indigenous yellow-feathered broilers under long-term heat stress. Animals 9: 506.

Liu, Z., Shen, H. and Zhang, J., 2009. Clinical effect of silymarin combined with levocarnitine on non-alcoholic fatty liver disease. China Modern Medicine 7: R575.

Lovelace, E.S., Wagoner, J., MacDonald, J., Bammler, T., Bruckner, J., Brownell, J., Beyer, R.P., Zink, E.M., Kim, Y.M., Kyle, J.E. and Webb-Robertson, B.J.M., 2015. Silymarin suppresses cellular inflammation by inducing reparative stress signaling. Journal of Natural Products 78: 1990-2000.

Malayeri, M.R.M., Tehrani, A.D. and Rezaei, A., 2014. Preventive effects of silymarin extract on carbon tetrachloride-induced hepatotoxicity in broilers. Journal of Veterinary Clinical Pathology 8: 445-459.

Metzler-Zebeli, B.U., Eklund, M. and Mosenthin, R., 2009. Impact of osmoregulatory and methyl donor functions of betaine on intestinal health and performance in poultry. World's Poultry Science Journal 65: 419-442.

Mousa, M.A. and Osman, A.S., 2016. The implication of L-carnitine and silymarin supplementation on growth performance and some blood parameters of broilers. Assiut Veterinary Medical Journal 62: 132-138.

Muhammad, D., Chand, N., Khan, S., Sultan, A. and Mushtaq, M., 2012. Hepatoprotective role of milk thistle (*Silybum marianum*) in meat type chicken fed aflatoxin B1 contaminated feed. Pakistan Veterinary Journal 32: 443-446.

Nagashimada, M. and Ota, T., 2019. Role of vitamin E in nonalcoholic fatty liver disease. IUBMB life 71: 516-522.

Ore, A. and Akinloye, O.A., 2019. Oxidative stress and antioxidant biomarkers in clinical and experimental models of non-alcoholic fatty liver disease. Medicina 55: 26.

Ratriyanto, A. and Mosenthin, R., 2018. Osmoregulatory function of betaine in alleviating heat stress in poultry. Journal of Animal Physiology and Animal Nutrition 102: 1634-1650.

Ridell, C., 1991. Developmental, metabolic, and miscellaneous disorders. In: Calnek, B.W., Barnes, H.J., Beard, C.W., Reid, W.M. and Yofer Jr, H.W., (eds) Diseases of poultry. Iowa State University Press, Ames, IA, USA, pp. 827-862.

Ringseis, R., Keller, J. and Eder, K., 2018. Basic mechanisms of the regulation of L-carnitine status in monogastrics and efficacy of L-carnitine as a feed additive in pigs and poultry. Journal of Animal Physiology and Animal Nutrition 102: 1686-1719.

Roth, S.D., Schüttrumpf, J., Milanov, P., Abriss, D., Ungerer, C., Quade-Lyssy, P., Simpson, J.C., Pepperkok, R., Seifried, E. and Tonn, T., 2012. Chemical chaperones improve protein secretion and rescue mutant factor VIII in mice with hemophilia A. PloS One 7: e44505.

Saeed, M., Babazadeh, D., Naveed, M., Arain, M.A., Hassan, F.U. and Chao, S., 2017. Reconsidering betaine as a natural anti-heat stress agent in poultry industry: a review. Tropical Animal Health and Production 49: 1329-1338.

Saeed, M., Babazadeh, D., Arif, M., Arain, M.A., Bhutto, Z.A., Shar, A.H., Kakar, M.U., Manzoor, R. and Chao, S., 2017. Silymarin: a potent hepatoprotective agent in poultry industry. World's Poultry Science Journal 73: 483-492.

Spurlock, M.E. and Savage, J.E., 1992. Research note: antioxidant activity of Japanese quail liver cytosol in the absence and presence of reduced glutathione. Poultry Science 71: 928-931.

Squires, E.J. and Leeson, S., 1988. Aetiology of fatty liver syndrome in laying hens. British Veterinary Journal 144: 602-609.

Surai, P.F., 2015a. Silymarin as a natural antioxidant: an overview of the current evidence and perspectives. Antioxidants 4, 204-247.

Surai, P.F., 2015b. Carnitine enigma: from antioxidant action to vitagene regulation. Part 1. Absorption, metabolism and antioxidant activities. Journal of Veterinary Science & Medicine 3, 2: 14.

Surai P.F., 2015c. Carnitine enigma: from antioxidant action to vitagene regulation. Part 2. Transcription factors and practical applications. Journal of Veterinary Science & Medicine 3, 2: 17.

Surai, P.F., 2015d. Antioxidant action of carnitine: molecular mechanisms and practical applications. EC Veterinary Science 2.1: 66-84.

Surai, P.F., 2018. Taurine and carnitine in poultry production: from vitagene activation to chicken health maintenance. Ptakhivnitstvo.ua (Ukrainian Poultry Science, Kiev) 1-2: 12-17.

Surai, P.F., 2019. From healthy liver to general health and high productivity of poultry. Ptitzevodstvo.ua (Ukrainian Poultry Science, Kiev) 1-2: 4-7

Surai, P.F. and Fisinin, V.I., 2015. Antioxidant-prooxidant balance in the intestine: applications in chick placement and pig Weaning. Journal of Veterinary Science & Medicine 3: 16.

Surai, P.F., Kochish, I.I. and Fisinin, V.I., 2017. Antioxidant systems in poultry biology: nutritional modulation of vitagenes. European Journal of Poultry Science 81: 1612-9199.

Surai, P.F., Kochish, I.I., Fisinin, V.I., Grozina, A.A. and Shatskikh, E.V., 2018. Molecular mechanisms of gut health support in poultry: role of microbiota. Agricultural Technologies, Moscow, Russia.

Surai, P.F., Kochish, I.I., Fisinin, V.I. and Kidd, M.T., 2019. Antioxidant defence systems and oxidative stress in poultry biology: an update. Antioxidants 8, 7: 235.

Tedesco, D., Steidler, S., Galletti, S., Tameni, M., Sonzogni, O. and Ravarotto, L., 2004. Efficacy of silymarin-phospholipid complex in reducing the toxicity of aflatoxin B1 in broiler chicks. Poultry Science 83: 1839-1843.

Wang, Y.W., Ning, D., Peng, Y.Z. and Guo, Y.M., 2013. Effects of dietary L-carnitine supplementation on growth performance, organ weight, biochemical parameters and ascites susceptibility in broilers reared under low-temperature environment. Asian-Australasian Journal of Animal Sciences 26: 233-240.

Wen, C., Chen, Y., Leng, Z., Ding, L., Wang, T. and Zhou, Y., 2019. Dietary betaine improves meat quality and oxidative status of broilers under heat stress. Journal of the Science of Food and Agriculture 99: 620-623.

Wolford, J.H. and Polin, D., 1972. Lipid accumulation and hemorrhage in livers of laying chickens. A study on fatty liver-hemorrhagic syndrome (FLHS). Poultry Science 51: 1707-1713.

Xing, J., Kang, L. and Jiang, Y., 2011. Effect of dietary betaine supplementation on lipogenesis gene expression and CpG methylation of lipoprotein lipase gene in broilers. Molecular Biology Reports 38: 1975-1981.

Xu, D., Xu, M., Jeong, S., Qian, Y., Wu, H., Xia, Q. and Kong, X., 2019. The role of Nrf2 in liver disease: novel molecular mechanisms and therapeutic approaches. Frontiers in Pharmacology 9: 1428.

Zaefarian, F., Abdollahi, M.R., Cowieson, A. and Ravindran, V., 2019. Avian liver: the forgotten organ. Animals 9: 63.

Zhan, X.A., Li, J.X., Xu, Z.R. and Zhao, R.Q., 2006. Effects of methionine and betaine supplementation on growth performance, carcase composition and metabolism of lipids in male broilers. British Poultry Science 47: 576-580.

Zhan, X., 2001. Effects of betaine on immune responses of Newcastle disease in broiler chicks. Chinese Journal of Veterinary Science 21: 99-102.

Zhong, S., Fan, Y., Yan, Q., Fan, X., Wu, B., Han, Y., Zhang, Y., Chen, Y., Zhang, H. and Niu, J., 2017. The therapeutic effect of silymarin in the treatment of nonalcoholic fatty disease: a meta-analysis (PRISMA) of randomized control trials. Medicine 96: e9061.

Chapter 16
Vitagenes in gut health and immunity

All is well that ends well

16.1 Introduction

In previous chapters the vitagene concept was described in details and special attention has been paid to possible applications of the concept in the development of the effective strategy for preventing negative consequences of commercially relevant stresses in poultry production. In fact, nutritional modulation of vitagene expression and activities is considered to be a new promising direction in nutritional sciences. It is important to note that vitagenes operate in the body in close relationship with various transcription factors, including Nrf2 and NF κB. Therefore, vitagene activation withing the framework of cell signalling have a great variety of additional applications, including gut health and immunity. Elucidating the cellular factors and signalling pathways that maintain intestinal integrity, epithelial proliferation and differentiation is vital to our understanding of gut adaptation to various stresses (Li *et al.*, 2020).

This chapter presents recent research data related to possible molecular mechanisms of regulatory roles of vitagenes in gut health and immunocompetence maintenance in stressful commercial conditions of industrial poultry production.

16.2 Role of vitagenes in the gut defence

Nutritional modulation of vitagenes described in Chapters 9-15 is of great importance for gut health maintenance. In fact, vitagenes are expressed in the gut and defensive antioxidant compounds coded by vitagene represent major protective mechanism in the gut. This is especially important in conditions of prohibition of feed antibiotics.

16.2.1 Superoxide dismutase

Detailed information related to SOD in the gut can be found in the Chapter 4. In fact, data on SOD expression and activity in the intestine are quite limited. In general, SOD in the small intestine and colon of the rat includes both MnSOD and Cu/Zn-SOD with cytosolic form of the enzyme to be characterised by increased expression in comparison to MnSOD (Frederiks and Bosch, 1997; Table 16.1).

There is a significant difference in total SOD activity between different segments of the chicken gut (Figure 16.1). Based on the SOD activity the laying hen intestinal segments can be placed in the following descending order: duodenum>jejunum=

DOI 10.3920/978-90-8686-906-0_16, © Wageningen Academic Publishers 2020

Table 16.1. Localisation of superoxide dismutase (SOD) activity in tissues of 3-month-old male Wistar rats (adapted from Frederiks and Bosch, 1997).[1]

Tissue		SOD activity	Cu/Zn-SOD	Mn-SOD
Small intestine	Enterocytes	+	++	+
	Goblet cells	+	++	+
Colon	Resorptive cells	++	++	+
	Goblet cells	+	++	+

[1] + = moderate; ++ = high.

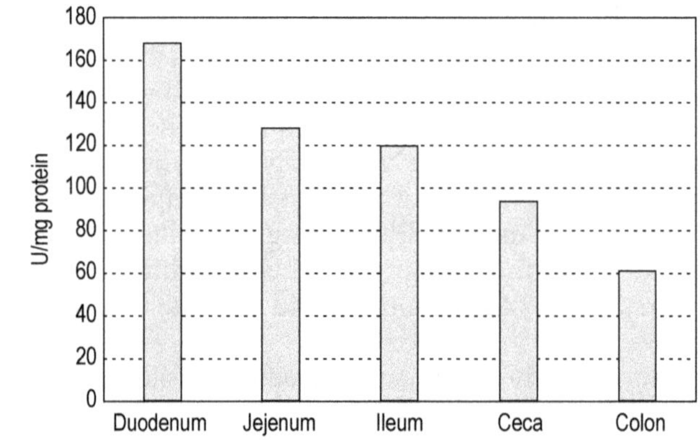

Figure 16.1. Superoxide dismutase activity in chicken gut (adapted from Surai *et al.*, 2018a).

ileum>>ceca>colon. On the one hand, the total SOD activity in the duodenum was significantly (by 40%, $P<0.05$) higher than that in jejunum or ileum. On the other hand, the total SOD activity in the ceca was shown to be significantly (by 32%, $P<0.05$) lower than that in jejunum or ileum, but substantially higher (by 24%, $P<0.05$) in comparison to that in the colon. Interestingly, lipid peroxidation (MDA) was found to be highest in jejunum and the order of the intestinal segments in accordance with MDA level (jejunum>duodenum>ileum>ceca=colon) is not the same as for SOD, reflecting importance of other protective mechanisms in the intestine (beyond SOD) responsible for antioxidant defences.

Indeed, there is a need for more detailed evaluation of the antioxidant system of the chicken gut depending on age, nutrition and stress. It would be also particularly important to study a relationship between antioxidant defences and microbiota in the chicken gut to understand molecular mechanisms of the maintenance of healthy gut in stress conditions.

16.2.2 Heat shock protein 70

Recently, new mechanisms of adaptive defences of the gastrointestinal mucosa at the intracellular level have been characterised. One of these responses, known as the heat shock response, is considered to be a universal fundamental mechanism necessary for cell survival under a variety of unfavourable conditions (Santoro, 2000). As mentioned above, intestinal cells are challenged with a great variety of potentially toxic compounds and their protection is a vital part of the strategy to maintain animal/poultry health. In mammalian/avian cells, the induction of the heat shock response requires the activation and translocation to the nucleus of one or more heat shock transcription factors, which control the expression of a specific set of genes encoding cytoprotective heat shock proteins (Santoro, 2000). Indeed, HSPs have a broad range of functions related to their major role in cellular homeostasis and protect cells against apoptotic cell death. The expression of HSP27, heat shock cognate 70 (HSC 70), HSP70 and HSP90 along the GIT of young pigs and the effect of weaning on this expression were studied (David *et al.*, 2002). There was a site specificity in HSP expression in the gut. For example, the expression of HSP27 and HSP70 was increased in the stomach and duodenum between 6 and 12 h post-weaning and between 24 and 48 h in the mid-jejunum, ileum and colon. At the same time, their expressions were transiently decreased in the ileum. Indeed, in normal porcine GI tract HSP expression is gut region- and cell type-specific in response to dietary components, microbes, and microbial metabolites to which the mucosa surface is exposed (Liu *et al.*, 2014).

Therefore, HSPs function as molecular chaperones in regulating cellular homeostasis and promoting survival. However, if the stress is too high, a signal that leads to programmed cell death, apoptosis, is activated, thereby providing a finely tuned balance between survival and death (Kopecek *et al.*, 2001). In addition to extracellular stimuli, several non-stressful conditions induce HSPs during normal cellular growth and development. In particular, the HSP family is activated under oxidative stress and provides an important protection against protein denaturation and modifications by capping and refolding or drives damaged proteins into appropriate proteolytic pathways (Yenari, 2002). In fact, HSPs have been assigned to multiple subcellular sites and implicated in multiple functions ranging from stress response, intracellular trafficking, antigen processing, control of cell proliferation, differentiation, and tumorigenesis (Wadhwa *et al.*, 2002). Therefore, in response to environmental or physiological stresses cells increase synthesis of HSP (Tsukimi and Okabe, 2001). ROS-mediated damage has been implicated in the pathophysiology of the gastrointestinal mucosa and HSPs are suggested to play an important role in cytoprotection against oxidative stress-induced injury (Prabhu and Balasubramanian, 2002). For example, the mammalian intestinal epithelial cells respond to heat stress by producing heat shock proteins that provide protection in stress conditions, which would otherwise lead to cell damage or death. The protective effects of HSPs are seen in heat stress, infection, and inflammation (Malago *et al.*, 2002). The molecular mechanisms of heat shock response-induced cytoprotection were described in Chapter 5. In fact, they involve inhibition of proinflammatory cytokine production and induction of cellular proliferation for restitution of the damaged epithelium (Malago *et al.*,

2002). It is interesting to note that HSP72, the stress-inducible form of HSP70, was detected in samples from rat distal colon, proximal colon, and terminal ileum, but was not found in proximal small bowel or other organs (liver, kidney, spleen, heart, and brain) of unstressed animals (Beck *et al.*, 1995). HSPs play an important role in gastric mucosal defence under conditions of stress. For example, exposure of rats to restraint and water-immersion stress caused rapid HSP70 mRNA expression and HSP70 accumulation in gastric mucosa and the extent of HSP70 induction inversely correlated to the severity of mucosal damage (Rokutan, 1999). Therefore, HSP70 is involved in repair of partially damaged proteins and substantially contributes to protection of the gastrointestinal mucosa against various necrotising factors (Tsukimi and Okabe, 2001). Exposure of mice to thermal stress was shown to result in the rapid induction and expression of HSP70 in the intestine and other organs (the liver, pancreas, heart, lung and adrenal cortex) not constitutively expressing HSP70 (Huang *et al.*, 2001). HSP70 is reported to maintain barrier function as a result of stabilisation of the tight junctions between intestinal epithelial cells (Kojima *et al.*, 2003; Liedel *et al.*, 2011; Much *et al.*, 1999). Indeed, mother's milk-induced HSP70 expression was shown to preserve intestinal epithelial barrier function in an immature rat pup model (Liedel *et al.*, 2011). Geranylgeranylacetone, a clinically used antiulcer drug, was reported to increase expression of HSP70 in the small intestine and suppress indomethacin-induced lesions (e.g. inflammation) of the small intestines in wild-type mice (Asano *et al.*, 2009). Intestinal HSP70 was shown to play a protective role as a result of re-establishing the balance between the intestinal post-inflammatory and anti-inflammatory cytokines in a post-infectious irritable bowel syndrome mouse model (Lan *et al.*, 2016). In patients with inflammatory bowel diseases, HSP70 expression was found to be significantly increased during the active stage of the disease and overexpression of HSP70 was reported to protect against the development of inflammation in the large intestinal mucosa provoked by various damaging factors (Samborski and Grzymisławski, 2015). The authors suggested that microbiota can affect HSP70 expression in the gut and it seems likely that augmentation of HSP70 expression in the gut could be considered as an important strategy to deal with various gut integrity-related problems. In fact, co-localised HSP70 and tight junction protein zona occludens-1 (ZO-1) reported recently, suggests physical interaction of HSP70 and tight junction proteins to protect tight junction function (Rentea *et al.*, 2018).

16.2.3 Heme oxygenase 1

It is accepted that HO-1, known as HSP32, can be induced by various stresses. Indeed, HO-1 induction and the maintenance of its appropriate activity is critical in protecting the intestinal epithelial cells from oxidative injury (Fujii *et al.*, 2003). It is interesting that in the aforementioned experiment HO-1 was markedly induced following LPS treatment in the mucosal epithelial cells in the upper intestine (duodenum and jejunum) but not in the lower intestine (ileum and colon). It seems likely, that there is a delicate interaction between HSPs and other antioxidant defence mechanisms to maintain mucosal integrity and repair of acute mucosal damage. It was suggested that activation of HO-1 could be an important natural defensive mechanism to alleviate inflammation and tissue injury in the gastrointestinal tract (Guo *et al.*, 2001). Increased expression of HO-1 was shown to be a protective mechanism behind alleviation of the intestinal injury by hydrogen gas in a model of severe sepsis in mice, (Li *et al.*, 2015) or due to pharmacological preconditioning with vitamin C after haemorrhagic shock in rats (Zhao *et al.*, 2014), ischemic preconditioning in the late phase of IR injury of the intestine (Mallick *et al.*, 2010) or haemorrhagic shock-induced intestinal tissue injury in rats (Inoue *et al.*, 2008; Umeda *et al.*, 2009). Induction of HO-1 was suggested to be a mechanism of protective action of hypothermia (Attuwaybi *et al.*, 2003), hemin (Attuwaybi *et al.*, 2004) or a somatostatin analogue, octreotide (Takano *et al.*, 2012) against ischemia/reperfusion-induced impairment of intestinal integrity in rats. Similarly, induction of HO-1 was indicated to improve impaired intestinal transit after burn injury in rats (Gan and Chen *et al.*, 2007). The HO-1 system is believed to provide gut cytoprotection and decreasing ischemia/reperfusion injury by inhibiting inflammation, oxidation, and apoptosis (Liao *et al.*, 2013). A great number of publications indicated that HO-1/CO system is characterised by potent antioxidant, antiapoptotic, anti-inflammatory and cytoprotective activities against I/R injury. This was demonstrated by genetic overexpression of HO-1 cDNA or pharmacological induction with drugs/nutrients (Cheng and Rong, 2017). In fact, resveratrol (Res) was shown to protect oxidative stress-induced intestinal epithelial barrier dysfunction by upregulating HO-1 and SOD expression. It is of note that protective effects of Res were abolished by the HO-1 inhibition or HO-1 knockdown by siRNA (Wang *et al.*, 2016). Similarly, HO-1 was found to serve as an effector of the anti-inflammatory action of hydrogen sulphide donor (NaHS-PC) in postischemic murine small intestine (Zuidema *et al.*, 2011). Fish oil dietary supplementation was shown to induce HO-1 expression in the mouse intestine associated with increased concentration of 4-HHN, a product of n-3 PUFAs peroxidation (Nakagawa *et al.*, 2014). Furthermore, in endotoxic shock model, intestinal preconditioning was able to prevent inflammatory responses by modulating HO-1 expression (Tamion *et al.*, 2007). HO-1 is shown to be important player in the anti-inflammatory activities of M-CSF-polarised M2 macrophages (Sierra-Filardi *et al.*, 2010). In fact, HO-1 production by intestinal $CX3CR1^+$ macrophages in mouse was found to help resolving gut inflammation (Marelli *et al.*, 2017). Nutritional modulation of the function of intestinal macrophages to restore the immunological homeostasis associated with upregulated HO-1 activity and rebalance the enteric commensal flora is considered as a promising strategy to deal with gut inflammation (Ju *et al.*, 2018). Furthermore, HO-1 has been considered as an immunomodulator playing a key role in the homeostasis maintenance in the gastrointestinal tract. Indeed, the protective role of HO-1 in the control of the intestinal inflammation is associated with its connection with the gut microbiota (Marelli and Allavena, 2020). Low dose of DON was reported to trigger low-grade inflammation in liver and changes in gut microbiota in mice. Interestingly, HO-1 was shown to exert protective effect in DON-induced hepatotoxicity, which was suggested to be associated with microbiota modulation by HO-1 (Peng *et al.*, 2019). Interestingly, a food-grade bacterium genetically modified to deliver bioactive HO-1 *in situ* was reported to possess a protective effect against intestinal mucosal injury in rats with endotoxemia as a result of modulation of the immune system (Pang *et al.*, 2008). The central role of HO-1 in modulating the immune system is associated with decreasing secretion of inflammatory cytokines, such as TNF, IL-1β, IL-6 and IL-17 and increased levels of anti-inflammatory mediators such as IL-10 or IL-22 due to increased production of CO (Marelli and Allavena, 2020). In addition, as a result of the low levels of HO-1, the processes of phagocytosis were impaired and the dysbiosis occurred leading to chronic inflammation in the gut (Marelli and Allavena, 2020).

16.2.4 Thioredoxin system

Thioredoxin and TrxR are considered to be also important members of vitagene family playing important roles in the gut. For example, Takaishi *et al.* (2003) identified rat Trx as a growth-promoting factor for intestinal epithelial cells, while Higashikubo *et al.* (1999) showed that cellular oxidative stress caused an increase in the activity of thioredoxin, which is involved in the defence mechanism against oxidative stress. In particular, H_2O_2 was cytotoxic to the small intestine epithelial cell line, IEC-6 and the glutathione S-transferase and thioredoxin reductase activities and SH content decreased dose-dependently with H_2O_2 treatment, while Trx activity increased at low H_2O_2 concentrations. is not clear at present, but clearly, they participate in maintenance of the redox balance in the gut. It is important to note that Trx expression is particularly high in the small intestinal mucosa and colon (Gasdaska *et al.*, 1996), while TrxR expression in small intestine is substantially higher than in colon. All three thioredoxin reductases are expressed in the intestine at least at the mRNA level, which is relatively unaffected by marginal Se deficiency and in the case of TrxR2 and TrxR3 rather increased when Se becomes limited (Kipp *et al.*, 2009). The presence of a full complement of Trx/TrxR proteins in stomach, duodenum, jejunum, ileum and colon suggesting their function in antioxidant defence and redox regulation in the intestinal tract (Godoy *et al.*, 2011). Furthermore, the selenoproteins GPx2, TrxR2 and TrxR3 in the gut are regulated by the Wnt pathway (Kipp *et al.*, 2012). It has been also shown that unstimulated lamina propria T lymphocytes exhibited high expression of Trx which involved in the regulation of intracellular redox homeostasis in these cells (Sido *et al.*, 2005). The authors suggested that Trx may play a key role in the specialised intestinal microenvironment in amplifying immediate immune responses. In fact, membrane-bound Trx converts human β-defensin 1 to a potent antimicrobial peptide *in vivo* (Jaeger *et al.*, 2013). Recently *Caenorhabditis elegans* Trx-3, the first metazoan thioredoxin with a tissue-specific expression pattern restricted to intestine has been characterised (Jiménez-Hidalgo *et al.*, 2014). Its role in the gut redox balance maintenance and search for avian orthologs await investigation.

16.2.5 Glutathione system

The importance of cellular GSH/GSSG redox in modulating cell transitions and their oxidative susceptibility has been clearly demonstrated (Jefferies *et al.*, 2003). In particular, it was shown that the intestinal epithelium is capable of transporting luminal GSH to augment intracellular peroxide catabolism and suggested that the loss of mucosal redox balance is a driving force in etiology of chronic gut pathologies. (Aw, 2005). The gut contains internally-originated antioxidant enzymes SOD, GPx and CAT and they represent an important mechanism of the enterocyte defence from oxidative damage. A specific gastrointestinal GPx2 has been described in 1993

(Chu *et al.*, 1993). GPx2 activity was expressed in both the villus and crypt regions of the rat mucosal epithelium and its activity nearly equalled that of classical GPx throughout the small intestine and colorectal segments (Esworthy *et al.*, 1998). When enzymatic activity in liver, small intestine, cecum, and colon of rats was compared, GPx2 was highest in large bowel (Meinl *et al.*, 2009). In fact, GPx2 was detected in proximal, transverse and distal colon being located in the luminal epithelium and lymphatic tissue of rats (Drew *et al.*, 2005). GPx2 could be considered to be a barrier against hydroperoxide resorption (Brigelius-Flohé, 1999; Brigelius-Flohé and Maiorino, 2013). Furthermore, in the gastrointestinal tract there are at least three more selenoproteins, including plasma GPx, selenoprotein P and TrxR (Mork *et al.*, 1998). GSH and GSH-dependent enzymes contribute significantly towards intestinal antioxidant defences. In fact, an important peroxide detoxification pathway in the intestine is based on the GSH redox system (LeGrand and Aw, 1998). In this system GPx reduces peroxides at the expense of GSH oxidation. Oxidised glutathione is reduced back to the active form by glutathione reductase utilising reducing potential of NADPH which is produced in the pentose phosphate pathway. Various studies in mouse models of colon cancer and selenoprotein gene deletion studies indicated that selenoproteins play a pivotal role in the maintenance of gut homeostasis. In particular, GPx has been extensively studied for its redox regulation, antioxidant and anti-inflammatory roles in preventing chronic intestinal inflammation (Reeves and Hoffman, 2009; Speckmann and Steinbrenner, 2014). Indeed, GPx1 is shown to be expressed in all cell types of the gut, whereas GPx2 is predominantly expressed in the epithelial cells, including the paneth cells, of the gastrointestinal mucosa (Brigelius-Flohé *et al.*, 2001; Esworthy *et al.*, 1998; Florian *et al.*, 2001) and GPx4 is found to be expressed in epithelial cells and the lamina propria of the intestine (Speckmann *et al.*, 2011; Takahashi *et al.*, 1987).

Up-regulation by the Nrf2/Keap1 system indicates that GPx2 is an antioxidant and anti-inflammatory enzyme (Brigelius-Flohé and Flohé, 2020). In fact, an increased apoptotic cell proportion in the crypts of the colon was reported as a major pathological observation in gpx2$^{-/-}$ mice under unstressed condition (Florian *et al.*, 2010). However, in a model of inflammation-mediated colon carcinogenesis, inflammatory scores were shown to be increased in gpx2$^{-/-}$ than in WT mice (Krehl *et al.*, 2012). It is of note an existence of a cooperative action of various GPxs in the gut. On the one hand, a double knock-out of GPx2 and GPx1 caused ileo-colitis (Esworthy *et al.*, 2001). On the other hand, in the colon of gpx2$^{-/-}$ mice a compensatory dramatic overexpression of gpx1 at the transcriptional and translational level were observed (Florian *et al.*, 2010). Interestingly, in gpx2$^{-/-}$ mice the total GPx activity in the gut was shown to be significantly increased in comparison to wild type mice (Müller *et al.*, 2013) and unique role of GPx2 in preventing ectopic apoptosis in the intestine awaits further investigation (Brigelius-Flohé and Flohé, 2020). Loss of epithelium specific GPx2 was reported to lead to aberrant cell fate decisions during intestinal differentiation in mice (Lennicke *et al.*, 2017). A significant increase in SOD1, SOD3, Prx6, GPx2, GPx7 expression during I/R-induced damage of the small intestine was observed in rats. Interestingly, injection of exogenous Prx6 prior to induced ischemia showed a significant protective effect in the intestine leading to minimisation of oxidative injury with reducing necrosis and apoptosis simultaneously with normalisation of expression of antioxidant enzymes (Gordeeva *et al.*, 2015).

Interestingly, protective effect of *Lactobacillus plantarum* (Wang *et al.*, 2018) and *Bifidobacterium bifidum* (Din *et al.*, 2020) in dextran sodium sulphate-induced colitis in mice was associated with up-regulation of ROS-scavenging enzymes, including GPx2. Similarly, gene expression of SOD2 in jejunum tissue of weaned piglets was increased after oral administration of *Lactobacillus reuteri* in the form of a liquid preparation (Zhang *et al.*, 2017). When porcine epithelial cells were treated with zearalenone (10 μM for 24 h), microarray results identified 190 genes significantly and differentially expressed, of which 70% were up-regulated, including GPx6, GPx2, GPx1(Taranu *et al.*, 2015). It seems likely that interactions between various antioxidants within the antioxidant defence network is of great importance in the animal/chicken gut. For example, vitamin E (200 mg/kg BW) was shown to reduce toxic effect of phoxim (organophosphate pesticides) on intestinal structure, alleviated the oxidative stress in intestinal tissue, decreased the level of proinflammatory TNF-α and increased expression of GPx2 and SOD in rats (Sun *et al.*, 2018). Gut microbiota was shown to affect host amino acid and GSH metabolism in mice (Mardinoglu *et al.*, 2015). Under oxidative stress conditions, imposed by peroxidised oil dietary consumption in weaned piglets, jejunal integrity is shown to be dependent on the local and hepatic GSH redox system which is responsible for the elimination of luminal peroxides, and thereby protecting duodenal barrier function (Degroote *et al.*, 2019). Interestingly, GPx8 is shown to protect against colitis by negatively regulating caspase-4/11 activity and decreasing inflammation. Indeed, mice lacking GPx8 (the oxidative stress sensor) are characterised by increased susceptibility to colitis and endotoxic shock and disturbed gut microbiome (Hsu *et al.*, 2020).

16.2.6 Sirtuins

It is generally accepted that SIRTs are involved in the intestine protection in stress conditions. For example, mice with an intestinal specific SIRT1 deficiency (Sirti1int$^{-/-}$) were characterised by dysregulated intestinal cell differentiation with altered gut microbiota (Lo Sasso *et al.*, 2014a). Indeed, SIRT1 is shown to have anti-ianflammatory effects in the intestine via regulating the gut microbiota (Wellman *et al.*, 2017). SIRT1 deficiency in the intestine was reported to elevate the number of secretory cells, including intestinal Paneth and goblet cells (Wellman *et al.*, 2017). SIRT1 was reported to alleviate endoplasmic reticulum stress-mediated apoptosis of intestinal epithelial cells in an ulcerative colitis model system (Ren *et al.*, 2019). Curcumin and resveratrol were shown to exert their protective effects on DSS-induced colitis in mice partially through regulating SIRT1/mTOR signalling leading to suppression of the intestinal inflammation (Zhang *et al.*, 2019). Resveratrol was shown to have protective effects against radiation-induced intestinal injury at least partly via activation of SIRT1 (Zhang *et al.*, 2017). In general, defective expression of SIRT1 is believed to contribute to activation of inflammatory pathways in the human gut (Caruso *et al.*, 2014). Interestingly, other SIRTs are believed to be also involved in gut integrity maintenance. For example, compared to wild type mice,

SIRT2 deficient mice showed exaggeration of cellular adhesion with sepsis in the small intestine, while SIRT2 overexpression showed protective effects (Buechler *et al.*, 2017). Interestingly, under basal condition, SIRT2 deficiency did not affect the basal phenotype and intestinal morphology, whereas SIRT2 deficiency ($Sirt2^{-/-}$) was shown to promote dextran sodium sulphate-induced inflammatory responses. Interestingly, anti-inflammatory action of SIRT2 was shown to be associated with regulation of NF-κB acetylation and macrophage polarisation (Lo Sasso *et al.*, 2014b). SIRT2 was found to be expressed in small intestinal villi and its levels were shown to be dramatically increased in differentiated intestinal cells compared with undifferentiated cells and it was clearly shown that SIRT2 to be essential for maintenance of normal intestinal homeostasis (Li *et al.*, 2020). Indeed, deletion of SIRT2 was reported to result in decreased enterocyte and goblet cell differentiation, whereas SIRT2 deficiency increased the Paneth cell lineage and it seems likely that SIRT2 regulates intestinal cell differentiation through regulation of Wnt/β-catenin signalling (Li *et al.*, 2020). SIRT3 deficiency was shown to cause an impaired intestinal permeability and inflammation in high fat fed mice, which was attenuated by sodium butyrate (Chen *et al.*, 2019). Furthermore, SIRT3 was reported to be anti-inflammatory factor interacting with the gut microbiota during colon tumorigenesis (Zhang *et al.*, 2018). It is believed that PRDX3 is a key protective factor for intestinal I/R injury, and SIRT3-mediated PRDX3 deacetylation was shown to alleviate intestinal I/R-induced mitochondrial oxidative damage and apoptosis (Wang *et al.*, 2020). Importantly, SIRT6 was found to be predominantly expressed in epithelial cells in intestinal crypts and its expression was decreased in colitis in both mice and humans (Liu *et al.*, 2017). Protective effects of SIRT6 to intestine was confirmed in a study showing increased susceptibility of the cells to injurious insults after knockdown of SIRT6 expression, while YAMC cells with SIRT6 overexpression were characterised by increased resistance to injurious insult. In fact, intestinal epithelial-specific *Sirt6* ($Sirt6^{IEC\text{-}KO}$) knockout mice were indicated to show increased susceptibility to dextran sulphate sodium (DSS)-induced colitis (Liu *et al.*, 2017). The authors showed that protective effects of SIRT6 on intestinal epithelial cells challenged with inflammatory injury was related to preservation of R-spondin-1 (a critical growth factor for intestinal epithelial cells) levels in the cells. By targeting SIRT6 and MAPK13, microRNA-351-5p was shown to aggravate I/R injury associated with increased intestinal mucosal oxidative stress, inflammation, and apoptosis (Hu *et al.*, 2018).

16.3 Gut redox balance and microbiota

It is believed that under physiological conditions, the gut cell can tolerate a certain level of ROS due to its antioxidant capacity, which is critical for intestinal homeostasis (Tian *et al.*, 2017). However, in stress conditions an excessive ROS production can break an existing antioxidant-prooxidant (redox) balance and enhance membrane permeability, alter the inflammatory response, and cause lipid and protein modifications, DNA damage and apoptosis (Tian *et al.*, 2017). The avian gastrointestinal tract harbours trillions of commensal microorganisms, collectively known as the microbiota (Bhat and Kapila, 2017). For example, chicken intestinal tract is composed of duodenum, jejunum, ileum, cecum, and colon, and there are significant differences in microbiota concentration and composition between the aforementioned gut sections (Xiao *et al.*, 2017). Interestingly, cecum is characterised by the most complex microbial community dominated by the phyla *Firmicutes*, *Bacteroidetes*, *Actinobacteria* and *Proteobacteria* (Sergeant *et al.*, 2014). On the other hand, at the genus level, the major microbial genera across all gut sections were shown to be *Lactobacillus*, *Enterococcus*, *Bacteroides*, and *Corynebacterium* (Xiao *et al.*, 2017). Furthermore, *Bacteroides* was shown to be the dominant group in cecum, while *Lactobacillus* was predominant in the small intestine sections (duodenum, jejunum and ileum; Xiao *et al.*, 2017). The microbiota's various bacterial members are involved in a physiological network of cooperation and competition within the gut (Stecher *et al.*, 2015). On one hand, microbiota is shaped by environmental factors. On the other hand, gut environment including redox balance is shaped by microbiota. Redox signalling in the chicken gut regulates many physiological functions, including self-renewal, proliferation, migration and differentiation of epithelial cells (Perez *et al.*, 2017). By modulating NADPH oxidases commensal microbiota substantially contribute to this signalling. In fact, normal commensal microbiota is responsible for so called colonisation resistance by creating hostile/unfavourable conditions for colonisation of enteric pathogens (Stecher *et al.*, 2015). Interestingly, the same nutritional e.g. (antibiotics and other drugs) or environmental (changes in diet, disease challenge) factors can affect antioxidant-prooxidant balance (Surai and Fisinin, 2015) and disrupt microbiota in the gut (Stecher *et al.*, 2015). For example, experimental infection with *Cryptosporidium parvum* in immunocompromised Swiss albino mice caused an increase in lipid peroxidation and decrease in GSH, CAT and SOD at the peak of infection in the intestine and liver (Phagat *et al.*, 2017). An important consequence of redox balance and microbiota disturbances is inflammatory host responses and activation of immune cells resulting in further production of ROS in the gut and deepen the problem. Furthermore, under inflammatory conditions, the microbial community could shift from obligate to facultative anaerobes (Rigottier-Gois, 2013). For example, the gut microbiota is shown to contribute to the control of *Campylobacter jejuni* colonisation and could prevent lesion development (Han *et al*, 2017).

Of note, intestinal microbiota is shown to play an important role in mucosal immunity and dysbiosis is associated with the pathogenesis of inflammatory diseases (Shi *et al.*, 2017). In the complex intestinal ecosystem, there could be a range of microbes able to facilitate oxidation/reduction reactions and maintenance of the redox balance. Indeed, a redox based response within cells is emerging as an important and conserved element of host cell and symbiotic microbe interaction (Jones and Neish, 2017). It has been suggested that in the gut commensal microflora, in particular *Lactobacillus* species, are able to stimulate rapid, non-pathogenic levels of ROS production which would oxidise reactive cysteine residues within proteins controlling cell signalling pathways. Indeed, proteins that harbour reactive cysteine residues, for example Keap1 or IκB, could function as redox sensors and transducers of signalling initiated by increased ROS concentrations (Jones and Neish, 2017). It has been shown that ROS production by intestinal epithelial cells is mediated by receptors and enzymatic processes similar to those employed by phagocytic cells to induce microbial death

(Neish *et al.*, 2013). Indeed, reversible oxidation of cysteine residues is responsible for adaptive response to fluctuating levels of ROS. In fact, ROS signalling in the gut is suggested to represent an evolutionary ancient form of host cell and microbe cross-talk (Jones *et al.*, 2012; Neish and Jones, 2014). Free radical production in the gut due to commensal microbiota could also oxidise Trx and GSH, changing redox balance and inducing the transcription factors such as the Nrf2 and NF-κB. This could lead to modulation of gut inflammation and other cellular processes. It seems likely that the aforementioned microbiota-host interactions represent a universal mechanism used by bacterial communities to affect a variety of signalling and homeostatic processes in the host (Lee, 2008). Of interest, it was shown that the loss of mucus due to pathogenic bacteria is associated with significantly increased ROS production (Alam *et al.*, 2016) further aggravating disturbances in the redox balance of the gut. Therefore, two important balances in the gut namely antioxidants/prooxidants and commensal/pathogenic microbiota works in synergy. They maintain gut redox balance via transcription factors (e.g. Nrf2 and NF-κB) and vitagene modulation and are main protective mechanisms of the healthy gut (Figure 16.2).

Interestingly, probiotics may modulate the redox status of the gut via their metal ion chelating ability, antioxidant systems modulation, regulating signalling pathways, enzyme producing ROS, and intestinal microbiota (Wang *et al.*, 2017). However, the authors raised several questions as to efficacy of probiotics in modulation of gut health. For example, incapability of probiotic bacteria to colonise the gut and their elimination shortly after their introduction substantially restrict their long-term action. Interestingly, *Bacillus* amyloliquefaciens SC06 was shown to alleviate the oxidative stress of intestinal porcine epithelial cells via modulating Nrf2/Keap1 signalling pathway and decreasing ROS production (Wang *et al.*, 2017a). Clearly there is a need for more detailed elucidation of the microbial population in poultry gut depending

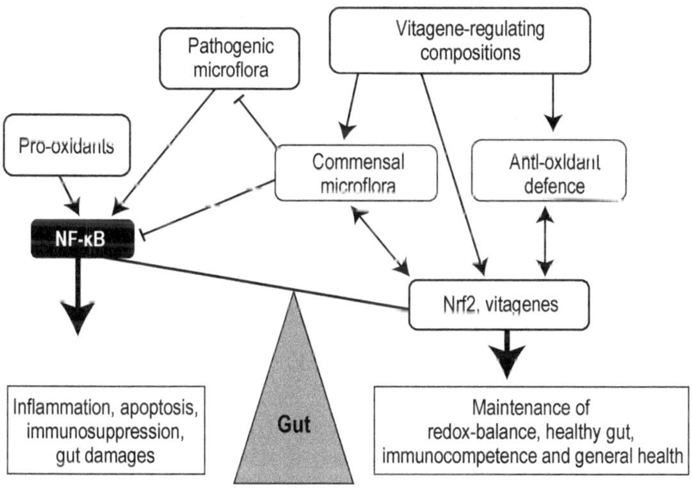

Figure 16.2. Antioxidant-prooxidant balance and microbiota in the gut (adapted from Surai, 2018).

on the diet, feed supplements, including pre- and probiotics and vitagene-modulating nutrients. The characterisation of the stress factors leading to intestinal dysbiosis and the identification of the microbial taxa contributing to pathological effects present a new direction of future research to better understand the impact of the microbiota on health and disease (Weiss and Hennet, 2017). Therefore, interactions between RONS, vitagene-regulated antioxidants and microbiota represent an important mechanism of the gut integrity/health maintenance in stressful conditions of commercial poultry production.

16.4 Vitagenes and immunity

Animal defence against various diseases depends on the efficacy of the immune system responsible for elimination of foreign substances (e.g. parasites, bacteria, moulds, yeast, fungi, viruses and various macromolecules) or the creation of specific inhospitable/hostile conditions within the host for a wide range of antigens. This protective capacity is based on the effective immune system which is considered to be a major determinant of animal health and wellbeing. Therefore, a remarkable ability of components of the immune system to distinguish between self and non-self is a great achievement of animal evolution.

16.4.1 Immunity in poultry production

Invading pathogens are controlled/destroyed by the natural and adaptive branches of the immune system. It is well established that in poultry natural immunity is responsible for recognition of invading pathogens by specific receptors. Binding of pathogen to those receptors induces the production of RONS, pro-inflammatory cytokines as well as communicating molecules which are responsible for sending regulatory signals to the adaptive immunity (Surai, 2018). Adaptive immunity is based on activity of B- and T-lymphocytes, which produce antibodies to specific non-self substances (B-lymphocytes) or directly attach to them (T-lymphocytes) and remove them from the cell (Juul-Madsen *et al.*, 2008). The adaptive immunity is characterised by high plasticity to recognise a great number (up to 10^{11}) distinct structures and is tightly regulated to turn on or off a response aiming in eradication of pathogens but not destruction of self (Surai, 2018). In the healthy animal/poultry resistance to infection relies on a balance between the natural and adaptive immunity. Regulation of the immune system is extremely complex. We are only starting to understand how the immune system co-ordinates the body's response to a disease or invading pathogen. It seems likely, that communication between immune cells is a crucial factor of immunocompetence (Surai, 2002, 2006, 2018). Interaction between the different immune cell types that make up these components of host defence is carried out by the relative mix of cytokines, hormone-like proteins, as well as other communicating molecules (Castle, 2000). The innate response, including its inflammatory component, reacts initially to the stimulus, acting directly to eliminate it by the activities of complement or phagocytosis. Cytokines produced by monocytes and macrophages regulate this response and also act on the liver, skeletal muscle, adipose and brain changing their metabolism and stimulating various responses. The cytokines also interact with T-lymphocytes (Calder, 2001).

Commercial poultry production is based on modern genetics, balanced diets and optimised environmental conditions. However, it is very difficult to avoid various stresses which are responsible for immunosuppression and increased susceptibility to various diseases leading to decreased productive and reproductive performance of poultry. In such situations immunomodulating properties of certain macro- and micronutrients are of great importance for the poultry industry. In fact, almost all nutrients in the diet play important roles in maintaining an 'optimal' immune response, and both insufficient and excessive nutrient intakes can bring negative consequences in terms of the immune status and susceptibility to a variety of pathogens. Indeed, immunocompetence can be presented as a 'violin music'. One can have a best violoncellist in the world with a Stradivari violin in the hands, however, before the violin is finely tuned, there would be no music, just a noise. On the one hand, that is exactly what is happening with an immunocompetence when the immune system is overreacted. This is associated with allergy, autoimmune diseases, food intolerance and other immune conditions. Furthermore, to maintain such an increased immune reactivity important nutrients are used. Such a redistribution of internal resources/nutrients would lead to decreased productive and reproductive performance. On the other hand, an underreacting immune system is also a problem, since it is not able to adequately protect the body from invaders, including microbes, viruses, etc. Therefore, only well-tuned immune system can do a proper job of the optimal protection without compromising performance of poultry.

16.4.2 Antioxidants, vitagenes and immunomodulation

Information has been actively accumulated for the last 20 years indicating that antioxidants are among major immunomodulating agents and their requirement for such action could be higher than that for animal growth and development (Surai, 2002, 2006, 2018). Banning feed grade antibiotics in Europe has made immune system competence the major factor determining efficiency of poultry production. Molecular immunology is developing very quickly and mechanisms of immunocompetence have received substantial attention and nutritional modulation of resistance to infectious diseases (Colitti *et al.*, 2019; Venter *et al.*, 2020) is a frontline for future research.

Importance of the vitagene network for immunity is related to several biological features (Surai, 2018):

- Phagocyte cells produce RONS in physiological conditions and use them as a weapon to kill pathogens (Figure 16.3).
- Immunocommunication between various immune cells is considered to be key element of effective immunocompetence. Since many immune receptors are redox-sensitive, antioxidant defence and adaptive redox homeostasis represent major mechanisms of immune system regulation.
- Redox equilibrium in the immune cells is related to their activity (activation, inactivation and apoptosis).

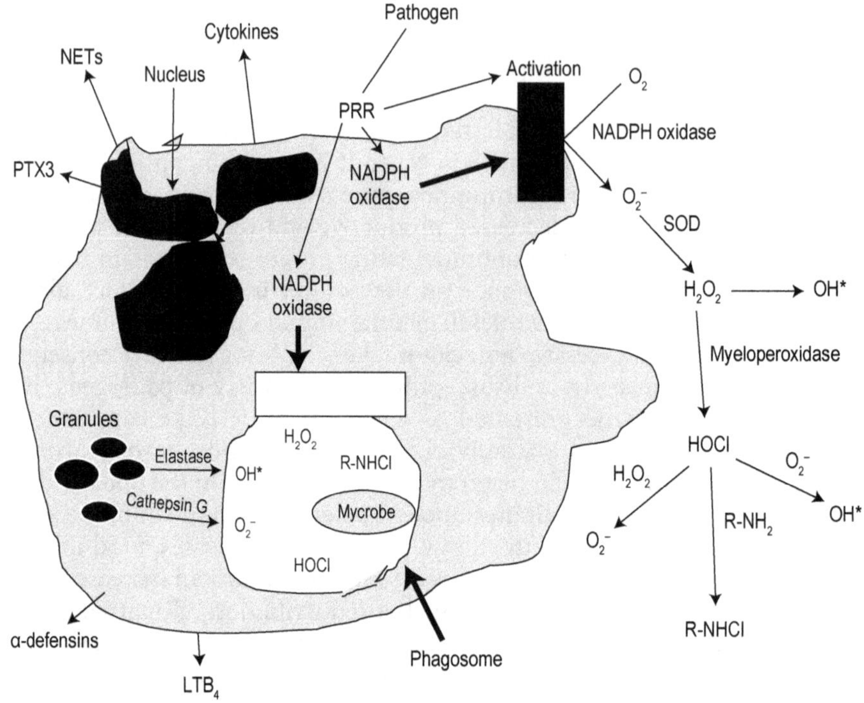

Figure 16.3. Respiratory burst in neutrophils (adapted from Surai, 2006, 2018).

- Disease challenge represent major stress associated with oxidative stress.
- Vaccination is associated with post-vaccinal stress and with some features of oxidative stress.

Neutrophil activation and phagocytosis of foreign particles are regularly accompanied by a so-called 'respiratory burst', an increase in the production of reactive oxygen and nitrogen species (RONS), exerted by the enzyme complex NADPH oxidase. Therefore neutrophils as well as other phagocyte leukocytes (e.g. macrophages, monocytes and eosinophils) can synthesise toxic oxygen metabolites such as superoxide anion (O_2^-), hydroxyl radical (OH*), singlet oxygen ($^1O_2$), hydrogen peroxide (H_2O_2), nitric oxide (NO), peroxynitrite (ONOO-), hypochlorous acid (HOCl) and chloramines during the respiratory burst (Zhao *et al.*, 1998). For example, a bacterium coming into contact with the plasma membrane is enclosed in a plasma membrane vesicle containing NADPH oxidase and exposed to an intensive flow of superoxide radical (Gille and Sigler, 1995).

Superoxide radical can disproportionate to H_2O_2 which penetrates into the bacterium with a production of hydroxyl radical, which is ultimately a deadly weapon able to damage any biological molecules. In general, the production of RONS is a characteristic for both mammalian and avian macrophages (Qureshi *et al.*, 1998). It is interesting that, people whose phagocytes possess no functional NADPH oxidase, are shown to suffer from chronic infections of the skin, lung, liver and bones leading to a premature death (Halliwell and Gutteridge, 2015).

Therefore, natural immunity is dependent on the efficient function of phagocytic cells, namely neutrophils (heterophils in avian species) and macrophages. These cells are equipped with an array of microbicidal weapons, such as proteases, enzymes that hydrolyse proteins and disrupting membranes. This weaponry is stored in granules in the cytoplasm. Furthermore, these cells have a powerful system for generating large amounts of RONS and they use them as an effective chemical weapon to kill pathogen. However, on escape from the phagosome the same free radicals become dangerous and can damage immune cells, compromising phagocyte function and reducing adaptive immunity. Phagocytes also produce communication molecules (eicosanoids, cytokines, etc.), that are used for effective communications between various immune cells and their production in many cases is redox-sensitive process.

16.4.3 Immunocommunication and receptors

If we imagine that immune system is an army fighting against invaders (microorganisms, viruses, etc.) than we would expect them to use signalling devices (similar to mobile phones) to receive and send signals to each other. Remarkably enough, major immune cells (macrophages, neutrophils, dendrites, NK cells, T- and B-lymphocytes) have on their surface something like 'mobile phones' called receptors. Those receptors are extremely sensitive to communicating molecules, but they are also sensitive to redox balance and to damages by free radicals and their expression and integrity can be easily compromised/damaged. In such a situation without proper communication all those huge armies of immune cells would become useless. Without proper communications, those cells can start fighting each other as well as eventually destroying immunocompetence. Furthermore, if we present immune cells (phagocytes) as 'soldiers' using chemical weapon to kill enemy, then special ammunition protecting them from their own weapon would be a crucial element for effective battle. In the case of immune cells such ammunition is represented by natural antioxidants with HSP, HO-1, SOD, thioredoxin and glutathione systems and sirtuins being major defences. Indeed, if not properly protected, phagocyte/macrophage functions could be compromised including initial overproduction of free radicals with consecutive damages to specific enzymatic systems resulting in decreasing efficiency of oxidative burst and apoptosis. In fact, redox balance/homeostasis is major regulator of the phagocyte functions and in the case of compromised antioxidant defences those functions would be compromised. Based on the presented model it is clear that the vitagene network, responsible for the antioxidant defence, signalling and adaptive homeostasis is a crucial factor of immune defence in the body.

For example, selenoprotein deficiency in T cells led to an inability of these cells to suppress RONS production, which in turn affected their ability to proliferate in response to T cell receptor stimulation (Carlson *et al.*, 2010; Shrimali *et al.*, 2008). It was shown that spleen oxidative stress induced by a high-fat diet (HFD) in mice induced the decreased expression of genes associated with antioxidant defence, as well as Fc receptor and an antioxidant (lipoic acid) supplementation attenuated the aforementioned changes (Cui *et al.*, 2012). It has been learned a lot recently about the innate receptors, such as Toll-like receptors, that activate antigen-presenting cells (APCs) in response to microbial products and initiate an immune response. It is well established that the immune response is regulated by a very complex interactions between immune cells via cytokine system as well as via other signalling molecules. In fact, there are many various cytokines responsible for immune cell communications. The most difficult question remains to answer is 'how do cells interpret these signals to modulate the immune response?'. A key aspect of the answer lies at the receptor and the multiple modifications and regulatory proteins that collectively shape a cytokine response (Delgoffe *et al.*, 2011). For example, for an optimal and appropriate immune response, T cells require activation through the T cell receptor (TCR), which recognises specific antigen presented in the context of major histocompatibility complex (MHC) (Williams and Kwon, 2004). This recognition also confers the ability of T cell responses to distinguish between 'self' and 'nonself'. Engagement of the TCR, throughout the lifetime of the T cell, controls the survival, proliferation, and/or differentiation of T cells. Thus, signalling through the TCR has important consequences for proper immune response and the effectiveness of that response. In fact, receptor expression can be dynamically modulated by the cell, based on cell type, stimulation, and cytokine activity. Some receptors are constitutively expressed, while others require additional signals in order to be upregulated and properly expressed. In addition, a lot of cytokine receptors are formed from multiple chains that have distinct modes of regulation. Furthermore, cells can temporally modulate the expression of certain receptors thereby altering the cytokines to which a particular cell is responsive (Delgoffe *et al.*, 2011). As a result of cytokine stimulation, cells can become activated and differentiate and most of the physiologic outcomes observed in the immune system are mediated by cytokines. It seems likely that T-cell activation requires two distinct signals: the first signal (recognition) relies on the interaction between the T-cell receptor (TCR) and a MHC-peptide complex, while the second signal is provided by the cross-linking between co-stimulatory molecules on activated T cells and compromise or loss of one or more of such regulatory and/or co-stimulatory molecules can be detrimental for immunocompetence (Pizzi *et al.*, 2016).

The immune response to pathogens is characterised by a biphasic response: the initial non-specific innate immune response and the subsequent pathogen-specific adaptive immune response. It is well established that the innate immune response is mediated mainly by macrophages, dendritic cells and neutrophils/heterophils, and occurs rapidly after these cells encounter a pathogen, while the adaptive immune response is controlled by T and B cells, which takes place several days after pathogen invasion (Kingeter and Lin, 2012). It is necessary to underline that the innate immune response is crucial for the development of the adaptive immune response. First, phagocytic cells such as dendritic cells and/or macrophages perform antigen presentation functions. They engulf extracellular pathogens, or infected host cells, and present pathogen-associated molecular patterns (PAMPs) to T cells. In this way, the innate immune system indicates to the cells of the adaptive immune system presence of pathogen and the appropriate pathogen-specific adaptive lymphocytes are activated (Kingeter and Lin, 2012). Therefore, signal transduction pathways are triggered resulting in the activation of various transcription factors, such as NF-κB, responsible for the production of pro-inflammatory cytokines and chemokines, which are involved in further regulation of the immune response. Indeed, immune cell receptor expression and integrity are major determinants of poultry immunocompetence. In fact, we can consider immunocompetence as an effective communication between all major cells of the immune system. It should be mentioned that both the innate and adaptive immunity are closely related and there is a complex interplay between innate and adaptive responses. In fact, the innate immune system can be considered as an effector arm of the adaptive response. For example, interactions between T cells and macrophages are of great importance. Indeed, T cell responses are associated with the production of interferon gamma (IFNγ), a cytokine possessing the ability to activate macrophages, increasing their capacity to phagocytose and kill invading microorganisms. Similarly, macrophages express Fc receptors that bind immunoglobulin (produced by B cells) and enhance recognition and engulfment of foreign pathogen-derived material. There is also a close interactions between T cells and goblet cells resulting in increased secretion of defensins and altered mucus composition (Juul-Madsen *et al.*, 2008). Indeed, initial encounter of pathogens with the innate system leads not only to the destruction of the pathogens but also initiates a cascade of important immunological events, including recruitment of various immune components, as well as induction and modulation of the adaptive immune system. In such interactions immune receptors are of great importance and their integrity and protection in stress conditions is a key for effective immunocompetence. For example, in the cellular innate immune system up to 100 different receptors are expressed at a relatively high frequency (Juul-Madsen *et al.*, 2008) and each receptor is usually expressed on millions of innate and adaptive immune cells.

Based on common structural features, cytokine receptors can be grouped into several families including the class I (haematopoietin) cytokine receptor family, the class II (IFN/IL-10) cytokine receptor family, the TNFRSF, the IL-1 receptor family, the TGF-β receptor family and the chemokine receptors (Kaiser and Staheli, 2008). Indeed, optimal redox status and antioxidant defences are needed for increase lymphocyte proliferation, expression of the high-affinity IL-2R, cytolytic T lymphocyte efficacy/activity, and NK-cell function, enhancing resistance to infections through modulation of interleukin production and subsequently the Th1/Th2 response. In fact, it seems likely that vitagene-regulating nutrients can upregulate the expression or protect from disruptions/damages by RONS receptors for IL-2 and other cytokines on the surface of activated lymphocytes and NK cells. This event favours the interaction of this cytokine with its respective receptors (Puertollano *et al.*, 2011; Surai, 2002, 2006, 2018).

16.4.4 Reactive oxygen and nitrogen species, redox homeostasis and immunocommunication

Over the last decade crucial roles of RONS and redox balance in regulating T- and B-cell functions was well recognised. In particular, the redox balance is believed to be regulated by dithiol/disulphide equilibrium and additional components of the antioxidant defence system encoded by vitagenes including thioredoxin, thioredoxin reductase, NADPH, SOD, catalase, peroxiredoxins, glutaredoxins, etc. (Patwardhan *et al.*, in press). Indeed, the redox balance in general was shown to be able to affect/ determine the phenotype and effector functions of T cells (Checker *et al.*, 2010; Gostner *et al.*, 2013; Mak *et al.*, 2017; Sena *et al.*, 2013). In fact, T cell activation is believed to depend on the redox balance at the site of their action and it is associated with ROS production in mitochondria. Furthermore, defects in mitochondria ETC are associated with impaired TCR-dependent RONS production and defects in antigen specific proliferation (Patwardhan *et al.*, in press). In particular, RONS are considered to be essential for T lymphocyte activation, expansion and effector function (Gambhir *et al.*, 2019; Moro-García *et al.*, 2018). In general, T cell activation is associated with a range of important changes. For example, during T cell activation mitochondria are characterised by elevated membrane potential (Wherry, 2011), metabolic re-programming (Wang *et al.*, 2011) and located to the immunological synapse (Quintana *et al.*, 2007). On the one hand, macrophage derived RONS are shown to induce Tregs in an NADPH oxidase dependent fashion (Kraaij *et al.*, 2010). On the other hand, there is also an activation of NADPH oxidase in T cells (Jackson *et al.*, 2004) with Duox1 (NOX family member) being responsible for TCR-mediated H_2O_2 production and involved in TCR signalling (Kwon *et al.*, 2010) with simultaneous changes in antioxidant defence mechanisms (Gambhir *et al.*, 2019; Sukumar *et al.*, 2016). It seems likely that GSH plays a special regulatory role in this process (Mak *et al.*, 2017), since GSH depletion by different means was shown to impair inflammatory responses (Checker *et al.*, 2010; Gambhir *et al.*, 2014; Ulrich and Jacob, 2019). In fact, GSH is considered to be a central metabolic integrator in immune responses mediated by T cells (Patwardhan *et al.*, in press) with the redox balance in T cells to control metabolic reprogramming (Almeida *et al.*, 2016). Indeed, both excess and insufficient levels of ROS was suggested to impair T cell metabolic reprogramming and functional activity (Patwardhan *et al.*, in press).

It should be mentioned that beyond GSH/GSSG involvement in the redox balance maintenance in immune cells a special role of thioredoxins, low molecular weight (12-kDa), ubiquitously expressed, and encoded by a vitagene molecules deserve more attention. Indeed, by reducing proteins as a result of cysteine thiol-disulphide exchange, thioredoxins are considered to be major players in redox balance maintenance and immune cell activation. It seems likely that thioredoxins are responsible for integrity and functional activity of TCRs. For example, Trx1 is shown to play an essential role in maintaining surface thiol density/activity of Tregs thereby offering them enhanced tolerance to oxidative stress and regulating their proliferation (Mougiakakos *et al.*, 2011). Indeed, Trx1 was found to exert antioxidant effect controlling ROS accumulation and redox status leading to anti-inflammatory effects and downregulating T cell responses in various model systems (Sofi *et al.*, 2019). The Trx1 system is believed to be essential for controlling DNA synthesis during T-cell metabolic reprogramming and proliferation (Muri *et al.*, 2018). Furthermore, Trxs are involved in NF-κB (Lorenzen *et al.*, 2017) and AP1 signalling (Patwardhan *et al.*, in press) in immune cells. Interestingly, the Trx1 and GSH/Grx1 systems are shown to redundantly fuel murine B-cell development and responses (Muri *et al.*, 2019).

The interaction of cell surface receptors on immune cells, including lymphocytes and phagocytes, with their cognate ligands on interacting cells or such soluble mediators as cytokines, chemokines and growth factors is a vital process responsible for generating and controlling an appropriate immune response including specific cell activation, proliferation and/or migration to sites of infection (Stegmann *et al.*, 2018). It is well appreciated that interacting proteins in intact three-dimensional structures are necessary to present their complementary binding interfaces for mediating receptor-ligand interactions. In fact, post-translation protein modifications are considered to be important mechanisms controlling receptor-ligand interactions and downstream events of immune response (Stegmann *et al.*, 2018).

16.4.5 Thiols and immunocommunication

Disulphide bonds are known to be important determinants of protein structure and function. In particular, they can sense and control changes in the redox environment. They also regulate protein folding, structural stabilisation, activity, localisation and interactions and therefore disturbances of protein disulphide bonds lead to various detrimental consequences to cellular/body homeostasis, including immune system dysregulation. In particular, oxidative stress-induced disulphide formation and thiol-disulphide exchange-mediated reduction of allosteric disulphides could affect immune cell structure and functions (Bechtel and Weerapana, 2017). Protein disulphide bonds serve not only structural role, but they also control the function of the mature protein in which they reside. On the one hand, structural disulphides are shown to be formed during protein folding and they are involved in the folding process and responsible for stability of the resultant protein architecture. On the other hand, functional disulphides can be divided into two groups, namely catalytic disulphides present in specific enzymes (e.g. protein-disulphide isomerase; PDI) or antioxidant compounds (e.g. thioredoxin), and responsible for manipulation of disulphides in other proteins and allosteric disulphides inducing changes in the tertiary structure and controlling the function of the protein in which they are located (Chiu and Hogg, 2019). Indeed, allosteric disulphide bond formation or cleavage can switch the structure and function of a protein (Kuroi *et al.*, 2020). In particular, these so called 'allosteric disulphides' allowing post-translation modification of proteins, have been shown to control various physiological processes, including the immune response (Chiu and Hogg, 2019). Indeed, modification of these disulphide bonds is believed to cause conformational changes of the protein and altering its activity and function, processes observed as a result of the entry of several enveloped viruses into their host cells, e.g. HIV, hepatitis C virus and Newcastle disease virus (Stegmann *et al.*, 2013). These labile disulphide bonds are common, with several classes of proteins being identified including integrins, receptors, transporters and cell-cell recognition proteins (Metcalphe *et al.*, 2011). The importance of disulphide bonds and central role for redox chemistry at the cell surface in creating and adequate immune response have been reviewed and analysed by Metcalfe *et al.* (2012) summarising that:

- disulphide bonds in integrins maintain their low affinity state and functional activity and their activities can be modulated by reducing agents;
- anti-microbial activity of defensins can be increased by reduction of disulphide bonds;
- IL-4 contains a disulphide bond susceptible to mild reduction;
- there are some redox-sensitive ligands for cell surface receptors;
- vaccination is associated with increase in the number of free thiols in lymph nodes;
- activation of T cells by dendritic cells is associated with secretion of Trx into the extracellular space, where it can catalyse reduction of disulphide bonds on the surface of immune cells, including T cells;
- Trx1 is shown to reduce disulphide bonds in the TNF receptor family member CD30 leading to altered cytokine binding and signalling;
- the Cys183–Cys232 disulphide bond in mouse CD132 (an important part of the receptor for several cytokines, including IL-4, IL-21, IL-7, IL-9, IL-15) is shown to be susceptible to reduction by enzymes such as gamma interferon-inducible lysosomal thiol reductase, protein disulphide isomerase and antioxidant thioredoxin, which are commonly secreted during immune activation;
- the Cys183–Cys232 disulphide bond is indicated to be reduced in an *in vivo* LPS-induced acute model of inflammation;
- reduction of the Cys183–Cys232 disulphide in CD132 was reported to inhibit IL-2 binding to the receptor complex (Metcalfe *et al.*, 2012).

Therefore, the activity of a range of cytokine receptors is shown to be affected by redox sensitive disulphide bonds and this could be an important mechanism controlling the levels of responses of cells to exogenous cytokines (Metcalfe *et al.*, 2012), associated with redox balance in the cell reflecting efficacy of antioxidant defence and adaptive vitagene-related signalling (Surai *et al.*, 2019). In fact, disulphide bond redox changes could be considered as a universal mechanism of immune receptor regulation. For example, the cluster of differentiation 4 (CD4), an integral membrane glycoprotein expressed on the surface of specific leukocytes, including T-cells, is shown to be a co-receptor stabilising T-cell receptor interactions with MHC class II molecules on antigen presenting cells (Owen *et al.*, 2018). Therefore, CD4 is suggested to play a vital role in the immune response and redox exchange of the CD4 disulphides was shown to regulate its functions. Indeed, each human CD4 monomer was indicated to comprise four immunoglobulin like ectodomains (D1-D4), a transmembrane segment, and a cytoplasmic tail and 3 domains are characterised by presence of a single disulphide bond located in their hydrophobic cores (Owen *et al.*, 2018). In fact, ablation of the allosteric disulphide bond in domain 2 was shown to lead to both a favourable structural collapse and an increase in the stability of CD4 (Owen *et al.*, 2016).

Interestingly, blocking cell surface redox exchanges with both a membrane-impermeable sulfhydryl blocker (DTNB) and specific antibody inhibitors of Trx1 was shown to induce translocation of CD4 into detergent-resistant membrane domains, while Trx1 inactivation did not affect the localisation of the chemokine receptor CCR5 (Moolla *et al.*, 2016). Furthermore, the αβ TCR is known to be a multimeric transmembrane complex consisting of a disulphide-linked antigen binding heterodimer associated with three signal-transducing CD3 subunits (Kim *et al.*, 2012). The TCR is known to mediate recognition of antigenic peptides bound to MHC molecules (pMHC), while the CD3 molecules transduce activation signals to the T cell (Mariuzza *et al.*, 2020). The CD3 delta chain was identified as having a labile disulphide bond and all three of the CD3 delta chains of the heterodimeric ectodomains of the TCR was shown to contain a conserved CysXXCys motif in their membrane proximal stalk. Those cysteines are believed to be essential for TCR-mediated T cell activation and form a disulphide bond which needs to be oxidised for signalling to occur (Brazin *et al.*, 2014).

CD44, the primary leukocyte cell surface receptor for hyaluronic acid, was also shown to have labile disulphide bond formed by Cys77 and Cys97 being susceptible to chemical and enzymatic reducing agents (Kellett-Clarke *et al.*, 2015). Immune cell function can be controlled by modulating the structure of either the receptor or the ligand (Stegmann *et al.*, 2018). In fact, the authors identified, quantified and monitored a reduction of labile disulphide bonds in primary cells during immune activation. Indeed, a reduction of labile disulphide bonds was shown to be thiol oxidoreductase-dependent and to affect activatory (e.g. CD132, SLAMF1) and adhesion (CD44, ICAM1) molecules (Stegmann *et al.*, 2018).

Increased levels of oxidative stress associated with several diseases are counteracted by the integrated antioxidant defence system which includes activities of various oxidoreductase enzymes, such as Trx, which are effective reductases of allosteric disulphide bonds in proteins (Gurjar *et al.*, 2019).

The importance of disulphide bonds management in stress conditions was demonstrated in experiments with monoclonal antibodies (mAbs). In particular, it was shown that Trx can effectively reduce the interchain disulphide bonds of the mAbs altering their function: increasing antigen-binding capacity and decreasing the Fc receptor binding with significant loss in both complement-dependent cytotoxicity and antibody-dependent cellular cytotoxicity activity (Gurjar *et al.*, 2019). The immune cells possess an important capacity to produce ROS and various inflammatory compounds as a part of their functional activities. However, as a result of oxidative stress leading to redox disturbance of cell signalling/communications can be compromised causing altered immune response and over-inflammation, a situation observed in adult mice with premature aging (Garrido *et al.*, 2019). In carp head kidney lymphocyte models, TCRγ knockdown was shown to significantly increase the mRNA expression of IFN-γ, IL-1β, IL-8, IL-10, Nrf2 and NF-κB. It simultaneously decreased the mRNA expression of SOD and CAT and reduced the activities of GPx,

T-AOC, CAT and SOD and increased the content of MDA and H_2O_2 (Yang *et al.*, 2020).

Interestingly, RONS are also involved in BCR signalling. Indeed, BCR stimulation was shown to result in extracellular ROS generation by NADPH oxidase enzymatic complex with following participation in a range of cellular events leading to the B cell activation and signalling (Bertolotti *et al.*, 2012). Furthermore, it was calculated that a single plasma cell secreting $\geq 10^3$ IgM per second, each with $\geq 10^2$ disulphides, has to form $\geq 10^5$ bonds per second during Ig production (Bertolotti *et al.*, 2010; Anelli *et al.*, 2015). In fact, disulphide bond formation is associated with formation of H_2O_2, an important signalling molecules which can be converted into a range of various RONS if not under antioxidant system control.

It was suggested that BCR, TCR and other Ig family members evolved the H_2O_2-generating activity to facilitate and amplify their signalling output (Reth, 2002).The authors suggested that the second messenger H_2O_2 is the molecule that communicates between BCR and TLR receptor systems, while BCR coreceptors could also be involved by increasing H_2O_2 production around the activated BCR. There is also a possibility that activated macrophages can function as APCs for T cell priming as a result of H_2O_2 diffusing from the macrophage to the T cell during their close interaction phase (Reth, 2002).

16.4.6 Vitagene-immunity interactions

The general scheme of vitagene-immunity interactions is depicted in the Figure 16.4.

Vitagene are responsible for synthesis of a range protective molecules, including HSP70, HO-1, SOD, elements of Trx-system and GSH-systems and sirtuins. As it was clearly shown in previous chapters these protective molecules provide optimal redox balance for cell signalling, including immunocommunication. Furthermore, vitagene-encoded molecules prevent receptor damages/altered expression and provide environment for their optimal expression. Therefore, increased expression of vitagenes in stress conditions is considered to be an important mechanism of maintaining high immunocompetence in stress conditions. Details of controlling functions of vitagenes on immunity are shown in Figure 16.5.

It is important to mention that vitagenes act in concert providing optimal conditions for cell signalling and stress adaptation. This helps maintain high immunocompetence in stress conditions. First of all, SODs are responsible for regulation of concentrations two major signalling molecules namely H_2O_2 and O_2^-. At the same time SOD could play a role of a transcription factor responsible for maintaining AO defences (see Chapter 4 for details). Secondly, Trx- and GSH-systems maintain redox signalling and redox homeostasis, necessary conditions for integrity maintenance and optimal immunoreceptor expression and immune cell signalling. Thirdly, HSP70 is responsible for proteostasis and integrity of immunoreceptors in stress conditions. Fourthly, HO-1 is involved in AO defence and cell signalling providing additional defence for

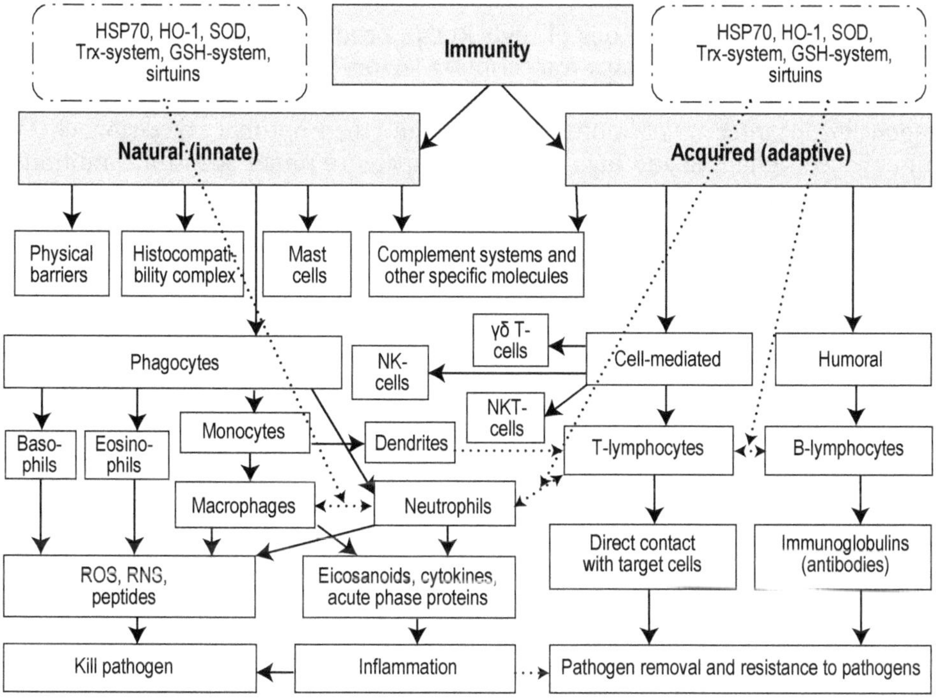

Figure 16.4. Vitagene-immunity interactions (adapted from Surai *et al.*, 2018).

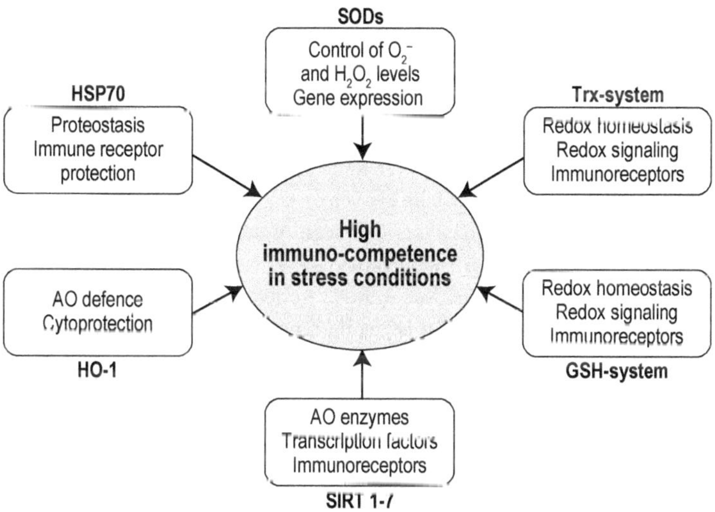

Figure 16.5. Regulating roles of vitagenes in immunocompetence.

immune cells under conditions of high RONS production. Finally, SIRTs regulate expression and activity of many transcription factors and AO enzymes orchestrating optimal AO defence, cell signalling and immunocompetence. Indeed, vitagenes are responsible for integrity of immune receptors and their optimal expression, effective immune cell signalling and high immunocompetence under stressful conditions of commercial poultry production.

16.5 Conclusions

The future applications of the vitagene concept are gut health and immunomodulation. Indeed, in both cases improved antioxidant defences and adaptive homeostasis are important targets for vitagene activation. In fact, a complex antioxidant network in the gut is responsible for gut integrity determining effective nutrient absorption and providing barrier function to prevent pathogen penetration into the body. Furthermore, maintaining optimal immune-communications between various cell types is key for high immunocompetence under various stress conditions. A new frontiers in understanding of the effect of microbiota on antioxidant-prooxidant (redox) balance in the gut as well as interactions between gut redox status, microbiota and gut immunity await further investigation. Indeed, vitagene future looks bright.

References

Alam, A., Leoni, G., Quiros, M., Wu, H., Desai, C., Nishio, H., Jones, R.M., Nusrat, A. and Neish, A.S., 2016. The microenvironment of injured murine gut elicits a local pro-restitutive microbiota. Nature Microbiology 1: 15021.

Almeida, L., Lochner, M., Berod, L. and Sparwasser, T., 2016. Metabolic pathways in T cell activation and lineage differentiation. Seminars in Immunology 28: 514-524.

Anelli, T., Sannino, S. and Sitia, R., 2015. Proteostasis and 'redoxtasis' in the secretory pathway: tales of tails from ERp44 and immunoglobulins. Free Radical Biology and Medicine 83: 323-330.

Asano, T., Tanaka, K.I., Yamakawa, N., Adachi, H., Sobue, G., Goto, H., Takeuchi, K. and Mizushima, T., 2009. HSP70 confers protection against indomethacin-induced lesions of the small intestine. Journal of Pharmacology and Experimental Therapeutics 330: 458-467.

Attuwaybi, B.O., Hassoun, H.T., Zou, L., Kozar, R.A., Kone, B.C., Weisbrodt, N.W. and Moore, F.A., 2003. Hypothermia protects against gut ischemia/reperfusion-induced impaired intestinal transit by inducing heme oxygenase-1. Journal of Surgical Research 115: 48-55.

Attuwaybi, B.O., Kozar, R.A., Moore-Olufemi, S.D., Sato, N., Hassoun, H.T., Weisbrodt, N.W. and Moore, F.A., 2004. Heme oxygenase-1 induction by hemin protects against gut ischemia/reperfusion injury. Journal of Surgical Research 118: 53-57.

Aw, T.Y., 2005. Intestinal glutathione: determinant of mucosal peroxide transport, metabolism, and oxidative susceptibility. Toxicology and Applied Pharmacology 204: 320-328.

Bechtel, T.J. and Weerapana, E., 2017. From structure to redox: the diverse functional roles of disulfides and implications in disease. Proteomics 17: 10.1002.

Beck, S.C., Paidas, C.N., Mooney, M.L., Deitch, E.A. and De, A.M., 1995. Presence of the stress-inducible form of hsp-70 (hsp-72) in normal rat colon. Shock 3: 398-402.

Bertolotti, M., Sitia, R. and Rubartelli, A., 2012. On the redox control of B lymphocyte differentiation and function. Antioxidants & Redox Signaling 16: 1139-1149.

Bertolotti, M., Yim, S.H., Garcia-Manteiga, J.M., Masciarelli, S., Kim, Y.J., Kang, M.H., Iuchi, Y., Fujii, J., Vené, R., Rubartelli, A. and Rhee, S.G., 2010. B-to plasma-cell terminal differentiation entails oxidative stress and profound reshaping of the antioxidant responses. Antioxidants & Redox Signaling 13: 1133-1144.

Bhagat, M., Sood, S., Yadav, A., Verma, P., Manzoor, N., Chakraborty, D., Katoch, R. and Sangha, N., 2017. Alterations in oxidative stress parameters and its associated correlation with clinical disease on experimental Cryptosporidium parvum infection in Swiss albino mice. Journal of Parasitic Diseases 41: 707-712.

Bhat, M.I. and Kapila, R., 2017. Dietary metabolites derived from gut microbiota: critical modulators of epigenetic changes in mammals. Nutrition Reviews 75: 374-389.

Brazin, K.N., Mallis, R.J., Li, C., Keskin, D.B., Arthanari, H., Gao, Y., Wu, S.L., Karger, B.L., Wagner, G. and Reinherz, E.L., 2014. Constitutively oxidized CXXC motifs within the CD3 heterodimeric ectodomains of the T cell receptor complex enforce the conformation of juxtaposed segments. Journal of Biological Chemistry 289: 18880-18892.

Brigelius-Flohé, R., 1999. Tissue-specific functions of individual glutathione peroxidases. Free Radical Biology and Medicine 27: 951-965.

Brigelius-Flohé, R. and Flohé, L., 2020. Regulatory phenomena in the glutathione peroxidase superfamily. Antioxidants & Redox Signaling 33: 498-516. https://doi.org/10.1089/ars.2019.7905

Brigelius-Flohé, R. and Maiorino, M., 2013. Glutathione peroxidases. Biochimica et Biophysica Acta 1830: 3289-3303.

Brigelius-Flohé, R., Muller, C., Menard, J., Florian, S., Schmehl, K. and Wingler, K., 2001. Functions of GI-GPx: lessons from selenium-dependent expression and intracellular localization. BioFactors 14: 101-106

Buechler, N., Wang, X., Yoza, B. K., McCall, C. E. and Vachharajani, V., 2017. Sirtuin 2 Regulates Microvascular Inflammation during Sepsis. Journal of Immunology Research 2017: 2648946.

Calder, P.C., 2001. Polyunsaturated fatty acids, inflammation, and immunity. Lipids 36: 1007-1024.

Cao, Y.Z., Maddox, J.F., Mastro, A.M., Scholz, R.W., Hildenbrandt, G. and Reddy, C.C., 1992. Selenium deficiency alters the lipoxygenase pathway and mitogenic response in bovine lymphocytes. The Journal of Nutrition 122: 2121-2127.

Carlson, B.A., Yoo, M.H., Shrimali, R.K., Irons, R., Gladyshev, V.N., Hatfield, D.L. and Park, J.M., 2010. Role of selenium-containing proteins in T-cell and macrophage function. Proceedings of the Nutrition Society 69: 300-310.

Caruso, R., Marafini, I., Franzè, E., Stolfi, C., Zorzi, F., Monteleone, I., Caprioli, F., Colantoni, A., Sarra, M., Sedda, S. and Biancone, L., 2014. Defective expression of SIRT1 contributes to sustain inflammatory pathways in the gut. Mucosal Immunology 7: 1467-1479.

Castle, S.C., 2000. Clinical relevance of age-related immune dysfunction. Clinical Infectious Diseases 31: 578-585.

Checker, R., Sharma, D., Sandur, S.K., Subrahmanyam, G., Krishnan, S., Poduval, T.B. and Sainis, K.B., 2010. Plumbagin inhibits proliferative and inflammatory responses of T cells independent of ROS generation but by modulating intracellular thiols. Journal of Cellular Biochemistry 110: 1082-1093.

Chen, M., Hui, S., Lang, H., Zhou, M., Zhang, Y., Kang, C., Zeng, X., Zhang, Q., Yi, L. and Mi, M., 2019. SIRT3 deficiency promotes high-fat diet-induced nonalcoholic fatty liver disease in correlation with impaired intestinal permeability through gut microbial dysbiosis. Molecular Nutrition & Food Research 63: e1800612.

Cheng, Y. and Rong, J., 2017. Therapeutic potential of heme oxygenase-1/carbon monoxide system against ischemia-reperfusion injury. Current Pharmaceutical Design 23: 3884-3898.

Chiu, J. and Hogg, P.J., 2019. Allosteric disulfides: sophisticated molecular structures enabling flexible protein regulation. Journal of Biological Chemistry 294: 2949-2960.

Chu, F.F., Doroshow, J.A. and Esworthy, R.S., 1993. Expression, characterization, and tissue distribution of a new cellular selenium-dependent glutathione peroxidase, GSHPx-GI. Journal of Biological Chemistry 268: 2571-2576.

Colitti, M., Stefanon, B., Gabai, G., Gelain, M.E. and Bonsembiante, F., 2019. Oxidative stress and nutraceuticals in the modulation of the immune function: Current knowledge in animals of veterinary interest. Antioxidants 8: 28.

Cui, J., Xiao, Y., Shi, Y.H., Wang, B. and Le, G.W., 2012. Lipoic acid attenuates high-fat-diet-induced oxidative stress and B-cell-related immune depression. Nutrition 28: 275-280.

David, J.C., Grongnet, J.F. and Lalles, J.P., 2002. Weaning affects the expression of heat shock proteins in different regions of the gastrointestinal tract of piglets. Journal of Nutrition 132: 2551-2561.

Degroote, J., Wang, W., Vergauwen, H., De Smet, S., Van Ginneken, C. and Michiels, J., 2019. Impact of a dietary challenge with peroxidized oil on the glutathione redox status and integrity of the small intestine in weaned piglets. Animal 13: 1641-1650.

Delgoffe, G.M., Murray, P.J. and Vignali, D.A., 2011. Interpreting mixed signals: the cell's cytokine conundrum. Current Opinion in Immunology 23: 632-638.

Din, A.U., Hassan, A., Zhu, Y., Zhang, K., Wang, Y., Li, T., Wang, Y. and Wang, G., 2020. Inhibitory effect of Bifidobacterium bifidum ATCC 29521 on colitis and its mechanism. The Journal of Nutritional Biochemistry 79: p.108353.

Drew, J.E., Farquharson, A.J., Arthur, J.R., Morrice, P.C. and Duthie, G.G., 2005. Novel sites of cytosolic glutathione peroxidase expression in colon. FEBS Letters 579: 6135-6139.

Esworthy, R.S., Aranda, R., Martín, M.G., Doroshow, J.H., Binder, S.W. and Chu, F.F., 2001. Mice with combined disruption of Gpx1 and Gpx2 genes have colitis. American Journal of Physiology-Gastrointestinal and Liver Physiology 281: G848-G855.

Esworthy, R.S., Swiderek, K.M., Ho, Y.S. and Chu, F.F., 1998. Selenium-dependent glutathione peroxidase-GI is a major glutathione peroxidase activity in the mucosal epithelium of rodent intestine. Biochimica et Biophysica Acta 1381: 213-226.

Florian, S., Krehl, S., Loewinger, M., Kipp, A., Banning, A., Esworthy, S., Chu, F.F. and Brigelius-Flohé, R., 2010. Loss of GPx2 increases apoptosis, mitosis, and GPx1 expression in the intestine of mice. Free Radical Biology and Medicine 49: 1694-1702.

Florian, S., Wingler, K., Schmehl, K., Jacobasch, G., Kreuzer, O.J., Meyerhof, W. and Brigelius-Flohé, R., 2001. Cellular and subcellular localization of gastrointestinal glutathione peroxidase in normal and malignant human intestinal tissue. Free Radical Research 35: 655-663.

Frederiks, W.M. and Bosch, K.S., 1997. Localization of superoxide dismutase activity in rat tissues. Free radical Biology & Medicine 22: 241-248.

Fujii, H., Takahashi, T., Nakahira, K., Uehara, K., Shimizu, H., Matsumi, M., Morita, K., Hirakawa, M., Akagi, R. and Sassa, S., 2003. Protective role of heme oxygenase-1 in the intestinal tissue injury in an experimental model of sepsis. Critical Care Medicine 31: 893-902.

Gambhir, L., Checker, R., Thoh, M., Patwardhan, R.S., Sharma, D., Kumar, M. and Sandur, S.K., 2014. 1, 4-Naphthoquinone, a pro-oxidant, suppresses immune responses via KEAP-1 glutathionylation. Biochemical Pharmacology 88: 95-105.

Gambhir, L., Sharma, V., Kandwal, P. and Saxena, S., 2019. Perturbation in cellular redox homeostasis: decisive regulator of T cell mediated immune responses. International Immunopharmacology 67: 449-457.

Gan, H.T. and Chen, J.D.Z., 2007. Induction of heme oxygenase-1 improves impaired intestinal transit after burn injury. Surgery 141: 385-393.

Garrido, A., Cruces, J., Ceprián, N., Vara, E. and De la Fuente, M., 2019. Oxidative-inflammatory stress in immune cells from adult mice with premature aging. International Journal of Molecular Sciences 20: 769.

Gille, G. and Sigler, K., 1995. Oxidative stress and living cells. Folia Microbiologica 40: 131-152.

Gordeeva, A.E., Temnov, A.A., Charnagalov, A.A., Sharapov, M.G., Fesenko, E.E. and Novoselov, V.I., 2015. Protective effect of peroxiredoxin 6 in ischemia/reperfusion-induced damage of small intestine. Digestive Diseases and Sciences 60: 3610-3619.

Gostner, J.M., Becker, K., Fuchs, D. and Sucher, R., 2013. Redox regulation of the immune response. Redox Report 18: 88-94.

Guo, X., Shin, V.Y. and Cho, C.H., 2001. Modulation of heme oxygenase in tissue injury and its implication in protection against gastrointestinal diseases. Life Sciences 69: 3113-3119.

Gurjar, S.A., Wheeler, J.X., Wadhwa, M., Thorpe, R., Kimber, I., Derrick, J.P., Dearman, R.J. and Metcalfe, C., 2019. The impact of thioredoxin reduction of allosteric disulfide bonds on the therapeutic potential of monoclonal antibodies. Journal of Biological Chemistry 294: 19616-19634.

Halliwell, B. and Gutteridge, J.M.C., 2015. Free radicals in biology and medicine. Oxford University Press, New York, NY, USA.

Han, Z., Willer, T., Li, L., Pielsticker, C., Rychlik, I., Velge, P., Kaspers, B. and Rautenschlein, S., 2017. Influence of the gut microbiota composition on *Campylobacter jejuni* colonization in chickens. Infection and Immunity 85: e00380-17.

Hsu, J.L., Chou, J.W., Chen, T.F., Hsu, J.T., Su, F.Y., Lan, J.L., Wu, P.C., Hu, C.M., Eva, Y., Lee, H.P. and Lee, W.H., 2019. Glutathione peroxidase 8 negatively regulates caspase-4/11 to protect against colitis. EMBO Molecular Medicine 12: e9386.

Hu, Y., Tao, X., Han, X., Xu, L., Yin, L., Sun, H., Qi, Y., Xu, Y. and Peng, J., 2018. MicroRNA-351-5p aggravates intestinal ischaemia/reperfusion injury through the targeting of MAPK13 and Sirtuin-6. British Journal of Pharmacology 175: 3594-3609.

Huang, L., Mivechi, N.F. and Moskophidis, D., 2001. Insights into regulation and function of the major stress-induced hsp70 molecular chaperone *in vivo*: analysis of mice with targeted gene disruption of the hsp70. 1 orhsp70. 3 gene. Molecular and Cellular Biology 21: 8575-8591.

Inoue, K., Takahashi, T., Uehara, K., Shimuzu, H., Ido, K., Morimatsu, H., Omori, E., Katayama, H., Akagi, R. and Morita, K., 2008. Protective role of heme oxygenase 1 in the intestinal tissue injury in hemorrhagic shock in rats. Shock 29: 252-261.

Jackson, S.H., Devadas, S., Kwon, J., Pinto, L.A. and Williams, M.S., 2004. T cells express a phagocyte-type NADPH oxidase that is activated after T cell receptor stimulation. Nature Immunology 5: 818-827.

Jefferies, H., Bot, J., Coster, J., Khalil, A., Hall, J.C. and McCauley, R.D., 2003. The role of glutathione in intestinal dysfunction. Journal of Investigative Surgery 16: 315-323.

Jiménez-Hidalgo, M., Kurz, C.L., Pedrajas, J.R., Naranjo-Galindo, F.J., González-Barrios, M., Cabello, J., Sáez, A.G., Lozano, E., Button, E.L., Veal, E.A. and Fierro-González, J.C., 2014. Functional characterization of thioredoxin 3 (TRX-3), a *Caenorhabditis elegans* intestine-specific thioredoxin. Free Radical Biology and Medicine 68: 205-219.

Jones, R.M. and Mercante, J.W. and Neish, A.S., 2012. Reactive oxygen production induced by the gut microbiota: pharmacotherapeutic implications. Current Medical Chemistry. 19: 1519-1529.

Jones, R.M. and Neish, A.S., 2017. Redox signaling mediated by the gut microbiota. Free Radical Biology and Medicine 105: 41-47.

Ju, S., Ge, Y., Li, P., Tian, X., Wang, H., Zheng, X. and Ju, S., 2018. Dietary quercetin ameliorates experimental colitis in mouse by remodeling the function of colonic macrophages via a heme oxygenase-1-dependent pathway. Cell Cycle 17: 53-63.

Juul-Madsen, H.R., Viertlboeck, B., Smith, A.L. and Gobel, T.W.F., 2008. Avian innate immune responses. In: Davison, F., Kaspers, B. and Schat, K.A. (eds) Avian immunology. Elsevier, Amsterdam, the Netherlands, pp. 129-158.

Kaiser, P. and Staheli, P., 2008. Avian cytokines and chemokines. In: Davison, F., Kaspers, B. and Schat, K.A. (eds.) Avian immunology. Elsevier, Amsterdam, the Netherlands, pp. 203-222.

Kellett-Clarke, H., Stegmann, M., Barclay, A.N. and Metcalfe, C., 2015. CD44 binding to hyaluronic acid is redox regulated by a labile disulfide bond in the hyaluronic acid binding site. PloS One 10: e0138137.

Kim, S.T., Shin, Y., Brazin, K., Mallis, R.J., Sun, J., Wagner, G., Lang, M. and Reinherz, E.L., 2012. TCR mechanobiology: torques and tunable structures linked to early T cell signaling. Frontiers in Immunology 3: 76.

Kingeter, L.M. and Lin, X., 2012. C-type lectin receptor-induced NF-κB activation in innate immune and inflammatory responses. Cell Molecular Immunology 9: 105-112.

Kojima, K., Musch, M.W., Ren, H., Boone, D.L., Hendrickson, B.A., Ma, A. and Chang, E.B., 2003. Enteric flora and lymphocyte-derived cytokines determine expression of heat shock proteins in mouse colonic epithelial cells. Gastroenterology 124: 1395-1407.

Kopeček, P., Altmannová, K. and Weigl, E., 2001. Stress proteins: nomenclature, division and functions. Biomedical Papers 145: 39-47.

Kraaij, M.D., Savage, N.D., Van der Kooij, S.W., Koekkoek, K., Wang, J., Van den Berg, J.M., Ottenhoff, T.H., Kuijpers, T.W., Holmdahl, R., Van Kooten, C. and Gelderman, K.A., 2010. Induction of regulatory T cells by macrophages is dependent on production of reactive oxygen species. Proceedings of the National Academy of Sciences 107: 17686-17691.

Krehl, S., Loewinger, M., Florian, S., Kipp, A.P., Banning, A., Wessjohann, L.A., Brauer, M.N., Iori, R., Esworthy, R.S., Chu, F.F. and Brigelius-Flohé, R., 2012. Glutathione peroxidase-2 and selenium decreased inflammation and tumors in a mouse model of inflammation-associated carcinogenesis whereas sulforaphane effects differed with selenium supply. Carcinogenesis 33: 620-628.

Kuroi, K., Kamijo, M., Ueki, M., Niwa, Y., Hiramatsu, H. and Nakabayashi, T., 2020. Time-resolved FTIR study on the structural switching of human galectin-1 by light-induced disulfide bond formation. Physical Chemistry Chemical Physics. 22: 1137-1144.

Kwon, J., Shatynski, K.E., Chen, H., Morand, S., De Deken, X., Miot, F., Leto, T.L. and Williams, M.S., 2010. The nonphagocytic NADPH oxidase Duox1 mediates a positive feedback loop during T cell receptor signaling. Science Signaling 3: ra59.

Lan, C., Sun, X.N., Zhou, X.C., Yang, B., Huang, B.L., Deng, T.Z., He, Z.T. and Han, X.Y., 2016. Preinduced intestinal HSP70 improves visceral hypersensitivity and abnormal intestinal motility in PI-IBS mouse model. Asian Pacific Journal of Tropical Medicine 9: 302-305.

Lee, W.J., 2008. Bacterial-modulated signaling pathways in gut homeostasis. Science Signaling 1: pe24.

Legrand, T.S. and Aw, T.Y., 1998. Chronic hypoxia alters glucose utilization during GSH-dependent detoxication in rat small intestine. American Journal of Physiology-Gastrointestinal and Liver Physiology 274: G376-G384.

Lennicke, C., Rahn, J., Wickenhauser, C., Lichtenfels, R., Müller, A.S., Wessjohann, L.A., Kipp, A.P. and Seliger, B., 2018. Loss of epithelium-specific GPx2 results in aberrant cell fate decisions during intestinal differentiation. Oncotarget 9: 539-552.

Li, C., Zhou, Y., Rychahou, P., Weiss, H.L., Lee, E.Y., Perry, C.L., Barrett, T.A., Wang, Q. and Evers, B.M., 2020. SIRT2 contributes to the regulation of intestinal cell proliferation and differentiation. Cellular and Molecular Gastroenterology and Hepatology 10(1): 43-57.

Li, Y., Li, Q., Chen, H., Wang, T., Liu, L., Wang, G., Xie, K. and Yu, Y., 2015. Hydrogen gas alleviates the intestinal injury caused by severe sepsis in mice by increasing the expression of heme oxygenase-1. Shock 44: 90-98.

Liao, Y.F., Zhu, W., Li, D.P. and Zhu, X., 2013. Heme oxygenase-1 and gut ischemia/reperfusion injury: a short review. World Journal of Gastroenterology 19: 3555-3561.

Liedel, J.L., Guo, Y., Yu, Y., Shiou, S.R., Chen, S., Petrof, E.O., Hu, S., Musch, M.W. and Claud, E.C., 2011. Mother's milk-induced Hsp70 expression preserves intestinal epithelial barrier function in an immature rat pup model. Pediatric Research 69: 395-400.

Liu, F., Bu, H.F., Geng, H., De Plaen, I.G., Gao, C., Wang, P., Wang, X., Kurowski, J.A., Yang, H., Qian, J. and Tan, X.D., 2017. Sirtuin-6 preserves R-spondin-1 expression and increases resistance of intestinal epithelium to injury in mice. Molecular Medicine 23: 272-284.

Liu, H.Y., Dicksved, J., Lundh, T. and Lindberg, J.E., 2014. Expression of heat shock proteins 27 and 72 correlates with specific commensal microbes in different regions of porcine gastrointestinal tract. American Journal of Physiology – Gastrointestinal and Liver Physiology 306: G1033-1041.

Lo Sasso, G., Menzies, K.J., Mottis, A., Piersigilli, A., Perino, A., Yamamoto, H., Schoonjans, K. and Auwerx, J., 2014b. SIRT2 deficiency modulates macrophage polarization and susceptibility to experimental colitis. PloS One 9: e103573.

Sasso, G.L., Ryu, D., Mouchiroud, L., Fernando, S.C., Anderson, C.L., Katsyuba, E., Piersigilli, A., Hottiger, M.O., Schoonjans, K. and Auwerx, J., 2014a. Loss of Sirt1 function improves intestinal anti-bacterial defense and protects from colitis-induced colorectal cancer. PloS One 9: e102495.

Lorenzen, I., Mullen, L., Bekeschus, S. and Hanschmann, E.M., 2017. Redox regulation of inflammatory processes is enzymatically controlled. Oxidative Medicine and Cellular Longevity 2017: 8459402.

Mak, T.W., Grusdat, M., Duncan, G.S., Dostert, C., Nonnenmacher, Y., Cox, M., Binsfeld, C., Hao, Z., Brüstle, A., Itsumi, M., Jäger, C., Chen, Y., Pinkenburg, O., Camara, B., Ollert, M., Bindslev-Jensen, C., Vasiliou, V., Gorrini, C., Lang, P.A., Lohoff, M., Harris, I.S., Hiller, K. and Brenner, D., 2017. Glutathione primes T cell metabolism for inflammation. Immunity 46: 675-689.

Malago, J.J., Koninkx, J.F. and Van Dijk, J.E., 2002. The heat shock response and cytoprotection of the intestinal epithelium. Cell Stress Chaperones 7: 191-199.

Mallick, I.H., Winslet, M.C. and Seifalian, A.M., 2010. Ischemic preconditioning of small bowel mitigates the late phase of reperfusion injury: heme oxygenase mediates cytoprotection. The American Journal of Surgery 199: 223-231.

Mardinoglu, A., Shoaie, S., Bergentall, M., Ghaffari, P., Zhang, C., Larsson, E., Bäckhed, F. and Nielsen, J., 2015. The gut microbiota modulates host amino acid and glutathione metabolism in mice. Molecular Systems Biology 11: 834.

Marelli, G. and Allavena, P., 2020. The good and the bad side of heme-oxygenase-1 in the gut. Antioxidants & Redox Signaling 32: 1071-1079.

Marelli, G., Erreni, M., Anselmo, A., Taverniti, V., Guglielmetti, S., Mantovani, A. and Allavena, P., 2017. Heme-oxygenase-1 production by intestinal CX3CR1+ macrophages helps to resolve inflammation and prevents carcinogenesis. Cancer Research 77: 4472-4485.

Mariuzza, R.A., Agnihotri, P. and Orban, J., 2020. The structural basis of T-cell receptor (TCR) activation: An enduring enigma. Journal of Biological Chemistry 295: 914-925.

Meinl, W., Sczesny, S., Brigelius-Flohé, R., Blaut, M. and Glatt, H., 2009. Impact of gut microbiota on intestinal and hepatic levels of phase 2 xenobiotic-metabolizing enzymes in the rat. Drug Metabolism and Disposition 37: 1179-1186.

Metcalfe, C., Cresswell, P. and Barclay, A.N., 2012. Interleukin-2 signalling is modulated by a labile disulfide bond in the CD132 chain of its receptor. Open Biology 2: 110036.

Metcalfe, C., Cresswell, P., Ciaccia, L., Thomas, B. and Barclay, A.N., 2011. Labile disulfide bonds are common at the leucocyte cell surface. Open Biology 1: 110010.

Moolla, N., Killick, M., Papathanasopoulos, M. and Capovilla, A., 2016. Thioredoxin (Trx1) regulates CD4 membrane domain localization and is required for efficient CD4-dependent HIV-1 entry. Biochimica et Biophysica Acta (BBA)-General Subjects, 1860: 1854-1863.

Mörk, H., Lex, B., Scheurlen, M., Dreher, I., Schutze, N., Köhrle, J. and Jakob, F., 1998. Expression pattern of gastrointestinal selenoproteins – targets for selenium supplementation. Nutrition and Cancer 32: 64-70.

Moro-García, M.A., Mayo, J.C., Sainz, R.M. and Alonso-Arias, R., 2018. Influence of inflammation in the process of T lymphocyte differentiation: proliferative, metabolic, and oxidative changes. Frontiers in Immunology 9: 339.

Mougiakakos, D., Johansson, C.C., Jitschin, R., Böttcher, M. and Kiessling, R., 2011. Increased thioredoxin-1 production in human naturally occurring regulatory T cells confers enhanced tolerance to oxidative stress. Blood, The Journal of the American Society of Hematology 117: 857-861.

Müller, M.F., Florian, S., Pommer, S., Osterhoff, M., Esworthy, R.S., Chu, F.F., Brigelius-Flohé, R. and Kipp, A.P., 2013. Deletion of glutathione peroxidase-2 inhibits azoxymethane-induced colon cancer development. PLoS One 8: e72055.

Muri, J., Heer, S., Matsushita, M., Pohlmeier, L., Tortola, L., Fuhrer, T., Conrad, M., Zamboni, N., Kisielow, J. and Kopf, M., 2018. The thioredoxin-1 system is essential for fueling DNA synthesis during T-cell metabolic reprogramming and proliferation. Nature Communications 9: 1851.

Muri, J., Thut, H., Heer, S., Krueger, C.C., Bornkamm, G.W., Bachmann, M.F. and Kopf, M., 2019. The thioredoxin-1 and glutathione/glutaredoxin-1 systems redundantly fuel murine B-cell development and responses. European Journal of Immunology 49: 709-723.

Musch, M.W., Sugi, K., Straus, D. and Chang, E.B., 1999. Heat-shock protein 72 protects against oxidant-induced injury of barrier function of human colonic epithelial Caco2/bbe cells. Gastroenterology 117: 115-122.

Nakagawa, F., Morino, K., Ugi, S., Ishikado, A., Kondo, K., Sato, D., Konno, S., Nemoto, K.I., Kusunoki, C., Sekine, O. and Sunagawa, A., 2014. 4-Hydroxy hexenal derived from dietary n-3 polyunsaturated fatty acids induces anti-oxidative enzyme heme oxygenase-1 in multiple organs. Biochemical and Biophysical Research Communications 443: 991-996.

Neish, A.S. and Jones, R.M., 2014. Redox signaling mediates symbiosis between the gut microbiota and the intestine. Gut Microbes 5: 250-253.

Neish, A.S., 2013. Redox signaling mediated by the gut microbiota. Free Radical Research 47: 950-957.

Owen, G.R., Channell, J.A., Forsyth, V.T., Haertlein, M., Mitchell, E.P., Capovilla, A., Papathanasopoulos, M. and Cerutti, N.M., 2016. Human CD4 metastability is a function of the allosteric disulfide bond in domain 2. Biochemistry 55: 2227-2237.

Owen, G.R., Le, D., Stoychev, S., Cerutti, N.M. and Papathanasopoulos, M., 2018. Redox exchange of the disulfides of human two-domain CD4 regulates the conformational dynamics of each domain, providing insight into its mechanisms of control. Biochemical and Biophysical Research Communications 497: 811-817.

Pang, Q., Ji, Y., Li, Y., Bermúdez-Humarán, L.G., Hu, G. and Zeng, Y., 2008. Intragastric administration with recombinant Lactococcus lactis producing heme oxygenase-1 prevents lipopolysaccharide-induced endotoxemia in rats. FEMS Microbiology Letters 283: 62-68.

Patwardhan, R.S., Singh, B., Pal, D., Checker, R., Bandekar, M., Sharma, D. and Sandur, S.K., in press. Redox regulation of regulatory T-cell differentiation and functions. Free Radical Research. https://doi.org/10.1080/10715762.2020.1745202

Peng, Z., Liao, Y., Chen, L., Liu, S., Shan, Z., Nüssler, A. K., Yao, P., Yan, H., Liu, L. and Yang, W., 2019. Heme oxygenase-1 attenuates low-dose of deoxynivalenol-induced liver inflammation potentially associating with microbiota. Toxicology and Applied Pharmacology 374: 20-31.

Pérez, S., Taléns-Visconti, R., Rius-Pérez, S., Finamor, I. and Sastre, J., 2017. Redox signaling in the gastrointestinal tract. Free Radical Biology and Medicine 104: 75-103.

Pizzi, M., Boi, M., Bertoni, F. and Inghirami, G., 2016. Emerging therapies provide new opportunities to reshape the multifaceted interactions between the immune system and lymphoma cells. Leukemia 30: 1805-1815.

Prabhu, R. and Balasubramanian, K.A., 2002. Heat preconditioning attenuates oxygen free radicalmediated alterations in the intestinal brush border membrane induced by surgical manipulation. Journal of Surgical Research 107: 227-233.

Puertollano, M.A., Puertollano, E., De Cienfuegos, G.Á. and De Pablo, M.A., 2011. Dietary antioxidants: immunity and host defense. Current Topics in Medicinal Chemistry 11: 1752-1766.

Quintana, A., Schwindling, C., Wenning, A.S., Becherer, U., Rettig, J., Schwarz, E.C. and Hoth, M., 2007. T cell activation requires mitochondrial translocation to the immunological synapse. Proceedings of the National Academy of Sciences 104: 14418-14423.

Qureshi, M.A., Hussain, I. and Heggen, C.L., 1998. Understanding immunology in disease development and control. Poultry Science 77: 1126-1129.

Reeves, M.A. and Hoffmann, P.R., 2009. The human selenoproteome: recent insights into functions and regulation. Cellular and Molecular Life Sciences 66: 2457-2478.

Ren, M.T., Gu, M.L., Zhou, X.X., Yu, M.S., Pan, H.H., Ji, F. and Ding, C.Y., 2019. Sirtuin 1 alleviates endoplasmic reticulum stress-mediated apoptosis of intestinal epithelial cells in ulcerative colitis. World Journal of Gastroenterology 25: 5800-5813.

Rentea, R.M., Guo, Y., Zhu, X., Musch, M.W., Chang, E.B., Gourlay, D.M. and Liedel, J.L., 2018. Role of intestinal Hsp70 in barrier maintenance: contribution of milk to the induction of Hsp70. 2. Pediatric Surgery International 34: 323-330.

Reth, M., 2002. Hydrogen peroxide as second messenger in lymphocyte activation. Nature Immunology 3: 1129-1134.

Rigottier-Gois, L., 2013. Dysbiosis in inflammatory bowel diseases: the oxygen hypothesis. The ISME journal 7: 1256-1261.

Rokutan, K., 1999. Molecular stress response in the stomach. Nippon Yakurigaku Zasshi 114: 265-272.

Samborski, P. and Grzymisławski, M., 2015. The role of HSP70 heat shock proteins in the pathogenesis and treatment of inflammatory bowel diseases. Advances in Clinical and Experimental Medicine 24: 525-530.

Santoro, M.G., 2000. Heat shock factors and the control of the stress response. Biochemical Pharmacology 59: 55-63.

Sena, L.A., Li, S., Jairaman, A., Prakriya, M., Ezponda, T., Hildeman, D.A., Wang, C.R., Schumacker, P.T., Licht, J.D., Perlman, H. and Bryce, P.J., 2013. Mitochondria are required for antigen-specific T cell activation through reactive oxygen species signaling. Immunity 38: 225-236.

Sergeant, M.J., Constantinidou, C., Cogan, T.A., Bedford, M.R., Penn, C.W. and Pallen, M.J., 2014. Extensive microbial and functional diversity within the chicken cecal microbiome. PloS One 9: e91941.

Shi, N., Li, N., Duan, X. and Niu, H., 2017. Interaction between the gut microbiome and mucosal immune system. Military Medical Research 4: 14.

Shrimali, R.K., Irons, R.D., Carlson, B.A., Sano, Y., Gladyshev, V.N., Park, J.M. and Hatfield, D.L., 2008. Selenoproteins mediate T cell immunity through an antioxidant mechanism. Journal of Biological Chemistry 283: 20181-20185.

Sierra-Filardi, E., Vega, M.A., Sánchez-Mateos, P., Corbí, A.L. and Puig-Kröger, A., 2010. Heme oxygenase-1 expression in M-CSF-polarized M2 macrophages contributes to LPS-induced IL-10 release. Immunobiology 215: 788-795.

Sofi, M.H., Wu, Y., Schutt, S.D., Dai, M., Daenthanasanmak, A., Voss, J.H., Nguyen, H., Bastian, D., Iamsawat, S., Selvam, S.P. and Liu, C., 2019. Thioredoxin-1 confines T cell alloresponse and pathogenicity in graft-versus-host disease. The Journal of Clinical Investigation 129: 2760-2774.

Speckmann, B. and Steinbrenner, H., 2014. Selenium and selenoproteins in inflammatory bowel diseases and experimental colitis. Inflammatory Bowel Disease 20: 1110-1119.

Speckmann, B., Bidmon, H.J., Pinto, A., Anlauf, M., Sies, H. and Steinbrenner, H., 2011. Induction of glutathione peroxidase 4 expression during enterocytic cell differentiation. Journal of Biological Chemistry 286: 10764-10772.

Stecher, B., 2015. The roles of inflammation, nutrient availability and the commensal microbiota in enteric pathogen infection. Microbiology Spectrum 3: 10.1128.

Stegmann, M., Barclay, A.N. and Metcalfe, C., 2018. Reduction of leucocyte cell surface disulfide bonds during immune activation is dynamic as revealed by a quantitative proteomics workflow (SH-IQ). Open Biology 8: 180079.

Stegmann, M., Metcalfe, C. and Barclay, A.N., 2013. Immunoregulation through membrane proteins modified by reducing conditions induced by immune reactions. European Journal of Immunology 43: 15-21.

Sukumar, M., Liu, J., Mehta, G.U., Patel, S.J., Roychoudhuri, R., Crompton, J.G., Klebanoff, C.A., Ji, Y., Li, P., Yu, Z. and Whitehill, G.D., 2016. Mitochondrial membrane potential identifies cells with enhanced stemness for cellular therapy. Cell Metabolism 23: 63-76.

Sun, Y., Zhang, J., Song, W. and Shan, A., 2018. Vitamin E alleviates phoxim-induced toxic effects on intestinal oxidative stress, barrier function, and morphological changes in rats. Environmental Science and Pollution Research 25: 26682-26692.

Surai, P.F., 2002. Natural antioxidants in avian nutrition and reproduction. Nottingham University Press, Nottingham, UK.

Surai, P.F. 2006. Selenium in nutrition and health. Nottingham University Press, Nottingham, UK.

Surai, P.F., 2015. Silymarin as a natural antioxidant: an overview of the current evidence and perspectives. Antioxidants 4: 204-247.

Surai, P.F., 2018. Selenium in poultry nutrition and health. Wageningen Academic Publishers, Wageningen, the Netherlands.

Surai, P.F., 2020. Antioxidants in poultry nutrition and reproduction: an update. Antioxidants 9, 2: 105.

Surai, P.F. and Fisinin, V.I., 2015. Antioxidant-prooxidant balance in the intestine: applications in chick placement and pig weaning. Journal of Veterinary Science and Medicine 3: 16.

Surai, P.F., Kochish, I.I., Fisinin, V.I., Grozina, A.A. and Shatskikh E.V., 2018. Molecular mechanisms of chicken gut health maintenance: role of microbiota. Agricultural Technologies, Moscow, Russia.

Surai, P.F., Kochish, I.I. and Fisinin, V.I., 2017. Antioxidant systems in poultry biology: nutritional modulation of vitagenes. European Journal of Poultry Science 81: 1612-9199.

Surai, P.F., Kochish, I.I., Fisinin, V.I. and Kidd, M.T., 2019. Antioxidant defence systems and oxidative stress in poultry biology: an update. Antioxidants 8, 7: 235.

Surai, P.F., Kochish, I.I. and Kidd, M.T., 2020. Taurine in poultry nutrition. Animal Feed Science and Technology 260: 114339.

Surai, P.F., Kochish, I.I. and Shapovalov, S.O., 2018a. Superoxide dismutase activity in chicken gut. British Poultry Abstracts 11-12.

Takahashi, K., Avissar, N., Whitin, J. and Cohen, H., 1987. Purification and characterization of human plasma glutathione peroxidase: a selenoglycoprotein distinct from the known cellular enzyme. Archives of Biochemistry and Biophysics 256: 677-686.

Takano, T., Yonemitsu, Y., Saito, S., Itoh, H., Onohara, T., Fukuda, A., Takai, M. and Maehara, Y., 2012. A somatostatin analogue, octreotide, ameliorates intestinal ischemia-reperfusion injury through the early induction of heme oxygenase-1. Journal of Surgical Research 175: 350-358.

Tamion, F., Richard, V., Renet, S. and Thuillez, C., 2007. Intestinal preconditioning prevents inflammatory response by modulating heme oxygenase-1 expression in endotoxic shock model. American Journal of Physiology-Gastrointestinal and Liver Physiology 293: G1308-G1314.

Taranu, I., Braicu, C., Marin, D.E., Pistol, G.C., Motiu, M., Balacescu, L., Neagoe, I.B. and Burlacu, R., 2015. Exposure to zearalenone mycotoxin alters *in vitro* porcine intestinal epithelial cells by differential gene expression. Toxicology Letters 232: 310-325.

Tian, T., Wang, Z. and Zhang, J., 2017. Pathomechanisms of oxidative stress in inflammatory bowel disease and potential antioxidant therapies. Oxidative Medicine and Cellular Longevity 2017: 4535194.

Tsukimi, Y. and Okabe, S., 2001. Recent advances in gastrointestinal pathophysiology: role of heat shock proteins in mucosal defense and ulcer healing. Biological and Pharmaceutical Bulletin 24: 1-9.

Ulrich, K. and Jakob, U., 2019. The role of thiols in antioxidant systems. Free Radical Biology and Medicine 40: 14-27.

Umeda, K., Takahashi, T., Inoue, K., Shimizu, H., Maeda, S., Morimatsu, H., Omori, E., Akagi, R., Katayama, H. and Morita, K., 2009. Prevention of hemorrhagic shock-induced intestinal tissue injury by glutamine via heme oxygenase-1 induction. Shock 31: 40-49.

Venter, C., Eyerich, S., Sarin, T. and Klatt, K.C., 2020. Nutrition and the immune system: a complicated tango. Nutrients 12: 818.

Wadhwa, R., Taira, K. and Kaul, S.C., 2002. An Hsp70 family chaperone, mortalin/mthsp70/PBP74/Grp75: what, when, and where? Cell Stress & Chaperones 7: 309-316.

Wang, N., Han, Q., Wang, G., Ma, W.P., Wang, J., Wu, W.X., Guo, Y., Liu, L., Jiang, X.Y., Xie, X.L. and Jiang, H.Q., 2016. Resveratrol protects oxidative stress-induced intestinal epithelial barrier dysfunction by upregulating heme oxygenase-1 expression. Digestive Diseases and Sciences 61: 2522-2534.

Wang, R., Dillon, C.P., Shi, L.Z., Milasta, S., Carter, R., Finkelstein, D., McCormick, L.L., Fitzgerald, P., Chi, H., Munger, J. and Green, D.R., 2011. The transcription factor Myc controls metabolic reprogramming upon T lymphocyte activation. Immunity 35: 871-882.

Wang, Y., Guo, Y., Chen, H., Wei, H. and Wan, C., 2018. Potential of Lactobacillus plantarum ZDY2013 and Bifidobacterium bifidum WBIN03 in relieving colitis by gut microbiota, immune, and anti-oxidative stress. Canadian Journal of Microbiology 64: 327-337.

Wang, Y., Wu, Y., Wang, Y., Fu, A., Gong, L., Li, W. and Li, Y., 2017a. *Bacillus amyloliquefaciens* SC06 alleviates the oxidative stress of IPEC-1 via modulating Nrf2/Keap1 signaling pathway and decreasing ROS production. Applied Microbiology and Biotechnology 101: 3015-3026.

Wang, Y., Wu, Y., Wang, Y., Xu, H., Mei, X., Yu, D., Wang, Y. and Li, W., 2017. Antioxidant properties of probiotic bacteria. Nutrients 9: 521.

Wang, Z., Sun, R., Wang, G., Chen, Z., Li, Y., Zhao, Y., Liu, D., Zhao, H., Zhang, F., Yao, J., and Tian, X., 2020. SIRT3-mediated deacetylation of PRDX3 alleviates mitochondrial oxidative damage and apoptosis induced by intestinal ischemia/reperfusion injury. Redox Biology 28: 101343.

Weiss, G.A. and Hennet, T., 2017. Mechanisms and consequences of intestinal dysbiosis. Cellular and Molecular Life Sciences 74: 2959-2977.

Wellman, A.S., Metukuri, M.R., Kazgan, N., Xu, X., Xu, Q., Ren, N.S., Czopik, A., Shanahan, M.T., Kang, A., Chen, W. and Azcarate-Peril, M.A., 2017. Intestinal epithelial sirtuin 1 regulates intestinal inflammation during aging in mice by altering the intestinal microbiota. Gastroenterology 153: 772-786.

Wherry, E.J., 2011. T cell exhaustion. Nature Immunology 12: 492-499.

Williams, M.S. and Kwon, J., 2004. T cell receptor stimulation, reactive oxygen species, and cell signaling. Free Radical Biology and Medicine 37: 1144-1151.

Xiao, Y., Xiang, Y., Zhou, W., Chen, J., Li, K. and Yang, H., 2017. Microbial community mapping in intestinal tract of broiler chicken. Poultry Science 96: 1387-1393.

Yang, J., Gong, Y., Cai, J., Zheng, Y., Liu, H. and Zhang, Z., 2020. Chlorpyrifos induces redox imbalance-dependent inflammation in common carp lymphocyte through dysfunction of T-cell receptor γ. Journal of Fish Diseases 43: 423-430.

Yenari, M.A., 2002. Heat shock proteins and neuroprotection. Advances in Experimental Medicine and Biology 513: 281-299.

Zhang, D., Shang, T., Huang, Y., Wang, S., Liu, H., Wang, J., Wang, Y., Ji, H. and Zhang, R., 2017. Gene expression profile changes in the jejunum of weaned piglets after oral administration of Lactobacillus or an antibiotic. Scientific Reports 7: 15816.

Zhang, H., Yan, H., Zhou, X., Wang, H., Yang, Y., Zhang, J. and Wang, H., 2017. The protective effects of Resveratrol against radiation-induced intestinal injury. BMC Complementary and Alternative Medicine 17: 410.

Zhang, L., Xue, H., Zhao, G., Qiao, C., Sun, X., Pang, C. and Zhang, D., 2019. Curcumin and resveratrol suppress dextran sulfate sodium-induced colitis in mice. Molecular Medicine Reports 19: 3053-3060.

Zhang, Y., Wang, X.L., Zhou, M., Kang, C., Lang, H.D., Chen, M.T., Hui, S.C., Wang, B. and Mi, M.T., 2018. Crosstalk between gut microbiota and sirtuin-3 in colonic inflammation and tumorigenesis. Experimental & Molecular Medicine 50: 21.

Zhao, B., Fei, J., Chen, Y., Ying, Y.L., Ma, L., Song, X.Q., Wang, L., Chen, E.Z. and Mao, E.Q., 2014. Pharmacological preconditioning with vitamin C attenuates intestinal injury via the induction of heme oxygenase-1 after hemorrhagic shock in rats. PloS One 9: e99134.

Zhao, W., Han, Y., Zhao, B., Hirota, S., Hou, J. and Xin, W., 1998. Effect of carotenoids on the respiratory burst of rat peritoneal macrophages. Biochimica et Biophysica Acta 1381: 77-88.

Zuidema, M.Y., Peyton, K.J., Fay, W.P., Durante, W. and Korthuis, R.J., 2011. Antecedent hydrogen sulfide elicits an anti-inflammatory phenotype in postischemic murine small intestine: role of heme oxygenase-1. American Journal of Physiology. Heart and Circulatory Physiology 301: H888-H894.

Chapter 17
Looking ahead

'The future belongs to those who believe

in the beauty of their dreams.'

– Eleanor Roosevelt –

17.1 Introduction

The vitagene concept has been developed and successfully applied for poultry sciences and commercial egg and meat production. In fact, the concept helps explain molecular mechanisms of poultry adaptation to various stresses. Several redox-sensitive genes were shown to be involved in stress sensing and synthesis of a range of protective substances involved in redox balance maintenance, antioxidant defences and adaptive homeostasis in stressed birds. To finalise this book, it would be important to try to have a look at future developments in this exciting area of research.

Indeed, this chapter presents my view on future development in vitagene-related research and applications.

17.2 Integrated antioxidant defence network

Taking into account all previous information provided above (see Chapters 1-16), an update on the vitagene-regulated antioxidant defence mechanisms and their involvement in stress adaptation can be presented as follows (Figure 17.1). Mitochondria and phagocyte cells are major sources of RONS. There are also various stress conditions in poultry production increasing RONS production. It is well appreciated that RONS play important roles as signalling molecules; however, when their concentration is above threshold level, they can cause damage to main biological molecules, including lipids, proteins, and DNA/RNA. As a result of RONS excess, a stress response program (Oxidative stress response, OSR) is activated.

Since oxidative stress can cause a range of damages to various molecules, in addition to OSR, other stress response programs, including heat shock response (HSR), unfolded protein response (UPR), hypoxia-induced response (HIR), and DNA damage response, are also activated. This leads to activation of various transcription factors, including HSF1, Nrf2, NF-κB, FOXO, HIF, p53, and others. As a result of the upregulation of transcription factors, various genes, including vitagenes, are activated. In fact, activation of HSF increased production of HSP70, Nrf2 activation

Peter F. Surai **Vitagenes in avian biology and poultry health**
DOI 10.3920/978-90-8686-906-0_17, © Wageningen Academic Publishers 2020

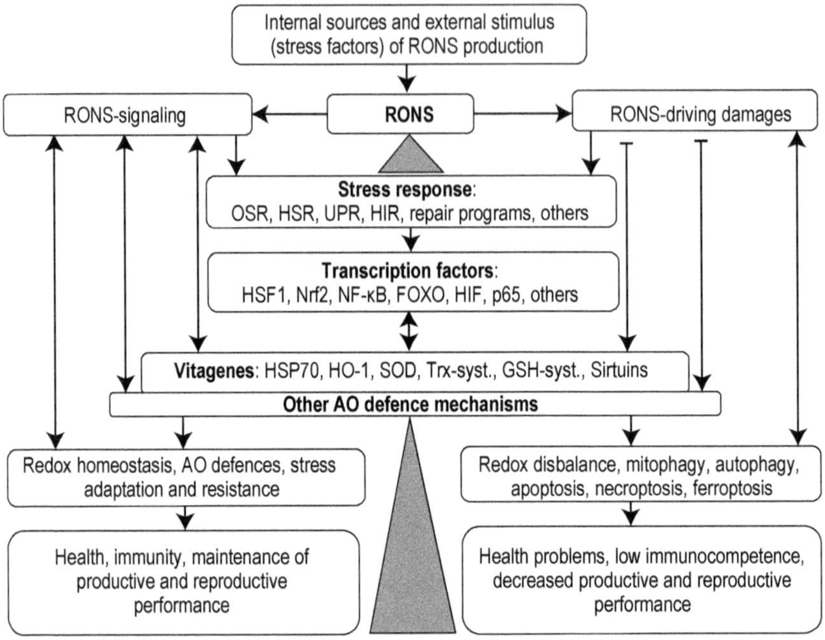

Figure 17.1. Antioxidant defence network in poultry (adapted from Surai, 2020).

increased synthesis of SOD, HO-1, elements of thioredoxin and glutathione systems; and FOXO activation would affect expression of sirtuins. There is a complex system of interplay between vitagenes and transcription factors. In fact, some vitagenes, like SOD, are affected by several transcription factors including Nrf2, NF-κB, p53, etc. Since SOD is responsible for production of H_2O_2, major signalling RONS, control of its concentration is of paramount importance. Furthermore, products of vitagene activation, e.g. sirtuins, would affect expression and activity of some transcription factors, including Nrf2, NF-κB, FOXO, etc. Some vitagenes can be activated directly without transcription factor involvement. This includes transcriptional regulation of SOD in response to RONS as well as activation of sirtuins by changes in NAD+/NADH ratio. In general, redox homeostasis plays an important role in the regulation of antioxidant defences. RONS are also responsible for adaptive production/activation of other antioxidants, which are not included into vitagene family (e.g. CoQ, catalase, various selenoproteins, etc.) and they all responsible for redox balance maintenance, stress resistance and adaptation leading to good health, high immunocompetence, high productive and reproductive performance of poultry. However, when the antioxidant defence system, together with the vitagene network, are not able to prevent or repair damages imposed by RONS to biological molecules, other protective mechanisms including mitophagy, autophagy, apoptosis, necroptosis, and ferroptosis are dealing with terminally damaged molecules, organelles or cells. As a result of disrupted redox balance and accumulation of damages in cells/tissues, health-related problems, including low immunocompetence, appear. In addition, decreased productive and reproductive performance can cause heavy economic losses for poultry industry.

17.3 Vitagenes and stress adaptation

A hypothetical scheme showing effects of vitagene-modulating nutrients/compositions on stress adaptability is shown in Figure 17.2.

It seems likely that low levels of stress can be considered as pre-conditioning action to activate stress response (SR) pathways to upregulate a range of redox-sensitive protective genes, including vitagenes, leading to improvement of antioxidant defences and adaptation and resistance to the stressful conditions. In fact, hormetic action of various chemicals/nutrients deserve more attention. Furthermore, usage of vitagene-modulating nutrients in poultry/animal nutrition appeared as a new direction of nutritional sciences.

Since it is almost impossible to avoid stresses in commercial poultry production, a search for nutritional means of antioxidant system modulation, including usage of increased doses of vitamin E (Surai *et al.*, 2019a), selenium (Surai, 2018; Surai and Kochish, 2019), taurine (Surai *et al.*, 2020) and polyphenolics (Hu *et al.*, 2019; Surai, 2015), is on the agenda of many research groups world-wide. It seems likely that vitagene modulation by nutritional means (e.g. carnitine, betaine, silymarin, taurine, vitamin and minerals (Surai *et al.*, 2017, 2019) is a new strategy to prevent commercially relevant stresses and maintain high productive and reproductive performance of commercial poultry (Surai *et al.*, 2018).

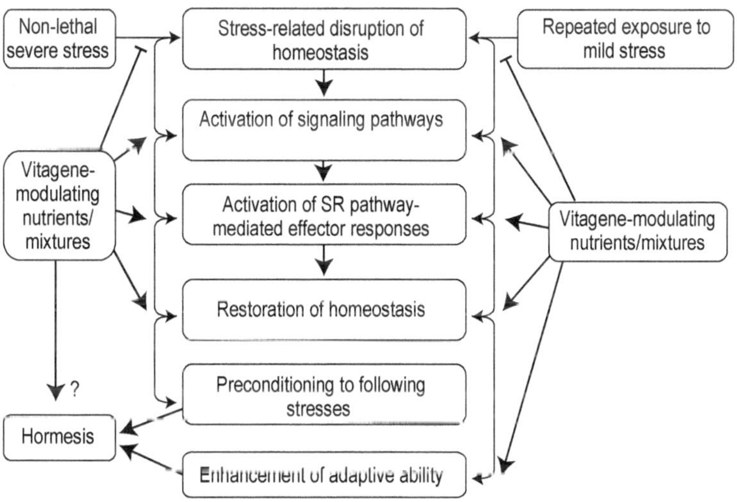

Figure 17.2. Vitagenes and stress adaptation (adapted from Bhattacharaya and Rattan, 2019; Sies and Jones, 2020; Surai, 2020; Surai *et al.*, 2019).

17.4 Future prospects

The term 'vitagene' was introduced in 1988 by Professor Suresh Rattan at Aarhus University (Denmark) and the vitagene concept in medical sciences has been developed by a group of Italian scientists at the University of Catania under the lead of Professor Vittorio Calabrese (for details and references see Chapter 3). We modified the concept, adapted it in relation to protection against various stresses and successfully transferred it into poultry sciences. The next steps in the development and application of the concept could be considered as follows:

- Vitagenes and epigenetics mechanisms. Understanding molecular mechanisms of vitagene involvement into epigenetic regulation of various physiological processes awaits future research.
- Vitagenes and maternal programming. Effect of maternal vitagene modulation on the antioxidant defences, redox homeostasis and adaptive ability to stress in progeny in human and animals is of paramount importance in biology, medicine and poultry/animal sciences.
- Effect of various drugs, including antibiotics, on the vitagene network in human and poultry/animals would help designing optimised programs of disease prevention/treatment.
- The vitagene concept application to farm animals (pigs, cattle, sheep, etc.), companion animals (cats, dogs, etc.), fur animals, fish, etc. await investigations.
- Application of the vitagene concept to ecology and evolutionary biology in relation to wild birds/animals would contribute to deeper understanding molecular mechanisms of stress adaptation.
- More research related to a relationship between hormesis and the vitagene network would help to expand search for vitagene-modulating compounds/mixtures to use them in medical practises, veterinary medicine and animal/poultry nutrition.

17.5 General conclusions

The vitagene concept of fighting stresses in avian biology and poultry production has been developed and successfully tested in commercial meat and egg production systems. Indeed, nutritional modulation of vitagenes is considered to be a new direction in nutritional sciences. In fact, by upregulating vitagenes it is possible to improve adaptive ability of poultry in stress conditions and maintain poultry health/ immunity, productive and reproductive performance of commercial birds under commercially relevant stress conditions. The applications of vitagene regulation were related to general stress fighting as well as improving eggshell gland and liver health.

Summarising information presented in the previous 16 chapters it could be concluded:

- Oxidative stress is the major mechanism of detrimental consequences of various stresses in poultry production.
- Low concentrations of RONS (mainly H_2O_2) are used as signalling molecules to upregulate a range of protective pathways and improve adaptive ability of poultry/animals to various stresses.
- A range of redox-sensitive genes called VITAGENES are an important and essential link between RONS, antioxidants, redox signalling, redox homeostasis and stress adaptation.
- The vitagene network can be upregulated via different transcription factors, including Nrf2, NF-κB, FOXO, HSF, etc. or directly by stress signals.
- Vitagene upregulation is an evolutionary mechanism responsible for stress adaptation and stress resistance.
- There is a range of nutrients having vitagene-modulating properties and they can be used in optimal combinations in poultry/animal diets as feed additive or water supplements to improve their adaptive ability to various stresses. This could decrease detrimental consequences of various stresses and prevent economic losses.
- The vitagene concept is an essential part of an integrated system including molecular mechanisms of the antioxidant defence network, redox signalling in cells/body and adaptive homeostasis and antioxidant/vitagene modulating feed additives.
- There is a range of publications proving an efficacy of the vitagene-regulating compositions in preventing/mitigating detrimental consequences of various stresses in commercial poultry and pig production.
- In future, further development of the vitagene concept would bring new knowledge into various areas of biological sciences with possible practical applications in medicine, veterinary medicine, and nutritional sciences.

Where there's a will, there's a way

References

Bhattacharya, S. and Rattan, S.I., 2019. Primary stress response pathways for preconditioning and physiological hormesis. In: Rattan, S. and Kyriazi, M. (eds) The science of hormesis in health and longevity. Academic Press, Cambridge, MA, USA, pp. 35-51.

Hu, R., He, Y., Arowolo, M.A., Wu, S. and He, J., 2019. Polyphenols as potential attenuators of heat stress in poultry production. Antioxidants 8: 67.

Sies, H. and Jones, D.P., 2020. Reactive oxygen species (ROS) as pleiotropic physiological signalling agents. Nature Reviews. Molecular Cell Biology 21: 363-383. https://doi.org/10.1038/s41580-020-0230-3

Surai, P.F., 2015. Silymarin as a natural antioxidant: an overview of the current evidence and perspectives. Antioxidants 4: 204-247.

Surai, P.F., 2018. Selenium in poultry nutrition and health. Wageningen Academic Publishers, Wageningen, the Netherlands.

Surai, P.F., 2020. Antioxidants in poultry nutrition and reproduction: an update. Antioxidants 9, 2: 105.

Surai, P.F. and Kochish, I.I., 2019. Nutritional modulation of the antioxidant capacities in poultry: the case of selenium. Poultry Science 98: 4231-4239.

Surai, P.F., Kochish, I.I., Fisinin, V.I., Grozina, A.A. and Shatskikh E.V., 2018. Molecular mechanisms of chicken gut health maintenance: role of microbiota. Agricultural Technologies, Moscow, Russia.

Surai, P.F., Kochish, I.I. and Fisinin, V.I., 2017. Antioxidant systems in poultry biology: nutritional modulation of vitagenes. European Journal of Poultry Science 81: 1612-9199.

Surai, P.F., Kochish, I.I., Fisinin, V.I. and Kidd, M.T., 2019. Antioxidant defence systems and oxidative stress in poultry biology: an update. Antioxidants 8, 7: 235.

Surai, P.F., Kochish, I.I., Romanov, M.N. and Griffin, D.K., 2019a. Nutritional modulation of the antioxidant capacities in poultry: the case of vitamin E. Poultry Science 98: 4030-4041.

Surai, P.F., Kochish, I.I. and Kidd, M.T., 2020. Taurine in poultry nutrition. Animal Feed Science and Technology 260: 114339.

www.ingramcontent.com/pod-product-compliance
Lightning Source LLC
LaVergne TN
LVHW080801011026
849160LV00015B/1505